Human Genetic Diversity

Contents

Preface

There is a danger when the pace of change is so fast, and the discoveries so remarkable, that we can lose sight of lessons learned in the past and the broader context of a field of scientific research. For human genetics, advances over the last twenty years since I studied as a medical student have been nothing short of revolutionary, changing what had been regarded as a sleepy backwater in the medical curriculum into a leading light, radically advancing our understanding of the basis of disease. The ramifications of contemporary genetic research extend far beyond medicine however, providing fundamental insights into biology and human origins, as well as important questions about the individual and society. This book was written as an introduction to the nature and functional consequences of human genetic variation, aiming to convey some of the excitement associated with recent advances by reviewing cutting-edge research while also providing a broad biological and historical context by considering some of the many landmark papers in the field.

The book begins with a review of the pioneering work into variation involving the genes encoding haemoglobin which has provided many fundamental insights into this field of research. Chapter 1 also serves to provide a primer in molecular genetics with examples given of the different types of genetic variation found at the globin gene loci, and more widely across the genome through the remarkable efforts to sequence the human genome. The major classes of genetic variation are then reviewed in more detail, ranging from the cytogenetically visible structural genomic variation seen at a microscopic chromosomal level (Chapter 3) to smaller scale submicroscopic structural variation which is increasingly recognised as copy number variation among healthy individuals as well as

contributing to common multifactorial diseases (Chapter 4) and genomic disorders (Chapter 5). The origins and role of segmental duplications in evolution and structural variation are described (Chapter 6) as well as the remarkable insights and application of research into tandem repeats (Chapter 7) and mobile DNA elements (Chapter 8). Sequence level diversity is then described including the efforts to catalogue and define the genomic architecture of such variation with implications for understanding susceptibility to common disease (Chapter 9) and selective pressures operating in our evolutionary and recent past (Chapter 10). Approaches to dissecting the genetic basis of classical mendelian diseases as well as common multifactorial traits are described, ranging from linkage and positional cloning to current genome-wide association studies (Chapters 2 and 9). Much remains to be understood about how genetic variants may be acting at a molecular level to modulate the nature or function of the protein encoded by a gene or the levels of expression of the gene itself, a topic explored in Chapter 11. Many of the successes and current roadblocks in our understanding of the nature and consequences of human genetic diversity are then highlighted by the extreme diversity found at the major histocompatibility complex on chromosome 6 (Chapter 12) while the past and ongoing battles seen in major parasitic diseases such as malaria (Chapter 13) and human immunodeficiency virus (Chapter 14) show how human genetic diversity reflects in part a genomic battlefield where specific allelic variants may change in frequency dependent on particular selection pressures.

The text is deliberately focused on exploring specific examples to illustrate the field rather than aiming to be comprehensive, and I hope that the references and reviews cited will serve as a jumping point for further reading. The glossary explains some of the more technical subject matter, with terms highlighted in bold on first mention in the text.

I hope that this will serve to make the book accessible and informative, of relevance to the practicing clinician, as well as to the specialist researcher, and to medical students and others studying at an advanced undergraduate and graduate level. The subject matter encompasses the many disciplines impacted by human genetic diversity, ranging from medicine to evolutionary biology, biological anthropology and molecular biology. These are times of dramatic scientific advances in human genetics that will impact on us all as individuals and as a society, and important choices lie ahead in how we choose to use such knowledge. If this book can go a small way towards conveying the excitement felt in contemporary research into human genetic variation, and the breadth of its application, then it will have succeeded, and done so by dint of the remarkable nature of the subject matter more than by my own efforts.

Acknowledgements

I am very grateful for advice and comments on many of the chapters from colleagues including Sunil Ahuja, Duncan Campbell, Chris Conlon, Doug Higgs, Mark Hirst, David Keeling, Samantha Knight, Anthony Monaco, Sreeram Ramagopalan, Kirk Rockett, Emanuela Volpi and Andrew Wilkie. I am also very grateful to the skilled editorial and other staff at Oxford University Press including Jane Andrew, Carol Bestley, Helen Eaton and Ian Sherman. The work in my lab is funded by the Wellcome Trust who have been instrumental in supporting many of the key programmes of research described in this book which have advanced our understanding of human genetic variation. Finally, I would like to thank my family for their patience and unwaiving support during the preparation of this book. This book is dedicated to my wife Marian and three daughters Jess, Pip and Kitty.

CHAPTER 1

Lessons from haemoglobin

1.1 Introduction

As individual human beings we are unique, shaped by our environment and experiences but also by the specific **DNA** sequences we have inherited. The extent and nature of variation in our DNA is remarkable with consequences ranging from physical appearance to risk of disease. With the rapid pace of advances in genetics over the last 20 years the amount of information related to human genetic diversity is at once overwhelming, and any review rapidly out of date. The intention of this book is not therefore to be exhaustive but rather to take a broader view using specific examples that illustrate the historical context of research in this area, and how it encompasses a range of scientific disciplines which have led to remarkable progress over a relatively short space of time. Just as the traveller with a round-the-world air ticket will not see the world in its entirety but will hopefully gain some appreciation of its extent and diversity, this book seeks to provide an overview of research into human genetic variation and some insight into its functional implications for health and disease.

This first chapter aims to provide a framework for understanding human genetic diversity by giving a detailed review of different types of genetic variation involving the **genes** encoding haemoglobin (Hb), the iron-containing protein found in red blood cells which is responsible for oxygen transport around the body. At first glance, this may seem too narrow a focus. However, by doing so we are entering regions of our **genome** remarkable for both the extent of their diversity, and the exhaustive research programmes over many years that have so elegantly delineated the nature and consequences of such variation. At the globin genes and neighbouring regions we can find examples of almost all known forms of human genetic diversity, many with dramatic consequences for human health. A discussion of the molecular basis of inherited disorders affecting haemoglobin (Box 1.1) will also allow an introduction to a number of fundamental concepts in human molecular genetics and a review of the often complex terminology used to describe genetic variation.

The pioneering work which has been carried out to understand genetic diversity at the globin genes (Fig. 1.1),

---

Box 1.1   Haemoglobinopathies

The haemoglobinopathies are inherited disorders affecting the structure or synthesis of haemoglobin. They are remarkably common, with an estimated 7% of the world population being **carriers** (Weatherall 2000). Haemoglobinopathies include structural variants of haemoglobin such as sickle haemoglobin (Hb S) and disorders in which the synthesis of one or more globin molecules is reduced or absent (the thalassaemias) (Weatherall 2001).

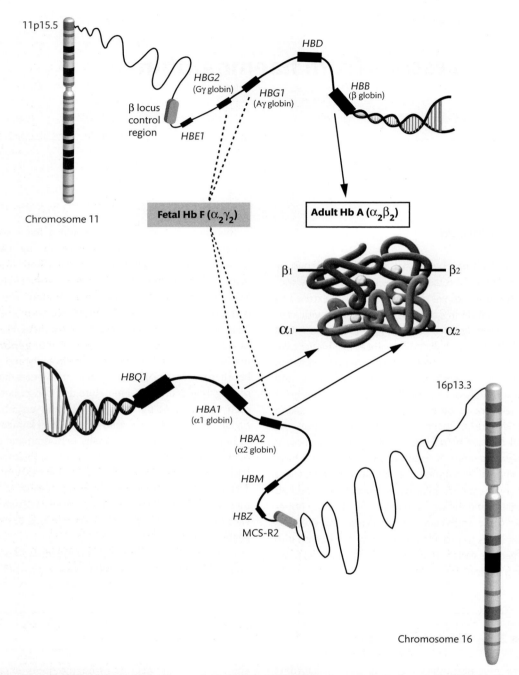

Figure 1.1 Genes encoding haemoglobin. Haemoglobin is a tetrameric molecule comprised of two pairs of identical polypeptides encoded by genes in the α globin and β globin gene clusters on chromosomes 16 and 11, respectively. Fetal haemoglobin (Hb F) ($\alpha_2\gamma_2$) comprises two α globin chains and two γ globin chains; after birth this is replaced by adult Hb A ($\alpha_2\beta_2$) and very small amounts of Hb A2 ($\alpha_2\delta_2$). In Hb S ($\alpha_2\beta^s_2$), the β globin polypeptide encoded by the *HBB* gene contains a glutamine to valine substitution.

and its implications for disease, has led to a succession of remarkable discoveries that have been a paradigm for much of our current understanding of human genetics. It has been a treasure trove of discovery to which many investigators have devoted their scientific careers, an investment in detailed research which has been repaid many times over and continues to reveal new discoveries to this day.

1.2   Genetic variation and a molecular basis for disease

Our journey begins with the example of sickle cell disease, a common clinically important genetic disorder (Box 1.2) (Ingram 2004; Frenette and Atweh 2007). Research into sickle cell disease was pivotal in demonstrating a molecular basis for disease, illustrating how DNA sequence variation (the **genotype**) can have profound functional consequences, in this case at a structural level in the encoded protein, resulting in a specific disease (the observed **phenotype**) (Box 1.3).

Research into the molecular basis of sickle cell disease was however unusual in some respects, as in this disease it was observed variation in a specific protein that allowed the underlying variation at the DNA level to be defined. This contrasts with the majority of inherited diseases in which a phenotype was mapped to a genetic locus on a **chromosome** (Box 1.4), and the particular genes (Box 1.5) and causative **mutations** within them identified without knowledge of the encoded protein (**linkage** and positional cloning are described in detail in Chapter 2).

1.2.1 A difference at the protein level between haemoglobin molecules

In 1910 Herrick described an unusual case of a patient with anaemia associated with lung symptoms in whom abnormal 'sickle-shaped' red blood cells were seen (Fig. 1.4) (Herrick 1910). This change in red cell morphology was shown to be dramatically modulated by low levels of oxygen (hypoxia) and led to the hypothesis that an abnormality in haemoglobin may be responsible (Hanh and Gillespie 1927; Scriver and Waugh 1930). In 1949

Box 1.2   Sickle cell disease

Sickle cell disease is an inherited structural disorder of haemoglobin. The disorder specifically involves the β globin subunit of the haemoglobin molecule such that instead of normal adult haemoglobin Hb A ($\alpha_2\beta_2$), Hb S results ($\alpha_2\beta^S_2$) (Fig. 1.1). Sickle cell disease is recognized to occur in particular ethnic groups, notably individuals of African, Asian or more rarely Mediterranean ancestry in whom the variant **allele** is present at a relatively high frequency. It is one of the most common inherited diseases known in man, with an estimated **incidence** in African Americans of 1 in 625 and a carrier frequency of 8%; carrier frequencies of up to 20% have been reported in Uganda and Kenya (Young 2005).

Box 1.3   Genotype and phenotype

Genotype refers to the hereditary or genetic constitution of an individual, either as a whole or for a specific **locus** within the genome. Phenotype describes an observable characteristic, which may range from appearance to a structural, biochemical, physiological or behavioural character. The phenotype is often considered as being the product of both genotype and environmental factors.

Box 1.4   Chromosomes

A chromosome consists of a very long molecule of DNA with associated proteins found in the nucleus of the cell. There are 46 chromosomes in a human **diploid** cell bearing the normal complement of chromosomes, made up of 22 pairs of **autosomes** (numbered 1 to 22) and two sex chromosomes (either XX or XY). Many genes are found on a given chromosome. For autosomal genes (i.e. those residing on chromosomes 1 to 22), a given cell has two copies of each gene, one on each pair of autosomes. Chromosomes constitute a discrete unit of the genome and vary in size. Chromosomes become visible on microscopy as distinct nuclear bodies during cell division (**mitosis**) when they are highly condensed, notably during **metaphase** (the point in mitosis when the condensed chromosomes align with each other) when particular banding patterns can be seen after staining with specific dyes (Fig. 1.2). During metaphase the **centromere** is visible as a constriction in the chromosome, seen as the point where the two sister **chromatids** are held together. The centromere is pivotal to the process of cell division and control. 'Chromatids' refers to each of the two copies of a replicated chromosome during the process of cell division (mitosis or **meiosis**), specifically at the time they are joined at the centromere; on separation (anaphase) the two are said to be 'daughter chromosomes'. The nature of chromosomes and the structural genetic variation that can be seen at a microscopic level are explored in Chapter 3.

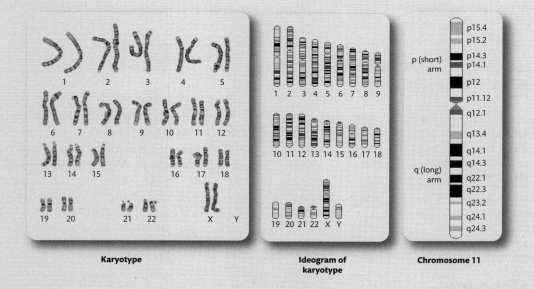

Karyotype   **Ideogram of karyotype**   **Chromosome 11**

Figure 1.2 Chromosomes and karyotypes. Human chromosome complement (karyotype) for a human female (using G-banding). Image provided by Dr Ros Hastings, UKNEQAS for Clinical Cytogenetics, Oxford. The karyotype can be shown in an ordered diagrammatic form as an ideogram with chromosome bands numbered according to position on the short (p) or long (q) arm of the chromosome. Where the arms are of equal length the chromosome is said to be metacentric, where unequal, submetacentric. For chromosomes 13, 14, 15, 21, and 22 the p arm is very short, and these chromosomes are said to be acrocentric. Ideogram of karyotype and details of chromosome 11 prepared with permission using screen shots from the Ensembl Genome Browser (http://Jul2008.archive.ensembl.org/Homo_sapiens/mapview?chr=11).

Box 1.5   Genes

The definition of a gene as a discrete unit of heredity has developed since the work of Mendel, from the recognition of a gene as a blueprint for a protein, to a transcribed code in the nucleic acid sequence, to a DNA sequence possessing particular characteristics allowing widespread annotation of the genome (Fig 1.3). The 'protein-centric' view of a gene has required adaptation with the recognition that there are many sites of **transcription** in the genome that lead to RNA but do not result in **translation** into a protein ('noncoding RNA'). Such transcriptionally active regions are involved in a diverse array of predominantly regulatory functions. The very extensive use of **alternative splicing** (Section 11.6) and the presence of many transcriptional start sites further complicate our view of a 'gene'. Gerstein and colleagues note that a gene must encompass the concept that there is a genomic sequence encoding a functional RNA or protein product: where a number of functional products share overlapping genomic sequences a coherent union of all is considered (Gerstein *et al.* 2007).

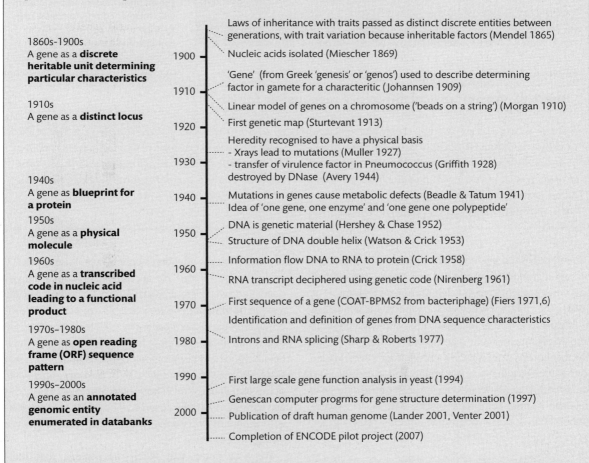

1860s-1900s
A gene as a **discrete heritable unit determining particular characteristics**

1910s
A gene as a **distinct locus**

1940s
A gene as **blueprint for a protein**

1950s
A gene as a **physical molecule**

1960s
A gene as a **transcribed code in nucleic acid leading to a functional product**

1970s–1980s
A gene as **open reading frame (ORF) sequence pattern**

1990s–2000s
A gene as an **annotated genomic entity enumerated in databanks**

1900 — Laws of inheritance with traits passed as distinct discrete entities between generations, with trait variation because inheritable factors (Mendel 1865)

Nucleic acids isolated (Miescher 1869)

1910 — 'Gene' (from Greek 'genesis' or 'genos') used to describe determining factor in gamete for a characteritic (Johannsen 1909)

Linear model of genes on a chromosome ('beads on a string') (Morgan 1910)

1920 — First genetic map (Sturtevant 1913)

Heredity recognised to have a physical basis
- Xrays lead to mutations (Muller 1927)

1930 — - transfer of virulence factor in Pneumococcus (Griffith 1928) destroyed by DNase (Avery 1944)

1940 — Mutations in genes cause metabolic defects (Beadle & Tatum 1941) Idea of 'one gene, one enzyme' and 'one gene one polypeptide'

DNA is genetic material (Hershey & Chase 1952)

1950 — Structure of DNA double helix (Watson & Crick 1953)

Information flow DNA to RNA to protein (Crick 1958)

1960 — RNA transcript deciphered using genetic code (Nirenberg 1961)

1970 — First sequence of a gene (COAT-BPMS2 from bacteriophage) (Fiers 1971,6)

Identification and definition of genes from DNA sequence characteristics

1980 — Introns and RNA splicing (Sharp & Roberts 1977)

1990 — First large scale gene function analysis in yeast (1994)

Genescan computer progrms for gene structure determination (1997)

2000 — Publication of draft human genome (Lander 2001, Venter 2001)

Completion of ENCODE pilot project (2007)

Figure 1.3 Changing views of a 'gene'. A timeline highlighting evolving concepts of a gene is shown together with dates of key advances. An open reading frame (ORF) refers to a sequence of bases that could potentially encode a protein. Adapted with permission from Gerstein *et al.* (2007).

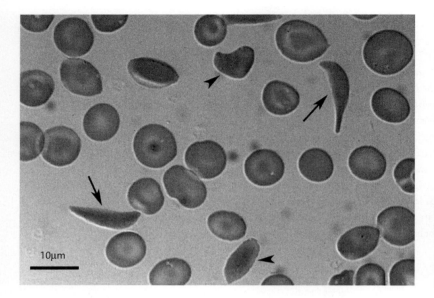

10μm

Figure 1.4 Peripheral blood smear from a patient with sickle cell disease. Sickle-shaped red blood cells are shown (indicated by arrows) together with misshaped cells (arrowheads). Reprinted with permission from Frenette and Atweh (2007).

Box 1.6   An amino acid difference responsible for Hb S

A single amino acid substitution from glutamic acid to valine results in Hb S. Glutamic acid is negatively charged while valine is hydrophobic: the amino acid change in each of the β globin chains in the haemoglobin molecule promotes hydrophobic contacts with alanine, phenylalanine, and leucine residues in adjacent molecules such that there is reversible association in conditions of deoxygenation forming 14 stranded polymers which can crosslink, the long fibres stretching and deforming the red blood cells (Vekilov 2007).

Pauling published evidence of a difference in behaviour of haemoglobins from affected and unaffected individuals when subjected to gel electrophoresis (Pauling *et al.* 1949). The difference between unaffected and affected individuals was further refined by analysis of polypeptides resulting from enzymatic digestion of haemoglobin using electrophoresis and partition chromatography. This led to the discovery that a single amino acid, a change from glutamic acid to valine, was responsible for the difference between normal adult haemoglobin (Hb A) to haemoglobin S (Hb S, sickle variant haemoglobin) (see Fig. 1.1; Box 1.6) (Ingram 1957, 1958, 1959).

1.2.2 Mendelian inheritance, alleles and traits

Among individuals whose red blood cells are observed to sickle in particular conditions, only a minority were noted to have the severe phenotype of sickle cell disease (Box 1.7) while the others had no apparent pathological consequence and were described as having sickle cell trait. Initially a single **dominant** gene with variable expression was proposed as being responsible but in 1947 Neel hypothesised that within an affected population, there were individuals heterozygous and homozygous for the condition such that sickle cell anaemia was only seen in **homozygotes** while sickle cell trait occurred in **heterozygotes** (Neel 1949). This was consistent with the observed inheritance in families involving parents who were unaffected, had sickle cell trait or sickle cell disease (Fig 1.5).

We now know that Neel was correct, that individuals who have the genetic variant leading to Hb S on both copies of the gene encoding β globin (*HBB* on chromosome 11), denoted 'Hb SS', have the severe haematological disorder sickle cell disease. These

Box 1.7   Phenotype of sickle cell disease

The clinical manifestations of sickle cell disease are severe and potentially life threatening (Ashley-Koch *et al.* 2000). The disease becomes symptomatic as β globin expression replaces γ globin during the first 6 months of life, with anaemia and jaundice as a result of increased haemolysis of the red blood cells. The red blood cells have an irregular sickled appearance in low oxygen states (see Fig. 1.4) and tend to adhere to blood vessel walls with reduced blood flow and risk of complete vascular occlusion. Symptomatically, sickle cell disease is characterized by acute painful crises affecting the musculoskeletal system; a pneumonia-like illness described as acute chest syndrome; stroke; and increased susceptibility to bacterial infection. A longitudinal prospective **cohort** study (the Cooperative Study of Sickle Cell Disease) highlighted the high incidence of severe pneumococcal infection in infancy which led to routine neonatal screening for the disorder and use of prophylactic penicillin (Gaston and Rosse 1982; Gaston *et al.* 1986). Sickle cell disease is associated with premature death, with 11% of patients having a stroke by age 20 and 24% by age 45 years (Platt *et al.* 1994; Ohene-Frempong *et al.* 1998). The use of hydroxyurea which is associated with an increase in the level of Hb F has proved a very important therapeutic intervention in sickle cell disease, reducing sickling, the frequency of painful crises, and mortality (Letvin *et al.* 1984; Charache *et al.* 1995; Steinberg *et al.* 2003).

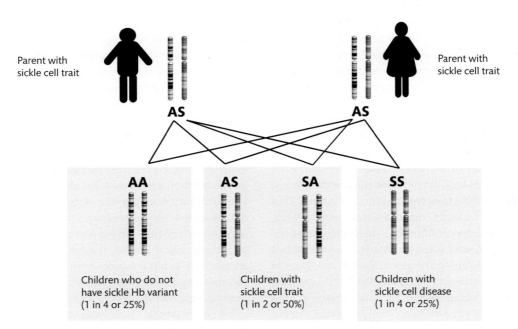

Figure 1.5 Inheritance of alleles for sickle haemoglobin. Illustration of two parents who have one allele bearing the sickle variant of β globin (S) and one with the normal variant (A) which shows the likelihood of their offspring having different combinations of the alleles. Individuals with sickle cell trait are carriers of the disease-associated allele: having two parents who are carriers ('AS') results in a 25% chance that a child will have two disease alleles ('SS') and develop sickle cell disease; 50% chance of a child being a carrier ('AS'); and 25% chance of a child having two alleles without the disease-causing variant ('AA')

Box 1.8 Alleles

Mendelian inheritance is concerned with the transfer of genetic information from parents to children. The alternate forms of the same gene are known as alleles. When an individual has two identical alleles, they are described as homozygous, when the alleles differ, as heterozygous. A child receives one allele from each of their parents for a given autosomal gene. If there is a genetic difference between the alleles, having one 'normal' allele (often denoted 'A') and one 'variant' allele ('a') may be sufficient for a particular phenotype or trait to become manifest. The individual is heterozygous for the alleles ('Aa'),

and inheritance of the character (also described as the trait or phenotype) is described as autosomal dominant. One copy of the variant allele is sufficient, for example, to result in disease, as seen in Huntington's disease where an unstable repeat expansion in the DNA sequence causes a dramatic change in the encoded protein which is toxic to the cell (Box 7.13). Sometimes the character will only be manifest if an individual inherits two alleles with the variant (homozygous 'aa'). The phenotype is described as autosomal recessive and examples include cystic fibrosis (Section 2.3.1).

---

individuals are homozygous for this allele (Box 1.8). By contrast, possession of one chromosome with the variant encoding Hb S while having a second normal chromosome which continues to encode adult haemoglobin, Hb A, results in sickle cell trait ('Hb AS'). This has minimal adverse effect except under conditions of more severe hypoxia such as underwater diving or high altitude. In most conditions therefore, heterozygous individuals with sickle cell trait are phenotypically normal and the inheritance of sickle cell disease can be described as autosomal recessive. However as heterozygotes will express Hb S as well as Hb A, when the altered form of haemoglobin is considered as the observed phenotype, it is inherited as a co-dominant trait.

This illustrates how the mode of inheritance refers to the specific trait or phenotype under consideration. Thus the presence of a single copy of an allele with the sickle variant can also be considered co-dominant in terms of susceptibility to sickling at high altitude, or overdominant in terms of conferring the phenotype of resistance to malaria. Here heterozygotes (those with sickle cell trait) have a marked advantage in terms of protection from malarial infection due to *Plasmodium falciparum* without the cost of sickling crises and other pathology seen in homozygotes with sickle cell disease (Allison 1964). This selective advantage is thought to have been responsible for the relatively high allele frequency of the genetic

variant responsible for sickle haemoglobin in sub-Saharan Africa and parts of India (Section 13.2.3).

It has also become apparent that sickle cell disease can arise if an individual has a copy of the Hb S variant together with some other sequence variant involving the *HBB* gene (such as resulting in Hb C) (Section 1.3.1). It was also notable that among patients homozygous for Hb S there is clinical heterogeneity in the severity of the disease phenotype observed, ranging from early death to disease with few complications. Genetic modifiers include α-thalassemia and variants determining the levels of Hb F. The latter were known to include rare deletions in the β globin gene cluster and point mutations in γ-globin genes resulting in hereditary persistence of fetal haemoglobin, but more recent work has identified variants near *HBB* and in other chromosomal regions which together are associated with determining nearly half of the variance in levels of Hb F among non-anaemic populations (Higgs and Wood 2008a). The functional mechanisms remain unclear but the variants on chromosome 2 and chromosome 6 may involve the oncogene *BCL11A* and haemopoietic transcription factor *MYB* respectively.

For multifactorial traits and diseases (sometimes described as 'complex' or '**polygenic** diseases') such as malaria (Box 13.1) or rheumatoid arthritis (Section 12.2.3), genetic factors play a role but are not inherited in a simple mendelian manner. Here, multiple genetic loci and variants are important in determining disease susceptibility together with environmental factors. The lines of division between such diseases and classical 'mendelian' disorders are however becoming increasingly blurred. Indeed, the growing awareness of the complexity of the genetic and other determinants of sickle cell disease have promoted the proposal that rather than considering this disease as a simple monogenic condition it should be 'considered as a complex multigenic disorder' (Higgs and Wood 2008a).

1.2.3 Sequencing the HBB gene and defining the variant responsible for Hb S

DNA was recognized as the 'heritable material' in 1944 by Avery, Macleod, and McCarty (Avery *et al.* 1944); its double helical structure was elucidated by Watson and Crick in 1953 (Watson and Crick 1953); and the nature of the genetic code solved by Nirenberg, Khorana, and Holley in the early 1960s (Box 1.9) (Fig 1.6 and Fig 1.7) (Nirenberg 1963). The determination of the DNA sequence for the globin genes had its basis in the development of groundbreaking new chemical and enzymatic methods for DNA sequencing (Box 1.10). Initial sequencing studies focused on determining the partial and later full sequence of globin **RNA.** For example the β globin messenger RNA sequence was determined using the Maxam and Gilbert technique by synthesizing double-stranded DNA from the RNA, and this was found to agree with predictions based on the amino acid sequence and earlier partial sequencing of RNA (Efstratiadis *et al.* 1977). In the same year, sequences of the noncoding region of human β globin RNA were published using the 'plus/minus' method of Sanger (Proudfoot 1977).

Advances in molecular cloning techniques (Cohen *et al.* 1973) allowed isolation and amplification of the β globin gene that was localized to the short arm of chromosome 11 (Messing et al. 1977; Wilson et al. 1977; Sanders-Haigh et al. 1980); subsequently the full nucleotide sequence was determined using the Maxam Gilbert sequencing method (Lawn et al. 1980).

Variation in DNA sequence of the *HBB* gene (see Fig. 1.1), encoding β globin, was found to be responsible for Hb S. The *HBB* gene is located on chromosome 11p15.4 and comprises three **exons** (Fig. 1.9). The linear flow of information from DNA to RNA to amino acid

chain through transcription and translation are illustrated with reference to the *HBB* gene (Box 1.11 and Box 1.12) (Fig 1.10 and Fig 1.11) (Strachan and Read 2004). The sequence variant resulting in Hb S is found near the start of the first exon of *HBB* and comprises an A to T nucleotide substitution in the non-template strand, which alters the RNA codon from 'GAG' to 'GUG', resulting in a change in amino acid residue from glutamic acid to valine (Fig. 1.12) (Kan and Dozy 1978; Frenette and Atweh 2007).

How should the variation responsible for Hb S be described? A number of different approaches have been taken and illustrate some of the complexities of defining and describing DNA sequence diversity (Fig. 1.13) (Beutler 1993; Beutler *et al.* 1996). Historically, an amino acid-based designation for describing variants was used as sequences were first available at the protein level, as was the case for haemoglobin, in advance of knowledge of the DNA code (Beutler 1993). A numbering system based on the amino acid sequence was possible with names beginning with a letter, for example E6V (glutamic acid for valine substitution at position 6) (Beaudet and Tsui 1993; AHCMN 1996). This system described the protein phenotype rather than the genotype and had the advantage of relative simplicity and insights into biological effect. However a number of problems with an amino acid-based approach were noted, not least that a particular amino acid change may result from a number of different nucleotide changes due to degeneracy of the genetic code (Beutler 1993; Beutler *et al.* 1996). For example, a histidine to glutamine substitution may result from a change in the codon from CAU, to CAA or CAG. It is therefore not always possible to deduce the DNA sequence variant from the amino acid change.

Furthermore there was controversy in amino acid notation in terms of the starting point. Early literature based on protein sequence considered the processed protein in which methionine is co-translationally cleaved at the point the amino acid sequence is about 25 amino acids long such that valine is the first amino acid and the Hb S variant denoted E6V (Glu6Val). However, current recommendations refer to the unprocessed protein, the primary translation product in which methionine is amino acid +1 so that the Hb S change would now be designated E7V. Ambiguities also arise in terms of whether the native, partly processed,

Box 1.9   DNA structure

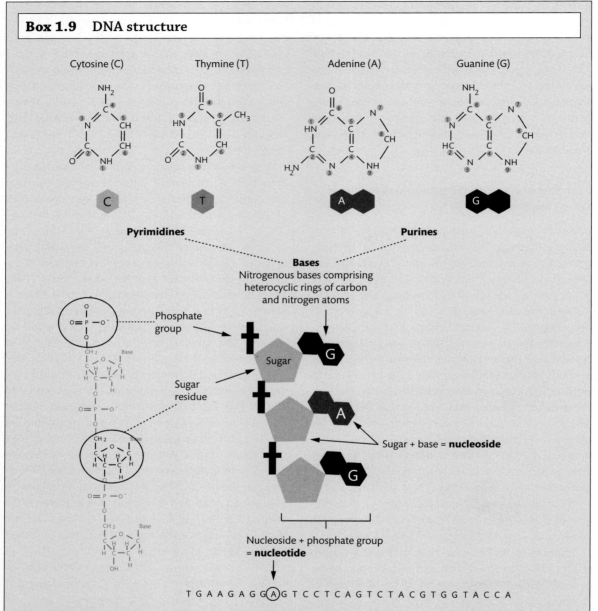

Figure 1.6 Bases, nucleosides, and nucleotides. The basic repeating units of DNA are nucleotides, shown here as a DNA sequence (from the *HBB* gene, encoding β haemoglobin) within which one nucleotide is circled ('A'). Nucleosides comprise a sugar residue (deoxyribose, a five carbon sugar), which is covalently linked to a nitrogenous base. In DNA there are four types of nitrogenous base: adenine (A), guanine (G), cytosine (C), and thymine (T); these are classified as either purines (A and G; designated as 'R') or pyrimidines (C and T; designated as 'Y'). Purines consist of two interlocked heterocyclic rings of carbon and nitrogen atoms, pyrimidines of one heterocyclic ring. Sugar residues are linked by covalent phosphodiester bonds, from the carbon atom 3′ of a sugar to the carbon atom 5′ of the next sugar residue. Nucleotides are nucleosides with a phosphate group attached.

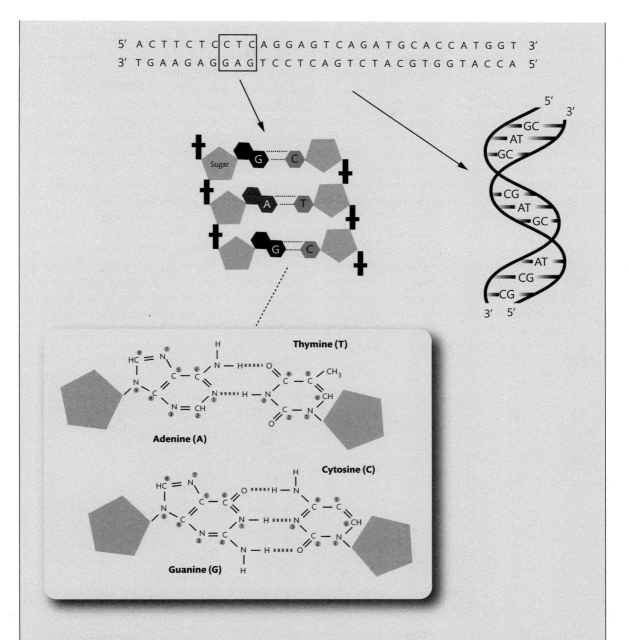

5′ A C T T C T C CTC A G G A G T C A G A T G C A C C A T G G T 3′
3′ T G A A G A G GAG T C C T C A G T C T A C G T G G T A C C A 5′

Thymine (T)

Adenine (A)

Cytosine (C)

Guanine (G)

Figure 1.7 DNA strands. Two DNA strands are held together by relatively weak hydrogen bonds between base pairs (bp); A specifically binds to T, and C to G to give rise to a DNA duplex or double helix structure. Most eukaryotic DNA is in a B-DNA right-handed helical structure of 10 bp per turn. The so called 5′ end of the DNA strand has a terminal carbon atom number 5 without a phosphodiester bond while the at the 3′ end of the strand there is carbon atom number 3. The two DNA strands always anneal together such that the 5′ to 3′ direction of one strand is opposite to, and complimentary in sequence to, the other strand. The sequence is normally quoted only for one strand in the 5′ to 3′ direction, either as 5′pGpApG-OH 3′ (p is the phosphodiester bond, -OH is the terminal OH group at the 3′ end of the strand) or in abbreviated form as 5′ GAG 3′.

Box 1.10 DNA sequencing

In 1977 Maxam and Gilbert published a DNA sequencing method based on different chemical modifications of DNA (such as dimethyl sulphate) which then allowed cleavage of the DNA only at specific bases (Maxam and Gilbert 1977). The DNA was radio labelled and visualized by electrophoresis on a polyacrylamide gel, generating a ladder of fragments of different lengths terminating for example at 'A' nucleotides. Depending on the specific chemical modification used, other bases could be preferentially cleaved, for example **purines** (giving a G+A ladder), or **pyrimidines** (C+Ts).

The 'plus/minus' method of Sanger involved enzymatic synthesis of a new DNA strand from a single-stranded template using DNA polymerase and a specific **primer.** The synthesis terminated in the absence of a particular **nucleoside** in the reaction mix, the plus reaction for example containing one nucleoside with the minus reaction containing the other three (Sanger and Coulson 1975).

Subsequently the chain terminator or dideoxy method was established using all four deoxynucleotide triphosphates (dNTPs) and a small amount of a specific dideoxynucleotide (ddNTP) which on random incorporation into the extending DNA strand would terminate its synthesis (as they lack the necessary 3′ OH needed to form a phosphodiester bond between nucleotides) (Sanger *et al.* 1977) (Fig. 1.8). Radioactive labelling of the primer allowed sequence determination on size separation using a denaturing polyacrylamide gel.

In a short space of time following development of these sequencing technologies the human mitochondrial genome was sequenced (16.5 **kilobases** (kb) in length) followed by a series of viral genomes. The use of fluorophores for primer labelling in enzymatic sequencing in 1986 (Smith *et al.* 1986) (or to label each ddNTP) set the stage for automation in detection and high throughput sequencing, culminating in publication of the draft human genome sequence in 2001 (Section 1.4.2) (Lander *et al.* 2001; Venter *et al.* 2001).

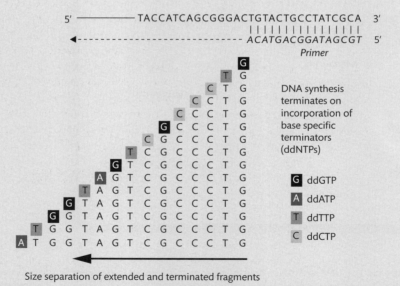

Size separation of extended and terminated fragments

Figure 1.8 Dideoxy DNA sequencing. Chain termination occurs on incorporation of a specific dideoxynucleotide into the nascent complementary DNA strand synthesized from the 3′ end of the annealed primer; different sequencing reactions are set up with one of four ddNTPs and all four dNTPs. Depending on where the ddNTP is incorporated, specific fragment lengths result. For a ddTTP reaction in this example, fragments of 2, 8, 11, and 14 nucleotides plus the length of the primer are observed indicating a 'T' at these positions in the sequence. Sequence read from +1 after 3′ end of primer: GTCCCGCTGATGGTA

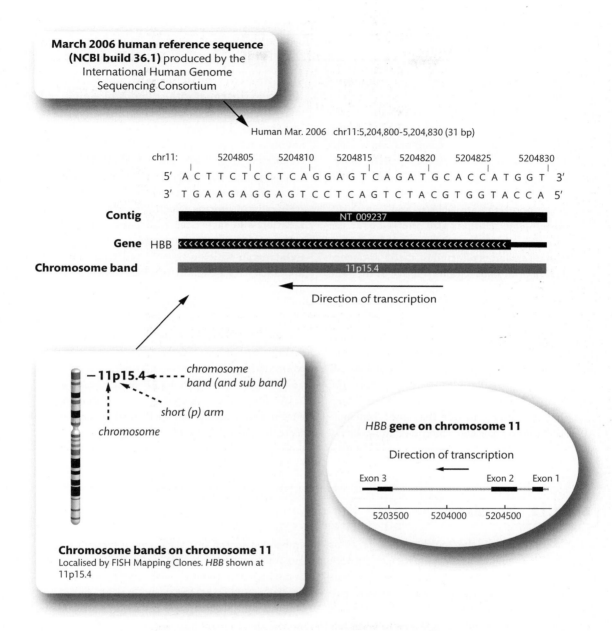

Figure 1.9 DNA sequence from *HBB* gene showing location and identifiers. Genomic DNA sequence from chromosome 11, located at 5 204 800 bp from the end of the p arm is shown. This is the March 2006 human reference sequence (NCBI build 36.1), which is a 'finished' highly accurate sequence with less than one error on average per 10 000 bases. The reference sequence used here is 'contig NT_009237'; a contig is a contiguous stretch of DNA sequence without gaps assembled using direct sequencing information. Of note, the *HBB* gene is orientated and transcribed in a centromeric to telomeric direction which is in the opposite direction to the numbering of the reference human genome sequence. The portion of the figure showing sequences and genomic location is adapted from a screenshot of the UCSC Genome Browser (Kent *et al.* 2002) (http://genome.ucsc.edu/) (Human March 2006 Assembly).

Box 1.11   Transcription

Transcription refers to the process of synthesizing RNA from a DNA template. This involves the unwinding of double-stranded DNA to allow DNA sequence information to be transcribed from the template DNA strand into complementary RNA molecules (primary transcripts) by **RNA polymerase**: the linear sequence (5´ to 3´) of the RNA molecule will be the same sequence and direction as the non-template DNA strand (Fig. 1.10). The non-template strand is also referred to as the 'sense strand', and is the sequence usually shown; the template strand is referred to as the 'antisense strand'. Orientation of regions relative to the gene sequence is with respect to the sense strand, for example the 5´ untranslated region is at the 5´ end of the sense strand. Genes are found in both orientations throughout the genome such that genes may be transcribed in opposite orientations. Sets of three DNA nucleotides ('base triplets', for example 'ATG') are decoded to RNA where the sequence of three nucleotides (a **codon**) define a particular amino acid (for the codon sequence 'AUG' the amino acid is methionine). The primary RNA transcript is 'spliced' to remove internal sequences (**introns**) with rejoining of protein coding exonic sequences: this complex process generates increased protein diversity for most genes by having a number of possible alternatively spliced transcripts (Fig. 1.10). The 5´ end of the primary RNA transcript is capped by linkage of a methylated nucleoside to the first 5´ nucleotide of the transcript, a process which protects against degradation and facilitates transport and **splicing**. Other RNA processing events include **polyadenylation** in which a poly(A) tail (approximately 200 adenylate residues) is added at the 3´ cleavage site at the end of the transcript, about 15–30 nucleotides (nt) downstream of the AAUAAA polyadenylation signal sequence. The poly(A) tail is also involved in RNA transport, stability, and facilitating translation. Although traditionally viewed as occurring in a stepwise manner, transcription and RNA processing are highly dynamic and interactive processes such that there is a complex and extensively coupled network of gene expression factories (Maniatis and Reed 2002).

Box 1.12   Translation

Translation involves the synthesis of proteins using the RNA template. Processed messenger RNA (mRNA) molecules are exported from the nucleus to the cytoplasm where the process of translation to polypeptide chains occurs (Fig. 1.11). This involves large RNA–protein complexes (ribosomes) with successive groups of three nucleotides (codons) coding particular amino acids. This specificity is achieved by transfer RNA (tRNA) molecules which have a particular trinucleotide sequence (anticodon) complementary to the RNA codon, and which bind a particular amino acid. The RNA sequence 'AUG' is recognized as the start or initiation codon when embedded in the initiation codon recognition sequence (for example GCCPuCCAUGG, where Pu is purine). The 'genetic code' relates codons in the RNA to amino acids in the protein. Each amino acid is specified on average by three different codons: overall there are 64 possible codons (4^3) and about 30 types of cytoplasmic tRNA (Fig. 1.11). This relates to degeneracy in codon–anticodon pairing most often involving the third base of the codon, such that either of two possible purine or pyrimidine bases will be recognized (for example AAA or AAG for lysine) or any of the four possible bases (ACA, ACG, ACC, ACU for threonine).

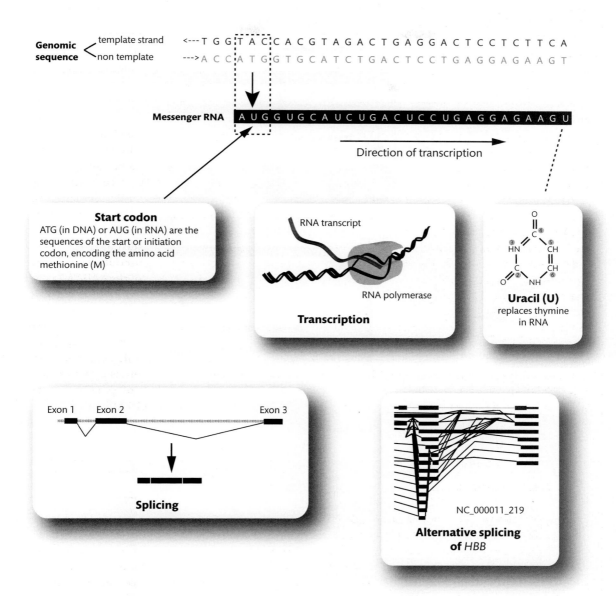

Figure 1.10 Transcription and RNA processing. The relationship between messenger RNA and the genomic DNA sequence at the start codon for the *HBB* gene is shown, together with the structure of uracil, the base which replaces thymine in RNA. A schematic of the process of transcription is shown. For clarity the *HBB* sequence is orientated with respect to the direction of transcription. Alternative splicing for *HBB* is summarized in the alternative splicing graph reproduced from the UCSC Genome Browser (www.genome.ucsc.edu/cgi-bin/hgTracks).

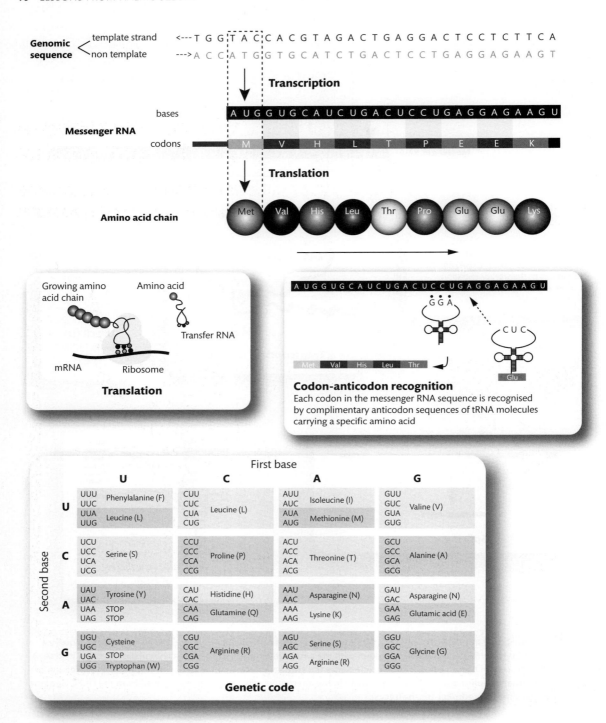

Figure 1.11 Translation. An overview of aspects of the process of translation is illustrated for *HBB* together with a summary of the genetic code. For clarity the *HBB* sequence is orientated with respect to the direction of transcription.

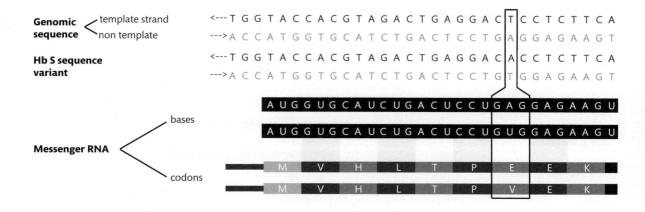

Genomic sequence ⟨ template strand / non template

Hb S sequence variant

Messenger RNA ⟨ bases / codons

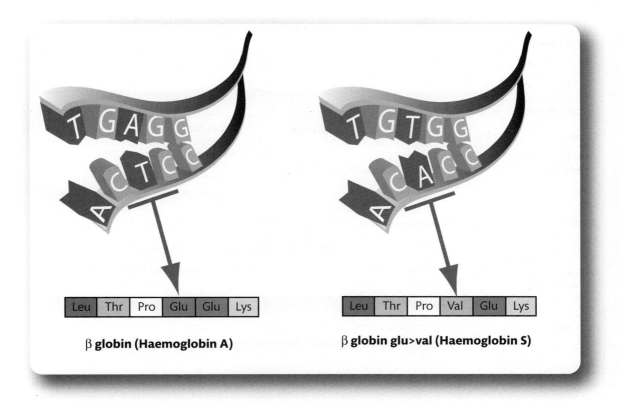

β **globin (Haemoglobin A)**   β **globin glu>val (Haemoglobin S)**

Figure 1.12 Molecular basis for sickle haemoglobin. An A to T substitution at the DNA level results in a codon change from GAG to GUG, and substitution of glutamic acid to valine. The altered β globin polypeptide results in Hb S. For clarity the *HBB* sequence is orientated with respect to the direction of transcription.

Descriptors for HbS sequence variant

Nicknames or common names

HbS mutation

Amino acid level nomenclature

E6V
Glu6Val
HBB: p.Glu7Val

Glutamic acid to valine substitution at amino acid position 6 or 7 (counting methionine as +1)

DNA level nomenclature

HBB: c.20A>T

A to T substitution in coding DNA sequence
Substitution at nucleotide 20; nucleotide '1' is 'A' of 'ATG' start codon

1		10		20	

A T G G T G C A T C T G A C T C C T G (A) G G A G A A G T

chr 11:5204808-5204808

Nucleotide position
Position in nucleotide sequence counting from p arm telomere of chromosome 11 in March 2006 human reference sequence (NCBI Build 36.1) produced by the International Human Genome Sequencing Consortium

Unique identifiers

rs334

dbSNP assigned reference SNP ID (rs)
http://www.ncbi.nlm.nih.gov/SNP

Entrez SNP
A repository maintained by National Centre for Biotechnology Information (NCBI) for single nucleotide substitutions and short deletion and insertion polymorphisms

ss79088884

dbSNP assigned submitter SNP ID(ss)

OMIM 141900.0243

Haemoglobin S, allelic variant at beta haemoglobin locus [HBB, Glu6Val]
Listing in Online Mendelian Inheritance in Man (OMIM)

HbS beta 6 (A3) Glu>Val
Hbvar ID 226

HbVar: a locus specific database of human haemoglobin variants and thalassemias
http://globin.bx.psu.edu/hbvar

VAR_002863

Swiss-Prot variant http://us.expasy.org/cgi-bin/get-sprot-variant.pl?VAR_002863

Figure 1.13 Nomenclature used to describe the DNA sequence variant and associated amino acid change underlying Hb S. A diverse range of terminology and identifiers are associated with specific sequence variants. Web addresses are given for the HbVar relational database of haemoglobin variants and thalassaemias (Patrinos *et al.* 2004) and the Swiss-Prot database of curated protein sequences and associated amino acid variants (Yip *et al.* 2004).

Box 1.13   **Nomenclature to describe sequence variants using Human Genome Variation Society (HGVS) recommendations**

Variants are described at the DNA level, with reference to a coding DNA or reference DNA sequence. The coding sequence should be the major and largest transcript of a gene. For the coding sequence, nucleotide '1' is 'A' of the 'ATG' translation initiation codon; the nucleotide immediately 5′ to this is denoted '−1' and the nucleotide 3′ of the translation stop codon is denoted '*1'. The first base of an intron is denoted 'c.27+1A', '27' being the number of the nucleotide in the last exon; the base at the end of an intron is given as 'c.28−1A', '28' being the nucleotide number in the next exon. When the exact position of the sequence change is not known, the possible range can be given in brackets. For example an insertion in a coding sequence somewhere between nucleotides 27 and 30 is given

as: 'c.(27–30)'. For an insertion/deletion, the positions of the nucleotides that flank the insertion are given. Genomic sequence should be from RefSeq (www.ncbi.nlm.nih.gov/projects/RefSeq/) with listing of database accession and version number. For genomic sequence, numbering starts with 1 as the first nucleotide of the database reference file and the variant is preceded by "g.". The genomic sequence should be complete. When referring to protein level changes, this should refer to the primary translation product, not a processed mature protein, with methionine as '+1'; the described variant is preceded by "p.". A tryptophan to cysteine substitution at position 26 would be denoted 'p.Trp26Cys' or 'p.W26C'; a change to a stop codon as p.Trp26X.

---

or processed protein is considered. It was also not possible to describe many observed variants on the basis of amino acids, for example insertions and deletions, and variants in noncoding DNA such as introns and promoter regions.

Several variants, notably those described in the earlier literature, were based on a 'nickname' or 'common' name derived from a particular amino acid change, a **restriction enzyme** site or patient name, or base number in the coding DNA or genomic DNA sequence. The growing need for a more systematic approach was recognized and in particular the realization that this should be based at the DNA level, either in terms of coding DNA or genomic DNA. There are a number of issues with a coding DNA-based approach, notably in terms of introns and alternative splicing, but it was favoured initially given the much later availability of genomic DNA sequence in which errors had to be accommodated and with variable starting points for numbering. For Hb S, the sequence variant can be described in terms of the coding DNA sequence based on nucleotide 1 as the 'A' of the 'ATG' start codon: this resolves to a substitution of nucleotide 20, written in shorthand as HBB: c.20A>T ("c" indicating the reference sequence used is coding DNA). The current availability of a curated non-redundant reference sequence database for genomic DNA (www.ncbi.nlm.nih.gov/RefSeq/)

has greatly facilitated such nomenclature (Section 1.4.2) (Pruitt *et al.* 2007). The variant nucleotide position has a unique position in the human reference sequence (chr 11:5204808) and, with respect to the forward or sense strand, is an A to T substitution.

A complementary approach was to consider unique identifiers for a particular sequence variant. An internationally accepted repository of **single nucleotide polymorphisms (SNPs)** and other short sequence variants is Entrez SNP (**dbSNP**) and the Hb S DNA sequence variant has a unique identifier in terms of a 'reference SNP ID', the **rs number** (rs334) and 'submitter SNP ID' (ss79088884) (Fig. 1.14). Unique identifiers are also found within other large curated databases and repositories such as Online Mendelian Inheritance in Man (**OMIM**) (OMIM 141900.0243) (Fig. 1.15), protein sequence databases such as Swiss-Prot (VAR_002863) (Fig. 1.13), and locus-specific databases such as HbVar (HbVar ID 226) (Fig. 1.13) which describe the specific variant and its context. A number of recommendations have been proposed for how sequence variants should be described in the literature or databases, for example those proposed by the Human Genome Variation Society (www.hgvs.org/mutnomen/recs.html) (Box 1.13; Fig. 1.16) (Antonarakis 1998; den Dunnen and Antonarakis 2000).

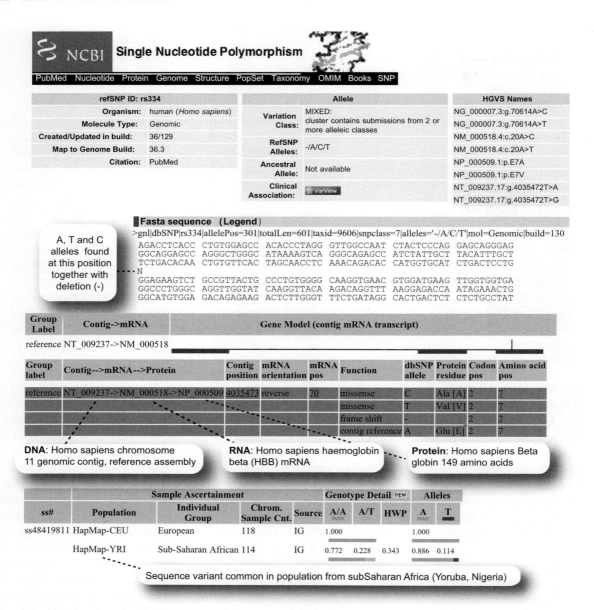

Figure 1.14 The dbSNP repository and the Hb S DNA sequence variant. In 1988 the National Centre for Biotechnology Information (NCBI) was set up by the National Institutes of Health to generate information systems for use in molecular biology (www.ncbi.nlm.nih.gov/) (Wheeler *et al.* 2008b) Among the many database resources now available, dbSNP is a repository of single nucleotide substitutions and short deletions and insertions containing over 12 million human SNPs (www.ncbi.nlm.nih.gov/SNP) (Sherry *et al.* 1999, 2001; Wheeler *et al.* 2008b). Some of the data available for rs334 is shown as an illustration using screenshots from dbSNP (www.ncbi.nlm.nih.gov/SNP/snp_ref.cgi?type=rs&rs=rs334). A number of the genome project sequence resources are indicated allowing the nucleotide substitution to be resolved in a particular contig, mRNA, or amino acid sequence. Validation status together with population-specific allele frequencies can be determined. Data from the International HapMap Project (www.hapmap.org) is shown for various population groups, highlighting how the Hb S sequence variant is common in a sub-Saharan population from Nigeria.

Sickle cell anaemia phenotype http://www.ncbi.nlm.nih.gov/entrez/dispomim.cgi?id=603903

NCBI | OMIM Online Mendelian Inheritance in Man | Johns Hopkins University | My NCBI [Sign In]

All Databases | PubMed | Nucleotide | Protein | Genome | Structure | PMC | OMIM

Search OMIM for Go Clear

Limits | Preview/Index | History | Clipboard | Details

Display Detailed Show 20 Send to

#603903

Clinical Features
Diagnosis
Gene Therapy
Molecular Genetics
Population Genetics
Animal Model

References
Contributors
Creation Date
Edit History

Gene map

Entrez Gene
Nomenclature

GenBank

UniGene

#603903 GeneTests,

SICKLE CELL ANEMIA

Gene map locus 11p15.5

TEXT

A number sign (#) is used with this entry because sickle cell anemia is the result of mutant beta globin (HBB; 141900 in which the mutation causes sickling of hemoglobin rather than reduced amount of beta globin which causes beta-thalassemia.

The most common cause of sickle cell anemia is Hb S (141900.0243), with SS disease being most prevalent in Africans.

CLINICAL FEATURES

Scriver and Waugh (1930) reported detailed studies of a 7-year old child with sickle cell anemia. Her main complaints were cough, night sweats, vague pains in the legs and joints, occasional abdominal pain, poor appetite, and increasing

HBB gene and allelic variants http://www.ncbi.nlm.nih.gov/entrez/dispomim.cgi?id=141900

+141900

HEMOGLOBIN--BETA LOCUS; HBB

Alternative titles; symbols

BETA-THALASSEMIAS, INCLUDED
METHEMOGLOBINEMIA, BETA-GLOBIN TYPE, INCLUDED
ERYTHREMIA, BETA-GLOBIN TYPE, INCLUDED

Gene map locus 11p15.5

The alpha and beta loci determine the structure of the 2 types of polypeptide chains in adult hemoglobin, Hb A. Mutant beta globin that sickles causes sickle cell anemia (603903). Absence of beta chain causes beta-zero-thalassemia. Reduced amounts of detectable beta globin causes beta-plus-thalassemia. For clinical purposes, beta-thalassemia is divided into thalassemia major (transfusion dependent), thalassemia intermedia (of intermediate severity), and thalassemia minor (asymptomatic).

.0243 HEMOGLOBIN S [HBB, GLU6VAL] dbSNP

SICKLE CELL ANEMIA, INCLUDED
MALARIA, RESISTANCE TO, INCLUDED

The change from glutamic acid to valine in sickle hemoglobin was reported by Ingram (1959). Ingram (1956) had reported that the difference between hemoglobin A and hemoglobin S lies in a single tryptic peptide. His analysis of this peptide, peptide 4, was possible by the methods developed by Sanger for determining the structure of insulin and Edman's stepwise degradation of peptides.

Figure 1.15 Online Mendelian Inheritance in Man (OMIM) and Hb S. Mendelian Inheritance in Man (MIM) is a definitive reference for human genes and genetic disorders started in 1966 by Dr Victor McKusick and made available online in 1987 as OMIM (www.ncbi.nlm.nih.gov/omim/) (McKusick and Antonarakis 1998; Hamosh *et al.* 2005). OMIM is described as a 'knowledgebase' with extensive full text describing particular genes and genetic phenotypes with 12 277 genes of known sequences described together with 2319 phenotypes with a known molecular basis and 10 990 loci (www.ncbi.nlm.nih.gov/Omim/mimstats.html; 11 May 2008). Portions of entries for sickle cell anaemia and *HBB* gene with allelic variants including Hb S are shown; OMIM also includes extensive links to Entrez Gene, external resources and databases. The figure was prepared with permission using screenshots from http://www.ncbi.nlm.nih.gov/omim/

(A)

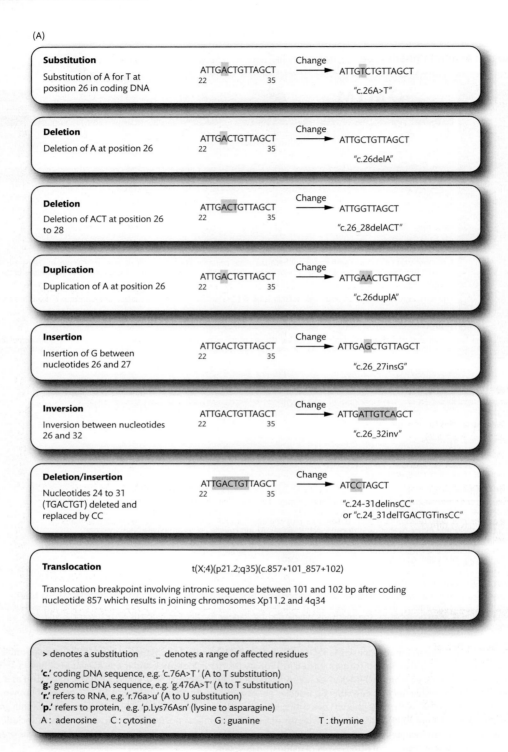

Substitution

Substitution of A for T at position 26 in coding DNA

ATTGACTGTTAGCT
22   35

Change →

ATTGTCTGTTAGCT

"c.26A>T"

Deletion

Deletion of A at position 26

ATTGACTGTTAGCT
22   35

Change →

ATTGCTGTTAGCT

"c.26delA"

Deletion

Deletion of ACT at position 26 to 28

ATTGACTGTTAGCT
22   35

Change →

ATTGGTTAGCT

"c.26_28delACT"

Duplication

Duplication of A at position 26

ATTGACTGTTAGCT
22   35

Change →

ATTGAACTGTTAGCT

"c.26duplA"

Insertion

Insertion of G between nucleotides 26 and 27

ATTGACTGTTAGCT
22   35

Change →

ATTGAGCTGTTAGCT

"c.26_27insG"

Inversion

Inversion between nucleotides 26 and 32

ATTGACTGTTAGCT
22   35

Change →

ATTGATTGTCAGCT

"c.26_32inv"

Deletion/insertion

Nucleotides 24 to 31 (TGACTGT) deleted and replaced by CC

ATTGACTGTTAGCT
22   35

Change →

ATCCTAGCT

"c.24-31delinsCC"
or "c.24_31delTGACTGTinsCC"

Translocation

t(X;4)(p21.2;q35)(c.857+101_857+102)

Translocation breakpoint involving intronic sequence between 101 and 102 bp after coding nucleotide 857 which results in joining chromosomes Xp11.2 and 4q34

> denotes a substitution   _ denotes a range of affected residues

'c.' coding DNA sequence, e.g. 'c.76A>T' (A to T substitution)
'g.' genomic DNA sequence, e.g. 'g.476A>T' (A to T substitution)
'r.' refers to RNA, e.g. 'r.76a>u' (A to U substitution)
'p.' refers to protein, e.g. 'p.Lys76Asn' (lysine to asparagine)
A : adenosine   C : cytosine   G : guanine   T : thymine

Figure 1.16 *Continued*

(B)

Sequence change by allele

To specify an allele, [ ] are used.

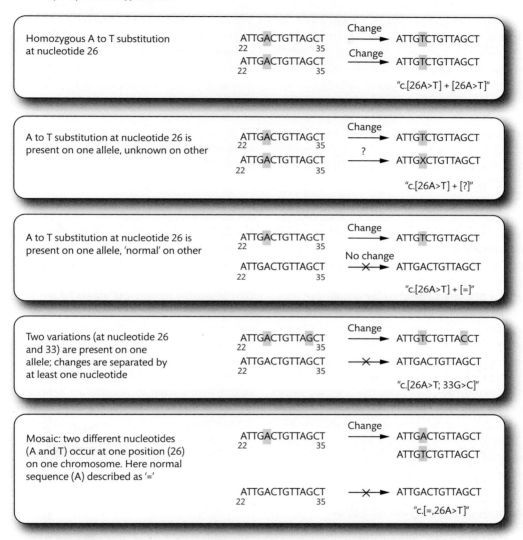

Figure 1.16 Description of sequence variants based on Human Genome Variation Society (HGVS) recommendations. Examples are given to illustrate the nomenclature recommended by the HGVS for description of sequence variants in scientific literature and databases. (A) A number of different sequence variants are shown including substitution (>), deletion (del), duplication (dup), insertion (ins), inversion (inv), and deletion/insertion (delins). (B) Allele-specific sequence variation involving one or both alleles, more than one variant, and a mosaic. The figure is based on examples and description given at www.hgvs.org/mutnomen (Antonarakis 1998; den Dunnen and Antonarakis 2000).

Box 1.14   Mutation and polymorphism

The word 'mutation' derives from the Latin '*mutatio*' meaning change or alteration (Marshall 2002). In the context of genetics, the term 'mutation' refers to a 'permanent structural change in the DNA' (www.genome.gov/glossary) (Cotton and Scriver 1998) and may involve changes in individual base pairs of DNA or larger structural changes in DNA. In some contexts and disciplines, 'mutation' has become associated with a 'disease-causing change' in the DNA, notably in terms of clinical genetics, while in the public perception there has been an increasingly negative association with the term 'mutation' (Condit *et al.* 2002). By contrast 'polymorphism' was regarded as a 'non-disease-causing change' or 'neutral' DNA variant (Cotton and Scriver 1998). The term polymorphism is used to refer to 'common variation in the sequence of DNA among individuals' (www.genome.gov/glossary) and more specifically to refer to sequence variants present at a frequency of 1% or more in a population, without reference to a particular phenotypic effect (Cotton and Scriver 1998).

A further question in terms of terminology when describing DNA sequence variants relates to the terms 'mutation' or '**polymorphism**' (Box 1.14). Sequence variants at their outset involve a *de novo* mutational event in the DNA of a single individual, and in general the term mutation is used in the context of a rare variant causing disease, but variants ('mutations') may rise to high allele frequencies in particular populations as seen with Hb S. Moreover, while the vast majority of common sequence variants described as 'polymorphisms' have no functional consequence, some are recognized to play an important role in common multifactorial diseases. When considering fine scale sequence changes or diversity, the phrase 'DNA sequence variant' carries no assumption about disease association or frequency. The occurrence of the variant should be validated and carefully assigned in terms of location and nature using specific terminology and unique identifiers as discussed.

1.2.4 Methods of detecting the Hb S DNA sequence variant

A **restriction fragment length polymorphism** (**RFLP**) was first found for a specific site 5 kb downstream of the *HBB* gene recognized by the restriction enzyme *HpaI*, which showed variation in the length of the digested restriction fragments between individuals, specifically among those of African descent in whom additional bands (7 and 13 kb in length) were recognized in addition to the expected size of 7.6 kb (Kan and Dozy 1978). Significantly, the 13 kb bands were associated with individuals who had sickle haemoglobin. Kan and Dozy noted in their paper that this 'may be useful for the prediction of the sickle cell gene in prenatal diagnosis' and that 'polymorphism in a restriction enzyme site could be considered as a new class of genetic marker and may offer a new approach to linkage analysis and anthropological studies' – views that were to be substantially validated although more specific restriction sites for the Hb S variant and other methods of detection were to develop.

Restriction sites spanning the specific single nucleotide variant were utilized to discriminate between DNA sequence for the two alleles, for example the enzymes *DdeI* or *MstII* (Geever *et al.* 1981; Chang and Kan 1982). In the presence of the A to T substitution the recognition site is lost and the enzyme no longer cuts, allowing those individuals with sickle cell trait and sickle cell disease to be identified (Fig. 1.17). Initial work relied on Southern blotting to detect DNA sequences. In this technique, invented by Edwin Southern, restriction enzyme digested DNA is separated on the basis of size by agarose gel electrophoresis, transferred to a membrane and then hybridized with a labelled (usually radioactive) probe to enable detection (Southern 1975).

The development of the **polymerase chain reaction** (**PCR**) made this process significantly faster and easier, with very much less DNA required (Fig. 1.18) (Mullis and Faloona 1987). One of the earliest applications of PCR was to detect the Hb S DNA sequence variant. DNA was amplified by PCR using the Klenow fragment of

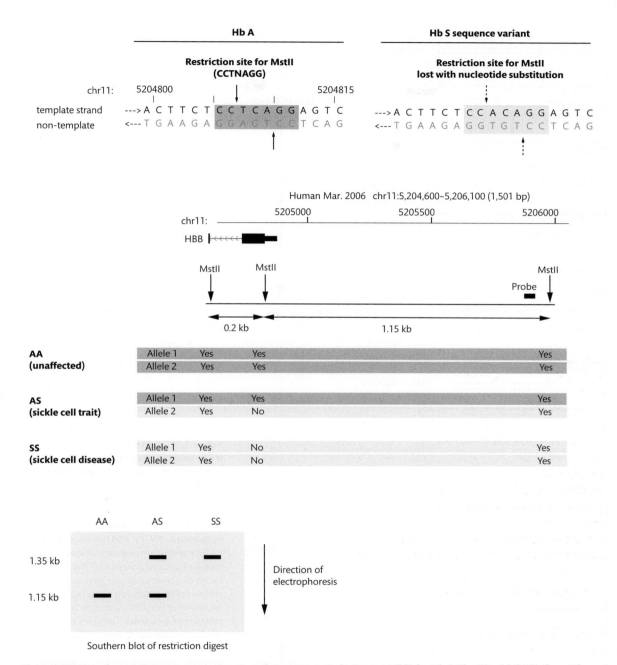

Figure 1.17 Use of restriction enzymes to genotype the sequence variant responsible for Hb S. The T to A substitution at the DNA level results in loss of the restriction enzyme recognition site for *MstII*. Restriction enzymes are highly specific such that they cut particular nucleic acid sequences: in the case of *MstII*, 5′-CCTNAGG-3′ where N is any nucleotide. In the presence of the A to T substitution the recognition site is lost (CCTGTGG). Restriction enzymes are derived from bacteria, for example *MstII* originates from *Microcoleus* species. Southern blotting using a probe hybridizing to DNA sequence 5′ to *HBB* gene results in a Southern blot where particular sizes of fragments allow individuals with sickle cell trait or sickle cell disease to be identified versus unaffected individuals.

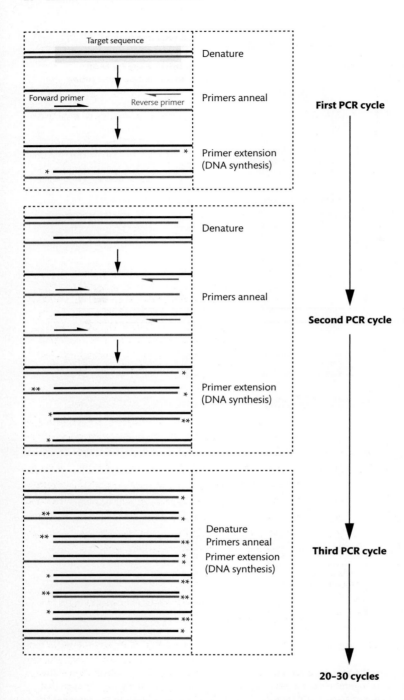

First PCR cycle

Second PCR cycle

Third PCR cycle

20–30 cycles

Figure 1.18 Polymerase chain reaction (PCR) amplification. Exponential amplification of a target DNA sequence can be achieved by using a pair of forward and reverse primers complementary to the region of interest, a heat-resistant DNA polymerase enzyme such as Taq polymerase, dNTPs, and specific buffer including divalent and monovalent cations (magnesium and potassium). Repeated cycles of denaturation, primer annealing, and template extension allow rapid and exponential amplification; typically 20–30 cycles are performed. * denotes newly synthesized strand; ** denotes a new strand of delimited length (by locations of primers). Redrawn from Young (2005), by permission of Oxford University Press.

Escherichia coli DNA polymerase I enzyme and variants detected by restriction enzyme digestion using *Dde*I (Saiki *et al.* 1985). This paper demonstrated how exponential amplification of DNA was achieved (some 220 000-fold increase after 20 cycles of PCR) and that 200 times less DNA was required: 20 nanograms versus two micrograms for Southern blotting.

As an alternative to restriction enzyme digestion, allele-specific detection was developed in which an **oligonucleotide** probe with or without the sequence

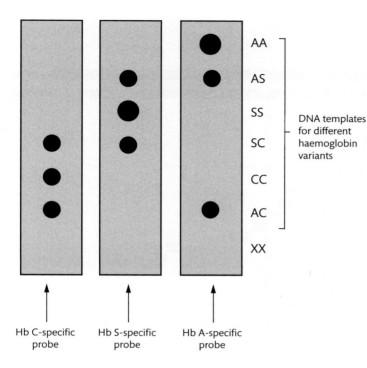

DNA templates
for different
haemoglobin
variants

AA
AS
SS
SC
CC
AC
XX

Hb C-specific
probe

Hb S-specific
probe

Hb A-specific
probe

Figure 1.19 Detection of β globin sequence variants by allele-specific oligonucleotide (ASO) hybridization. PCR amplified DNA from clinical blood samples of differing β globin type was detected by ASO hybridization using radiolabelled probes. Reprinted by permission from Macmillan Publishers Ltd: Nature (Saiki *et al.* 1986), copyright 1986.

variant could be labelled and hybridized to the PCR amplified DNA: under specific conditions, the single base pair mismatch was sufficient to prevent probe binding. The approach was applied to sickle cell and other β globin sequence variants. Allele-specific oligonucleotide (ASO) hybridization ('dot blot') allowed detection from as little as one nanogram genomic DNA and was highly specific (Fig. 1.19) (Saiki *et al.* 1986). DNA-based diagnosis can be particularly important for sickle cell screening during pregnancy with antenatal diagnosis using PCR-based methodologies including restriction enzyme digestion, ASO dot blot hybridization, and amplification refractory mutation system (ARMS) (a further allele-specific approach); these may involve analysing samples from amniocentesis or chorionic villus sampling (Old 2007).

1.3 Genetic diversity involving the globin genes

The variant responsible for Hb S is only one of a remarkable number of differences in DNA sequence identified within *HBB* (Fig. 1.20) and move broadly across the α globin and β globin gene clusters on chromosomes 16 and 11 (see Fig. 1.1). Such variation underlies not only structural variants of haemoglobin but also differences in globin synthesis leading to thalassaemia. Analysis of genetic diversity in the globin loci has served to define many of the different classes of variation now recognized across the genome, from single nucleotide changes to major structural and chromosomal rearrangements.

1.3.1 Structural variants of haemoglobin and the thalassaemias

Structural variants of haemoglobin are incredibly diverse with over 980 described (http://globin.cse.psu.edu/hbvar/menu.html) (Giardine *et al.* 2007). Mainly they arise due to single amino acid substitutions in the globin protein sequence, with a few examples of lengthening or shortening of the globin chain (Fig. 1.21) (Weatherall 2001; Old 2006). Hb S is an example of a structural haemoglobin variant common in sub-Saharan Africa, the Mediterranean, Middle East, and India. Other variants

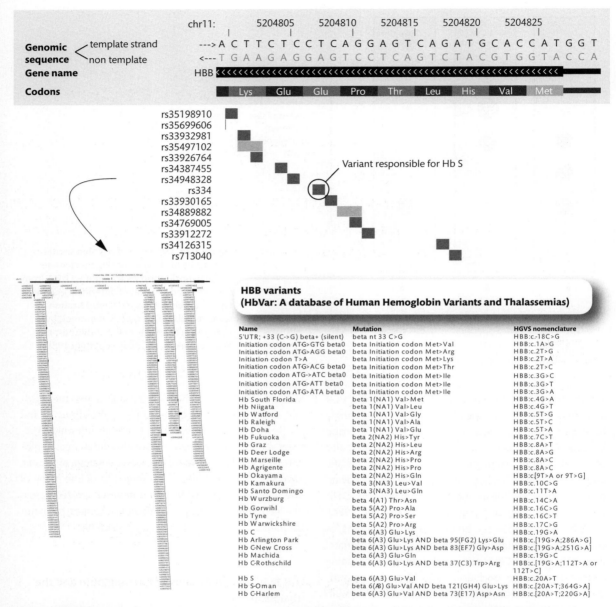

Figure 1.20 Sequence variation in *HBB*. The DNA sequence of *HBB* has been found to be remarkably diverse, with many nonsynonymous variants resulting in amino acid changes. Single amino acid substitutions have been reported in at least 138 out of 146 codons in β globin with over 335 different single nucleotide changes as well as concurrent variants, insertions, and deletions (Huisman *et al.* 1996). The portion of the figure showing sequences, genomic location, and SNPs was adapted from a screenshot of the UCSC Genome Browser (Kent *et al.* 2002) (http://genome.ucsc.edu/) (Human March 2006 Assembly). Screenshot of the Database of Hemoglobin Variants and Thalassemias (http://globin.bx.psu.edu/hbvar/menu.html) (Patrinos *et al.* 2004), reprinted with permission.

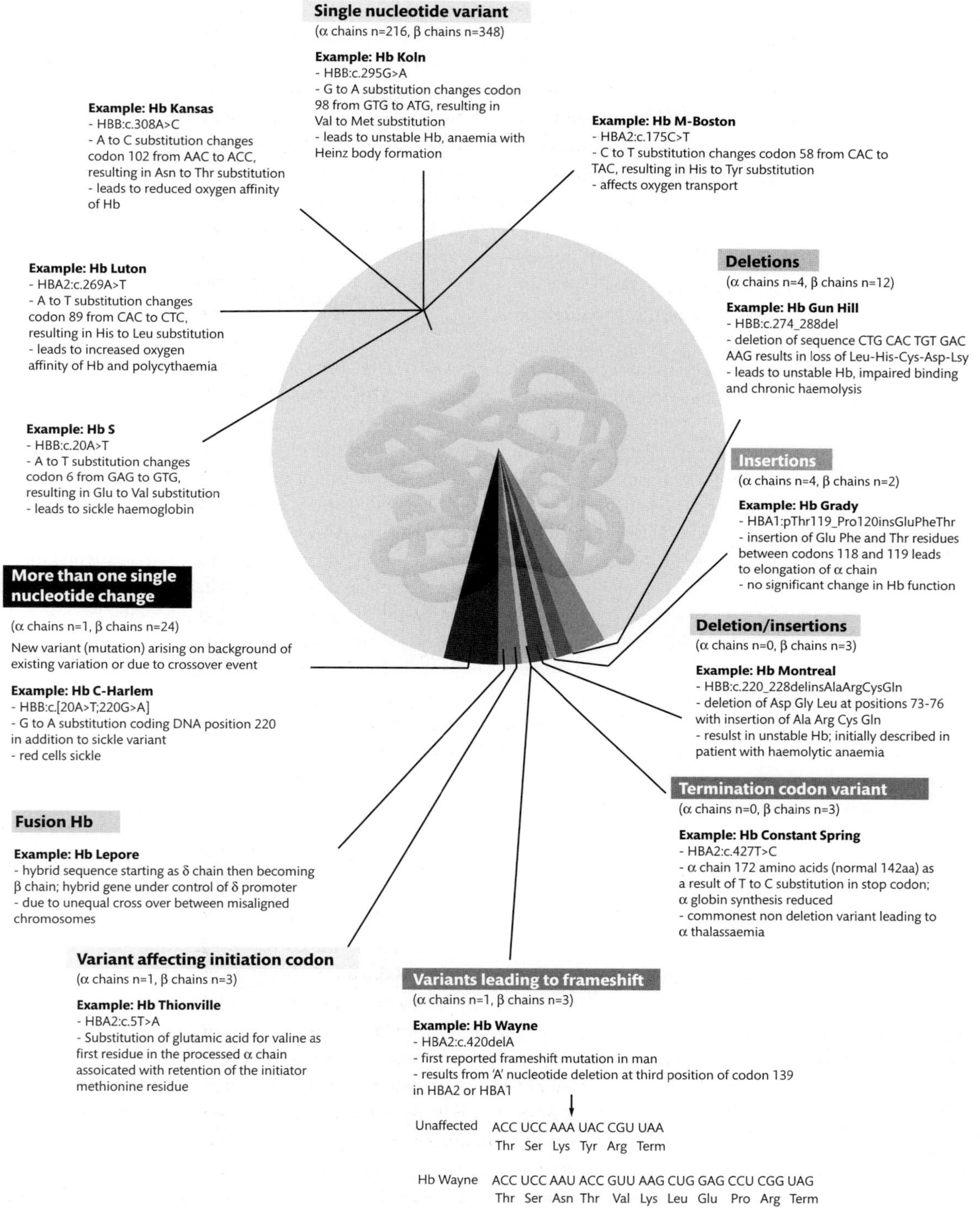

Single nucleotide variant
(α chains n=216, β chains n=348)

Example: Hb Koln
- HBB:c.295G>A
- G to A substitution changes codon 98 from GTG to ATG, resulting in Val to Met substitution
- leads to unstable Hb, anaemia with Heinz body formation

Example: Hb Kansas
- HBB:c.308A>C
- A to C substitution changes codon 102 from AAC to ACC, resulting in Asn to Thr substitution
- leads to reduced oxygen affinity of Hb

Example: Hb M-Boston
- HBA2:c.175C>T
- C to T substitution changes codon 58 from CAC to TAC, resulting in His to Tyr substitution
- affects oxygen transport

Example: Hb Luton
- HBA2:c.269A>T
- A to T substitution changes codon 89 from CAC to CTC, resulting in His to Leu substitution
- leads to increased oxygen affinity of Hb and polycythaemia

Deletions
(α chains n=4, β chains n=12)

Example: Hb Gun Hill
- HBB:c.274_288del
- deletion of sequence CTG CAC TGT GAC AAG results in loss of Leu-His-Cys-Asp-Lsy
- leads to unstable Hb, impaired binding and chronic haemolysis

Example: Hb S
- HBB:c.20A>T
- A to T substitution changes codon 6 from GAG to GTG, resulting in Glu to Val substitution
- leads to sickle haemoglobin

Insertions
(α chains n=4, β chains n=2)

Example: Hb Grady
- HBA1:pThr119_Pro120insGluPheThr
- insertion of Glu Phe and Thr residues between codons 118 and 119 leads to elongation of α chain
- no significant change in Hb function

More than one single nucleotide change

(α chains n=1, β chains n=24)

New variant (mutation) arising on background of existing variation or due to crossover event

Example: Hb C-Harlem
- HBB:c.[20A>T;220G>A]
- G to A substitution coding DNA position 220 in addition to sickle variant
- red cells sickle

Deletion/insertions
(α chains n=0, β chains n=3)

Example: Hb Montreal
- HBB:c.220_228delinsAlaArgCysGln
- deletion of Asp Gly Leu at positions 73-76 with insertion of Ala Arg Cys Gln
- resulst in unstable Hb; initially described in patient with haemolytic anaemia

Fusion Hb

Example: Hb Lepore
- hybrid sequence starting as δ chain then becoming β chain; hybrid gene under control of δ promoter
- due to unequal cross over between misaligned chromosomes

Termination codon variant
(α chains n=0, β chains n=3)

Example: Hb Constant Spring
- HBA2:c.427T>C
- α chain 172 amino acids (normal 142aa) as a result of T to C substitution in stop codon; α globin synthesis reduced
- commonest non deletion variant leading to α thalassaemia

Variant affecting initiation codon
(α chains n=1, β chains n=3)

Example: Hb Thionville
- HBA2:c.5T>A
- Substitution of glutamic acid for valine as first residue in the processed α chain assoicated with retention of the initiator methionine residue

Variants leading to frameshift
(α chains n=1, β chains n=3)

Example: Hb Wayne
- HBA2:c.420delA
- first reported frameshift mutation in man
- results from 'A' nucleotide deletion at third position of codon 139 in HBA2 or HBA1

Unaffected ACC UCC AAA UAC CGU UAA
            Thr Ser Lys Tyr Arg Term

Hb Wayne ACC UCC AAU ACC GUU AAG CUG GAG CCU CGG UAG
          Thr Ser Asn Thr Val Lys Leu Glu Pro Arg Term

Figure 1.21 Overview of structural variants of haemoglobin and their molecular basis at the DNA level. Numbers of haemoglobin variants known for α and β chains are shown on the central pie chart and under major headings classifying types of variation. The vast majority are single nucleotide changes. Numbers of variants and selected examples derived from Old (2006).

such as Hb C (common in West Africa and areas of the Mediterranean) and Hb E (India, South East Asia) were named on the basis of letters of the alphabet, while more recently described variants have names derived from the place of discovery (Hb Leidin, Hb Lepore).

An online database of sequence variants leading to haemoglobin variants, thalassaemias, and haemoglobinopathies 'Human Hemoglobin Variants and Thalassemias' is available (http://globin.bx.psu.edu/hbvar) (Giardine *et al.* 2007). In most cases haemoglobin variants are not associated with disease, however some can affect the normal functioning of the haemoglobin molecule in a variety of ways (Fig. 1.21). The variant haemoglobin may be unstable, precipitating within the red blood cells and leading to haemolytic anaemia as seen with Hb Koln (Carrell *et al.* 1966), or oxygen affinity may be reduced (Hb Kansas) (Reissmann *et al.* 1961) or increased (Hb Luton) (Williamson *et al.* 1992). Major physiological changes can result from single amino acid substitutions, as seen with Hb M Boston in which a critical amino acid involved in haem binding within a pocket created by the globin chain is changed from histidine to tyrosine due to a C to T nucleotide substitution (HBA2:c.175C>T). The iron becomes locked in an oxidized state and unable to participate in oxygen transport (Gerald *et al.* 1957; Pulsinelli *et al.* 1973). Clinically this is manifest as methaemoglobinaemia and cyanosis. While the examples discussed to date result from **single nucleotide substitutions**, more rarely other DNA sequence diversity can lead to structural haemoglobin variants. These include multiple single nucleotide changes (Hb C-Harlem), nucleotide deletions (Hb Gun Hill) or insertions (Hb Grady), or fusion events combining different globin chains (for example Hb Lepore) (Fig. 1.21).

Thalassaemias result from a defect or imbalance in synthesis of one or more of the globin molecules (Weatherall 2001, 2004a, 2004b; Weatherall and Clegg 2001). The mechanisms whereby changes in DNA may result in altered globin synthesis and thalassaemia are diverse, involving the processes of transcription, RNA processing and translation. α thalassaemia can arise from a variety of different genetic causes, ranging from single base changes in the DNA sequence of genes encoding α globin (*HBA1* and *HBA2*) to small

and large scale deletions; over 80 such events have been identified with large deletions most common (Box 1.15). Single base changes (point mutations) and small deletions or insertions most commonly account for β thalassaemia, with over 200 different DNA sequence variants (mutations) identified in the β globin gene *HBB*. In terms of global health, β thalassaemia represents the greater burden of disease and public health challenge (Box 1.16).

1.3.2 HBB sequence diversity and sickle cell disease

Sequence diversity in the β globin cluster plays an important part in the observed clinical heterogeneity in sickle cell disease for affected individuals when an individual inherits the Hb S variant on one allele and a different β globin variant on the other. Thus while possessing two copies of Hb S is associated with the most severe disease, having a copy of Hb S combined with a chromosome with a non-functional *HBB* gene (β⁰ thalassaemia) may be of comparable severity. Much less severe disease is associated with having a chromosome encoding Hb S together with a chromosome bearing a partially active *HBB* gene (β⁺ thalassaemia) although this varies between geographical regions, being more severe for example in variants of Mediterranean rather than African origin (Serjeant *et al.* 1973, 1979).

Hb C is a structural haemoglobin variant resulting from a nucleotide substitution at a position adjacent to that leading to Hb S. In Hb S, an A to T substitution (rs334, HBB:c.20A>T) changes the codon from GAG to GUG and the resulting amino acid residue from glutamic acid to valine; in Hb C, a G to A substitution at chr 11:5204809 (rs33930165, HBB:c.19G>A) alters the same codon but from GAG to AAG, resulting in a glutamic acid to lysine change (HBB Glu6Lys) (Itano and Neel 1950). Having both Hb S and Hb C (so-called sickle haemoglobin C disease) may result in relatively mild disease (Platt *et al.* 1994) and, indeed, Hb C is thought not to be associated with adverse consequences but rather to be a recently arising variant which significantly protects against malaria and is rising to high allele frequencies in some African populations (Section 13.2.3) (Agarwal *et al.* 2000; Modiano *et al.* 2001).

Box 1.15   Alpha thalassemia (OMIM 141800)*

The phenotype associated with α thalassemia variants varies from being clinically silent to fatal before birth. In unaffected individuals there are four copies of the α globin genes (αα/αα), one copy of *HBA1* on each of two chromosomes, similarly *HBA2*. For individuals with loss (deletion) of one copy of a single α globin gene (αα/α-) the result is clinically silent: the blood picture and function is normal but such individuals are carriers of the variant with potential consequences for their children. Individuals with loss (deletion) of two copies (α-/α- or αα/- -) are usually asymptomatic and detected on routine blood testing with mild anaemia and red cells appearing small. Severe anaemia results from loss of three copies of α globin genes (α-/- -) in which Hb H disease is seen: in affected individuals red cells appear small and irregular, and are prone to haemolysis; the spleen is enlarged. The lack of α globin chains mean that the excess β globin chains are associated with each other as a tetramer (Hb H) ($β_4$), the molecule cannot function and accelerates red cell destruction through damage to the cell membrane. When all four copies of the α globin genes are lost (- -/- -) affected individuals usually die before birth with a disorder called Hb Bart's hydrops fetalis. Here characteristic Hb molecules are seen in which groups of four γ chains associate together. The high incidence of α thalassemia in malarial regions relates to a selective advantage seen for carriers of the variant (Section 13.2.2). In parts of South East Asia, the carrier frequency is up to one in five individuals.

* Over the course of this book, reference identifiers for diseases and variants will be given relating to Online Mendelian Inheritance in Man at www.ncbi.nlm.nih.gov/omim/. For α thalassemia see OMIM 141800.

Box 1.16   Beta thalassemia (OMIM 141900)

The diverse array of sequence variants causing β thalassemia either act to reduce *HBB* gene expression ($β^+$ type) or suppress it completely ($β^0$). A variety of phenotypes are observed with classification into β thalassemia minor, intermedia, and major. Individuals with β thalassemia minor (trait) have one normal copy of *HBB* and one affected ($β^0$ or $β^+$). Such people are usually diagnosed on routine testing with reduced red cell indices. With β thalassemia intermedia (often $β^+/β^+$), patients have a significant dyserthyropoietic anaemia as the excess of α globin chains impair red cell maturation, while those red cells that do manage to enter the blood stream are removed early by the spleen. Increased susceptibility to infection, iron overload, bone deformities and fractures, and enlargement of the spleen are seen. In β thalassemia major ($β^+/β^0$ or $β^0/β^0$), disease usually presents around six months of age when β globin chains would normally replace γ globin, often with increased susceptibility to infection or failure of infants to thrive. Patients are dependent on blood transfusions and require chelation therapy with agents such as desferroxamine to avoid iron overload; bone marrow transplantation may be curative but is infrequently available. β thalassemia is common in the Mediterranean with carrier frequencies of up to 20% in some areas. The disease is also common in North Africa, the Middle East, India, and South East Asia. The high allele frequency is associated with a selective advantage seen with protection from malarial infection (Section 13.2.2).

1.3.3 Transitions versus transversions

Nucleotide substitutions comprise either **transversions** or **transitions** (Freese 1959). The A to T substitution responsible for Hb S is an example of a transversion: transversions involve substitutions of **pyrimidines** for **purines**, or purines for pyrimidines (A↔C, A↔T, C↔G, G↔T). In contrast, transitions involve substitution of a pyrimidine for a pyrimidine (C for T and vice versa, C↔T), or a purine for a purine (A↔G). Based on the number of possible combinations, a ratio of 2 : 1 in the occurrence of transversions versus transitions would be expected for randomly occurring substitutions. However, transition nucleotide substitutions predominate across a range of species. A number of reasons have been proposed for the excess of transition over transversions (Strachan and Read 2004). Transition from C to T is more common than expected: this is associated with the high rate of methylation of cytosine residues, the 5-methylcytosines are in turn susceptible to spontaneous deamination to thymine (Box 1.17). In coding DNA, transitions are usually better tolerated than transversions. There is also differential repair of mispaired bases by proofreading activities of DNA polymerases which may favour transitions over transversions.

Box 1.17   CpG dinucleotides and C to T transitions

CG dinucleotides are denoted as 'CpG', this refers to cytosine followed by guanine separated by a linking phosphate, and avoids confusion with cytosine in a base pair with guanine. CpG dinucleotides are much rarer than expected in the genome (only one-fifth the expected frequency). This relates to the fact that 60–90% of CpGs are methylated at the 5′ position in the cytosine ring (5-methylcytosines) (Fig. 1.22) (Bird 1986). This methylated dinucleotide is hypermutable because of the failure of DNA repair mechanisms to recognize deamination of the 5-methylcytosines. In 1990, Cooper and Krawczak noted the excess of point mutations responsible for human genetic disease which comprised CG to TG and CG to CA transitions; these comprised 32% of disease-causing point mutations, some 12-fold greater than expected (Cooper and Krawczak 1990).

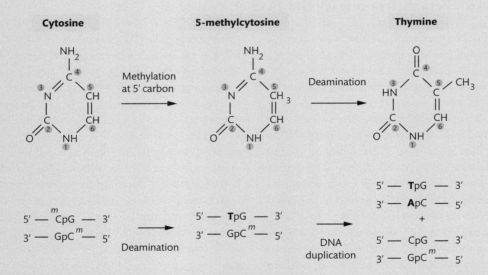

Figure 1.22 CpG dinucleotides have a high rate of mutation. Cytosine methylation and spontaneous deamination to thymine is illustrated. Adapted with permission from Strachan and Read (2004).

1.3.4 Synonymous versus nonsynonymous changes

The single base substitutions in coding DNA that result in Hb S and Hb C are examples of non synonymous changes in which a DNA sequence variant results in a change in the RNA codon and alters the encoded protein. In contrast, a synonymous change results in a new codon but not a change in amino acid; across the genome such events are the most common variants seen in coding DNA. Within *HBB*, an example of a synonymous DNA variant is HBB:c.33C>A (Fig. 1.23A). With this C to A substitution, the third base of the codon is altered (codon changes from GCC to GCA) but the encoded amino acid remains alanine. The position of the base change in the RNA codon is significant: changes in the codon sequence at the third base often do not change the encoded amino acid (for example GCU, GCC, GCA, and GCG all encode alanine). This is the most common site of synonymous changes and is sometimes referred to as 'third base wobble'.

By contrast, changes involving the first or second base often result in structurally dissimilar amino acids. 'Non-degenerate' sites refer to positions where all possible substitutions are non synonymous: substitution rates are very low and such sites are most frequent in the first or second codon base position (Box 1.18). Nonsynonymous sequence variants may be classified into missense or nonsense changes. Missense variants may be either conservative (the amino acid change has minimal consequence for protein function) or non-conservative (a significant structural change). Hb S (HBB:c.20A>T) is an example of a non-conservative missense variant with dramatic consequences for Hb structure and function. By contrast, Hb Graz (HBB:c.8A>T) is an example of a conservative missense variant (Fig. 1.23B). Here the A to T nucleotide substitution changes the codon from CAU to CUU, changing the amino acid from histidine to leucine without significant consequences for the structure or function of the haemoglobin molecule. The variant was identified by accident on cation exchange high performance liquid chromatography in apparently healthy unrelated adults (Liu *et al.* 1992).

A nonsense change involves a DNA sequence variant resulting in a codon change from encoding a particular amino acid to being a stop codon: such events typically have dramatic consequences for gene function and are rare. The first **nonsense mutation** identified in man was in the *HBB* gene (Chang and Kan 1979). Some Chinese patients in whom no β globin synthesis was observed (having β⁰ thalassaemia) were found to have an A to T nucleotide substitution (HBB:c.52A>T) which changed the RNA codon from AAG to UAG (Fig. 1.23C). The consequences were dramatic as instead of encoding a lysine at amino acid 18 (17 if valine is counted as +1), a termination signal is shown and synthesis of the amino acid chain ended prematurely.

There are rare examples of single nucleotide substitutions disrupting the ATG initiation codon leading to loss of gene expression. For example in a number of Sardinian patients with α thalassaemia, a T to C substitution at the start of *HBA2* changes the ATG to ACG (HBA2:c.2T>C) (Pirastu *et al.* 1984). Similarly in *HBA1*,

---

> ### Box 1.18   Codon position and degeneracy
>
> Sites where all possible substitutions are synonymous are also known as 'four-fold degenerate' sites and are relatively common in frequency compared to other coding changes such that their frequency is similar to substitution rates within introns. 'Two-fold degenerate' sites refer to positions where one of three potential base substitutions is synonymous. Non-degenerate, four-fold degenerate, and two-fold degenerate sites make up 65%, 16%, and 19% respectively of the base positions in human codons. The nature of the genetic code promotes conservative missense substitutions as the sequence of codons specifying similar amino acids are themselves similar (for example asparagine (GAU, GAC) and glutamic acid (GAA, GAG)). The 'mutability' of specific amino acids is seen to vary, from highly mutable (for example threonine) to rarely mutated (cysteine).

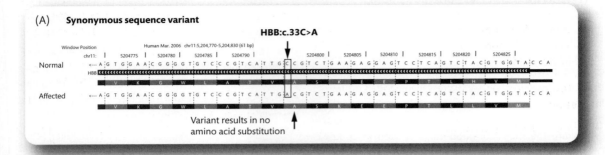

(A) **Synonymous sequence variant**

HBB:c.33C>A

Variant results in no amino acid substitution

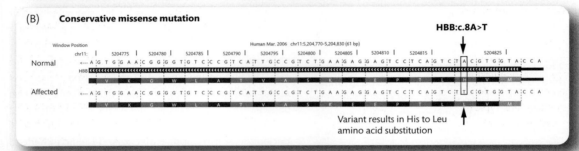

(B) **Conservative missense mutation**

HBB:c.8A>T

Variant results in His to Leu amino acid substitution

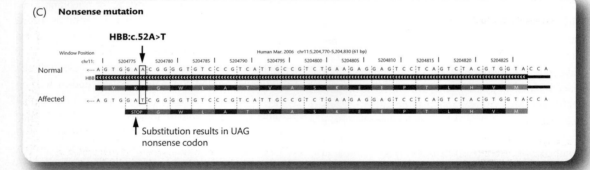

(C) **Nonsense mutation**

HBB:c.52A>T

Substitution results in UAG nonsense codon

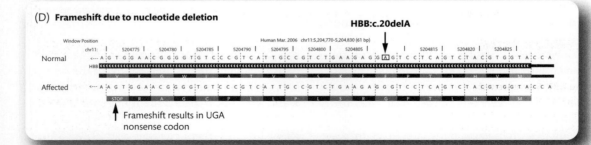

(D) **Frameshift due to nucleotide deletion**

HBB:c.20delA

Frameshift results in UGA nonsense codon

Figure 1.23 *Continued*

an A to G substitution was found to change ATG to GTG (HBA1:c.1A>G) (Moi *et al.* 1987). The phenotypic consequences of the former were more severe due to the more prominent contribution of *HBA2* versus *HBA1* to normal α globin expression.

1.3.5 Insertions or deletions may result in frameshift events

Insertions or deletions in coding DNA typically have dramatic consequences due to the nature of the genetic code. Codons are defined from a fixed starting point (the ATG start codon) and are non-overlapping sets of three nucleotides. A nucleotide insertion or deletion will disrupt the 'reading frame' of the codons causing a 'frameshift'. The consequences are carried through the remaining sequence, and may often manifest in a termination sequence being revealed or a stop codon not being read; the phenotypic consequences are typically severe. For example, a single nucleotide deletion in *HBB*, HBB:c.20delA, results in

a frameshift and alters the amino acids encoded downstream of the sequence variant until a nonsense codon UGA is generated at codon 19 (Fig. 1.23D) (Kazazian *et al.* 1983). The result is loss of β globin synthesis and β⁰ thalassemia. Remarkably, the deleted nucleotide is the same one in which a substitution results in Hb S but rather than a nucleotide substitution from GAG to GTG, the nucleotide is deleted resulting in GG (G_G).

Up to this point the focus of this chapter has been on single nucleotide changes but a diverse array of other variants is found at the globin loci and elsewhere in the genome. An early example of fine scale sequence diversity at *HBB* was the recognition of Hb Leiden, which results from deletion of the three nucleotides GAG encoding glutamic acid (Glu) at amino acid position 7 or 8 (counting Met as +1) (De Jong *et al.* 1968). Here the coding sequence variant, denoted HBB:c.22_24delGAG, results in loss of glutamic acid from the amino acid chain which in combination with β⁰ thalassemia may lead to severe

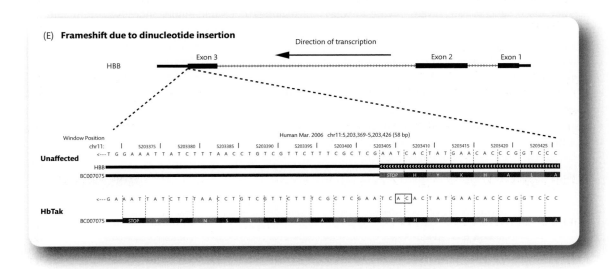

Figure 1.23 Sequence variation involving coding DNA. **(A)** An example of a synonymous DNA variant: C to A nucleotide substitution at ch11:5204795 (HBB:c.33C>A) (rs35799536) in *HBB* exon 1 coding sequence. **(B)** An example of a conservative missense mutation: Hb Graz results from an A to T substitution (HBB:c.8A>T) (OMIM 141900.0429; HbVar ID 222). **(C)** An example of a nonsense mutation: A to T nucleotide substitution (HBB:c.52A>T) (OMIM 141900.0311; rs33986703; HbVar ID 800). **(D)** An example of a frameshift event due to a nucleotide deletion, HBB:c.20delA (HbVar ID 784). **(E)** Hb Tak results from a CA insertion leading to a frameshift and amino acid chain elongation (modified C-terminal sequence: (147)Thr-Lys-Leu-Ala-Phe-Leu-Leu-Ser-Asn-Phe-(157)Tyr-COOH). Nomenclature used to describe this variant includes HBB:c.441_442, HbVar710 (OMIM 141900.0279, rs33999427, beta 147(+AC). The portion of the figure showing sequences and genomic location is adapted from a screenshot of the UCSC Genome Browser (Kent *et al.* 2002) (http://genome.ucsc.edu/) (Human March 2006 Assembly).

haemolytic anaemia. Other nomenclatures and database references for this deletion include HbVar ID 720, OMIM 141900.0156, and rs34948328.

A two nucleotide insertion (CA) at the end of the *HBB* gene coding sequence was found to result in a frameshift and extension of the β globin chain by 11 amino acids (Fig. 1.23E) (Flatz *et al.* 1971). The CA insertion at codon 146 changes the sequence from CAC TAA to CAC ACT, abolishing the normal stop codon 147 (UAA) and allowing chain elongation until a stop codon is encountered (now at position 157). This elongation event results in Hb Tak, which is associated with increased oxygen affinity and polycythaemia.

1.3.6 Deletions, duplications, and copy number variation

Larger scale deletions may involve all or part of a gene, a gene region, or indeed an entire chromosome. The first demonstration of a gene deletion being responsible for a **monogenic** disorder was in patients with α thalassemia, specifically those patients with severe disease in which no α globin was synthesized (see Box 1.15). In the most severely affected individuals dying of the disease before birth with Hb Barts hydrops fetalis syndrome, the α globin proteins were found to be completely absent (Weatherall *et al.* 1970). A relative imbalance of mRNA for α and β globin proteins was recognized among patients with α and β thalassemia (Housman *et al.* 1973) and subsequent analysis using **complementary DNA** probes showed that among

patients homozygous for α thalassemia, all or large parts of the genes encoding α globin were deleted (Ottolenghi *et al.* 1974; Taylor *et al.* 1974). Loss of different numbers of copies of the α globin genes was found to associate with disease severity; moreover the clinical heterogeneity in α thalassemia was reflected in molecular diversity in terms of specific deletions (Embury *et al.* 1979; Orkin *et al.* 1979). Deletions result from unequal **crossing over** at meiosis (**non-allelic homologous recombination**) involving very similar sequences within the gene cluster (Section 5.2), however the particular regions involved are highly heterogeneous, with a diverse array of deletions reported to cause α thalassemia (Higgs *et al.* 1989).

Defective expression of one copy of an α globin gene (α[+] thalassaemia) may arise from loss of the gene due to deletion, or to a variety of variants (mutations) affecting processing of RNA, translation, or the stability of the resulting protein chain (Higgs *et al.* 1981, 1989; Old 2006). The most frequently occurring deletions involve loss of a 3.7 kb region ($-\alpha^{3.7}$) found among individuals from Africa, the Mediterranean, Middle East, and India; or of a 4.2 kb region ($-\alpha^{4.2}$) found among individuals from South East Asia and the Pacific.

α[0] thalassaemia occurs with loss of both copies and most commonly involves deletions of both α globin genes (Fig. 1.24) (Higgs *et al.* 1989; Old 2006). More rarely there are deletions involving the α globin regulatory regions which may lead to α thalassaemia. In these deletions the α globin genes themselves remain intact (for example αα[RA]) (Section 1.3.9) (Hatton *et al.* 1990).

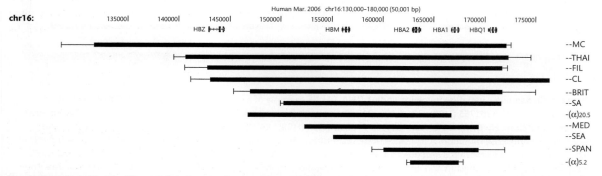

Figure 1.24 Deletions leading to α[0] thalassaemia. A variety of deletions have been shown to result in α[0] thalassaemia (indicated by black bars) for example --[SEA] (found in South East Asia), --[THAI] (Thailand), --[FIL] (Philippines), --[MED] (Mediterranean), and --[BRIT] (England). Regions of uncertainty in breakpoints are shown by error bars. Adapted with permission from Higgs *et al.* (1989).

A variety of different deletions have been identified involving *HBB* in patients with β^0 thalassemia although they account for only a small proportion of variants responsible for β thalassaemia. These range from portions of the *HBB* gene, to the entire gene or gene region. For example, a 0.6 kb deletion at the 3′ end of the gene was identified in Indian patients with β^0 thalassemia (Orkin *et al.* 1979), while a 1.35 kb deleted region at the 5′ end of the gene was found in an African American with Hb S/β^0 thalassemia and his uncle with β^0 thalassemia trait (Padanilam *et al.* 1984). Larger deletions of 4.2 kb involving upstream regions and *HBB* up to the second intron are described (Popovich *et al.* 1986); or involving the entire *HBB* gene in β^0 thalassemia patients of Thai (Sanguansermsri *et al.* 1990) or Dutch descent (Gilman *et al.* 1984).

The α globin gene cluster containing two highly **homologous genes** *HBA1* and *HBA2* also highlights the occurrence of gene duplication. Gene duplication has played a key role in the evolution of the globin gene families from an ancestral globin gene some 800 million years ago with divergence of α and β globin an estimated 450–500 million years ago. The process has involved a complex set of events including gene duplication, inactivation, fusion, and conversion specific to different mammalian lineages (Section 6.4.1) (Ingram 1961; Aguileta *et al.* 2004).

Individuals have been identified with an increased number of α globin genes from the normal complement of four copies ($\alpha\alpha/\alpha\alpha$). In a Welsh family, a father and his two sons were found to have five copies ($\alpha\alpha\alpha/\alpha\alpha$); there was no apparent haematological abnormality or final globin chain imbalance, although a 20% increase at the mRNA level in ratio of α versus β was seen (Higgs *et al.* 1980). Individuals from other ethnic backgrounds have also been reported having one chromosome with the triplicated α globin loci (Goossens *et al.* 1980). This was thought to be due to unequal crossing over, and was anticipated to occur given the occurrence of deletions at the same locus. Quadruplication of α globin has also been reported ($\alpha\alpha\alpha\alpha$) (Gu *et al.* 1987).

Such **structural genomic variation** involving the α globin locus, involving duplication or deletion of the *HBA* genes, provides examples of **copy number variation** (Box 1.19) – an important and common cause of structural variation in the human genome.

1.3.7 Gene fusion

The *HBD* (δ globin) and *HBB* (β globin) genes (Fig 1.1) show a high degree of sequence homology, and misalignment can lead to unequal crossing over at meiosis. Normally *HBD* is found upstream of *HBB* and both are transcribed from the same strand. Unequal crossing over leads to the deletion of approximately 7 kb of sequence between the two genes and results in formation of a *HBD/HBB* fusion gene (Baglioni 1962; Flavell *et al.* 1978). Expression of the encoded haemoglobin, named Hb Lepore (Gerald and Diamond 1958), is controlled by the original *HBD* **promoter** and is thus expressed at relatively low levels leading to the clinical manifestation of thalassaemia.

The precise crossover points between the two genes have been shown to vary with three different Hb Lepore variants described comprising different proportions of the protein encoded by *HBD* and *HBB*. For example, Hb Lepore-Hollandia (OMIM 142000.0021), originally described in patients with β thalassaemia major from Papua New Guinea, comprises amino acids encoded by *HBD* codon 1–22, then amino acids encoded by

---

Box 1.19   Copy number variation

Copy number variation refers to structural variation involving DNA segments larger than 1 kb in which there is relative gain or loss of copy number relative to a reference genomic sequence (Scherer *et al.* 2007). This may involve multiple copies and has been found to occur much more commonly than expected within phenotypically normal individuals as well as being an important cause of disease. The nature and biology of copy number variation is reviewed in Chapter 4.

HBB codon 50 onwards (Barnabas and Muller 1962); Hb Lepore-Baltimore (OMIM 142000.0019), found for example in Brazil and Portugal, derives from *HBD* up to codon 50, then *HBB* from codon 86 (Ostertag and Smith 1969); and Hb Lepore-Boston-Washington (OMIM 142000.0020), found in Italy and Spain, derives from *HBD* to codon 87, then *HBB* from codon 116 (Labie *et al.* 1966). By contrast, Hb Miyada (OMIM 141900.0179) was found to be the result of a fusion whereby the reverse configuration is seen with the fusion gene comprising *HBB* amino acids up to codon 12, then amino acids encoded by *HBD* from codon 22 (Kimura *et al.* 1984).

1.3.8 Sequence variation, RNA splicing, and RNA processing

RNA splicing (Box 1.20) may be modulated by DNA sequence diversity in a number of ways including by changes in regulatory sequences (for example exonic splicing **enhancers**), splice donor and acceptor sequences, or through activation of cryptic splice sites (Section 11.6) (Pagani and Baralle 2004; Wang and Cooper 2007). A notable example of this is seen with a G to A nucleotide substitution at position 1 of the second intron of *HBB* (intervening sequence 2, IVS2) in which the conserved dinucleotide GT at the 5′ splice site is changed to AT (Fig. 1.25A). When such an event was analysed based on DNA sequence from a fetus with β^0 thalassemia, significant changes in alternative splicing were seen with insertion of the initial 47 bp of IVS2 between exons 2 and 3, together with a less abundant RNA in which exon 1 was spliced directly onto exon 3 (Treisman *et al.* 1982).

Hb E is a structural variant of haemoglobin that is very common in South East Asia (with estimates of 30 million carriers in the region) in which there is a G to A substitution (HBB:c.79G>A) resulting in an amino acid substitution, Glu26Lys, due to the change in codon sequence from GAG to AAG (Fig. 1.25B). Hb E was found to be associated with a slight reduction in amounts of β globin synthesis and production of small amounts of an alternatively spliced transcript (Traeger *et al.* 1980; Orkin *et al.* 1982b). The β^E allele had a 'thalassaemic nature': the basis for this lay in alternative splicing as the nucleotide substitution activated a cryptic donor site in exon 1. The intronic sequence IVS1 was seen to be excised more slowly than normal and a small amount of alternatively spliced transcript involving the cryptic donor site was observed (Fig. 1.25B). This issue becomes significant if an individual inherits a β thalassaemia allele together with Hb E when severe transfusion-dependent disease may result.

A further example involving splicing is provided by a pentanucleotide deletion (TGAGG) at the 5′ donor sequence at the start of intron 1 in *HBA2* which was found to result in α thalassaemia due to effects on splicing (Fig. 1.25C) (Orkin *et al.* 1981; Felber *et al.* 1982). This variant significantly disrupted the normal donor sequence, notably the invariant GT sequence, abolishing use of that site and leading to activation of a new donor consensus site 49 nt upstream within exon 1. This results in truncation of the mRNA such that a normal globin chain cannot be encoded; the alternative splice site itself, however, functions well with no unspliced or alternatively spliced RNA apparent.

Other DNA sequence variants leading to α^+ thalassaemia are recognized that involve RNA processing and translation. The sequence AAUAAA on the RNA molecule some 11–30 nt from the terminal poly(A) tract plays an important role in 3′ RNA processing, evidence for which was provided by an A to G nucleotide substitution in *HBA2* (HBA2:c*+94A>G) that changes the DNA

Box 1.20   Splicing

Splicing is the process whereby coding RNA sequences are identified and joined together; alternative splicing produces diversity in this process such that multiple gene products arise from a single coding sequence.

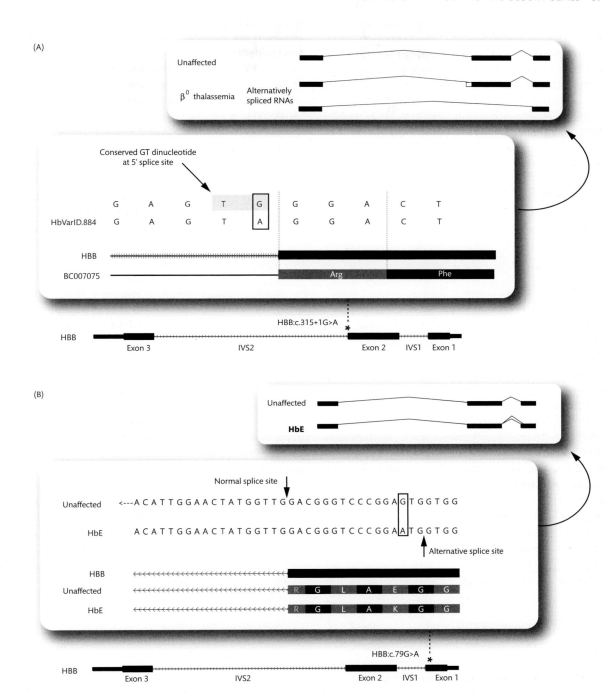

Figure 1.25 *Continued*

sequence from AATAAA to AATAAG (Higgs *et al.* 1983). Accumulation of mRNA was reduced and the transcripts were noted to fail to terminate normally, continuing to 'read through' past the normal poly(A) addition site and revealing decoupling of the normal processes of termination of transcription and RNA 3′ processing (Whitelaw and Proudfoot 1986).

1.3.9 Sequence diversity in noncoding DNA modulating gene expression

A diverse array of regulatory events control the process of transcription ranging from protein–DNA interactions involving the proximal promoter region upstream of the transcriptional start site to more distant regulatory regions. In the *HBB* gene promoter region 'CACCC' box elements have been shown to be important, interacting with specific **transcription factors** such as erythroid Kruppel-like factor (KLF). A proximal CACCC element has been shown to be the site of a number of single nucleotide variants that disrupt the consensus binding site for the interacting proteins leading to reduced gene expression and β⁺ thalassaemia.

An early example was a C to G nucleotide substitution 87 nt upstream of the transcriptional start site (HBB:c.-137C>G) which changed the CACCC box to

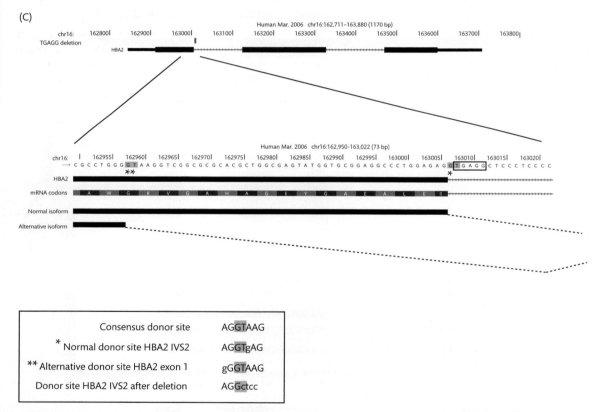

Figure 1.25 Sequence diversity and alternative splicing. (A) A variant resulting in β thalassaemia modulates alternative splicing: the G to A nucleotide substitution at IVS position 1 in *HBB* results in two abnormally spliced RNAs (HbVar ID 884; HBB:c.315+1G>A; OMIM 141900.0348). (B) A structural haemoglobin variant, Hb E, in which a G to A substitution activates a cryptic donor site and both mildly reduces β globin synthesis and results in alternative splicing (HbVar ID 227; HBB:c.79G>A; OMIM 141900.0071). (C) A pentanucleotide deletion modulates splicing of *HBA2*. The normal exonic structure of *HBA2* is shown together with sequence spanning the 3′ half of exon 1 and start of intron 1. A pentanucleotide deletion (TGAGG) (HBA2:c.95+2_95+6deltGAGG) abolishes the normal donor site at the exon/intron boundary and leads to activation of an alternative site within exon 1.

CACGC and was associated with a ten-fold reduction in gene expression (Orkin *et al.* 1982a; Treisman *et al.* 1983). The same nucleotide position was found to be the site of a C to T substitution (HBB:c.-137C>T) which reduced gene expression by half (Kulozik *et al.* 1991); a change in the neighbouring nucleotide from C to T (HBB:C.-138C>T) similarly reduced gene expression (Orkin *et al.* 1984).

Analysis of gene regulation within the globin loci has provided many important fundamental insights into the molecular basis for the control of gene expression, notably the role of more distant upstream regulatory elements (Higgs and Wood 2008b). For example, patients with thalassaemia were identified in whom the structural genes were intact but upstream elements deleted. Hatton and colleagues described a 62 kb deletion (ααRA) linked to an α thalassaemia phenotype which was associated with a marked reduction in α globin gene expression (Fig. 1.26) (Hatton *et al.* 1990). At least 14 similar deletions have been described within the α globin locus of variable size but involving upstream elements, described as 'multispecies conserved sequences' (MCSs) as they are conserved across species; co-localization with erythroid specific **DNase I hypersensitive sites** was also noted (Higgs and Wood 2008b).

DNase I hypersensitive sites are regions of DNA more susceptible to cleavage by the enzyme DNase I due to the more open **chromatin** conformation associated with regulatory elements. One MCS in particular, MCS-R2 (also known as HS-40), was shown to enhance α globin expression on its own; deletion of this site reduced expression to less than 5% of normal, while if all MCSs are deleted there is complete silencing of α globin gene transcription. Notably, deletions or insertions of intervening sequence between the upstream elements and gene promoters did not affect expression. MCSs appear to play a critical role in recruitment of the RNA polymerase enzyme and general transcription factors. Similar effects were noted in the β globin locus when particular upstream DNase I hypersensitive sites were lost due to deletions (Kioussis *et al.* 1983; Driscoll *et al.* 1989).

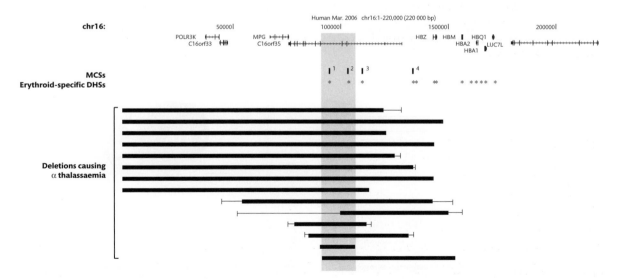

Figure 1.26 Deletions removing regulatory elements leading to α thalassaemia. Schematic of α globin cluster and flanking region showing multispecies conserved sequences (MCSs) and DNase hypersensitive sites (DHSs) together with deletions affecting regulatory elements that result in α thalassaemia. The shortest region of overlap between the deletions is shown shaded and includes MCS-R1 and 2. Adapted with permission from Higgs and Wood (2008b).

Single nucleotide substitutions at a distance from a gene can have dramatic consequences as seen with a T to C substitution at chr16:149,709 (dubbed 'SNP 195') between *HBZ* (ζ) and *HBM* (α^D). De Gobbi and colleagues analysed individuals from Melanesia with α thalassaemia and defined through **resequencing** and systematic functional analysis that this particular variant was associated with gain of function, such that an intergenic region became transcriptionally active (De Gobbi *et al.* 2006). Remarkably, this single nucleotide change led to creation

of a promoter-like element associated with recruitment of the RNA polymerase enzyme Pol II and a new peak of active chromatin (Fig. 1.27). At a sequence level, the T to C nucleotide substitution defined a new GATA-1 binding site. The downstream α globin genes (*HBA2*, *HBA1*) showed reduced gene expression leading to the α thalassaemia phenotype: the new promoter being thought to outcompete the downstream promoters through interacting with the upstream MCS elements first, 'stealing' transcriptional activity.

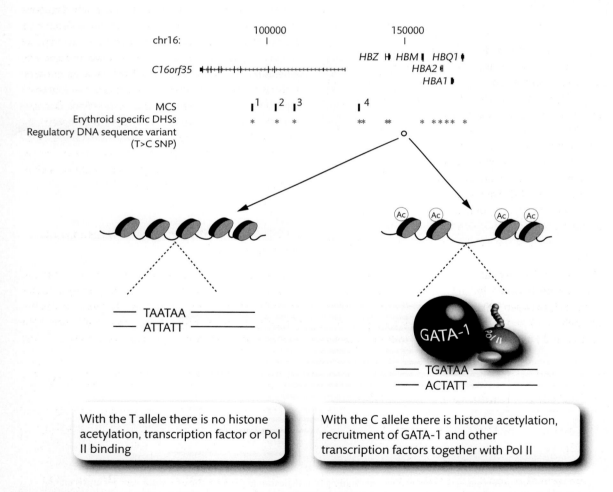

Figure 1.27 A regulatory SNP creates a new promoter. Schematic of α globin cluster and flanking region showing multispecies conserved sequences (MCSs) and DNase hypersensitive sites (DHSs) together with location of regulatory sequence variant (de Gobbi *et al.* 2006).

1.3.10 Tandem repeats

DNA sequence within the α and β globin loci was recognized to vary between individuals irrespective of a particular disease phenotype. Specific DNA sequences recognized by particular restriction enzymes were found to be polymorphic, providing genetic markers for early linkage studies. Repetitive DNA sequences are seen ranging from simple dinucleotide repeats to more complex repeating units spanning hundreds of nucleotides. Repeat units are tandemly arranged 'head to tail' without intervening sequence, constituting '**tandem repeats**' which may be highly polymorphic within and between populations (Box 1.21; reviewed in detail in Chapter 7).

'Microsatellites' comprise arrays less than 100 base pairs (bp) in length made up of simple repeats up to 6 bp in length. Microsatellites are common, constituting an estimated 3% of the human genome with over one million such loci (Lander *et al.* 2001; Ellegren 2004). Sequencing of the terminal two **megabases** (Mb) (1000 kb) of chromosome 16p revealed 322 simple repeats, constituting 1.8% of the sequence and some 35 625 bp (Daniels *et al.* 2001). A variety of microsatellites are recognized in the α and β globin loci, notably dinucleotide repeats such as (AC) n (Poncz *et al.* 1983; Shen *et al.* 1993). Use of microsatellite markers including dinucleotide, trinucleotide, and tetranucleotide repeats allowed **linkage maps** (Section 2.2.2) of chromosome 16 to be established (Kozman *et al.* 1995).

Longer tandem repeats are also seen which can be hypervariable between individuals. Such 'minisatellites' are typically between 100 bp and 20 kb in size with repeat units between 7 and 100 bp. A hypervariable region was recognized 8 kb downstream of *HBA1* in both patients with thalassaemia and unaffected individuals, which comprised between 70 and 450 repeats of a 17 bp sequence 'ACACGGGGGGAACAGCG' (Higgs *et al.* 1981; Jarman *et al.* 1986). Some 100 kb upstream of the α globin gene cluster a further minisatellite is seen in which there are between five and 55 repeats of a 57 bp sequence (Jarman and Higgs 1988). These and other minisatellites, together with restriction enzyme fragment length polymorphisms in the globin loci, were found to be highly variable among diverse populations providing markers for genetic population studies (Higgs *et al.* 1986).

'Satellite' DNA comprises very large arrays spanning hundreds to thousands of kilobases of DNA. Such sequences are commonly encountered across the genome, notably at centromeres (a region visible as a constriction during **metaphase** typically in the middle of chromosomes; see Box 1.4) as well as pericentromeric and telomeric regions (repetitive DNA found at the end of chromosomes seen as simple repeats of the sequence TTAGGG; Box 7.3) where DNA is typically in a compressed transcriptionally inactive state (**heterochromatin**). Satellite DNA is recognized in the terminal region of chromosome 16p but it is in the pericentromeric region of chromosome 16p11 that large tracts of alpha satellite DNA have been found (Martin *et al.* 2004).

1.3.11 Mobile DNA elements and chromosomal rearrangements

Mobile DNA elements are segments of DNA that can transport or duplicate themselves (transpose) to other genomic regions. They are commonly found across the human genome although almost exclusively such DNA segments now represent a genomic 'fossil record' of past

---

Box 1.21   Tandemly repeated DNA

Tandemly repeated DNA sequences in a head to tail configuration occur commonly across the genome and they are typically polymorphic in nature due to expansion or contraction of the number of repeating units. In some cases they show extreme levels of variation allowing useful approaches such as **DNA** **fingerprinting** to uniquely identify a particular individual (Section 7.4.1) (Jeffreys *et al.* 1985). Classification is possible on the basis of increasing size into **microsatellite, minisatellite** and **satellite DNA**.

events with only a very small number of full length elements remaining competent for transposition. The biology of mobile DNA elements is reviewed in detail in Chapter 8 where the major classes, **retrotransposons** (which transpose via an RNA copy) and DNA **transposons** are discussed. Retrotransposons include long interspersed elements (LINEs) and short interspersed elements (SINEs) of which some are autonomous, such as the LINE 1 family, while others such as **Alu elements** are dependent on active L1 elements encoding the proteins required for retrotransposition. Alu elements are characteristic DNA sequences some 300 bp in length which are very common, constituting some 10% of human genomic sequence with an estimated one million copies present. They are polymorphic and have proved highly informative in population genetic analysis and studies of evolutionary history.

Sequencing of the terminal 2 000 000 bp of chromosome 16, which includes the α globin gene cluster, revealed a very gene-rich region with 100 confirmed and 20 predicted genes. The sequence also contained a high proportion of Alu repeats (nearly 20%) with 1442 such elements identified, constituting the majority of the SINEs identified (Daniels *et al.* 2001). LINEs over this region were present less commonly than the genomic average, with 279 described (5% of the sequence); elsewhere within the region, 126 long terminal repeat (LTR) retrotransposons were found (2.5% of sequence).

The high sequence homology of Alu repeats predisposes to unequal homologous recombination leading to a range of chromosomal rearrangements including deletions, duplications, and **translocations**. Such an event was recognized approximately 105 kb from the 16p subtelomeric region leading to an interstitial rearrangement (deletion or translocation) which led to loss of the MCS-R2 (HS-40) upstream regulatory element and α thalassaemia (Flint *et al.* 1996). Homologous recombination due to a crossover event between Alu elements is also recognized, leading for example to a 62 kb deletion and α thalassaemia (denoted ααRA) (Nicholls *et al.* 1987). Indeed, Alu elements are found to be present at many of the breakpoints around the α globin gene locus (Nicholls *et al.* 1987).

1.3.12 Monosomy and trisomy of the terminal end of chromosome 16p

Transfer of chromosomal regions between two non-homologous chromosomes as a result of breakage and reattachment results in chromosomal rearrangements, a structural change described as a translocation (Box 1.22).

Study of the thalassaemias has provided examples of specific translocation events leading to partial monosomy or trisomy. This is illustrated by the case of a 3-year-old child who presented with moderate global developmental delay and a mild hypochromic

---

Box 1.22   Translocations

Translocations may result after breaks involving two chromosomes, for example during meiotic recombination involving mispaired chromosomes, with part of one chromosome becoming detached and reattaching to another non-homologous chromosome. There may be no net gain or loss of genetic material, in which case the translocation is said to be balanced, or there may be, in which case it is unbalanced. Reciprocal translocations between non-homologous chromosomes are relatively common and usually spontaneous leading to

stable balanced exchanges. **Robertsonian translocations** are the most common recurrent type of translocation and specifically involve the acrocentric chromosomes (chromosomes 13, 14, 15, 21, and 22) in which breaks in the very short p arm lead to fusion of the remaining long arms (Section 3.4.4). Most often this involves exchanges between chromosomes 13 and 14, and 14 and 21 – for the individual there is little phenotypic consequence but for their offspring there is a risk of **monosomy** or **trisomy**. Translocations are reviewed in Section 3.4.

microcytic anaemia (Buckle *et al.* 1988). She was found to have α thalassaemia trait with a moderate degree of mental retardation together with mildly dysmorphic features. Her father had no cytogenetic abnormalities detected (his chromosomal complement of 23 chromosome pairs denoted 46,XY), however her mother had a balanced reciprocal translocation involving the long arm of chromosome 10 (10q26.13) and short arm of chromosome 16 (16p13.3). This is denoted t(10,16) (q26.13;p13.3) giving the nomenclature 46,XX,t(10,16) (q26.13;p13.3). For her mother there was no apparent phenotypic consequence as there was no net gain or loss or genetic material. However, the affected child inherited a 'derived' copy of chromosome 16 lacking the terminal region of 16p beyond band 13.3 (denoted 16p13.3→pter) and having in its place 10q26.13→qter, together with a normal copy of chromosome 16 from her father, meaning that she had partial monosomy for the terminal region of 16p beyond 16p13.3 which includes the α globin gene cluster (Fig. 1.28). DNA studies confirmed that the affected child did not inherit the maternal copy of chromosome 10 bearing the terminal region of chromosome 10q; rather she inherited two normal copies of chromosome 10 which meant that in combination with the terminal portion of 10q present on the rearranged chromosome 16 she inherited, the child had partial trisomy for chromosome 10q26.13→qter. This study provided evidence supporting the physical location of the *HBA* genes within the terminal region of 16p13.3→pter.

Other examples involving partial monosomy for the terminal portion of chromosome 16p have been reported. A Nigerian family was described in which the mother had a balanced translocation involving the subtelomeric regions of the short arms of chromosomes 1 and 16 which were cytogenetically invisible but had significant consequences. Her son inherited the derived copy of chromosome 16 leading to partial monosomy and α thalassaemia; his sister inherited a normal copy of chromosome 16 but the derived copy of chromosome 1 (Lamb *et al.* 1989). Both children had some dysmorphic features and mental retardation (Box 1.23).

Trisomy involving the terminal end of chromosome 16p is also seen, for example trisomy distal to 16p12 was found to be present in a 7-week-old infant presenting with a number of congenital malformations including cleft palate, talipes (club foot), and hypospadias (a urinary tract abnormality) (Wainscoat *et al.* 1981). This resulted from a maternal reciprocal translocation involving the short arms of chromosomes 14 and 16 (14p11 and 16p12) denoted t(14;16)(p11;p12). As would be predicted with additional copies of the *HBA* genes being present, an excess of α globin chain production was seen.

The consequences of possession of an abnormal number of chromosomes (**aneuploidy**) or portions of chromosomes (segmental aneusomy) are reviewed in more detail in Chapter 3. Loss of a single chromosome (monosomy) involving autosomal chromosomes

---

Box 1.23   Alpha thalassaemia and mental retardation

The relationship between α thalassaemia and mental retardation was recognized some years earlier in some patients with Hb H disease (Weatherall *et al.* 1981) and is now classified as α thalassaemia/mental retardation, deletion type (ATR16) (OMIM 141750). Wilkie and colleagues reported a series of patients with α thalassaemia, mental retardation, and a range of dysmorphic features – all had deletions involving the terminal end of 16p (16p13.3)

but of variable extent with four cases due to unbalanced translocations (Wilkie *et al.* 1990). Among the *de novo* truncations, one case involved a substantial 2 Mb terminal deletion which was found to be stabilized at the chromosomal breakpoint by the addition of telomeric (TTAGGG)n repeats (Lamb *et al.* 1993), a mechanism found to be present in a series of similar truncations of chromosome 16 involving 16p13.3 (Flint *et al.* 1994).

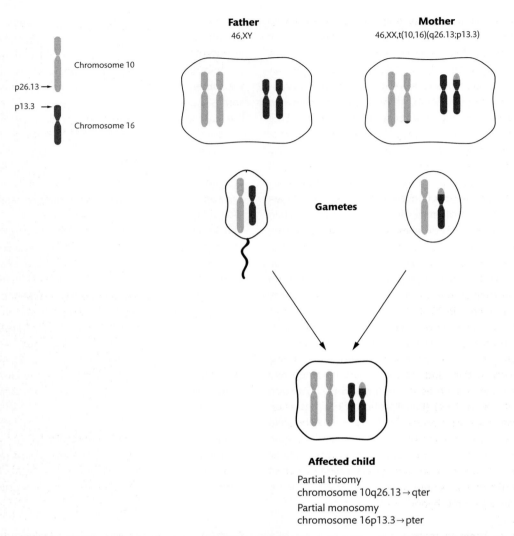

Figure 1.28 Example of translocation leading to partial trisomy and monosomy involving the α globin gene cluster. A balanced translocation affecting a mother between chromosome 16p13.3 and chromosome 10q26.13 leads to a child with partial monosomy of 16p13.3→pter and partial trisomy 10q26.13→qter; the child had α thalassaemia trait (Buckle *et al.* 1988).

is lethal during development and therefore monosomy of chromosome 16 is not seen among live born infants. However, individuals with loss of all or part of an X chromosome may survive, leading to Turner syndrome (Box 3.5). Trisomy of certain chromosomes may be compatible with survival, notably chromosomes 21 (Down syndrome) (Box 3.3), 18 (Edward syndrome), and 13 (Patau syndrome); possession of additional sex chromosomes can lead to specific syndromes such as Kleinfelter (XXY) (Box 3.4).

1.4 Diversity across the genome

1.4.1 Classifying genetic variation

Our journey across the variable landscape of the globin genes has served to highlight many different classes of genetic variation and the functional consequences this may have. Research into human genetic variation has a clear historical context which has been reflected in prevailing views about the relative importance of different forms of diversity that have been observed (Feuk *et al.* 2006a). Early reports involved large 'microscopic' structural genomic changes at a chromosomal level typically at least 3 Mb in size which changed the quantity or structure of chromosomes; such events were relatively rare but often associated with dramatic consequence for the individual concerned in terms of a clear observed phenotype (described in Chapter 3). Subsequently, DNA sequence level variation was highlighted with nucleotide substitutions, deletions, or insertions recognized, notably those that modulated coding DNA sequences to change the structure or function of encoded proteins.

The extent of such diversity, and of repetitive elements, became clearer as more sequencing information was generated with technological advances and increasingly automated high throughout collaborative studies culminating in the publication of the draft human genome sequence in 2001 (Lander *et al.* 2001; Venter *et al.* 2001). Appreciation of the extent of 'simple' nucleotide variation has continued apace with sophisticated analytical approaches and comprehensive cataloguing of diversity within and among human populations through projects such as the **International HapMap Project** (Section 9.2.4) (Frazer *et al.* 2007). There has also been a growing appreciation of the importance and frequency of intermediate scale structural genomic variation greater than 1 kb in size, in particular the high frequency of copy number variation through the development of techniques utilizing **microarrays** to perform comparative genomic hybridization (Section 4.2) (Redon *et al.* 2006).

To try to begin to make sense of the richness and nature of human genetic variation, clear definitions and nomenclature are essential together with some form of systematic approach to classification. Over the course of this chapter we have gone from single nucleotide changes through to major structural events involving chromosomal regions. This broadly follows a classification based on size described by Scherer and Lee (Scherer *et al.* 2007) which serves as an important framework within which to understand the different classes of variation (Fig. 1.29).

The coming chapters will serve to illustrate and describe these different classes through detailed discussion of a number of specific examples. In Chapter 3,

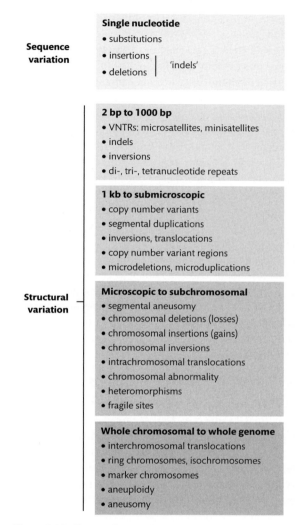

Figure 1.29 Classes of genomic variation. Redrawn and reprinted by permission from Macmillan Publishers Ltd: Nature Genetics (Scherer *et al.* 2007), copyright 2007.

chromosomal level variation is reviewed including gain or loss of whole chromosomes, translocations and chromosome rearrangements, **inversions**, and other structural variation detectable at a microscopic (cytogenetic) level. Submicroscopic structural variation is then discussed in terms of copy number variation among healthy individuals and its role in susceptibility to disease (Chapter 4) before considering pathogenic copy number variation and **genomic disorders** (Chapter 5). **Segmental duplications** are reviewed in Chapter 6 with diverse implications for our understanding of population genetics, evolution, and disease risk. Tandemly repeated DNA is reviewed in Chapter 7, including satellite, minisatellite, and microsatellite repeats. Mobile DNA elements are discussed in Chapter 8 before a review of sequence level diversity including single nucleotide polymorphisms in Chapter 9. Further examples of these and other variants are then considered in the remaining chapters of the book focused on evidence of selection (Chapter 10), effects on gene expression (Chapter 11), and specific genomic regions (Chapter 12) or diseases (Chapters 13 and 14). Before concluding this chapter, this seems an appropriate point to conclude a story begun earlier when DNA sequencing technologies were introduced (Box 1.10). The fruits of such work were to enable the remarkable feat of sequencing the human genome, and in so doing to provide a reference sequence for comparing and annotating genetic variation, as well as uncovering many new sequence variants which served to highlight the extent of diversity within our genomes.

1.4.2 Sequencing the human genome

In 2001 draft sequences for the human genome were published (Lander *et al.* 2001; Venter *et al.* 2001). We now have a detailed route map of our genome which is publically available and finished to a high degree of accuracy and coverage (IHGSC 2004), although work remains ongoing to close the final gaps in the sequence (Cole *et al.* 2008). In terms of understanding human genetic variation, such studies have been of fundamental importance. They established a reference human genome sequence to which sequence variants can be mapped and compared, most recently updated as the February 2009 human reference sequence (GRCh37) produced by the Genome Reference Consortium, (http://genome. ucsc.edu/cgi-bin/hgGateway?org=Human&db=hg19). However, this is not the sequence of a single individual but rather a composite of DNA sequence derived from different people reflecting the hierarchical approach to mapping and sequencing used in the **Human Genome Project**.

The Human Genome Project was launched in 1990, with publication of the draft sequence in 2001 (Lander *et al.* 2001). It was the result of an international consortium involving 20 scientific laboratories from the United States, United Kingdom, Japan, France, Germany, and China; a remarkable collaborative effort that resulted in the successful sequencing of the human genome, the first vertebrate genome to be extensively sequenced. Over the course of the project sequence data were publically available and updated daily. In the same year a draft sequence was published by Celera Genomics, a private company (Venter *et al.* 2001).

Both sequences were based on **'shotgun' sequencing** using the Sanger dideoxy method (Box 1.10): 'shotgun' refers to the fact that a library of clones is sequenced prepared from randomly fragmented genomic DNA. Celera adopted a whole genome shotgun sequencing strategy while the International Human Genome Sequencing Consortium used a hierarchical shotgun sequencing approach in which clones containing large inserted fragments of genomic DNA are selected to generate an overlapping set of mapped segments of DNA (Fig. 1.30). This facilitates final assembly of sequence and was felt particularly important given the number of repeats, but carried increased financial cost. A given large insert clone was derived from a single **haplotype** (the combination of genetic markers or alleles found in a specific region of a single chromosome of a given individual); the need to sequence overlaps between clones generated significant amounts of data on sequence diversity given the multiple individuals from whom libraries and hence clones were derived. The introduction of large insert cloning systems such as **bacterial artificial chromosomes** (**BACs**) was important in enabling such work (Shizuya *et al.* 1992).

Sequencing of a complete chromosome from yeast in 1992 (Oliver *et al.* 1992) and an extensive sequence from the nematode worm in 1994 (Wilson *et al.* 1994) highlighted the feasibility of large scale sequencing. Pilot projects to assess feasibility and define approaches to sequencing the human genome were completed in 1999, together with some 15% of the human genome sequence. This led on to a 'full scale production' phase: the sequence of the two smallest chromosomes in the genome, chromosomes 22 and 21 were published in 1999 and 2000, respectively (Dunham *et al.* 1999; Hattori *et al.* 2000); the draft sequence for the entire genome was published in 2001 (Lander *et al.* 2001).

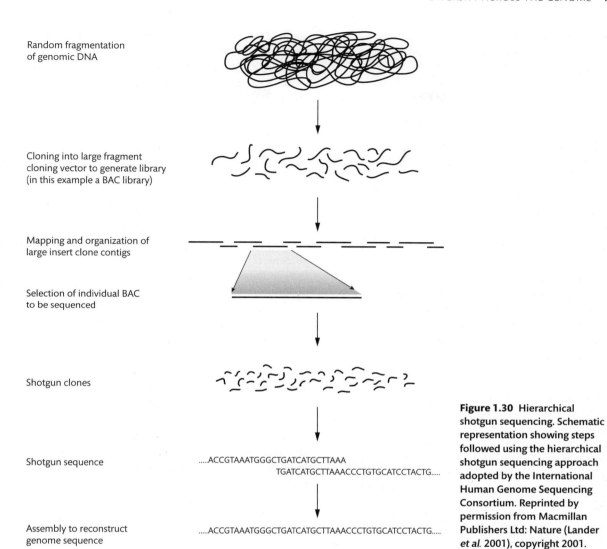

Random fragmentation
of genomic DNA

Cloning into large fragment
cloning vector to generate library
(in this example a BAC library)

Mapping and organization of
large insert clone contigs

Selection of individual BAC
to be sequenced

Shotgun clones

Shotgun sequence

.....ACCGTAAATGGGCTGATCATGCTTAAA
TGATCATGCTTAAACCCTGTGCATCCTACTG.....

Assembly to reconstruct
genome sequence

.....ACCGTAAATGGGCTGATCATGCTTAAACCCTGTGCATCCTACTG.....

Figure 1.30 Hierarchical
shotgun sequencing. Schematic
representation showing steps
followed using the hierarchical
shotgun sequencing approach
adopted by the International
Human Genome Sequencing
Consortium. Reprinted by
permission from Macmillan
Publishers Ltd: Nature (Lander
et al. 2001), copyright 2001.

Based on draft human genome sequences, the size of the human genome was estimated at 3.2 Gb (3 2000 000 000 bp). Sequencing efforts had focused on the 2.95 Gb of the genome comprised of **euchromatin** (light-staining regions of chromosomes), in contrast to the regions of heterochromatin rich in repetitive satellite DNA seen as darkly staining chromosomal regions composed of highly condensed, transcriptionally inactive DNA at centromeres, pericentromeres, and **telomeres**. Even

within euchromatin, some 150 000 gaps in the sequence remained with 10% of the sequence omitted (Baltimore 2001). Nearly complete coverage with very high accuracy in sequence was achieved over the subsequent three years with 99% of the euchromatic genome sequenced and an error rate of 1 in 100 000 bases. Data were now available on 2 851 330 913 nucleotides with an estimated 20 000–25 000 protein coding genes in the human genome, thought to be 3.08 Gb in size (IHGSC 2004).

1.4.3 Repetitive DNA sequences are common

At a structural level, repetitive DNA sequences were found to be remarkably common, constituting over half of our DNA sequence with mobile DNA elements making up 45% of the human genomic sequence (Lander *et al.* 2001). Breaking this down into different classes of mobile DNA elements, 20% of the sequence was made up of LINEs (on average 6 kb in length with an estimated 850 000 copies), 13% was made up of SINEs (1 500 000 copies) of which Alu elements were most common (some 10% of human genomic DNA sequence), 8% was LTR retrotransposons (450 000 copies), and 3% DNA transposons (300 000 copies). This varied across the genome: we have seen how the terminal 2 Mb of chromosome 16p was rich in Alu elements (nearly 20% of the sequence) (Section 1.3.11), while at chromosome Xp11 a 525 kb region was noted where 89% of the sequence was transposable elements. By contrast apparently highly conserved and functionally important gene regions such as four homeobox gene clusters (HOXA–D) critical during development were almost completely repeat-free.

In terms of other classes of repeats, an estimated 3% of the human genome involves short repetitive DNA sequences such as micro- and minisatellites with di-nucleotide repeats most common (0.5%) (Lander *et al.* 2001). Of these, 'AC' repeats were most frequent (50%), then 'AT' (35%), 'AG' (15%), and 'GC' (0.1%) which were very rare. Segmental duplications of genomic sequence between 1 and 200 kb in size to another location were common, particularly in pericentromeric and subtelomeric regions. Both inter- and intrachromosomal segmental duplications are seen; the former between non-homologous chromosomes while the latter occur within a chromosome or chromosomal arm. In relative terms, 1.5% of genomic sequence was thought to involve interchromosomal segmental duplications and 2% intrachromosomal dupl ications. Pending a finalized finished sequence, an estimated 5% of genomic sequence was thought to involve large recent segmental duplications.

1.4.4 Whose genome was sequenced?

For the Human Genome Project volunteers of diverse ethnic background were recruited and samples anonymized, with immortalized cell lines established and individual genomic libraries prepared for sequencing. Of the eight main libraries used to generate clones for sequencing, 66.3% of the reference sequence derives from one BAC library; 32.1% of the reference sequence from an additional 41 BAC, P1-derived artificial chromosome (PAC), cosmid, and fosmid libraries; and 1.9% from 706 non-standard clone sources (Feuk *et al.* 2006a). For the draft sequence generated by Celera Genomics, a composite derived from five donors of diverse ethnic backgrounds was generated (Venter *et al.* 2001).

The fact that the DNA that was sequenced originated from a diverse array of individuals allowed a very large amount of fine scale DNA sequence diversity to be defined through analysis of regions of sequenced clone overlap, notably single nucleotide polymorphisms (SNPs) with more than 1.4 million SNPs reported in 2001 as the result of the International Human Genome Sequencing Consortium and SNP Consortium (Sachidanandam *et al.* 2001). The International HapMap Project subsequently led to the establishment of dense maps of the extent and nature of nucleotide diversity within and between diverse human populations (www.hapmap.org/; Section 9.2.4) (Consortium 2003, 2005; Frazer *et al.* 2007). The number of SNPs has continued to grow at a remarkable rate, providing a fundamental resource as genetic markers for population genetic studies such as **genome-wide association studies** to define disease susceptibility for a broad range of important common human diseases (Section 9.3). In 'dbSNP' (www.ncbi.nlm.nih.gov/projects/SNP/) (Fig 1.14) the major public domain SNP database of nucleotide sequence variation, over 18 million SNPs are listed (NCBI dbSNP Build 129, April 2008). The utility, extent, and biology of SNPs are reviewed in Chapters 9 to 11.

1.4.5 Resequencing diploid human genomes

In 2007 the sequence of the diploid genome of one individual was published, that of J. Craig Venter, based on sequencing using Sanger dideoxy methodology (Box 1.10) (Levy *et al.* 2007). This was followed in 2008 by publication of the sequence of James Watson using next generation sequencing technologies (Box 1.24) (Wheeler *et al.* 2008a). This has allowed an approximation of the extent of diversity in a given individual. Comparison with the reference sequence assembly revealed that the Venter sequence had 4.1 million DNA variants, comprising some 12.3 Mb of sequence, of which 1.3 million variants were novel (this high number reflecting the fact that variant discovery by deep resequencing remains at an early stage) (Levy *et al.* 2007). There were 3 213 401 SNPs; 58 823 nucleotide substitutions involving 2–206 bp of sequence; 292 102 heterozygous insertion/deletions (**indels**) (ranging from 1 to 571 bp) and 559 473 homozygous indels (1 to 82 711 bp); 90 inversions; multiple segmental duplications and copy number variable regions (Levy *et al.* 2007). SNPs comprised the majority of individual variations (78%), but only 26% of variable bases, illustrating the importance of structural genomic variation such as **copy number polymorphism**. Given that the diploid genome was sequenced, this analysis also provides an estimate that for two **homologous chromosomes** in a given individual, there was an estimated 99.5% similarity.

Use of next generation sequencing (Margulies *et al.* 2005) provides the potential to radically increase capacity for resequencing as illustrated by the publication of the Watson genome. The extent of diversity was broadly similar to the Venter genome, both being of European ancestry, with 3.3 million SNPs identified on comparison to the reference genome. Comparison of the Venter and Watson genomes showed for example that they differed by 7648 nonsynonymous SNPs. The increased capacity for high throughput sequencing at significantly reduced cost offered by next generation technologies is driving a number of large scale private and publically funded projects that will allow the resequencing of a large number of individuals. The '1000 genomes' project (www.1000genomes.org), for example, is a publically funded international consortium aiming to sequence at least 1000 people of African, Asian, and European ancestry such that rare human genetic variation (less than 1% minor allele frequency) can be extensively catalogued (Siva 2008). The estimated cost of this work was up to $50 million; the drive for cheaper sequencing remains with a challenge to produce the first '$1000' genome (Mardis 2006).

Both the Watson and Venter genomes were also analysed by **microarray-based comparative genome hybridization** (**array CGH**) (Box 4.2), a technique to determine the extent of copy number variation. For the Watson genome, 23 apparent copy number variable regions were seen ranging in size from 26 kb to 1.6 Mb with nine regions showing gains and 14 showing losses in DNA which would be predicted to affect 34 genes (Wheeler *et al.* 2008a). A higher number were reported with the Venter genome with 62 copy number variable regions, 32 losses and 30 gains (Levy *et al.* 2007), although

---

Box 1.24   Next generation sequencing

When a pyrosequencing-based method for sequencing by synthesis was scaled up into a massively parallel system, a remarkable 25 million bases of DNA were reported to be sequenced with 99% accuracy in a 4-hour run (Margulies *et al.* 2005). Here when a nucleotide is added to a DNA strand, pyrophosphate is released. In the presence of firefly luciferase enzyme, this leads to a flash of light tracked by a computer capable of recording hundreds of thousands of synthesizing DNA molecules simultaneously. Reaction volumes are minute, a picolitre scale reaction. Initial fragmentation of the genome is carried out with capture of DNA on beads to allow clonal amplification of individual fragments. Sequencing can be achieved much faster (the Watson genome was reported to have been completed over a 2-month period) and at a fraction of the cost of capillary-based electrophoresis methods: figures quoted were of less than US$1 million, compared with an estimated cost of $100 million for the Venter genome (Wheeler *et al.* 2008a).

this varied depending on the array platform used with limitations in coverage and sensitivity. This makes accurate estimation of the extent of such diversity difficult, particularly in the light of a lack of standard reference for comparison when considering copy number variation (Box 4.3).

1.5 Summary

Research into the molecular genetics of the globin genes has been a remarkable feat of scientific endeavour which has yielded a succession of novel and fundamental insights which underpin much of our current understanding of human genetics. This chapter has highlighted how all major classes of human genetic variation can be found within or involving the globin loci, many with profound functional consequences in terms of structural variants of haemoglobin and the thalassaemias. This story continues to advance with new insights into the complexities of gene regulation and how it may be modulated by genetic variants being uncovered, and new methodologies for such studies being described. As a paradigm for human genetics, and in particular our understanding of the nature and consequences of genetic variation, research involving the globin genes has been in many ways unparalleled while also having profound translational implications for patient care. The availability of specific cell types across developmental stages has allowed context specific effects to be defined and the complexities of regulation uncovered. The clinical importance and biological interest of these gene clusters has meant that many new technologies have been applied first to these regions while in turn this research involving globin has itself been a source of many new approaches.

Looking beyond globin, completion of the sequencing of the euchromatic portion of the human genome has set the stage for ongoing investigations of the nature of human genetic variation. The sequencing of model organisms such as the yeast *Saccharomyces cerevisiae*, the nematode worm *Caenorhabditis elegans* (Consortium 1998), and the fruit fly *Drosophila melanogaster* (Adams *et al.* 2000) allowed fundamental insights to be gained and comparative studies to be initiated (Rubin *et al.* 2000). The more recent sequencing of the mouse (Waterston *et al.* 2002), chimpanzee (Box 10.1) (CSAC 2005), and

rhesus macaque (Box 10.2) (Gibbs *et al.* 2007) genomes has allowed further comparative genomic analyses which have provided dramatic insights into the nature and origins of genetic diversity. Such studies are discussed through the course of this book.

The major classes of genetic variation and their functional consequences are described over the course of Chapters 3 to 9. Before this, Chapter 2 provides a review of approaches used to find genes and specific variation within them associated with phenotypic traits, illustrating many general principles fundamental to our understanding of human genetic diversity and its implications for health and disease.

1.6 Reviews

Reviews of subjects in this chapter can be found in the following publications:

Topic	References
DNA structure and biology; transcription and translation	Lewin 2004; Strachan and Read 2004
DNA sequencing	Hutchison 2007
Sickle cell disease	Ingram 2004; Frenette and Atweh 2007
Thalassaemia	Weatherall, 2001, 2004b, Weatherall and Clegg 2001
α globin gene regulation	Higgs 2004; Higgs and Wood 2008b
Genetic diversity at globin loci	Young 2005; Old 2006
OMIM (www.ncbi.nlm. nih.gov/omim)	Hamosh *et al.* 2005
dbSNP (www.ncbi.nlm. nih.gov/projects/SNP)	Sherry *et al.* 2001
UCSC Genome Browser (genome.UCSC.edu)	Kent *et al.* 2002
Microsatellites	Ellegren 2004
Structural genomic variation	Feuk *et al.* 2006a, Freeman *et al.* 2006; Scherer *et al.* 2007
Splicing	Wang 2007; Kiim 2008
Mobile DNA elements	Kazaziam 2004; Xing 2007

CHAPTER 2

Finding genes and specific genetic variants responsible for disease

2.1 Introduction

In this chapter approaches to defining the genetic basis of disease are explored. Genetic mapping using DNA sequence variants as markers to look for correlation of segregation with phenotype in family pedigrees (linkage analysis) has proved a powerful approach to identify specific genes and particular variants responsible for relatively rare diseases showing a mendelian pattern of inheritance with high **penetrance**. More recently, association studies in populations using many hundreds of thousands of common single nucleotide polymorphisms (SNPs) to perform genome-wide association studies has begun to dissect the genetic contribution to common multifactorial diseases.

Genetic variation has played an essential role as a tool to help localise particular regions of our genome involved in a given disease. Within such loci, specific causative genetic variants have been identified in some monogenic diseases, for example in coding DNA sequence with dramatic functional consequences for the structure or function of the encoded protein. By contrast, much more modest but still significant effects may result from variation in non-coding DNA, changing levels of gene expression. Indeed in multifactorial traits, a number of different genetic variants and several genes may be involved, contributing to a particular phenotype in combination with other important environmental factors. In this chapter some of the basic principles underpinning research to define the role of genetic variation in disease are explored, in particular the role of linkage and **linkage disequilibrium**. A number of examples of specific traits are described

to illustrate approaches which have been used and the interested reader is also referred to some of the many excellent reviews relating to linkage analysis and genetic association studies (Section 2.8). The population based approaches involving genetic association are explored in greater detail in Chapter 9 where the large scale identification and characterisation of SNPs and recent dramatic successes achieved through genome-wide association studies in common disease are reviewed. Other examples of the successful application of linkage and of genetic association are illustrated elsewhere over the course of the book.

2.2 Linkage analysis

2.2.1 Defining linkage

Linkage describes the coinheritance or cosegregation of different genetic variants or alleles between parents and offspring. If alleles at two loci are inherited together more commonly than expected based on independent inheritance, they are said to be in linkage. The reason that alleles on a given parental chromosome do not remain in linkage is because at meiosis (Box 2.1) homologous recombination or crossing over occurs between chromatids derived from homologous chromosomes (Box 2.2). This means that alleles at any two genetic loci present on a particular parental chromosome may become separated rather than inherited together, the further apart they are the greater the chance of recombination occurring.

Linkage analysis allowed the relative positions of genetic markers to be defined based on how they segregate through family **pedigrees**; it also allowed analysis

---

Box 2.1   Meiosis

Meiosis is a specialized form of cell division occurring in reproductive tissues generating sperm and egg cells. The diploid germ cells undergo DNA replication and two successive rounds of cell division. The genetic complement of the resulting **haploid** cells comprise one of each pair of homologous chromosomes and is made genetically unique by the independent assortment of chromosomes originally of paternal or maternal origin, and by the reshuffling of genetic material due to homologous recombination, also described as 'crossover events' (Box 2.2). Independent assortment and recombination mean that a person can produce highly variable and distinct individual gametes.

Box 2.2   Homologous recombination (crossing over)

Early in meiosis the cell contains four copies of each chromosome (duplicate pairs of homologous chromosomes) described at this time as chromatids. Homologous recombination results from the process of breakage of non-sister chromatids (one paternal and one maternal chromatid) of a pair of homologues and the rejoining of the fragments to generate new recombinant strands such that there are equal exchanges between allelic sequences at the same positions within the alleles (Fig. 2.1). Homologous recombination is a critical means of generating genetic diversity.

---

of mendelian characters to map specific disease or trait loci based on segregation within a pedigree of a particular trait and a polymorphic marker (Fig. 2.1). In order to interpret the significance of linkage results, the logarithm of the odds or '**lod score**' is used (Chotai 1984; Dawn Teare and Barrett 2005). This describes the recombination fraction (Box 2.3) between a genetic marker and disease locus in terms of a likelihood ratio, in which the null hypothesis is of no linkage between the marker and disease loci (Box 2.4).

In order to analyse cosegregation of genetic loci in family pedigrees, parametric linkage analysis is used. Calculation of a lod score requires a particular genetic model to be specified for the disease under investigation. In fully penetrant mendelian diseases the model would involve the mode of inheritance of the disease and disease allele frequency; models become more complex with incomplete penetrance. Two-point analysis involves estimating linkage between individual genetic markers and a disease locus; multipoint analysis involves at least two markers and is typically used in a small genomic region to localize a disease locus relative to a fixed map of markers.

Non-parametric linkage analysis is used in multifactorial traits without a clear mode of inheritance and does not rely on a particular disease model. In this situation sibling pairs are often analysed looking for excess sharing of alleles that are **identical by descent** among affected sib pairs more often than would be expected by chance. This approach was used for example to identify linkage to the **major histocompatibility complex** (**MHC**) on chromosome 6p21 and other genomic loci for type 1 diabetes (Davies *et al.* 1994).

2.2.2 Genetic markers

Prior to the availability of genetic markers, a range of other polymorphic characters have been studied including blood group and human leukocyte antigen (HLA) type. Historically association with ABO blood group (OMIM 110300) was defined for a number of diseases including peptic ulcer disease and cancer (Clarke 1959), while for the MHC at chromosome 6p21 the advent of HLA serological testing provided important new evidence of association, notably with autoimmune disease (Chapter 12). However, it was the advent of quantifiable genetic diversity in the form of genetic markers in the 1980s that saw the power of linkage become manifest, in particular when highly polymorphic markers were available in the framework of a genetic map (Botstein *et al.* 1980; Donis-Keller *et al.* 1987; NIH/CEPH 1992; Weissenbach *et al.* 1992; Dib *et al.* 1996; Broman *et al.* 1998).

Initial work based on genetic markers used restriction fragment length polymorphisms (RFLPs), typically single nucleotide differences resulting in the gain or loss of a specific sequence recognized by a restriction enzyme (Section 1.2.4) (Botstein *et al.* 1980). These biallelic markers were increasingly replaced by highly polymorphic minisatellites, then bi-, tri-, or tetranucleotide microsatellites which had the advantage of being relatively easy to genotype and highly informative due to the number of different alleles observed in a population for a given marker (Chapter 7) (Weber and

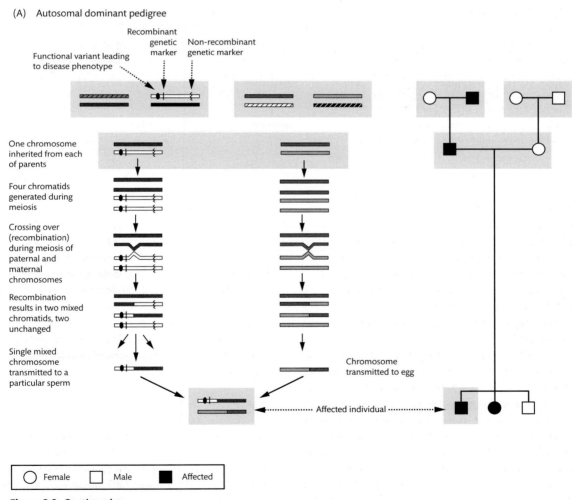

Figure 2.1 *Continued*

(B)   Autosomal recessive pedigree

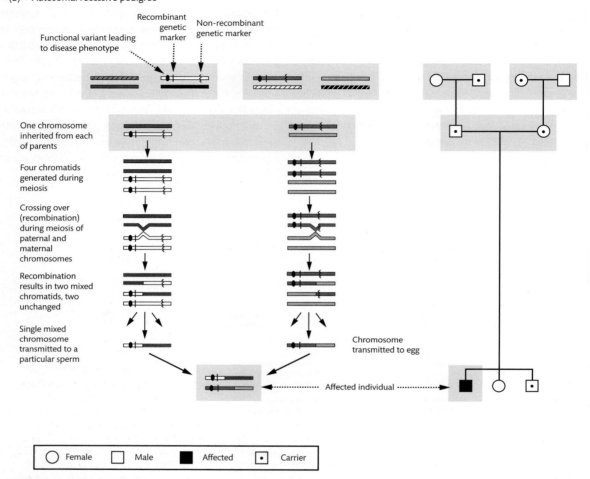

Figure 2.1 *Continued*

May 1989). The construction of linkage maps to assign a genomic location for genetic markers enabled the rapid growth in successful application of linkage analysis to mendelian traits. More recently the availability of dense panels of SNPs, which can be genotyped at ultrahigh throughput, has led to these markers being available for such studies.

Typically sets of polymorphic DNA markers 5–10 cM apart have been used to look for evidence of linkage with denser sets of markers targeted on regions showing linkage (Botstein and Risch 2003). However, the limit

of resolution with linkage mapping is relatively large (1–10 cM) as it is determined by the number of meioses in which crossover events may take place.

2.3   Application of linkage analysis and positional cloning to mendelian diseases

In this section some examples are provided of diseases where linkage analysis has been successfully employed

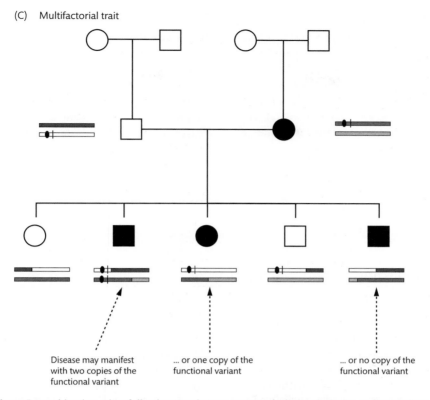

(C)   Multifactorial trait

Disease may manifest
with two copies of the
functional variant

... or one copy of the
functional variant

... or no copy of the
functional variant

Figure 2.1 Linkage. Recombination arises following crossing over events during meiosis. Examples are shown for simple family pedigrees showing autosomal dominant (A) or autosomal recessive (B) inheritance. In the autosomal dominant pedigree, a single copy of the functional variant (mutation) is sufficient to cause disease. Transmission of the functional variant occurs with a closely linked genetic marker; in many linkage analyses polymorphic microsatellites are used as genetic markers. Recombination between chromatids leads to reshuffling of DNA and this can be informative for linkage analysis when a set of different genetic markers are used, the marker showing strongest evidence of linkage (from the lod score) localizing the likely site of the functional variant. In the figure, two genetic markers are shown, one recombinant (solid line, lying close to functional variant) and one non-recombinant (serrated line). The resolution achieved by linkage analysis is, however, modest and typically requires finer scale mapping. (C) In a common multifactorial disease such as asthma, the occurrence of a functional variant is neither sufficient nor necessary for disease to occur but may be an important determinant of disease susceptibility; typically several variants involving different gene loci are involved, each with a relatively modest magnitude of effect. This significantly limits the power of applying a linkage-based analysis.

to define the genetic basis of relatively rare, highly penetrant diseases showing a minimally ambiguous phenotype and clear evidence of familial segregation, in most cases showing dominant, codominant, recessive, or X-linked mendelian pattern of inheritance (Botstein and Risch 2003). Integral to the success of linkage analysis was the use of positional cloning (Box 2.5) to identify specific genes within genomic regions resolved by linkage analysis, and of DNA sequencing among affected

and unaffected individuals to identify specific causal mutations.

The combination of linkage analysis and a positional cloning approach is illustrated by the pioneering studies in cystic fibrosis and Treacher Collins syndrome. Other examples of application to specific diseases are described elsewhere in the book including haemochromatosis (Section 12.6), Duchenne muscular dystrophy (Box 3.7), and Huntington's disease (Box 7.13).

Box 2.3   Recombination fraction and genetic distance

The recombination fraction describes the probability of recombination between two loci at meiosis. Each recombination event or crossing over involves two of the four chromatids such that a single crossover event can lead to only 50% recombinants. Thus even for very widely spaced loci on a chromosome, the maximum recombination fraction is 50%. Genetic distance can be defined based on the probability of a crossover event. One **centimorgan (cM)** describes a region in which a crossover event is expected to occur once in every 100 meioses and approximates in physical distance to 1 million bases (1 Mb), but this varies depending on genomic location as the recombination rate differs between regions of the genome.

Box 2.4   Lod score

The lod score is a measure of linkage between loci. The score uses the observed recombination fraction in order to derive the likelihood ratio in comparison with the null hypothesis of no linkage being present. Mathematically the lod score, $z\theta$, can be defined as log base 10 of the likelihood ratios between observed linkage (with a recombination fraction θ) and that of no linkage ($\theta = 0.5$):

$$z\theta = \log_{10}[p(r;\theta)/p(r;0.5)]$$

where $p(r;\theta)$ is the probability of data r when the true recombination fraction is θ (Chotai 1984; Dawn Teare and Barrett 2005).

A higher positive lod score provides evidence of linkage; when above 3 this is regarded as significant, and equivalent to a P value of less than 0.0001. For genome-wide significance, a threshold of 3.3 is proposed for parametric linkage analysis to give a genome-wide **type I error** (risk of a false positive result or rejecting a true null hypothesis of no linkage) of 0.05 (Dawn Teare and Barrett 2005). A lod score of less than −2 provides significant evidence against linkage.

Box 2.5   Positional cloning

Positional cloning refers to the cloning or identification of a gene based on its chromosomal location rather than identification based on knowledge of the encoded protein. It has also been described as 'reverse genetics' and contrasts with the situation seen with sickle cell anaemia where knowledge of the haemoglobin protein allowed the amino acid change to be defined (Section 1.2), or other instances where knowledge of the amino acid sequence or availability of antibodies to a protein has allowed screening of complementary DNA (cDNA) libraries and identification of a specific gene. cDNA refers to DNA generated by reverse transcriptase from a single strand of mature fully spliced mRNA.

2.3.1 Cystic fibrosis and the delta-F508 mutation

The identification of the role of the *CFTR* gene in cystic fibrosis (Box 2.6) by positional cloning was a landmark in the field, demonstrating the power of this approach (Kerem *et al.* 1989; Riordan *et al.* 1989; Rommens *et al.* 1989). A substantial body of work from many independent groups had established linkage to chromosome 7q31 (Buchwald *et al.* 1989), including a large collaborative study of more than 200 families (Beaudet *et al.* 1986). Cloning the DNA sequence for this linkage region and looking for candidate gene sequences was a very considerable undertaking at the time this work was done. Rommens and colleagues adopted a chromosome walking and jumping approach to assemble a contiguous 280 kb region spanning the putative gene locus (Rommens *et al.* 1989). Chromosome walking was based on overlapping the ends of clone segments, jumping to bypass unclonable regions, then beginning bidirectional walks after each jump.

The limited resources available at the time meant that the identification of possible gene sequences was based on several analyses including cross species hybridization to indicate evolutionary conservation (zoo blots), detection of **CpG islands**, RNA hybridization experiments, isolation of cDNA, and DNA sequencing (Rommens *et al.* 1989). cDNA screening of many different libraries using specific DNA segments as probes allowed the first exon of the putative cystic fibrosis gene to be identified from a cDNA library originating from epithelial cells of a sweat gland. Further screening with additional clones allowed the mRNA of the putative gene to be resolved. Expression of a 6.5 kb transcript was seen in a number of tissues, notably the lung, colon, and sweat glands, which was consistent with tissues known to be involved in the disease process (Riordan *et al.* 1989). The investigators defined a gene spanning some 250 kb with 24 exons based on hybridization to genomic DNA; specific motifs suggested membrane association and adenosine triphosphate (ATP) binding (Riordan *et al.* 1989). This is reflected in the current nomenclature referring to the gene as the cystic fibrosis transmembrane conductor regulator gene, *CFTR*, which encodes an ATP-binding cassette (ABC) transporter protein.

In order to identify associated mutations, the cDNA sequences from cystic fibrosis patients and unaffected individuals were compared. A 3 bp deletion was noted on two clones derived from a cystic fibrosis cDNA library which was predicted to result in loss of the amino acid phenylalanine at position 508 of the encoded polypeptide (Riordan *et al.* 1989). Kerem and colleagues found the deletion was present in 145 out of 214 (68%) cystic fibrosis chromosomes derived from individuals in a general clinic population (Kerem *et al.* 1989). Further studies have confirmed that the 'delta-F508 mutation', a three nucleotide deletion in exon 10 of the *CFTR* gene on chromosome 7q31.2 denoted 'p.F508del', is the most common mutation leading to cystic fibrosis. The deletion results in retention and degradation of the CFTR protein within the endoplasmic reticulum. It is particularly associated with pancreatic insufficiency, which

---

Box 2.6   Cystic fibrosis (OMIM 219700)

Cystic fibrosis is an autosomal recessive disorder in which specific mutations of the *CFTR* gene result in dysfunction of epithelial secretion, obstructing ducts in specific organs such as the lung, liver, and bowel. Cystic fibrosis transmembrane conductor regulator (CFTR) protein functions as a calcium channel and is also involved in regulation of other cellular transport channels. Airway obstruction and chronic bronchopulmonary infection are characteristic of cystic fibrosis, notably persistent infection with *Pseudomonas aeruginosa*. Pancreatic insufficiency, intestinal obstruction, and biliary cirrhosis may occur, together with infertility; all patients have abnormally high levels of sweat electrolytes. Cystic fibrosis is one of the commonest genetic diseases in Caucasian populations affecting one in 2500 live births in the United Kingdom (Dodge *et al.* 2007); it is rare in other populations.

occurs in most patients with cystic fibrosis: pancreatic insufficiency is present in 99% of patients homozygous for p.F508del, 72% of those heterozygous for p.F508del, and 36% of cystic fibrosis patients with other mutations of the gene.

Worldwide, p.F508del accounts for an estimated 66% of cystic fibrosis chromosomes and the frequency of the deletion correlates with the incidence of cystic fibrosis (Bobadilla *et al.* 2002). Analysis of microsatellite makers in European populations suggest a single origin for the deletion, at least 52 000 years ago (Morral *et al.* 1994). Frequencies of the deletion among cases range along a risk gradient increasing from southeast to northwest, with values of 24% in Turkey to 88% in Denmark and 100% in the Faroe Islands. This may relate to the migration of early farmers across Europe carrying the mutation, or to a selective advantage of p.F508del accounting for its persisting high frequency in Caucasian populations. A number of other *CFTR* variants associated with cystic fibrosis are thought to have ancient origins, for example a single nucleotide substitution in exon 11, p.G542X, is associated with sites of occupation by ancient Phoenicians (Loirat *et al.* 1997).

There are currently 1601 *CFTR* mutations listed on the cystic fibrosis mutation database (http://www.genet.sickkids.on.ca/cftr/StatisticsPage.html; date of access 20 August 2008) acting through a number of molecular mechanisms, including mutations resulting in defective protein production, processing, regulation, and conductance (Welsh and Smith 1993).

2.3.2 Treacher Collins syndrome

The steps involved in positional cloning are also well illustrated by work identifying the gene and specific mutations responsible for Treacher Collins syndrome (Box 2.7), a disorder of craniofacial development showing autosomal dominant inheritance. Linkage analysis of 12 affected families found significant linkage to chromosome 5q31-q34 for three out of five RFLP markers with a maximum lod score of 9.1 (Dixon *et al.* 1991) and in an independent study to 5q31.3-q33.3 (Jabs *et al.* 1991). As we saw with cystic fibrosis, the first step in positional cloning involves defining the candidate region and, because at the time this work was done the genetic and physical maps of the genome were limited, a high resolution map of the 5q31-q33 region was prepared by Loftus and colleagues that combined genetic mapping derived from analysis of **CEPH** pedigrees with a physical map from radiation hybrid mapping (Loftus *et al.* 1993). The candidate region was further refined by linkage analysis using more informative polymorphic microsatellite markers to 5q32-q33.1 (Dixon *et al.* 1993).

The next step involved establishing a **contig** of genomic clones across the region of interest, in this case the contig being an ordered arrangement of cloned fragments from yeast artificial chromosomes to collectively contain the underlying DNA sequence for this region (Dixon *et al.* 1994). The current availability of the human genome sequence now allows direct downloading of assembled sequences, but in the early days of positional cloning this was a major obstacle to be overcome.

Box 2.7 Treacher Collins–Franceschetti syndrome (OMIM 154500)

This disease was described in 1900 by E. Treacher Collins and is thought to relate to abnormal development involving the first and second brachial arches, leading to the observed phenotype which may include conductive hearing loss (the external ear, auditory canal, and inner ear ossicles are malformed) and a characteristic facial appearance with downsloping palpebral fissures, hypoplasia of the mandible, and cleft palate. The phenotype is, however, variable and the disease is diagnosed in one in 50 000 live births (Fazen *et al.* 1967). In 60% of cases there is no family history and the disease has arisen as a *de novo* mutation; extended family pedigrees have been described including for example 14 affected individuals over five generations in a family from Kentucky (Rovin *et al.* 1964).

Armed with contig information and additional short tandem repeat markers, this allowed Dixon and colleagues to narrow the candidate region to 840 kb (Dixon *et al.* 1994). A transcriptional map was then generated for the region from exon amplification clones, resolving at least four known and two novel candidate genes in the region (Loftus *et al.* 1996).

The Treacher Collins Syndrome Collaborative Group then successfully completed the positional cloning process. Firstly they used new polymorphic short tandem repeat markers to analyse segregation in families showing recombination involving the critical region to define overlapping recombination events in two unrelated individuals. This indicated that the transcription map needed to be extended with additional cosmids used to screen cDNA libraries. Positive plaques from this screening were cloned and sequenced leading to identification of a novel gene initially named Treacle but now given the nomenclature *TCOF1* (Treacher Collins–Franceschetti syndrome 1). The gene had no known homology but was shown to be conserved, and to show widespread expression among different tissues by northern blotting.

The team defined the exon structure and now completed the final step in the positional cloning process by screening for mutations among affected individuals. PCR amplified exonic DNA from one individual in each of 33 families was screened by single-strand conformation analysis, leading to the identification of five mutations: all cosegregated with disease and were not found among 100 ethnically matched controls (TCSCG 1996). The mutations identified included single base insertions leading to frameshifts and premature termination of the encoded protein as well as single nucleotide substitutions. By 2005 more than 118 different pathogenic mutations in 165 patients had been reported involving the coding regions of *TCOF1*, in two-thirds of cases involving small deletions leading to frameshifts (www.genoma.ib.usp.br/TCOF1_database/) (Splendore *et al.* 2005). *TCOF1* is a large gene with 28 exons; a third of mutations are reported in exons 23 and 24, which may be mutational **hotspots** (Splendore *et al.* 2002). The pathophysiology remains to be resolved but is thought to involve **haploinsufficiency**. *TCOF1* encodes a nucleolar protein involved in transcription of ribosomal DNA.

2.3.3 Linkage disequilibrium mapping and mendelian disease

Positional cloning to identify disease genes proved highly successful for rare diseases inherited in a mendelian fashion but was often a very challenging and time consuming approach. This was facilitated in many cases by other genomic features helping to identify the gene, such as chromosomal deletions, translocations, or repeat expansions. Linkage disequilibrium mapping has also been used to provide a finer level of resolution of disease loci (Lehesjoki *et al.* 1993; Hastbacka *et al.* 1994); in genetically isolated populations it has also been used at a genome-wide level to map disease gene loci (Houwen *et al.* 1994; Puffenberger *et al.* 1994). Linkage disequilibrium mapping is based on the association that exists between the genotyped variants (markers) and causative variants at the disease locus. Linkage disequilibrium and haplotypic structure (Box 2.8) are key concepts in the study of human genetic variation. They are described in greater detail later in this chapter in the context of common disease, and elsewhere in this book with examples of many applications in human genetics and evolutionary biology.

The application of linkage disequilibrium mapping has been most effective for recently founded, isolated populations. An elegant example of the successful application of the approach is provided by work to localize the gene associated with diastrophic dysplasia (DTD), a disease affecting the skeletal system which is relatively common in Finland (Box 2.9). It was thought that the approach would be most successful if an **ancestral haplotype** could be defined from a set of closely linked genetic markers within which there had been some degree of decay in linkage disequilibrium by recombination, such that the disease-associated region could be resolved to a high degree of resolution. The fact that the Finnish population has remained genetically isolated for over 2000 years since the immigration of a small number of settlers into the southwest of the country was ideal for such an analysis (Hastbacka *et al.* 1992).

The search for the gene and disease causing mutation(s) responsible for DTD began with a conventional linkage screen using a set of genome-wide genetic markers (Hastbacka *et al.* 1990). These markers had been published 3 years earlier as a linkage map comprising 403

Box 2.8   Linkage disequilibrium and haplotypes

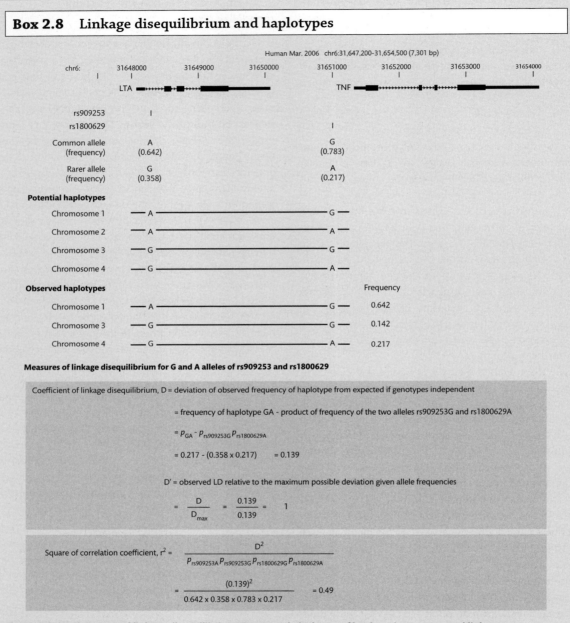

Measures of linkage disequilibrium for G and A alleles of rs909253 and rs1800629

Coefficient of linkage disequilibrium, D = deviation of observed frequency of haplotype from expected if genotypes independent

= frequency of haplotype GA - product of frequency of the two alleles rs909253G and rs1800629A

$= p_{GA} - p_{rs909253G} p_{rs1800629A}$

$= 0.217 - (0.358 \times 0.217)$ $= 0.139$

D' = observed LD relative to the maximum possible deviation given allele frequencies

$= \dfrac{D}{D_{max}}$ $= \dfrac{0.139}{0.139}$ =   1

Square of correlation coefficient, $r^2 = \dfrac{D^2}{p_{rs909253A} p_{rs909253G} p_{rs1800629G} p_{rs1800629A}}$

$= \dfrac{(0.139)^2}{0.642 \times 0.358 \times 0.783 \times 0.217} = 0.49$

Figure 2.2 Haplotypes and linkage disequilibrium. An example is shown of haplotypic structure and linkage disequilibrium for two biallelic SNPs in the *TNF* locus on chromosome 6p21. The two SNPs are identified by their rs numbers from dbSNP (www.ncbi.nlm.nih.gov/projects/SNP/) and are found in the first intron of *LTA*, encoding lymphotoxin alpha (rs 909253) and in the promoter region of *TNF*, encoding tumour necrosis factor (rs 1800629). The number of possible haplotypes for biallelic SNPs is 2^n, where n is equal to the number of SNPs; meaning that in this case there are four (2^2) possible haplotypes. However, when analysed in a population of European geographic ancestry (the CEU cohort of the International HapMap Project; Box 9.1) only three haplotypes are observed with frequencies as shown. Measures of linkage disequilibrium are shown including D, D', and r^2. For the coefficient of linkage disequilibrium, D, values can range from –0.25 to 0.25 depending on allele frequencies with $D = 0$ indicating linkage equilibrium; in this example a value of 0.139 was found. To allow for comparison between pairs of SNPs, a normalized value D' is used based on the ratio of D to the maximum possible value given the allele frequencies; in this case D' was 1 indicating at least one of the possible haplotypes (in this case the A-A haplotype) was not found in the population analysed. The square of the correlation coefficient, r^2, provides a further measure of linkage disequilibrium and varies from 0 (perfect equilibrium) to 1 (two alleles perfectly correlated). In this example, r^2 would be 1 if only two of the four potential haplotypes was found. Portion of figure showing genomic location of *LTA* and *TNF* adapted from screenshot of UCSC Genome Browser (Kent *et al.* 2002) (http://genome.ucsc.edu/) (Human March 2006 Assembly).

Linkage disequilibrium refers to the non-random association of alleles at two or more loci (Slatkin 2008). The term derives from population genetics (Lewontin and Kojima 1960) and is a measure of statistical association, in other words whether the observed frequency of two alleles occurring together is greater than that expected based on their allele frequencies (Fig. 2.2). The combination of alleles found on a single chromosome represents the **haplotype** (Fig. 2.2). In principle this could be the whole chromosome but the term is generally used to describe linked alleles in a much smaller region, typically less than 100 kb in size, although there are examples of particular loci where consideration of 'extended' haplotypes has been of value such as at the MHC on chromosome 6p (Section 12.3.2). Across successive generations a haplotype will degenerate as recombination occurs and further diversity arises leading to new haplotypes. This is reflected in a reduction in the coefficient of linkage disequilibrium between loci over time, depending not only on recombination and mutation, but also on selection, gene flow (transfer of alleles between populations), and **genetic drift** (changes in allele frequency in a population arising by chance) (Slatkin 2008).

polymorphic loci; the vast majority of markers were RFLPs which included single nucleotide substitutions that generated a site recognized by a specific restriction enzyme. Hastbacka and colleagues analysed 13 families with two or three affected siblings; a total of 84 individuals of whom 29 were affected by DTD (Hastbacka *et al.* 1990). This demonstrated linkage to the distal end of chromosome 5q with significant pairwise lod ratios (up to 7.4 for a marker D5S72 with no recombination). The following year, in 1991, the same group used a dense set of 16 polymorphic markers to further refine linkage to the region 5q31-q34 (Hastbacka *et al.* 1991).

Finer resolution was to come from combining this approach with linkage disequilibrium mapping that could be done with families having a single affected child, which would not be informative for linkage analysis. The team resolved the haplotypic structure using a number of markers and found very different haplotypic frequencies between DTD chromosomes and non-DTD chromosomes (Fig. 2.3A). A particular haplotype '1–1' defined by the markers *StyI* and *EcoRI* was found to account for almost all of the disease chromosomes but was otherwise rare, present at a frequency of 3% in the unaffected population. This indicated that the disease-causing mutation had occurred on this ancestral haplotype with relatively little decay of linkage disequilibrium over the estimated 2000 years since the founding of the Finnish population (Hastbacka *et al.* 1992).

What of the disease gene itself? Aided by the fine mapping, a region 100 kb proximal to the *CSF1R* gene (in which the haplotypic markers had been found) was found to show amino acid homology to a recently cloned

---

Box 2.9   Diastrophic dysplasia (OMIM 222600)

Diastrophic dysplasia is a disorder affecting the skeletal system (an osteochondroplasia), first described in 1960 by Lamy and Maroteaux, which may cause severe physical handicap. The disease is characterized by short limbs and short stature, spinal deformities such as scoliosis, together with specific joint abnormalities such as flexion limitation of proximal finger joints, 'hitch hiker thumb', and club foot (Walker *et al.* 1972). An autosomal recessive pattern of inheritance is seen. The disease is very rare outside Finland where it is the commonest form of dwarfism, and constitutes one of the commonest autosomal recessive diseases, the carrier frequency being 1–2% (Hastbacka *et al.* 1999).

---

rat gene which encoded a sulphate transporter protein. A large RNA was then identified in a range of tissues using the clone as a probe, but importantly the specific RNA was absent in an affected patient (Hastbacka *et al.* 1994). The DTD sulphate transporter gene (*DTDST*, now known as *SLC26A2*) was identified but an initial search for disease-causing sequence variants by direct sequencing failed to identify the major pathogenic variants, although a small number of 'non-Finnish' mutations were resolved including two **frameshift mutations** and one splice acceptor mutation (Hastbacka *et al.* 1994).

The full extent of the *SLC26A2* transcriptional unit was subsequently resolved and sequencing of two individuals homozygous for the ancestral 1–1 DTD haplotype identified a single nucleotide substitution at the start of intron 1 (Hastbacka *et al.* 1999). This mutation had occurred in the essential 'GT' sequence of the 5′ splice donor site, changing it to 'GC' (a T to C substitution denoted c.26-+2T>C), which reduced mRNA expression by 95% (Fig. 2.3B). The sequence variant was identified in almost all Finnish DTD chromosomes and no parental non-DTD chromosomes; intriguingly the small number of DTD chromosomes without this substitution had either a C to T substitution resulting in an amino acid substitution (p.R279W) (Fig. 2.3C) or a 3 bp deletion resulting in loss of the amino acid valine (p.V340del). The fact that the commonest variant abolishing the splice donor site also led to loss of the restriction enzyme recognition site for *Hph*I provided a convenient test for carrier and prenatal screening (Fig. 2.3D) (Hastbacka *et al.* 1999). More work needs to be done to define the precise functional basis

for this disease but *SLC26A2* is known to encode a sulphate transporter protein. Deficiency of intracellular sulphate pools within chrondrocytes is thought to result in undersulphated proteoglycans and to the specific developmental abnormalities affecting the skeletal system seen in DTD.

2.3.4 Allelic and genetic heterogeneity in mendelian diseases

The remarkable successes of linkage analysis and positional cloning in mendelian diseases have defined many genes and mutations involved in human disease (Botstein and Risch 2003). These mutations are typically very rare in populations and in most cases act to change the structure and function of the encoded protein with only a very small minority of disease causing mutations located in regulatory regions. The situation is complex as while a single gene, and specifically a particular mutation involving that gene, may cause the major phenotypic effect observed for a given individual and family pedigree, for the population of individuals affected by the disease typically a large number of rare mutations in a given gene or gene locus can be responsible for the observed phenotype. This 'allelic heterogeneity' is apparent when we consider the many different genetic variants that can lead to thalassaemia (Section 1.3), or other mendelian characters described in this chapter with over 1600 mutations of the *CFTR* gene listed on the cystic fibrosis mutation database (Section 2.3.1) and 118 mutations involving the *TCOF1* gene in Treacher Collins syndrome (Section 2.3.2). Indeed

(A)

StyI-EcoRI haplotype	Non-DTD chromosomes	DTD chromsomes
1-1	4	144
1-2	28	1
2-1	7	0
2-2	84	7

(B)

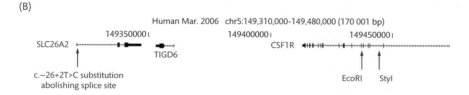

Human Mar. 2006   chr5:149,310,000-149,480,000 (170 001 bp)

(C)

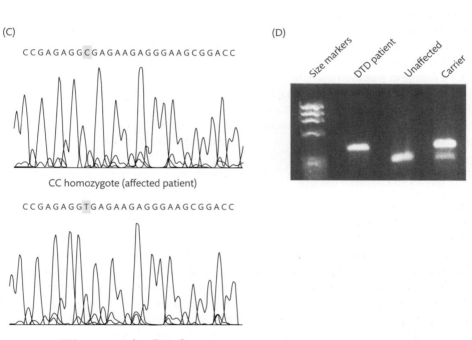

CC homozygote (affected patient)

TT homozygote (unaffected)

(D)

Figure 2.3 Finding sequence variants responsible for diastrophic dysplasia (DTD). (A) Linkage disequilibrium mapping showed that almost all DTD chromosomes carried an ancestral haplotype 1–1 based on restriction enzyme digestion. (B) *SLC26A2* gene showing the site of T to C substitution abolishing splice site in intron 1 together with restriction sites for *StyI* and *EcoRI* in the adjacent *CSF1R* gene. (C) Sequencing traces illustrating the nucleotide substitution responsible for almost all Finnish cases of DTD. (D) Restriction digest using *HphI* illustrating the loss of restriction site in presence of T to C substitution for an affected patient; in the carrier both T and C alleles are present, leading to two bands. Redrawn and reprinted by permission from Macmillan Publishers Ltd: Nature Genetics (Hastbacka *et al.* 1992), copyright 1992; European Journal of Human Genetics (Hastbacka *et al.* 1999), copyright 1992. Panel B adapted from screenshot of UCSC Genome Browser (Kent *et al.* 2002) (http://genome.ucsc.edu/) (Human March 2006 Assembly).

the finding of allelic heterogeneity is seen as strengthening the evidence of a causal relationship between variation in the gene and the phenotype (Risch 2000).

Variation at different loci can also lead to the same phenotype (described as genetic heterogeneity) as illustrated by early onset familial Alzheimer's disease, which is discussed in more detail later in Section 2.5.1. In this disease, rare variants at three different genes (*APP*, *PSEN1*, and *PSEN2*) have been shown to cause the observed phenotype. Within families the relationship is, however, specific between a given gene variant and the phenotype, which allows linkage and positional cloning approaches to be applied. In this example a common pathway leading to Alzheimer's disease based on amyloid accumulation provides a unifying mechanism within which variation at these different genes may lead to the same phenotype.

Other genetic and environmental variation may also contribute to the phenotypic heterogeneity observed in some mendelian disorders. For example, marked variation in penetrance is seen for the iron storage disorder haemochromatosis, which in most cases arises due to homozygosity for a **missense mutation** in the *HFE* gene (Section 12.6). Among the proposed environmental and genetic modifiers is a common SNP of the *BMP2* gene (encoding bone morphogenetic protein 2) which modulates iron burden (Milet *et al.* 2007).

2.3.5 Linkage analysis and common disease

For mendelian diseases, linkage studies have proved a very robust approach with a low false positive rate (Risch 2000). By contrast, much less success has been achieved using linkage-based approaches for common multifactorial traits. Here it has been increasingly recognized that diversity in many genes is likely to be involved, each with individually modest effect sizes and further modulated by environmental factors to a much greater extent than with rare diseases showing Mendelian inheritance. Critically, for polygenic diseases there is usually no clear pattern of inheritance within a pedigree.

Linkage approaches have been applied in common multifactorial diseases but in only a minority of cases has this been highly informative. There have been some striking successes such as the mapping of a Crohn's disease susceptibility locus to chromosome 16q by genome-wide

linkage analysis (Section 9.5.2). Crohn's disease is a common debilitating inflammatory bowel disorder in which genetic factors have been extensively investigated (Section 9.5). Positional cloning and linkage disequilibrium mapping helped to refine the inflammatory bowel disease locus (IBD1) on chromosome 16 and led to the identification of specific mutations of the *NOD2* gene conferring significantly increased risk of disease (Section 9.5.2). Other loci resolved by linkage analysis in Crohn's disease include regions of chromosome 6p (IBD3) and chromosome 5q31 (IBD5), however many other IBD loci reported from linkage scans have not been convincingly replicated (Section 9.5.3).

Linkage analysis has also been applied successfully to study particular subtypes of some common diseases where individuals display a mendelian pattern of inheritance as seen with early onset Alzheimer's disease (Section 2.5.1) and maturity onset diabetes of the young (OMIM 606391). The latter is caused by mutations in a number of different genes including *HNF4A* (encoding hepatocyte nuclear factor-4-alpha) on chromosome 20q12-q13.1 (MODY type 1) (Yamagata *et al.* 1996).

2.4 Genetic association studies and common disease

Genetic association studies have been extensively used to try to define genetic variation associated with disease susceptibility, notably in the context of common multifactorial traits. The approach was advocated by Risch and Merikangas in 1996 as more powerful than linkage analysis for such diseases where a modest effect size was likely, such that application of linkage analysis would require an unfeasibly high number of families (Risch and Merikangas 1996). It was only 10 years later, however, that our ability to genotype hundreds of thousands of SNPs as common genetic markers in the context of a publically available finished human genome sequence and improved understanding of the genetic architecture of human diversity that would allow the successful application of genome-wide association studies (Section 9.3).

Prior to this, association studies adopting a candidate gene approach based on biological plausibility were extensively used across a very wide range of common diseases for which there were varying levels of evidence to support a role for genetic factors. There were notable successes but also increasing scepticism as in many cases initial studies failed to be replicated. There are many reasons proposed for this but in essence what was not clear at that time was the small size of relative risks associated with possession of the majority of individual alleles and the nature of the underlying allelic architecture. We now know that common multifactorial traits are likely to involve several genes and multiple variants of individually modest magnitude of effect, which are neither necessary nor sufficient for disease to occur.

Many studies did not have sufficient statistical **power** to find an association or to replicate it, and there were issues with the significance thresholds chosen, how to correct for multiple comparisons, and the potential bias from underlying **population stratification** (Section 2.4.3). The effects of linkage disequilibrium were difficult to dissect, particularly as often only a very limited number of genetic markers were selected for analysis. Issues with phenotype definition, overestimation of the magnitude of initial association, testing of multiple hypotheses, publication bias, population-specific differences in underlying linkage disequilibrium, and gene–gene and gene–environment interactions have all been raised as further reasons for failure to replicate genetic association studies (Cardon and Bell 2001; Hirschhorn et al. 2002; Healy 2006). A number of these factors are considered in more detail below and elsewhere in the context of specific diseases or traits.

2.4.1 A small number of robustly demonstrated associations?

In a comprehensive literature review of association studies published between 1986 and 2000, a total of 603 different gene disease associations involving 238 genes and 133 common diseases were analysed by Hirschhorn and colleagues (2002). These excluded the very many associations reported with the MHC on chromosome 6 and those with blood group antigens, and were restricted to variants with a minor allele frequency of at least 1% in or close to known genes. For 166 associations in which

three or more publications were available to review, only six associations were judged to be reproducible with a high level of statistical confidence (Hirschhorn et al. 2002). This is raising the bar to a high level and should be regarded as a minimal set of robust associations with many other reported associations likely to be informative.

The six associations noted by Hirschhorn and colleagues included factor V Leiden and venous thrombosis (reviewed in Section 2.6.1); possession of the APOE $\varepsilon 4$ allele and late onset Alzheimer's disease (reviewed in Section 2.5.2); a nonsynonymous SNP of CLTA4 and Graves' disease (an autoimmune disease involving the thyroid gland) (Donner et al. 1997); a 32 bp deletion of the CCR5 gene and HIV-1 infection (Section 14.2.1); a tandem repeat upstream of the INS gene and type 1 diabetes (Box 7.5); and a nonsynonymous SNP of the prion protein gene PRNP on chromosome 20p13 associated with sporadic Creutzfeldt–Jakob disease (Palmer et al. 1991). Despite this very low number of apparently robust associations, in a subsequent paper Hirschhorn and colleagues highlighted how while false positive results are common among initial reports, a substantial number of real associations of modest effect are likely to be present which become significant on **meta analysis** (Lohmueller et al. 2003). Such analysis showed significant replication among follow-up studies for eight out of 25 studies selected from the initial set of 166 frequently studied associations (Lohmueller et al. 2003). This and other studies illustrate the power of meta analysis in this context (Ioannidis et al. 2001).

2.4.2 Study design and statistical power

For any genetic study of human disease, whether linkage or association based, a clearly defined phenotype using specific diagnostic criteria is essential. There will often be variation within a phenotype related to aetiology or other factors, and minimizing such heterogeneity is important to maximize the chances of success.

The levels of resolution achieved by genetic association studies are potentially very different from linkage analysis. With linkage analysis, sets of markers are used to map chromosomal location by statistical analysis of observed informative recombination events. The resolution will be coarse as there are relatively few opportunities for such events within the generations of the pedigrees studied, and the detail achieved can be seen as large blocks of inherited haplotypes with a 'disease region' that may span hundreds of genes. This contrasts with genetic association studies. Here an association will usually arise due to linkage disequilibrium between the genetic marker and functional variant unless the causative variant has been included in the set of genetic markers analysed. As it is the population which is being sampled rather than a limited number of generations of a family, the scale of resolution will be much finer as the observed linkage disequilibrium at a given chromosomal region will reflect human ancestry and the multiple complex recombination and mutational events which will have occurred (Fig. 2.4) (Cardon and Bell 2001). Despite this, fine mapping disease association studies remains a formidable challenge and our ability to resolve causative functional variants, particularly for noncoding changes, still represents a major road block.

Both case–control and family-based study designs have been extensively used in genetic association studies, the case–control design looking for evidence of statistically significant association between the frequency of a given genetic variant among cases of the disease compared to controls. The selection of gene regions for candidate gene association studies was driven by biological plausibility with well characterized genes, such as *TNF* encoding the cytokine tumour necrosis factor, the subject of many reported disease association studies ranging from infectious and autoimmune disease to cancer (Section 2.4.4).

Often, single genetic markers (usually SNPs) were selected for analysis in a case–control design, while later studies sought to increase SNP coverage for a gene region as awareness of the extent of diversity, haplotype structure, and linkage disequilibrium increased. Given that there are very few instances where a causative variant had been identified, candidate gene association studies relied on either successfully genotyping the latter or finding association with SNPs in linkage disequilibrium with the causative variant. The low prior probability of finding a true association when testing a small number of markers in a candidate gene is likely to have been a major factor in the large number of false positive studies reporting an initial association (Risch 2000). In Chapter 12 the nature and consequences of

genetic diversity in the MHC are reviewed: relative to other genomic loci many more robust disease associations were found in this region, which Risch suggested was due to the much higher prior probability of a functional or linked variant being analysed (Risch 2000).

Study power is extremely important in association studies, with many reported studies underpowered to detect a given minimal magnitude of effect and allele frequency. Large sample sizes are required for more modest effect sizes and rarer allele frequencies. Power is also influenced by local patterns of linkage disequilibrium and other alleles at the same or a different locus which lead independently to the same disease (allelic and genetic heterogeneity, respectively). Cardon and Bell describe how in 2001 reported sample sizes for **case–control** studies had been modest to that time, with 100 or less cases and equal numbers of controls.

For the modest effect sizes typically seen with common multifactorial disease traits, insufficient statistical power can lead to false negative results on initial reporting of association testing or with replication studies. This was illustrated by work looking at genetic variation in the *PPARG* gene at chromosome 3p25 encoding a specific transcription factor 'peroxisome proliferator-activated receptor gamma', important in adipocyte differentiation and gene expression. A specific nonsynonymous SNP, rs1801282, in which a C to G nucleotide substitution leads to substitution of proline for alanine at amino acid position 12 (p.P12A), was initially reported as showing a strong association with type 2 diabetes with an **odds ratio (OR)** of 4.35 ($P = 0.03$) for individuals homozygous for the C allele (Deeb *et al.* 1998). Four of five subsequent studies failed to find a significant association with diabetes although a modest elevated risk was present and the studies were thought to have insufficient sample sizes to reliably detect the association (Altshuler *et al.* 2000a). A larger study of 3000 individuals comprising different family-based and case–control **cohorts** did find significant association for the C allele encoding proline although the odds ratio was modest (OR 1.25, $P = 0.002$) (Altshuler *et al.* 2000a), highlighting that initial reports may be overestimates of an association and that replication studies should be well powered (Hirschhorn *et al.* 2002). This association also serves to illustrate that it is not necessarily possession of the rarer allele which is associated with disease risk; the high frequency of the

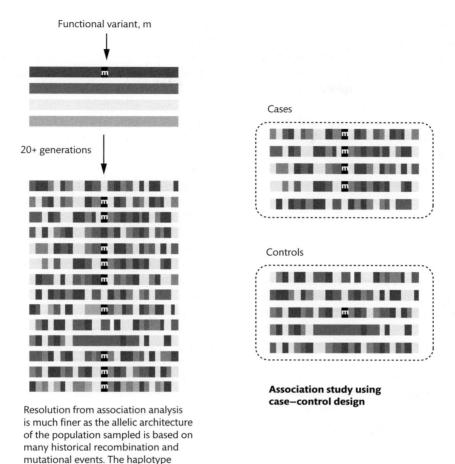

Functional variant, m

20+ generations

Cases

Controls

Association study using case–control design

Resolution from association analysis is much finer as the allelic architecture of the population sampled is based on many historical recombination and mutational events. The haplotype on which the functional mutation arose breaks down over time.

Figure 2.4 Genetic association analysis. Genetic association studies sample from a population in which many historical recombination and mutational events will have occurred allowing increased resolution from genotyped markers. Linkage disequilibrium between the genotyped genetic marker and functional variant(s) allows association with disease even if the latter is not directly genotyped. Adapted by permission from Macmillan Publishers Ltd: Nature Reviews Genetics (Cardon and Bell 2001), copyright 2001.

allele also emphasizes that while the effect for the individual on disease susceptibility is modest, the attributable risk may be much higher.

The need for much larger genetic association studies involving thousands of individuals which more recently become possible through collaboration between research groups, allowing genome-wide association studies and replication studies to become a reality (WTCCC 2007). Large ongoing prospective cohort collections such as UK Biobank (www.ukbiobank.ac.uk), which aims to recruit

500 000 people aged 40–69 years, will prove very exciting resources for future genetic studies with a wealth of associated epidemiological data; power calculations suggest for example that after 6 years of recruitment the study will be powered to detect an odds ratio of 1.3 or higher for type 2 diabetes (Palmer 2007).

In Chapter 9 the development and application of genome-wide genetic association studies are reviewed in detail (Section 9.3). A number of different factors enabled such studies to become a reality, not least the availability of large cohorts of carefully phenotyped cases of different diseases. The remarkable efforts to catalogue SNPs and the underlying haplotypic diversity across populations (Section 9.2) were to dramatically improve our understanding of allelic architecture and allow the selection of informative common SNP markers for use in genome-wide association studies. These factors, together with the capacity for high throughput genotyping, the availability of the human genome sequence, and advances in statistical analysis, were to set the stage for dramatic improvements in our ability to interrogate the genetic basis of common multifactorial disease.

2.4.3 Genetic admixture and association with disease

A major potential confounder of case–control disease association studies is the effect of genetic admixture or population subdivision. This can arise when a population containing two or more ethnic groups or subgroups is studied. The frequency of many alleles varies between ethnic groups or 'segments' of a population relating to genetic drift or **founder effects**, as may the **prevalence** of disease (Slatkin 1991; Cavalli-Sforza et al. 1994; Pritchard and Rosenberg 1999). If there has not been careful matching of ethnic groups between cases and controls, apparent disease associations may arise. Thus if disease prevalence in a given group is higher than the others, that group may be overrepresented among the cases compared to the controls and any genetic markers that are present at a high frequency in that particular ethnic group will appear to be associated with disease (Hirschhorn et al. 2002). A number of conditions vary in prevalence between ethnic groups, such as hypertension, which is more common among African Americans than Caucasians; study of a mixed population will run the risk of spurious association with disease for any marker allele more frequent in the African American population (Reich and Goldstein 2001).

The issues relating to genetic admixture and disease association are well illustrated by a study of type 2 diabetes among the Pima and Papago Native American tribes of southern Arizona (Knowler et al. 1988). A large longitudinal study over 20 years allowed the association with disease of a particular immunoglobulin heavy chain haplotype (Gm[3;5,13,14]) to be analysed among a cohort of

4920 individuals with a high prevalence of diabetes. The association appears dramatic in a well powered study with a clear protective effect associated with possession of a specific Gm haplotype and type 2 diabetes (Fig. 2.5A) (Knowler et al. 1988). However, this result arises due to Caucasian admixture among the Native American population studied leading to population stratification. Among those of full Pima and Papago Indian heritage, type 2 diabetes has a much higher prevalence compared to Caucasians or those with lower fractions of Indian heritage; conversely the particular Gm haplotype studied has a very low frequency among those of full Indian heritage (0.006) compared to Caucasians in the United States (0.665) (Fig. 2.5B). Thus when the data are stratified based on Caucasian admixture (expressed as the fraction of full Indian heritage), the prevalence of diabetes is the same among those who do or do not possess the Gm haplotype (Fig. 2.5C) (Knowler et al. 1988). The risk of diabetes does vary dependent on the degree of Caucasian admixture but this has nothing to do with the Gm haplotype studied – that haplotype is simply a very good marker of the degree of Caucasian admixture in Native Americans (Williams et al. 1986).

Careful matching of cases and controls is important to avoid the effects of significant population stratification but even so 'cryptic stratification' may persist. Family-based methodologies, notably the transmission disequilibrium test, circumvent the problem by analysing affected children and their parents. Actual transmission of alleles to offspring can then be compared to expected transmission, provided that at least one parent is heterozygous for a given allele allowing testing for linkage and association between marker and disease (Spielman et al. 1993; Ewens and Spielman 1995). Alleles showing disease association will be transmitted more often than the expected 50 : 50 transmission based on mendelian inheritance. Recruitment of family members is, however, more demanding than for unrelated individuals, particularly for late onset diseases, and requires additional genotyping (Hirschhorn et al. 2002). Unlinked genetic markers have also been used to both define the extent of population stratification and as a means of statistical correction (Pritchard and Rosenberg 1999; Reich and Goldstein 2001). More recent genome-wide association studies using many hundreds of thousands of genetic markers allow powerful statistical approaches to defining hidden

(A)

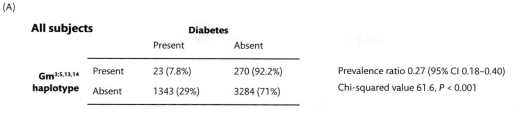

		Diabetes	
		Present	Absent
Gm³;⁵,¹³,¹⁴ haplotype	Present	23 (7.8%)	270 (92.2%)
	Absent	1343 (29%)	3284 (71%)

All subjects

Prevalence ratio 0.27 (95% CI 0.18–0.40)

Chi-squared value 61.6, $P < 0.001$

(B)

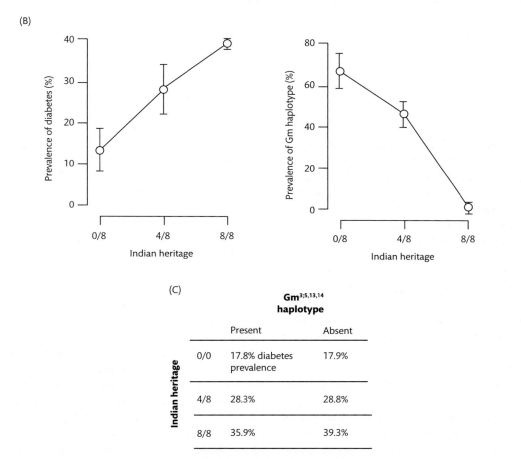

(C)

	Gm³;⁵,¹³,¹⁴ haplotype	
Indian heritage	Present	Absent
0/0	17.8% diabetes prevalence	17.9%
4/8	28.3%	28.8%
8/8	35.9%	39.3%

Figure 2.5 Genetic admixture and potential confounding of disease association. **(A)** Disease association for Gm³;⁵,¹³,¹⁴ haplotype with type 2 diabetes among all subjects. **(B)** Prevalence of diabetes and of the Gm haplotype varies with degree of Indian heritage. **(C)** Stratification based on Indian heritage shows no difference in diabetes prevalence in the presence or absence of the haplotype for a given level of Indian heritage. Reproduced with permission from Knowler *et al.* (1988) .

population structure and taking account of it, as illustrated by the Wellcome Trust Case Control Consortium study of seven common diseases (Section 9.3.2) (WTCCC 2007).

As with any disease association study, replication of results is extremely important within and between populations. Where the causative allele(s) are not definitively identified, differences in genetic architecture between populations, for example patterns of linkage disequilibrium, can lead to different results in replication studies carried out in different ethnic groups. This is commonly seen in case–control studies among populations of African ancestry where linkage and haplotype structure are often very different. This can, however, help to narrow down the region of association as illustrated by work on narcolepsy (Section 12.4.1) where analysis in African American individuals helped define the role of HLA DQB1*0602, which in Caucasians, but not African Americans, is in strong linkage with DRB1*1501 making refining the association problematic when analysing Caucasians alone (Mignot *et al.* 1994).

2.4.4 TNF and candidate gene association studies

The *TNF* gene on chromosome 6p21 encodes a pleiotropic cytokine with key roles in the immune and inflammatory response. There is evidence that genetic factors are important determinants of interindividual variation in levels of TNF expression with specific genetic variants, notably in the promoter region upstream of *TNF*, postulated to modulate gene expression (Knight and Kwiatkowski 1999; Bayley *et al.* 2004). Dysregulation of TNF production is important in the pathogenesis of a number of different infectious and autoimmune diseases, notably septic shock (Tracey and Cerami 1993) and severe malaria (Kwiatkowski *et al.* 1990) while anti-TNF therapies are now extensively used in conditions such as rheumatoid arthritis and inflammatory bowel disease.

Genetic variation involving the *TNF* locus has been extensively studied by investigators adopting a candidate gene approach given its plausibility in disease pathogenesis. Early successes such as the dramatic association with severe malaria of homozygosity for a *TNF* promoter SNP (rs1800629) 308 nt upstream of the transcriptional start ('TNF-308'; Section 13.2.6) (McGuire *et al.* 1994) have been followed by a remarkable number of reported case–control studies, over 90 relating to this promoter SNP alone by 2004 (Bayley *et al.* 2004). The list of reported associations for *TNF* continues to grow. Infectious disease associations include with malaria, leishmaniasis, leprosy, scarring trachoma, meningococcal disease, HIV-1 infection, hepatitis B and C, melioidosis,

and septic shock. Other reported autoimmune disease associations include with multiple sclerosis, coeliac disease, diabetes, erythema nodosum, common variable immunodeficiency, and fibrosing alveolitis. Further details of *TNF* locus variants associated with disease are listed in the Cytokine Gene Polymorphism in Human Disease online database (www.nanea.dk/cytokinesnps/) (Hollegaard and Bidwell 2006).

Unfortunately many of these studies have failed to replicate and illustrate many of the issues relating to study design and analysis described previously. More recent studies adopting genome-wide approaches have also found association with the diversity at the *TNF* locus, for example in the neighbouring gene *LTA* (encoding lymphotoxin alpha) SNPs have been associated with myocardial infarction in a Japanese population (Ozaki *et al.* 2002) and with leprosy (Alcais *et al.* 2007). The *TNF* locus lies within the MHC class III region (Fig. 12.1) and genetic diversity at this locus shows extensive linkage disequilibrium across the MHC such that assigning association to specific SNPs is problematic. Disease associations for example with rheumatoid arthritis are likely to be the result of coinherited variants elsewhere in the region (Newton *et al.* 2004a), while even within the *TNF* locus, linkage between the TNF promoter SNP rs1800629 and an intronic SNP of *LTA* rs909253 ('LTA+252') (Fig 2.2) has led to considerable debate over the causative variant. There is some evidence to support a functional role for both SNPs that may be highly context specific, although there is also evidence that the functional variant(s) may lie elsewhere within the locus or be operating at a distance (Section 13.2.6) (Knight *et al.* 2003; Knight *et al.* 2004).

2.5 Alzheimer's disease

Alzheimer's disease is the commonest cause of dementia among European and North American populations and has been the subject of intense research aiming to understand the underlying pathophysiology and **risk factors** for developing the disease (Box 2.10).

Early onset Alzheimer's disease before the age of 65 years makes up only a small minority of all cases of Alzheimer's disease (between 1 and 6%) but is

---

Box 2.10   Alzheimer's disease (OMIM 104300)

Alzheimer's disease is a neurodegenerative disorder characterized by an insidious onset of memory problems, which progresses to global cognitive decline. A progressive dementia is seen with confusion, disorientation, poor judgement, language problems, impaired reasoning, agitation, and hallucinations among the many clinical manifestations. The disease course is usually for 8–10 years before death, most commonly as a result of pneumonia, cardiovascular disease, cachexia, and dehydration (Attems *et al.* 2005; Koopmans *et al.* 2007). Alzheimer's disease is a devastating disease for the individual patient and their family, as well as representing a very significant public health burden owing to the disease being very common among the growing elderly populations of Europe and North America. In the year 2000, for example, there were estimated to be 4.5 million people with Alzheimer's disease in the United States (Hebert *et al.* 2003). Alzheimer's disease is age-related, with a reported incidence of 2.8 per 1000 person-years among individuals between 65 and 69 years of age, rising to 56.1 among individuals over 90 years of age (Kukull *et al.* 2002).

Imaging of affected individuals with advanced disease reveals marked cortical atrophy and ventricular enlargement in the brain while neuropathological studies at post mortem show characteristic lesions associated with accumulation of amyloid beta (Aβ) protein. Extracellular deposits of Aβ form the characteristic amyloid plaques present in high numbers among affected individuals. Intracellular deposits of neurofibrillary tangles comprising hyperphosphorylated microtubule-associated protein tau are also seen, together with vascular deposits. A higher incidence of dementing illness among relatives of patients with Alzheimer's disease was noted by Heston and colleagues, suggesting a genetic link (Heston *et al.* 1981). Twin studies demonstrate high **heritability**, noted in a large study using the Swedish Twin Registry to be between 58% and 79% (Gatz *et al.* 2006). Alzheimer's disease is a complex and heterogeneous disorder, with environmental and genetic factors as causal or modifying risk factors. Genetic factors have been the subject of several comprehensive reviews (Cruts and Van Broeckhoven 1998a; Tanzi and Bertram 2005; Waring and Rosenberg 2008) and online resources include the Alzheimer Disease and Frontotemporal Dementia Mutation Database (www.molgen.ua.ac.be/ADMutations) (Cruts and Van Broeckhoven 1998b, 1998a) and the 'AlzGene' database of genetic association studies in Alzheimer's disease maintained by the Alzheimer Research Forum (www.alzgene.org) (Bertram *et al.* 2007a).

---

often familial with an autosomal dominant pattern of mendelian inheritance seen in some families (Bird 2008). Research into the genetic basis of such cases has been highly informative in advancing our understanding of disease pathogenesis and its potential modulation. Rare variants have been identified in three genes (*APP*, *PSEN1*, and *PSEN2*), which encode the Aβ precursor protein (APP) and specific proteases responsible for its cleavage into a particular form of amyloid beta, Aβ42. Accumulation of Aβ42 in the brain is central to

disease pathogenesis according to the amyloid hypothesis (Hardy and Selkoe 2002) with characteristic plaque formation, neurofibrillary tangles, synaptic loss, and neuronal death (Fig. 2.6).

Familial disease is, however, most commonly seen in patients with late onset disease, comprising between 15% and 25% of cases of Alzheimer's disease (Bird 2008). Among this group of patients, possession of the *APOE* ε4 allele is a major risk factor for age of disease onset. Like the rare variants in *APP*, *PSEN1*, and *PSEN2* associated

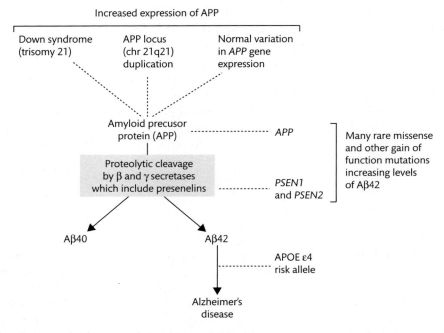

Figure 2.6 Factors influencing the accumulation of Aβ42, thought to be important in the pathogenesis of Alzheimer disease based on the amyloid hypothesis. Adapted by permission from Macmillan Publishers Ltd: Nature Genetics (Hardy 2006), copyright 2001.

with early onset disease, the *APOE* ε4 allele is thought to play a key role in the accumulation of Aβ protein. A small number of cases, estimated at less than 1%, are caused by trisomy 21 as part of Down syndrome (Box 3.3), the characteristic neuropathological changes of Alzheimer's disease being seen among almost all individuals with Down syndrome after 40 years.

This review falls into two sections. Firstly, a discussion of early onset familial Alzheimer's disease, which illustrates many of the issues and approaches inherent to investigating a mendelian trait. Secondly, a consideration of late onset Alzheimer's disease and the role of the *APOE* ε4 allele, which illustrates the success of a candidate gene approach in defining genetic determinants of common multifactorial disease.

2.5.1 Early onset familial Alzheimer's disease: rare variants underlying a mendelian trait

The prevalence of early onset familial Alzheimer's disease in the French city of Rouen was estimated to be

41.2 per 100 000 people at risk aged between 41 and 60 years; of these 13% showed clear cut autosomal inheritance (Campion *et al.* 1999). Study of such families showing segregation consistent with autosomal dominant inheritance was highly informative for both linkage analysis and identification of specific mutations. The link with Down syndrome was highly suggestive that a risk locus may be present on chromosome 21. Aβ was thought to play a key role in disease pathogenesis and the determination of the peptide sequence from patients with Alzheimer's disease and with Down syndrome (Glenner and Wong 1984a, 1984b; Masters *et al.* 1985), including material purified from senile plaques, allowed cloning of the *APP* gene (Goldgaber *et al.* 1987; Kang *et al.* 1987; Robakis *et al.* 1987; Tanzi *et al.* 1987). The gene was mapped to chromosome 21q21 and, together with initial evidence of linkage among families with early onset familial Alzheimer's disease, further heightened interest in a potential disease susceptibility locus on chromosome 21 dubbed 'familial Alzheimer disease 1', AD1 (OMIM 104300).

The first pathogenic sequence variation of *APP* was reported among Dutch families with a rare autosomal dominant disorder involving amyloid protein, called hereditary cerebral haemorrhage with amyloidosis (OMIM 609065). Levy and colleagues sequenced the exons encoding the Aβ domain of the protein in two patients and defined a G to C transversion which resulted in the substitution of glutamic acid to glutamine (p.E693Q) (Levy *et al.* 1990).

Among patients with early onset familial Alzheimer's disease showing autosomal inheritance, Goate and colleagues used additional markers on the long arm of chromosome 21 to further define linkage to chromosome 21 which included *APP* (Goate *et al.* 1991). Sequencing of exon 17, involved in encoding the Aβ domain, showed a G to A transition cosegregating with disease in the family kindred analysed; two individuals with the same variant were defined in a second affected family. The nucleotide substitution was a further missense mutation, changing the amino acid sequence in the encoded protein from valine to isoleucine at position 717 (c.2149G>A, p.V717I) (Goate *et al.* 1991).

A total of 29 pathogenic mutations have now been identified at the *APP* gene, the majority being missense mutations, and are listed at the Alzheimer Disease and Frontotemporal Dementia Mutation Database (www.molgen.ua.ac.be/ADMutations) (Cruts and Van Broeckhoven 1998a, 1998b). Cumulatively, however, *APP* sequence variants account for a minority of early onset familial disease cases (estimated at 10–15%) and overall a very small number of Alzheimer's disease cases. They appear to be **gain of function mutations**, resulting in increased levels of the Aβ42 protein thought to be central to the pathogenesis of Alzheimer's disease (Scheuner *et al.* 1996).

A **gene dosage** effect is suggested by the neuropathology seen in Down syndrome, with trisomy of chromosome 21 thought to result in increased lifetime accumulation of Aβ. It is striking that gene duplication events involving *APP* have now been identified among families with early onset familial Alzheimer's disease showing that overexpression of the normal protein can lead to disease (Rovelet-Lecrux *et al.* 2006). Among 65 families, individuals from five families were defined with duplications ranging in size from 0.58 to 6.37 Mb, encompassing five to 12 known genes including *APP*;

the affected families included individuals with evidence of cerebral amyloid angiopathy suggesting a potentially more complex phenotype.

It was clear that not all families with early onset Alzheimer's disease showed linkage to chromosome 21q21 (St George-Hyslop *et al.* 1990). A number of groups found evidence instead of linkage to chromosome 14q24.3 (Schellenberg *et al.* 1992; Van Broeckhoven *et al.* 1992; Campion *et al.* 1995), leading to definition of the early onset familial Alzheimer disease-3 (AD3) subtype (OMIM 607822) due to rare mutations affecting the *PSEN1* gene. Analysis of short tandem repeat markers using the lod score method resolved strong linkage for 14q24.3. A positional cloning approach was then used by Sherrington and colleagues to define a novel gene containing specific disease-associated sequence variants (Sherrington *et al.* 1995). Additional markers helped resolve the region of linkage with haplotypic analysis defining recombination boundaries. The authors hybridized cDNA generated from human brain mRNA to a series of genomic DNA fragments to perform transcription mapping, then looked for sequence differences. Analysis of affected family members from tightly linked pedigrees versus normal brain samples resolved a novel gene encoding the 'S182 protein' with five nonsynonymous sequence variants specific to affected individuals (p.M146L, p.A246E, p.286V, p.C410Y) (Sherrington *et al.* 1995).

The novel gene now called *PSEN1*, at chromosome 14q24.3, encodes presenilin 1 which is part of the gamma secretase complex involved in cleavage of the amyloid precursor protein to form Aβ protein. At least 168 different pathogenic sequence variants have been reported at *PSEN1*, the vast majority being missense mutations (Alzheimer Disease and Frontotemporal Dementia Mutation Database). Overall, mutations at AD3 involving *PSEN1* account for 20–70% of cases of early onset familial Alzheimer's disease. The *PSEN1* variants are thought to be gain of function mutations, increasing production of the Aβ42 peptide (Scheuner *et al.* 1996; De Strooper *et al.* 1998). Reports suggest the p.M146L variant has been associated with calcium dysregulation affecting Aβ processing (Cheung *et al.* 2008) while p.V97L was associated with increased Aβ protein through an effect on insulin degrading enzyme (Qin and Jia 2008).

A further gene locus associated with early onset familial Alzheimer's disease was reported in 1995 by two independent groups taking different approaches (Levy-Lahad *et al.* 1995; Rogaev *et al.* 1995). Rogaev and colleagues used their finding of the novel 'S182' gene (*PSEN1*) to identify homologous genes. They screened cDNA libraries and identified a 2.3 kb transcript which they mapped to chromosome 1. The new gene (now called *PSEN2*) encoded a protein with 63% amino acid identity to presenilin-1 and the authors sought to identify underlying variants associated with disease among 23 family pedigrees where variation at *APP* or *PSEN1* had been excluded. Two missense mutations were identified: an A to G substitution leading to a methionine to valine substitution at codon 239 (p.M239V) was found in all four affected members of an Italian origin extended pedigree; and a further mutation in three of four pedigrees of Volga German origin (p.N141I) (Rogaev *et al.* 1995).

By contrast, Levy-Lahad and colleagues adopted a linkage-based approach to identify the same gene locus by analysing seven related families of Volga German ancestry, which included 94 affected people (Levy-Lahad

et al. 1995). Volga Germans are thought to have settled in the Volga region of Russia in the 1760s but to have remained genetically isolated, emigrating to the United States between 1870 and 1920. A founder effect had been proposed within this population group who show a number of families with early onset familial Alzheimer's disease inherited in an autosomal dominant manner (Fig. 2.7), the disease manifesting on average at 55 years of age (in contrast to AD3 at 45 years of age) (Bird *et al.* 1988, 1996). Analysis of 162 markers showed linkage to chromosome 1q31-q34 (Levy-Lahad *et al.* 1995), the early onset familial Alzheimer locus 'AD4' (OMIM 606889) in which *PSEN2* was identified by Rogaev and colleagues. Mutations of *PSEN2* are rare and account for a very small proportion of cases of early onset familial Alzheimer's disease; to date ten missense mutations of *PSEN2* have been identified and associated with Alzheimer's disease among 18 families, some with incomplete penetrance (Alzheimer Disease and Frontotemporal Dementia Mutation Database). Like *PSEN1* and *APP* mutations, the consequences are thought to relate to increased Aβ42 accumulation.

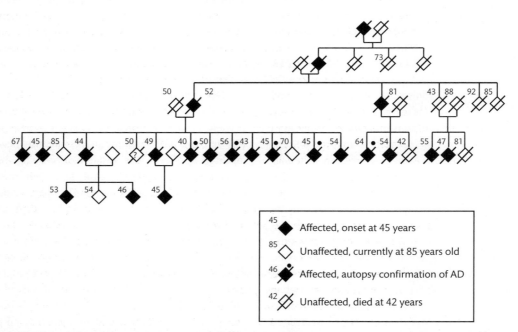

Figure 2.7 Family pedigree of Volga German ancestry showing autosomal dominant segregation for early onset familial Alzheimer disease (AD). Redrawn from Levy-Lahad *et al.* (1995), reprinted with permission from AAAS.

2.5.2 APOE ε4 and late onset Alzheimer's disease

The association of a particular **isoform** of the ApoE protein with age of onset in late onset Alzheimer's disease is a remarkable story. In the context of common multifactorial disease traits, it was an early example of the potential power of a candidate gene approach yet has proved subsequently in many respects to be unusual: this is a common variant associated with a large effect size whose association with disease susceptibility has been robustly replicated. Biological plausibility as a candidate gene combined with some evidence of linkage to the region of chromosome 19q where the *APOE* gene is found (19q13.2) led to reports in 1993 of association between the ε4 allele and late onset Alzheimer's disease. Pericak-Vance and colleagues had reported in 1991 evidence that familial late onset Alzheimer's disease was linked to the proximal region of the long arm of chromosome 19, in contrast to early onset familial disease for which they found evidence of linkage to chromosome 21. Linkage studies in late onset disease were more problematic due to the advanced age of cases and lack of clear mendelian segregation, but they were able to derive some data using affected pedigree member linkage analysis looking for differences in independent segregation between markers and disease (Pericak-Vance *et al.* 1991).

The linkage data were suggestive but the findings that ApoE protein was a significant component of amyloid plaques, neurofibrillary tangles, and vascular amyloid highlighted the biological plausibility of diversity involving the *APOE* gene in disease pathogenesis. *APOE* had been found to be strongly expressed in the brain by astrocytes, with increased levels following injury or in chronic neurodegenerative disease including in Alzheimer's disease. Strittmatter and colleagues found that when one patient from each of 30 predominantly late onset families with Alzheimer's disease was analysed, the frequency of a particular *APOE* allele was significantly increased among cases compared to 91 age-matched controls (Strittmatter *et al.* 1993). Amino acid sequencing had previously determined three major **isoforms** of ApoE, ε3, ε4 and ε2, which differed by particular amino acid substitutions (Fig. 2.8) (Mahley 1988). These could be determined by genotyping using different methodologies including

restriction enzyme digestion with the *HhaI* restriction enzyme (Hixson and Vernier 1990). It was the ε4 allele that was significantly increased in frequency among cases (0.5 ± 0.06) versus controls (0.16 ± 0.03) ($P = 0.01$) (Strittmatter *et al.* 1993).

As well as familial late onset Alzheimer's disease, the APOE ε4 allele was found to be associated with sporadic cases of the disease, including an early report detailing 176 cases of sporadic disease where diagnosis was confirmed at post mortem and the ε4 allele frequency was found to be 0.40 ± 0.03 ($P = 0.00001$) (Saunders *et al.* 1993). No association was found with early onset familial Alzheimer's disease. The disease association was robustly replicated for different populations by numerous independent researchers, showing similar results in men and women. These studies, together with very many other published disease association studies in Alzheimer's disease, are described in the 'AlzGene' online database maintained by the Alzheimer Research Forum (www.alzforum.org) (Bertram *et al.* 2007a). The strongest association is found when a family history of dementia is present and with two copies of the ε4 allele (ε4/ε4). There is also some evidence of a protective effect associated with possession of the *APOE* ε2 allele (Corder *et al.* 1994). Analyses of polymorphic variants in the region of *APOE* showed no evidence that the observed associations were due to linkage disequilibrium with other genetic variation in the locus (Strittmatter and Roses 1996).

A meta analysis published in 1997 included data for 5930 patients with Alzheimer's disease and 8607 controls, predominantly for populations of European ancestry (Fig. 2.9) (Farrer *et al.* 1997). Among Caucasians from clinics or with post mortem proven disease, possession of a copy of the ε4 allele (ε4/ε3) versus the common haplotype ε3/ε3 carried an odds ratio of 3.2 (2.8–3.8) for Alzheimer's disease, and possession of two copies (ε4/ε4) versus ε3/ε3 an OR of 14.9 (10.8–20.6); there was also evidence that possession of the ε2 allele was associated with some protection (ε2/ε3) versus ε3/ε3 with an OR of 0.6 (0.5–0.8) (Farrer *et al.* 1997). A more recent study among African Americans demonstrated similar age-related disease risk with ORs versus ε3/ε3 of 2.6 (1.8–3.7) for ε4/ε3 and 10.5 (5.1–21.8) for ε4/ε4 (Graff-Radford *et al.* 2002); among Hispanics the association also appears robust with a strong association with

	ε3 allele	ε4 allele	ε2 allele
Population frequency	77%	15%	8%
rs429358 c.388T>C	T	C	T
p.C130R	Cys	Arg	Cys
Flanking sequence	GTG[T]GCGG	GTG[C]GCGG	GTG[T]GCGG
Hha digest (cleaves GCGC)	No	Yes	No
rs7412 c.526C>T	C	C	T
p.R176C	Arg	Arg	Cys
Flanking sequence	AAG[C]GCCT	AAG[C]GCCT	AAG[T]GCCT
Hha digest (cleaves GCGC)	Yes	Yes	No

Figure 2.8 APOE alleles. Schematic illustration of SNP alleles and amino acid substitutions found with specific alleles.

late onset familial Alzheimer's disease reported in 2002 (Romas *et al.* 2002).

A gene dosage relationship for the ε4 allele was described by Corder and colleagues in 1993 who found that among 42 late onset disease families (which included 95 affected individuals of whom 90% were autopsy proven) the risk of disease increased 2.8-fold with each extra allele (Corder *et al.* 1993). More fundamentally, possession of the ε4 allele was found to be associated with younger age of onset such that the mean age of onset with no copies of the ε4 allele was 84.3 years; one copy of the ε4 allele was associated with a mean age of onset of 75.5 years; and two copies with a mean age of onset at 68.4 years (Corder *et al.* 1993). This strong association with age of onset was clearly demonstrated by several independent investigators (Blacker *et al.* 1997; Meyer *et al.* 1998), with a large prospective study showing that while possession of the ε4 allele accelerated the age of onset of disease, there was no difference in the 100-year lifetime incidence (Khachaturian *et al.* 2004).

The *APOE* ε4 allele is strongly associated with earlier age of onset of disease but is neither necessary nor sufficient for Alzheimer's disease to occur: 42% of people with Alzheimer's disease do not possess a copy of the ε4 allele; while possession of two copies of the ε4 allele, the strongest disease association seen, does not equate to certain development of the disease (Bird 2008). Indeed for a young asymptomatic individual, the ε4/ε4 genotype is estimated as being associated with a 30% lifetime risk of developing Alzheimer's disease (Breitner 1996). This should not detract from the strength of association seen with the *APOE* ε4 allele – it is remarkably strong for a common disease variant and is found across diverse populations. It is notable, for example, that of all known disease susceptibility loci, James Watson only requested that gene information relating to APOE was not publically released on completion of the resequencing of his genome (Section 1.4.5) (Wheeler *et al.* 2008a).

Early reports investigating amyloid deposition in sporadic late onset Alzheimer's disease showed that when

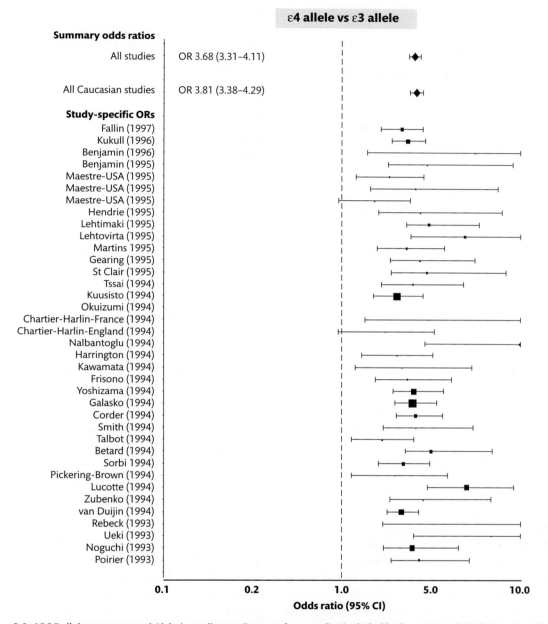

Figure 2.9 APOE allele genotype and Alzheimer disease. Data are from studies included in the meta analysis by Farrer and colleagues (1997) at the AlzGene Database (left) and genotype summaries for Alzheimer patients and controls from case–control studies categorized by ethnic group are shown on the right. Pooled odds ratio estimates are for ε4 versus ε3 allele. Accessed at www.alzgene.org on 19 September 2008 (Bertram *et al.* 2007b), redrawn with permission.

analysed in terms of *APOE* alleles, the ε4 allele was associated with increased Aβ deposition in plaques and cerebral vessels (Schmechel *et al.* 1993) and it is considered that the mode of action of this variant involves increased aggregation and reduced clearance of Aβ (Bird 2008). As such, like trisomy 21 and localized duplication of the *APP* gene locus, the *APOE* ε4 allele may be acting to accelerate the disease process involving Aβ accumulation and associated neuropathology leading to earlier presentation when present in conjunction with other (largely unknown) genetic and environmental factors.

There remain many more genetic factors to be defined. A candidate gene approach adopted by numerous studies has highlighted many loci for which further replication and analysis is warranted (such disease association studies are collated in the AlzGene online database www.alzforum.org/res/com/gen/alzgene) while genome-wide association studies are already revealing new associations (Waring and Rosenberg 2008). Reassuringly, Coon and colleagues found the strongest association was with rs4420638, a SNP in linkage disequilibrium with the ε4 variant (the latter was not part of the 502 000 Affymetrix SNP panel genotyped) (OR 4.01, $P = 5.3 \times 10^{-34}$) (Coon *et al.* 2007). It is worth noting however that had this single SNP not been genotyped, the association with *APOE* would have been missed highlighting the imitations inherent in SNP coverage in genome-wide association studies (Section 9.3). Analysis taking into account ε4 genotype may prove productive as highlighted by the association

with *GAB2* encoding GRB-associated binding protein 2 (Reiman *et al.* 2007). Variation in *GAB2* may modify disease risk associated with possession of the *APOE* ε4 allele as altered *GAB2* expression affected phosphorylation of the tau protein, a key component of neurofibrillary tangles and the pathology of Alzheimer's disease.

2.6 Common and rare genetic variants associated with venous thrombosis

Venous thrombosis is a common and potentially fatal disease of significant public health importance, in which the occurrence of recurrent events in an individual or family has long been recognized. The development of venous thrombosis has been associated with a diverse array of acquired and inherited risk factors (Box 2.11). The study of the role of genetic variation in determining disease risk illustrates the complex interplay between genetic and environmental factors, as well as the success which has been achieved in adopting largely a candidate gene approach based on knowledge of the underlying pathophysiology of the coagulation cascade and disease process. The high proportion of the disease risk attributable to genetic factors, which has now been defined by specific common and rare variants, is unusual for a multifactorial disease trait, as are the high relative risks associated with possession of particular variants, and the robustness with which these findings have been replicated. More work

---

Box 2.11   Thrombophilia (OMIM 188050) and venous thrombosis

Thrombophilia describes a tendency or propensity to venous thrombosis. Venous thrombosis is a potentially life threatening condition in which there is localized coagulation or clotting of blood within the vascular system. The disease is reported as having an annual incidence of one in 1000 adults (Cushman 2007), thrombosis most commonly occurring in the lower limbs with the risk of embolism to the pulmonary vasculature and death in 1–2% of cases (Rosendaal 1999). A major risk factor is increasing age, while many other inherited and acquired factors contribute to the risk of venous thrombosis, often occurring together. Thus an environmental trigger such as recent surgery, in a patient having a specific genetic variant predisposing to a hypercoagulable state, may lead to venous thrombosis. Acquired risk factors include immobilization, hospitalization, pregnancy, the oral contraceptive pill, female hormone treatments, surgery, trauma, the presence of specific antiphospholipid antibodies (lupus anticoagulant, anticardiolipin antibodies), and underlying malignancy.

remains to be done to resolve remaining genetic determinants. Adopting a non-hypothesis-driven approach based on genome-wide association scans is likely to yield additional variants in potentially unexpected genes whose magnitude of effect may be smaller, but whose importance for our ability to define risk and understand disease pathogenesis is likely to be significant.

In the normal physiological state a balance exists between pro-coagulant and anticoagulant mechanisms, allowing extravascular blood to clot while maintaining blood flow within the circulation. Central to this process is thrombin, which promotes haemostasis when generated by vascular injury with physiological control of bleeding through vasoconstriction, platelet aggregation and coagulation, leading to fibrin polymerization and clot formation. However on binding to the endothelial cell membrane protein thrombomodulin, thrombin promotes anticoagulant effects through activation of protein C, which cleaves and inactivates specific coagulation factors Va and VIIIa to limit clot formation. This protease activity of protein C is dependent on a cofactor, protein S.

Rare deficiencies of specific inhibitors of the pro-coagulant system (the natural anticoagulants) have been recognized for many years among families with a strong history of recurrent thrombotic events. In 1965, Egeberg and colleagues identified deficiency of antithrombin as a cause of thrombophilia (Egeberg 1965). Antithrombin is an inhibitor of thrombin and thought to be the most important physiological inhibitor of the coagulation pathway. Multiple rare variants have been described in the *SERPINC1* (previously known as *AT3*) gene encoding antithrombin at chromosome 1q23-q25.1 including missense, nonsense, and deletion variants affecting the function or levels of antithrombin protein. Homozygosity for antithrombin deficiency is thought to be lethal; heterozygotes have a ten-fold increased risk of thrombosis and are thought to be present among one in 2000 individuals. Protein C deficiency (Griffin *et al.* 1981) and protein S deficiency (Schwarz *et al.* 1984) are further conditions due to many different specific, very rare variants associated with risk of thrombosis in the heterozygous state but whose effects are typically modulated by other genetic and environmental factors. Unlike antithrombin deficiency they are found in the homozygous form, presenting with severe thrombosis soon after birth as neonatal purpura fulminans. Cumulatively, variation in the genes encoding antithrombin, protein C, and protein S are thought to account for only a small proportion of the genetic factors underlying thrombophilia, for example representing less than 5% of patients with familial forms of disease.

2.6.1 Factor V Leiden

A major breakthrough in this field was the discovery of a common variant involving the *F5* gene encoding coagulation factor V, which conferred resistance to cleavage by activated protein C, and was found to account for a substantial proportion of the genetic contribution to risk of venous thrombosis in the general population (Bertina *et al.* 1994). The discovery had its origins in the recognition that a patient with a recurrent history of venous thrombosis showed a poor anticoagulant response to activated protein C (Dahlback *et al.* 1993). This led to the development of a new assay to screen for this phenotype, 'activated protein C resistance', that was shown in the Leiden thrombophilia case–control study to occur in 5% of healthy individuals but in 21% of unselected consecutive patients with a first episode of deep vein thrombosis (Koster *et al.* 1993) and in over 40% of those with a recurrent or family history of thrombosis (Griffin *et al.* 1993; Svensson and Dahlback 1994).

Family studies suggested that resistance to activated protein C was inherited in an autosomal dominant fashion and the molecular basis for this was found to involve coagulation factor V (Bertina *et al.* 1994; Dahlback and Hildebrand 1994). Linkage analysis using microsatellite markers implicated chromosome 1q21-q25, specifically the *F5* gene locus, with activated protein C resistance. Resequencing of gene regions encoding the proposed activated protein C binding site and cleavage site in the coagulation factor V protein resolved a specific SNP in exon 10 of *F5*. The SNP, a G to A transition altered the codon from CGA to CAA leading to an amino acid substitution from arginine to glutamine at position 506 (p.R506Q) (c.1517G>A) (rs6025) (Bertina *et al.* 1994). The protein variant was named factor V Leiden after the Dutch city of Leiden where the study was based. Complete segregation was found in the affected family pedigree between heterozygosity for the SNP and activated protein C resistance.

Analysis of the Leiden thrombophilia case–control cohort showed that of the 64 patients and six controls with a phenotype of activated protein C resistance, 56 possessed the allelic variant while none of those without activated protein C resistance had the rarer A allele (Bertina *et al.* 1994). The variant modulated an activated protein C cleavage site, rendering the factor V Leiden protein significantly less sensitive to degradation and inactivation, and promoting a hypercoagulable state (Aparicio and Dahlback 1996). What was particularly striking was that this was a common SNP, originally reported as present in 2% of the Dutch population, with subsequent studies showing it to be common in those of European ancestry, with a prevalence of up to 15%, but rare or absent in African, East Asian, or other population groups (Rees *et al.* 1995; Dahlback 2008). Possession of one copy of the A allele (being heterozygous for factor V Leiden) is associated with a five-fold increased risk of venous thrombosis; having two copies leads to a 50-fold increased risk (Dahlback 2008). All those with the variant allele have the same haplotype at *F5* with evidence that this was a recent mutational event, estimated as occurring 21 000 years ago, after the proposed migration out of Africa (Zivelin *et al.* 2006). Whether there was any selective advantage to possession of this variant or associated alleles is unclear; reduced risk of severe bleeding after child birth has been proposed (Lindqvist *et al.* 1999; Dahlback 2008).

2.6.2 Genetic diversity and thrombophilia: insights and applications

Soon after this discovery, a candidate gene approach led to the identification of a further common variant associated with risk of venous thrombosis. Here the candidate was prothrombin, encoded by the *F2* gene at chromosome 11p11-q12, which is a precursor of thrombin. Poort and colleagues sequenced the 5′ and 3′ regions of the *F2* gene, together with the exons, and found a G to A transition in the 3′ untranslated region of the gene at nucleotide 20210, which was present in five out of 28 individuals with a personal or family history of venous thrombosis but none of the five controls (Poort *et al.* 1996). Analysis of the Leiden thrombophilia study case–control cohort, which included 424 unselected patients with deep vein thrombosis and 474 controls, showed 6.2% of cases and 2.3% of controls were heterozygous for the SNP (Poort *et al.* 1996). Possession of the SNP was associated with an

OR of 2.8 (1.4–5.6) for venous thrombosis and increased circulating levels of prothrombin, suggesting this or a linked polymorphism was acting as a **regulatory variant** modulating gene expression (Poort *et al.* 1996). Like the factor V Leiden SNP, in certain populations such as in southern Europe, the prothombin SNP is present at a relatively high frequency (2–4% of healthy individuals) but is rare in populations of non-European ancestry, with evidence of being a relatively recent mutation arising some 24 000 years ago (Zivelin *et al.* 2006).

The manifestation of disease in those with particular genetic variants is, however, highly variable such that some patients may never have a venous thrombosis while others will do so at a young age and have recurrent events. This reflects the multifactorial nature of the disease, with gene–gene and gene–environment interactions being important. A large meta analysis of 2130 cases and 3204 controls from eight case–control studies showed that possession of the factor V Leiden SNP was associated with an OR of 4.9 (4.1–5.9) for venous thrombosis, while the prothrombin SNP had an OR of 3.8 (3.0–4.9); possession of a copy of the variant allele for both SNPs was associated with an OR of 20 (11.1–36.1) for venous thrombosis and a significantly earlier age of having a first thrombotic event, suggesting a multiplicative rather than additive effect (Emmerich *et al.* 2001). Use of the oral contraceptive pill was associated with an odds ratio for venous thrombosis of 2.3 (1.7–3.0) which increased to 10.2 (5.7–18.4) in the presence of the factor V Leiden variant, and to 7.1 (3.4–165.0) with the prothrombin variant.

Research in this area has given us important insights into pathophysiology. Indications for thrombophilia screening are however currently controversial. It has been suggested that screening should be undertaken in those presenting with venous thrombosis at a young age (less than 50 years of age) or those presenting without apparent cause, or with a history of recurrent thrombotic events (Whitlatch and Ortel 2008). Others have argued that for the most common mutations, management is not altered and so screening rarely alters clinical care. Patients heterozygous for the factor V Leiden SNP or the prothrombin SNP alone do not have a clinically significantly increased risk of recurrence (Ho et al. 2006; Marchiori et al. 2007). Screening might however help in the counselling of asymptomatic relatives in terms of preventative measures or use of the oral contraceptive pill.

Those individuals possessing a copy of the risk allele for both the common SNPs, or being homozygous for factor V Leiden, or having antithrombin deficiency, will be rare but may be candidates for lifelong anticoagulation, similarly those with a history of recurrent or life threatening thrombotic events, antiphospholipid antibodies, or underlying cancer (Whitlatch and Ortel 2008).

2.7 Summary

The search for genetic determinants of disease has required remarkable perseverance and scientific endeavour, a task which has been facilitated by recent advances in human genetics, notably the completion of sequencing of the human genome and our understanding of the nature and extent of human genetic variation. Prior to this, the availability of increasingly polymorphic genetic markers and associated genetic and physical maps set the stage for linkage analysis and positional cloning to unlock the secrets of many rare monogenic diseases showing a clear phenotype with a mendelian pattern of inheritance. From the 1980s onwards over 1200 genes involved in such diseases have been defined (Botstein and Risch 2003). In this chapter a series of different diseases has been reviewed to highlight the theoretical basis of linkage analysis and positional cloning, and different issues that have arisen in their application. Pioneering work in cystic fibrosis, for example, highlighted the very considerable hurdles that had to be overcome to define the candidate region, identify and assemble clone DNA fragments, and define, prioritize, and interrogate transcribed regions before resolving a specific novel transcript in which it was possible to identify and screen for specific mutations. In most diseases showing mendelian inheritance investigated by linkage and positional cloning, a series of rare mutations were found to underlie the observed disease phenotype, usually altering the structure and function of the encoded protein. The situation was unusual for cystic fibrosis where a 3 bp deletion (the delta-F508 deletion) was found to account for 66% of cystic fibrosis chromosomes worldwide, although there are over 1600 other rare mutations of the *CFTR* gene identified.

For more common multifactorial traits not showing a clear mendelian segregation of inheritance, linkage analysis was less successful. Here it was becoming clearer

that several genes were likely to be involved and many variants, with an individually modest magnitude of effect. Genetic association studies were extensively employed using a candidate gene approach although only a relatively small number of examples of robustly replicated associations were described. There have been significant limitations to many such studies relating to definition of the disease phenotype, a low prior probability of success, underlying population stratification, the choice of a limited number of markers for analysis, and underpowered study design.

The cataloguing of the extent, nature, and coinheritance of genetic diversity is continuing to advance our ability to define the genetic basis for common disease, notably in terms of SNP diversity with high throughput genotyping technologies and the availability of large panels of clinical samples and controls enabling genome-wide association studies as reviewed in Chapter 9. The distinction and similarities of using genetic markers to define disease regions in linkage analysis and association studies have been reviewed, the former utilizing a limited number of informative recombination events in the family pedigrees analysed to have a relatively broad resolution of disease regions. This contrasts with association studies where historical recombination and mutational events, and varying linkage disequilibrium between variants, provide increased resolution on analysing individuals sampled from a population.

In this chapter several different diseases have been reviewed in detail and other examples are found elsewhere in this book. Alzheimer's disease provides an elegant example of a complex multifactorial trait in which a proportion of cases with early onset familial disease show a mendelian inheritance that has allowed rare variants in at least three different genes (*APP*, *PSEN1*, and *PSEN2*) to be defined by linkage. Gain of function mutations in these three genes appear to lead to common mechanisms involving accumulation of Aβ protein. Gene dosage effects involving the normal protein are also apparent for *APP* at chromosome 21q21 as seen with trisomy 21 in Down syndrome and gene duplication events involving *APP*. In late onset Alzheimer's disease a common variant, the ε4 allele, is significantly associated with age of onset of disease. This involves increased aggregation and reduced clearance of Aβ, but inheritance of this allele is neither necessary nor sufficient for disease to occur.

Research into the genetic basis of monogenic diseases showing mendelian inheritance has enabled the development of clinical genetics as a medical speciality, with application across medical disciplines to facilitate diagnosis and allow clear delineation of genetic risk. By contrast, research to determine specific genetic factors contributing to common multifactorial traits is in many ways less advanced. New insights into underlying disease pathogenesis have been defined but it is a complex picture, such that application to personalized medicine and precise delineation of individual risk associated with inheritance of particular alleles remains a significant hurdle to be overcome. Over the course of Chapters 3–9 different classes of genetic diversity are reviewed, set in the context of population and molecular genetics with examples relating to human disease and evolution as well as diversity in other species. We return specifically to the theme of defining the genetics of common disease in detail in Chapter 9, when SNP diversity is considered in greater detail and the steps which have led to genome-wide association studies are reviewed.

2.8 Reviews

Reviews of subjects in this chapter can be found in the following publications:

Topic	References
Linkage and positional cloning	Collins 1991, 1992, 1995; Ballabio 1993; Risch and Merikangas 1996; Botstein and Risch 2003; Strachan and Read 2004; Dawn Teare and Barrett 2005
Genetic association studies in complex disease	Risch and Merikangas 1996; Kruglyak 1999; Risch 2000; Cardon and Bell 2001; Ioannidis *et al.* 2001; Hirschhorn *et al.* 2002; Botstein and Risch 2003; Carlson *et al.* 2004; Healy 2006; Chanock *et al.* 2007; Kruglyak 2008; Manolio *et al.* 2008; McCarthy *et al.* 2008; Altshuler *et al.* 2008
Linkage disequilibrium	Slatkin 2008
Transmission disequilibrium test	Ewens and Spielman 1995
Cystic fibrosis genetics	Buchwald *et al.* 1989
Epidemiology and risk factors for venous thrombosis including genetic basis	Rosendaal 1999; Cushman 2007; Dahlback 2008; Whitlatch and Ortel 2008
Alzheimer's disease	Heston *et al.* 1981; Gatz *et al.* 2006
Genetics of Alzheimer's disease	Cruts and Van Broeckhoven 1998a; Bird 2008; Waring and Rosenberg 2008
TNF polymorphisms, biology and relation to disease	Tracey and Cerami 1993; Bayley *et al.* 2004; Knight 2005
Cytokine polymorphism in human disease	Hollegaard and Bidwell 2006

CHAPTER 3

Cytogenetics and large scale structural genomic variation

3.1 Introduction

Structural genomic variation is an integral part of our variable genomes, both in health and disease. Structural variation has been defined as involving segments of DNA more than 1 kb in size. This ranges from very large 'microscopic' structural variants (visible down a microscope and typically more than 3 Mb in size) to smaller 'submicroscopic' structural variation (Feuk *et al.* 2006a; Scherer *et al.* 2007). In this chapter the focus is on large scale microscopic structural variation, which was the subject of early research into human genetic diversity through cytogenetic approaches. As technology has advanced, our ability to resolve submicroscopic structural variation has dramatically improved, demonstrating that this class of structural genomic variation is much more common than previously thought (reviewed in Chapters 4 and 5). For both microscopic and submicroscopic structural variation, diversity may arise through copy number changes involving deletions, insertions, or duplications, or structural variation resulting in a positional change, for example balanced translocations, or an alteration in orientation, as seen with inversions (Fig 1.29) (Scherer *et al.* 2007).

In this chapter major classes of chromosomal variation are described ranging from gain or loss of whole chromosomes to specific chromosomal rearrangements and resulting structural variation such as deletions, duplications, inversions, and translocations. The impact of such diversity on human health is illustrated by discussion of a number of important associated clinical syndromes. For further information, a number of excellent reviews have been published on this subject (Section 3.7) and the interested reader is also referred to the diverse array of databases and other electronic resources that have been established to try and catalogue human structural variation and the phenotypes that may be associated with them (Box 3.1). A detailed description of terminology used to describe human chromosomes is also available, published as recommendations of the International Standing Committee on Human Cytogenetic Nomenclature (Shaffer and Tommerup 2005).

3.2 A historical perspective on cytogenetics

The study of human chromosome number and variations in chromosome structure has provided the basis for cataloguing and investigating human genetic diversity. Although human chromosomes were first described in the 1880s by Flemming and Arnold, it was only in 1956 that Tjio and Levan at the University of Lund in Sweden correctly reported that we possess 46 chromosomes within a normal cell rather than the long-held view that there were 48 (Fig. 3.1) (Tjio and Levan 1956). This work, combined with other conceptual and technological advances at the time, led to a very rapid period of development in the study of chromosomes and associated chromosomal abnormalities. This established human cytogenetics as a scientific discipline that has continued to play a key role in clinical medicine and the study of human genetics (Trask 2002). Some of the important discoveries and advances relating to cytogenetics are summarized in Fig. 3.2.

The careful association of chromosomal abnormalities with observed phenotypes in specific human disorders

---

Box 3.1   Electronic resources and databases of human structural genomic variation

A number of different databases and resources are available, in many cases including both microscopic and submicroscopic variation.

Large scale (microscopic) variation

- *Chromosomal Anomaly Collection* (www.som. soton.ac.uk/research/geneticsdiv/Anomaly%20 Register/). Database including large scale copy number variation in phenotypically normal individuals. Unbalanced chromosomal abnormalities without apparent phenotype are described, including duplications and deletions. Also includes large scale cytogenetically visible copy number variants.
- *Chromosome Abnormality Database* (www.ukcad. org.uk/cocoon/ukcad/index.html). Database containing constitutional and acquired abnormal karyotypes reported by UK regional cytogenetics centres.
- *DECIPHER: DatabasE of Chromosomal Imbalance and Phenotype in Humans using Ensembl Resources* (www.sanger.ac.uk/PostGenomics/decipher/). Database covering submicroscopic chromosomal imbalances (chromosomal microdeletions, duplications, insertions, translocations, and inversions). Clinical information and genomic location.
- *ECARUCA: European Cytogeneticists Association Register of Unbalanced Chromosome Aberrations* (http://agserver01.azn.nl:8080/ecaruca/ ecaruca.jsp). Database relating to rare chromosomal disorders including microdeletions and microduplications.

- *Human Structural Variation Project* (http://paralogy. gs.washington.edu/structuralvariation/). This database catalogues large and intermediate scale structural variation together with copy number polymorphism.
- *OMIM: Online Mendelian Inheritance in Man* (www.ncbi.nlm.nih.gov/omim/). An extensive and frequently updated summary of human genes, genetic variation, and genetic disorders written and edited by Johns Hopkins University and distributed by the National Center for Biotechnology Information (NCBI) (Hamosh *et al.* 2005).

Smaller scale (submicroscopic) variation

- *Copy Number Variation Project* (www.sanger. ac.uk/humgen/cnv/). Data for copy number variation within HapMap samples hosted at the Wellcome Trust Sanger Institute, UK.
- *Database of Genomic Variants* (http://projects. tcag.ca/variation/). A catalogue for data from healthy control subjects of structural variation (greater than 1 kb in size) and indels (100 bp to 1 kb).
- *dbSNP* (www.ncbi.nlm.nih.gov/projects/SNP/). A catalogue of human genome variation maintained by the NCBI (Sherry *et al.* 2001).
- *Human Gene Mutation Datatabase* (www.hgmd. org). Database of mutations causing or associated with human inherited disease (Stenson *et al.* 2003). Restricted to germline mutations and nuclear genes.

---

has proved extremely informative. In 1959, gain or loss of specific chromosomes was associated with a number of disorders, including Down syndrome due to possession of an extra copy of chromosome 21 (Lejeune *et al.* 1959), while gain of an X chromosome was found to cause Klinefelter syndrome (XXY) (Jacobs and Strong 1959), and its loss Turner syndrome (XO) (Ford *et al.* 1959). In

1963 the first inherited syndrome due to a chromosomal deletion was reported named 'cri du chat', a syndrome characterized by affected babies having a high pitched, cat-like cry due to deletion of some or all of the short arm of chromosome 5 (Box 5.9) (Lejeune *et al.* 1963). It was also quickly appreciated that the majority of spontaneous abortions could be attributed to chromosomal

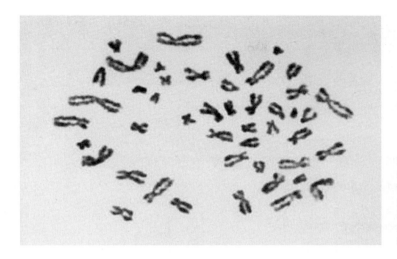

Figure 3.1 Human cells contain 46 chromosomes. Photomicrograph showing the normal human 46,XY chromosome number by Joe-Hin Tjio, December 1955. Reprinted with kind permission from Springer Science + Business Media: Human Genetics (Harper 2006).

abnormalities (Clendenin and Benirschke 1963), while development of amniocentesis allowed for the screening and identification of fetal chromosomal abnormalities. The key role of chromosomal defects in human cancer was also made clear by a number of studies, including the discovery in 1960 of the Philadelphia chromosome as a cause of chronic myeloid leukaemia (Nowell and Hungerford 1960), later shown to be the result of a translocation between chromosomes 9 and 22 (Rowley 1973). Here, as elsewhere in this book, our focus is on constitutional human genetic diversity, while detailed discussion of the essential and expanding role of cytogenetics in cancer is beyond the scope of this chapter.

A series of key technological developments have characterized advancement of the field of cytogenetics. Essential among these was the establishment of banding technologies such that not only could individual chromosomes be specifically identified, but also locations within them (Caspersson *et al.* 1968). Staining of chromosomes revealed dark and light bands, a 'barcode' with which deletions, inversions, insertions, and other rearrangements could be resolved and studied (Fig 1.2). This was later taken to progressively higher states of resolution. Approaches such as somatic cell hybridization allowed the mapping of genes and markers to specific chromosomes (Weiss and Green 1967; Ruddle *et al.* 1971). Flow cytometry and sorting proved to be very useful techniques in characterizing human chromosomal variation, both quantitatively and in separating normal and abnormal chromosomes (Carrano

et al. 1979). Hybridization of specific labelled DNA probes proved highly informative, particularly after the development of fluorescent labels, which proved safer and to give higher resolution than radioactive labelling (Landegent *et al.* 1985). **Fluorescence in situ hybridization (FISH)** techniques have grown in sensitivity to remarkable levels and have been used in the discovery of a number of key chromosomal rearrangements including the remarkable **imprinting** disorders Prader–Willi (Box 5.6) and Angelman syndromes (Box 5.7) (Knoll *et al.* 1989). The development of chromosome-specific paints and use of a combination of fluorochromes has allowed multiplex FISH or spectral karyotyping with increased automation in chromosome analysis.

Comparative genome hybridization (CGH) (Kallioniemi *et al.* 1992) has proved a very powerful genome-wide approach in cytogenetics to detect gain or loss of chromosomal material, notably with clinical application in cancer genetics. Here the hybridization of test and reference samples of genomic DNA (labelled with different fluorochromes) to sets of normal human chromosomes can be compared. The use of microarray technology in combination with comparative genome hybridization (array CGH) (Box 4.2) (Pinkel *et al.* 1998) has dramatically improved our ability to define and quantify structural genomic variation, notably submicroscopic structural variation. This has lead to an appreciation of the extent of copy number variation among healthy individuals (discussed in detail in Section 4.2).

Figure 3.2 Major advances related to human cytogenetics. Selected overview of advances leading to the development of modern cytogenetics. References can be found in the bibliography or within the text of the published reviews from which this figure was prepared (Trask 2002; Pearson 2006).

3.3   Chromosomal diversity involving gain or loss of complete chromosomes

3.3.1 Constitutional and somatic variation in chromosome number

Variation in the number of chromosomes in a cell is a common occurrence. When this affects all cells of the body, the abnormality is constitutional and is present from very early in development, originating from anomalies in the sperm, egg, fertilization, or very early embryogenesis. An estimated 10–30% of fertilized human eggs have an abnormal number of chromosomes, most often gain or loss of a single chromosome (trisomies and monosomies) (Hassold and Hunt 2001) (Box 3.2). Such events are usually lethal with about one-third of spontaneous abortions found to be aneuploid; a few are compatible with survival such that aneuploidy is the leading genetic cause of congenital birth defects and mental retardation (Hassold and Hunt 2001).

Somatic variation in chromosome number also occurs. Within a given individual, variation in the number of chromosomes is seen for particular cells or tissues. This somatic variation may be part of normal cellular physiology or a cancerous growth. Most human cells have two copies of the chromosome set (diploidy or 2n chromosomes), while sperm and egg cells have a single chromosome set and are haploid (n chromosomes). Some normally functioning cells within everyone are polyploid: these are usually found in highly differentiated tissues involving for example up to eight copies of the chromosome set (8n) in liver cells (hepatocytes) and heart muscle cells (cardiomyocytes), while the giant bone marrow megakaryocyte cells (precursors of platelets) can show extreme levels of polyploidy, typically with 16 but sometimes 64 copies (16n to 64n). Megakaryocytes achieve this extreme level of polyploidy through a variation on the normal cell cycle and by doing so are able to generate thousands of platelets from fragmentation of a single highly **polyploid** megakaryocyte cell (Ravid et al. 2002). Terminally differentiated cells such as circulating red blood cells, platelets, and keratinocytes (cells forming the outer layer of the skin) are nulliploid as they have no nucleus.

The occurrence of mitotic errors after conception and during embryogenesis may result in chromosomal **mosaicism** in which two or more genetically distinct cell lines are seen, the cell lines are genetically identical apart from the specific chromosomal difference resulting from the mitotic error. Post-zygotic non-disjunction for example may lead within a normal conceptus (46,N) to a cell line monosomy 21 (which is lost) while another is trisomy 21; the normal cell line also persists and the individual has a karyotype 46,N/47,+21. This may manifest in a specific phenotype if occurring early in embryogenesis, as a greater fraction of the tissue involved is likely have the chromosomal abnormality. Any consequence will critically depend on the tissue involved. Gonadal mosaicism is usually only recognized after the same *de novo* abnormality is found in two or more children.

Chromosomal missegregation leading to an abnormal number of individual chromosomes has been estimated to occur on average once every 10^4–10^5 cell divisions in mammals. It has long been recognized that almost all cancer cells are aneuploid and it is postulated that the cancerous state may arise through having extra copies of an oncogene or by losing a tumour suppressor gene (Lengauer *et al.* 1998). How such aneuploidy arises is

---

Box 3.2   Aneuploidy and polyploidy

Aneuploidy refers to the state of having an abnormal number of individual chromosomes. This may involve loss of a chromosome pair (**nullisomy**) or single chromosome (monosomy); or gain of a chromosome pair (tetrasomy) or single chromosome (trisomy). Segmental aneusomy involves a portion of a chromosome. Polyploidy involves having more than the normal complement of two chromosome sets (diploidy, 2n). Cells may, for example, be triploid (3n) or tetraploid (4n). There are 22 pairs of autosomes, numbered 1 to 22 from largest to smallest in size, and one pair of sex chromosomes, giving 23 chromosome pairs in a normal cell, which is denoted 46,XX in a female and 46,XY in a male (Fig 1.2).

unclear, but tumourigenesis is thought to involve abnormalities of the mitotic checkpoint that normally act to prevent missegregation of chromosomes.

3.3.2 Chromosomal abnormalities and development

There are very high rates of pregnancy wastage before or during implantation. Among those pregnancies that do become clinically recognizable, an estimated further 15% of conceptions are lost as spontaneous abortions between the 6th and 28th week of pregnancy. There are many reasons why an embryo may not survive, ranging from genetic to infectious, anatomical, immunological, or blood clotting aetiologies. Study of spontaneous abortions shows about half have evidence of chromosomal abnormalities with the vast majority involving numerical abnormalities in chromosomal number (Hassold et al. 1980; Eiben et al. 1990). For example, a cytogenetic study of 750 spontaneous abortions up to 25 weeks' gestation, of whom the majority were before the 12th week, showed 50% had an abnormal chromosomal complement (karyotype). Of these 62.1% were trisomies (most commonly involving chromosomes 16, 22, and 21), 12.4% showed triploidy, 10.5% had monosomy of the X chromosome, 9.2% had tetraploidy, and 4.7% had structural chromosomal anomalies such as translocations (Eiben et al. 1990).

The incidence of aneuploidy varies dramatically during development (Fig. 3.3) (Hassold and Hunt 2001). Among newborns, about 0.3% are aneuploid, most commonly trisomy of chromosome 21 (Down syndrome) or of the sex chromosomes (for example 47,XXX, 47,XXY, or 47,XYY).

The incidence is about ten-fold higher among still births (those fetuses that die between 20 weeks' gestation and term), at about 4%, while among spontaneous abortions about 35% of all conceptions are trisomic (chromosomes 16, 21, and 22) or monosomic (45,X, comprising 10% of spontaneous abortions). Among gametes, aneuploidy rates in sperm have been estimated at 2%, and in eggs at 20–25%; the latter derive mainly from studies of oocytes at in vitro fertilization (IVF) clinics (Jacobs 1992) but similar rates are reported from analysis of oocytes from unstimulated ovaries (Volarcik et al. 1998). Karyotyping spare diploid embryos from IVF or gamete intrafallopian transfer (GIFT) procedures showed nearly 20% are aneuploid (Jamieson et al. 1994).

3.3.3 Polyploidy

Triploidy is one of the commonest chromosomal abnormalities seen in humans and is usually rapidly lethal during development. Rarely triploidy is compatible with survival to term but not beyond a year: an infant has been reported with a 69,XXY chromosomal constitution who survived to nearly 11 months (Sherard et al. 1986). A detailed analysis of 91 triploid spontaneous abortions of less than 20 weeks' gestation showed that 66% (60/91) were of paternal origin, 37 arising from two sperm fertilizing an egg (dispermy), five from diploid (2n) sperm, one post-meiotic error, and 17 undetermined (Zaragoza et al. 2000). Among the 30% of maternal origin, most were because of a diploid egg arising from failure of meiosis. Tetraploidy (4n) is rarer than triploidy and always lethal. Typically the DNA has replicated to give four chromosomes but cell division failed to take place normally.

Gestation (weeks)			0 ———————— 6-8———— 20 ———— 40				
	Sperm	Oocytes	Pre-implantation embryos	Pre-clinical abortions	Spontaneous abortions	Still births	Live births
Incidence	1–2%	20%	20%	?	35%	4%	0.3%
Most common aneuploides	Various	Various	Various	?	45,X; +16; +21; +22	+13; +18 +21	+13; +18; +21 XXX; XXY; XYY

Figure 3.3 Aneuploidy during development. Overview of the incidence and nature of aneuploidy in reproductive cells and at different stages of gestation. Reprinted by permission from Macmillan Publishers Ltd: Nature Reviews Genetics (Hassold and Hunt 2001), copyright 2001.

3.3.4 Trisomy

Whole chromosome trisomies resulting from meiotic or mitotic non-dysjunction events are relatively common and if they involve the autosomes this is usually lethal for the developing embryo or fetus. Trisomies are present in many spontaneous abortions with trisomy of chromosome 16 occurring in approximately one in 13 of such cases. Trisomies of chromosomes 13, 18, or 21 are found among infants surviving to term, of which trisomy of chromosome 21 (Down syndrome) is the most common and compatible with survival to adulthood (Fig. 3.4; Box 3.3).

It is likely that several genes are involved in Down syndrome and the occurrence of rare individuals with partial trisomy 21 has allowed resolution of specific regions associated with some Down syndrome phenotypes such as congenital heart disease (Barlow *et al.* 2001). A gene dosage effect is postulated with 1.5-fold overexpression of the unbalanced gene (Amano *et al.* 2004). The sequencing of chromosome 21, the smallest of the autosomes, revealed 127 known genes (Hattori *et al.* 2000) and facilitated the hunt for genetic determinants of Down syndrome. Recent evidence implicates *DSCR1* and *DYRK1A* genes in the 'Down syndrome critical region' on chromosome 21 (Arron *et al.* 2006). These genes encode proteins that are key regulators of phosphorylation of NFATc (nuclear factor of activated T cells), a family of proteins involved in calcium signalling that activate expression of many genes involved in immunity and development. Allelic variability may contribute to the great phenotypic variation seen in this syndrome with 137 000 probable single nucleotide polymorphisms (SNPs) identified on chromosome 21.

Diagnosis at birth of Edwards syndrome (trisomy of chromosome 18) (Edwards *et al.* 1960) is rare, being present in one out of 8000 births and associated with growth failure, multiple congenital anomalies, and mental retardation. The major cause of death is congenital heart disease and less than 10% of infants survive beyond the first year of life. Patau syndrome (Patau *et al.* 1960), trisomy of chromosome 13, is even rarer, found in one per 12 000 live births and such infants usually die within days of cardiopulmonary complications. Why should trisomy of only these three autosomes be compatible with survival to term? The answer remains unclear but interestingly these chromosomes have relatively low numbers of genes.

In contrast to the autosomes, having an additional sex chromosome – as found in syndromes such as Klinefelter syndrome (47,XXY) (Box 3.4), triple X syndrome (47,XXX), and the rare Y polysomy (47,XYY) – cause relatively few problems and are compatible with a normal lifespan.

The characteristic clinical phenotypes associated with monosomies and trisomies are thought to relate to a relatively small number of major gene effects together with a larger number of smaller effects during development. Having an extra copy of a chromosome has been shown to be associated with higher gene expression. However, there is evidence that cells compensate for this with greater protein turnover, which requires more energy and exerts a selective disadvantage on those cells with an antiproliferative effect (Segal and McCoy 1974; Torres *et al.* 2007). The most significant risk factor for trisomy is maternal age: the striking relationship between increasing maternal age and Down syndrome was reported as long ago as 1933 (Penrose 1933). Overall the incidence of trisomy in clinically recognized pregnancies is about 2% for women under the age of 25 years and 35% for those over 40 years of age (Hassold and Hunt 2001).

3.3.5 Monosomy

Autosomal monosomies are lethal during early embryonic development. Similarly, loss of a sex chromosome is highly deleterious. Embryos 45,Y do not survive while those 45,X almost all spontaneously abort, but those who do survive have relatively few problems and features of Turner syndrome (Box 3.5).

In most cases of Turner syndrome the single X chromosome is maternally derived with the paternal X or Y chromosome lost either in meiosis or early in embryogenesis (Jacobs *et al.* 1997). The phenotypes seen in Turner syndrome are thought to arise due to having half the normal gene dosage (haploinsufficiency) of X-linked genes that escape X inactivation (Zinn and Ross 1998). Early on in normal female embryogenesis one X chromosome is inactivated, but in fact about 15% of the X chromosome genes remain variably active on both chromosomes (Carrel and Willard 2005). At least one such gene, the *SHOK* homeobox gene has been associated with short

Box 3.3 Down syndrome (OMIM 190685)

The clinical entity of Down syndrome has been described for about 150 years and was associated with trisomy of chromosome 21 in 1959 (Lejeune *et al.* 1959). Down syndrome occurs in about one in every 700 live births and is characterized by marked hypotonia (low muscle tone), a constellation of characteristic facial and other physical features, cognitive impairment ranging from mild learning difficulties to severe mental disability, and a high frequency of congenital heart disease and other major congenital abnormalities including gastrointestinal tract obstruction or dysfunction (for reviews see Antonarakis *et al.* 2004; Antonarakis and Epstein 2006). The occurrence of particular phenotypes and their severity are highly variable between individuals with Down syndrome and manifest at different times. For example, congenital heart disease is present in 40% of individuals while a degree of cognitive impairment is always seen.

(A)

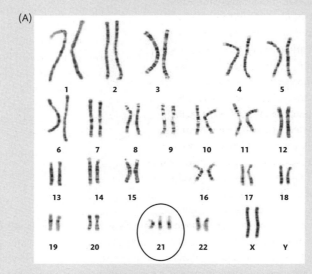

(B)

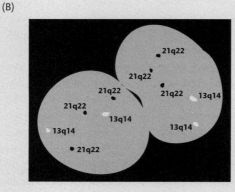

Figure 3.4 Trisomy 21. (A) Three copies of chromosome 21 are seen on a G-banded karyotype of a female with trisomy 21. (B) Schematic representation of results of fluorescent in situ hybridization (FISH) for interphase nuclei from a fetus with trisomy 21 using probes specific for the 13q14 and 21q22.13-q22.2 chromosomal regions. Redrawn and reprinted by permission from Macmillan Publishers Ltd: Nature Reviews Genetics (Antonarakis *et al.* 2004), copyright 2004.

Box 3.4 Klinefelter syndrome and sex chromosome aneuploidy

Klinefelter syndrome (47,XXY) is the most common disorder affecting the sex chromosomes described in humans and occurs in one of every 500 male births. In 1942 Harry Klinefelter described a series of nine men with testicular abnormalities who failed to produce sperm and had enlargement of the breast (Klinefelter *et al.* 1942). This was found in 1959 to be the result of having an extra copy of the X chromosome (Jacobs and Strong 1959). Most affected individuals are infertile and have some reduction in speech and language ability. Other karyotypes have more recently been identified including 48,XXYY (one in 17 000 live male births) and 48,XXXY (one in 50 000) (Visootsak and Graham 2006). A translocation of Y material, which includes the key sex determining region, to the X chromosome during paternal meiosis means that the rarely occurring 46,XX males have normal male sexual development (Ferguson-Smith 1966).

---

stature in Turner syndrome (Rao *et al.* 1997). *SHOK* was identified by positional cloning (Box 2.5) among individuals with short stature and sex chromosome abnormalities. It is found in the pseudoautosomal region at the tips of the short arm of chromosomes X and Y, a region of sequence identity within which recombination and genetic exchange between sex chromosomes is restricted. Genes in this region are expressed in diploid dosage in males and females, making it a candidate region for genes underlying the Turner syndrome phenotype. Interestingly, while haploinsufficiency is associated with short stature, gene overdose seen in sex chromosome polyploidy is associated with tall stature (Ogata *et al.* 2001).

3.4 Translocations

Chromosomal breakage and rearrangement may result in a translocation event in which there is transfer of chromosomal regions between non-homologous chromosomes. This may be constitutional, involving for example recombination between mispaired chromosomes in the meiotic steps of gamete formation; or be acquired and arise in somatic cells secondary to exogenous or endogenous agents causing double-stranded DNA breaks. Such agents include ionizing radiation, oxygen free radicals, and enzymes of DNA metabolism. Somatic cell translocations may result in cancer through upregulation of proto-oncogenes or downregulation of tumour suppressor genes.

Constitutional translocations may be classified into reciprocal translocations, occurring between any two non-homologous chromosomes, or Robertsonian translocations, which are specific to whole arm exchanges between a subset of chromosomes distinguished by having as their normal state tiny short arms (chromosomes 13, 14, 15, 21, and 22) (Shaffer and Lupski 2000; Strachan and Read 2004) (Fig. 3.5). There may be a net loss or gain of genetic material making the translocation unbalanced, or there may be no overall loss or gain making a balanced translocation. Carriers of balanced translocations typically have no adverse effects except relating to reproduction where infertility, recurrent spontaneous abortion, and risk of chromosomal imbalance among offspring occur.

3.4.1 Reciprocal translocations

Reciprocal translocations are common, being present in about one in 625 people in the general population (Van Dyke et al. 1983), and may result from single breaks in any two non-homologous chromosomes. While the exchange of two acentric fragments leads to a stable reciprocal translocation, exchange of centric and acentric fragments leads to chromosomes with two or no centromeres which are lost (Fig. 3.5A). For those carrying a balanced reciprocal translocation, there are usually no phenotypic consequences for the individual but it may lead to reproductive problems for their offspring. After fertilization with a normal gamete, a normal baby,

---

Box 3.5 Turner syndrome

The characteristic clinical features were first described by Henry Turner in 1938 as a syndrome of infantilism, congenital webbed neck, and a deformity of the elbow (cubitus valgus) (Turner 1938). In 1959 Ford and colleagues described a female patient with Turner syndrome in which there was sex chromosome monosomy, in other words only one copy of the X chromosome (Ford *et al.* 1959).

Turner syndrome is the phenotype associated with the absence of all or part of one X chromosome. Like Down syndrome, it is a complex and variable phenotype which includes short stature, ovarian failure, specific neurocognitive deficits, and anatomical abnormalities. The syndrome is found on average in one out of every 2500–3000 live female births.

---

a balanced carrier, or partial trisomies and monosomies may result (Fig. 3.6).

Translocation events involve the egg, sperm, or very early embryo and are almost always spontaneous, 'private' non-recurring rearrangements found only in a particular individual or family. Palindromic AT-rich repeat (PATRR) sequences are often found at breakpoints in translocations, including a recurrent reciprocal translocation involving chromosomes 11q23 and 22q11 (Box 3.6) (Inagaki *et al.* 2008). Elsewhere homologous clusters of olfactory receptor genes have been found on chromosomes 4p16 and 8p23 to contain breakpoints leading to t(4;8)(p16;p23) translocations, predisposed by mothers of cases having inversion polymorphisms in both the 4p and 8p regions (Section 5.5.1) (Giglio *et al.* 2002).

Rarely, reciprocal translocations can result in disease. For example, recurrent *de novo* reciprocal translocations have been reported involving the short arm of the X chromosome and a variety of different autosomes (including chromosomes 1, 3, 5, 6, 9, 11, 21) which result in the X-linked recessive disease Duchenne muscular dystrophy (Box 3.7) occurring in females where the normal X chromosome is inactivated, for example t(X;5)(p21.2;q31.2) (Greenstein *et al.* 1977; Nevin *et al.* 1986). This unusual scenario was seen to always involve the Xp21 region of the X chromosome and helped to define the likely site of the gene involved in causing the disease. A translocation involving chromosome 21, for example, allowed an elegant study in which probes could be designed based on the known ribosomal gene sequence found in the region of the translocation on chromosome 21 to clone the region spanning the breakpoint on the X chromosome (Ray *et al.* 1985).

3.4.2 Robertsonian translocations

Robertsonian translocations are the most common recurrent chromosomal rearrangement found in humans, occurring in about one in 1000 of the general population (Hamerton *et al.* 1975), and are restricted to chromosomes 13, 14, 15, 21, and 22. These 'acrocentric' chromosomes have very small short arms which comprise repeats of **satellite** DNA (very large arrays of tandemly repeated noncoding DNA) (Section 7.2) and ribosomal RNA genes (Page *et al.* 1996). In a Robertsonian translocation, breaks in the short arms near to the centromere leads to fusion of the remaining two long arms (Fig. 3.5B). This typically involves two different acrocentric chromosomes, or very rarely the same numbered chromosome, with one or both centromeres. As they are very close together, the two centromeres are functionally one, and the chromosome can segregate normally. There is a net loss of genetic material as the acentric fragment from the two short arms is lost, however this has little phenotypic effect and therefore it is regarded as a balanced translocation. The consequences of being a carrier of a balanced Robertsonian translocation can, however, be severe for future reproduction with the possibility, after fertilization with a normal gamete, of trisomy or monosomy of the chromosomes involved in the translocation (Fig. 3.7).

Approximately 85% of all Robertsonian translocations involve whole arm exchanges between chromosomes 13 and 14, or 14 and 21, which are denoted as rob(13q14q) and rob(14q21q) (Therman *et al.* 1989). For these translocations there is evidence that breakpoints recur in specific regions: between specific satellite DNA repeats for 14p, and between a satellite DNA and ribosomal DNA

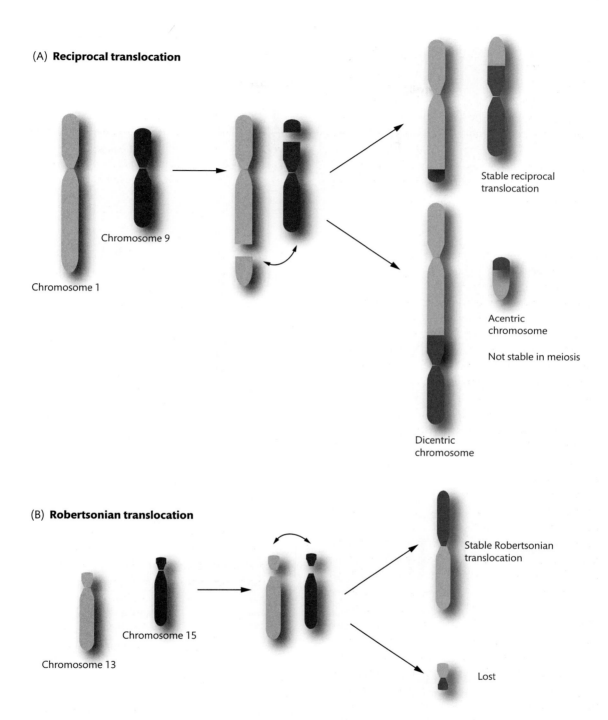

Figure 3.5 Reciprocal (A) and Robertsonian (B) translocations. Schematic representation of the processes involved in the two major classes of translocation. Redrawn and adapted with permission from Strachan and Read (2004).

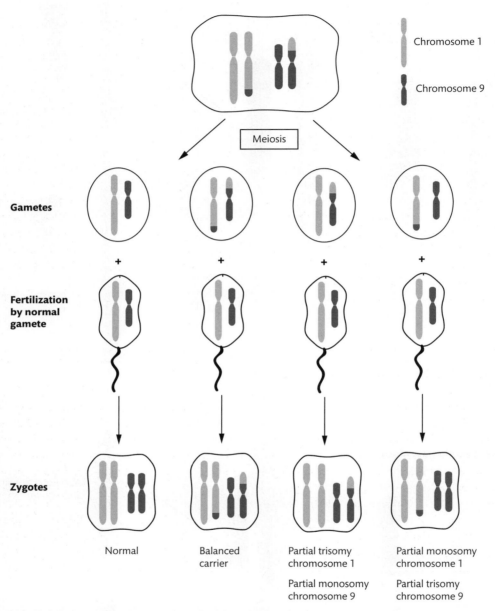

Figure 3.6 Meiosis in a carrier of a balanced reciprocal translocation. Partial monosomy or partial trisomy may result from unbalanced gametes of a carrier of a balanced reciprocal translocation. Redrawn and adapted with permission from Strachan and Read (2004).

Box 3.6   Palindromic AT-rich repeats and recurrent reciprocal translocation

The most common recurrent reciprocal translocation found in man involves breakpoints on chromosomes 11q23 and 22q11. This translocation is denoted t(11;22)(q23;q11) and has been reported in more than 160 unrelated families. The propensity for such recurrent events was resolved to specific PATRR sequences found at the chromosomal breakpoints resulting in unstable secondary structures (Edelmann *et al.* 2001; Kurahashi and Emanuel 2001; Kurahashi *et al.* 2004, 2007). PATRRs associated with translocation comprise several hundred bases of DNA of a near perfect **palindrome** (sequence reading the same in either direction) with an AT-rich centre and another flanking AT-rich region; they can self pair leading to hairpin or cruciform structures which are unstable and hotspots for recombination. Among healthy individuals, the size of the PATRR at 11q23 varies significantly and was an important determinant of rates of *de novo* translocations in sperm (Kato *et al.* 2006). The risk for balanced carriers of the t(11;22)(q23;q11) translocation is of the supernumerary der(22)t(11;22) syndrome or Emanuel syndrome (OMIM 609029) (Zackai and Emanuel 1980) occurring in their children due to a 3 : 1 meiotic non-disjunction event (Shaikh *et al.* 1999). The risk of such an unbalanced event occurring is highest for female heterozygotes (estimated at 10% risk). Affected individuals show a characteristic syndrome of mental retardation, craniofacial abnormalities, and congenital heart defects. Chromosome 22q11 is a region prone to non-allelic homologous recombination and recurrent genomic rearrangements (reviewed in Section 5.2.2).

Box 3.7   Duchenne muscular dystrophy (OMIM 310200)

This devastating neuromuscular disorder affects one in 3300 male live births with a progressive muscular dystrophy. The mean age at diagnosis is between 4 and 6 years, the ability to walk is generally lost at 8–10 years of age, with death in the late teens or early twenties through respiratory and cardiac failure (Biggar *et al.* 2002). Skeletal and cardiac muscle is affected with defective membrane stability; necrosis and regeneration of muscle fibres is seen (Deconinck and Dan 2007). About 30% of cases arise due to spontaneous mutations, the remainder show X-linked recessive inheritance. Affected individuals lack dystrophin, in about 70% of cases due to deletions of one or multiple exons of the dystrophin (*DMD*) gene at Xp21; about 20% are due to small deletions, insertions, or point mutations (Hoffman *et al.* 1987; Koenig *et al.* 1987; Davies *et al.* 1988; Aartsma-Rus *et al.* 2006). *DMD* is a remarkably large gene spanning 2.3 Mb with 79 exons and at least seven promoters.

(rDNA) for 13p and 21p (Page *et al.* 1996; Bandyopadhyay *et al.* 2002). It may be that the particular sequence or genomic architecture in these regions predisposes to this, with translocations then resulting from meiotic mispairing between two non-homologous chromosomes and recombination involving specific highly homologous sequences found on the short arms (Bandyopadhyay *et al.* 2002). Such events occur predominantly during oogenesis; overall 93% of Robertsonian translocations are maternal in origin (Bandyopadhyay *et al.* 2002).

3.5 Chromosomal rearrangements

3.5.1 Large scale structural variation resulting from intrachromosomal rearrangements

Within a single chromosome, or homologous pair of chromosomes, a number of cytogenetically visible variants have been identified. These may be classified into interstitial and terminal deletions, interstitial and terminal duplications, **marker chromosomes**, inversions, and **isochromosomes** (Fig. 3.8) (Shaffer and Lupski 2000). A deletion or duplication is considered interstitial if there is exchange within a chromosomal arm, and the original telomere is retained. Interstitial deletions and duplications are thought to occur in at least one in 4000 of the

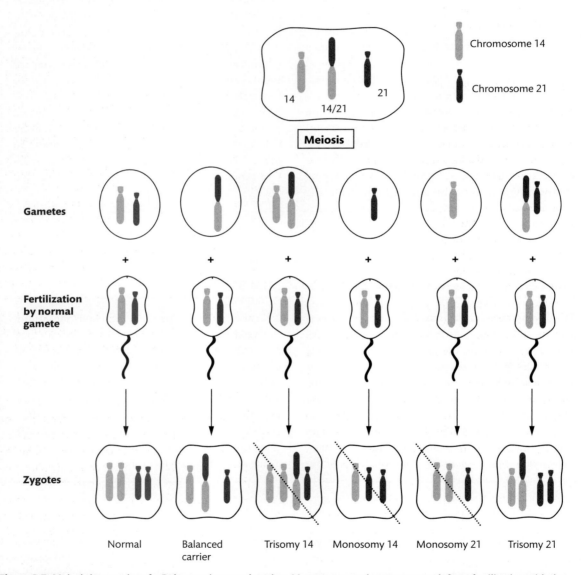

Figure 3.7 Meiosis in a carrier of a Robertsonian translocation. Monosomy or trisomy may result from fertilization with the unbalanced gametes of a carrier of a Robertsonian translocation. Redrawn and adapted with permission from Strachan and Read (2004).

general population and have been associated with specific disorders (Fig. 3.9). There are certain genomic regions that appear more or less susceptible to deletions and duplications. Some regions have never been observed to undergo such events and this may be because of the critical role of the gene, or of a change in gene dosage in development such that those embryos are lost. In general there appears to be greater tolerance of trisomy than haploinsufficiency, with only 2.1% of possible autosomal bands never seen to be involved in any duplication compared with 11% in any deletion (Shaffer and Lupski 2000).

3.5.2 Genomic disorders

Genomic disorders are a diverse group of genetic diseases in which constitutional genomic rearrangements result in gain, loss, or disruption of dosage sensitive genes (Lupski 1998). Individually, genomic disorders are rare, but collectively they account for a large proportion of birth defects (0.7% of live births). Most genomic disorders include association with mental retardation and congenital malformations, but there is great diversity in observed phenotypes for genomic disorders, specific to the dosage sensitive gene(s) involved in a particular disease. Some are associated with congenital heart disease as part of their phenotype such as DiGeorge and velocardiofacial syndrome, which may result from a 3 Mb deletion at 22q11 (Box 5.2), while a 1.5 Mb deletion at 7q11.23 involving the elastin gene results in Williams–Beuren syndrome and abnormalities of connective tissue and the nervous system (Box 5.3). In other cases genomic disorders may involve imprinted genes, as seen at 15q11–13 where a 4 Mb deletion involving the paternally imprinted genes (including *SNRPN*) is an important cause of Prader–Willi syndrome (Box 5.6), while deletion involving the maternal allele, notably the imprinted gene *UBE3A*, leads to Angelman syndrome (Box 5.7) (Ledbetter *et al.* 1981; Knoll *et al.* 1989).

Genomic disorders typically arise due to chromosomal rearrangements occurring recurrently as *de novo* events at specific genomic locations where the genomic architecture predisposes to such events, for example due to high sequence homology at segmental duplications (Section 5.2.1) (Inoue and Lupski 2002). Large scale rearrangements involving microscopically visible regions of DNA imbalance may result in genomic disorders. However, it is now clear that most cases arise due to submicroscopic events and genomic disorders are considered in greater detail in Section 5.2.

3.5.3 Marker chromosomes

Marker chromosomes are small, structurally abnormal chromosomes that are found in addition to the normal complement of 46 chromosomes (Liehr and Weise 2007). They occur in about one in 2000 of the general population. Typically they have no phenotypic consequences with 70% of individuals possessing marker chromosomes reported to be clinically normal. Most occur *de novo* and an estimated 0.04% of newborn infants have a marker chromosome: worldwide there are 2.7 million carriers of such marker chromosomes. About 30% originate from chromosome 15; other common origins are the X chromosome and chromosome 22. Clinical syndromes are associated with such events, the cat eye syndrome, for example, is associated with a supernumerary small chromosome 22 (Box 3.8). Marker chromosomes of chromosome 15 origin comprise two copies of the short arm in an inverted orientation which may be associated with mental retardation. About 10% of marker chromosomes are found to be in a circular conformation (Blennow *et al.* 1994). These small supernumerary ring chromosomes have a highly variable correlation with phenotype and about 40% are thought to be inherited.

3.5.4 Isochromosomes

Isochromosomes result from the duplication of one chromosome arm. They may arise when the centromere splits transversely rather than longitudinally: the resulting chromosomes will thus have either two identical long arms of the original chromosome but no short arms, or two short arms and no long arms. They may also result from a U-type exchange between sister **chromatids** in the short arm giving a dicentric and an acentric fragment (Fig. 3.8B).

The most common isochromosomes seen in humans involves the long arm of the X chromosome at a rate

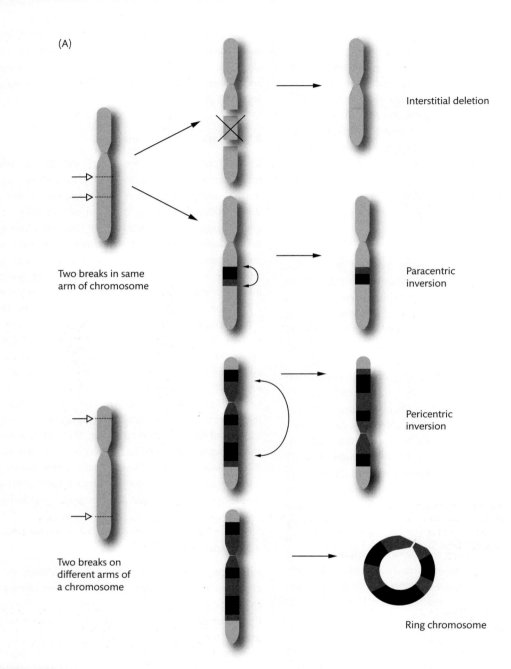

(A)

Two breaks in same arm of chromosome

Two breaks on different arms of a chromosome

Interstitial deletion

Paracentric inversion

Pericentric inversion

Ring chromosome

Figure 3.8 *Continued*

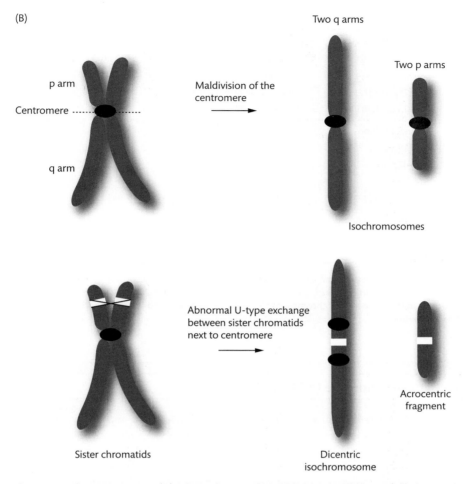

(B)

p arm

Centromere

q arm

Maldivision of the
centromere

Two q arms

Two p arms

Isochromosomes

Sister chromatids

Abnormal U-type exchange
between sister chromatids
next to centromere

Dicentric
isochromosome

Acrocentric
fragment

Figure 3.8 Intrachromosomal rearrangements. **(A)** Schematic examples are shown of some large scale chromosomal rearrangements involving either two breaks in the same arm of a chromosome or two breaks on different arms of a chromosome. Redrawn and adapted with permission from Strachan and Read (2004). **(B)** Generation of isochromosomes. Redrawn and adapted from Shaffer and Lupski (2000) with kind permission of Annual Reviews.

of about one in 13 000 people. Among those individuals with Turner syndrome, the percentage with an isochromosome of Xq is high at 15%. The isochromosomes are characteristically dicentric with the breakpoint in the proximal short arm of the X chromosome, and may be of maternal or paternal origin (James *et al.* 1997). Any of the acrocentric chromosomes have also been found to form isochromosomes and these underlie a significant proportion of rearrangements seen in spe-

cific syndromes, for example Patau (chromosome 13), Prader–Willi or Angleman (chromosome 15), and Down syndrome (chromosome 21) (Shaffer and Lupski 2000).

3.6 Summary

Cytogenetics has been fundamental to our understanding of human genetic variation. At a microscopic level,

Chromosomal anomaly	Syndrome or disorder	Estimated frequency
Interstitial deletions		
· del(7)(q11.23q11.23)	Williams	1 in 20 000–50 000
· del(8)(q24.1q24.1)	Langer–Giedion	–
· del(15)(q12q12)	Prader–Willi or Angleman	1 in 20 000
· del(22)(q11.2q11.2)	DiGeorge/velocardiofacial	1 in 4000
Terminal deletions		
· del(1)(p36.3)	Monosomy 1p	1 in 10 000
· del(4)(p16)	Wolf–Hirschhorn	1 in 50 000
· del(5)(p15)	Cri du chat	1 in 50 000
· del(17)(p13.3)	Rubinstein–Taybi	1 in 125 000
Interstitial duplications		
· dup(7)(p12p13)	Russell–Silver	–
· dup(17)(p12p12)	Charcot–Marie–Tooth type 1A	1 in 2500

Figure 3.9 Human chromosomal rearrangements associated with specific genetic disorders. Examples of some different deletions (denoted del) and duplications (dup) are shown, for example del(7)(q11.23q11.23) which is an interstitial deletion arising from two breaks in the same arm of chromosome 7 at q11.23. Reprinted from Shaffer and Lupski (2000) with kind permission of Annual Reviews.

Box 3.8   Cat eye syndrome (OMIM 115470)

This syndrome is associated with an additional marker chromosome 22, which typically has two centromeres and is structurally abnormal with an inversion duplication involving 22.q11. The characteristic ocular features and absence of an anal opening were reported in 1965 by Schachenmann and colleagues (1965). Cat eye syndrome has since been found to be a rare and highly diverse syndrome that includes abnormality of the iris (coloboma, due to failure of fusion of the intraocular fissure), anal atresia with fistula, down slanting openings of the eyes between the eye lids (palpebral fissures), tags of ear tissue usually just in front of the ear (preauricular tags), and frequent heart and renal abnormalities.

---

an appreciation of structural genetic diversity was rapidly established in terms of gain or loss of whole chromosomes, or major chromosomal rearrangements. As cytogenetic techniques have improved in sensitivity and resolution, our appreciation of the extent and diversity of smaller scale structural variation at a submicroscopic level has more recently advanced. This has brought a clearer understanding of the nature and mechanisms underlying chromosomal rearrangements and, where these involve dosage sensitive genes, recognition of a class of genetic diseases dubbed genomic disorders (Lupski 1998).

Aneuploidy is commonly seen in eggs and to a lesser extent sperm; among spontaneous abortions about a third of conceptions have a gain or loss of a single chromosome; while in still births the incidence is about 4%, and among newborns 0.3% (Hassold and Hunt 2001). The latter is most commonly trisomy of chromosome 21 (Down syndrome) and the sex chromosomes, or absence of all or part of the X chromosome leading to Turner syndrome. The relatively low number of genes on the particular autosomes (chromosomes 21, 18, and 13) where trisomy is compatible with survival to term is thought to be important but what determines the phenotypic diversity seen within particular syndromes associated with aneuploidy remains unclear. Gene dosage effects are important and specific genes/gene regions have been implicated

Chromosomal rearrangements within and between chromosomes have been described over the course of this chapter, including an overview of translocations and an introduction to genomic disorders. Over the coming chapters we will explore the extent and diversity of submicroscopic and fine scale structural variation found among normal individuals, and those associated with diverse phenotypic consequences including severe disease.

3.7   Reviews

Reviews of subjects in this chapter can be found in the following publications:

Topic	References
Historical perspective of cytogenetics	Trask 2002; Pearson 2006
Chromosome abnormalities	Gardner and Sutherland 2004
Aneuploidy	Hassold and Hunt 2001
Down syndrome	Antonarakis and Epstein 2006
Klinefelter syndrome	Visootsak and Graham 2006
Turner syndrome	Jacobs et al. 1997; Zinn and Ross 1998
Duchenne muscular dystrophy	Biggar et al. 2002
Structural genomic variation	Feuk et al. 2006a
Mechanisms of chromosomal rearrangements and genomic disorders	Shaffer and Lupski 2000; Stankiewicz and Lupski 2002; Lupski and Stankiewicz 2005

CHAPTER 4

Copy number variation in health and susceptibility to disease

4.1 Introduction

Submicroscopic structural genomic variation includes genomic alterations involving segments of DNA more than 1 kb in size and typically less than 3 Mb, which result in a change in DNA dosage, referred to as 'copy number variants' (Box 4.1), together with chromosomal rearrangements leading to a change in position or orientation (Scherer *et al.* 2007). In this chapter the particular focus is on copy number variants that include deletions, duplications, or insertions, and such genomic loci may be biallelic or multiallelic. Biallelic copy number variable loci involving deletions have a diploid copy number of none, one, or two whereas those involving duplications typically have two, three, or four copies. Where duplications and deletions involve the same locus, multiallelic copy number variation is seen, ranging for example from none, one, two, three, or four diploid copies for the gene *FCGR3B* (Section 4.5.2) (Aitman *et al.* 2006; Redon *et al.* 2006); from one to six copies of the chemokine gene *CCL3L1* among most individuals (Section 4.5.3) (Gonzalez *et al.* 2005); while at *TSPY* gene array on the Y chromosome, copy number ranges from 23 to 64 in men (Repping *et al.* 2006).

In this chapter copy number variation is described, demonstrating how the nature and extent of such variation within normal human populations has only relatively recently been appreciated and shown to be widespread (Iafrate *et al.* 2004; Sebat *et al.* 2004). The impact of such variation on gene expression is reviewed before a detailed discussion of the impact of such variation on common disease, metabolism, and heritable traits such as colour vision and rhesus blood group. This subject is continued in Chapter 5 where pathogenic copy number variation and other classes of submicroscopic structural variation are described, including a detailed review of the causes and consequences of genomic disorders resulting from chromosomal rearrangements that change gene copy number.

4.2 Surveys of copy number variation

The process of defining and mapping copy number variation has been greatly facilitated and enabled by

---

Box 4.1 Copy number variation and polymorphism

Scherer and colleagues define **copy number variation** as DNA segments greater than 1 kb in size in which a comparison of two or more genomes reveals gains (by insertion or duplication) or losses (by deletions or null genotypes) of genomic copy number relative to a designated reference genome sequence; **copy number polymorphism** is present when such variation is present in more than 1% of the reference or general population (Scherer *et al.* 2007).

technological advances to detect submicroscopic structural variation at a genome-wide level. These include microarray-based comparative genomic hybridization (array CGH) (Box 4.2) using oligonucleotides (Sebat *et al.* 2004; Hinds *et al.* 2005) and bacterial artificial chromosome (BAC) clones (Iafrate *et al.* 2004; Sharp *et al.* 2005; Redon *et al.* 2006; Wong *et al.* 2007), comparing clone paired-end sequence to the reference human assembly sequence (Tuzun *et al.* 2005) or human genome assemblies (Khaja *et al.* 2006) and detection of deletions and duplications based on single nucleotide polymorphism (SNP) mapping (Conrad *et al.* 2006; McCarroll *et al.* 2006) (Fig. 4.1). These techniques have bridged the gap between classical cytogenetic techniques using microscopy to detect structural variation greater than 3–15 Mb in size (depending on the banding pattern of the chromosome), and molecular approaches for the detection of

small scale variation including targeted fluorescence in situ hybridization (FISH) (25–300 kb), multiplex ligation-dependent probe amplification (MLPA), quantitative polymerase chain reaction (PCR), and DNA sequencing (1–700 bp) that can detect down to the resolution of single nucleotide changes.

4.2.1 Copy number variation is common within normal populations

In 2004 two landmark papers were published which demonstrated that among apparently phenotypically normal individuals there was a much higher level of large scale copy number variants than previously appreciated (Iafrate *et al.* 2004; Sebat *et al.* 2004). Both research teams used microarray technology to hybridize DNA from panels of healthy people and identify copy number variation based

---

Box 4.2   Using DNA microarrays to analyse copy number variation

Array CGH has proved a powerful method of identifying copy number variation (reviewed in Carter 2007). Conventional metaphase CGH has its origins in tumour biology, in particular analysis of metaphase chromosomes (a stage in mitosis when chromosomes are condensed and centrally aligned prior to separation into two daughter cells) (Kallioniemi *et al.* 1992). Test and reference DNA samples were labelled with different fluorochromes and analysed by FISH. The relative intensities of fluorescence for the two fluorochromes analysed along the chromosomes allowed the detection of regions of gain or loss but afforded only low resolution, down to 5–10 Mb and often less, at telomeres. Arrays spotted with large insert clone DNAs originally prepared for sequencing as part of the Human Genome Project were subsequently used and provided much improved resolution, with test and reference DNAs hybridized together to the array after fluorescent labelling each with different dyes (Pinkel *et al.* 1998). Iafrate and colleagues, for example, used one BAC clone every 1 Mb (Iafrate *et al.* 2004). Using BACs provides genome-wide coverage and

low noise but the resolution is still limited (maximum theoretical resolution is the size of one BAC, ~100–300 kb, but ~2–3 Mb in practice); this can be improved with fosmid and cosmid clones and cDNA clones, and PCR products have also been used. However, to date the highest resolutions are achieved with oligonucleotide arrays, theoretically as low as 5 kb using highest density arrays containing for example 2 million probes (Nimblegen HD2) (Carter 2007). There are limitations of oligo arrays in terms of signal to noise ratios and coverage in repeat regions rich in **low copy repeats** (**LCRs**) and segmental duplications. Genotyping (SNP) arrays have also been used successfully, based on hybridizing a single DNA sample and determining relative copy number variation compared with a reference set of individuals (Redon *et al.* 2006). The challenge remains how to identify copy number variation in the range 500 bp to 5 kb and this may be facilitated by lower cost high throughput sequencing by synthesis on arrays, for example the Roche-Nimblegen 454 and Illumina (Solexa) systems (Bentley 2006).

Reference	Coverage	Analysis	Number of individuals	Number of events or regions	Size range (bp)	Average size (bp) Median size (bp)	Total bp
Mills 2006	16 million whole genome shotgun traces	Alignment of sequence traces from SNP Consortium resequencing	36	415 434	1–9 989	20 / 2	8 360 235
Conrad 2006	1.3 million SNPs	HapMap SNP genotyping data mining*	180 (CEU and YRI)	609	25–993 000	34 996 / 17 217	21 313 127
McCarroll 2006	1.3 million SNPs	HapMap SNP genotyping data mining**	269 (CEU, YRI, and CHB+JPT)	538	96–745 418	16 874 / 6887	9 078 084
Hinds 2006	100 million to 200 million bp	Oligonucleotide array hybridization	24	1000	72–8 001	1379 / 947	137 912
Tuzun 2005	8 coverage fosmid library	Paired-end sequencing	1	297	700–1 944 156	55 706 / 25 230	14 984 826
Iafrate 2004	5264 BACs	BAC array-CGH	55	246	19 597–337 967	146 189 / 150 395	35 962 540
Sharp 2005	1986 BACs	BAC array-CGH	47	124	29 514–410 301	170 019 / 164 704	21 082 320
Sebat 2004	85 000 oligonucleotides	ROMA CGH	20	76	754–1 698 859	350 670 / 199 800	25 248 203
Wong 2007	26 363 BACs	BAC array-CGH	105	1365	50 459–1 037 332	185 504 / 175 314	253 212 685
Redon 2006	26 574 BACs	BAC array-CGH	269 (CEU, YRI, and CHB+JPT)	913	2639–7378760	349 880 / 227 889	319 440 476
Redon 2006	500 000 SNPs	Affyx 500K SNP array analysis	269 (CEU, YRI, and CHB+JPT)	980	1033–3605436	165 996 / 63 140	162 675 683
All variations	NA	NA	NA	323 573	1–7 442 054	1901 / 2	615 095 095
All variations >1 kb	NA	NA	NA	4131	1004–7442053	148 578 / 93 356	613 774 371

Figure 4.1 Surveys of structural variation in the human genome 2004–2007. Populations from the International MapMap Project analysed include CEU (Utah residents with ancestry from northern and western Europe), YRI (Yoruba in Ibadan, Nigeria), and CHB+JPT (Han Chinese in Beijing, China and Japanese in Tokyo). * HapMap SNP genotyping data mining based on mendelian inconsistencies; ** based on null genotypes, mendelian inconsistencies, and deviations from Hardy–Weinberg. BAC, bacterial artificial chromosome; CGH, comparative genomic hybridization; NA, not applicable; ROMA, representational oligonucleotide microarray analysis; SNP, single nucleotide polymorphism. References can be found in the bibliography or within the text of the published review from which this figure was prepared. Reprinted by permission from Macmillan Publishers Ltd: Nature Genetics (Cooper *et al.* 2007), copyright 2007.

on comparative gains or losses of DNA. Iafrate and colleagues used large insert clone arrays to cover about 12% of the genome with a detection size of approximately 50 kb, while Sebat and coworkers used much smaller probes (oligonucleotides) with on average one probe every 35 kb (detection size 105 kb). Both technologies allowed only limited resolution and genomic coverage, and represented only a fraction of the likely variation across the genome. However, the unexpected scale of copy number variation that was found promoted great research interest and further high resolution analyses.

Iafrate and colleagues analysed 55 unrelated individuals of whom 16 had chromosomal imbalances (the latter group were used to assess the sensitivity and specificity of the approach: all expected abnormalities were found) (Iafrate *et al.* 2004). DNA from this panel of individuals was hybridized to the array and compared to pooled male or female genomic DNA from karyotypically and phenotypically normal individuals. In total 255 copy number variants were found (Fig. 4.2), an average of 12.4 per individual, with 102 occurring in more than one individual, 24 in more than 10% of people, and six in more than 20%. The most common copy number polymorphism identified, present in nearly half of people analysed, involved a 150–425 kb region spanning *AMY1A* and *AMY2A* (amylase alpha loci) with gain or loss of the region in an equal proportion of those bearing the polymorphism. Entire genes were spanned by 67 of the 255 copy number variants. The authors believed that most of the variation represented tandem copy number changes and established the Database of Genomic Variants as a repository for structural genomic variation (http://projects.tcag.ca/variation/).

Sebat and colleagues (2004) analysed a smaller number of individuals (20) and identified 76 unique copy number variants, only five of which were previously known. They found that among the individuals studied, on average any two individuals differed by 11 such variants (a similar figure to Iafrate and colleagues) with an average length of 465 kb. Cytogenetic analysis confirmed a high proportion of variants tested. In this study the copy number variants were again widely distributed across chromosomes although clusters of three or more variants were noted at regions of chromosomes 6, 8, and 15, suggestive of hotspots of variation. A six-fold excess of segmental duplications was noted among deleted

regions, in duplicated regions a 12-fold excess was noted (see Section 4.2.6 for a discussion of the role of segmental duplications in copy number variation).

4.2.2 Towards a global map of copy number variation

In contrast to these initial studies, Tuzun and colleagues in 2005 published the results of a sequence-based approach to try to produce a 'fine scale map' of structural variation across the genome (Tuzun *et al.* 2005). They hoped to assess a much greater proportion of the human genome and at a higher level of precision than the array CGH approaches reported by Sebat and Iafrate, so as to detect deletions and insertions, together with inversions and smaller insertion/deletion polymorphisms (indels). Tuzun and coworkers compared sequence data for a North American female donor individual (NA15510) with that of the reference genome, which is more than 70% derived from a single individual (RPC1–11). This involved mapping 'paired end sequence data from the fosmid DNA genomic library' for NA15510, analysing 581 Mb of sequence, and identifying structural variants greater than 8 kb in size.

Overall 297 sites of variation were found, 139 insertions, 102 deletions, and 56 inversion breakpoints; 112 structural variants were validated and when subsequently 57 sites were analysed in a panel of 47 individuals, 28% showed copy number variation by array CGH (Tuzun *et al.* 2005). The data of Tuzun and colleagues showed a ten-fold enrichment of structural variation at segmentally duplicated regions of the genome. In common with other surveys of structural variation they also noted overrepresentation of structural variation involving genes involved in drug detoxification, innate immunity and inflammation, surface integrity, and antigens.

In 2006 Redon and colleagues published a first-generation map of copy number variation across the human genome for a large panel of ethnically diverse individuals (Redon *et al.* 2006). The team studied 270 **lymphoblastoid cell lines** from the International HapMap Project, established from individuals of African, European, and Asian origin (Box 9.1). An important question with such immortalized cell lines is whether somatic artefacts would confound analysis of germline copy number variation. To address this, the investigators

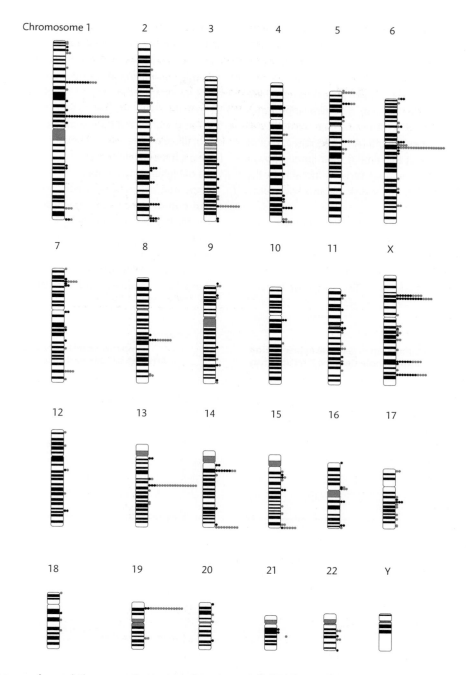

Figure 4.2 Copy number variation across the genome. An unexpectedly high level of variation in copy number was found on array CGH analysis (Iafrate *et al.* 2004). Circles shown to the right of each chromosome indicate the number of individuals with copy gains (black) and losses (grey) for each clone among 39 unrelated, healthy control individuals. Redrawn and reprinted by permission from Macmillan Publishers Ltd: Nature Genetics (Iafrate *et al.* 2004), copyright 2004.

performed extensive karyotyping and removed aberrant chromosomes from analysis. Two platforms were used for detecting copy number variation, comparative analysis of hybridization intensities on SNP arrays and comparative genomic hybridization (CGH) using Whole Genome TilePath array (Fig. 4.3). The two approaches were complementary, the former facilitating detection of smaller copy number variants. An important caveat of this and other studies at this time is their limited power to detect smaller copy number variants (notably in the range 1–20 kb) such that many such variants have yet to be identified.

Redon and colleagues found 1447 copy number variable regions among the 270 individuals, spanning a total of 360 Mb of sequence (Redon *et al.* 2006). This equates to 12% of the human genome and accounts for more nucleotide content than SNPs. Two-thirds of variants

were replicated, predominantly using the second technological platform. Within parent–child trios, biallelic copy number variants were found to be heritable. A heterogeneous chromosomal distribution was found for copy number variants, the proportion of any given chromosome involved ranged from 6% to 19% (Fig. 4.4). For most variants the ancestral state could not be determined: the minor allele was denoted as being derived, with deletions having a minor allele of lower copy number and duplications of higher copy number.

Copy number variable regions were more commonly found outside genes and ultraconserved elements, with a greater bias away from genes for deletions compared to duplications (Redon *et al.* 2006). This is thought to be due to deletions being under stronger selective pressure (purifying selection) for the removal of deleterious variants from the population (Brewer *et al.* 1998, 1999;

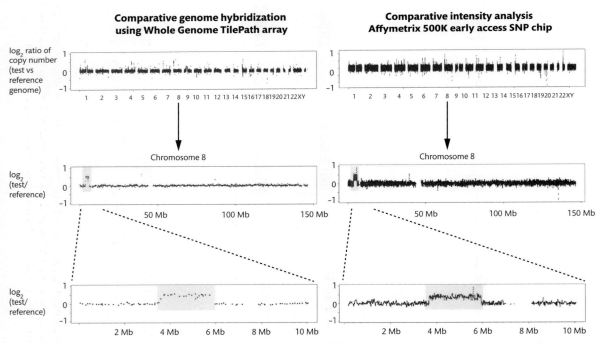

Figure 4.3 Technology platforms used to detect copy number variation. Redon and colleagues used a CGH Whole Genome TilePath array with large insert clones covering 93.7% of the euchromatic human genome, and a comparative intensity analysis using a high density single nucleotide polymorphism (SNP) array. In this example, two male genomes were compared for all chromosomes (top panels), an individual chromosome (middle panels), or a window on chromosome 8 (bottom panels) showing in detail a 2 Mb duplication in one genome. Redrawn and reprinted by permission from Macmillan Publishers Ltd: Nature (Redon *et al.* 2006), copyright 2006.

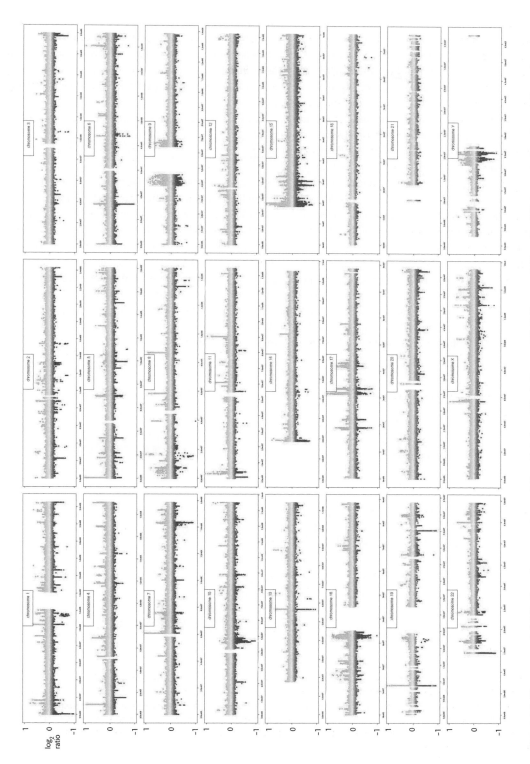

Figure 4.4 Chromosomal distribution of copy number variation. Data from Whole Genome TilePath array platform for 269 dye-swap experiments. For a given chromosome, log₂ ratios for each individual are shown superimposed. Gaps correspond to centromeric regions where data are not available. Reprinted by permission from Macmillan Publishers Ltd: Nature (Redon et al. 2006), copyright 2006.

Locke *et al.* 2006). There was, however, no difference in the observed frequency of deletions compared to duplications as might have been expected if this were the case, although deletions were three times shorter than duplications (43 kb versus 120 kb). In terms of genes which were involved in copy number variation, many were involved in cell adhesion and perception of smell and chemical stimuli. Nearly half of the regions known to be involved in genomic disorders such as DiGeorge syndrome (Box 5.2) and Williams–Beuren syndrome (Box 5.3) showed high levels of copy number polymorphism in this population of apparently healthy individuals (Redon *et al.* 2006).

Redon and colleagues also analysed copy number variation for evidence of population differentiation and clustering to suggest recent positive selection (Fig. 4.5) (Redon *et al.* 2006). A number of specific examples were found including *CCL3L1*, a chemokine multicopy gene previously shown to be associated with protection from human immunodeficiency virus type 1 (HIV-1) infection (Section 14.2.4) (Gonzalez *et al.* 2005). A further striking example the authors found was a duplication near to the

MAPT gene present only in the population of European ancestry: this region of chromosome 17 is implicated in neurodegenerative disease and is associated with a large inversion polymorphism subject to recent positive selection in Europeans (Section 5.5.2) (Stefansson *et al.* 2005).

4.2.3 Finding deletions across the genome within normal human populations

The extent of deletions in the human genome has only recently been investigated for phenotypically normal individuals. The availability of high density genotyping data for common SNPs across the genome from the International HapMap Project (Section 9.2.4) has allowed investigators to identify deletions (Conrad *et al.* 2006; McCarroll *et al.* 2006) and inversions (Bansal *et al.* 2007). Other investigators have used an oligonucleotide array comparative hybridization approach comparing data from individuals to the reference human genome to identify deletions (Hinds *et al.* 2006).

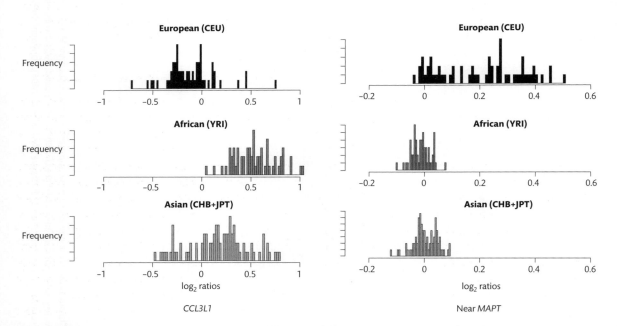

Figure 4.5 Examples of population differentiation in copy number variation indicating positive selection. Frequency histograms for individuals in different HapMap populations are shown for two examples, encompassing *CCL3L1* and a region near to *MAPT*. Redrawn and reprinted by permission from Macmillan Publishers Ltd: Nature (Redon *et al.* 2006), copyright 2006.

McCarroll and colleagues describe how 'footprints' in the contiguous SNP data allow for identification of deletions: a deletion is likely if a cluster of SNPs are seen that deviate from the expected mendelian inheritance within a family trio; or there are a cluster of SNPs showing apparent deviation from Hardy–Weinberg equilibrium; or there are a cluster of null genotypes (Fig. 4.6) (McCarroll et al. 2006). There is a high risk of false positives in such an approach due to artefact or genotyping errors, but by looking for clusters and using appropriate statistical thresholds, the authors were able to analyse 1.3 million SNPs among 269 individuals of four ethnic groups from the Phase I HapMap (Box 9.1). This led to the identification of 541 candidate deletion variants with a median size of 7 kb (range 1–745 kb), of these 278 occurred in multiple, unrelated individuals.

McCarroll and colleagues validated 90 predicted variants using a range of FISH and PCR-based approaches. The approach allowed the identification of ten genes relatively commonly deleted among the populations studied, although the frequencies of the deleted alleles varied considerably between populations (Fig. 4.7) and present in some cases as homozygous nulls. The observed variation in gene expression for GSTT1 and GSTM1, encoding glutathione S-transferases involved in detoxification by conjugation with glutathione (Section 4.4.3), and UGT2B17, encoding uridine diphosphate glucuronosyl-transferases involved in steroid hormone metabolism, was largely explained by the differences in gene dosage between individuals (Fig. 4.7).

Other investigators have also mined the dense genotyping data from HapMap to identify deletions. Conrad and colleagues looked for mendelian inconsistencies among trios to identify deletions transmitted to the child and performed extensive validation using a custom tiling path array for comparative genome hybridization to find a false discovery rate of 14% (Conrad et al. 2006). A total of 345 predicted deletions were found among 30 individuals of European descent (CEU) with a median size of 10.6 kb (range 0.3–404 kb), while those 30 individuals of African origin (YRI) had 590 predicted deletions (median 8.5 kb, range 0.5–1200 kb) consistent with greater genetic diversity in African populations. The deletions included 267 known or predicted genes of which 92 were completely deleted. Overall, the authors suggested a typical individual was **hemizygous** for between 30 and 50

deletions of greater than 5 kb in size (lower and upper range derived from European and African HapMap populations, respectively) (Conrad et al. 2006).

Finally, Hinds and colleagues used an array CGH approach to analyse 24 unrelated individuals and find evidence of deletions based on comparison with the reference human genome (Hinds et al. 2006). The investigators screened 100–200 Mb of DNA and found 215 deletions ranging in size from 70 bp to 7 kb (median size of 0.75 kb); 100 of these deletions were validated (Hinds et al. 2006).

Combining the three studies by McCarroll, Conrad, and Hinds to screen for deletions among apparently normal individuals, 1000 deletions were identified. There was surprisingly little overlap between the studies in term of the deletions identified, illustrating differences in approach and the likely large number of predominantly smaller deletions still to be identified (Eichler 2006). Greater diversity in deletions was noted for those of African descent, deletions found were smaller than perhaps expected, and they were underrepresented on the X chromosome and for coding exons. In comparison with SNPs, more extreme selective pressure was noted for deletions with a greater number of rare deletions in comparison with SNPs (Eichler 2006).

4.2.4 Integrating surveys of structural variation

Between 2004 and 2007, at least 12 surveys were published investigating the extent of human genomic structural variation among individuals with apparently normal phenotypes (summarized in Fig. 4.1). These studies were reviewed by Scherer and colleagues (2007), who noted the significant heterogeneity in many aspects of these and other studies of structural variation. These included the number of genomes studied, the tissue sources of DNA analysed, the different reference samples used for genome comparisons (Box 4.3), the range of technologies and approaches used for discovery (having very different resolutions and abilities to detect variants of particular sizes), the experimental quality controls used, and the extent to which putative structural variants were validated.

The difficulties of integrating current surveys of structural variation lead Scherer and colleagues to call for improved standardization in the field with respect to

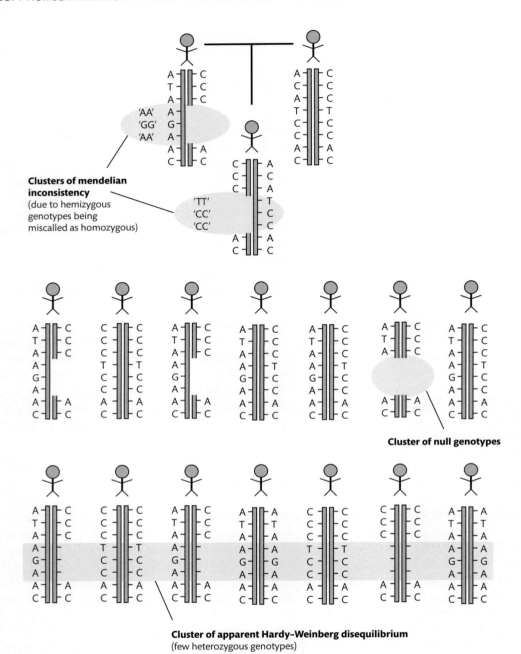

Figure 4.6 Discovery of segregating deletions using SNP genotyping data. Cluster patterns within SNP genotyping data showing apparent mendelian inconsistency for a family trio, clusters of null genotypes, and clusters of apparent Hardy–Weinberg disequilibrium, provide a 'footprint' suggestive of segregating deletions. Redrawn and reprinted by permission from Macmillan Publishers Ltd: Nature Genetics (McCarroll *et al.* 2006), copyright 2006.

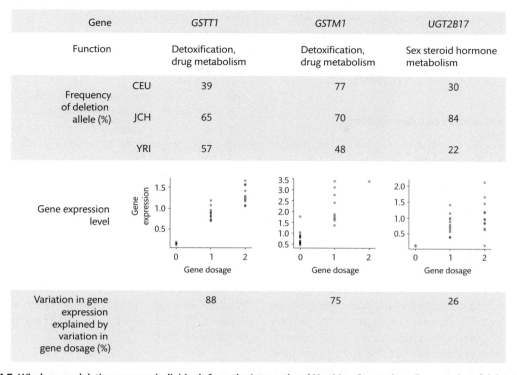

Gene		GSTT1	GSTM1	UGT2B17
Function		Detoxification, drug metabolism	Detoxification, drug metabolism	Sex steroid hormone metabolism
Frequency of deletion allele (%)	CEU	39	77	30
	JCH	65	70	84
	YRI	57	48	22
Variation in gene expression explained by variation in gene dosage (%)		88	75	26

Figure 4.7 Whole gene deletions among individuals from the International HapMap Consortium. Frequencies of deleted alleles among 269 individuals from the three HapMap populations are shown: CEU (European geographic ancestry), JCH (Japanese and Chinese), and YRI (African). Gene expression levels are for lymphoblastoid cell lines established from individuals in HapMap populations. Redrawn and reprinted by permission from Macmillan Publishers Ltd: Nature Genetics (McCarroll *et al.* 2006), copyright 2006.

Box 4.3   Reference DNA and copy number variants

To identify structural variation requires comparison with a reference DNA source, dataset, or genomic sequence. In 2007, the human genome reference assembly release from the US National Center for Biotechnology Information Build 36 was a mosaic of 708 different sources (Feuk *et al.* 2006a) with 302 known gaps and incomplete coverage: this may confound comparative analysis and, for unannotated segments, lead to structural variation being missed (Scherer *et al.* 2007). Moreover, the use of multiple different DNAs or pools of DNA as reference controls significantly complicates the analysis of copy number differences and database standardization; adoption of a standardized reference control DNA would be of significant benefit (Scherer *et al.* 2007).

---

terminology, complete reporting of sample descriptions and experimental methodologies, quality control, and annotation of structural variants (Scherer *et al.* 2007). Only recently has the terminology for describing structural variants become more uniform (Redon *et al.* 2006) but differences in terminology still underlie much of the heterogeneity currently observed between studies of structural variation. The different technologies currently in use differ significantly in sensitivity and specificity, and the extent of smaller copy number variants remains largely unknown because of the limitations of current detection methods. Moreover, lack of resolution means the boundaries or breakpoints of variants remain poorly resolved.

4.2.5 Extent of copy number variation

With these caveats in mind, the number of copy number variants reported in specific surveys (see Fig. 4.1) ranges from 76 (Sebat *et al.* 2004) to 3654 (Wong *et al.* 2007), with the latter thought to have a high proportion of false positives. In total 17 641 copy number variants are recorded at the Database of Genomic Variants (http://projects.tcag.ca/variation/; date of access September 2008), which combined data from 49 studies. Using this database, Scherer and coauthors noted that structural variation covered 18.8% of the euchromatic genome (538 Mb), based on analysis of less than 1000 genomes of people without a known disease phenotype (Scherer *et al.* 2007). Copy number variants are found genomewide but show evidence of clustering in pericentromeric and subtelomeric regions – regions known to be rich in segmental duplications (Section 6.2.3). As well as segmental duplications, copy number variants show strong correlations with exons and mobile DNA elements such as Alu repeats (Section 8.4). At present there is considerable difficulty in defining the precise locations and specific DNA sequences of copy number variants within copy number variable regions (the 'variant breakpoints') as the resolution of current techniques is generally poor and based on, for example, the coordinates of the BAC probe.

In order to 'complete the map of human genetic variation', a project has been launched from the National Human Genome Research Institute by the Human Genome Structural Variation Working Group which aims to sequence large insert clones from many different phenotypically normal individuals of African, European, and Asian ancestry so as to systematically identify and resolve structural variants (Eichler *et al.* 2007). The increased capacity of new sequencing technologies such as the 454 system employing highly parallel array-based pyrosequencing will facilitate such an effort (Box 1.24). Korbel and colleagues published work in 2007 showing

how structural variation down to a 3 kb resolution could be defined by paired-end mapping using the 454 sequencing system for two individuals, a female thought to be of European descent previously analysed by Tuzun and colleagues (NA15510) and a female from the Yoruba population in Nigeria (NA18505) (Korbel *et al.* 2007). This work, which involved sequencing more than 10 million and 21 million paired ends for the two individuals, respectively, fine mapped 853 deletions, 322 insertions, and 122 inversions. Based on full genomic coverage, 761 structural variants of more than 3 kb in size were predicted for NA15510 compared to the reference genome, and 887 for NA18505. Overall, 45% of structural variants were shared between the two individuals. In terms of alleles, 23% and 15–20% of structural variants were homozygous for NA15510 and NA18505, respectively. The authors noted that two genomic regions known to be associated with genomic disorders were hotspots for structural variation with 13 structural variants in an 8 Mb region at 22q11.2 and 29 in an 18 Mb region at 7q11 (Korbel *et al.* 2007).

4.2.6 Segmental duplications and identifying copy number variation

The presence of highly homologous, large segmental duplications flanking a region predisposes it to recurrent chromosomal rearrangements and genomic disease (Section 5.2.1). Sharp and colleagues investigated the role such duplications might play in generating structural variation within normal populations by designing a CGH array targeted at regions flanked by highly homologous intrachromosomal duplications (Sharp *et al.* 2005). One hundred and thirty such regions were identified, which the authors described as 'potential hotspots of recombination' or 'regions of potential genomic instability'. The authors analysed 47 ethnically diverse, phenotypically normal individuals and found 119 regions of copy number polymorphism, 73 of which were previously unknown. Copy number polymorphisms were found in 39% of hotspots, a four-fold enrichment in regions flanked by or containing large highly homologous segmental duplications. In terms of copy number polymorphism, equal numbers of deletions and duplications were found.

In a follow-up study of larger numbers of normal individuals, Locke and colleagues analysed 269 individuals from the International HapMap Project compared to a single reference individual using their targeted CGH array. This again showed a clear association with copy number variation, 84 out of 130 hotspot regions showed copy number differences (Locke et al. 2006). The presence of parent–child trios within the individuals analysed also allowed heritability to be assessed with a mendelian pattern of inheritance for copy number polymorphisms observed. These studies provided important further evidence to support a key role for segmental duplications in mediating chromosomal rearrangements, leading to both genomic disease and structural variation within normal human populations. The approach also proved an efficient way of identifying new likely pathogenic microdeletions and duplications among individuals with mental retardation (Section 5.4).

4.2.7 Structural versus nucleotide diversity

Surveys of structural variation have highlighted how for any two individuals in a population there is greater difference at the structural level than at the level of nucleotide diversity (Sebat 2007). Copy number variation between individuals has been conservatively estimated at 4 Mb of genetic difference in comparison with 2.5 Mb for SNPs; the contribution of copy number variation is likely to be significantly higher given our limited ability to detect small copy number variants with available technologies. The total genomic variability between people has been estimated at a difference of at least 0.2%, with more than 0.12% at a structural level and 0.08% at the nucleotide level (Sebat 2007). Structural variation also contributes significantly more to genetic diversity between species than single nucleotide substitutions (Cooper et al. 2007). Analysis of sequence divergence between humans and chimpanzees (*Pan troglodytes*) is ~1.2% from 35 million fixed single nucleotide substitutions; this increases to ~5% when approximately 5 million structural variants involving gain or loss of DNA between species are considered (Fig. 4.8) (Cheng et al. 2005; Feuk et al. 2005; Newman et al. 2005).

4.3 Copy number variation and gene expression

Copy number variation may modulate levels of gene expression through effects on gene dosage in which there is loss or gain of functional gene copies (McCarroll et al. 2006), or through disruption of the gene or noncoding DNA sequences involved in control of gene expression (Kleinjan and van Heyningen 2005; Lee et al. 2006). The latter proved surprisingly common when a genome-wide analysis of the association between copy number variation and gene expression was published in 2007 (Stranger et al. 2007a). Stranger and colleagues analysed gene expression in resting lymphoblastoid cell lines established from 210 unrelated individuals in the International HapMap Project (Section 9.2.4) from four populations. They were able to analyse the association between expression levels of 14 925 transcripts from 14 072 genes and copy number variation; for this collection of cell lines, data on

Type	Size	Number of events	Mb
Substitution	1 bp	35 000 000	35
Structural	<80 bp	4 93 000	18.1
	80 bp–15 kb	70 000	48.9
	>15 kb	1000	21
	Segmental duplications (lineage-specific)	940	46
	Segmental duplications (shared showing quantitative differences)	590	26
Total structural	>1 bp	5 000 000	160

Figure 4.8 Genetic variation between human and chimpanzee genomes. Reprinted by permission from Macmillan Publishers Ltd: Nature Genetics (Cooper et al. 2007), copyright 2007.

genome-wide, large scale copy number variation were available from earlier study data generated by Redon and colleagues (Section 4.2.2) (Redon *et al.* 2006). The authors found significant association for copy number variation and gene expression for all classes of copy number variation – namely deletions, duplications, deletions/duplications at a given locus, multiallelic, and complex (as classified by Redon and colleagues (2006)).

In a conservative analysis of 'high confidence' copy number variants, Stranger and coworkers reported association with 99 non-redundant genes across the four populations, of which 34% were significantly associated in at least two populations and 7% in all four populations (Stranger *et al.* 2007a). The authors estimated that 50% of effects were due to copy number variation modulating regulatory sequences within genes or in noncoding DNA, in some cases acting at a considerable distance from the gene. In the same study, the association of expression with SNP diversity was analysed (Section 11.4.1). For 700 000 SNPs, approximately four times as many expression phenotypes were found to be associated, although there was little overlap with those associated with copy number variants. Overall, Stranger and colleagues estimated that up to 17.7% of variation in heritable gene expression was explained by copy number variation. Given that the majority of copy number variants are small and thought to remain undetected

using the detection technology employed, this figure is likely to be much greater.

4.4 Copy number variation, diet, and drug metabolism

4.4.1 Duplication of the salivary amylase gene and high starch diet

The gene encoding salivary amylase, *AMY1*, lies within a highly variable region of chromosome 1p21.1 and is thought to have undergone tandem duplication events within the human lineage (Groot *et al.* 1989; Iafrate *et al.* 2004). It was the most common large scale copy number variation identified in a genomic screen by Iafrate and colleagues, present in half of individuals studied with relative gains and losses in equal numbers of people (Iafrate *et al.* 2004).

Perry and coworkers found that among a cohort of European Americans, copy number variation for *AMY1* significantly correlated with levels of salivary amylase protein (Perry *et al.* 2007). They hypothesized that copy number variation in *AMY1* may vary between those populations with higher starch consumption (for example agricultural societies and specific hunter gatherer groups relying on starch-rich tubers such as the Hadza) compared to those with low starch diets (rainforest

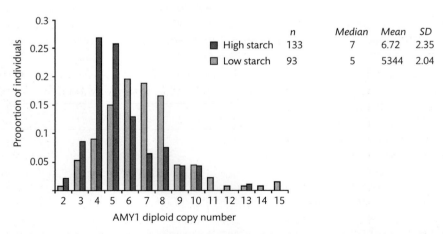

Figure 4.9 Copy number variation of *AMY1* and traditional dietary starch intake. Comparison of copy number variation for *AMY1* between populations traditionally having diets with high and low levels of starch. Reprinted by permission from Macmillan Publishers Ltd: Nature Genetics (Perry *et al.* 2007), copyright 2007.

hunter gatherers such as the Biaka and Mbuti groups from central Africa, and pastoralist fishing societies such as the Yakut from Siberia). The authors compared copy number variation between these different population groups and found that those with high starch diets did indeed have more *AMY1* copies (Fig. 4.9). Based on this analysis and that of other copy number variants and microsatellites, Perry and colleagues proposed that there had been positive or directional selection for copy number variation among these individuals, while in populations with low starch diets there had been **genetic drift** and the locus had evolved neutrally (Perry *et al.* 2007).

4.4.2 Copy number variation and drug metabolism: role of *CYP2D6*

The consequences of genetic diversity for interindividual variation in drug metabolism can be profound, both in terms of drug toxicity and response. Genetic variation involving enzymes of drug metabolism has therefore been of great research interest, notably for *CYP* genes encoding cytochrome P450 enzymes (Box 4.4), which are highly polymorphic with at least 350 different functional alleles (www.cypalleles.ki.se) (Ingelman-Sundberg 2004, 2005; Eichelbaum *et al.* 2006; Ingelman-Sundberg *et al.* 2007). These include deletions, duplications, specific mutations, and copy number variation. Among these, *CYP2D6* has 63

alleles and *CYP2A6* has 22 alleles; copy number variation of both can have significant effects on drug metabolism and response. Individuals who possess more than two active functional copies of a specific *CYP* gene may show increased drug metabolism and absence of response at ordinary drug dosages, classified as 'ultrarapid metabolizers', while 'poor metabolizers' lack functional enzyme due to defective or deleted genes.

The *CYP2D6* gene is found at chromosome 22q13.1 near two pseudogenes. In 1993, marked copy number variation of *CYP2D6* was described that was stably inherited, with individual alleles bearing between zero and 13 copies of the gene (Johansson *et al.* 1993). Since then 63 polymorphic alleles have been recognized resulting in complete loss, reduced, normal, or ultrarapid enzyme activity (reviewed in Ingelman-Sundberg *et al.* 2007). Such genetic diversity is very important: CYP2D6 is involved in the metabolism of an estimated 20–25% of all marketed drugs and genetic polymorphism of *CYP2D6* is responsible for much of the observed variation between individuals in enzyme activity (Ingelman-Sundberg *et al.* 2007).

In western Europe, 7% of the population are poor metabolizers, while up to 5.5% are ultrarapid metabolizers (Ingelman-Sundberg 2005; Eichelbaum *et al.* 2006). Within Europe, the distribution of ultrarapid metabolizers varies from 1–2% in northern Europe to 10% in Mediterranean regions including Spain and Italy (Ingelman-Sundberg

---

Box 4.4   *CYP* genes encode cytochrome P450 enzymes

The name 'cytochrome P450' refers to coloured (chrome) cellular (cyto) proteins that were noted to have a peak light absorbance near 450 nm when the haem iron found at the active site of the enzyme was reduced. Cytochrome P450 enzymes are responsible for the oxidative metabolism of many different endogenous (such as steroids and fatty acids) and exogenous (such as toxins, drugs, and carcinogens) molecules. They are found conserved across species, ranging from mammals to fish, plants, fungi, and bacteria. In humans, the liver and mucosa of the gastrointestinal tract are major sites of action

of these enzymes, which are responsible for about 80% of phase I drug metabolism (Eichelbaum *et al.* 2006). Cytochrome P450 enzymes are encoded by *CYP* genes, with over 57 genes and 59 **pseudogenes** identified in humans including 18 families and 43 subfamilies. The nomenclature comprises CYP, followed by a numeral to indicate the gene family, a capital letter for the subfamily, followed by another numeral for the individual gene. The *CYP2* family, which includes *CYP2D6* discussed in the main text, is involved in drug and steroid metabolism with at least 13 subfamilies and 16 genes.

Box 4.5   Consequences of *CYP2D6* duplication: a case report

A case report published in the *Lancet* in 2006 highlights the potentially fatal consequences of altered drug metabolism associated with polymorphism of *CYP2D6* (Koren *et al.* 2006). The case describes the death of an infant 13 days after birth who had lethal levels of morphine. The mother had been breast feeding while taking high doses of codeine for pain relief following an episiotomy. Morphine, the active metabolite of codeine, was

subsequently found at 70 ng/ml in the infant's blood (normal levels found in neonates of breast feeding mothers taking codeine are 0–2.2 ng/ml). High levels of morphine were also found in stored breast milk. Investigations showed the mother was an ultrarapid metabolizer and heterozygous for the CYP2D6*2 allele, having a gene duplication of *CYP2D6*.

2004). Ultrarapid metabolizers are particularly common in populations from northeast Africa and Oceania. The consequences of such differences are seen in a number of important drug classes, notably antidepressants, pain relieving analgesics (Box 4.5), anti-emetics (controlling nausea and vomiting), anti-arrhythmics (for disorders of heart rhythm), neuroleptics (drugs treating psychotic disorders such as schizophrenia), and anti-cancer drugs such as tamoxifen.

The allele frequencies and functional consequences of the different *CYP2D6* alleles have been extensively investigated (reviewed in Eichelbaum *et al.* 2006; Ingelman-Sundberg *et al.* 2007). CYP2D6*4 and CYPD6*5 are **null alleles** in which the encoded enzyme has no activity: a splicing defect is present in CYP2D6*4 while the *CYP2D6* gene is deleted in CYP2D6*5. CYP2D6*4 is common among Caucasians with an allele frequency up to 21%, but is less frequent in other populations (1–4%). The gene deletion seen with CYPD6*5 has a similar allele frequency across ethnic groups of about 5%.

Alleles associated with significantly reduced enzymatic activity include CYPD6*10, CYPD6*17, and CYPD6*41. CYPD6*10 encodes an unstable enzyme due to impaired folding resulting from specific mutations; this allele is very common in Asian populations with allele frequencies of over 50%, and less frequent in other groups (1–2% in Caucasian populations, 6% in African), however given the high frequency in Asia, worldwide it is the most common CYPD6 allele. The CYPD6*17 allele is common in sub-Saharan Africans and carries mutations altering the

affinity of the enzyme for specific substrates; CYPD6*41 is associated with a splicing defect.

Alleles showing copy number variation include deletions (CYPD6*5), duplications, and multiduplications leading to alleles with none, one, two, three, four, five, and 13 copies. Such duplications may or may not be functional genes but when the duplicated genes are functional, ultrarapid enzymatic activity is seen with markedly increased levels of metabolites (up to 30-fold) which may be toxic, as well as reduced drug levels including loss of therapeutic response. CYPD6*2 is the commonest duplicated allele in Caucasians (Sachse *et al.* 1997) although this is at a low frequency (1–5%) compared to some populations, notably north African populations. A study of healthy Ethiopians from Addis Ababa showed 28% of individuals had duplication or multiduplications of CYPD6 genes (Aklillu *et al.* 1996), while duplications were found in 21% of Saudi Arabian subjects tested (McLellan *et al.* 1997b). In Oceania, the CYP2D6*1 duplicated allele is common at a frequency of 22%. In other populations the frequency of *CYP2D6* duplications is lower, up to 4% among sub-Saharan Africans (Griese *et al.* 1999), 0–2% among Asians (Garcia-Barcelo *et al.* 2000; Nishida *et al.* 2000), and absent among Aborigines from Western Australia (Griese *et al.* 2001). A recent large study of African Americans showed a much higher rate of duplications compared to Caucasians (5.75% versus 1.3%) with many novel duplications among the African American cohort (Gaedigk *et al.* 2007).

4.4.3 Whole gene deletions of glutathione S-transferase enzymes, catalytic activity, and cancer risk

Genes encoding members of the glutathione S-transferase (GST) families have been shown to exhibit striking levels of polymorphism, notably *GSTM1* and *GSTT1* for which whole gene deletions are commonly found in a Caucasian population (reviewed in Hayes and Strange 2000). Proteins with GST activity provide protection against oxidative stress and toxic chemicals, and have been implicated in cancer pathogenesis.

GSTM1 lies on chromosome 1p13.3 within a GST gene cluster (Seidegard *et al.* 1988). An unequal crossing over event about 5 kb downstream of the gene is thought to have led to an 18 kb deletion that includes the entire *GSTM1* gene. About 50% of Caucasians are homozygous for the deleted allele and do not express the encoded protein (Board *et al.* 1990). These individuals are '*GSTM1* null' and have lost catalytic activity, they are more sensitive to specific toxic chemicals (for example carcinogenic epoxides), and a number of studies have shown increased cancer risk often associated with smoking (Hayes and Strange 2000). By contrast a study from Saudi Arabia revealed a small number (3% of the population phenotyped) who had 'ultrarapid' GSTM1 activity; these individuals were found to have a duplication of the *GSTM1* gene leading to overexpression of the encoded protein (McLellan *et al.* 1997a).

Detoxification of monohalomethanes in red blood cells by glutathiones was recognized to be clearly delineated within the human population into 60–70% of people who could conjugate and the remainder who could not (Peter *et al.* 1989). Pemble and colleagues found that this phenotype was explained by a deletion of the *GSTT1* gene on chromosome 22q11.23 (Pemble *et al.* 1994). Like *GSTM1*, *GSTT1* null has been associated with a wide range of cancers (Hayes and Strange 2000).

4.5 Copy number variation and susceptibility to common multifactorial disease

4.5.1 Psoriasis risk and β defensin gene copy number

Chromosome 8p23.1 is a genomic region often involved in chromosomal rearrangements for which a high level of multi-allelic copy number polymorphism has been described. The majority of individuals were found to have two to six copies, but in some individuals there were up to a maximum of 12 copies present. Levels of gene expression strongly correlated with copy number variation (Hollox *et al.* 2003).

A recent study has highlighted the role of copy number variation involving the 300 kb repeat region in susceptibility to psoriasis (Hollox *et al.* 2008) (Box 4.6). The region spans seven genes of the β defensin gene family. β defensins are small secreted antimicrobial peptides with cytokine-like properties which are found expressed at high levels in psoriatic plaques; the genes encoding them constitute plausible candidate genes for psoriasis (de Jongh *et al.* 2005). In their study published in 2008,

---

Box 4.6   Psoriasis (OMIM 177900)

Psoriasis is a chronic inflammatory disease affecting the skin. It is common, affecting 2% of the population in developed countries, difficult to treat, and associated with significant physical and psychological morbidity. The name psoriasis comes from the Greek word '*psora*', to itch, and the disease is characterized by red, scaling, elevated plaques found on the elbows, knees, and trunk, often associated with severe arthritis. Genetic factors, notably the *PSOR1* locus spanning HLA-C on chromosome 6, together with environmental triggers such as β haemolytic streptococcal infection, are important in early onset psoriasis (Griffiths and Barker 2007). While *PSOR1* is thought to account for 35–50% of the heritability of psoriasis (Trembath *et al.* 1997), a number of other genetic susceptibility loci have been described.

Copy number	Population frequency	Relative risk	95% confidence interval
2	3.50%	0.31	0.12–0.77
3	21.30%	0.51	0.37–0.71
4	42.60%	0.84	0.67–1.05
5	24.40%	1.08	0.83–1.40
6 or more	8.10%	1.69	1.16–2.48

Figure 4.10 Beta defensin copy number polymorphism and relative risk of psoriasis. Data from case–control studies in Dutch and German populations. Reprinted by permission from Macmillan Publishers Ltd: Nature Genetics (Hollox *et al.* 2008), copyright 2008.

Hollox and colleagues investigated the relationship between copy number variation and risk of psoriasis in a case–control cohort from the Netherlands and replicated their result in a German population (Hollox *et al.* 2008). The results were dramatic: when more than two copies were present, each extra copy increased the relative risk of disease by 34% (95% CI 25–43%) (Fig. 4.10). The effect may be due to dosage of one or several of the β defensin genes within the 300 kb repeat (including *DEFB4*, *DEFB103*, and *DEFB104*). Intriguingly, low copy number at this locus has been associated with inflammatory bowel disease, in particular colonic Crohn's disease (Fellermann *et al.* 2006).

4.5.2 Copy number variation of *FCGR3B* and susceptibility to autoimmune disease

In autoimmune diseases involving the kidney, the tiny blood vessels responsible for filtering blood can become inflamed and progressively blocked, leading to a condition known as glomerulonephritis and eventually renal failure (Nadeau and Lee 2006). A rat model of glomerulonephritis allowed Aitman and colleagues to map two major **quantitative trait loci** (**QTLs**) (Section 11.3) and determine that within one region of linkage, copy number of one of a family of genes encoding immune receptors was important (Aitman *et al.* 2006). Here, the gene *Fcgr3* was found in single copies among susceptible rat strains, while resistant strains had a second copy of the gene, *Fcgr3-related sequence* (*Fcgr3-rs*). *Fcgr3* encodes a transmembrane receptor involved in inflammation, usually found on macrophages. It is thought that the second

gene copy arose through a duplication event. Differences have subsequently arisen, including a single nucleotide deletion in *Fcgr3-rs* affecting the size of the cytoplasmic domain, and an extra 226 bp at the 3′ untranslated region of *Fcgr3* thought to modulate expression.

In humans, copy number variation of *FCGR3B* (encoding the Fc fragment of IgG, low affinity IIIb receptor) on chromosome 1q23 was associated with glomerulonephritis in patients with the autoimmune disease systemic lupus erythematosus (SLE). Low copy number (less than two, the healthy control population median copy number) significantly increased disease risk (Aitman *et al.* 2006; Fanciulli *et al.* 2007). An association was also found for SLE regardless of renal involvement ($P = 2.7 \times 10^{-8}$) and two other systemic autoimmune diseases, Wegner's granulomatosis and microscopic polyangiitis (Fanciulli *et al.* 2007). However, no association was reported for Graves' disease and Addison's disease, organ specific autoimmune diseases affecting the thyroid and adrenal glands respectively (Fanciulli *et al.* 2007).

4.5.3 CCL3L1, HIV, and autoimmunity

Copy number variation in the chemokine gene *CCL3L1* on chromosome 17q was found to be significantly associated with both susceptibility to HIV-1 and rate of disease progression (Gonzalez *et al.* 2005) (Section 14.2.4). CCL3L1 inhibits infection of cells by HIV-1 strains using the CC chemokine receptor 5 (CCR5). Most people possess one to six copies of the *CCL3L1* gene. Among individuals with a higher number of copies than the median for the ancestral population, risk was significantly reduced while lower copy numbers were associated with faster disease progression. High copy number of *CCL3L1* in combination with specific genetic variants of *CCR5* has also been associated with risk of Kawasaki disease, a childhood vasculitis (Burns *et al.* 2005). Chemokines including CCR5 play an important role in autoimmune disease and there is evidence to support a role for copy number variation in *CCL3L1* and the risk of rheumatoid arthritis (McKinney *et al.* 2007) and SLE (Mamtani *et al.* 2007).

4.5.4 Copy number and complement genes

Extreme genetic diversity, including copy number variation, has been described for the complement component C4 and its two major isotypes, C4A and C4B, encoded in the major histocompatibility complex (MHC) on chromosome 6 (Section 12.7). Increased risk of the autoimmune disease SLE was found for individuals with lower numbers of copies of total C4 and C4A, while risk is significantly reduced among individuals with higher numbers of copies (Yang *et al.* 2007).

4.6 Summary

As technology has advanced and our views of human genetic variation evolved, the extent of structural variation has been increasingly recognized. Over the course of Chapters 3 and 4 the nature and consequences of microscopically visible and submicroscopic structural variation have been reviewed. Such variation contributes to greater diversity between humans, and between humans and chimpanzees, than that found at a nucleotide level (Sebat 2007).

The recognition of copy number variation involving DNA segments greater than 1 kb in size as common within apparently phenotypically normal individuals has only become apparent since 2004 with the landmark papers of Sebat and Iafrate and colleagues (Iafrate *et al.* 2004; Sebat *et al.* 2004). Since that time there have been several surveys of copy number variation with increasing coverage across the genome and with improved resolution, although it is recognized that the limitations in resolution currently mean the majority of smaller copy number variants remain undetected. Array CGH has provided

the tool for unlocking many of the secrets of copy number variation with low cost, high throughput sequencing technologies likely to provide major advances over the coming years.

In one of the largest surveys involving 270 individuals by Redon and colleagues, 1447 copy number variable regions were identified involving 12% of the human genome (Redon *et al.* 2006). Paired-end mapping with high throughput sequencing of two individuals, females of probable European (NA15510) and African ancestry (NA18505), fine mapped 853 deletions, 322 insertions, and 122 inversions; the authors predicted that with full genome coverage there would be 761 structural variants more than 3 kb in size for NA15510 compared to the reference genome, and 887 for NA18505 (Korbel *et al.* 2007). By September 2008, 17 641 copy number variants were recorded in the Database of Genomic Variants, one of the largest repositories for this class of genetic variation.

Copy number variation is common and found across the genome with some notable hotspots of variation identified that include known sites of recurrent genomic disorders such as 22q11.2 and 7q11. Consideration of copy number variation is an essential part of disease mapping including genetic association studies of susceptibility to common multifactorial disease. Such work will be greatly facilitated by the maps and databases of copy number variation being established, and the technologies available for detection, including strategies based on SNP genotyping analysis. The contribution of this class of variation to common complex disease traits has only recently begun to be appreciated. Gene dosage effects through copy number variation have been implicated in infectious and autoimmune disease, notably *CCL3L1* and HIV-1 (Gonzalez *et al.* 2005), *FCGR3* and glomerulonephritis/SLE (Aitman *et al.* 2006), and β defensins and psoriasis (Hollox *et al.* 2008). The story also extends to selective pressures of high starch diet and copy number variation of the gene encoding salivary amylase, *AMY1*, seen to vary across populations dependent on dietary intake of starch (Perry *et al.* 2007). Perhaps the most dramatic effects are seen in drug metabolism including copy number variation of *CYP2D6* resulting in poor and ultrarapid drug metabolizers (Ingelman-Sundberg *et al.* 2007) with potentially serious consequences for effective drug dosage and toxicity.

This chapter has highlighted the extent and import-ance of copy number variation. Copy number variation is a major player in determining gene expression (Stranger *et al.* 2007a) and underlies a number of common multifac-torial common multifactorial traits. In the coming chap-ter the role of pathogenic copy number variation leading to specific mendelian traits is reviewed, focusing on the diverse group of diseases dubbed genomic disorders. As our ability to detect copy number variation at higher reso-lution improves, the importance of this class of genetic variation is only likely to increase.

4.7 Reviews

Reviews of subjects in this chapter can be found in the following publications:

Topic	References
Structural genomic variation including copy number variation	Eichler 2006; Feuk *et al.* 2006a; Carter 2007; Cooper *et al.* 2007; Scherer *et al.* 2007; Sebat 2007
Pathogenesis and clinical features of psoriasis	Griffiths and Barker 2007
Pharmacogenomics	Eichelbaum *et al.* 2006
Cytochrome P450 and CYP2D6	Ingelman-Sundberg 2005; Ingelman-Sundberg *et al.* 2007
Glutathione S-transferase polymorphisms	Hayes and Strange 2000

CHAPTER 5

Submicroscopic structural variation and genomic disorders

5.1 Introduction

The subdivision of structural genomic variation based on size, or more specifically method of detection, is to a certain extent arbitrary. It allows us to divide structural variants into microscopic and submicroscopic based on whether variants can or cannot be detected using conventional karyotyping. The focus of this chapter is on genomic disorders involving chromosomal rearrangements in which a change in gene copy number results in diverse but specific clinical syndromes (Box 5.1; Section 3.5.2). In some individuals, specific disorders arise due to events that are large enough to be cytogenetically detectable but, in most cases of a given genomic disorder, disease results from submicroscopic deletions, duplications, or other events. In the classification used here distinction has also been drawn between structural variants arising among healthy individuals or contributing to susceptibility to multifactorial traits (which are described in Chapter 4) versus those that are directly pathogenic. Again there is overlap in terms of underlying mechanism and use of terminology.

The concept of genomic disorders is relatively new (Lupski 1998) but has been extremely helpful in advancing our understanding of the basis of genetic disease due to structural genomic variation and our ability to discover such events and diagnose specific clinical syndromes. Remarkable progress has been made as our ability to detect copy number variants has improved. In this chapter, a number of genomic disorders are explored in detail to illustrate some of the common themes and principles underlying this diverse and fascinating group of diseases. For further information the interested reader is referred to the many excellent reviews available on this topic (Section 5.7).

5.2 Genomic disorders

5.2.1 Segmental duplications and genomic disorders

The specific genomic architecture associated with recurrent chromosomal rearrangements and genomic disorders was highlighted by analysis of recombination breakpoints. This showed that segmental duplications or LCR sequences (Box 5.1) showing high levels of sequence homology flanked recurrently recombining segments,

---

Box 5.1   Genomic disorders

Genomic disorders are a diverse group of genetic diseases that involve gain, loss, or disruption of dosage sensitive genes and which arise due to specific features of the genomic architecture, not-

ably highly homologous low copy repeats (LCRs), which predispose to chromosomal rearrangements through non-allelic homologous recombination (Lupski 1998; Lupski and Stankiewicz 2005).

with breakpoints often clustered in small regions of near-perfect sequence identity. These results suggested that rearrangements were mediated by non-allelic homologous recombination (also known as unequal crossover) between the highly homologous repeats (Inoue and Lupski 2002). Deletions and duplications may result from unequal crossover where there is a tandem or direct orientation relationship between the LCRs, while inversions will result if there is a reverse orientation relationship between the homologous regions. Rarely, non-allelic homologous recombination will involve LCRs on different chromosomes leading to reciprocal translocations.

The probability of non-allelic homologous recombination occurring was shown to be proportional to the degree of sequence identity (Waldman and Liskay 1988). Larger and more homologous duplicated regions are associated with more frequent rearrangements: the commonest known microdeletion syndromes, velocardiofacial syndrome and DiGeorge syndrome (see Box 5.2), involve duplicated blocks more than 300 kb long with greater than 99.7% sequence homology (Edelmann et al. 1999a; Shaikh et al. 2000). Non-allelic homologous recombination mainly involves intrachromosomal segmental duplications. Breakpoints are typically associated with particular repeat sequences or AT-rich sequences, with Alu repeats (Section 8.4) thought to be particularly significant in this process.

These mechanisms are illustrated by a number of different diseases. In Williams–Beuren syndrome (see Box 5.3), a recurrent common deletion of a 1.5 Mb region has been found to involve specific tandemly arranged segments within two complex 320 kb LCRs that flank the deleted region (Perez Jurado et al. 1996). A similar arrangement is seen among the 5–22% of cases of neurofibromatosis where deletion of a 1.5 Mb region spanning the NF1 gene occurs: the region is flanked by 85 kb LCRs arranged in a direct orientation (Dorschner et al. 2000). Neurofibromatosis (OMIM 162200) is an autosomal dominant condition characterized by café-au-lait spots and neurofibromas, patients with this specific deletion characteristically develop the skin neurofibromas at an early age. In haemophilia A (see Box 5.12), 25–30% of patients have a recurrent 400 kb inversion involving F8 (encoding factor VIII) at Xp28 mediated by LCRs present in inverted orientations to each other (Lakich et al. 1993).

5.2.2 Recurrent rearrangements involving chromosome 22q11

Chromosome 22q11 is an example of a genomic region particularly susceptible to chromosomal rearrangements, with the observed disease phenotype dependent on gene dosage and the specific sequences involved (Fig. 5.1) (McDermid and Morrow 2002). LCRs are found in this region of chromosome 22 ('LCR22s') setting the stage for non-allelic homologous recombination events and recurrent chromosomal rearrangements. The LCR22s are found at breakpoints identified in the different genomic disorders (Fig. 5.2).

Possessing one copy of the region due to a hemizygous deletion leads to haploinsufficiency and DiGeorge/velocardiofacial syndrome (Box 5.2). In most patients with DiGeorge syndrome, a 3 Mb deletion results from an unequal crossover event between two 240 bp LCRs, denoted LCR22-2 and LCR22-4, with homologous modules sharing more than 99% sequence identity (de la Chapelle et al. 1981; Edelmann et al. 1999b; Shaikh et al. 2000) (Fig. 5.2). Chromosome 22q11 deletions are the second commonest cause of congenital heart disease after Down syndrome (Box 3.3). A variety of other phenotypes are associated with deletions in this region for which the acronym CATCH has been proposed: cardiac abnormalities, T cell deficits, clefting, and hypocalcaemia. The developmental basis for this involves a disturbance in cervical neural crest migration into derivatives of the pharyngeal arches and pouches. Within the deleted region, haploinsufficiency for the TBX1 gene (encoding a T-box containing transcription factor) is thought to be particularly important in DiGeorge syndrome with evidence from mouse studies showing Tbx1 is responsible for development of the cardiovascular phenotype seen in humans (Lindsay et al. 2001; Merscher et al. 2001).

More rarely, individuals are found who possess additional copies of chromosome 22q11 (Fig. 5.1). Three copies may be found in people with Emanuel syndrome, also known as supernumerary der(22) syndrome. This arises from non-disjunction events in carriers of the constitutional t(11;22) translocation (Box 3.6) leading to partial trisomy of both 11q23 and 22q11. Finally, four copies of the region may occur as a result of having a supernumerary marker chromosome leading to cat eye syndrome (Box 3.8).

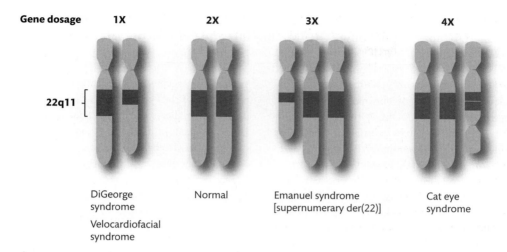

Figure 5.1 Chromosomal rearrangements and chromosome 22q11. Differences in gene dosage involving 22q11 result in genomic disorders: interstitial hemizygous deletions are associated with DiGeorge/velocardiofacial syndrome; partial trisomy of 22pter-q11 is seen in supernumerary der(22) (Emanuel syndrome); and partial tetrasomy is seen in cat eye syndrome where an additional marker chromosome 22 containing an inversion duplication involving 22.q11 is seen. Redrawn and adapted from McDermid and Morrow (2002), copyright 2002, with permission from Elsevier.

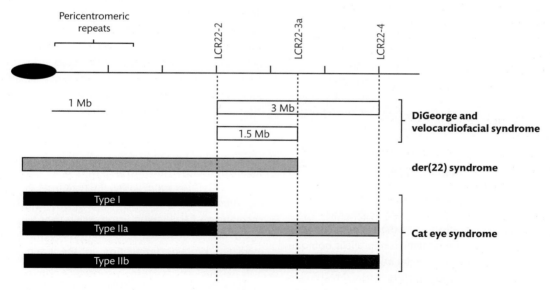

Figure 5.2 Low copy repeats and chromosomal rearrangements at chromosome 22q11. The three breakpoint regions involved in DiGeorge/velocardiofacial syndrome, der(22) syndrome, and cat eye syndrome involve low copy repeats (dubbed 'LCR22') within which complex inverted and direct orientations of highly homologous sequences are seen predisposing to the different chromosomal rearrangements. Hemizygous deletions (leaving one copy of the region) of 3 Mb (and more rarely 1.5 Mb) associated with DiGeorge and velocardiofacial syndrome are shown as white boxes; partial trisomies (three copies) are shown in grey; and partial tetrasomies (four copies) in black. The constitutional t(11; 22) translocation disrupts LCR22–3a. Redrawn and adapted from McDermid and Morrow (2002), copyright 2002, with permission from Elsevier.

Box 5.2   DiGeorge syndrome (OMIM 188400) and velocardiofacial syndrome (OMIM 192430)

A variety of clinical phenotypes have been found to result from deletion of chromosome 22q11.2. DiGeorge syndrome (DiGeorge 1968) is characterized by tetany and seizures in infancy due to low blood calcium (hypocalcaemia resulting from parathyroid gland hypoplasia), susceptibility to infection due to a T cell deficit (thymic hypoplasia), and defects involving the outflow tract of the heart. There is clinical overlap with the velocardiofacial

syndrome described by Shprintzen, which is a common cause of cleft palate; other features include cardiac abnormalities, a typical facial appearance, and learning disabilities (Shprintzen *et al.* 1981). The deletion can also be associated with isolated cardiac outflow tract defects such as the tetralogy of Fallot. Together, DiGeorge and velocardiofacial syndrome comprise the commonest known genomic disorder with an estimated frequency of one in 4000.

Box 5.3   Williams–Beuren syndrome (OMIM 194050)

This syndrome was thought to be relatively rare, affecting one in 20 000 people but more recent estimates suggest a prevalence of one in 7500 people (Stromme *et al.* 2002). The syndrome involves abnormalities of connective tissue and the central nervous syndrome. Congenital vascular and cardiovascular anomalies such as supravalvular aortic stenosis occur, while infantile hypercalcaemia, mental retardation, and a characteristic appearance, personality, and cognitive profile are all reported.

Supravalvular aortic stenosis when present alone is recognized as an autosomal dominant disorder (OMIM 185500) caused by mutations restricted to the *ELN* elastin gene (Ewart *et al.* 1994; Li *et al.* 1997); the same gene is part of the 1.5 Mb region which is deleted in almost all patients with Williams–Beuren syndrome, although the region includes at least 27 other genes (Fig. 5.3). Haploinsufficiency of these genes is believed to cause the syndrome with its diverse phenotypic manifestations.

5.2.3 Reciprocal genomic disorders

In some genomic disorders gene duplications and deletions are recognized involving the same genomic region but with contrasting effects, in others the phenotypes may be similar. A common theme is non-allelic homologous recombination with parental inversions implicated as an underlying predisposing chromosomal rearrangement. For example, a large 1.5 Mb deletion at 7q11.23 was found to be responsible for over 95% of cases of Williams–Beuren syndrome (Box 5.3) (Ewart *et al.* 1993). The deletion region includes at least 28 genes (Fig. 5.3) with haploinsufficiency of a number of different genes including *ELN*, the elastin gene, which is responsible for

the cardiovascular phenotype (Ewart *et al.* 1993; Pober *et al.* 2008). Three genes within the region are part of the tripartite motif (TRIM) protein family, including *TRIM50* encoding an E3 ubiquitin ligase which may play a role in the observed mental retardation through ubiquitin-mediated protein degradation in the brain (Micale *et al.* 2008).

The deleted region is flanked by repeating units of DNA which are highly homologous and present in the same and opposite orientations. These 400 kb segmental duplications are postulated to lead to the deletion through non-allelic homologous recombination (Baumer *et al.* 1998). The same 1.5 Mb region has also been found

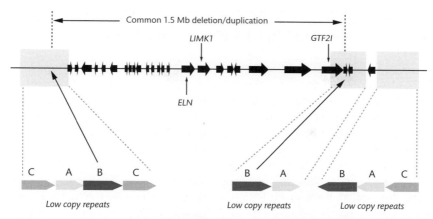

Figure 5.3 Region of chromosome 7q11.23 involved in Williams–Beuren syndrome. A cluster of genes (denoted by black arrows) is contained within a 1.5 Mb region deleted in this syndrome; individuals with duplication of the region have also been reported. *ELN* (elastin), *LIMK* (LIM domain kinase 1), and *GTF2I* (general transcription factor II, I) have been associated with Williams–Beuren syndrome. Flanking low copy repeats are shown, each composed of highly homologous blocks of DNA (denoted A, B, C). Almost all deletions occur between the homologous B blocks. Redrawn from Osborne and Mervis (2007), with permission from Cambridge University Press.

to be inverted in some individuals; again this is thought to involve the highly homologous DNA repeats through misalignment and unequal crossover of repeat units in an inverted orientation (Osborne *et al.* 2001). Such individuals are very rare but found among atypical cases of Williams–Beuren syndrome and in parents of children with the syndrome. Recently, a case has been reported involving tandem duplication of the same region: in contrast to Williams–Beuren syndrome, the affected individual with dup(7)(q11.23) was found to have severe delay of expressive speech (Somerville *et al.* 2005). This is now recognized as Williams–Beuren region duplication syndrome (OMIM 609757) and is characterized by severe language problems and autistic features (Berg *et al.* 2007).

A further example of reciprocal genomic disorders is provided by the *PMP22* gene encoding a myelin protein: having three copies (trisomy) of the gene associated with duplication leads to a nerve conduction deficit due to demyelination (Charcot–Marie–Tooth disease (CMT) type 1A) (Lupski *et al.* 1992), while deletion results in having only one copy (monosomy) leads to nerve conduction block (hereditary neuropathy with pressure palsies) (Chance *et al.* 1993). CMT is the commonest known inherited disorder of the peripheral nervous system (Box 5.4). Research into this disorder has

proved a remarkable story which has significantly contributed to our understanding of the disease and also served as a paradigm of how the copy number of a gene can influence phenotype and cause disease. CMT type 1A is a common and specific type of the disorder in which there is loss of the protective myelin sheath surrounding nerve fibres (demyelination). Family studies allowed investigators to map CMT type 1A to genetic markers on chromosome 17 by linkage analysis (Vance *et al.* 1989). Remarkably for an autosomal dominant disorder, complete linkage and association with CMT type 1A was then found for a tandem 1.5 MB duplication of a region of chromosome 17 in seven multigenerational family pedigrees and isolated patients (Lupski *et al.* 1991).

An estimated 30–50 genes are present in this region but an important clue as to which gene was important in causing the disease was provided by mouse studies. In 1992, a mouse with the autosomal dominant 'trembler mutation' in the *pmp22* gene on mouse chromosome 11 had been identified with a similar phenotype to CMT (Suter *et al.* 1992). *pmp22* encodes peripheral myelin protein-22, a protein found in peripheral myelin. A number of groups in the same year then reported that the human *PMP22* gene mapped to the region duplicated in CMT type 1A (Matsunami *et al.* 1992; Patel *et al.*

Box 5.4 Charcot–Marie–Tooth disease (OMIM 118220)

In 1886, Charcot and Marie described a slowly progressive muscular atrophy that manifested as weakness in the feet and legs followed by the hands (Charcot and Marie 1886); in the same year Tooth described the same disease although he correctly attributed it to a neuropathy (Tooth 1886). CMT is the most common inherited disorder of the peripheral nerves in man, affecting one in 2500 individuals (Skre 1974). Autosomal dominant, autosomal recessive, and X-linked forms of the disease occur. The disease can be divided into a demyelinating form, CMT type 1, which accounts for about 70% of cases and is most often transmitted as autosomal dominant; and an axonal form, CMT type 2. CMT type 1 can be further classified into CMT1A and CMT1B and they have different genetic aetiologies. CMT is characterized by slowly progressive weakness and wasting of foot and leg muscles followed by the hands, impaired sensation, and loss of tendon reflexes. Onset is usually in the first or second decade, and clinical features vary.

1992; Timmerman *et al.* 1992; Valentijn *et al.* 1992b). *PMP22* is a dosage sensitive gene and increased levels of expression of the gene were found in the peripheral nerves of patients with the CMT type 1A duplication (Yoshikawa *et al.* 1994). More severe disease is seen in those with four copies of the gene. Overall, about 70% of individuals with CMT type 1A have the duplication and where this occurs *de novo* it is usually of paternal origin. Remarkably, among patients without the duplication, investigators found that the same leucine to proline substitution occurred as had been seen in the Trembler-J mouse (Valentijn *et al.* 1992a).

Duplications and point mutations in the *PMP22* gene can result in CMT type 1A disease; deletion of the gene is associated with another neurological condition involving demyelination called hereditary neuropathy with pressure palsies (HNPP; OMIM 162500) (Chance *et al.* 1993). HNPP is an autosomal dominant disorder in which a focal demyelinating neuropathy is seen after minor trauma to peripheral nerves. Family pedigree studies showed association with a large interstitial deletion mapping to the same breakpoints as the duplication seen in CMT type 1A. There is evidence that the CMT type 1A duplication and the HNPP deletion are the reciprocal products of an unequal crossover event. The 1.5 Mb duplicated/deleted region is flanked by complex low copy number repeat sequences, which if misaligned may result in the reciprocal recombination event (Reiter *et al.* 1996).

The examples described to date illustrate how gene dosage effects are integral to the observed phenotype with contrasting effects between duplication and deletion in the 'reciprocal' genomic disorders. In other genomic disorders, gene duplications can lead to an overexpression phenotype which is similar to that seen for disorders associated with haploinsufficiency for the same region. For example, mental retardation associated with progressive spasticity has been reported in men as a result of duplications affecting *MECP2* (encoding methyl CpG-binding protein 2) (Van Esch *et al.* 2005); the same gene is involved in Rett syndrome (OMIM 312750), a neurodegenerative disorder affecting women but here the defect is due to haploinsufficiency of the gene due to mutations (Amir *et al.* 1999) or large deletions (Archer *et al.* 2006). Elsewhere, a 3.7 Mb interstitial deletion at 17p11.2 is associated with mental retardation and multiple congenital abnormalities (Smith–Magenis syndrome; OMIM 182290); a similar but milder phenotype is associated with duplications at the same locus (Potock–Lupski syndrome; OMIM 610883) (Potocki *et al.* 2000).

Finally, there are examples of reciprocal events where the deletion is pathogenic but the duplication may be benign. For example on chromosome 16, reciprocal deletion and duplication events 1.65 Mb in size were defined at 16p13 involving chromosome 16-specific LCRs by a microarray-based comparative genome hybridization (array CGH) screen of patients with mental retardation and/or major congenital abnormalities (Hannes *et al.* 2008). Intriguingly the duplication, but not the deletion, was found among five of 1682 phenotypically normal individuals.

5.2.4 Non-recurrent genomic disorders

Non-recurrent rearrangements can also lead to genomic disorders, often with highly complex patterns of deletion and duplication rearrangements. This is seen for the dosage sensitive proteolipid protein 1 gene (*PLP1*) at Xq22 associated with Pelizaeus–Merzbacher disease (OMIM 312080) (Sistermans *et al.* 1998; Inoue *et al.* 1999; Lee *et al.* 2007). Most cases of this X-linked demyelinating disease are due to non-recurrent duplications, while more rarely non-recurrent deletions or point mutations are the cause. Other genomic disorders due to non-recurrent rearrangements include a progressive neurodevelopmental syndrome found in males (Lubs X-linked mental retardation syndrome; OMIM 300260) due to complex *MECP2* duplications and triplications (del Gaudio *et al.* 2006); and rare familial forms of Parkinson's disease due to duplication (Chartier-Harlin *et al.* 2004) and triplication (Singleton *et al.* 2003) of the α-synuclein gene locus (*SNCA*) at 4q21, and of Alzheimer's disease due to duplication of *APP* (Rovelet-Lecrux *et al.* 2006), which are described in more detail in the remainder of this section. Non-recurrent chromosomal rearrangements may arise through non-homologous end joining or complex replication-based mechanisms as proposed for duplications and deletions involving *PLP1* with a model of 'Fork Stalling and Template Switching' described by Lee *et al.* (2007).

Most cases of Parkinson's disease are sporadic, with some rare familial forms of the disease described (Box 5.5). In 1997, a missense mutation was identified in the *SNCA* gene in autosomal dominant Parkinson's disease in a number of different families (Polymeropoulos *et al.* 1997). *SNCA* encodes a presynaptic protein α-synuclein,

which was found in the same year to be a major constituent of Lewy bodies, characteristic proteinaceous lesions seen in the disease (Spillantini *et al.* 1997). In a large family kindred with early onset parkinsonism associated with dementia, triplication of a 1.6–2 Mb region spanning *SNCA* was then identified (Singleton *et al.* 2003). *SNCA* triplication was found to have occurred independently in a Swedish American family where the predicted doubling of *SNCA* expression (there being three copies of the gene on the disease-linked chromosome and one on the normal chromosome, giving a total of four copies) was demonstrated in brain messenger RNA and soluble protein levels (Farrer *et al.* 2004).

Gene duplications involving *SNCA* have also been found in three families from France and Italy (Chartier-Harlin *et al.* 2004; Ibanez *et al.* 2004). Here three copies of the gene are present, the disease has a later age of onset, and dementia is rare. Strikingly, genetic variation involving *SNCA* has been associated with sporadic Parkinson's disease risk, notably specific haplotypes (Farrer *et al.* 2001) and a mixed dinucleotide repeat about 10 kb upstream of the gene (Maraganore *et al.* 2006) together with other mutations (Simon-Sanchez *et al.* 2007). The dinucleotide repeat is thought to act as a negative regulator of gene expression, dependent on allelic length (Chiba-Falek *et al.* 2005).

A further example of the phenotypic consequences of gene duplication events is provided by Alzheimer's disease (Section 2.5.1). Duplication of *APP*, the gene encoding amyloid beta precursor protein (APP), was found in patients with autosomal dominant, early onset Alzheimer's disease. These patients had a progressive dementia with severe cerebral amyloid angiopathy in brain tissue due to overexpression of APP, leading to

---

Box 5.5   Parkinson's disease (OMIM 168600)

Parkinson's disease is a common neurodegenerative disorder characterized by parkinsonism, in which there is tremor at rest, movements become difficult to initiate and are slower, with rigidity and postural instability. It is a common, devastating, and progressive disease found in 1% of the population at age 65 years and 4–5% by age 85 years. In addition

to movement disorders, depression and dementia are common. Pathologically, neuronal dysfunction and cell death occur with loss of neuromelanin-containing monoamine neurones (notably containing the neurotransmitter dopamine) and characteristic Lewy bodies (proteinaceous intracellular inclusions) seen in the brainstem (Fahn 2003; Farrer 2006).

accumulation of amyloid beta peptides (Rovelet-Lecrux *et al.* 2006). *APP* is found on chromosome 21q21; trisomy 21 will also lead to overexpression of this locus and patients with Down syndrome have also been recognized to develop Alzheimer's disease associated with cerebral amyloid angiopathy.

Triplication has also been implicated in other rare familial disorders, notably hereditary pancreatitis (OMIM 167800) (Le Marechal *et al.* 2006). A 605 kb region spanning the *PRSS1* gene (encoding cationic trypsinogen) on chromosome 7 was found to be triplicated in five French families with hereditary pancreatitis; a gain of function missense mutation of the same gene had previously been shown to cause the same disorder (Whitcomb *et al.* 1996).

5.2.5 Genomic disorders and control of gene expression

Chromosomal rearrangements leading to genomic disorders may alter gene dosage through duplication or deletion (haploinsufficiency) of regions spanning entire genes, or alter gene expression through smaller variants involving parts of genes or **cis-acting** regulatory regions. Modulation of gene expression leading to the observed phenotype may involve more complex 'position effects' often operating at a distance (Kleinjan and van Heyningen 2005; Feuk *et al.* 2006b). Separation of a gene from its control elements through translocation, inversion, deletion, or duplication events may dramatically alter regulation of gene expression, or bring a gene into a new relationship with existing or distant control elements to alter gene expression.

An example of this is seen for a rearrangement involving the α globin gene cluster on chromosome 16p13.3 where an 18 kb deletion was found to abolish α globin gene expression leading to α thalassaemia (Barbour *et al.* 2000). Here the deletion did not structurally involve the globin gene or a major regulatory element but rather a positional effect that changed the chromatin make up of the region silencing the gene, the CpG island in the promoter becoming densely methylated. This contrasts with the more than 80 structural deletions involving the *HBA1* and *HBA2* genes encoding α globin genes and regulatory elements that have

been shown to result in α thalassaemia (together with a small number of point mutations) (Section 1.3).

5.2.6 Genomic disorders showing parent of origin effects

Deletions of a critical region of the proximal long arm of chromosome 15 result in particular phenotypes depending on whether the maternal or paternal copy of the region is deleted (Knoll *et al.* 1989). This relates to the remarkable phenomenon of genomic imprinting (Box 11.5) in which only the copy of the gene derived from a particular parent is expressed in the normal state, with the other copy inactivated due to **epigenetic** modifications such as hypermethylation (Driscoll *et al.* 1992). Imprinting is observed in a small number of autosomal genes, of which a cluster is found in the deleted region 15q11-q13. Normally both chromosomes are differentially imprinted over this region depending on their parental origin, and both are required in normal development. Prader–Willi syndrome (Box 5.6) results from loss of function of paternally imprinted genes including *SNRPN*, while Angelman syndrome (Box 5.7) occurs if there is loss of the maternally imprinted *UBE3A* gene. These syndromes are a further example of reciprocal genomic disorders as duplications also occur that can lead to chromosome 15q11-q13 duplication syndrome (OMIM 608636), which includes features of autism (Fig. 5.4) (Clayton-Smith *et al.* 1993; Baker *et al.* 1994). Indeed individuals may be found with one (deletion), two (normal), four (duplication), and six (triplication) copies of 15q11-q13.

5.3 Terminal deletions and subtelomeric disease

The most common terminal deletion in the human genome involves the distal short arm of chromosome 1 leading to monosomy of a contiguous region and a recognized clinical syndrome (Box 5.8). A number of chromosomal rearrangements may result in deletion of this region. These include a pure terminal deletion (52% cases), interstitial deletion (29%), unbalanced translocations (7%),

---

Box 5.6   Prader–Willi syndrome (OMIM 176270)

Prader–Willi syndrome is characterized before birth by reduced fetal movements, and after birth by obesity, reduced muscle tone, mental retardation, short stature, and reproductive problems. In 75% of cases the disorder results from a 4 Mb *de novo* deletion of the 15q11-q13 region. It can also occur in 20% of cases due to loss of the paternal chromosome 15 with two maternal copies of the chromosome (so-called maternal uniparental disomy) – this may occur because of trisomy in the early embryo, which is rescued by paternal chromosomal loss (Cassidy *et al.* 1992). In 5% of cases there are translocations or other structural rearrangements,

while in 1% there is a microdeletion in a specific 'imprinting centre' (Buiting *et al.* 1995). Although very rare, such events provide a remarkable insight into control of imprinting across the critical region (regulating chromatin structure, DNA methylation, and gene expression) for the paternal state. Such microdeletions also account for recurrent cases seen in families. The critical region for Prader–Willi syndrome includes a number of paternally imprinted genes, critical among which is the *SNRPN* gene, which encodes small nuclear ribonucleoprotein, a ribosome-associated protein important in control of splicing (Ozcelik *et al.* 1992).

Box 5.7   Angelman syndrome (OMIM 105830)

Angelman syndrome is characterized by mental retardation, absent speech, seizures, and motor dysfunction. Like Prader–Willi syndrome the disorder involves a critical region on chromosome 15 but here it is loss of the allele derived from the mother that is important. In particular, failure to express the maternally imprinted gene *UBE3A* (encoding ubiquitin protein ligase E3A) appears critical. Normally

the maternal allele is transcribed in the hippocampus and cerebellum, but in Angelman syndrome gene expression is lost (Albrecht *et al.* 1997). This most often occurs because of a 5q11-q13 deletion (60% cases) (Knoll *et al.* 1989), paternal uniparental disomy (3–5%) (Malcolm *et al.* 1991), imprinting centre mutations (10%), or mutations in *UB3EA* resulting in failure of the gene to be expressed (Fang *et al.* 1999).

---

and complex rearrangements (12%) (Shaffer *et al.* 2006; Ballif *et al.* 2007). A range of different deletion sizes are recognized with a common minimal region. The terminal deletion syndrome is thought to result from haploinsufficiency of particular genes in the common region, although these remain to be clearly defined. Such telomeric regions of chromosomes are typically gene dense. Several candidate genes have been proposed, notably *KCNAB2* associated with seizures, *SK1* with cleft lip and palate, *MMP23* with cranial suture closure, and *GABRD* with neuropsychiatric and developmental abnormalities (Gajecka *et al.* 2007).

The first reported inherited deletion syndrome was the striking cri du chat syndrome which has since been

associated with terminal and interstitial deletions of the short arm of chromosome 5 (Box 5.9). The specific genomic regions and genes responsible for the clinical phenotypes seen in this syndrome are now being resolved.

The extensive segmental duplication found in subtelomeric regions between chromosome-specific sequences, and the arrays of telomeric repeats (Box 7.3) found at the tips of chromosomes, predisposes to chromosomal rearrangements. Many subtelomeric deletion syndromes are now recognized and associated with specific phenotypes, notably mental retardation (Section 5.4.2) (Knight and Regan 2006). These include chromosome 1q43-q44 deletion syndrome (OMIM 612337) (van Bon *et al.* 2008),

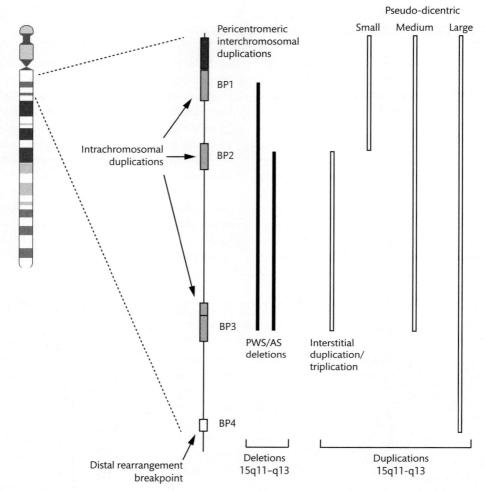

Figure 5.4 Segmental duplications and chromosomal rearrangements at 15q11-q13. Schematic representation showing major interchromosomal and intrachromosomal duplications associated with breakpoints (BP) and chromosomal rearrangements. Common deletions and duplications are shown. AS, Angelman syndrome; PWS, Prader–Willi syndrome. Adapted and reproduced from Locke *et al.* (2004), copyright 2004, with permission from BMJ Publishing Group Ltd.

chromosome 3q29 microdeletion syndrome (OMIM 609425) (Willatt *et al.* 2005), chromosome 9q34.3 deletion syndrome (OMIM 610253) (Harada *et al.* 2004; Stewart *et al.* 2004b), and chromosome 22q13.3 deletion syndrome (OMIM 606232) (Flint *et al.* 1995; Wong *et al.* 1997). Detection of subtelomeric genomic imbalance can be achieved by a number of different approaches, such as fluorescence in situ hybridization (FISH) and array CGH, the latter including focused subtelomeric arrays (Knight and Regan 2006).

Other genetic diseases may also arise due to chromosomal rearrangements in subtelomeric regions, notably involving a tandem repeat at 4q35 in the subtelomeric region of chromosome 4q. This was found to cause a specific autosomal dominant disorder, fascioscapulohumeral muscular dystrophy (OMIM 158900), a neuromuscular disorder characterized by an asymmetrical weakness of facial, shoulder, and upper arm muscles (Wijmenga *et al.* 1992). Affected patients were found to have only one to ten 3.3 kb repeat units compared with normal individuals

Box 5.8   Terminal deletion of chromosome 1p36 syndrome (OMIM 607872)

Deletion at the telomeric end of the short arm of chromosome 1 occurs in about one in 5000 of the population. Deletion of 1p36 results in a characteristic syndrome including specific physical characteristics, psychomotor retardation, fits, delayed growth, and mental retardation (Shapira *et al.* 1997). Overall the syndrome is thought to account for 0.5–1.2% of idiopathic mental retardation.

Box 5.9   Cri du chat syndrome (OMIM 123450)

This syndrome was first described by Lejeune in 1963 (Lejeune *et al.* 1963). The incidence of the syndrome is between one in 20 000 and one in 50 000 births (Niebuhr 1978). Affected newborn babies have a diagnostic high pitched, cat-like cry. Young children show characteristic facial features, speech delay, and mental retardation and die early in childhood. The syndrome arises due to terminal, or less commonly interstitial, deletions in the short arm of chromosome 5. These can range from very small (5p15.2) to the whole short arm of the chromosome. Most arise *de novo*, while about 12% arise from unbalanced segregation of translocations or recombination events. The critical region underlying the characteristic cry was mapped to 5p15.3 and other features of the syndrome to 5p15.2 (Overhauser *et al.* 1994). More recent work has implicated haploinsufficiency of a specific gene involved in telomerase activity (*TERT*, encoding telomerase reverse transcriptase) located at 5p15.33 (Zhang *et al.* 2003). Within the critical region, the *SEMA5A* gene encoding a semaphorin protein, and the catenin gene *CTNND2,* are linked with cortical development and may play a role in the mental retardation seen among patients with this syndrome (Cerruti Mainardi 2006). Use of array CGH has allowed high resolution mapping and is facilitating genotype–phenotype associations with specific regions (Fig. 5.5).

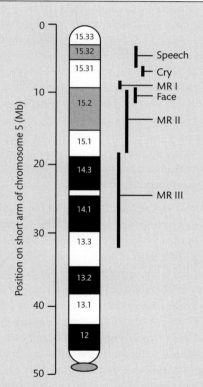

Figure 5.5 Cri du chat and chromosome 5p. Mapping of genomic regions of the short arm of chromosome 5 associated with specific phenotypes. Three regions associated with mental retardation (MR) are shown. Redrawn from Zhang *et al.* (2005b), copyright 2005, with permission from Elsevier.

having ten to several hundred tandemly repeated units (Section 7.2.2; Box 7.2).

5.4 Pathogenic copy number variation, mental retardation, and autism

The analysis of copy number variation has been highly informative in efforts to define the role of genetic factors in mental retardation (also termed learning disability). This is a very common condition affecting 1–3% of the population (Roeleveld et al. 1997). In approximately half of cases the cause is unknown but genetic factors are often postulated (Flint and Wilkie 1996). Large scale microscopically visible chromosomal abnormalities are associated with mental retardation and detectable using cytogenetic approaches, for example in Down syndrome (Box 3.3). Submicroscopic structural variation also plays a very important role, although only comparatively recently has the full extent of this become apparent through technological advances allowing the detection of progressively smaller sizes of pathogenic copy number variation.

5.4.1 Subtelomeric rearrangements and idiopathic mental retardation

Early success was achieved by resolution of subtelomeric rearrangements among individuals with idiopathic mental retardation. Flint and colleagues studied a group of 99 such individuals in whom no major chromosomal abnormalities were apparent using standard karyotype analysis. A combination of approaches resolved and confirmed monosomy in subtelomeric regions at 22q in two cases and 13q in one case (28 chromosome ends examined among 23 chromosomes) (Flint et al. 1995). Analysis of subtelomeric regions using different approaches, including FISH or more recently array CGH, has defined pathogenic subtelomeric copy number variation in about 5% of cases of idiopathic mental retardation – of which 50% were thought to occur de novo (Knight and Flint 2004). Analysis of subtelomeric genomic imbalances forms an important tool for clinical geneticists and paediatricians investigating idiopathic mental retardation and developmental delay.

5.4.2 Copy number variation among cases of mental retardation

Considerable success has been achieved through analysis of interstitial copy number variation to detect novel submicroscopic structural variants responsible for specific phenotypes among cases of idiopathic mental retardation. Such variants and associated syndromes had been previously unknown. Advances in our ability to detect copy number variants now allows for large scale screening of patient cohorts and clinical application through the use of arrays. A number of studies using array CGH at relatively low resolution (1 Mb) indicated that copy number variation could explain 10% of idiopathic mental retardation among individuals who also had dysmorphic features, a subset of cases likely to be enriched for underlying genetic abnormalities (Knight and Regan 2006). Higher resolution genome-wide tiling arrays comprising 32 447 bacterial artificial chromosomes (BACs) were used by de Vries and colleagues to demonstrate that among 100 individuals with idiopathic mental retardation and normal karyotype/subtelomeric analysis, seven cases had deletions and three had duplications, varying in size from 540 kb to 12 Mb and occurring de novo based on parental analysis (de Vries et al. 2005).

The use of BAC array CGH targeted at regions flanked by highly homologous segmental duplications was discussed in the context of phenotypically normal individuals (Section 4.2.6). The same array was also used by Sharp and colleagues to study a large cohort of 290 patients with idiopathic mental retardation (Sharp et al. 2006). Their earlier study (Sharp et al. 2005) allowed them to analyse in detail only copy number variation not found in the 'normal' population. Their approach was an efficient way of identifying new likely pathogenic microdeletions and duplications, which they found to be present in 16 individuals. These included a previously unrecognized 17q21.31 microdeletion syndrome (OMIM 610443) (Box 5.10), 15q24 microdeletion syndrome, and 15q13.3 microdeletion syndrome (Amos-Landgraf et al. 1999; de Vries et al. 2005).

Further advances were achieved through higher resolution oligonucleotide arrays allowing definition of very small sizes of copy number variation and mapping of breakpoints with greater precision (Shaikh 2007). For example using Affymetrix GeneChip Human Mapping

Box 5.10   De novo deletion at 17q21.31 and mental retardation

Sharp and colleagues identified four individuals with a novel deletion at 17q21.31 among 290 cases of idiopathic mental retardation. This genomic region has highly complex architecture with several large LCRs and segmental duplications (Fig. 5.6) that, as noted for other chromosomal rearrangements and genomic disorders, provide a substrate for non-allelic homologous recombination (Lupski 2006). Moreover, this region has been

shown to include a 900 kb inversion polymorphism which is common among Europeans (Fig. 5.6; Section 5.5.2) (Stefansson *et al.* 2005). Using higher resolution oligonucleotide arrays, Sharp and coworkers mapped the breakpoints of the deletion to flanking LCRs 38 kb in size, present in the same orientation and sharing 98% sequence identity (Sharp *et al.* 2006). There was evidence of a novel genomic disorder occurring sporadically *de novo*

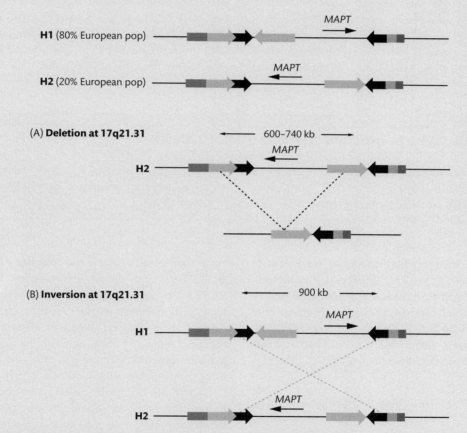

Figure 5.6 Genomic architecture at chromosome 17q21.31. Schematic representation of the 17q21.31 region showing two common haplotypes comprising low copy repeat regions (large block arrows) with shaded regions of the arrows indicating repeat subunits. (A) A deletion rearrangement spanning ~600–740 kb (includes the *MAPT* gene and five other genes not shown) is thought to occur through non-allelic homologous recombination on haplotype H2. (B) A common inversion spanning 900 kb, which includes the *MAPT* gene (and five other genes not shown), is present on haplotype H2. Adapted and reprinted by permission from Macmillan Publishers Ltd: Nature Genetics (Lupski 2006), copyright 2006.

with a reasonably specific clinical phenotype; indeed Sharp and colleagues were able to identify a fifth individual with mental retardation based on their clinical features who also had the deletion at 17q21.31. In the same issue of *Nature Genetics* where their paper was published, two other papers reported the same deletion as being associated with idiopathic mental retardation and a similar clinical phenotype, which included characteristic facial features, hypotonia, friendly or amicable behaviour, and structural brain abnormalities on imaging. Koolen *et al.* used array CGH to screen 360 individuals with idiopathic mental retardation and found a person with the 600 bp deletion; screening an additional 480 affected individuals revealed one person with the same 600 bp deletion and one with a smaller 100 bp deletion but having

one common breakpoint with the larger deletion (Koolen *et al.* 2006). Shaw-Smith and colleagues found three cases with the deletion and overall it is thought that this relatively large submicroscopic deletion at 17q21.31 could account for 1% of cases of mental retardation (Lupski 2006). This deletion has since been extensively characterized among 22 individuals, establishing 'chromosome 17q21.31 microdeletion syndrome' as a clearly defined genomic disorder (OMIM 610443) with an estimated prevalence of one in 16 000 (Koolen *et al.* 2008). An inversion event in a parent is necessary for the deletion to occur and within the proximal breakpoint a rearrangement hotspot was noted to lie in a mobile DNA element, an L2 long interspersed element (LINE) motif (Koolen *et al.* 2008).

Box 5.11   Autism spectrum disorders (OMIM 209850)

Autism spectrum disorders are characterized by difficulties with language and social interactions together with restricted and repetitive behaviour with onset by 3 years of age. Such disorders are common, with an estimated prevalence of one in 166.

100K arrays, Friedman and colleagues were able to resolve deletions as small as 178 kb in size: in 100 children with idiopathic mental retardation, eight had *de novo* deletions and two had duplications (Friedman *et al.* 2006). More widespread application of array CGH technology in clinical genetic testing is proposed as a cost effective approach (Wordsworth *et al.* 2007).

5.4.3 *De novo* copy number mutations and autism

There has been great research interest in the genetic basis of autism (Box 5.11), with strong evidence to support a role for inherited variation but marked genetic heterogeneity and multiple genetic loci reported (Klauck 2006). Highly diverse cytogenetic abnormalities are found in more than 5% of autistic children, occurring across all

chromosomes, with notable examples being duplication at 15q11-q13 and 16p13.1, and deletions at 2q37 and 22q13.3. However, the extent of submicroscopic structural variation remains unclear (Vorstman *et al.* 2006).

Sebat and colleagues sought to address this using high resolution CGH arrays, although the authors noted that they were probably failing to detect the vast majority of copy number variants because of their small size (Sebat *et al.* 2007). With this caveat in resolution imposed by available technology, the results were still striking. A significantly higher rate of *de novo* copy number mutation among patients with sporadic autism was found compared to healthy controls (10% versus 1%, $P = 0.0005$) (Sebat *et al.* 2007). The investigators were careful to exclude cases of syndromic autism (severe mental retardation or other congenital abnormalities) and patients with known cytogenetic abnormalities. Among

195 patients with autism spectrum disorders, 14 *de novo* copy number mutations were found. The authors estimated that once cytogenetically visible abnormalities were taken into account, the frequency of *de novo* copy number variation was 15% at current detection thresholds. The true figure may prove very much higher, with a diverse array of rare, highly penetrant sporadic copy number mutations postulated to underlie much of the sporadic nature of autism (Sebat *et al.* 2007).

5.5   Inversions in health and disease

5.5.1 Inversions may cause severe disease

Inversions are reported to occur on all chromosomes and may involve two breaks on different arms (pericentric inversion) or the same arm (paracentric inversion) (Fig. 3.8A). Many inversions may not be associated with any phenotypic consequences, for example an inversion on chromosome 9 inv(9)(p11q13) is found in 1–3% of the population without any known clinical significance (Hsu *et al.* 1987; Yamada 1992). There are, however, some notable examples of inversions leading to severe disease, illustrated by inversions involving intron 22 and intron 1

of the *F8* gene encoding factor VIII which are associated with severe haemophilia A (Box 5.12).

An inversion at the *IDS* gene in the same region of the X chromosome, Xq28, has been associated with another X-linked recessive disorder, Hunter syndrome (Bondeson *et al.* 1995). Here a deficiency of idunorate 2-sulphatase, encoded by *IDS*, can lead to severe disease with progressive damage to the brain and liver, often causing death by the age of 15 years.

Inversions of segmentally duplicated olfactory receptor genes at 4p16 and 8p23 were found to occur commonly among control subjects of European descent at 12.5% and 26%, respectively, with 2.5% of individuals heterozygous for both inversions (Giglio *et al.* 2001, 2002). This latter state, while carrying no phenotype for the affected individual, has been associated with having offspring carrying a recurrent translocation for the two loci denoted t(4;8)(p16;p23) (Section 3.4). Thus having two inversion polymorphisms in a heterozygous state on non-homologous chromosomes can lead to interchromosomal rearrangements. The translocation, in some individuals, is associated with Wolf–Hirschhorn syndrome (OMIM 194190). This syndrome is characterized by severe growth retardation, mental retardation, a characteristic facial appearance, and closure defects (for example cleft lip or palate, coloboma of the eye, and cardiac septal defects).

Box 5.12   Haemophilia A (OMIM 306700)

This is one of the commonest X-linked diseases affecting one in 5000 males. It is caused by a deficiency of coagulation factor VIII, which leads to easy bruising, haemorrhage into joints and muscles, and prolonged bleeding. Approximately 50% of people have a severe phenotype with less than 1% residual factor VIII, while 10% have moderate disease (2–5% residual factor VIII) and 40% have mild disease (5–30% residual factor VIII). The condition results from genetic variation at the *F8* gene encoding factor VIII: in those with severe disease, half of cases arise from an inversion involving intron 22 and a homologous region telomeric to the *F8* gene

(Lakich *et al.* 1993). A further 5% of severe cases were reported to arise from a second inversion, this time involving intron 1 (Bagnall *et al.* 2002). In both situations, inversions have arisen due to intrachromosomal recombination between homologous regions in introns 1 or 22, and regions telomeric to the *F8* gene. A large gene deletion is thought to account for 5% of severe cases while the remainder, together with all moderate and mild cases, arise due to point mutations and small insertions/deletions of the *F8* gene (Castaldo *et al.* 2007). These are thought to be newly occurring mutations in approximately one-third of cases.

5.5.2 Inversion and deletion at 17q21.31 with evidence of selection

Large inversions are thought to often arise from non-allelic homologous recombination involving highly homologous low copy repeat regions present in an inverted orientation (Shaw and Lupski 2004). Such a situation is seen in the complex genomic architecture on the long arm of chromosome 17 at 17q21.31 where both deletions and a large inversion have been identified (see Fig. 5.6) (Stefansson *et al.* 2005; Lupski 2006).

Stefansson and colleagues found a large inversion polymorphism that was common among Europeans (frequency of inversion 21%), but considerably rarer in Africans (6%) and almost absent among Asians (1%) (Stefansson *et al.* 2005). The inversion spans a region of particular interest in neurological diseases, with two highly divergent haplotypes previously identified involving *MAPT* (encoding microtubule-associated protein tau), designated H1 and H2, and shown to be associated with progressive supranuclear palsy (Baker *et al.* 1999) and Parkinson's disease (Skipper *et al.* 2004). Analysis of 60 microsatellite markers showed that the H2 haplotype was structurally distinct, and that a 900 kb segment was inverted compared to H1 and the reference assembly for the region. Analysis of human and chimpanzee (*Pan troglodytes*) clones suggested very great mutational differences and a very ancient divergence between the H1 and H2 lineages, such that the inversion polymorphism may be 3 million years old (Stefansson *et al.* 2005). Moreover, analysis in an Icelandic population showed that the H2 lineage was in fact undergoing positive selection, with carrier females of the inversion polymorphism having more children and higher recombination rates (Stefansson *et al.* 2005).

5.5.3 Finding inversions across the human genome

High throughput methods for detecting inversion polymorphisms are not currently available, however approaches using high density single nucleotide polymorphism (SNP) genotyping data have been proposed. For example, Bansal and colleagues developed a statistical method for the detection of large inversions based on unusual patterns of linkage disequilibrium indicative of inversions using the HapMap dataset (Bansal *et al.* 2007). When such patterns were present in a majority of chromosomes from the population, and were in an inverse orientation to the reference human genome sequence, this was taken as evidence of a candidate inversion. One hundred and seventy-six such events were reported, ranging in size from 200 kb to several megabases, however only a minority were likely to be real.

Comparison of a second genome sequence based on **fosmid** paired-end sequencing data with the reference human assembly has also been informative, with 56 likely inversion breakpoints greater than 8 kb in size identified (Tuzun *et al.* 2005). This approach utilized a human fosmid DNA genomic library generated from an anonymous North American female (donor GM15510) to identify orientation discrepancies, and the authors validated a large proportion of those identified.

Comparison of human and chimpanzee genome sequences has also allowed definition of potential sites of inversions between the two species (Feuk *et al.* 2005; Szamalek *et al.* 2006). Feuk and colleagues identified 1576 putative inversions in such regions, ranging in size from 23 bp to 62 Mb, with 33 inversions greater than 100 kb and the highest count seen on the X chromosome (Feuk *et al.* 2005). Of 27 putative inversions tested, 23 were experimentally validated including a novel 4.3 Mb inversion at 7p14. Three of these inversions were found to be polymorphic among a panel of ten individuals of European descent: a 730 kb inversion at 7p22 (5% minor allele frequency), a 13 kb inversion at 7q11 (30%), and a 1 kb inversion at 16q24 (48%).

Among apparently healthy human subjects, 182 inversions greater than 1 kb in size had been reported by December 2007 (Database of Genomic Variants, http://projects.tcag.ca/variation/) (Zhang *et al.* 2006). As described earlier, pericentric inversions on chromosome 9, inv(9)(p11q12)/inv(9)(p11q13), are found in about 2% of the normal population and are regarded as polymorphic variants without significant consequences: homologous regions in the long and short arms of chromosome 9 around the breakpoints may facilitate recombination and inversion (Park *et al.* 1998). Other large inversions have been found. For example, on chromosome Xq28, mispairing between two large inverted repeats was found to lead to a 48 kb inversion of a region containing the

FLN1/EMD genes in 18 out of 108 human X chromosomes analysed (Small *et al.* 1997).

5.6 Summary

This chapter has highlighted the importance of pathogenic copy number variation, in particular how this relates to the growing list of genomic disorders. Here genomic architecture, most notably highly homologous stretches of DNA, is highly significant in modulating chromosomal rearrangements and genomic disease (Inoue and Lupski 2002). Examples of hotspots of recurrent rearrangements and disease include chromosome 22q11 (DiGeorge syndrome and velocardiofacial syndrome; Box 5.2) and the proximal short arm of chromosome 17 (CMT type 1A; Box 5.4). A number of different genomic disorders have been described to illustrate the basis and nature of recurrent and non-recurrent disease. Sites involving reciprocal genomic disorders were noted such as duplication, deletion, and inversions at 7q11.23 (Williams–Beuren syndrome); or at 17p11.2 involving the dosage-sensitive gene peripheral myelin protein-22 (*PMP22*) leading to demyelinating peripheral neuropathies (CMT type 1A associated with duplications, and deletions with HNPP) (Lupski *et al.* 1991; Chance *et al.* 1993). Other genomic disorders have been described showing parent of origin effects such as Prader–Willi and Angelman syndromes.

The importance of recent advances in the analysis of submicroscopic structural variation have also been emphasized by progress that has been achieved in the diagnosis and understanding of mental retardation. The analysis of copy number variation has led, for example, to the identification of previously unrecognized genomic disorders such as 17q21.31 microdeletion syndrome, which may account for 1% of cases of idiopathic mental retardation (Lupski 2006). Many other new syndromes and underlying structural causes have been defined by high resolution mapping techniques to define pathogenic copy number variation, representing major advances of significant clinical importance. Array CGH techniques developed for a research setting are now being implemented for clinical diagnostic use, highlighting the utility of such advances to direct patient care.

Terminal deletions and subtelomeric events have also been reviewed, together with structural variation involving a change in orientation in the form of inversions. Inversions are recognized to occur in all chromosomes and can lead to severe disease, as seen with inversions involving the *F8* gene and haemophilia A, or reach high frequencies in particular populations associated with positive selection (Lakich *et al.* 1993; Stefansson *et al.* 2005). Systematic surveys to detect inversions within human populations are now becoming feasible, as they are for other classes of submicroscopic variation.

The role of segmental duplications and in particular low copy repeats in genomic disorders has been described. In the next chapter the nature and importance of segmental duplications are described taking a broad view of the field, which allows analysis of their extent and evolutionary significance. Further examples involving copy number variation and other genetic variation are also described in Chapter 6 through review of the genetics of colour vision and rhesus blood groups.

5.7 Reviews

Reviews of subjects in this chapter can be found in the following publications:

Topic	References
Mechanisms of chromosomal rearrangements and genomic disorders	Lupski 1998; Shaffer and Lupski 2000; Inoue and Lupski 2002; Shaw and Lupski 2004; Lupski and Stankiewicz 2005; Gu *et al.* 2008
Clinical and genetic aspects of Parkinson's disease	Fahn 2003; Farrer 2006
Charcot–Marie–Tooth disease	Murakami *et al.* 1996
Williams–Beuren syndrome	Osborne and Mervis 2007
Cri du chat syndrome	Niebuhr 1978
Chromosomal rearrangements at chromosome 22q11	McDermid and Morrow 2002
Genetics of autism spectrum disorder	Klauck 2006
Molecular genetics of haemophilia A	Castaldo *et al.* 2007

CHAPTER 6

Segmental duplications and indel polymorphisms

6.1 Introduction

In this chapter the theme of structural genomic variation is continued, in particular focusing on segmental duplications. Segmental duplications (Box 6.1) are blocks of DNA ranging in size from one to several hundred kilobases which map to more than one location in the genome. They can be found in tandem or at interspersed locations, inter- or intrachromosomally and are estimated to comprise 5% of the human genome (Cheung *et al.* 2001, 2003a; Bailey *et al.* 2002). There is evidence that the majority of segmental duplications have arisen relatively recently in our evolutionary past, notably in the last 35 million years (Hattori *et al.* 2000; Lander *et al.* 2001; Bailey *et al.* 2002; Cheung *et al.* 2003a). The study of duplications has been of particular interest to evolutionary biologists as a substrate for adaptive evolution and this theme is explored over the course of this chapter. At the end of the chapter, smaller scale insertions and deletions involving one or more contiguous nucleotides, collectively described as 'indels', are discussed. Such diversity occurs more commonly than was originally thought and can have significant phenotypic consequences.

6.2 Nature and extent of segmental duplications

6.2.1 Segmental duplications are common in the human genome

The publication of the human genome sequence in 2001 highlighted the abundance of segmental duplications within the human genome (Fig. 6.1) (Lander *et al.* 2001, 2001). The extent of segmentally duplicated sequence varies significantly between chromosomes, estimated at between 1% and 14% among all 24 chromosomes (Fig. 6.2) (Zhang *et al.* 2005a). A number of databases catalogue segmental duplications across the human genome, for example as part of the Human Genome Structural

---

Box 6.1 Terminology relating to duplication events

Segmental duplications are continuous portions of DNA greater than 1 kb in size that occur in two or more copies per haploid genome with the copies sharing more than 90% sequence identity (Scherer *et al.* 2007). In tandem duplications the duplicated segment of DNA is found immediately adjacent to the ancestral copy in the same orientation. **Low copy repeats** (**LCRs**) refers to sequences of 10 kb or more in size, with at least 95% sequence identity, which are separated by 50 kb to 10 Mb of intervening sequence (Stankiewicz and Lupski 2002). Segmental duplications can be LCRs. A **duplicon** is a duplication traceable to an ancestral or donor location (Bailey and Eichler 2006).

Variation Project (http://humanparalogy.gs.washington.edu/) and in the Database of Genomic Variants (http://projects.tcag.ca/variation/).

An estimated 50% of duplications are strictly intrachromosomal. In comparison with interchromosomal duplications, intrachromosomal duplications tend to be larger (18.5 kb versus 14.8 kb) and a higher proportion contain complete genes (6.2% versus 1.3% for known genes) (Zhang *et al.* 2005a). Within the long arm of chromosome 22, for example, 9.1% of the sequence has been identified as being involved in segmental duplications: 3.9% occurred between non-homologous chromosomes (interchromosomal) while 6.4% was intrachromosomal (Fig. 6.3) (Bailey *et al.* 2001).

Analysis of the finished sequence of chromosome 15 showed 8.8% of the euchromatic sequence was composed of segmental duplications of which 50% were purely intrachromosomal, 20% interchromosomal, and 30% both intra- and interchromosomal (Zody *et al.* 2006). The segmental duplications were broadly clustered into two regions, on the proximal and distal ends of 15q. The 15q11-q13 region is particularly prone to chromosomal rearrangements with recombination among segmental duplications, and associated with a variety of deletions, duplications, triplications, inversions, translocations, and marker chromosomes (Fig. 5.4; Section 5.2.6).

Approximately one-third of duplicated genes are arranged in tandem with a heterogeneous distribution among chromosomes (Fig. 6.4) (Shoja and Zhang 2006). There are thought to be 902 tandemly arranged gene clusters in the human genome, of which 68% have two genes in the tandem array; together, tandemly arranged genes make up 14–17% of all genes in the genome (Shoja and Zhang 2006).

6.2.2 Pericentromeric and subtelomeric regions are hotspots for segmental duplications

In terms of chromosomal regions, pericentromeric regions contain a large proportion of segmental duplications: within 5 Mb of the centromere, 22.7% of bases are segmentally duplicated and thought to account for about one-third of all human duplicated sequence (47.2 out of 152 Mb) (She *et al.* 2004). A gradient is seen towards the centromere of increasing numbers of interchromosomal duplications and reduced transcriptional diversity. The rate of duplicative transposition of segments of DNA towards pericentromeric regions has been estimated at six to seven events per million years of primate evolution (She *et al.* 2004).

A two step process is postulated for the generation of such segmental duplications with an initial duplication event to a pericentromeric region, followed by

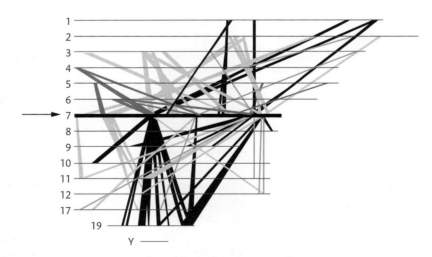

Figure 6.1 Interchromosomal duplications revealed from sequencing the human genome. An example is shown for chromosome 7 with lines to other chromosomes indicating homologous blocks, each comprising at least three genes. From Venter *et al.* (2001), reprinted with permission from AAAS.

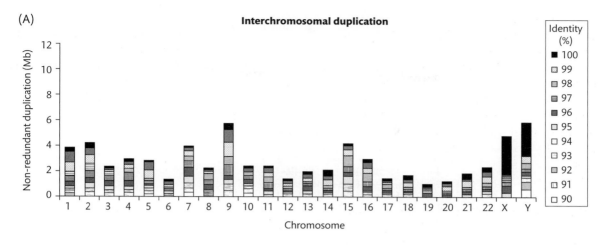

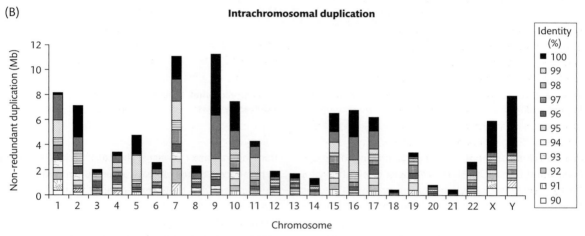

Figure 6.2 Chromosomal distribution of recent segmental duplication in the human genome. (A) Interchromosomal duplication. (B) Intrachromosomal distribution. Shading indicates percent sequence identity. Reproduced with permission from She *et al.* (2006).

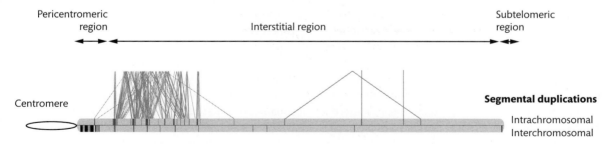

Figure 6.3 Segmental duplications of chromosome 22q. Segmental duplications of the long arm of chromosome 22 that are greater than 10 kb in size with more than 90% sequence identity are shown. Interchromosomal (displayed in black below the line) and intrachromosomal (in dark grey above the line) segmental duplications are given, with light grey lines joining homologous duplications. Figure redrawn and reprinted by permission from Macmillan Publishers Ltd: Nature Reviews Genetics (Bailey and Eichler 2006), copyright 2006.

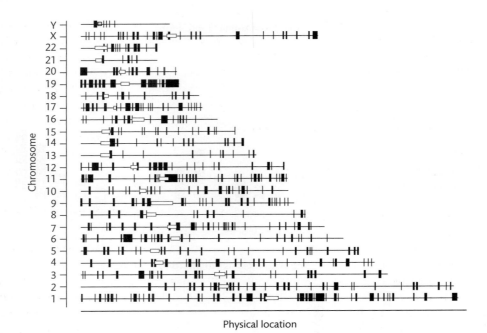

Figure 6.4 Tandemly arranged genes show variable distribution across chromosomes. Open boxes represent centromeres. Reproduced from Shoja and Zhang (2006), by permission of Oxford University Press.

duplication to nonhomologous pericentromeric regions (Bailey and Eichler 2006). For example, duplication of a 9.7 kb segment in the adrenoleukodystrophy ALD locus from the X chromosome to a pericentromeric region of chromosome 2, followed by duplication from 2p11 to 10p11, 16p11, and 22q11. These duplicated regions still have high sequence homology and are believed to have arisen over the last 5–10 million years (Eichler et al. 1997).

Subtelomeric regions are a second major 'hotspot' of duplication in which a high concentration of recent interchromosomal segmental duplications have been found (Linardopoulou et al. 2005). Subtelomeres are transition zones ranging in size from 10 to 300 kb, which are found near the tips of chromosomes, between chromosome-specific sequences and the arrays of telomeric repeats that cap each chromosome (Box 7.3) (Mefford and Trask 2002). Subtelomeres are remarkably dynamic and variable regions, comprising 25 gene families in a patchwork array of duplicated blocks showing high homology but great diversity in copy number and chromosomal location. Gene products include olfactory receptors and cytokines as well as

transcription factor proteins. The patchwork of segmental duplications is thought to have arisen due to complex double-stranded DNA breakage and repair leading to numerous repeated translocations at the ends of chromosomes (Linardopoulou et al. 2005). Multiple events have led to a mosaic of adjacent duplicons with maintenance of sequence orientation between copies. These interchromosomal duplications are thought to have occurred over very recent evolutionary time: 49% of known subtelomeric sequence is believed to have been generated after humans and chimpanzees diverged (Linardopoulou et al. 2005).

For such a small portion of the genome, subtelomeres account for a very high proportion of highly homologous duplications: 40% of all duplications in the sequenced genome with a sequence identity of 98.7% or more (Linardopoulou et al. 2005). The rate of gene duplication in subtelomeres is four times that of the genome-wide average with seven gene duplicates having arisen in human subtelomeres per million years. Polymorphism in the extent of duplication in healthy individuals is recognized, for example stable interindividual variation in the length of alleles involving the subtelomeric region on the short arm of chromosome 16 (Wilkie *et al.* 1991). The consequences can also be severe, with a number of genetic diseases associated with subtelomeric chromosomal rearrangements (Section 5.3).

6.2.3 Non-allelic homologous recombination, segmental duplications and genomic disorders

Non-allelic homologous recombination (section 5.2.1) and replication error are thought to be the most important mechanisms leading to tandem duplications (Bailey and Eichler 2006). Segmental duplications, and in particular LCRs, are themselves prone to non-allelic homologous recombination at meiosis which can lead to genomic rearrangements and genomic disorders (Section 5.2); such events in mitosis can result in a mosaic somatic cell population carrying genomic rearrangements which are associated with cancer or mosaic manifestations of genomic disorders (mechanisms of genomic rearrangements reviewed by Gu *et al.* 2008). If LCRs are present on the same chromosome and in a direct orientation with each other, non-allelic homologous recombination may result in duplications and deletions; if LCRs are present in opposite orientations, inversions. The efficiency of non-allelic homologous recombination is related to the distance between LCRs. Within LCRs, a minimal length of extremely homologous sequence is required, usually between 300 and 500 bp. Particular sequences are associated with double stranded DNA breaks (for example palindromes, minisatellites and mobile DNA elements) which are seen as 'hotspots' for non-allelic homologous recombination (Gu *et al.* 2008).

Specific chromosomal regions which have been the target of multiple independent duplication events are described as duplication hubs or acceptor regions (reviewed in Bailey and Eichler 2006). These can be associated with disease phenotypes such as DiGeorge syndrome, a genomic disorder caused by non-allelic homologous recombination between three highly identical (greater than 99%) duplication hubs on chromosome 22q11 (Box 5.2).

6.2.4 Alu elements and segmental duplications

Alu elements are a family of retrotransposons (class I mobile DNA elements), noncoding DNA sequences approximately 300 bp long which are found extensively across the human genome (Section 8.4). They are particularly associated with segmental duplications and low copy repeats. Alu insertions are seen to be enriched at the junctions of duplications, with 27% of segmental duplications found to terminate within an Alu repeat (Bailey *et al.* 2003). A burst of Alu retroposon activity has been identified in primates occurring 35–40 million years ago (Shen *et al.* 1991). Bailey and colleagues propose that this sensitized the ancestral genome to Alu-Alu-mediated recombination events and initiated an expansion of gene-rich segmental duplications (Bailey *et al.* 2003).

6.2.5 Segmental duplications in primates and other species

Segmental duplications have been found in a range of other genomes, ranging from the worm (*Caenorhabditis elegans*) (Mounsey *et al.* 2002) and the fly (*Drosophila melanogaster*) (Fiston-Lavier *et al.* 2007), to the mouse (*Mus musculus*) (Cheung *et al.* 2003b; Bailey *et al.* 2004), rat (*Rattus norvegicus*) (Tuzun *et al.* 2004), and dog (*Canis familiaris*) where an estimated 2–4% of the genome is duplicated. Segmental duplications among mammalian genomes are larger than those seen in the worm or fly. Differences are seen between mammals, with interchromosomal segmental duplications more common in humans (48%) compared with mice and rats (13% and 15%, respectively) (Bailey *et al.* 2004; Tuzun *et al.* 2004), while tandemly duplicated segments are less common (45% in humans versus 70–90% in mice, chickens, and rats) (She *et al.* 2006)).

Publication of the chimpanzee (*Pan troglodytes*) genome sequence (Box 10.1) highlighted significant differences in terms of segmental duplications with one third of human duplications (showing more than 94% sequence identity) not found in the chimpanzee genome (Cheng *et al.* 2005). These were remarkable differences, highlighting the extent of recent segmental duplications in primate evolution. Cheng and colleagues estimated that differences involving duplicated segments accounted for much more of the sequence difference between humans and chimpanzees than fine scale sequence diversity (2.7% compared to 1.2% at the level of single base pair differences). Even within shared duplications, there was evidence of significant copy number variation. Further insights came from analysis of the genome sequences of other primates. The rhesus macaque (*Macaca mulatta*) is an Old World Monkey thought to share a common ancestor with humans some 25 million years ago (Fig 6.5). Sequencing of the macaque genome (Box 10.2) highlighted much lower levels of segmental duplication, comprising 2.3% of the genome in comparison with humans and chimpanzees where 5 to 6% of the genome consists of segmental duplications (Gibbs *et al.* 2007).

A recent analysis of duplications among four primate genomes (human, chimpanzee, orang-utan and macaque) resolved these differences further, highlighting how 80% of human segmental duplications arose after the divergence of the hominoid linages from Old World Monkeys (analyzing duplications more than 20 kb in size with more than 94% identity) (Marques-Bonet 2009). Indeed, Eichler and colleagues showed evidence of a highly significant increase in duplication activity in the common ancestor of humans and African great apes, and after the divergence of the gorilla and human-chimpanzee lineages - the authors note how this contrasts with the 'slowing' of other processes generating genetic diversity including point mutations and retrotransposon activity in the hominoid lineage (Marques-Bonet 2009).

6.3 Duplication and evolution

6.3.1 Whole genome duplications

Segmental duplications are thought to have had a less dominant evolutionary impact than ancient whole genome duplications, which were associated with major evolutionary changes involving evolutionary transitions and adaptive radiation of species (Maere *et al.* 2005). Current hypotheses suggest either two rounds of whole genome duplication early in vertebrate evolution (Dehal and Boore 2005) or a single round of whole genome duplication followed by gene family expansion through small scale segmental and tandem duplication 50–150 million years ago (Gu *et al.* 2002).

6.3.2 Gene creation

Segmental duplications are, however, thought to be one of the main mechanisms for the creation of new genes over evolutionary time, predominantly through duplications of entire genes and more rarely through exon shuffling and the creation of fusion transcripts (Box 6.2)

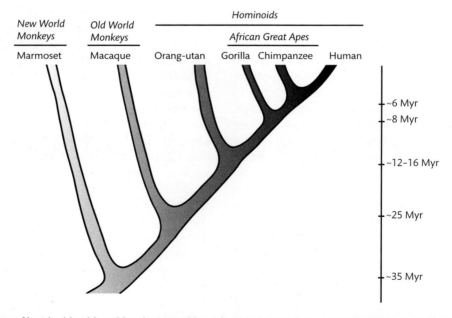

Fig 6.5 Phylogeny of hominoids, old world and new world monkeys. Estimated times of divergence indicated (Myr, million years ago). Figure adapted and reprinted by permission from MacMillan Publishers Ltd: Nature Reviews Genetics (Samonte and Eichler 2002), copyright 2002; and Nature (Marques-Bonet *et al.* 2009), copyright 2009.

(Taylor and Raes 2004; Bailey and Eichler 2006). Genes within segmental duplications have been noted to commonly (1) show an excess of structural (copy number) variation; (2) have strong positive **signatures of selection** (Section 10.2); and (3) encode proteins involved in the immune response, reproduction, nuclear function, olfactory reception, and drug detoxification (Bailey and Eichler 2006). Such features are consistent with an important role in primate and human adaptive evolution.

On the short arm of chromosome 16, segmental duplications are highly variable in copy number between species, with evidence of strong positive selection (Johnson *et al.* 2001). Segmental duplications account for more than 10% of the euchromatic sequence on the short arm of chromosome 16, with 20 elements described (denoted low copy repeat sequences on chromosome 16, LCR16a

to LCR16t) (Loftus *et al.* 1999). Of particular interest is LCR16a, a duplicated segment 20 kb in length with high sequence identity which is present as 15 duplicated copies in humans (Johnson *et al.* 2001). Johnson and colleagues found evidence of a major proliferation in the number of these duplicons between great apes and Old World monkeys, with 17 copies in gorillas and 25–30 copies in chimpanzees but only one or two copies of LCR16a found in all Old World monkeys.

A likely ancestral origin at 16p13.1 was found, with duplication to other chromosomes noted in orangutans and chimpanzees consistent with lineage-specific expansion. Strikingly, exonic regions within the repeats were found to be hypervariable (10% nucleotide divergence compared to 2% in intronic sequence). Analysis of average nucleotide substitution rates for nonsynonymous

---

Box 6.2   Juxtaposition of segmental duplications leading to new genes

A fusion of segmental duplications from two genes on the long arm of chromosome 17 were shown to lead to the creation of the oncogene USP6 (Tre2) (ubiquitin-specific protease 6) on the short arm of the same chromosome, at 17p13.2 (Fig. 6.6) (Paulding *et al.* 2003). This chimeric gene is expressed in a variety of cancers and is hominid-specific. It is thought to have arisen 21–33 million years ago from duplications of an ancient, highly conserved gene USP32 and TBC1D3 (TBC1 domain family member 3). TBC1D3 is thought to have itself arisen from a recent segmental duplication leading to rapid dispersal through the primate lineage.

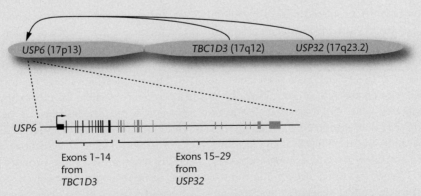

Figure 6.6 Gene creation through segmental duplication. Fusion of segmental duplications from *TBC1D3* and *USP32* result in a hominid-specific oncogene, *USP6*. Figure redrawn and reprinted by permission from Macmillan Publishers Ltd: Nature Reviews Genetics (Bailey and Eichler 2006), copyright 2006.

and synonymous changes (Ka and Ks respectively) within and between species showed evidence of strong positive selection. The differences between humans and Old World monkeys for example were very striking with Ka/Ks quotients of 13 (values above one are generally taken as evidence of selection). Further analysis suggested the major effect of positive selection was operating in a common ancestor to human and African apes, after the separation of human and chimpanzee lineages from the orangutan less than 12 million years ago (Johnson *et al.* 2001).

In their review, Conrad and Antonarakis describe a number of other factors influencing gene retention following duplication including the degree of conservation, the sensitivity of genes to dosage effects (with loss of fitness for example if only one copy were available), and the regulatory and architectural complexity of the gene (duplicated genes tending to encode longer proteins with more cis-regulatory domains) (Conrad and Antonarakis 2007). The other major mechanism for generating proteomic diversity, alternative splicing (Box 1.15; Section 11.6), shows a negative correlation with gene duplication when analysed among gene families of different sizes (Kopelman *et al.* 2005; Su *et al.* 2006). Duplicated genes show fewer alternatively spliced isoforms than single copy genes, particularly for recently duplicated genes.

6.3.3 Duplication rates over evolutionary timescales

Gene duplication has been considered to play a key role in evolution for a number of years. In 1970, Ohno proposed that gene duplication was very important in allowing genomes to grow and diversify (Ohno 1970). The subsequent availability of DNA sequence data from a diverse range of species allowed the rate of gene duplication and fate of duplicated genes to be assessed. The work of Connery and Lynch published in 2000 suggested that the rate of duplication was much higher than previously thought with, on average, a gene duplicating once every 100 million years (Lynch and Conery 2000). The authors compared the protein coding sequences available for human, mouse, chicken (*Gallus gallus*), worm, fly, rice (*Oryza sativa*), the flowering plant thale cress (*Arabidopsis thaliana*), and yeast (*Saccharomyces cerevisiae*). Dating of

duplication events from the number of silent nucleotide changes indicated that most duplicates were relatively young. Almost all were silenced by degenerative mutations within a few million years (the average half life for a gene duplicate was estimated to be 4 million years) while strong purifying selection was noted for the few surviving functional duplicates. The high rates of duplication were similar across species such that a genome of 15 000 genes was likely to acquire 60–600 duplicate genes over a million years.

In 2006, Dermuth and colleagues analysed a number of mammalian whole genome sequences and proposed that since the split from chimpanzees there have been 689 genes gained and 86 lost along the lineage leading to modern humans (Demuth *et al.* 2006). When the 689 genes gained in humans and the 729 lost in chimpanzees are combined, this showed that humans and chimpanzees differ by at least 6% in their complement of genes (1418 out of 22 000 genes).

6.3.4 Evolutionary fate of duplicated genes

In evolutionary terms, the most likely outcome for a duplicated gene will be loss of function (Ohno 1970). Duplication of a gene in the absence of selective pressure on the copy allows rapid divergence with deleterious mutations leading to loss of function, becoming a pseudogene (Box 6.3) and eventually disappearing from the genome.

By contrast, gain or change in function through mutation occurring in the duplicated copy will allow the mutation to become fixed in the population by natural selection and the gene duplication to persist in the genome (Force *et al.* 1999; Lynch and Conery 2000). For a small minority of duplicated genes such effects are seen: mutations may lead to selectively advantageous novel functions for the duplicated copy while the other copy retains its original function, a process described as 'neofunctionalization'. This may, for example, involve mutations in noncoding DNA leading to diversity in tissue or developmental specificity of gene expression, or more rarely a change in the coding sequence. Alternatively, both the original and duplicated copies of the gene may mutate and acquire new and complementary functions to those of the original gene, leading to 'subfunctionalization' (Lynch and Force 2000). A number of other models have been proposed

---

Box 6.3   Pseudogenes

Pseudogenes genes show a high degree of sequence homology to a non-allelic functional gene but are themselves non-functional, usually due to a lack of protein coding ability (Jacq *et al.* 1977). A pseudogene may be generated by nonsense mutation, frameshift mutation, or partial nucleotide deletion. There are many examples of pseudogenes across the genome, notably within the major histocompatibility complex (MHC) region.

---

for the duplicate gene copy, notably 'genetic robustness' whereby there is redundancy but the highly conserved copy acts as a backup in the event of deleterious mutations occurring in the original gene (Gu *et al.* 2003).

6.4   Gene duplication and multigene families

Gene duplication and conversion has led to the evolution of multigene families (Ohta 2000; Nei and Rooney 2005). Multigene families are thought to arise from a common ancestral gene, leading to a group of genes with similar functions and DNA sequence. In some cases 'supergene' families are seen composed of related multigene families.

6.4.1 Olfactory receptor and globin supergene families

The largest superfamily known in vertebrate genomes is the olfactory receptors, comprising 17 gene families (Glusman *et al.* 2001). In humans the family comprises about 800 genes found in clusters of tandem arrays on all chromosomes except 22 and Y; 42% of olfactory receptor genes are found on chromosome 11. The mean cluster size is 300 kb with 80% of clusters comprising six to 138 genes. This is a remarkable proportion of the genome, indeed it is described as the 'olfactory subgenome' comprising as it does of nearly 1% of the human genome and over 30 Mb of sequence. Relative to the mouse lineage, humans have similar numbers of gene clusters but appear to have lost many olfactory receptor genes, while mice have gained many (mice have about 1400 olfactory receptor genes of which about 1040 are functional)

(Niimura and Nei 2005). In humans only about 390 olfactory receptor genes are functional, the remainder being pseudogenes. The number of olfactory receptor genes is thought to have increased by tandem duplication and chromosomal rearrangements (Nei and Rooney 2005).

In the conclusion to his paper published in 1961 on gene evolution and the haemoglobins (Fig. 6.7), Vernon Ingram wrote:

> The suggestion is made that a single primitive myoglobin like haem protein is the evolutionary forerunner of all four types of peptide chain in the present day human haemoglobins, and of the corresponding peptide chains in other vertebrate haemoglobins. Such a scheme involves an increase in the number of haemoglobin genes from one to five by repeated gene duplications and translocations; the scheme may thus illustrate a general phenomenon in gene evolution. (Ingram 1961)

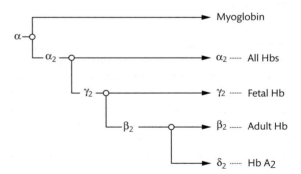

Figure 6.7 Evolution of the haemoglobin chains. This figure from Ingram's paper published in 1961 shows points of gene duplication followed by translocation (denoted by an open circle), the α chain is the ancestral peptide chain. Reprinted by permission from Macmillan Publishers Ltd: Nature (Ingram 1961), copyright 1961.

A large body of research has subsequently demonstrated that over the last approximately 800 million years a superfamily of genes have become established from an ancestral globin gene (Fig. 6.8). The encoded proteins, haemoglobin, myoglobin, neuroglobin, and cytoglobin, continue to share the ability to bind oxygen, although the process of evolution has led, for example, to tissue-specific expression (myoglobin in muscle, neuroglobin in neuronal tissues). Divergence of α and β globin is thought to have occurred 450–500 million years ago, with duplication within β globin 150–200 million years ago leading to the proto-β and proto-ε genes (Czelusniak *et al.* 1982; Goodman *et al.* 1987). A complex series of gene duplications, inactivation, fusion, and conversion events appear to have occurred specific to different mammalian lineages (Fig. 6.9) (Aguileta

et al. 2004). The consequences of genetic diversity at the human α and β globin gene clusters for disease were discussed in Chapter 1 and the evidence for selection, notably in relation to malaria, are discussed in Sections 10.2 and 13.2.

6.4.2 Models for the evolution of multigene families

Nei and Rooney have reviewed different models for the evolution of multigene families (Nei and Rooney 2005). Analysis of the genes encoding haemoglobin α, β, γ, δ, and myoglobin provided a paradigm for a divergent mode of evolution, in which phylogenetically related genes gradually diverged with the acquisition of new gene functions by duplicate genes (Ingram 1961). The

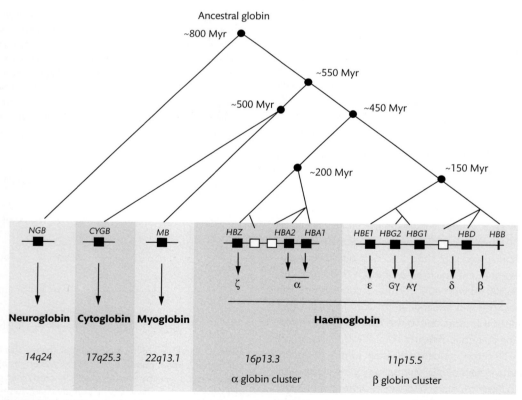

Figure 6.8 Evolution of the globin superfamily. Schematic representation of the likely evolution of the globin superfamily by gene duplications. Myr, million years ago. Redrawn with permission from Strachan and Read (2004).

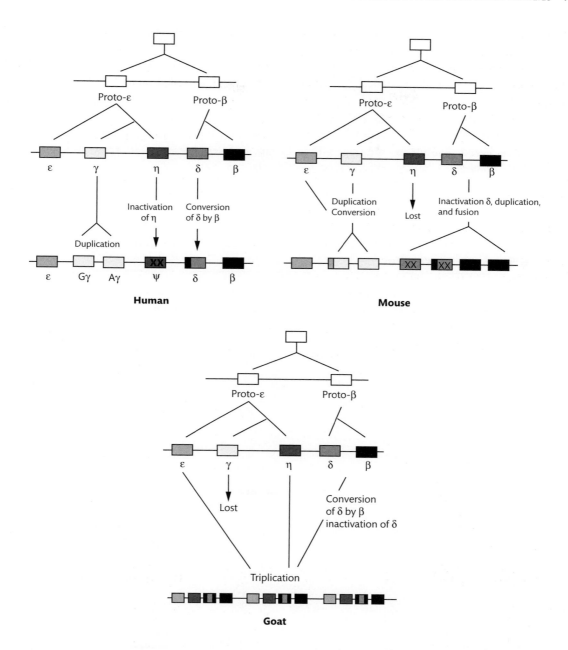

Figure 6.9 Evolution of β globin cluster. Complex gene duplications, conversions, and inactivation events have occurred in different mammalian lineages. Redrawn with permission from Strachan and Read (2004).

subsequent observation that for tandemly repeated ribosomal RNA genes (Box 6.4) the sequence between genes was more similar within a species than between two related species, led to a concerted model of evolution with all members of a multigene family assumed to evolve in a concerted manner rather than independently. Homogeneity between member genes is promoted by repeated unequal crossover events and gene conversion (Box 6.5) allowing a mutation to spread across the gene members of the family. The availability of sequence data

Box 6.4 Ribosomal RNA

In humans, ribosomal RNA (rRNA) is a tandemly arranged multigene family comprising about 500 repeats across the genome, found clustered on the short arms of the five acrocentric chromosomes. **Concerted evolution** is proposed to explain the remarkable homogeneity among the different copies: the rRNA coding gene regions 18S and 28S are virtually identical within and between hominid species, while greater diversity is seen in the intergenic spacer regions distal to the telomere within each array (Gonzalez and Sylvester 2001).

Box 6.5 Gene conversion

Gene conversion involves non-reciprocal transfer of sequence information between an acceptor and a donor sequence. The donor sequence remains unchanged but a copy of it replaces the sequence in the acceptor; this process involves heteroduplex formation and mismatch repair. Interallelic gene conversion is seen between pairs of allelic sequences, while non-allelic gene conversion between loci is promoted by high sequence homology (Fig 6.10).

Interallelic gene conversion

Interlocus gene conversion

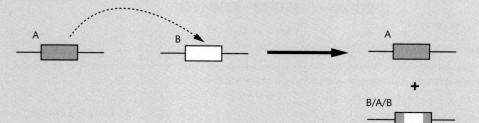

Figure 6.10 Gene conversion. Reproduced with permission from Strachan and Read (2004)

Box 6.6   *CASP12* gene duplication and selective advantage of an inactive pseudogene

A single nucleotide polymorphism (SNP) in exon 4 of the duplicated copy of the *CASP12* gene creates a pseudogene by converting an arginine codon (CGA) to a stop codon (TGA), truncating the protein to half its normal length and rendering it inactive. The ancestral state appears to be the active gene encoding the full length protein and is found in a range of primate and rodent species. Among present-day human populations, the active ancestral copy is found almost exclusively on chromosomes of African ancestry, its presence on chromosomes of non-African descent being less than 1%. Highest frequencies of the ancestral version are seen among present-day San (75%) and Mbuti (60%) pygmies; the average amongst present-day sub-Saharan

African populations is 28% (Xue *et al.* 2006). There is evidence that those individuals with the ancestral active form of *CASP12* produce lower amounts of cytokines on stimulation with lipopolysaccharide and are at greater risk of severe sepsis (Saleh *et al.* 2004). It is postulated that there was selection against the active form of the *CASP12* gene, seen at the level of an individual as resistance to sepsis in those carrying the protein-truncating mutation. It is thought that the mutation arose in Africa some 100 000–500 000 years ago, and that it was initially neutral or nearly neutral, and then subject to strong positive selection over the last 60 000–100 000 years which has driven the polymorphism to near fixation in non-African populations (Xue *et al.* 2006).

---

has more recently led to a model of 'birth and death' evolution by which duplicated genes may be maintained in the genome for a long time, or be deleted, or lose function through deleterious mutations (Nei and Hughes 1992).

The 'birth and death' evolution model appeared particularly applicable to immune genes such as immunoglobulin gene families (Section 6.4.3) and the MHC on chromosome 6 (Chapter 12). Mutations occurring in the duplicated copy may lead to deleterious consequences for the encoded gene but an overall selective advantage, as illustrated by a recent mutation affecting the *CASP12* gene, encoding caspase-12 (Box 6.6) (Saleh *et al.* 2004; Xue *et al.* 2006).

6.4.3 Immunoglobulin gene families

Gene families for the constant and variable regions of immunoglobulin light (λ and κ) and heavy chains are found clustered in the genome, with 50–100 genes identified in each of the variable region gene families. Immunoglobulin molecules comprise heavy and light chains joined by disulphide bonds leading to two variable region antigen binding sites and a constant region (C) (Fig. 6.11) (Janeway *et al.* 2005). Immunoglobulins on B cells are a critical component in our defence against infection, with diversity in

the variable regions essential to being able to respond to a great diversity of possible foreign antigens. Diversity in the variable region is achieved by a number of mechanisms, key to which is the fact that it is encoded by a number of different gene segments that are subject to somatic recombination in developing lymphocytes to randomly bring together a complete variable region sequence (Weigert *et al.* 1978; Tonegawa 1983; Janeway *et al.* 2005).

For light chains, there are V gene segments (comprising the bulk of the variable region domain) and J gene segments (joining segment); in heavy chains there are in addition diversity gene segments (D) (Fig. 6.11) (Early *et al.* 1980). A very high degree of potential combinatorial diversity can be achieved by such gene rearrangements: for the λ light chain locus at chromosome 22q11.2, 30 functional V_λ gene segments and four J_λ gene segments could generate 120 possible λ variable regions; for the κ light chain locus at chromosome 2p11.2, 200 variable regions are possible (40 V_κ × 5 J_κ). There is also evidence of a large inverted duplication spanning the V_κ cluster which is restricted to humans, thought to have occurred about 5 million years ago (Kawasaki *et al.* 2001). At the heavy chain locus on chromosome 14q32.3, 40 V_H segments combined with 25 D_H segments and six J_H segments provide in principle 6000 possible different variable heavy chain regions. Combinatorial diversity based on the 320 light chain peptides (120+200) with each of the ~6000 heavy chains gives a theoretical 1.9×10^6 different antibody specificities (Janeway *et al.* 2005).

Further somatic diversity is achieved through the joining process of somatic recombination: so-called junctional diversity involves both shortening of the gene segments (as joining ends is imprecise) and the insertion of one or several nucleotides between the segments during joining. Finally, once B cells have been activated by encountering antigen, somatic mutation of the rearranged variable region light and heavy chain genes occurs at a very high rate throughout the coding sequences (Weigert *et al.* 1970).

The remarkable diversity of immunoglobulin proteins is thus achieved by a series of different somatic events from

a starting point of distinct multigene families encoding segments of the variable region of the protein, which have arisen through gene duplication and been subject to 'birth and death' evolution (Ota and Nei 1994; Sitnikova and Nei 1998).

6.5 Segmental duplication, deletion, and gene conversion

6.5.1 Lessons from the study of the genetics of colour vision

Genes encoding the photoreceptor pigments found in photoreceptor cells are thought to have arisen by gene duplication events, with analysis of gene sequences

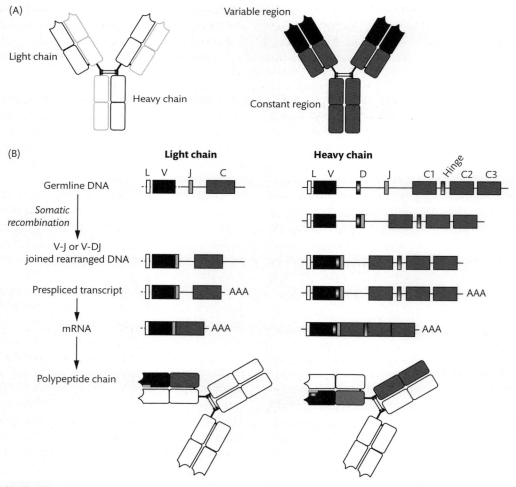

Figure 6.11 *Continued*

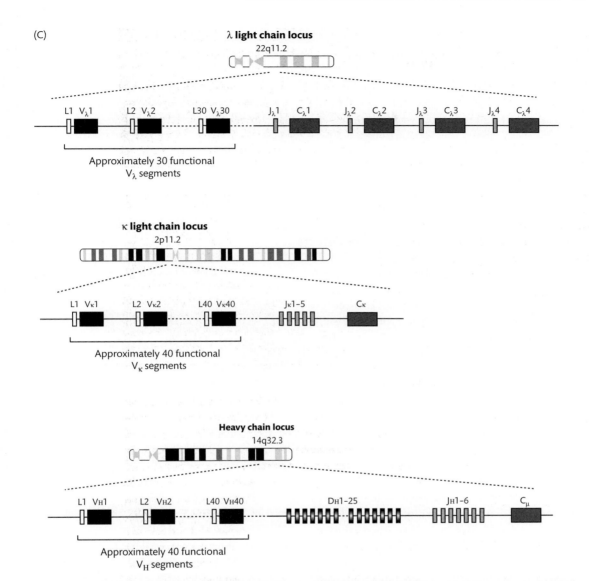

Figure 6.11 Immunoglobulin molecules and genes. (A) Schematic representation of an immunoglobulin molecule showing two heavy and two light chains joined by disulphide bonds, the heavy chains are in turn joined to a further pair of heavy chains to form the stalk of a Y configuration. The variable region containing antigen binding sites are shown in black with the constant region in dark grey. (B) Overview of how immunoglobulin gene segments come together to form a complete immunoglobulin gene in B lymphocytes. The variable region is assembled by gene rearrangement from the V and J gene segments (light chains) or V, D, and J gene segments (heavy chains). A leader sequence (L) directs the immunoglobulin polypeptide chains to the secretory pathways and is then cleaved (immunoglobulins being extracellular proteins). (C) Schematic of immunoglobulin gene clusters. Note this is a simplified representation and not to scale. Redrawn with permission from Janeway *et al.* (2005).

among fish, birds, and mammals suggesting an ancient initial event before the vertebrate radiation. *OPN1LW* (encoding the red photopigment 'opsin 1 long-wave-sensitive' found in cone photoreceptors) and *OPN1MW* (encoding the green photopigment, 'opsin 1 medium-wave-sensitive') are highly homologous and share 96% identity at the amino acid level. They are found in a head-to-tail tandem array on the X chromosome at Xq28 with

Box 6.7 Trichromatic colour vision

Normal human colour vision involves three classes of retinal cone with maximal light sensitivity at approximately 420 nm (short wave, blue), 530 nm (medium wave, green), and 560 nm (long wave, red).

A given photoreceptor cell will contain a single type of photopigment comprising opsin protein bound to chromatophore.

Box 6.8 Red–green colour vision defects

Among north Europeans approximately 8% of males and 0.5% of females have defective colour vision ranging from mild to very severe. Dichromatic vision is a severe defect in colour vision in which individuals have no functional red cones (protanopes, approximately 1% of males), green cones (deuteranopes, 1% of males), or blue cones (tritanopes, affecting less than one in 10 000 males due to a mutation in the *OPN1SW* gene on chromosome 7). Less severe defects are also common, involving red–green chimera genes; these conditions are described as anomalous trichromatic disorders, either protanomalous or deuteranomalous, affecting approximately 1% and 5% of north European males, respectively. Among females, homozygotes for protan or deutan gene arrays will have defective colour vision (0.5%); 16% are expected to be heterozygous carriers, and the majority have normal vision although extreme skewing of X inactivation can lead to defective colour vision. Enhanced colour vision has been found in some females who are carriers of anomalous trichromacy due to possession of four classes of cone photoreceptor (red, green, green-like, and blue) resulting from **X chromosome inactivation** (Jordan and Mollon 1993).

The true absence of colour discrimination, achromatopsia, is thought to have been first described in 1777 in a report describing a subject that

...could never do more than guess the name of any color; yet he could distinguish white from black, or black from any light or bright color ... He had 2 brothers in the same circumstances as to sight; and 2 brothers and sisters who, as well as his parents, had nothing of this defect' (Huddart 1777).

Congenital causes of achromatopsia include the absence of all sensitivity mediated by cone pigments (rod monochromacy) and individuals with rods and blue cones only (blue cone monochromacy). The study of the molecular genetics of blue cone monochromacy proved highly informative in understanding regulation of red/green pigment genes. The condition is very rare, found in only one in 100 000 people, but Nathan and colleagues were able to determine that among 12 affected families either a single visual pigment gene was present but inactivated by a point mutation, or in about half of the cases a deletion of variable size was present 5′ to the red/green pigment genes (Nathans *et al.* 1989). Remarkably, the common region present in all the deleted regions was found to span a highly conserved stretch of DNA which in transgenic mice revealed evidence of a locus control region present 3.1–3.7 kb 5′ to the red pigment gene, essential for expression of red and green pigment genes (Wang *et al.* 1992).

one copy of the gene encoding red pigment followed by one or more copies of the gene encoding green pigment (Fig. 6.12) (Nathans *et al.* 1986; Vollrath *et al.* 1988).

High sequence identity has predisposed to non-allelic homologous recombination: among male Caucasians, half carry two green pigment genes, one-quarter have a single copy, and one-quarter have three or more copies. However, only the *OPN1LW* (red) gene and the proximal *OPN1MW* (green) gene are expressed in the retina, and having additional copies of the *OPN1MW* pigment gene does not affect colour vision (Yamaguchi *et al.*

1997). The other photopigment-encoding genes found in humans are *RHO*, encoding rhodopsin found in rod photoreceptor cells, and *OPN1SW*, encoding 'opsin 1 short-wave-sensitive', the photopigment found in blue cones. These genes are found at 3q21-q24 and 7q31.3-q32, respectively, and share only 40–44% identity with the red and green pigment genes.

The long-wavelength-sensitive (red) and medium-wavelength-sensitive (green) cone pigment genes are believed to have arisen relatively recently by a duplication event within the Old World monkey lineage,

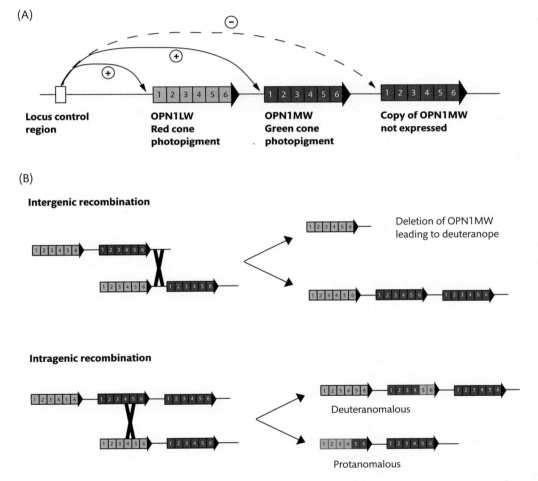

Figure 6.12 Molecular genetics of colour vision. (A) Tandem array of red and green cone photopigment genes on chromosome Xq28. Expression of a single photopigment gene in a given photoreceptor cell is controlled by a proximal locus control region. (B) Copy number variation through non-allelic homologous recombination in intergenic region can lead to deletion of green pigment encoding gene and dichromatic vision (deuteranope). Intragenic recombination can result in green–red hybrid genes (deuteranomalous colour vision) and red–green hybrid (typically protanomalous colour vision). Modified with permission from Deeb (2005).

Box 6.9   Rhesus blood group and disease

The Rh blood group system was discovered in the late 1930s (Levine and Stetson 1939; Landsteiner and Wiener 1940). The name relates to the experiments of Landsteiner and Weiner involving the rhesus macaque: when rabbits were immunized with rhesus monkey cells an antibody was produced that agglutinated human red blood cells (Landsteiner and Wiener 1940). Rh blood group was found to be of considerable clinical importance, in particular to be responsible for haemolytic disease of the newborn through fetomaternal incompatibilities (for review see Van Kim *et al*. 2006). In 1982, the proteins associated with Rh antigens were identified (Moore *et al*. 1982) and the *RHD*, *RHCE*, and *RHAG* genes encoding the Rh proteins and associated glycoprotein were subsequently cloned. Individuals who are RhD-negative do not express the RhD protein on their red blood cell membranes and produce allo anti-D antibodies on exposure to RhD-positive red blood cells. If an RhD-negative mother is exposed to red blood cells from an RhD-positive fetus (for example

due to transplacental haemorrhage, trauma, abortion, child birth, or medical procedures) then, in subsequent pregnancies, the resulting maternal anti-D antibodies may cross the placenta and destroy fetal RhD-positive red blood cells. The result is haemolytic disease of the newborn and potentially fetal death. Fortunately this can be prevented with anti-D immunoglobulin given just after delivery to prevent maternal development of anti-D antibodies, and can be treated with phototherapy and intravascular fetal transfusions. Fetal *RHD* genotyping from maternal plasma DNA shows excellent concordance with that gained from amniotic cells or neonatal blood, allowing antenatal anti-D immunoprophylaxis to be targeted to RhD-positive fetuses (Lo *et al*. 1998). Other examples of severe haemolytic reactions relating to Rh blood group include alloimmune transfusion reactions arising between donor and recipient following an Rh incompatible blood transfusion. Occasionally the RhCcEe antigen may underlie haemolytic reactions.

---

possibly 30 million years ago. Old World monkeys, great apes, and humans have trichromatic colour vision (Box 6.7) and possess two spectrally distinct cone pigment genes for mid and long wavelengths (in humans these are green and red pigments). This contrasts with most New World monkeys who have one gene for these wavelengths but can still possess trichromatic vision depending on allelic variation at that locus (Jacobs 1996). Convergent evolution appears to have led to the evolution of trichromatic vision independently among primates through powerful selective pressures including the need to detect coloured fruits (Shyue *et al*. 1995).

The high sequence homology and proximity of the red and green cone photopigment genes has led to frequent non-allelic homologous recombination and gene conversion events between X chromosomes during gamete formation in females leading to variation in gene number and hybrid genes, together responsible for most red–green colour vision defects (Box 6.8) (reviewed in

Deeb 2005). Specifically, intergenic unequal crossover events can result in variation in the number of green cone pigment genes, while intragenic events can lead to hybrid chimeras of red and green cone pigment genes in both orientations (Fig. 6.12). The red and green cone pigment genes each comprise six exons. Exon 5 accounts for most of the spectral difference between the pigments such that exchange of exon 5 will convert a red to a green pigment. Analysis of nucleotide diversity across diverse human populations has shown unusually high levels of recombination, notably within 169 bp of exon 3 of *OPN1LW*, consistent with gene conversion (Winderickx *et al*. 1993; Verrelli and Tishkoff 2004).

Among those with apparently normal colour vision, variation is seen in colour matching tests indicating more subtle differences in perception. A common SNP within the red pigment gene (resulting in a substitution of serine for alanine at amino acid residue 180 in the encoded protein) was found to correlate with higher sensitivity to red light (Winderickx *et al.* 1992). Among male Caucasians, 62% were found to have serine, and 38% alanine. *In vitro* experiments showed this related to a difference in wavelength of maximal absorption of 4–7 nm (Merbs and Nathans 1992). Possession of the polymorphism was also found to significantly modulate the severity of red–green colour vision defects. X chromosome inactivation in females means that for heterozygous carriers, half of red cones carry alanine and half serine. There is some evidence that this is associated with enhanced colour discrimination (Jameson *et al.* 2001).

6.5.2 Rhesus blood groups: genetic diversity involving duplication and deletion

Molecular studies of genetic variation determining Rh (rhesus) blood groups have shown how for a substantial proportion of Caucasian individuals, a gene arising from duplication during primate evolution has been since been deleted, leading to risk of severe haemolytic disease (Box 6.9) (for reviews see Wagner and Flegel 2004; Avent *et al.* 2006). A duplication event involving the ancestral *RHCE* gene is thought to have occurred in a common ancestor of humans, chimpanzees, and gorillas some 8.5 million years ago, leading to two homologous genes *RHCE* and *RHD* (Matassi *et al.* 1999; Wagner and Flegel 2002). These genes encode Rh proteins CcEe and D, respectively, which are found complexed with an Rh-associated glycoprotein in the red blood cell membrane. At least 49 different serologically defined Rh antigens have been described involving the RhD and RhCcEe proteins (www.uni-ulm.de/~fwagner/RH/RB/) and they determine an individual's Rh blood group. Clinically, the most important phenotype is that where individuals lack the D antigen (protein) on the surface of their red blood cells, denoted as D-negative. This phenotype is common among Caucasians, with 15–18% of the population RhD-negative, but rarer in those of African ancestry (3–7%) and very rare among individuals originating in the Far East (<1%) (Daniels 1995).

RHD and *RHCE* are found tandemly arranged in reverse orientations at chromosome 1p34.1–1p36 with the *SMP1* gene lying between the two Rh genes (Fig. 6.13) (Cherif-Zahar *et al.* 1991; Colin *et al.* 1991). Among Caucasians, the most common cause of having

the RhD-negative phenotype is deletion of the *RHD* gene. This deletion is thought to have arisen due to unequal crossing over involving two highly homologous regions flanking the *RHD* gene known as rhesus boxes (Fig. 6.13) (Wagner and Flegel 2000). Rhesus boxes are present in the same orientation and have 98.6% homology. The deletion haplotype is present in approximately 40% of Caucasians; those homozygous for the haplotype are RhD-negative, heterozygotes are RhD-positive but have only one functional *RHD* gene. In Caucasians, the D-negative phenotype can rarely arise due to gene rearrangements or point mutations (Wagner *et al.* 2001). In other populations, such as those of African ancestry, the D-negative phenotype is both rarer and arises from a more diverse set of causes. In a study of black Africans who were D-negative, only 18% had a deletion of *RHD*; 66% had an inactivated *RHD* gene, denoted *RHDψ*, which had a 37 bp insertion in intron 3 and exon 4 disrupting the open reading frame (ORF) (Fig. 6.13); and 15% had a *RHD-CE-D* hybrid gene (Singleton *et al.* 2000).

Hybrid genes are thought to arise due to gene conversion, made possible by the high homology between *RHD* and *RHCE*. A number of hybrid alleles are reported with evidence of selection acting to maintain antigen variation (Innan 2003). Other rare causes of antigen diversity include single base changes such as missense mutations changing the encoded protein, nonsense mutations resulting in premature stop codons and truncation of the proteins, splice site mutations, and frameshift mutations (reviewed in Wagner and Flegel 2004).

6.6 Insertion/deletion polymorphisms: 'indels'

The final section of this chapter describes small scale genetic variation involving insertions and deletions, collectively described as 'indels' (Box 6.10). The distinction in terms of size is, however, arbitrary in this context; such variants are typically considered in terms of being less than 1 kb in size but Scherer and colleagues would advocate no size restriction in use of the term indel (Scherer *et al.* 2007). When considering sequence level diversity, indels have until recently been somewhat overlooked in terms of their frequency and

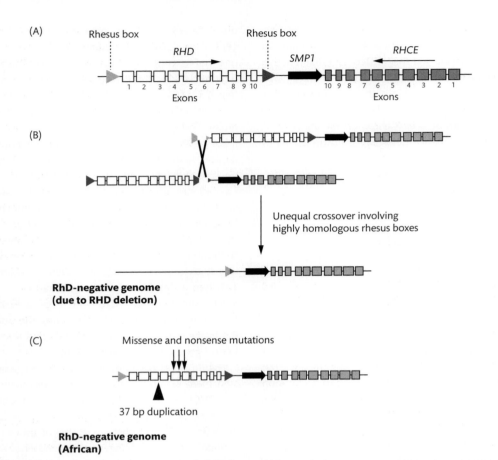

Figure 6.13 Gene arrangement of the human Rh locus and its diversity. (A) The tandem arrangement of the *RHD* and *RHCE* genes is shown in opposite orientations with *RHD* flanked by rhesus boxes. The RhD-negative phenotype may result from *RHD* deletion after unequal crossover involving highly homologous rhesus boxes (B) or by *RHD* gene inactivation associated with a 37 bp duplication and point mutations (C). Redrawn from Avent *et al.* (2006), with permission from Cambridge University Press; and with permission from Wagner and Flegel (2000).

Box 6.10   Indels

The term 'indel' encompasses genetic variation involving gain or loss of one or more contiguous nucleotides in a genomic sequence. This allows for ambiguous situations comparing genomes when it is not possible to determine whether there has been a gain or loss of nucleotides. For example, if sequence 1 has a run of three nucleotides (AAA), sequence 2 has a run of four nucleotides (AAAA), and a reference or ancestral sequence is unavailable, it will not be possible to determine if there has been a deletion event in sequence 1 or an insertion event in sequence 2. Typically, the term is used to describe small scale sequence variation and the most common indels involve a single base; they contrast with single nucleotide substitutions in which one nucleotide is replaced with another.

significance in comparison with single nucleotide substitutions. In fact, it is now apparent that indels are the second most frequently occurring type of genetic variation after single nucleotide substitutions and can have significant phenotypic consequences. Small (less than 20 bp) indels are estimated to account for 24% of disease-causing mutations in the Human Gene Mutation Database (www.hgmd. cf.ac.uk; accessed December 2008). Among apparently healthy subjects, 11336 indels in the size range 100 bp to 1 kb were reported in the Database of Genomic Variants (http://projects.tcag.ca/variation/; accessed December 2008) (Zhang *et al.* 2006).

6.6.1 Human-specific indels and selection

Completion of sequencing the common chimpanzee chromosome 22 allowed direct comparison with human chromosome 21 and the finding of nearly 68000 indels over the 33.3 Mb of sequence (Watanabe *et al.* 2004). Most indels were small, but a number of large human insertions/chimpanzee deletions were noted, up to 54000 bp in length. An increase in the number of indels around 300 bp in size was noted involving transposable elements.

In 2005, the draft chimpanzee whole genome sequence was published leading to evidence for 5 million indels on comparison with the human genome (Box 10.1) (CSAC 2005). Approximately one-third of these indels were noted to be repeats and one-quarter transposable elements. An excess of short transposable Alu sequence elements around 300 bases was found in the human compared to chimpanzee sequence (7000 versus 2300, respectively), while the number of long transposable elements (L1) was approximately the same.

Comparison of whole genome multiple alignments of human and chimpanzee DNA sequences, with the rhesus macaque genome sequence used as an outgroup species, has allowed investigators to define whether an indel is an insertion or deletion, and to resolve indels specific to the human branch since divergence from a common ancestor with the chimpanzee approximately 6 million years ago (Chen *et al.* 2007; de la Chaux *et al.* 2007; Messer and Arndt 2007). For example, in their analysis of short indels (less than 100 bp), Messer and colleagues identified 225744 insertions and 428048 deletions in the human branch of: their analysis suggested that tandem duplication was the major mechanism leading to generation of short insertions in the human genome over our recent evolutionary past (Messer and Arndt 2007).

Analysis of coding indels using a similar approach comparing human, chimpanzee, and rhesus whole genome multiple alignments showed coding indels to be significantly underrepresented (0.14% of all indels occurred in coding DNA, while 1.2% of the genome is coding) (de la Chaux *et al.* 2007). Insertions were found to involve small amino acids such as glycine and alanine more commonly than expected, while glutamic acid was overrepresented in deletions; hydrophobic, aliphatic, and aromatic amino acids were underrepresented in indels. Coding indels were noted by the authors to occur more often in genes where selective pressures were lower (as judged by the ratio of nonsynonymous to synonymous nucleotide substitution rates), and in regions of proteins not involving important structural domains (de la Chaux *et al.* 2007).

Chen and colleagues identified human-specific indels from a human/chimpanzee/mouse/rat/dog genomic sequence analysis and noted that human-specific traits may relate to changes in transcription and translation as human-specific coding indels were relatively enriched in such genes (Chen *et al.* 2007).

6.6.2 Mapping the extent of indel polymorphism

Indels appear common across a range of organisms: in the fly, indels comprised 16.2% of sequence polymorphisms when two common laboratory strains were analysed (Berger *et al.* 2001); in the worm, they comprised 25% of polymorphisms with most involving a single base pair (Wicks *et al.* 2001). Among humans, the completion of sequencing for chromosome 22 (Dunham *et al.* 1999) provided an early opportunity to assess on a large scale the occurrence of indels as nine libraries of different DNA were used, such that the sequence overlaps of the clones allowed potential variation to be resolved (Dawson *et al.* 2001). This analysis showed 18% of the variation seen was indels, of which just under two-thirds involved a single base change; over 90% of variants were subsequently confirmed in a panel of 92 individuals. A major limitation in the assessment of indels related to identification of the variants in the heterozygous state unless both homozygous individuals for the 'long' and 'short' forms were present in the sample.

Bhangale and colleagues (2005) were able to show that accurate identification could be achieved using a fluorescence-based sequencing approach. Resequencing of

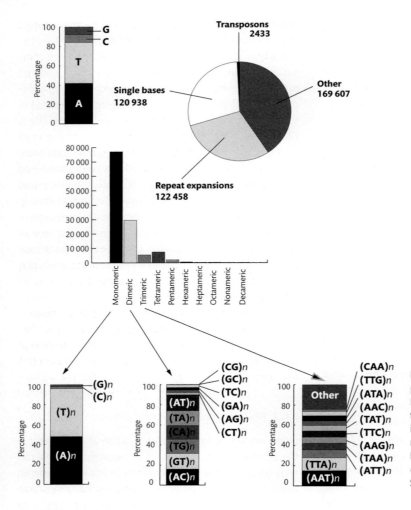

Figure 6.14 Insertion/deletion (indel) polymorphisms in the human genome. Data mining of resequencing traces from three diverse human populations involving 36 subjects revealed 415 436 non-redundant indels (Mills *et al.* 2006). Figure prepared using data from Mills *et al.* (2006), with permission from Cold Spring Harbor Laboratory Press.

330 candidate genes in 24 individuals of African descent and 23 individuals of European descent revealed 33 829 nucleotide substitutions and 2393 indels. On average, seven indels were found per gene, occurring once every 2714 bases and ranging in size from 1 to 543 bp in length; 46% of indels involved a single base pair and 84% were less than 5 bp. The allele frequencies and patterns of linkage disequilibrium for indels were noted by the authors to be similar to those of the nucleotide substitutions. In a later study, Bhangale and coworkers were able to apply their algorithm for automated detection and genotyping of indels from sequencing traces to regions of the genome functionally characterized in the **ENCODE** (ENCyclopedia Of DNA Elements) Project (Section 9.2.4) (Bhangale *et al.* 2006).

Mills and colleagues have recently produced a first map of indel variation across the human genome by using a computational strategy to mine DNA resequencing traces (Mills *et al.* 2006). These traces were generated by SNP discovery projects involving 36 diverse individuals across three diverse human population groups. Their map contains 415 436 unique indels (Fig. 6.14). About half could be mapped onto the chimp genome sequence to provide a reference ancestral sequence: based on this, 47% were insertions and 53% deletions. The indels ranged in size from 1 to 9989 bp in length. Approximately one-third of the indels involved a single base pair deletion or insertion, of these 84% were A:T or T:A. A further third of the indels were either monomeric base pair expansions or multi-base repeat expansions involving 2–15 bp repeat units.

Such repeats have been of great utility as genetic markers, for example (CA)*n* repeat expansions, as reviewed in Chapter 7; they may also be associated with significant phenotypes as seen with the trinucleotide (CGG)*n* repeat expansion of more than 200 repeats at the *FMR1* gene associated with fragile X mental retardation syndrome (Box 7.8) (Penagarikano *et al.* 2007). The remaining indels contained either random DNA sequence or, in a very small minority, transposon insertions.

The dataset from Mills and colleagues indicated an indel occurring on average once every 7.2 kb. Some genomic regions appeared to be 'hotspots' of diversity with a much higher frequency of indels (up to 24-fold); increased SNP diversity was found at the same sites. Overall, it was estimated that indels accounted for 15.6% of discovered polymorphisms and that human populations were likely to have approximately 1.5 million indels.

6.6.3 Functional consequences of indels

When indels occur in coding DNA, a change in the amino acid sequence will always occur, in contrast to single nucleotide substitutions which can be synonymous. The change may be to insert or delete an amino acid, or cause a frameshift and loss of protein function. Over the course of this book many examples of such events are described, including in relation to the globin genes (Section 1.3.5) and the delta-F508 mutation (p.F508del) in cystic fibrosis (Section 2.3.1). Other examples include an ACCC heterozygous deletion (c.989_992delACCC) in the *PAX8* gene on chromosome 2q12-q14, which results in a frameshift and premature stop codon, truncating the encoded protein and rendering it transcriptionally inactive. *PAX8* is an important transcription factor involved in thyroid cell proliferation and development and this deletion mutation is associated with thyroid dysfunction (de Sanctis *et al.* 2004). A 27 bp deletion in the *GPIBA* gene (encoding the platelet glycoprotein 1b receptor for von Willebrand factor) on chromosome 17 has been associated with a severe bleeding disorder, platelet-type von Willebrand's disease (Othman *et al.* 2005). As noted previously, there is evidence of strong selective pressures acting on indels occurring in coding DNA over time: since the divergence of humans and chimpanzees, occurrence of indels in coding compared to intergenic or intronic DNA has been found to be highly suppressed (Chen *et al.* 2007; de la Chaux *et al.* 2007; Messer and Arndt 2007).

When indels occur in the promoter and other regions of DNA important to regulation of gene expression, the consequences can also be marked. For example, a dramatic effect of a single base pair indel is seen at the *MMP1* (encoding matrix metalloproteinase 1) gene promoter: the presence of a G insertion (AAGAT to AAGGAT) 1607 nt upstream of the transcriptional start site created an erythroblast transformation-specific (Ets) transcription factor binding site (GGA) and was associated with increased transcriptional activity (Rutter *et al.* 1998). The '2G' indel was present at a high frequency in Caucasian individuals (allele frequency 0.5) and even higher among cancer cell lines where a copy of the 2G allele was present in seven out of eight lines tested. A number of studies have since demonstrated highly significant associations with cancer, notably ovarian (Kanamori *et al.* 1999), lung (Zhu *et al.* 2001), and colorectal cancer (Ghilardi *et al.* 2001; Hinoda *et al.* 2002).

In the promoter region of the *NFKB1* gene on chromosome 4q24, which encodes the **NFkB** transcription factor protein p105/p50 isoforms, an ATGG indel was found 94 nt from the transcriptional start site (Karban *et al.* 2004). This polymorphism modulated protein–DNA interactions and transcriptional activity in a **reporter gene** system and showed association with risk of the inflammatory bowel disease ulcerative colitis (Karban *et al.* 2004) and sporadic cancers (Lewander *et al.* 2007); other investigators showed no association with susceptibility to ulcerative colitis (Mirza *et al.* 2005; Oliver *et al.* 2005). Another example of indels modulating gene expression includes a 6 bp deletion in *CASP8*, which has been associated with loss of a binding site for the transcription factor stimulatory protein 1 (Sp1) and cancer risk (Sun *et al.* 2007), although this remains controversial (Frank *et al.* 2007; Haiman *et al.* 2008).

6.7 Summary

Deletion and duplication events have led to much of the diversity we see today in our genomic landscape and are thought to have been a major force in enabling evolutionary change. Segmental duplications are common, comprising an estimated 5% of the human genome, with an excess noted on particular chromosomes such as chromosomes 15 and 22, and within specific chromosomal regions, notably subtelomeric and pericentromeric regions (Bailey *et al.* 2001, 2002). Subtelomeric regions include tandemly arrayed members of the olfactory receptor gene superfamily, a remarkable set of 800 genes constituting 1% of the genome that are thought to have arisen by tandem duplication and chromosomal rearrangements (Glusman *et al.* 2001).

Comparative sequence analysis across species, notably with the availability of the chimpanzee genome sequence, has highlighted the recent nature of the majority of duplication events (Bailey and Eichler 2006). It is estimated that a genome comprising 15 000 genes will acquire 60–600 duplicated genes over a million years (Lynch and Conery 2000). The majority of duplicated genes lose function and are seen as pseudogenes, eventually disappearing from the genome, while a few may gain selectively advantageous mutations resulting in neo- or sub-functionalization or provide robustness to the organism in the event of deleterious mutations in the ancestral gene.

Segmental duplications and deletions may result in a change in gene dosage and a specific phenotype associated with altered gene expression. Highly homologous duplicated regions, notably LCRs, predispose to non-allelic homologous recombination and in turn chromosomal rearrangements and genomic disorders through generation of deletions, duplications, inversions, and translocations. The impact of duplications and deletions is, however, much broader, underlying a range of common traits. Work to define the genetic basis of colour blindness and the rhesus blood group system has illustrated how such events have occurred over evolutionary time and the consequences of later recombination events. For the photoreceptor pigment genes a series of duplication events has led to highly homologous genes with frequent unequal crossover and gene conversion leading to red–green colour vision defects of variable severity affecting up to 8% of male north Europeans (Deeb 2005). Specific rhesus blood groups, such as absence of the D antigen, carry significant risk including haemolytic disease of the newborn: 15% of Caucasians have a deletion of the *RHD* gene responsible for failure to express D antigen; this represents the loss of one of a pair of genes originally duplicated an estimated 8.5 million years ago (Wagner and Flegel 2000).

More generally the extent of insertion and deletion polymorphism is becoming apparent, with an estimated 1.5 million indels in human populations (Mills *et al.* 2006). Comparisons with primate species have allowed the ancestral states to be defined such that in the large scale survey by Mills and colleagues they found 47% of indels were insertions and 53% deletions; a third involved single base insertions or deletions (most commonly A:T or T:A) while a further third involved monomeric base pair expansions or multibase repeats. Small scale indel events may have dramatic consequences for the encoded protein or gene expression, as seen in the globin locus and elsewhere, with very important consequences for human health.

In Chapter 7 our focus switches to tandem repeats and the remarkable insights that study of this class of often highly variable repeats have provided in both health and disease.

6.8   Reviews

Reviews of subjects in this chapter can be found in the following publications:

Topic	References
Gene duplication, phenotypic diversity, and human disease	Conrad and Antonarakis 2007
Primate segmental duplications	Bailey and Eichler 2006
Introduction to evolution	Wood 2006
Evolution of multigene families	Ohta 2000; Nei and Rooney 2005
Evolution, recombination, and duplication in subtelomeric regions	Mefford and Trask 2002; Linardopoulou *et al.* 2005
Olfactory receptor gene superfamily	Glusman *et al.* 2001
Genetics of colour vision	Deeb 2005
Molecular biology and genetics of rhesus blood group	Wagner and Flegel 2004; Avent *et al.* 2006
Insertion/deletion (indel) polymorphisms	Mills *et al.* 2006

CHAPTER 7

Tandem repeats

7.1 Introduction

Repetitive DNA sequences, found in tandem arrays in a head to tail arrangement without sequences between repeating units, are a common feature of the human genome and have been classified on the basis of the size of the repeat array into satellites, minisatellites, and microsatellites (Fig. 7.1; Box 7.1). An important feature of tandemly repeated DNA is the capacity to vary in terms of length, based on the number of repeat units, which is a characteristic feature of this class of genetic diversity. Such polymorphism is notable among hypervariable minisatellites and also microsatellite DNA where repeat arrays are seen as the most unstable (variable) of any of the classes of satellite DNA.

7.2 Satellite DNA

Satellite DNA comprises very large arrays of tandemly repeated noncoding DNA, usually megabases in size, which form the major structural constituent of heterochromatin, a tightly packaged condensed state of transcriptionally suppressed DNA. Satellite DNA is the main component of functional centromeres, as well as heterochromatin in pericentromeric and telomeric regions, and in the short arms of the acrocentric chromosomes (chromosomes 13, 14, 15, 21, and 22) (Charlesworth et al. 1994; Schueler et al. 2001). Satellite DNA is recognized across eukaryotic genomes, in some species such as the kangaroo rat Dipodomys ordii comprising half of the total genomic DNA content (Hatch and Mazrimas 1974). The technical challenges of sequencing and analysing

such large repetitive DNA sequences has meant only limited coverage in genome assemblies, such that our understanding of the precise structure, organization, and functional importance of satellite DNA remains relatively incomplete. The unusual structure of satellite DNA may be the preferred state at centromeres and flanking regions. The specific composition of satellite DNA is variable and typically species-specific. Concerted evolution is thought to maintain sequence homogeneity of species-specific satellite arrays but allow for the rapid change in sequence or composition of repeats required during evolution (Plohl et al. 2008).

Early work using density gradient centrifugation resolved that specific 'satellite' bands could be separated from the bulk of genomic DNA due to differences in nucleotide content: three major classes were noted with different satellite subfamilies (Corneo *et al.* 1968; Prosser *et al.* 1981). Other types of satellite DNA cannot be easily separated by centrifugation and were initially identified by particular restriction enzymes recognizing motifs in the basic repeat unit. Satellite DNA varies in the nucleotide sequences involved, the complexity of the repeat structure, and genomic abundance. Many satellite monomers are in the size range 150–180 bp, the DNA length required to wrap around a **nucleosome**, or 300–360 bp (sufficient for two nucleosomes). It is now known that satellite DNA repeats include alpha satellite DNA (tandem repeats of a 171 bp monomer), beta satellite DNA (a 68 bp repeat), satellite family 1 DNA (AT rich with repeat units from 25 to 48 bp in length), and satellites families 2 and 3 (with much simpler 5 bp repeats). When it was first identified, satellite DNA was recognized to not encode specific proteins and was initially regarded

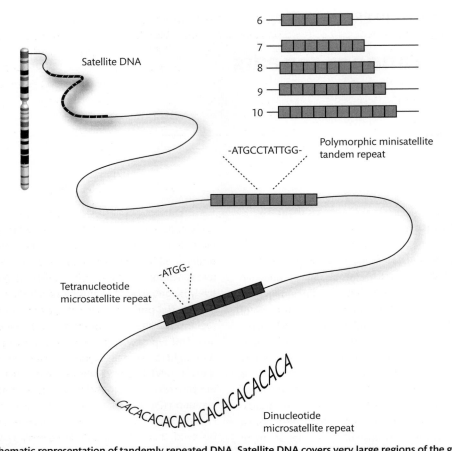

Figure 7.1 Schematic representation of tandemly repeated DNA. Satellite DNA covers very large regions of the genome, for example 171 bp monomers of alpha satellite DNA organized into units several hundred base pairs in length that may be repeated hundreds or thousands of times. A polymorphic minisatellite is shown comprised of between six and ten repeats of an ATGCCTATTGG sequence. A tetranucleotide (ATGG)n and dinucleotide (CA)n microsatellite are also shown.

Box 7.1   Satellites, minisatellites, and microsatellites

- *Satellite DNA* comprises very long arrays of tandem repeats typically 100 kb to several megabases in size, the repeat unit length varying between 5 and 171 bp.
- *Minisatellite DNA* arrays are of intermediate size and typically span between 100 bp and 20 kb, with each repeat unit between 6 and 100 bp in length.

- *Microsatellite DNA* comprises short arrays less than 100 bp in size, made up of simple tandem repeats 1–6 bp in length. There is, however, some blurring in the distinction between minisatellites and microsatellites in the repeat length between 6 and 12 bp in some classifications.

as 'junk' DNA (Ohno 1972) although others postulated a role in chromosome organization and pairing (John and Miklos 1979) which has been supported by more recent studies (Csink and Henikoff 1998).

Alpha satellite DNA is the best studied satellite DNA and constitutes the major structural component of all human centromeres as well as across different primate species. The basic 171 bp motif or monomer is AT rich and arranged in tandem with a head to tail arrangement; these monomers are organized in multimeric repeat units to form higher order repeat arrays spanning typically 3–5 Mb. At chromosome 17, for example, 16 monomers of alpha satellite DNA are organized into 2.7 kb units, tandemly repeated over 1000 times (Waye and Willard 1986). Irregularly interspersed monomeric alpha satellite DNA without a higher order structure is typically found flanking the repeat arrays, often with high levels of sequence divergence. The number of repeats may vary between individuals, as seen at the centromere of the X chromosome where alpha satellite length varied from 1380 to 3730 kb among 29 different X chromosomes; these differences were inherited in a mendelian fashion (Mahtani and Willard 1990). Specific organization and structure of alpha satellite arrays is found at the centromeres of different chromosomes, for example DXZ1 is X chromosome-specific (Willard and Waye 1987).

7.2.1 A functional role for satellite DNA?

The functional roles of satellite DNA remain unclear. In *Drosophila* and other species, transcriptionally active genes are found within alpha satellite DNA; such genes are not found in human centromeres. However, transcripts from satellite DNA in humans can generate small interfering RNA which has a critical role in heterochromatin formation (Grewal and Elgin 2007). Alpha satellite DNA contains specific binding sites for centromere associated proteins, notably a 17 bp sequence motif to which CENP-B binds; this protein plays a critical role in centromere formation in a highly regulated fashion (Okada *et al.* 2007). Satellite DNA is also thought to play a role in chromosomal stability and genome restructuring during development (Plohl *et al.* 2008).

7.2.2 Satellite repeats and disease

Polymorphism in satellite DNA has been associated with disease, a notable example being facioscapulohumeral muscular dystrophy (Box 7.2), which in almost all cases is associated with a reduced number of copies of a satellite repeat, dubbed D4Z4, at 4q35. Linkage studies had implicated the distal end of chromosome 4 in disease susceptibility and this was fine mapped to a subtelomeric region and specifically a region comprised of 3.3 kb repeats (Wijmenga *et al.* 1992). In unaffected individuals, between 11 and 100 copies of the repeat are found, while in patients between one and ten copies are observed on one of the two alleles. The phenotype is highly variable but, in general, having a lower number of repeat copies (between one and three) is associated with more severe disease occurring sporadically, while in familial cases four to ten copies are typically seen (Tawil and Van Der Maarel 2006). The molecular basis of the association remains unclear but is thought to involve epigenetic mechanisms (van Overveld *et al.* 2005) and a positional effect of the deletion modulating transcription of neighbouring genes (van Deutekom *et al.* 1996). Haplotypic analysis also suggests specific sequence differences may be important (Lemmers *et al.* 2007).

Box 7.2   Facioscapulohumeral muscular dystrophy (OMIM 158900)

This is an autosomal dominant inherited muscle disorder with an estimated prevalence of one in 20 000 (Tawil and Van Der Maarel 2006). There is a variable phenotype and age of onset. The disorder is characterized by sequential weakness of facial, shoulder, and upper arm muscles followed by trunk and lower extremity involvement. High frequency hearing loss and retinal telangiectasia are often associated.

7.3 Minisatellite DNA

Minisatellite DNA also comprises tandemly repeated DNA sequences but in this case the arrays are smaller than those seen within satellite DNA, typically between 100 and 20 000 bp, and each repeat unit is between 6 and 100 bp in length. Such tandem repeats are integral to our genetic makeup; simple repeats of TTAGGG, for example, constitute the DNA sequence found at telomeres (Box 7.3). Minisatellites are abundant in subtelomeric and centromeric regions, and are found at a higher density on certain chromosomes such as chromosome 19. Most classical minisatellites are GC rich in their nucleotide content.

7.3.1 Polymorphic minisatellites

A proportion of minisatellites show polymorphism in the length of tandem repeats due to unequal crossing over or gene conversion. Such polymorphic minisatellites are also described as **variable number tandem repeats** (**VNTRs**). A scan of the draft human genome suggests there are 157 549 minisatellites, of which 29 224 are predicted to be VNTRs (Naslund *et al.* 2005). In general when high copy numbers are seen, the minisatellite repeat units are small; large repeat units are associated with low copy number.

In some cases the level of polymorphism at minisatellites may be extremely high, seen in so-called 'hypervariable' minisatellite DNA. Such diversity found great utility in forensic analysis through the technique of **DNA fingerprinting**, in which a unique minisatellite diversity profile could be defined for a given individual (Section 7.4) (Jeffreys *et al.* 1985b, 1985c). Hypervariable minisatellites have also been used for linkage analysis and genetic mapping (Nakamura *et al.* 1987), but microsatellites and other genetic markers proved more powerful tools for this purpose. Minisatellites have been used in population genetics to provide evidence to support a recent African origin for modern humans (Box 8.5). Increased diversity in African populations was noted, for example, at a highly variable minisatellite MS205, which comprises up to 87 repeats of a 45–54 bp repeat unit (Armour *et al.* 1996). Certain minisatellites have been found to be

---

Box 7.3 Telomeres and tandem repeats

Telomeres, located at the ends of individual chromosomes, comprise long tracts of tandemly repeated DNA made up of simple repeats of the sequence TTAGGG (Moyzis *et al.* 1988; Blackburn 1991). This structure is found across vertebrates. In humans, telomeres are typically 3–20 kb in length. A complex of proteins is associated with the TTAGGG repeats, which protect chromosomal ends (de Lange 2005). The 'shelterin complex' includes proteins that directly recognize TTAGGG repeats: telomere repeat binding proteins 1 and 2 (TRF1 and TRF2) together with protection of telomeres 1 (POT1). Normal DNA polymerase is unable to copy to the ends of chromosomes. Telomeres can be elongated in certain cells, such as germline and stem cells, by a specific polymerase called telomerase. In most somatic cells, telomerase activity is very weak or undetectable; this contrasts with tumour and immortalized cells where high levels are seen. Loss of repeats at each cell division leads to telomere shortening, controlling cell proliferation in somatic cells as very short telomeres leads to senescence or apoptosis (Harley *et al.* 1990; Baird *et al.* 2003; Herbig *et al.* 2004). For a specific chromosome the number of repeats is seen to be maintained across tissues for a particular individual (Martens *et al.* 1998). Within a given age group, significant variation is seen in telomere length between people. This is at least partly genetically determined, as twin studies provide evidence of significant heritability in telomere size (Slagboom *et al.* 1994). It is thought that telomere length variation is highly specific to an individual (Gilson and Londono-Vallejo 2007). Telomere length is a critical component in the biological function of telomeres and plays an important role in diseases associated with aging and cancer.

hypermutable, such as CEB1, which has a mutation rate of 13% in the male germline (Vergnaud *et al.* 1991).

Minisatellites proved valuable as **biomarkers** for the effects of ionizing radiation after it was associated with increased minisatellite mutation rates in mice (Dubrova *et al.* 1993). Following the explosion at the Chernobyl nuclear power station in 1986, a two-fold increase in frequency of mutant alleles at hypermutable minisatellite loci was found in the exposed versus the control populations with a dose effect apparent (Dubrova *et al.* 1996, 1997).

7.3.2 Functional effects of minisatellites

The functional consequences of minisatellites can be diverse, ranging from differences in protein structure and function by altering open reading frames (ORFs), to effects on transcription (Nakamura *et al.* 1998). Gene transcription may be modulated by specific minisatellites, as seen at the *INS* gene (Section 7.3.3). Effects on transcription have also been found involving a minisatellite in an enhancer element 3.8 kb upstream from the transcription start site of the ABO blood group gene (Kominato *et al.* 1997). The minisatellite comprises four tandem copies of a 43 bp repeat to which the transcription factor CBF/NF-Y binds and modulates transcriptional regulation.

VNTRs have been found 1.2 kb upstream of the serotonin transporter gene *SLC6A4* (also known as *5FTT*, *SERT*) and in the second intron of the gene itself on chromosome 17q11. The latter has been of particular interest given disease associations with mood disorders (Cho *et al.* 2005). Nine, ten, or 12 copies of 16 or 17 bp repeat elements are found that have been related to differential gene expression (Fiskerstrand *et al.* 1999), including using a transgenic mouse model to define the effects of repeats on embryonic expression in the hind brain where this gene is known to be transcribed (MacKenzie and Quinn 1999). Y-box binding protein 1 (YB-1) and CTCF binding protein (CTCF) have been implicated in regulation at this VNTR (Klenova *et al.* 2004) with differential recruitment noted in the presence of lithium chloride, an important drug used in mood stabilization for mania and depression (Roberts *et al.* 2007).

Other mechanisms whereby minisatellites may modulate function include at the level of alternative splicing (Section 11.6). Similarities exist between the repeat unit of a minisatellite found at the interferon inducible gene IF16 on chromosome 1p35, and a mammalian splice donor consensus sequence. The minisatellite comprises 26 tandem repeats of 12 bp units; two of these repeats adjacent to a splice donor site in exon 2 function as additional splice donor sites, responsible for inserting four and eight additional amino acids each into the encoded protein (Turri *et al.* 1995). Only limited polymorphism of this minisatellite is seen within human populations, with most individuals having 26 repeats but some people having up to 29 repeats.

Altered activity of an enzyme involved in drug metabolism has been associated with a promoter VNTR in the *TPTT* gene encoding thiopurine methyltransferase (Spire-Vayron de la Moureyre *et al.* 1998, 1999; Yan *et al.* 2000). Significant interindividual variation in levels of enzymatic activity are seen that can determine therapeutic response or toxicity associated with specific drugs such as thiopurines: the consequences of polymorphism at the VNTR are important although the major determinants of variable enzymatic activity relate to nonsynonymous single nucleotide polymorphisms (SNPs) (Krynetski *et al.* 1995) and SNPs modulating alternative splicing (Otterness *et al.* 1998).

The functional consequences of minisatellites may also relate to mechanisms of genomic imprinting (Box 11.5), in which monoallelic expression of genes dependent on parental origin is seen. Tandem repeats were found to be enriched in CpG islands found at imprinted genes (Neumann *et al.* 1995; Hutter *et al.* 2006); more generally there is evidence that copy number of repeats plays a role in epigenetic silencing with tandem repetition of transgenes leading to gene silencing (Garrick *et al.* 1998).

7.3.3 Minisatellites and disease: examples from epilepsy and diabetes

As well as being used as genetic markers to map disease susceptibility, there are a number of examples where minisatellites have been directly implicated in disease. A specific inherited form of epilepsy, progressive myoclonic epilepsy (Box 7.4), was initially mapped by linkage analysis using polymorphic markers to the distal end of chromosome 21 (Lehesjoki *et al.* 1991), then fine mapped within an approximate 300 bp region of chromosome 21q22.3

Box 7.4   Progressive myoclonic epilepsy (OMIM 254800)

Myoclonic epilepsy of the Unverricht Lundborg type (also known as progressive myoclonic epilepsy (EPMI1) or Baltic myoclonic epilepsy) is an important autosomal recessive neurodegenerative disorder more common in Finland, where the incidence is one in 20 000 people, and the western Mediterranean. The disease onset is typically with seizures at the age of 6–13 years, followed by myoclonus (involuntary muscle jerking), cerebellar ataxia, mental deterioration, and dementia.

Box 7.5   *INS* variable number tandem repeat

A minisatellite approximately 600 bp upstream of the insulin *INS* gene transcriptional start site was found to be highly polymorphic (Bell *et al.* 1982). The VNTR comprised simple tandem repeats of a 14–15 bp sequence (ACAGGGGT(G/C)(T/C)GGGG) with at least three alleles recognized. Class I alleles were shortest, 570 bp on average, and comprised of 26–63 repeats; class III alleles were longest at 2470 bp on average with 141–210 repeats; and class II were intermediate, on average 1320 bp (Bell *et al.* 1984). Class I alleles were most common across ethnic groups, while class II were rare, except among individuals of African ancestry who showed significant diversity in allele length at this locus.

(Lehesjoki et al. 1993). Positional cloning identified *CSTB*, encoding a cysteine protease inhibitor cisplatin B, within the region which showed reduced expression in affected individuals in whom two specific mutations were identified (Pennacchio *et al.* 1996). In the majority of people affected by the disease, however, a minisatellite repeat was found to be significantly expanded, from two to three copies of a 12 bp repeat unit (CCCCGCCCCGCC) in unaffected individuals, to 30–80 copies in those with progressive myoclonic epilepsy (Lafreniere *et al.* 1997; Lalioti *et al.* 1997; Virtaneva *et al.* 1997). Individuals with the *CSTB* minisatellite expansion had lower levels of gene expression in the blood, which may relate to altered spacing of transcription factor binding sites (Lalioti *et al.* 1999); 'premutant alleles' were also described in some parents of affected individuals in whom 12-17 repeats were present (Lalioti *et al.* 1997).

A second example of a minisatellite directly implicated in disease comes from the genetics of susceptibility to type 1 diabetes. Type 1 or insulin-dependent diabetes is an autoimmune disorder in which there is destruction of pancreatic β cells and insulin deficiency: environmental and genetic factors are important with several genes implicated in disease susceptibility, notably within the major histocompatibility complex (MHC) (Section 12.4.3). Other loci are important (Box 12.8), including a polymorphic minisatellite upstream of the insulin *INS* gene on chromosome 11p15.5 (Box 7.5). When allele frequencies among cases of diabetes and controls were compared in a Caucasian population, possession of the shorter class I alleles found at this VNTR was associated with increased risk of disease (Bell *et al.* 1984).

This result was subsequently confirmed (Hitman *et al.* 1985) and intensive analyses mapped the association with type 1 diabetes to a 19 kb region spanning the *INS* gene and the neighbouring *IGF2* gene encoding insulin-like growth factor II (Julier *et al.* 1991). This was subsequently refined to a 4.1 kb region encompassing the VNTR and *INS* gene locus, which included ten SNPs (Lucassen *et al.* 1993). Bennett and colleagues used haplotypic analysis and linkage disequilibrium mapping to map disease susceptibility to within the VNTR and exclude these other SNPs (Bennett *et al.* 1995); later analysis using a much larger SNP and pedigree set found that two other SNPs in the region could not be excluded (Barratt *et al.* 2004). Overall homozygosity for class I alleles conferred the highest risk of type 1 diabetes (two- to five-fold), while possession of class III alleles was dominantly protective.

In studies involving *INS* expression in the pancreas, class III alleles were associated with lower expression relative to class I, for example when allele-specific RNA expression was analysed for fetal and adult pancreas tissues taken at post mortem (Bennett *et al.* 1995, 1996; Vafiadis *et al.* 1996). Further evidence was obtained from analysis of insulin secretion where possession of class III alleles was found to be associated with, on average, one-third the levels of glucose induced insulin secretion (Cocozza *et al.* 1988), although this remains controversial; and from reporter gene studies with evidence from a rodent pancreatic β cell line that class III alleles had lower expression than class I (Lucassen *et al.* 1995). This apparently contradictory situation of protective alleles being associated with lower insulin secretion may be partly explained by findings that the VNTR alleles were differentially expressed in the thymus with greater expression of the class III relative to the class I alleles seen here (Pugliese *et al.* 1997; Vafiadis *et al.* 1997). This may be important during development in establishing immune tolerance, with higher levels of expression associated with class III alleles leading to loss of insulin-specific T lymphocytes, which are thought to be important in disease pathogenesis.

7.4 Genetic profiling using mini- and microsatellites: DNA fingerprinting and short tandem repeats

7.4.1 DNA fingerprinting

In 1985, Jeffreys and colleagues published a series of papers highlighting the high level of polymorphism in the number of repeats found at certain minisatellites, and their utility in defining a unique pattern for a particular individual – an approach dubbed 'DNA fingerprinting' (Jeffreys *et al.* 1985b, 1985c). They found that a particular 33 bp repeat present in a short minisatellite of four tandem repeats within an intron of the myoglobin gene could be used to generate a probe that detected a number of other minisatellites scattered across the genome. This 'multilocus' probe hybridized simultaneously to multiple polymorphic minisatellites when used in a Southern blot of restriction enzyme digested DNA, giving

a highly specific pattern of multiple bands dependent on the numbers of repeats in particular minisatellites.

Jeffreys and coworkers noted a core sequence within the 33 bp repeat shared between the minisatellites the probe identified (GGGCAGGAXG). A given multilocus probe was highly specific: when a single probe (designated '33.15') was used the probability of a match between unrelated people was less than 3×10^{-11}; when two different probes ('33.15' and '33.6') detecting particular sets of minisatellites were used this fell to less than 5×10^{-19} (Jeffreys *et al.* 1985c). The approach radically altered forensic practice, being used initially in criminal investigations (Fig. 7.2) (Gill and Werrett 1987) and extensively in establishing paternity (Jeffreys *et al.* 1991) and immigration cases (Jeffreys *et al.* 1985a).

A limitation of the approach for analysing scene-of-crime material such as blood stains is the need for relatively large amounts of good quality DNA (several micrograms). The use of simpler single locus probes

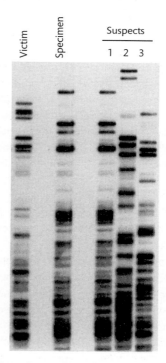

Figure 7.2 Forensic use of DNA fingerprinting. DNA fingerprinting was used in a rape case to identify the DNA fingerprint of suspect 1 as exactly matching that of the semen specimen from a vaginal swab of the rape victim. Reprinted with permission from Strachan and Read (2004).

(specific to a single minisatellite) allowed increased sensitivity with band detection down to 10 ng of DNA and avoided some of the difficulties of comparing between blots (Tamaki and Jeffreys 2005). The use of DNA fingerprinting and genetic profiling is now fundamental to forensic practice and utilizes a range of different types of human genetic variation and amplification/detection techniques, with a more recent focus on short tandem repeats (microsatellites) but also Y chromosome and mitochondrial polymorphisms, SNPs, and copy number variation.

7.4.2 Genetic profiling using a panel of short tandem repeats

The analysis of multiple microsatellite or short tandem repeat (STR) loci has dramatically improved levels of resolution in detecting picogram quantities of potentially degraded DNA. For example, six CA repeat microsatellite markers were used to identify a murder victim from DNA originating from 8-year-old skeletal remains (Hagelberg et al. 1991). Currently, tetranucleotide repeats are predominantly used in forensic DNA analysis. To facilitate data exchange and sharing, a core set of short tandem repeat loci are used in forensic DNA and identity testing with a number of commercial kits available (Butler 2006, 2007). In the USA, 13 STR markers have been used since 1997 as the core genetic markers in the Federal Bureau of Investigation Laboratory's Combined DNA Index System (CORDIS) (www.fbi.gov/hq/lab/html/codis1.htm); in the UK and much of Europe, ten core STR loci and eight recommended ones are used in human identity testing (Gill 2002). A database of STR loci, technologies, and associated information is maintained at www.cstl.nist.gov/biotech/strbase/.

Simultaneous amplification of multiple STRs is achieved with the polymerase chain reaction (PCR) using specific fluorescent dye labels and generating PCR products of different sizes (Fig. 7.3). The use of relatively small PCR products (typically 100–500 bp in length) has allowed small amounts of potentially degraded DNA to be analysed. The number of repeats in a given allele is resolved by size separation with a given amplicon fluorescently labelled to allow multiplexing. Forensic DNA analysis, paternity testing, missing person investigations, and mass disaster victim identification are some of the many current applications of STR technologies. The use

of additional loci, improvements in typing technologies, and the speed of amplification and analysis should further improve identity analysis based on STRs.

7.5   Microsatellite DNA

7.5.1 Short tandemly repeated DNA sequences are common and polymorphic

In the early 1980s a number of reports highlighted the occurrence, nature, and polymorphism of short tandemly repeated DNA in the human genome. Spritz reported the presence of a simple tandem repeat upstream of the β globin gene comprising four, five, or six copies of the repeat unit ATTTT (Spritz 1981). This work highlighted not only the presence of such a tandem repeat, but also that it varied between alleles, with some individuals gaining or losing a whole repeat unit to have four or six copies, and that variation within the tandem repeat array occurred much more frequently than polymorphism in flanking DNA. Elsewhere in the locus, Miesfield and coworkers described the occurrence of a tandem block of 17 TG dinucleotides (TGTGTGTGTGTGTGTGTGTGTGTGTGTGTGTGTG) between the β and δ globin genes that was highly conserved across species (Miesfeld et al. 1981).

The following year, Hamada and colleagues used probes specific for TG dinucleotide repeats to resolve by Southern blotting that such tandem repeats were found in a diverse array of organisms from yeast to humans, and were more common in higher eukaryotes, with approximately 50 000 copies in the human genome comprising typically ten to 60 repeat units (Hamada and Kakunaga 1982; Hamada et al. 1982). In the fourth intron of the human cardiac muscle actin gene, for example, 50 tandemly repeated TG dinucleotides were found (Hamada and Kakunaga 1982). Repeats of CG dinucleotides were also noted to occur very frequently in the human genome, but less commonly than TG repeats, and were not found in yeast (Hamada et al. 1982).

7.5.2 Classification and occurrence of microsatellites

Microsatellites, also referred to as short tandem repeats or simple sequence repeats, are a common class of genetic variation and are much more unstable than minisatellite

(A)

STR loci	Chromosomal location	Repeat Motif	Allele range	PCR product sizes
CSF1PO	5q33.1	TAGA	6–15	305–342 bp (6-FAM)
FGA	4q31.3	CTTT	17–51.2	215–355 bp (PET)
TH01	11p15.5	TCAT	4–13.3	163–202 bp (VIC)
TPOX	2p25.3	GAAT	6–13	222–250 bp (NED)
VWA	12p13.31	[TCTG] [TCTA]	11–24	155–207 bp (NED)
D3S1358	3p21.31	[TCTG] [TCTA]	12–19	112–140 bp (VIC)
D5S818	5q23.2	AGAT	7–16	134–172 bp (PET)
D7S820	7q21.11	GATA	6–15	255–291 bp (6-FAM)
D8S1179	8q24.13	[TCTA] [TCTG]	8–19	123–170 bp (6-FAM)
D13S317	13q31.1	TAT C	8–15	217–245 bp (VIC)
D16S539	16q24.1	GATA	5–15	252–292 bp (VIC)
D18S51	18q21.33	AGAA	7–27	262–345 bp (NED)
D21S11	21q21.1	[TCTA] [TCTG]	24–38	185–239 bp (6-FAM)
D2S1338	2q35	[TGCC] [TTCC]	15–28	307–359 bp (VIC)
D19S433	19q12	AAGG	9–17.2	102–135 bp (NED)
Amelogenin (sex typing)	Xp22.22 Yp11.2	Not applicable	Not applicable	X = 107 bp (PET) Y = 113 bp (PET)

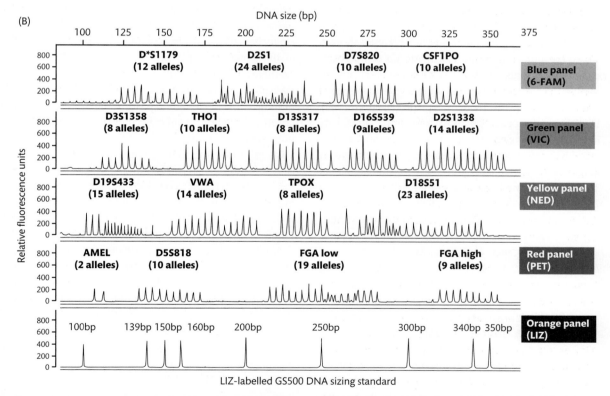

Figure 7.3 Use of multiple short tandem repeat genotyping in forensic analysis. (A) Characteristics of 15 STR loci (205 alleles) and a sex typing assay simultaneously amplified and analysed in AmpF*l*STR Identifiler kit (Applied Biosystems) (Collins *et al.* 2004). (B) DNA allele profiles with different colour panels based on fluorescent dye used. Reprinted with permission from Butler (2007).

or satellite DNA, leading to a high level of polymorphism in terms of gain or loss of repeat units. Microsatellites represent the shortest class of tandem repeats. Most authors consider the term 'microsatellites' to encompass repeat units between one and six bases in length (mono-, di-, tri-, tetra-, penta-, and hexanucleotide repeats), which form a tandem repeating sequence whose total length is less than 100 bp. There is debate over the minimum number of base pairs or repeat units required to constitute a microsatellite – whether, for example, CACA should be considered a $(CA)_2$ microsatellite – and the degree of degeneracy in the tandemly repeated sequence that can still allow classification as a microsatellite (Ellegren 2004).

Microsatellites can be classified into simple repeats (with a single 'perfect' repeating unit) and compound repeats (imperfect with variable composition among repeats). The nomenclature of microsatellites includes a number of pieces of information: 'D3S1266' refers to DNA ('D') of chromosome 3 ('3') where a short tandem repeat ('S') has the unique identifier '1266'.

Analysis of the draft human genome sequence showed 3% of the human genome was made up of microsatellite DNA (Lander et al. 2001). Over 1 million microsatellite loci are found, dependent on the parameters used in defining the tandem repeats (Ellegren 2004). Dinucleotide repeats are most frequent, followed by mono- and trinucleotide repeats, which are more common than tetranucleotide repeats – although if longer lengths of microsatellites are considered (greater than 12 bp in total), monomeric tracts predominate (Ellegren 2004) (Box 7.6). Many microsatellites were noted to be 'A' rich with early surveys of sequence databanks finding that three-quarters of microsatellites comprised, in order of frequency, of poly(A), AC, AAAN, AAN, and AG (where N is C, G, or T) (Beckman and Weber 1992). A-rich microsatellites are often located near to mobile DNA elements such as Alu repeats (Section 8.4), which may reflect their origin and mode of dispersal (Beckman and Weber 1992; Nadir et al. 1996).

In exonic DNA, strong selective pressures against frameshift mutations disrupting codon reading (which is present in three nucleotide units) is thought to have limited repeat expansion to trinucleotide or hexanucleotide repeats such that only these repeat lengths are enriched in coding DNA. This is in contrast to noncoding DNA where all microsatellite repeat lengths are found in excess (Metzgar et al. 2000; Borstnik and Pumpernik 2002). This difference is observed across species ranging from mammals to arthropods, worms, fungi, and yeast (Toth et al. 2000). Among dinucleotide repeats, $(CA)n$ ($=$ AC, GT, or TG) are most common and found on average once every 36 kb; $(AT)n$ ($=$ TA) occur every 50 kb; $(GA)n$ ($=$ AG, CT, TC) occur every 125 kb; and $(GC)n$ ($=$ CG) are least common, found every 10 Mb (Ellegren 2004; Strachan and Read 2004).

7.5.3 Generation and loss of microsatellite DNA

Buschiazzo and Gemmell (2006) reviewed the 'life cycle' of microsatellite DNA in eukaryotes, considering the processes whereby microsatellites may arise, expand in length, contract, and eventually become degenerate and die. Much remains unclear but it is thought that 'proto' microsatellites arise either de novo or through transposable DNA elements (Zhu et al. 2000; Wilder and Hollocher 2001). Single base substitutions may result in the generation of a two or three repeat locus, for example a G to A substitution altering the sequence from GAC<u>G</u>CACG to GAC<u>A</u>CACG and thereby creating a run of three 'AC' dinucleotide repeats. Alternatively an indel (insertion/deletion) event may generate a new dinucleotide repeat, for example GCAT becomes GCA<u>CA</u>T with a CA insertion. Such small repeats then serve as substrates for expansion. Messer and colleagues described the possible 'birth' of microsatellites within the η globin pseudogene based on primate sequence alignments, with single nucleotide substitutions leading to di- and tetranucleotide tandem repeats (Fig. 7.5) (Messier et al. 1996). The role of transposable elements (Chapter 8) may be to serve as both substrate, for example through reverse transcription errors involving poly(A) tracts at the 3′ ends of long interspersed elements (LINEs) and short interspersed elements (SINEs), and the means of dispersal of microsatellites.

Microsatellites undergo a high rate of 'mutation' in that they are observed to expand or contract in length, through gain or loss of single, or less commonly, multiple repeat units. The major mechanism for this instability involves replication slippage during DNA synthesis, resulting in

Box 7.6   Genome-wide survey of human microsatellites

In a genome-wide survey of simple sequence repeats, the highest density was found on chromosome 19 (Fig. 7.4) (Subramanian *et al.* 2003). Across the different chromosomes, mononucleotide repeats of poly(A) and poly(T) were found at a 300-fold higher density than poly(G) or poly(C) ones (Subramanian *et al.* 2003). Dinucleotide repeats were most common in intronic DNA, with AC and AT most frequent and GC rarest. Trinucleotide repeats were two times more common in exonic DNA with AAT, AAC, AAG, and ACG most common; and ACG, ACT, and CCG rare. Among tetranucleotide repeats AAAT, AAAG, AAAC, AAGG were most common; the highest densities were seen on chromosomes 7 and 22; and AGAT was the most frequent repeat found on the Y chromosome. The density of hexanucleotide repeats was two to three times that of trinucleotides in exonic DNA.

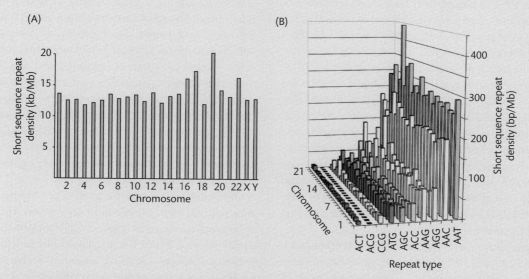

Figure 7.4 Genome-wide analysis of density of microsatellites. (A) Short tandem repeat density across human chromosomes. (B) Density of trinucleotide repeats for different human chromosomes. Reprinted with permission from Subramanian *et al.* (2003).

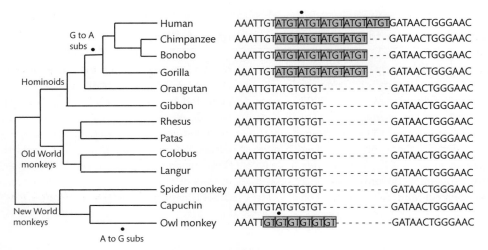

Figure 7.5 Birth of microsatellites. Analysis of primate sequences for the η globin pseudogene indicates a G to A substitution in the common ancestor of African apes and humans leading to four or five copies of a tetranucleotide repeat (ATGT); and an A to G substitution in the linage leading to owl monkeys resulting in six copies of a dinucleotide (TG) repeat. Reprinted by permission from Macmillan Publishers Ltd: Nature (Messier *et al.* 1996), copyright 1996.

variable numbers of repeat units (and hence the length of the microsatellite) between individuals (Levinson and Gutman 1987; Schlotterer and Tautz 1992). Transient dissociation and subsequent realignment of the elongating nascent DNA strand at a microsatellite sequence is often 'out of register' or misaligned, leading to a loop structure. When the loop is present in the newly synthesized DNA, the number of repeats is increased; when it occurs in the template strand, the number of repeats is reduced (Fig. 7.6) (Ellegren 2004). Such events occur very commonly at microsatellites *in vitro* but *in vivo* this is limited by the mismatch repair system. When mutations occur in yeast mismatch repair genes, for example, a 100–700-fold increase in instability was found (Strand *et al.* 1993). Unequal crossing over is much less important with micro- than minisatellites but may be involved in large scale gains and losses of repeat arrays.

The mutation rate of microsatellite DNA is estimated at between 10^{-3} and 10^{-6} per locus per generation (Weber and Wong 1993; Brinkmann *et al.* 1998; Kayser *et al.* 2000). The rate varies between microsatellites, with higher rates in disease-causing trinucleotide repeats (Section 7.6). Among non-disease-causing repeats, higher rates in dinucleotide than tetranucleotide repeats are reported, although other authors have found the converse to be the case (Weber and Wong 1993; Chakraborty *et al.* 1997;

Kruglyak *et al.* 1998). The expansion of microsatellites through replication slippage typically occurs when four or more repeats are in place (Buschiazzo and Gemmell 2006). Smaller length microsatellites are biased towards expansion, and longer microsatellites to contraction, with an exponential increase in mutation rate observed as the number of repeats increases (Lai and Sun 2003). Nucleotide content, genomic context, and internal architecture are important determinants of the course and rate of microsatellite mutation. Biological factors such as sex, age, and the environment are also important, with a higher mutation rate observed in older males, and particular selective forces found to be specific to a given species (Buschiazzo and Gemmell 2006).

Microsatellites will reach a point when the net balance switches from expansion to contraction. The rate of contraction increases exponentially with increasing length (Xu *et al.* 2000); the mechanisms underlying the switch to contraction remain unclear but involve selective pressures acting against very long alleles and the occurrence of point mutations limiting expansion by replication slippage. Likelihood-based simulation on 400 (CA)*n* microsatellites among 680 individuals suggested that once the number of repeats was greater than 20 a shift was observed towards contraction of microsatellite length (Whittaker *et al.* 2003). Over evolutionary

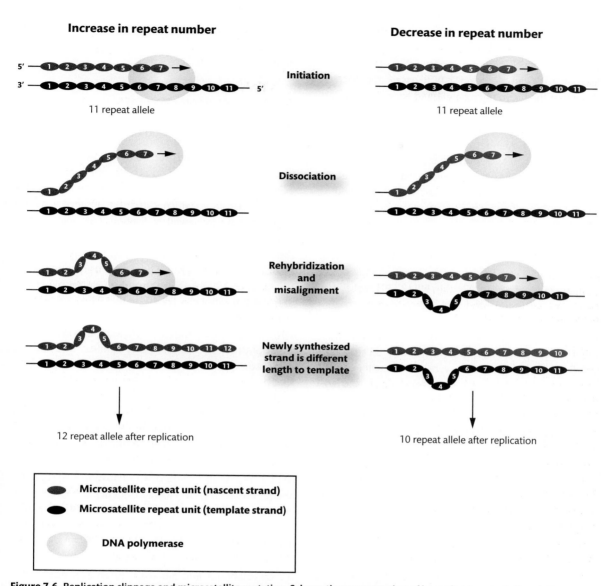

Figure 7.6 Replication slippage and microsatellite mutation. Schematic representation of how after transient dissociation, misalignment may result in gain or loss of a repeat unit in the nascent strand. Redrawn and reprinted by permission from Macmillan Publishers Ltd: Nature Reviews Genetics (Ellegren 2004), copyright 2004; and from Jobling *et al.* (2004).

time, interruptions and deletions lead to degeneracy of microsatellite sequences and 'scrambling' of the sequence, such that 'death' of the microsatellite is seen and the 'life cycle' of the microsatellite can be viewed as complete (Taylor *et al.* 1999; Buschiazzo and Gemmell 2006). The duration of the life cycle of a microsatellite may represent tens or hundreds of millions of years.

7.5.4 Utility of microsatellite markers

Simple mono-, di-, and trinucleotide repeats were found to vary within and between species, occurring five to ten times more commonly than random motifs and providing a major source of genetic diversity (Tautz *et al.* 1986). Microsatellites proved to be an extremely important tool as genetic markers: key to this was their diversity and ability to be assayed by PCR (Litt and Luty 1989; Tautz 1989; Weber and May 1989). Litt and colleagues, for example, used a PCR-based approach to show that the (TG)*n* microsatellite in exon 4 of the cardiac actin gene was highly polymorphic, with 12 different alleles among 37 unrelated individuals and evidence of codominant mendelian inheritance (Litt and Luty 1989). Weber and May found that all ten of the different (TG)*n* microsatellites they investigated showed length polymorphism between individuals, with PCR providing a rapid and sensitive assay for such diversity (Weber and May 1989).

The applications of polymorphic microsatellite markers have been wide-ranging. Microsatellites have played a fundamental role in generating linkage maps of the human genome. In 1992, the NIH/CEPH collaborative mapping group produced a high density map of the human genome comprising 1416 loci, of which the majority of polymorphic markers were restriction fragment length polymorphisms (RFLPs) and 339 of 1676 polymorphic markers used were microsatellites (NIH/CEPH 1992). In the same year Weissenbach and colleagues produced a high density linkage map also using eight large families from the CEPH resource, but here based on 792 CA repeat microsatellites of which 600 were noted to be highly informative and providing coverage of more than 90% of the human genome with 5 cM on average between microsatellites (Weissenbach *et al.* 1992). In 1996 a higher density linkage map was published using 5264 microsatellite markers comprising (CA/TG)*n* repeats to reduce the average size interval to 1.6 cM (Dib *et al.* 1996).

Microsatellites have also been hugely informative in studies of genetic diversity across and within different species and populations, from whales (Schlotterer *et al.* 1991) and birds (Ellegren 1991) to humans. Using microsatellites, information on relatedness and variation can be defined with important implications for our understanding of evolutionary biology, conservation, and population genetics (Bruford and Wayne 1993). Microsatellite typing across diverse human populations has provided important insights into our evolutionary past. The work of Bowcock and colleagues, for example, highlighted how microsatellite diversity was greatest among African populations, supporting an African origin for modern humans

(Fig. 7.7; Box 8.5) (Bowcock *et al.* 1994). Microsatellites have also been extensively used in forensic practice as discussed in Section 7.4.

7.5.5 Functional consequences of microsatellites

Short tandemly repeated sequences may significantly alter the control of gene expression or the structure and function of the encoded protein. The underlying mechanisms and pathophysiological consequences of microsatellite diversity have been best characterized for unstable repeat expansions, which can result in severe neurological disease (reviewed in Section 7.6), while microsatellite instability has also been associated with cancer (Box 7.7). Microsatellite dinucleotide repeats have been associated with stimulating homologous recombination (Wahls *et al.* 1990b); this is also observed for hypervariable minisatellite DNA (Wahls *et al.* 1990a). Local chromatin structure may also be modulated by microsatellite DNA: analysis of expanded CTG/CAG repeats in myotonic dystrophy (Box 7.15) showed they significantly enhanced nucleosome positioning creating a very stable structure that may repress transcription (Wang *et al.* 1994; Wang and Griffith 1995).

A range of different microsatellite repeats have been associated with differences in gene expression. Long tracts of (CA)*n* repeats were found to have enhancer activity, albeit to a relatively modest degree (Hamada *et al.* 1984), while analysis of a 170 bp long (CA) repeat in the rat prolactin gene showed evidence of repression of transcriptional activity (Naylor and Clark 1990). CA repeat number has been associated with levels of expression of the epidermal growth factor receptor (*EGFR*) gene. The CA repeat in intron 1 of *EGFR* varies between 14 and 22 dinucleotides and lies close to an enhancer element: higher repeat numbers are associated with reduced transcription (Gebhardt *et al.* 1999). Intriguingly, responsiveness to a small molecule inhibitor of EGFR tyrosine kinase is also related to the CA repeat number and preliminary evidence suggests an association with lung cancer risk (Amador *et al.* 2004; Han *et al.* 2007; Zhang *et al.* 2007). CA repeats in other genes have also been shown to modulate transcriptional activity, for example in the promoter of the acetyl CoA carboxylase gene (Tae *et al.* 1994) and the matrix metalloproteinase 9 gene (Shimajiri *et al.* 1999; Huang *et al.* 2003) .

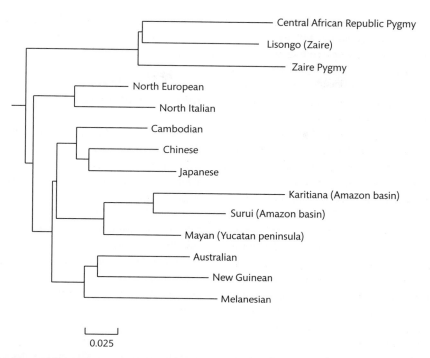

Figure 7.7 Use of microsatellites to define human population relationships. A neighbour joining tree for 30 microsatellites analysed for 148 individuals from 14 indigenous populations across five continents. Reprinted by permission from Macmillan Publishers Ltd: Nature (Bowcock *et al.* 1994), copyright 1994.

Box 7.7   Hereditary non-polyposis colorectal cancer (OMIM 120435) and microsatellite instability

Hereditary non-polyposis colorectal cancer (HNPCC; also known as Lynch syndrome) is the commonest hereditary colorectal cancer and is characterized by an autosomal dominant pattern of inheritance with early onset of predominantly proximal colonic cancer. The disease can be divided into Lynch syndrome I (site-specific colonic cancer) and Lynch syndrome II in which a wide range of extracolonic tumours occur involving the stomach, endometrium, biliary, pancreatic, and urinary tract. The genetic basis for the disease is reviewed by Rustigi (2007). Germline mutations in DNA mismatch repair enzymes are found, most commonly MSH2 and MLH1: in tumours the wild type allele is lost and widespread DNA replication errors occur, often involving dinucleotide repeats (Aaltonen *et al.* 1993; Fishel *et al.* 1993; Ionov *et al.* 1993). Microsatellite instability involving gain or loss of repeat units is seen with a panel of microsatellites recommended by the National Cancer Institute as part of guidelines on the identification of patients with HNPCC (Pinol *et al.* 2005).

Tetranucleotide repeats have also been associated with differences in transcriptional activity. Tyrosine hydroxylase is an important rate limiting enzyme in catecholamine synthesis encoded by the *TH* gene at chromosome 11p15.5. A microsatellite repeat (TCAT)n is found in the first intron of the gene with five to ten copies observed; in 30% of Caucasians the repeat is imperfect when present in ten copies [(TCAT)$_4$CAT(TCAT)$_5$]. Associations have been found between variation in this microsatellite and diseases where catecholamine transmission is thought to be important in disease pathogenesis including bipolar disorder (Meloni *et al.* 1995b), schizophrenia (Meloni *et al.* 1995a), and hypertension (Sharma *et al.* 1998), although the literature remains controversial. Twin studies demonstrate significant heritability in autonomic and catecholamine response and there is evidence that sequence diversity in the *TH* gene is an important determinant of this, although the precise role of the microsatellite remains unclear (Zhang *et al.* 2004b; Rao *et al.* 2007). Modulation of transcriptional enhancer and **silencer** activity has been reported depending on the genomic context and repeat length (Meloni *et al.* 1998; Albanese *et al.* 2001). In a further example, a (CCAT)n repeat in the promoter of the *TNFRSF8* gene, encoding a tumour necrosis factor receptor protein, was shown to repress transcription and there is evidence that this is mediated by a specific transcription factor Yin Yang (YY1) (Croager *et al.* 2000; Franchina *et al.* 2008).

A pentanucleotide repeat in the promoter region of the tumour protein p53 inducible protein 3 (*TP53I3*) gene was shown to be important in determining transcriptional activity dependent on the number of repeats, and that this involved binding by p53 directly to the TGYCC repeat unit (where Y = C or T) (Contente *et al.* 2002). Between ten and 17 repeats were observed in a human population with similar results in great apes; in Old and New World monkeys the number was reduced with more imperfect repeats and this was associated with loss of response to p53 (Contente *et al.* 2003). The relationship with cancer risk remains unclear; association was reported with invasive bladder cancer but not lung or breast cancer (Gorgoulis *et al.* 2004; Ito *et al.* 2006). In the inducible nitric oxide synthase 2A (*NOS2A*) gene a (CCTTT)n repeat in the promoter region was reported to also be associated with differences in transcriptional activity (Warpeha *et al.* 1999) and this may relate to susceptibility to autoimmune and infectious disease.

7.6 Unstable repeats and neurological disease

Repeat instability has been associated with more than 20 diverse neurological disorders, including neurodegenerative and neuromuscular disease (Gatchel and Zoghbi 2005; Pearson *et al.* 2005; Orr and Zoghbi 2007). Repeat expansions as a cause of disease was initially described in 1991 when specific unstable trinucleotide repeats were associated with fragile X syndrome (see Box 7.8) (Verkerk *et al.* 1991), and spinal and bulbar muscular atrophy (see Box 7.12) (La Spada *et al.* 1991). Expansions of trinucleotide repeats are responsible for the majority of disorders caused by unstable repeats. Other repeat lengths are also associated with disease, notably tetranucleotide repeats, for example CCTG repeats and dystrophia myotonica type 2 (see Box 7.15) (Liquori *et al.* 2001), and pentanucleotide repeats, for example ATTCT repeats and spinocerebellar ataxia type 10 (Fig. 7.8) (Matsuura *et al.* 2000). Instability or mutation of longer repeat lengths have also been found to cause disease, notably involving minisatellite DNA, as seen in progressive myoclonic epilepsy (see Box 7.4), and in satellite DNA where reduced numbers of a 3.3 kb repeat at 4q35 are associated with facioscapulohumeral dystrophy (see Box 7.2).

The study of the molecular basis of these 'unstable repeat' diseases has been of particular interest as the disease associated repeat mutations are dynamic, changing between generations and within individuals dependent on the specific tissues involved (Pearson *et al.* 2005; Orr and Zoghbi 2007). These were remarkable findings: that disease-causing mutations were not necessarily transmitted stably from parents to children and that the variation in disease phenotype could be related to the size of the repeat expansion. In most of these diseases, disease severity and age of onset is associated with increasing numbers of repeats, and the likelihood of repeat expansion increases with longer lengths of repeats. Across successive generations, as repeat expansion occurs, the disease severity in affected individuals is seen to increase with an earlier age of

Disease OMIM	Repeat unit	Gene name (protein)	Repeat length		Main clinical features
			Normal	Pathogenic	
Loss of function mechanism					
Fragile X syndrome 309550	(CGC)*n*	*FMR11* (FMRP)	6–60	>200 (full mutation)	Mental retardation, macroorchidism, connective tissue defects, behavioural abnormalities
Fragile X syndrome E 309548	(CCG)*n*	*FMR2* (FMR2)	4–39	200–900	Mental retardation
Friedreichs ataxia 229300	(GAA)*n*	*FRDA* (Frataxin)	6–32	200–1700	Sensory ataxia, cardiomyopathy, diabetes
Polyglutamine disorders caused by gain-of-function mechanism					
Huntingtons disease 143100	(CAG)*n*	*HD* (huntingtin)	6–34	36–121	Chorea, dystonia, cognitive deficits, psychiatric problems
Spinocerebellar ataxia 164400	(CAG)*n*	*SCA1* (ataxin 1)	6–44	39–82	Ataxia, slurred speech, spasticity, cognitive impairments
Spinocerebellar ataxia 183090	(CAG)*n*	*SCA2* (ataxin 2)	15–24	32–200	Ataxia, polyneuropathy, decreased reflexes
Spinocerebellar ataxia 3 109150	(CAG)*n*	*SCA3* (ataxin 3)	13–36	61–84	Ataxia, parkinsonism, spasticity
Spinocerebellar ataxia 6 183086	(CAG)*n*	*CACNA1A* (CACNA1$_A$)	4–19	10–33	Ataxia, dysarthria, nystagmus, tremors
Spinocerebellar ataxia 7 164500	(CAG)*n*	*SCA7* (ataxin 7)	4–35	37–306	Ataxia, blindness, cardiac failure in infantile form
Spinocerebellar ataxia 17 607136	(CAG)*n*	*SCA17* (TBP)	25–42	47–63	Ataxia, cognitive decline, seizures, and psychiatric problems
Spinal and bulbar muscular atrophy 313200	(CAG)*n*	*AR* (androgen receptor)	9–36	38–62	Motor weakness, swallowing, decreased fertility, gynaecomastia
Dentatorubral-pallidoluysian atrophy 125370	(CAG)*n*	*DRPLA* (atrophin)	7–34	49–88	Ataxia, seizures, choreoathetosis, dementia
RNA-mediated pathogenesis					
Dystrophia myotonica 1 160900	(CTG)*n*	*DMPK* (DMPK)	5–37	50–10,000	Myotonia, weakness, cardiac conduction, insulin resistance, cataracts, testicular atrophy, mental retardation (congenital form)
Dystrophia myotonica 2 602668	(CCTG)*n*	*ZNF9* (ZNF9)	10–26	75–11,000	Similar to DM1 but no congenital form
Fragile X tremor/ataxia syndrome 309550	(CGG)*n*	*FMR1* (FMRP)	6–60	60–200	Ataxia, tremor, Parkinsonism, and dementia
Unknown pathogenic mechanism					
Spinocerebellar ataxia 8 608768	(CTG)*n*	*SCA8*	16–34	>74	Ataxia, slurred speech, nystagmus
Spinocerebellar ataxia 10 603516	(ATTCT)*n*	Unknown	10–20	500–4500	Ataxia, tremor, dementia
Spinocerebellar ataxia 12 604326	(CAG)*n*	*PPP2R2B* (PPP2R2B)	7–45	55–78	Ataxia, seizures
Huntington disease-like 2 606438	(CTG)*n*	*JPH3* (junctophilin 3)	7–28	66–78	Similar to HD

Figure 7.8 Unstable repeat expansions and human disease. Overview of unstable repeat disorders classified based on knowledge of pathogenic mechanism. Adapted and reprinted by permission from Macmillan Publishers Ltd: Nature Reviews Genetics (Gatchel and Zoghbi 2005), copyright 2005; and with permission from Orr and Zoghbi (2007).

onset, a phenomenon described as **genetic anticipation** (Howeler *et al.* 1989). In myotonic dystrophy (see Box 7.15), for example, early observations by Greenfield (1911) and Fleischer (1918) described how among successive generations of affected families, disease manifestations included progressively earlier onset of cataracts, with later generations manifesting classical myotonic dystrophy with muscle weakness and presenile cataracts, and increasingly severe symptoms, mental retardation, infertility, and a high infant mortality rate. These observations were confirmed in later family studies (Howeler *et al.* 1989) and the molecular genetic basis for anticipation in this disease found to relate to dramatic expansions of a CTG repeat (Buxton *et al.* 1992).

In the following sections, some of the different disorders resulting from unstable repeat expansions are described to illustrate common themes in pathogenesis and specific mechanisms responsible for disease. In noncoding DNA, the expansion of unstable repeats may reduce gene expression resulting in loss of function as seen in fragile X syndrome (see Box 7.8) and Friedreich's ataxia (see Box 7.10); in transcribed but not translated DNA sequences disease may result through RNA-mediated mechanisms as seen with myotonic dystrophy (see Box 7.15). Expansions of unstable repeats in coding DNA sequence is seen with CAG trinucleotide repeats leading to polyglutamine diseases such as Huntington's disease through changes in protein conformation, with neuronal degeneration and loss (Fig. 7.8; Section 7.6.2). Other coding sequence trinucleotide repeat insertions and deletions may result in disease, notably GCG repeats encoding alanine resulting in rare congenital developmental and neuromuscular disorders through protein misfolding, aggregation, and degradation (Fig. 7.9) (Albrecht and Mundlos 2005). Such GCG expansions are, however, small and show low levels of polymorphism: the repeats are stable at meiosis or mitosis rather than dynamic, and are thought to arise mainly due to non-allelic homologous recombination at meiosis (Warren 1997).

7.6.1 Trinucleotide repeat expansion and loss of function: lessons from fragile X and Friedreich's ataxia

A significant proportion of the genetic causes of mental retardation are associated with the X chromosome, with over 217 conditions currently known (http://xlmr.interfree.it/home.htm; Chiurazzi et al. 2004). Fragile X mental retardation syndrome (Box 7.8) is responsible for up to 20% of cases of X-linked mental retardation. The pattern of inheritance of this syndrome suggested a complex underlying genetic origin with the existence of unaffected male carriers, affected female carriers, and disease risk dependent on position in a family pedigree. In 1969 Lubs described a family pedigree in which four mentally retarded males and three carrier females had 'unusual secondary constrictions' at the end of the long arm of the X chromosome (Fig. 7.10) (Lubs 1969). This fragile site was localized to Xq27.3 (Harrison et al. 1983) and a specific gene was identified (named fragile X mental retardation 1, FMR1) whose first untranslated exon contained a tandem repeat of CGG trinucleotides (Verkerk et al. 1991). This repeat expansion lay close to a CpG island which had been shown to be hypermethylated in affected individuals.

In normal individuals the repeat length is six to 54; in female and male carriers this increases to 55–200 repeats (the 'pre-mutation'; Box 7.9); while in affected individuals with fragile X mental retardation syndrome the 'full mutation' is greater than 200 repeats. In those with the full mutation, the repeat region, CpG island, and flanking DNA sequences are hypermethylated, silencing the FMR1 gene through inhibition of binding by the transcriptional machinery and specific transcription factor proteins (Pieretti et al. 1991). The likelihood of repeat expansion from pre- to full mutation increases with maternal transmission (Malter et al. 1997), increasing repeat length (Nolin et al. 2003), and the absence of (AGG) interruptions in the (CGG) repeat units (Eichler et al. 1994); typically such interruptions occur every nine to ten CGG units in normal alleles.

How does an absence of the fragile X mental retardation protein result in disease? The encoded FMRP protein is expressed in the brain and testes, and is an RNA

Box 7.8   Fragile X syndrome (OMIM 300624)

Fragile X mental retardation syndrome occurs in one in 4000 males (Turner et al. 1996) and is the commonest form of inherited mental retardation (pathophysiology is reviewed in Penagarikano et al. 2007). The syndrome is characterized by mental retardation, particularly affecting short term memory, visuospatial skills, and speech, enlarged testicles (macroorchidism), and a characteristic facial morphology.

Disease OMIM	Repeat unit	Gene name	Repeat length		Main clinical features
			Normal	Pathogenic	
Synpolydactyly 186000	GCG	HOXD13	15	22–29	Hand/foot malformation with syndactyly and polydactyly, brachydactyly, hypodactyly in homozygous individuals
Oculopharyngeal muscular dystrophy 164300	GCG	PABPN1	10	12–17	Progressive, late onset muscular weakness of oculopharyngeal muscles, nuclear inclusion bodies in affected tissues
Hand–foot–genital syndrome 140000	GCG	HOXA13	18	24–26	Hand/foot malformation with short thumbs/great toes, abnormal genitalia
Cleidocranial dysplasia 119600	GCG	RUNX2	17	27	Skeletal dysplasia with hypoplastic clavicles, open fontanelles, tooth abnormalities, short stature
Mental retardation with growth hormone deficiency 300123	GCG	SOX3	15	26	Combination of X-linked mental retardation and short stature caused by growth hormone deficiency
Mental retardation, epilepsy, West syndrome, Partington syndrome 300382	GCG	ARX	10–16	17–23	A spectrum of conditions including to variable extents of mental retardation, various forms of epilepsy, and dystonia
Congenital central hypoventilation, Haddad syndrome 603851	GCG	PHOX2B	20	25–33	Loss of ventilary response to high CO_2 and low O_2, also in combination with Hirschsprung disease (Haddad)
Holoprosencephaly 603073	GCG	ZIC2	15	25	Malformation of midline structures of the forebrain and facial cranium
Blepharophimosis-ptosis-epicanthus inversus syndactyly 110100	GCG	FOXL2	14	22–24	Blepharophimosis, ptosis, epicanthus inversus and ovarian failure

Figure 7.9 Polyalanine tract expansion disorders. Congenital malformation syndromes associated with (GCG)n expansions. The majority involve genes encoding transcription factor proteins. Adapted and reprinted from Albrecht and Mundlos (2005), copyright 2005, with permission from Elsevier.

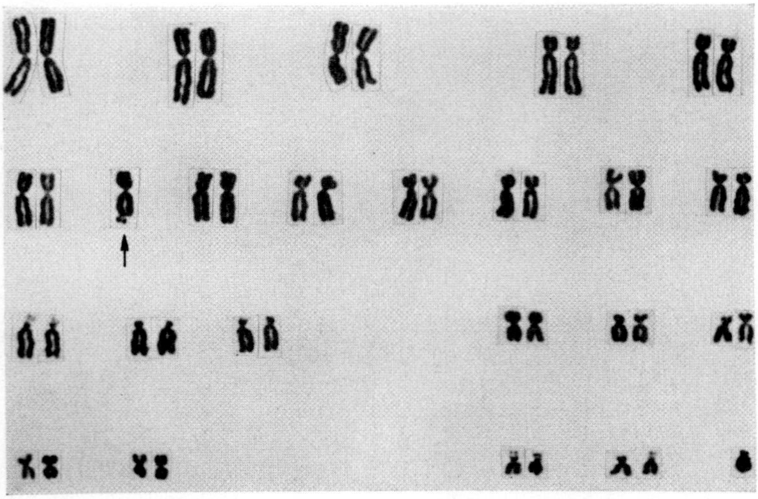

Figure 7.10 Fragile site in a marker X chromosome. Example of secondary constriction at the end of the long arm of X chromosome observed by Lubs in a mentally retarded male. Reprinted from Lubs (1969), copyright 1969, with permission from Elsevier.

Box 7.9   Fragile X tremor/ataxia syndrome (OMIM 30623)

Individuals possessing 55–200 repeats are now recognized to develop a specific neurodegenerative disorder called fragile X tremor/ataxia syndrome, which typically affects older males with intranuclear inclusions seen in neurones and astrocytes, and whose pathogenesis is thought to be RNA-mediated (Hagerman *et al.* 2001; Jacquemont *et al.* 2003).

---

binding protein that normally acts to suppress translation via a complex association with microRNA machinery, notably of specific mRNAs involved in synapse plasticity and neuronal maturation (Ashley *et al.* 1993; Weiler *et al.* 1997; Bear *et al.* 2004; Jin *et al.* 2004). Loss of the protein through gene silencing associated with hypermethylation critically affects these vital processes in the brain. Other deletions and point mutations in *FMR1* can also result in the fragile X phenotype if they result in a non-functional protein, and at least 15 such deletions have been described (Penagarikano *et al.* 2007).

Inhibition of gene expression associated with a trinucleotide repeat in noncoding DNA is also seen with expansion of a GAA repeat within an Alu element in the first intron of the *FXN* gene at 9q13 leading to Friedreich's ataxia (Box 7.10) (Campuzano *et al.* 1996). In unaffected individuals, lengths of six to 34 repeats are seen; affected individuals are homozygous for typically 200–1700 repeats; and those with 34–200 repeats are regarded as having highly unstable 'pre-mutation' alleles which can rapidly expand over a single generation. Larger repeat expansions are associated with more severe disease and earlier age of onset, and with more marked inhibition of *FXN* gene expression leading to lower levels of the encoded protein frataxin (Filla *et al.* 1996; Bidichandani *et al.* 1998; Ohshima *et al.* 1998). The somatic instability of repeat expansion progresses during life. Autopsy studies show progressive accumulation of large expansions in dorsal root ganglia, and blood samples from patients and carriers show increasing mutational load with age (De Biase *et al.* 2007a, 2007b).

Intronic GAA repeat expansion inhibits transcriptional elongation of *FXN*. This may involve formation of unusual DNA structures involving 'sticky DNA', leading to a very

---

Box 7.10 Friedreich's ataxia (OMIM 229300)

This autosomal recessive neurodegenerative disorder is the most common inherited ataxia in man, with a prevalence of one in 50 000 and a carrier frequency of one in 120 Caucasians (Cossee *et al.* 1997). The disease is rare in sub-Saharan Africa and not found in the Far East. Friedreich's ataxia is characterized by degenerative disease involving the central and peripheral nervous system, notably manifesting as incoordination and, in some cases, with heart involvement. Ataxia describes marked incoordination of muscle movements. In Friedreich's ataxia, symptoms are usually present before adolescence with a slowly progressive sensory ataxia affecting gait; loss of tendon reflexes, position, and vibration sense; and difficulty with articulating speech (dysarthria). Over time cardiac involvement is seen in most patients including cardiomyopathy and arrhythmias; diabetes mellitus develops in a minority of patients. The disorder is usually fatal by the age of 50 years. Degeneration is seen involving the dorsal root ganglia, together with axonal degeneration in specific sites – notably the posterior columns and spinocerebellar and corticospinal tracts.

---

stable transcriptionally inactive conformation (Sakamoto *et al.* 1999; Wells 2008). Frataxin is implicated in mitochondrial iron–sulphur cluster biosynthesis with dorsal root ganglia particularly sensitive to frataxin deficiency (Simon *et al.* 2004). The pathophysiology remains unclear but relates to iron dysregulation and oxidative stress (Orr and Zoghbi 2007).

7.6.2 Polyglutamine disorders

Expansion of an unstable translated CAG repeat found in different disease genes results in long tracts of polyglutamines in specific encoded proteins, and has been found to cause at least nine different neurodegenerative diseases including spinal and bulbar muscular atrophy, Huntington's disease, and six spinocerebellar ataxias. These diseases involve specific groups of neurones and have distinct clinical features (see Fig. 7.8). Disease is associated primarily with gain of function. For example, null mice for ataxin-1, the protein bearing a polyglutamine repeat expansion in spinocerebellar ataxia type 1 (SCA1) (Box 7.11), do not develop features of the disorder (Matilla *et al.* 1998), while transgenic and knock-in animal models expressing full length proteins containing long tracts of glutamine repeats do (Watase *et al.* 2002). Polyglutamine diseases are dominantly inherited and a 'toxic' change in the protein conferred by the glutamine expansion is postulated. A change in protein conformation and

accumulation and the formation of insoluble aggregates at nuclear and/or cytoplasmic locations are often seen. Overexpression of the protein without repeat expansion is noted to result in mild disease in some animal models, for example with wild type ATXN1 in the mouse and fly, consistent with the repeat expansion enhancing existing interactions and resulting in gain of function (Fernandez-Funez *et al.* 2000).

Expansion of the glutamine tract is necessary but not sufficient for pathology to result; other sequences are also important. In SCA1, for example, ataxin-1 containing the polyglutamine repeat expansion does not lead to cerebellar degeneration in the presence of a specific serine to alanine substitution (at residue 776) which prevents phosphorylation (Emamian *et al.* 2003). The situation is complex and may not be restricted to gain of function: in SCA1, there is evidence that the polyglutamine expansion in *ATXN1*, in the presence of phosphorylated serine 776, both enhances the formation of a protein complex with an RNA-binding motif protein 17 (RBM17) promoting neuropathology, and represses the formation and function of a protein complex containing the transcriptional repressor protein capicua (which has been associated with protection from toxicity associated with ataxin-1) (Lam *et al.* 2006; Lim *et al.* 2008). It may be, therefore, that SCA1 derives from both gain of function and partial loss of function, and that this is specific to particular protein interactions. The interaction with

---

Box 7.11   Spinocerebellar ataxia

Spinocerebellar ataxias are a group of dominantly inherited disorders resulting from a translated unstable CAG expansion, which manifest with symptoms of ataxia, tremor, and dysarthria together with specific clinical features for individual disease types (see Fig. 7.8). Marked cerebellar atrophy is observed. Presentation is usually in middle age, with premature death within 10–20 years, although juvenile

and late onset cases occur. The age of onset relates to the size of expansion, juvenile onset being associated with larger expansions. Spinocerebellar ataxia type 1 (OMIM 164400) is caused by repeat expansions in the *ATXN1* gene (Orr *et al.* 1993; Banfi *et al.* 1994). Expansions in the range 39–82 repeats result in a long polyglutamine tract in the encoded protein, ataxin-1, and are associated with disease.

RBM17 is regulated by the number of glutamine repeats in ataxin-1; repeat length in SCA1 is also associated with age of onset of disease, larger repeat expansions being associated with juvenile onset. Many of the features described in relation to SCA1 are shared with the other polyglutamine diseases, as will be illustrated in the following text reviewing spinal and bulbar muscular atrophy, and Huntington's disease.

The genetic locus responsible for the neurodegenerative disease spinal and bulbar muscular atrophy (Box 7.12) was localized by linkage analysis to the proximal long arm of the X chromosome (Fischbeck *et al.* 1986), and was found to be caused by the expansion of a CAG repeat in the first exon of the *AR* gene encoding androgen receptor (La Spada *et al.* 1991). Affected individuals have 38–62 repeats, while unaffected people have between nine and 36 repeats. Repeat length correlates with disease severity and age of onset although there is heterogeneity (Atsuta *et al.* 2006). The CAG repeat is translated into a polyglutamine tract in the N-terminal activation domain of the androgen receptor protein. Pathogenesis involves a change in protein conformation and altered cellular interactions, resulting in degeneration and loss of neurones; proteinaceous aggregates are seen. Neuronal dysfunction related to accumulation of the mutated protein is testosterone-dependent as androgen deprivation rescues animal models of the disease (Katsuno *et al.* 2002).

As a further example of a disease associated with the presence of long tracts of polyglutamine repeats, Huntington's disease (Box 7.13) illustrates how devastating the consequences of inheriting such variation can be. Huntington's disease is most often manifest in mid-adult life, the disease is fatal with an unremitting course, and

currently incurable. A clear familial basis to the disorder was originally reported by Huntington in 1872. In 1983 Gusella and colleagues reported evidence of linkage to the tip of the short arm of chromosome 4, using RFLPs as DNA markers to study affected families (Gusella *et al.* 1983).

This was a pioneering study, the first human disease locus to be mapped without *a priori* knowledge of its chromosomal location. Use of linkage disequilibrium and selective recombination data allowed resolution to a 500 kb region at 4p16.3. Within this region a new gene, *HD*, was identified by the Huntington's Disease Collaborative Research Group that contained a (CAG)n repeat strikingly polymorphic in length (HDCRG 1993). The number of repeats ranged from 11 to 34 copies in unaffected individuals, to more than 42 copies present in Huntington's disease chromosomes among 75 ethnically diverse, independent, affected families. A remarkable resource utilized in these and other studies of Huntington's disease are family kindreds from around Lake Maracaibo in Venezuela, the largest of which was reported in 2004 as comprising 14 761 individuals who were descended over ten generations from a woman who died of Huntington's disease in the early 1800s (Wexler *et al.* 2004). Repeat length is a critical determinant of age of onset of the disease but there is significant heterogeneity, with other genetic and environmental factors shown to be important by analysis of the Venezuelan kindreds (Wexler *et al.* 2004).

The expanded (CAG)n repeat is found in exon 1 of the *HD* gene encoding huntingtin. The analysis of repeat lengths has facilitated predictive testing for the disease (Box 7.14). Age of onset correlates inversely with the number of repeats; more than 70 repeats is associated

---

Box 7.12   Spinal and bulbar muscular atrophy (Kennedy's disease) (OMIM 313200)

This uncommon disorder affects males. It is characterized by adult onset of progressive muscle weakness and atrophy, often associated with gynaecomastia and infertility. An X-linked recessive pattern of inheritance is described (Kennedy *et al.* 1968) but in many cases there is no family history due to the late onset and lack of symptoms in carriers. The disease prevalence has been estimated at one in 40 000 people (Fischbeck 1997). Degeneration is seen in anterior horn cells, bulbar neurones, and dorsal root ganglia (Sobue *et al.* 1989); neurogenic atrophy and chronic denervation are observed.

Box 7.13   Huntington's disease (OMIM 143100)

In 1872 Huntington described the clinical features of this disease including movement disorder, personality change, and cognitive decline, together with the autosomal dominant pattern of inheritance (Huntington 1872). This is a neurodegenerative disorder with progressive and selective neural cell death, in most cases presenting in mid-adult life; life expectancy from presentation is usually 15–20 years. A small proportion of cases, approximately 5–7%, have a juvenile form of the disease, while late adult onset is also recognized.

Huntington's disease occurs with a frequency of between four and seven per 100 000 although this varies between geographical areas and populations groups, being lower for example in sub-Saharan Africa. Motor symptoms include clumsiness and chorea, where there are involuntary spasmodic movements affecting limbs and facial muscles. Other symptoms involve cognitive decline with deteriorating memory and development of dementia, changes in personality, and psychiatric disturbance.

---

with the juvenile onset form of the disease (Andrew *et al.* 1993; Duyao *et al.* 1993; HDCRG 1993; Telenius *et al.* 1993). The number of repeats changes between generations, with largest expansions seen in disease chromosomes that are paternally transmitted (Duyao *et al.* 1993; Telenius *et al.* 1993). It is recognized that juvenile cases predominantly inherit the disease from their fathers, with a significant difference in age of onset between parent and child ('genetic anticipation') (Ridley *et al.* 1988). Repeat number is also seen to vary between tissues. Dramatic increases in length are seen in areas of the brain showing neuropathological involvement (Telenius *et al.* 1994), in some cases involving increases of up to 1000 repeats at specific sites early in the course of the disease (Kennedy *et al.* 2003).

Much remains to be understood about the pathophysiology of Huntington's disease. Post mortem studies show particular regions of the brain such as the caudate and putamen are severely atrophied (Vonsattel *et al.* 1985). The presence of an expansion of $(CAG)n$ repeats in exon 1 of the *HD* gene is translated into long polyglutamine tracts in the N terminus of the encoded protein, huntingtin. Huntingtin is a large 348 kilodalton (kDa) protein expressed from early in development in many tissues including the central nervous system (Strong *et al.* 1993). The presence of multiple sites (HEAT repeats) for protein–protein interactions suggested a role as a scaffold protein and many different interacting partners have been identified (Andrade and Bork 1995; Harjes and Wanker 2003).

Animal models showed that a lack of huntingtin is associated with cell death and embryonic lethality (Duyao *et al.* 1995) while inactivation of the Huntington's disease gene in conditional knock-out mice demonstrated adult

---

Box 7.14   Predictive testing for Huntington's disease

Predictive testing has been available since 1987, initially based on linkage analysis (Meissen *et al.* 1988), and since 1993 by direct analysis of the (CAG)*n* repeat (Tibben 2007). Among unaffected individuals six to 34 repeats are found. Those with 40 or more copies of the repeat will progress to Huntington's disease; having 35–39 repeats leaves uncertainty about whether the disease will develop and to what degree; while those with 27–35 repeats have a small risk of their future children inheriting an allele bearing an expansion (Rubinsztein *et al.* 1996; Tibben 2007). International ethical guidelines are in place for predictive testing, and extensive pre- and post-test counselling and support are needed,

including neurological and psychiatric evaluation (Anonymous 1994). Longitudinal studies have shown that the majority of those at risk have not had predictive testing (Creighton *et al.* 2003). The decision to be tested carries very significant implications. Reasons for wanting predictive testing include relief of anxiety over whether the disease will develop, and clarity over future decision making, notably family planning. However, many people choose not to have predictive testing given that this is an incurable disease, concerns about family risk, consequences for health insurance, the costs of testing, the finality of test results, and how they as individuals and a family would cope with such knowledge.

---

neuronal degeneration (Dragatsis *et al.* 2000). The consequences of an abnormally long tract of glutamine residues has been studied using transgenic mice in which the presence of the human CAG repeat expansion resulted in a severe progressive neurological phenotype showing many of the features of the human disease (Mangiarini *et al.* 1996). Prior to onset of the phenotype, characteristic neuronal intranuclear inclusions were noted containing huntingtin and ubiquitin (Davies *et al.* 1997). These proteinaceous aggregates were remarkably similar to those seen in human brain tissue from affected patients with Huntington's disease where the extent of accumulation was related to repeat length (DiFiglia *et al.* 1997). A variety of mechanisms including transcriptional dysregulation, disruption of axonal transport, and altered mitochondrial function have been implicated (reviewed in Orr and Zoghbi 2007).

7.6.3 Disease resulting from RNA-mediated gain of function: myotonic dystrophy

Myotonic dystrophy is an autosomal dominant degenerative disorder affecting muscle fibres but with a complex phenotype including cataracts, cardiac conduction abnormalities, insulin resistance, sleep disorders,

testicular atrophy, and frontal balding (Box 7.15). The molecular basis for these seemingly unrelated manifestations involves altered RNA processing as a result of interactions between the repeat expansions in the encoded RNA, and specific RNA binding proteins. A change in the activity or level of these RNA binding proteins alters RNA splicing (Box 1.20) such that particular isoforms are expressed, with organ-specific consequences.

Expansion of (CTG)*n* repeats in the 3′ untranslated region (UTR) of the dystrophia myotonica-protein kinase (*DMPK*) gene at 19q13.3 was found to cause myotonic dystrophy in 1992 (Brook *et al.* 1992; Fu *et al.* 1992; Mahadevan *et al.* 1992). Affected individuals were found to have from 50 to several thousand copies of CTG repeats. It was subsequently recognized that a proportion of patients did not have this repeat and had some clinically distinct features, for example proximal rather than distal muscle weakness and involvement of type 2 rather than type 1 muscle fibres. These patients were classified as having myotonic dystrophy type 2 (DM2; OMIM 602668) and in 2001 the underlying defect was found to be a (CCTG)*n* repeat expansion comprising on average 5000 repeats in intron 1 of the zinc finger protein 9 (*ZNF9*) gene at 3q21 (Liquori *et al.* 2001).

Box 7.15 Myotonic dystrophy

Muscular dystrophies are inherited degenerative disorders affecting skeletal muscles leading to progressive weakness. The commonest adult muscular dystrophy is myotonic dystrophy type 1 (DM1) (OMIM 160900) with a prevalence of about one in 8000 people (Machuca-Tzili *et al.* 2005). Myotonic dystrophy is a specific type of muscular dystrophy in which myotonia is seen: abnormal electrical activity is found at muscle fibre membranes, manifest clinically with difficulty relaxing the grip, and also affecting talking, chewing, and swallowing.

The other distinct feature of myotonic dystrophies is the multisystem involvement that is manifest, involving brain (for example cognitive impairment, daytime sleepiness, psychological and personality traits), eyes (cataracts), heart (cardiac conduction abnormalities), and endocrine (insulin resistance, testicular atrophy) systems. For DM1, clinical anticipation is recognized with severely affected children born to mothers with minimal symptoms (Koch *et al.* 1991). Congenital, childhood, and adult onset forms of DM1 are recognized.

The (CTG)*n* and (CCTG)*n* repeat expansions found in the 3′ UTR of the *DMPK* gene, and intron 1 of *ZNF9*, results in RNA with long CUG and CCUG tracts, respectively. RNA was shown to be retained in the nucleus in DM1 (Davis *et al.* 1997). Moreover, the presence of CUG repeats in a transcript was sufficient to cause disease. In a transgenic mouse when a (CTG) expansion was inserted into the 3′ UTR of an unrelated gene, human skeletal actin, features of myotonia were seen (Mankodi *et al.* 2000). The myotonic dystrophy phenotype was also seen with transgenic mice producing *DMPK* RNA with at least 300 CUG repeats (Seznec *et al.* 2001).

Misregulation of alternative splicing for specific genes is found to occur in patients with myotonic dystrophy resulting in the presence of specific alternatively spliced isoforms, involving for example the muscle-specific chloride channel (CLCN1), insulin receptor (IR), and cardiac troponin T (cTNT) (Savkur *et al.* 2001; Charlet *et al.* 2002; Mankodi *et al.* 2002). The presence of particular isoforms has important consequences, for example reduced chloride channel expression and defective insulin signalling in skeletal muscle results in specific disease manifestations such as myotonia and insulin resistance. The basis for this effect on splicing was found to relate to the interaction of specific RNA binding proteins with the repeat expansions found in the *DMPK* and *ZNF9* RNA molecules. Musclebind-like (MBNL) protein and CUG binding protein (CUG-BP1) can bind CUG repeats and modulate splicing; the former interaction being dependent on repeat

length (Timchenko *et al.* 1996; Philips *et al.* 1998; Miller *et al.* 2000). The result is loss of function for MBNL and gain of function for CUG-BP1, altering RNA splicing with a diverse range of consequences depending on the specific proteins whose splicing is modulated (Fig. 7.11). A knock-out mouse model for MBNL leads to characteristic features of myotonic dystrophy (Kanadia *et al.* 2003) while overexpression of CUG-BP1 in skeletal and heart muscle (Ho *et al.* 2005b) reproduced features of the disease.

7.7 Summary

A large proportion of our genome is made up of tandemly repeated DNA. The study of this class of genetic variation has proved extremely important to our understanding of the human genome. Variation in number of repeat units provided genetic markers of diversity that were of fundamental importance in efforts to generate a map of the human genome, to try and identify the genetic basis of disease, and to understand human origins and evolutionary relationships. We have seen that classification is possible based on length of the tandemly repeated DNA, ranging from hundreds to thousands of kilobases in satellite DNA, to less than 100 base pairs in microsatellite DNA, with minisatellites of intermediate size.

The recognition that certain minisatellites were hypervariable in length, based on numbers of repeating units, and shared core motifs to enable recognition of multiple

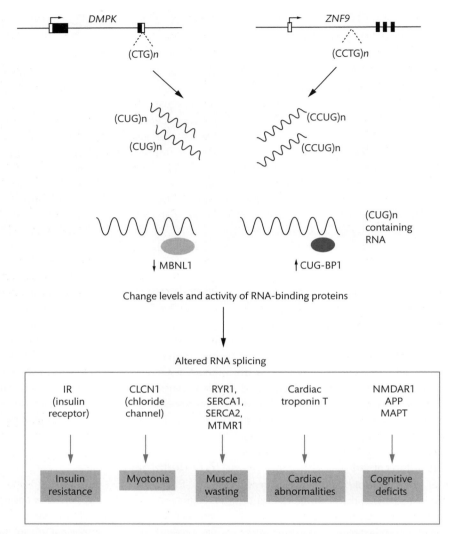

Figure 7.11 Myotonic dystrophy and modulation of alternative splicing. (CUG)n and (CCUG)n repeat expansions in RNA molecules from *DMPK* and *ZNF9* interact with RNA binding proteins MBNL1 and CUG-BP1, altering their levels and activity. The consequences of this are manifest across a range of tissues where specific alternatively spliced isoforms are found leading to particular clinical phenotypes observed in myotonic dystrophy.

minisatellites with a given single probe, revolutionized forensic practice. First came the technique of DNA fingerprinting, in which a unique pattern of bands for a given individual was recognized on a gel based on numbers of repeats in different minisatellites. This was superseded by genetic profiling based on simultaneous amplification of a panel of short tandem repeats (microsatellites) allowing much smaller quantities of DNA to be analysed (Jeffreys *et al.* 1985c; Butler 2007).

Microsatellites, by virtue of their short length and high degree of polymorphism, could be amplified by PCR and used as highly informative genetic markers to define linkage maps of the human genome (Dib *et al.* 1996). The high degree of mutation in microsatellites in which there is gain or loss of repeat units, involves replication slippage during DNA synthesis with misalignment and looping of the nascent or template strand. Mismatch repair enzymes act to control this: marked microsatellite instability is seen with germline mutations in specific repair enzymes and this is associated with hereditary non-polyposis colorectal cancer (Rustgi 2007). Unstable repeats, notably trinucleotide repeats in which marked expansions in numbers of repeats are seen, have been associated with a number of different neurological disorders (Orr and Zoghbi 2007). Dynamic changes in repeat number are observed between generations and within tissues: the consequences are typically severe and in some cases fatal.

Polyglutamine expansions in specific proteins resulting from translated CAG repeats are usually associated with gain of function, manifest as autosomal dominant disease with changes in protein conformation and function, for example Huntington's disease. In other diseases, repeat expansions have been associated with loss of function involving gene silencing associated with hypermethylation (fragile X syndrome) or inhibition of transcriptional elongation with adoption of particular transcriptionally inactive DNA conformations (Friedreich's ataxia). In myotonic dystrophy, interactions of the expanded CUG in RNA molecules transcribed from the *DMPK* gene with specific RNA binding proteins alters control of alternative splicing for a number of genes, leading to the specific and diverse clinical manifestations seen in this disease.

Polymorphism in minisatellite DNA can also have important implications for gene function and disease, notably reduced gene expression associated with repeat expansions in cisplatin B, a cysteine protease inhibitor, which is associated with progressive myoclonic epilepsy. One of the earliest and most significant disease associations with a minisatellite was found upstream of the insulin gene where shorter alleles were associated with increased risk of type 1 diabetes: there was evidence of a context-specific functional role for this minisatellite with increased expression seen for this class of alleles in the pancreas, and reduced expression in the thymus. Tandem repeats perform vital roles in telomeric DNA at the end of chromosomes where long tracts of TTAGGG repeats are found complexed with a number of different proteins; very large arrays of satellite DNA are seen at centromeres as well as heterochromatin in pericentromeric and telomeric regions. Satellite DNA has proved difficult to sequence and remains enigmatic but some secrets are starting to appear, notably a role in generating small interfering RNA critical for heterochromatin formation and in providing binding sites for centromere associated proteins; disease is also associated with diversity in repeat number for example at chromosome 4q35 associated with facioscapulohumeral dystrophy.

Tandem repeats have been, and will continue to be, extremely informative in our journey to understand human genetic diversity and its consequences. Variation in tandemly repeated DNA provides a tool for gene mapping and defining us as unique individuals but can also have profound consequences for disease. Study of tandemly repeated DNA has provided remarkable insights into disease pathogenesis and mechanisms of human diversity, a story which is only likely to continue to grow.

7.8   Reviews

Reviews of subjects in this chapter can be found in the following publications:

Topic	References
Satellite DNA evolution and use	Csink and Henikoff 1998
Facioscapulohumeral muscular dystrophy	Tawil and Van Der Maarel 2006
Telomere structure and function	Blackburn 1991
Genetic profiling and forensic DNA analysis	Tamaki and Jeffreys 2005; Butler 2007
Short tandem repeats	Fan and Chu 2007
Microsatellite biology, population genetics, and evolution	Bruford and Wayne 1993; Ellegren 2004; Buschiazzo and Gemmell 2006
Hereditary colon cancer	Rustgi 2007
Trinucleotide and unstable repeat disorders	Gatchel and Zoghbi 2005; Orr and Zoghbi 2007
Pathophysiology of fragile X	Penagarikano et al. 2007
Freidreich's ataxia	Wells 2008
Myotonic dystrophy	Machuca-Tzili et al. 2005

CHAPTER 8

Mobile DNA elements

8.1 Introduction

A remarkably high proportion of our genomic sequence is made up of mobile DNA elements (Box 8.1), estimated at 45% of the human genome (Lander *et al.* 2001). In non-human primates such as the chimpanzee *Pan troglodytes* (CSAC 2005) and rhesus macaque *Macaca mulatta* (Gibbs *et al.* 2007) between 40% and 50% of the genome is composed of mobile repetitive sequences (Fig. 8.1), with similar results in other mammals such as the mouse *Mus musculus* (37.5% of genome) (Waterston *et al.* 2002).

Two major classes of mobile DNA elements are recognized, each differentiated based on their method of transposition. Class I elements, or retrotransposons, transpose via an RNA copy which is then reverse transcribed into DNA and inserted into a new genomic location (Deininger and Batzer 2002). Retrotransposons include long interspersed elements (LINEs) and short interspersed elements (SINEs), the latter being less than 500 bp in length. Transposition may be achieved by replication machinery encoded by the DNA elements themselves, seen with so-called autonomous retrotransposons such as the LINE 1

(L1) retrotransposable family (Ostertag and Kazazian 2001); or be non-autonomous and depend on retrotransposition proteins encoded by active L1 elements, as seen with the Alu family (Batzer and Deininger 2002) (Fig. 8.2).

Class II elements are so-called DNA transposons; here transposition involves no RNA intermediate but rather DNA, in a 'cut and paste' process achieved by a transposase enzyme encoded by some transposons (Pace and Feschotte 2007). With the increasing amount of sequencing data becoming available from a diverse range of species, the extent and diversity of transposable elements is remarkable and a hierarchical classification system for transposable elements across different eukaryotic genomes has been devised (Wicker *et al.* 2007). This includes major classes, subclasses, orders, and superfamilies of transposable elements with the size of the target site duplication characteristic for most superfamilies.

Mobile DNA elements are found in almost all eukaryotic genomes and are seen as important drivers of genomic plasticity through their ability to transpose. Many promoters contain sequence derived from such events as do other *cis* regulatory elements (Jordan *et al.*

Box 8.1 Mobile DNA elements

Mobile DNA elements have been defined as segments of DNA that can transport or duplicate themselves to other regions of the genome (Xing *et al.* 2007). The process of moving or duplicating around the genome is described as transposition: this may be achieved directly by a 'cut and paste' type mechanism as seen with **DNA transposons**; or be achieved via an RNA intermediate to achieve duplication as seen with **retrotransposons**.

		Human	Chimpanzee	Macaque
DNA transposons		355 000	305 000	327 000
Long terminal repeat/ endogenous retroviruses (LTR/ERV)		506 000	453 000	432 000
Long interspersed elements (LINEs)	L1	572 000	558 000	531 000
	L2	363 000	315 000	298 000
Short interspersed elements (SINEs)	Alu	1 144 000	1 111 000	1 094 000
	Mammalian interspersed repeat (MIR)	584 000	553 000	539 000
SVA, a composite repetitive element including SINE and Alu components		3400	4400	150

Figure 8.1 Repeat elements in human, chimpanzee, and macaque genomes. Data from the human genome (version 18), common chimpanzee (*Pan troglodytes*) (version 2), and rhesus macaque (*Macaca mulatta*) (version 2). From Gibbs *et al.* (2007), reprinted with permission from AAAS.

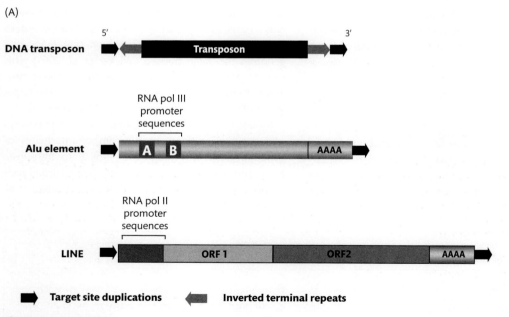

Figure 8.2 *Continued*

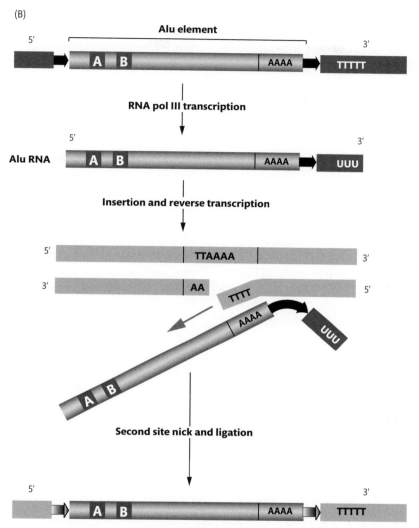

(B)

Figure 8.2 Structure of mobile DNA elements. (A) Active DNA transposons encode transposase; the transposon is flanked by inverted terminal repeats and duplicated target sequence as a result of integration. Alu elements do not encode any proteins and are transcribed by RNA pol III. Full length LINEs encode RNA binding protein and protein with endonuclease/reverse transcriptase activity. Adapted with permission from Xing *et al.* (2007). (B) Retrotransposition of an Alu element. Transcript from an Alu element runs into the 3′ flanking sequence as there is no RNA pol III termination signal coded in Alu – the signal is TTTT on the sense strand, and leads to UUU at the 3′ end of the majority of transcripts. L1 endonuclease is thought to usually make the first nick at the site of insertion at the TTAAAA consensus site. AAAAs at the 3′ end of the Alu element may anneal at the site of integration for target-primed reverse transcription. Note the new set of direct repeats flanking the newly inserted Alu element. Redrawn and reprinted by permission from Macmillan Publishers Ltd: Nature Reviews Genetics (Batzer and Deininger 2002), copyright 2002.

2003). Transposable elements are thought to have been important in the evolution of regulatory sequences and networks (Feschotte 2008). For example, a number of families of long terminal repeat elements (affili- ated to class I endogenous retroviruses) were found to be enriched for binding sites for p53 (Wang *et al.* 2007). These represented a third of p53 binding sites in the human genome previously identified by genome-wide *in vivo* binding studies using **chromatin immunoprecipitation (ChIP)** (Box 11.7) (Wei *et al.* 2006). Evolutionary analysis suggests dispersal by insertion and deletion of these long terminal repeat elements over 40 million years from a primate ancestor. This illustrates how a primate-specific regulatory network may arise with multiple genes controlled by a given transcription factor family, in this case the master regulatory factor p53 (Fig. 8.3) (Wang *et al.* 2007; Feschotte 2008).

Transposable elements may also result in novel functional sequences that are advantageous to the host (an example of exaptation or cooption), as illustrated by the origins of enhancers in SINE retrotransposons for the *ISL1* and *POMC* genes (Box 8.2) (Bejerano *et al.* 2006; Santangelo *et al.* 2007). Transposable elements are thought to directly modulate gene expression in many different ways involving both transcriptional and post-transcriptional mechanisms (Fig. 8.4). As we will see during the course of this chapter the consequences of mobile DNA elements are diverse and can be profound, contributing to the birth of new genes through fusion events, or can have deleterious consequences, for example disrupting open reading frames (ORFs) or predisposing to homologous recombination, and can result in a range of disease phenotypes.

8.2 DNA transposons: a fossil record in the genome

DNA transposons have only relatively recently been recognized in humans (Morgan 1995; Oosumi *et al.* 1995; Smit and Riggs 1996). They are thought to constitute 3% of the human genome (Lander *et al.* 2001). Active DNA transposon elements found in species such as *Drosophila* are characterized by encoding a transposase gene and having terminal inverted repeat sequences to which the transposase enzyme binds. Transposase catalyses transposition by cutting and pasting (van Luenen *et al.* 1994). The transposase excises the DNA transposon element and inserts it into a new genomic location after making

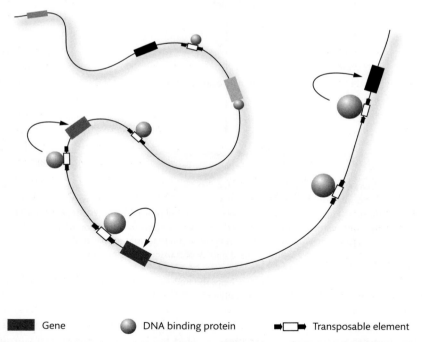

■■■ Gene   ● DNA binding protein   ■▭➤ Transposable element

Figure 8.3 Mobile DNA elements and wiring of a transcriptional network. Dispersal of a DNA binding site by multiple copies of a transposable element allows the potential for many genes to be simultaneously regulated (indicated by curved arrows) by binding of the same transcription factor to the DNA binding sites. Redrawn and reprinted by permission from Macmillan Publishers Ltd: Nature Reviews Genetics (Feschotte 2008), copyright 2008.

Box 8.2   Transposable elements and exaptation

There are now a number of examples where relics of transposable elements can acquire functions that are advantageous to their host genome. A neuronal enhancer of the proopiomelanocortin (*POMC*) gene is thought to have arisen from a CORE-SINE retrotransposon some 170 million years ago in the lineage leading to mammals (Santangelo *et al.* 2007). This group of SINEs lost transposable activity about 100 million years ago in placental mammals (where their estimated copy number is 300 000) but remained active until relatively recently in marsupials. A very ancient family of SINE elements, called the LF-SINE (lobe-finned fishes or 'living fossil' SINEs), is thought to have been active at least 410 million years ago in lobe-finned fishes and terrestrial vertebrates, and more recently in the 'living fossil' Indonesia coelacanth *Latimeria menadoensis*. LF-SINEs were identified from human sequences that are very strongly conserved across species (so-called ultraconserved elements) and 245 copies have been found to date in the human genome. One of these, some 500 kb upstream of the ISL LIM homeobox 1 (*ISL1*) gene (which encodes a transcription factor required for motor neurone development) acts as an enhancer and illustrates how mobile DNA elements can generate distal enhancers (Bejerano *et al.* 2006).

---

a break in the target site DNA sequence. At integration, a short and constant length duplication of the target site is created. This is characteristic of particular families of DNA transposons, visible in the genomic architecture as direct repeats flanking the inserted element (see Fig. 8.2). In *Drosophila* for example, the *Mariner* DNA transposon duplicates 2 bp (TA) on integration while the *Ac/hobo* elements duplicate 8 bp (Jacobson *et al.* 1986).

The work of Smit and Riggs in 1996 identified more than 100 000 copies of degenerate DNA transposon elements bearing characteristic short inverted terminal repeats and flanking 8 bp or TA duplications in the human genome. These included *mariner* elements and so-called *Tigger* elements similar to a *Drosophila* DNA transposon, *pogo* (Smit and Riggs 1996). Study of one of the two ancient *mariner* subfamilies found in humans, called *Hsmar1* transposons, has provided evidence of how transposons may have contributed to the birth of new genes through fusion events (Box 8.3). At least seven of the nine major eukaryotic superfamilies of DNA transposons are represented in the human genome, with 125 different families present at a copy number of greater than 100. These can be referenced through the database of repetitive elements maintained at the Genetic Information Research Institute (www.girinst.org/repbase/index.html) (Jurka *et al.* 2005).

In humans, DNA transposons represent 'genomic fossils': there is no evidence of any transposase activity for 37 million years through the anthropoid-specific lineage (Pace and Feschotte 2007). However, there is evidence of intense activity during the mammalian radiation and early in primate evolution. In fact 98 000 elements are believed to have been added to the primate genome over the last 80 million years, with over 40 transposon families active during the evolution of primates. A mass extinction of DNA transposons is then thought to have occurred in our anthropoid primate ancestor such that transposase activity was lost with no transposon elements younger than 37 million years (Pace and Feschotte 2007). DNA transposons have become fixed in human populations; in contrast, retrotransposons have remained active within the human genome and recent human-specific diversity within Alu and L1 elements have provided very important markers of genetic diversity for **phylogenetic** and population genetic analysis.

8.3   L1 retrotransposable elements

The L1 retrotransposable element family constitutes one-fifth of the human genome (Smit *et al.* 1995; Lander *et al.* 2001; Kazazian 2004). L1s are believed to have been established approximately 120 million years ago in our mammalian ancestors. L1 retrotransposons have been described as the 'most successful and enduring

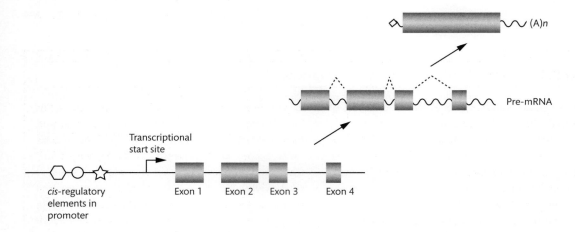

(A)n

Pre-mRNA

Transcriptional
start site

cis-regulatory
elements in
promoter

Exon 1   Exon 2   Exon 3   Exon 4

Consequences of insertion of a mobile DNA element for transcription:

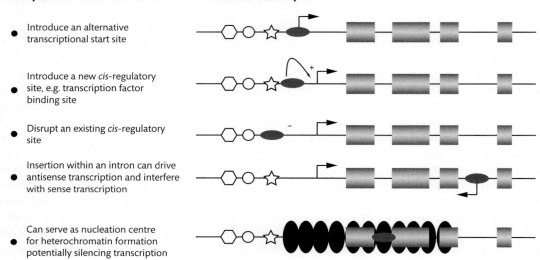

- Introduce an alternative transcriptional start site

- Introduce a new *cis*-regulatory site, e.g. transcription factor binding site

- Disrupt an existing *cis*-regulatory site

- Insertion within an intron can drive antisense transcription and interfere with sense transcription

- Can serve as nucleation centre for heterochromatin formation potentially silencing transcription

Consequences of insertion of a mobile DNA element for post-transcriptional events:

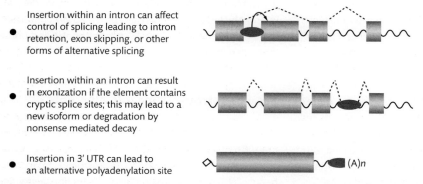

- Insertion within an intron can affect control of splicing leading to intron retention, exon skipping, or other forms of alternative splicing

- Insertion within an intron can result in exonization if the element contains cryptic splice sites; this may lead to a new isoform or degradation by nonsense mediated decay

- Insertion in 3′ UTR can lead to an alternative polyadenylation site

(A)n

Figure 8.4 Mobile DNA elements and gene expression. Schematic representation illustrating some of the ways transposable elements may directly affect local gene expression. Redrawn and reprinted by permission from Macmillan Publishers Ltd: Nature Reviews Genetics (Feschotte 2008), copyright 2008.

Box 8.3   SETMAR and gene fusion

Following transposition downstream of a SET gene (encoding a SET protein domain, characteristic of histone lysine methyltransferases and originally identified in *Drosophila*), the transposase gene from the **mariner mobile element** (MAR) was thought to have been captured and fused with the sequences encoding the SET domain leading to exonization and creation of a new intron. The result was the birth of a new chimeric gene in primates some 40–58 million years ago called *SETMAR* (Cordaux *et al.* 2006). In humans a copy is found on 3p26. In terms of function of the human SETMAR protein, this remains unclear but the SET domain has histone methyl transferase activity and the MAR domain is capable of binding to the many copies of the *Hsmar1* transposon found throughout the human genome which may serve to direct histone methylase activity (Liu *et al.* 2007).

---

self-replicating genomic parasite of the human genome' (Witherspoon *et al.* 2006). L1 elements encode their own replication machinery: full length active L1s have an RNA polymerase II promoter region, and two ORFs encoding an RNA binding protein, and a protein with **endonuclease** and **reverse transcriptase** activity – the former important for creating a nick at the point of genomic insertion (Mathias *et al.* 1991). This enables L1 elements to transpose through target-primed reverse transcription (Luan *et al.* 1993). There is a catch as completion of this process results in severe truncation of many L1 elements such that they lose the ability to catalyse their own replication. This means that there are only a small number of full length retrotransposon competent L1 copies in the human genome, so-called 'master' mobile elements approximately 6 kb long (Sassaman *et al.* 1997; Brouha *et al.* 2003). Over 500 000 copies of L1s are found in the human genome but more than 99.8% are inactive because of 5′ truncations, as well as internal rearrangements and mutations.

An average human is estimated to have between 80 and 100 retrotransposon competent L1s (Brouha *et al.* 2003). These competent L1s continue to replicate, which means within human populations the presence/absence of insertions remains polymorphic. An estimated 44% of competent L1s are reported to be polymorphic (Brouha *et al.* 2003). Human-specific L1s (L1Hs) are found, notably Ta (Transcribed, subset a). The preTa subfamily, for example, has an average age of 2.3 million years, with expansion after the divergence of humans and African apes (Boissinot *et al.* 2000; Myers *et al.* 2002; Badge *et al.* 2003; Salem *et al.* 2003b).

The consequences of L1 insertion events can be severe. Chen and colleagues reviewed L1 mediated retrotransposition events associated with human disease and identified 48 events (Chen *et al.* 2005). Overall, L1 mediated retrotransposition events are thought to account for 0.1% of known mutations leading to human genetic diseases (Chen *et al.* 2006). When 240 male patients with haemophilia A were screened for underlying mutations in the factor VIII gene, two unrelated cases were found to be due to large insertions in exon 14 of the gene (3.8 and 2.3 kb, respectively) (Kazazian *et al.* 1988). The sequences were consistent with L1 insertions, notably the 3′ portion of the L1 element including a poly(A) tract and target site duplications. In an analysis of cases of colon cancer, an L1 insertion event was found into the last exon of the tumour suppressor adenomatous polyposis coli (*APC*) gene (Miki *et al.* 1992).

Other examples include an L1 insertion in the dystrophin (*DMD*) gene as a rare cause of Duchene muscular dystrophy (Box 3.7). Two Japanese brothers have been reported in whom a 5′ truncated consensus L1 element was found inserted within exon 44 of the *DMD* gene which disrupts the process of splicing such that this exon is lost in the mRNA precursor (Narita *et al.* 1993). An L1 insertion in the retinitis pigmentosa 2 (*RP2*) gene at Xp11.4 has also been reported, leading to X-linked retinitis pigmentosa (OMIM 312600) (Schwahn *et al.* 1998). X-linked retinitis pigmentosa is a progressive disorder affecting the retina and is characterized by the constriction of visual fields and night blindness; there is significant loss of vision by the fourth decade of life and a range of mutations in the *RP* and neighbouring *RPGR* (retinitis pigmentosa GTPase regulator) genes have been identified (Pelletier *et al.* 2007). Deletions may also result from L1 insertions; 50 deletion events were identified in the human and chimpanzee genomes associated with L1 insertions, and during the primate radiation an estimated 7.5 Mb of sequence was deleted through such events (Han *et al.* 2005).

8.4 Alu elements: parasites of L1s

Alu elements are characteristic noncoding DNA sequences approximately 300 bp long and found in all primates. The name 'Alu element' derives from the fact that members of this family contain a recognition site specific to the *AluI* restriction enzyme (Houck *et al.* 1979). The origin of Alu elements is thought to involve duplication of the 7SL RNA gene some 65 million years ago, early in primate evolution (Ullu and Tschudi 1984). Other short interspersed elements are also thought to have their origins in small highly structural RNA genes (Okada and Ohshima 1993). Alu elements are considered parasites of L1 LINEs as Alu elements need the retrotransposition proteins encoded by active L1 elements in order to replicate (Kajikawa and Okada 2002; Dewannieux *et al.* 2003). In this, Alu elements have been extremely successful, with more than 1 million Alu insertions present in the human genome, equating to about one-tenth of the genomic sequence (Lander *et al.* 2001).

8.4.1 Extent and diversity of Alu elements

Alu elements constitute the largest family of mobile elements within the human genome. They are often found within introns or the 3′ untranslated regions of genes, or in intergenic regions (Batzer and Deininger 2002). Alu elements are transcribed by RNA polymerase III and contain specific promoter regions for this polymerase in their 5′ regions (denoted A and B) (see Fig. 8.2). The RNA polymerase III transcript originating from the Alu element is then reverse transcribed by necessary proteins encoded by L1s: the two groups of mobile DNA elements, L1 and Alu, compete for the same reverse transcription components. Short direct repeats are found flanking Alu elements that originate from the site of insertion. Within these, the 3′ end of the Alu element comprises a run of As (see Fig. 8.2).

Only a few Alu elements remain competent for retrotransposition (so-called 'master elements'), the remainder lack the necessary flanking sequences for activation (Ullu and Weiner 1985). Like other classes and families of transposable elements, the occurrence of Alu elements across the genome represents a fossil record of past insertions across evolutionary time. Decay of 'A'-rich Alu tails, and the high CpG content of Alu elements, minimizes retrotransposition activity even in the event of adjacent sequence activating an Alu (Batzer and Deininger 2002).

It has been possible to classify Alu elements into subfamilies based on elements accumulating mutations over time: older subfamilies have fewer subfamily-specific mutations and more random ones, in contrast to younger subfamilies. Recent Alu elements continue to replicate, generating subfamilies that show polymorphism within and between human populations (Batzer and Deininger 2002). Examples of evolutionarily young Alu subfamilies found almost exclusively in humans include Y, Yc1, Yc2, Ya5, Ya5a2, Ya8, Yb8, and Yb9. These subfamilies constitute about 5000 elements (~0.5% of all Alu repeats) and arose some 4–6 million years ago, after the separation of humans and African apes, and mostly before human migration from Africa. The majority are therefore seen to be **monomorphic** between human populations.

However, some very recently arising Alu family members are polymorphic between diverse human populations. These comprise about 1200 Alu elements for which polymorphism is seen in terms of the presence or absence of specific Alu insertions (Carroll *et al.* 2001; Roy-Engel *et al.* 2001; Batzer and Deininger 2002). Such Alu polymorphism may relate to a single individual for a *de novo* Alu insertion. In current human populations, one Alu insertion is thought to occur with every 200 new births. This contrasts with a much higher rate early in primate evolution, estimated at one insertion every birth (Deininger and Batzer 1999). An Alu insertion polymorphism may be specific to a population or be found in an ancestral population and particular subpopulations.

8.4.2 Consequences of Alu insertions

Most Alu element insertions are rapidly lost from a population due to genetic drift, particularly in larger populations. Rarely, the effects are beneficial through modulation of gene expression or alteration in coding DNA. More usually such effects are deleterious, for example through disruption of the ORF of a particular gene (Batzer and Deininger 2002). An estimated 0.1% of human genetic disorders arise due to Alu insertions (Fig. 8.5).

A striking example of the consequences of Alu insertions was found on screening 100 Japanese primary breast cancer patients: a 346 bp insertion was found in exon 22 of the *BRCA2* gene which resulted in exon skipping and early termination of translation (Miki *et al.* 1996). The insertion showed very high homology (278 out of 282 bases) to a conserved Alu subfamily with 8 bp flanking duplications of target site sequences.

An Alu insertion involving the *NF1* gene was found to be a rare cause of neurofibromatosis type 1 (OMIM 162200), a relatively common autosomal dominant disorder in which affected individuals present with cafe-au-lait spots and fibromatous skin tumours (Wallace *et al.* 1991). The Alu insertion resulted in splicing of an exon and a frameshift. A final example comes from Apert

syndrome (OMIM 101200), a rare disorder in which there is premature fusion of the cranial sutures (craniosynostosis) and digits of the hands and feet (syndactyly): two cases were found to be associated with Alu insertions in the fibroblast growth factor 2 (*FGFR2*) gene (Oldridge *et al.* 1999).

The high homology between Alu elements predisposes to non-allelic homologous recombination and an estimated further 0.3% of human genetic disease (Deininger and Batzer 1999). Such recombination may lead to deletions, duplications, or translocations (Section 5.2). Some 492 human-specific deletions have been identified as mediated by recombination between Alu elements (400 kb in total), which is responsible for much of the difference between the human and chimpanzee genomes (Sen *et al.* 2006). Specific examples of Alu elements underlying such recombination events include a deletion mediated by Alu elements in the promoter and first intron of the *HPRT1* gene at Xq26 and is a cause of Lesch–Nyhan syndrome (OMIM 300322), which is

Locus	Subfamily	Disease
CaR	Ya4	Hypocalciuric hypercalcaemia and neonatal severe hyperparathyroidism
Mlvi -2	Ya5	Associated with leukaemia
NF1	Ya5	Neurofibromatosis
PROGINS	Ya5	Linked with ovarian carcinoma
IL2RG	Ya5	XSCID
ACE	Ya5	Link with protection from heart disease
Factor IX	Ya5	Haemophilia
EYA1	Ya5	Brachio-oto-renal syndrome
FGFR2	Ya5, Yb8	Apert syndrome
Cholinesterase	Yb8	Cholinesterase deficiency
APC	Yb8	Hereditary desmoid disease
Btk	Y	X-linked agammaglobulinaemia
C1 inhibitor	Y	Complement deficiency
BRCA2	Y	Breast cancer
GK	Y	Glycerol kinase deficiency

Figure 8.5 Alu insertions and human disease. Alu insertion events have been associated with human disease. Reprinted from Deininger and Batzer (1999), with permission from Elsevier.

characterized by mental retardation, cerebral palsy, and urinary stones (Mizunuma *et al.* 2001). Other examples of disorders that can arise due to Alu mediated recombination include Tay–Sachs, familial hypercholestrolaemia, C3 deficiency, and α thalassemia (Deininger and Batzer 1999). Recently there has been evidence that Alu and L1 insertions can more directly lead to deletion events as part of the retrotransposition process (Callinan *et al.* 2005). In the human and chimpanzee genomes, 33 retrotransposition mediated deletion events were defined: over the course of primate evolution 3000 such deletion events are thought to have occurred leading to loss of a megabase of DNA.

Alu insertions are commonly associated with repeat sequences, notably microsatellites, and with single nucleotide polymorphisms (SNPs) (Batzer and Deininger 2002). Altogether an estimated 25% of simple repeat sequences are associated with Alu repeats, which are a rich substrate for such events (Jurka and Pethiyagoda 1995). The expansion of tandem repeats associated with Alu elements may lead to severe disease, a notable example being Freidreich's ataxia (Box 7.10). Over 95% of cases of Freidreich's ataxia arise due to a GAA repeat expansion within an Alu element in the first intron of the *FXN* gene at 9q13-q21.1 (Campuzano *et al.* 1996). The Alu element belongs to the AluSx subfamily and the repeat expansion reduces expression of the *FXN* gene at the mRNA level. Frataxin, encoded by *FXN*, is involved in mitochrondrial iron transport and its deficiency is associated with cellular stress. The higher incidence of SNPs in Alu insertions reflects the frequent occurrence of CpG dinucleotides in elements that are associated with a high rate of mutation (Batzer and Deininger 2002).

8.5 Mobile DNA elements and human population genetics

The discovery of recently occurring Alu insertions that are polymorphic across human populations has provided a very valuable tool to understand human population genetics and in particular our evolutionary history. Alu elements, as well as other SINEs and LINEs, have particular advantages for such studies over other forms of genetic variation such as microsatellites or SNPs. Notably, the ancestral state is known – the absence of the insertion – such that the root of a particular population tree can be determined (Batzer *et al.* 1994). With these markers it is also possible to infer identity by descent. If people share a particular Alu insertion, they have inherited it from a common ancestor, as the chances of a second Alu insertion having occurred independently at the same genomic location are extremely remote. The number of potential sites of insertion is huge and the rate of insertion low, moreover loss or reversion of an element is extremely rare (Ho *et al.* 2005a). For the study of human population genetics, Alu markers are considered to be 'homoplasy free': every observed insertion at a particular genomic locus being identical by descent to the insertion created by the original transposition event (Salem *et al.* 2003c; Witherspoon *et al.* 2006; Xing *et al.* 2007). A number of studies have investigated the diversity of recent Alu and L1 insertions in a range of different populations (Box 8.4).

8.5.1 Genetic diversity and human origins

In 1994 Batzer and colleagues analysed four human-specific polymorphic Alu insertions in 664 people from

---

Box 8.4   Diversity in recent Alu and L1 insertions

Recently integrated Ya5 and Yb8 Alu family members were found to occur almost exclusively in the human genome with 22% of family members found to be polymorphic for their presence or absence in different human populations (Carroll *et al.* 2001). In the Yg6 and Yi6 subfamilies, 9% were polymorphic (Salem *et al.* 2003a); while 12% of the human-specific Yd6 subfamily were polymorphic (Xing *et al.* 2003), and 20% of the Yb subfamily (Carter *et al.* 2004). For L1 elements, 45% of the human-specific Ta subfamily were found to be polymorphic for insertion or absence (Myers *et al.* 2002).

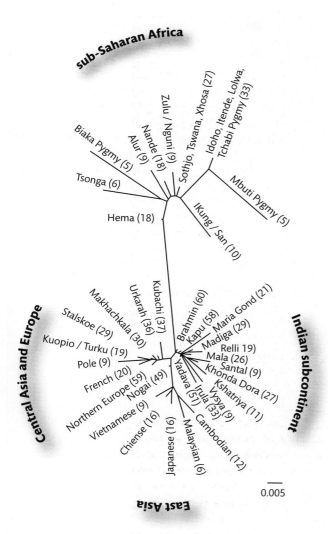

Figure 8.6 Genetic relationships between populations. A tree of 36 diverse human populations from Europe, sub-Saharan Africa, and Asia based on frequencies of 100 Alu insertions genotyped in 840 individuals (Bulayeva *et al.* 2003; Watkins *et al.* 2003). Reproduced with permission from Xing *et al.* (2007).

16 different human populations across the world (Batzer *et al.* 1994). This work supported an African ancestral origin for the polymorphic Alu insertions studied, consistent with earlier work using classical markers (Cavalli-Sforza *et al.* 1988; Nei and Roychoudhury 1993), restriction fragment length polymorphisms (RFLPs) (Wainscoat *et al.* 1986; Bowcock *et al.* 1991), microsatellites (Bowcock *et al.* 1994), and mitochondrial DNA types (Cann *et al.* 1987; Vigilant *et al.* 1991). Populations

which were geographically closer showed greater genetic similarity based on the polymorphic Alu insertions (Batzer *et al.* 1994); this was also seen in subsequent larger studies of Alu polymorphisms (Watkins *et al.* 2001, 2003) and L1 polymorphisms (Witherspoon *et al.* 2006). A study of 100 Alu insertions in 710 people from 31 different populations showed highest diversity in African populations, with the largest genetic distances among African populations, and between African and non-African

Box 8.5 Recent African Origins hypothesis

This hypothesis proposed the origins of anatomically modern humans as a small isolated ancestral population in Africa from which a number of individuals emigrated to colonize the rest of the world and replace, without significant genetic mixing, archaic (pre-modern) human populations such as the Neanderthals in Europe (Fig. 8.7) (Stringer and Andrews 1988). In terms of genetic diversity, the migrating individuals represented a limited subset of African genetic variation. Subsequently much of this diversity is thought to have been lost by genetic drift during repeated or severe transient reductions (**bottlenecks**) in population size. By contrast, the remaining African populations retained and continued to accumulate genetic diversity.

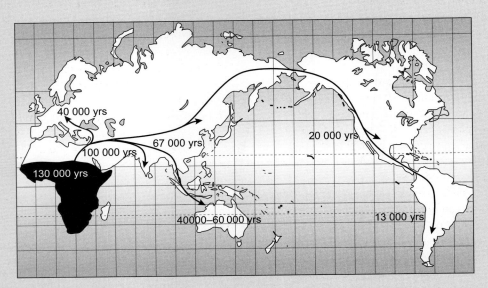

Figure 8.7 Dispersal of modern humans from sub-Saharan Africa. Earliest fossil and archaeological evidence (dates shown) supports genetic evidence for dispersal from Africa within the last 100 000 years. Reprinted by permission from Macmillan Publishers Ltd: Nature (Hedges 2000), copyright 2000.

populations (Watkins *et al.* 2003). Clustering of non-African populations is seen in such analyses, with separation from African populations by a long branch (Fig. 8.6) (Xing *et al.* 2007). The analysis of polymorphism in mobile DNA elements has provided important evidence supporting the Recent African Origins hypothesis (Box 8.5) (Xing *et al.* 2007).

The analysis of neutral human genetic diversity has provided important insights into our evolutionary past, complementing data from paleontological and archaeological research (Garrigan and Hammer 2006). Analysis of diversity in mitochondrial DNA and the Y chromosome provided significant early data to support a single origins model of anatomically modern humans in Africa (Box 8.5) (Cann *et al.* 1987; Stringer and Andrews 1988; Vigilant *et al.* 1991). More recent data analysing genetic diversity from the X chromosome, and among the different autosomes, suggest significant heterogeneity within the human genome in terms of evolutionary history and greater time to our most common ancestor (Harding *et al.* 1997; Harris and Hey 1999; Garrigan and Hammer 2006). Rather than completely replacing the archaic *Homo* populations without admixture, there may have been significant genetic contributions from these groups to anatomically modern humans.

Garrigan and Hammer (2006) describe how genetic diversity data may relate to different models for human origins. In contrast to a single origin model from a single geographically localized interbreeding population (**deme**), the multiple origins models propose gene flow among a number of ancestral demes before the transition to anatomically modern humans, or after the emergence of that population (Fig. 8.8). The unexpectedly distant origin of different haplotypes (Box 2.8) within the current human genome support the descent of anatomically modern humans from multiple archaic *Homo* subpopulations but this remains controversial. Many important questions remain as to the origins of anatomically modern humans and their subsequent global spread. The effective ancestral population size is thought to have been small, approximately 10 000 individuals (in contrast to great apes such as gorilla and orangutan). Fossil evidence supports an origin of anatomically modern humans in the last 200 000 years: hominin skulls from the Omo Valley in Ethiopia, for example, have been dated to 195 000 years ago (McDougall *et al.* 2005).

8.6 Summary

Mobile DNA elements within the human genome are in almost all cases inactive and represent a 'fossil record' in our DNA reflecting past transposition events. They constitute a remarkable 45% of human genomic sequence and have contributed greatly to the complexity and plasticity of our genome, and that of our primate and earlier ancestors. No evidence of transposase activity is found in the anthropoid lineage for 37 million years, but intense activity is believed to have taken place early in primate evolution (Pace and Feschotte 2007). The consequences of transposition can be profound, leading for example

to the generation of new genes through fusion events, as illustrated by the birth of the *SETMAR* gene in primates some 40–58 million years ago involving a mariner mobile element (Cordaux *et al.* 2006).

The dispersal and accumulation of transposable elements are believed to have had a dramatic influence on the evolution of eukaryotic genomes. They have the capacity to introduce significant diversity into the genome and the potential to serve as building blocks for coordinated regulatory networks, as illustrated by the dispersal of specific transcription factor binding sites within mobile elements for the master regulator p53 (Wang *et al.* 2007; Feschotte 2008). Elsewhere there is recent evidence of how specific SINEs may have generated important regulatory enhancer elements for the *POMC* and *ISL1* genes some 170 and 410 million years ago, respectively (Bejerano *et al.* 2006; Santangelo *et al.* 2007).

Mobile DNA elements are seen as genomic parasites, surviving as they replicate faster than the host genome in which they are found (Brookfield 2005). Within this diverse group, Alu elements are themselves parasites of LINEs, relying as they do on active L1 elements for the proteins they require to replicate. Over 99.8% of the more than 500 000 copies of L1 elements in our genome are inactive; polymorphism is recognized among human populations for the 80–100 competent L1s (Brouha *et al.* 2003). Similarly, only a small number of competent 'master' Alu elements are found among the more than 1 million Alu insertions in the human genome. A diverse range of human genetic diseases may result from either L1 or Alu element insertions – ranging from haemophilia (factor VIII gene) (Kazazian *et al.* 1988) to colon cancer (*APC* gene) (Miki *et al.* 1992) with L1s, and to breast cancer (*BRCA2*) (Miki *et al.* 1996), neurofibromatosis (*NF1*) (Wallace *et al.* 1991), and Apert syndrome (*FGFR2*) (Oldridge *et al.* 1999) with Alu insertions. Alu elements are also an important substrate for homologous recombination, which may lead to deleterious deletions, and for unstable repeats.

Polymorphism among recent Alu elements and other mobile DNA elements has been a very important tool in our understanding of human population genetics and evolutionary history, as illustrated by work on human origins and support for the Recent African Origins

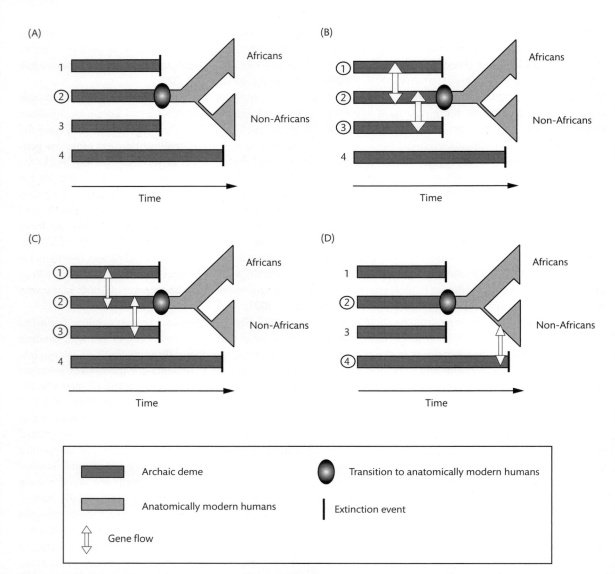

Figure 8.8 Models of human origins. Four archaic demes are shown labelled 1–4, with circles around those contributing genetic material to the anatomically modern human genome in a particular model. (A) In the single origin model, a single ancestral deme (2) gives rise to anatomically modern humans after which the African and non-African demes are shown to diverge, with bottlenecks in the latter depicted as a narrowed line followed by recent expansion. (B–D) Multiple ancestry models involve gene flow among demes either before (B and C) or after (D) the transition to anatomically modern humans. In the high migration model (B), several demes contribute equally to the anatomically modern human population and there is a high rate of gene flow between demes; in the low migration model (C) there is a single major archaic deme with others making a minor contribution through a low rate of gene flow. (D) The isolation and admixture model. Current evidence appears to favour model C. Redrawn and reprinted by permission from Macmillan Publishers Ltd: Nature Reviews Genetics (Garrigan and Hammer 2006), copyright 2006.

hypothesis (Batzer *et al.* 1994; Watkins *et al.* 2003). The ability to define the ancestral state (absence of an insertion) and infer identity by descent has been particularly helpful. Much remains to be understood about mobile DNA elements and their evolutionary history but there is no doubt that their influence on the human genome has been profound and continues to contribute to human genetic diversity.

8.7 Reviews

Reviews of subjects in this chapter can be found in the following publications:

Topic	References
Human origins and genomics	Hedges 2000; Garrigan and Hammer 2006
Mobile DNA elements and evolution	Kazazian 2004; Xing *et al.* 2007; Feschotte 2008
L1 retrotransposons	Ostertag and Kazazian 2001
Alu elements and genomic diversity	Roy-Engel *et al.* 2001; Batzer and Deininger 2002; Salem *et al.* 2003a
Alu elements and human disease	Deininger and Batzer 1999

CHAPTER 9

SNPs, HapMap, and common disease

9.1 Introduction

Over the course of this book the nature and consequences of genetic diversity have been explored. This journey has taken us from large scale structural genomic variation at a cytogenetic level, through different classes of submicroscopic structural variation, to fine scale sequence level diversity involving insertions and deletions of one or more contiguous nucleotides, or single nucleotide substitutions. In this chapter the extent and nature of genetic variation involving single nucleotide substitutions are reviewed in more detail, in particular the role of single nucleotide polymorphisms (SNPs) as genetic markers for association studies, a topic introduced in Chapter 2.

SNPs have been of great utility as genetic markers as they occur very commonly across the genome and can be genotyped using ultrahigh throughput technologies. As will be discussed over the course of this chapter, definition of the nature and genomic architecture of fine scale nucleotide diversity has allowed informative common SNPs to be employed at high density as genetic markers in genome-wide association studies. In some instances within these SNP panels, specific functional disease-associated variants will have been directly genotyped, but much more often the most strongly disease associated SNPs will be serving as genetic markers for additional variation in linkage disequilibrium (Box 2.8) with the genotyped variants. As we saw with the globin genes (Chapter 1), the potential functional consequences of sequence level diversity are very wide ranging, from effects on the structure or function of the encoded protein to differences in gene expression (reviewed in more detail in Chapter 11).

The successful application of genome-wide association studies to common multifactorial traits has radically advanced our ability to define the genetic determinants of common disease. In this chapter, the background to genome-wide association studies is described together with the insights and lessons which have been learned and the challenges that remain. An in depth review of two common diseases, age-related macular degeneration and Crohn's disease, is then presented which illustrate the historical and current approaches to dissecting genetic factors in multifactorial traits.

9.2 SNPs, association, and genetic susceptibility to common disease

The definition of a single nucleotide polymorphism or 'SNP' used here is of a single nucleotide substitution present in the human population with a frequency of both alleles of greater than 1%. These are 'common' genetic variants and as such were favoured for use as genetic markers, particularly as the vast majority of such single nucleotide changes are biallelic which facilitates high throughput genotyping. Fine scale DNA sequence variation also includes deletions or insertions of one or more nucleotides: in many cases it is not possible to determine whether there has been gain or loss of nucleotides and the variant is described as an insertion/deletion or 'indel' (Box 6.10). A DNA sequence variant may involve a single sequence, sometimes referred to as 'simple mutations', or rarely involve exchange between allelic or non-allelic sequences, as seen for example with unequal crossing over or gene conversion events common at segmental duplications (Chapter 6) and tandem repeats (Chapter 7).

Almost all SNPs are thought to have arisen as a unique mutation event given the very low rate of mutation (10^{-8} per site per generation) relative to the estimated 10 000 generations to a common ancestor of any two humans (TIHMP 2003). A newly occurring single nucleotide variant will not be found on a sequence devoid of variation, but rather will occur on a chromosome bearing many different variants reflecting the preceding history of mutations arising over time across generations and indeed species. The particular combination of variants (alleles) coinherited on a given chromosome or part of a chromosome is described as the haplotype (Box 2.8). Over time the haplotype will begin to break down as recombination occurs and further diversity arises, forming new haplotypes. The degree of association or correlation between any two variants (alleles) in a population is described by measures of linkage disequilibrium (Box 2.8). With increasing genomic distance between SNPs, the degree of association or linkage disequilibrium will typically reduce. Across the genome, linkage disequilibrium varies dependent on the particular chromosomal region and human population studied, but as meiotic recombination is concentrated in so-called **recombination hotspots** (Section 9.2.5), linkage disequilibrium between SNPs often persist.

For rare diseases showing a mendelian pattern of inheritance with a well defined phenotype arising due to rare variants with high penetrance, linkage analysis and positional cloning have been highly successful approaches (Section 2.3). By contrast, much less success was achieved using such approaches for common diseases without a clear pattern of inheritance such as diabetes or asthma, which involve to varying degrees both a genetic predisposition and environmental factors. In such diseases there was a growing awareness that multiple genetic loci were likely to be involved with a given disease phenotype that individually had a modest magnitude of effect.

Different strategic approaches for resolving genetic factors influencing common disease traits have been advocated, ranging from a 'direct' approach to catalogue and compare all diversity through genome-wide resequencing, to an 'indirect' approach using common informative genetic markers to capture the majority of variation and define regions of association (Collins *et al.* 1997; Botstein and Risch 2003; TIHMP 2003). The latter approach was adopted by the International HapMap Project (Section 9.2.4), and the analysis of common variants informative for underlying haplotypic structure has enjoyed considerable recent success in genome-wide association studies. However, the concept of 'common disease, common variant' has remained controversial and rare sequence variants – as well as structural genomic variants – are increasingly recognized to play an important role. The rapid advances in high throughput, low cost, 'next generation' sequencing technologies (Box 1.19) is enabling extensive resequencing studies that should advance still further our ability to discover genetic variation and understand its relationship with disease.

9.2.1 Strategic approaches

As already described, the successes of linkage-based approaches for the elucidation of genetic determinants of mendelian diseases was unfortunately not mirrored to the same extent in common multifactorial traits showing complex inheritance. Many linkage scans were performed predominantly using affected sibling pairs, but only relatively few reproducible loci were demonstrated. The power of linkage studies based on allele sharing by descent among affected relatives was thought to be low when effect sizes attributable to genetic variants were small, and the variants themselves were present at high frequency (Kruglyak 2008). This appeared to be the situation with common diseases, in contrast to mendelian disorders. Some affected individuals will not necessarily possess the risk allele and indeed have developed the disease as a result of other risk factors (including distinct genetic variants), while the relatively high frequency of the allelic variant (in contrast to the vary rare variants characteristically seen in mendelian disorders) allows for occurrence in a family through multiple founders and loss of the inheritance pattern (Kruglyak 2008).

The recognition that multiple genomic loci were likely to be involved in susceptibility to common multifactorial traits due to variants present at relatively high frequency with an individually small magnitude of effect led to proposals that association studies would be a more powerful approach than family-based linkage studies, and the development of the 'common disease, common variant' hypothesis (Lander 1996; Risch and Merikangas 1996). In 1997 Collins, Guyer, and Charkravarti described the power of a systematic cataloguing of DNA sequence variation for such proposed studies with the aim of achieving a genome-wide level of association (Collins *et al.* 1997). At the time the most likely source of functional variants was proposed to be in coding DNA and a direct approach was advocated to try and catalogue all such variants, anticipating that the causative functional variant would therefore be included in the association study. It was recognized that some functional variants would lie outside coding DNA, although it is only more recently that the extent to which this is true has become apparent and the myriad ways in which genetic diversity can have functional consequences.

In parallel, Collins and colleagues proposed an indirect approach based on establishing dense maps of SNPs and making use of the effects of linkage disequilibrium to indirectly find association with causative variants (Collins *et al.* 1997). The functional variant did not need to be directly genotyped as linkage disequilibrium allowed association with genotyped SNPs to be observed through linkage with the ungenotyped functional variant. Common sequence variants in the form of SNPs would provide effective markers to resolve such loci but this would require a much denser set of SNPs than was currently available. The power of linkage disequilibrium mapping in resolving specific disease genes had been previously demonstrated in mendelian diseases, including fine mapping of specific genes after conventional linkage analysis and directly mapping disease genes (Section 2.3.3). Indeed the use of linkage disequilibrium mapping for the analysis of complex traits had been proposed as early as 1986 by Lander and Botstein but required the establishment of an informative set of genomic markers (Lander and Botstein 1986). The proposal now was for an immediate large scale effort to catalogue sequence diversity, in particular SNPs, with an emphasis on coding variants.

It was envisaged that this would require a consortium effort to establish a dense map of at least 100 000 SNPs with data to be publically deposited and freely available. In 1999 the SNP Consortium was launched, followed in 2002 by the International HapMap Project (Section 9.2.4), a remarkable collaborative research effort to catalogue SNP diversity across the genome for a number of human populations, capturing patterns of common human sequence variation, and so informing and enabling future genetic studies of common human disease (Manolio *et al.* 2008). A number of other studies have contributed to generating the dense maps of common SNP diversity now available for genome-wide analysis of sequence diversity – notably from Perlegen Sciences, which in 2005 reported data on 1.6 million SNPs analysed among a panel of 71 unrelated Americans of European, African, and Asian ancestry (Hinds *et al.* 2005).

9.2.2 Surveying SNP diversity: lessons from the SNP Consortium and Human Genome Project

Prior to publication of the draft human genome sequence in 2001, a number of studies demonstrated that large scale identification of SNPs and their high throughput genotyping was possible. In 1998, Lander and colleagues published a large scale SNP analysis in which they demonstrated the utility of resequencing strategies and DNA 'chip' arrays for SNP discovery, while also showing how such arrays could be extended to allow high throughput genotyping (Wang *et al.* 1998). Resequencing over 279 kb of sequence using Sanger dideoxy methodology with fluorescently labelled primers (Box 1.10) among three individuals and a pool of ten individuals, revealed one SNP per kilobase with an excess of SNPs observed involving CpG dinucleotides, and transitions present twice as often as transversions. A total of 23 indels were also found. High density chips containing short DNA probes possessing the four possible nucleotide bases at their centre for a given genomic position were also used by the authors for SNP discovery. Screening 2 Mb of DNA for seven individuals yielded 2748 candidate SNPs (one per 721 bp) and was found to be highly accurate (Wang *et al.* 1998).

Importantly, this study also demonstrated how such DNA chips, designed to interrogate the two alternate SNP alleles, could be used for high throughput genotyping allowing the detection of allelic variants for 500 SNPs within a given DNA sample (Wang *et al.* 1998). Over the subsequent 10 years, such array-based approaches were to be scaled up massively with 1 million SNPs accurately genotyped on commercial genotyping arrays at a fraction of the original cost.

As the Human Genome Project advanced, increasingly large SNP surveys were published. The publication in 1999 of the sequence of chromosome 22 (Dunham *et al.* 1999) was followed by an annotated SNP map of that chromosome, comprising a total of 2370 SNPs (or one per 12 kb of sequence) (Mullikin *et al.* 2000). This study by Bentley and colleagues utilized a 'reduced representational shotgun strategy' and a 'genomic alignment strategy'. The former involved sequencing DNA clones from a library generated from DNA fragments originating from the genomic DNA of seven individuals subjected to restriction enzyme digestion (Mullikin *et al.* 2000). Repeated sampling and sequencing allowed comparisons of sequence differences to be resolved with 455 candidate SNPs identified (one per 4.8 kb of raw data). Genomic alignment of the results of such sequencing with the available finished genomic sequence further increased the efficiency of SNP detection, with 914 SNPs identified. Such genomic alignments using 'reads' from shotgun sequencing (Section 1.4.2) enriched the SNP set for chromosome 22, allowing annotation and construction of a SNP map containing 2370 SNPs (Mullikin *et al.* 2000). This work, together with other studies, set in progress the SNP Consortium with a goal of providing 300 000 SNPs as a publically available resource.

The utility of a reduced representational shotgun sequencing approach combined with genomic alignment at a genome-wide level was demonstrated by Altshuler and colleagues in the same year as part of the ongoing SNP Consortium (Altshuler *et al.* 2000b). From a panel of 24 ethnically diverse individuals, 20 libraries were prepared using four different restriction enzymes and 47 172 SNPs were identified (Altshuler *et al.* 2000b). By November 2000, the number of candidate SNPs identified by the SNP Consortium was 1 023 950. This SNP set, combined with the data available as a result of the International Human Genome Sequencing Consortium (some 971 077 SNPs) and a much smaller proportion of SNPs from other gene-based studies (about 5% of total), led to the publication of 1.42 million SNPs across 2.7 Gb of assembled genome sequence providing a SNP density map of one SNP per 1.9 kb (Sachidanandam *et al.* 2001).

The Human Genome Project allowed for SNP discovery through comparison of DNA sequence at sites of clone overlaps as the clones originated from a diverse panel of individuals (Section 1.4.4). Overall, the SNP density was similar across autosomes with lower diversity at the sex chromosomes. At the X chromosome this was postulated to relate to lower effective population sizes

and mutation rates; while at the Y chromosome markedly lower heterozygosity was seen as anticipated, notably the 'non-recombining Y region' where only 348 SNPs were mapped over 2 304 916 bp (Sachidanandam *et al.* 2001). The predicted sequence diversity for the Y chromosome compared to autosomes was 31%; the observed value was considerably lower at 20% (Sachidanandam *et al.* 2001). This contrasted with other genomic loci such as the major histocompatibility complex (MHC) on chromosome 6p21 where high heterozygosity was seen (Section 12.3.1). Within the 1.42 million SNPs, an estimated 60 000 were located in exonic regions.

9.2.3 Haplotype blocks and haplotype tagging SNPs

At this time data was published showing that the human genome showed regions of strong allelic association or linkage disequilibrium, interspersed with sites where association broke down, which correlated with hotspots of meiotic recombination (Daly *et al.* 2001; Jeffreys *et al.* 2001). This '**haplotype block structure**' was investigated among individuals from diverse population groups by Gabriel and colleagues in order to define in more detail the extent and nature of such variation, and whether it was possible to define it using common SNP markers (Gabriel *et al.* 2002). A total of 51 autosomal regions, each 250 kb in size, were successfully investigated among 275 individuals by genotyping 3728 SNPs. The regions were selected to be evenly spaced across the genome and to have an average density of one candidate SNP (discovered by the SNP Consortium) per 2 kb in their core region of 150 kb.

Two populations of African ancestry were analysed, 30 parent–child trios from Yoruba Nigeria and 50 unrelated African Americans; 93 individuals of European origin and 42 unrelated Japanese and Chinese individuals were also genotyped. Haplotype blocks were defined based on regions 'over which less than 5% of comparisons between informative SNP pairs showed strong evidence of recombination' (Gabriel *et al.* 2002). A total of 928 blocks were defined in the four population groups, with a mean block size of 11 kb among those of African ancestry (ranging from less than 1 to 94 kb) and of 22 kb in those of European or Asian ancestry (<1–173 kb). In terms of genomic coverage, half of the genome was thought to be made up of blocks of 22 kb or larger in those of African ancestry, and of blocks greater than 44 kb in those of European or Asian ancestry.

Within the haplotype blocks, only three to five common haplotypes were found (in this context, 'common' refers to a haplotype frequency of greater than 5%). This diversity was highest in those of African ancestry among whom, on average, five common haplotypes were present and was lowest among the Asian population studied (3.5 common haplotypes), with those of European ancestry having, on average, 4.2 common haplotypes. These numbers are considerably less than predicted based on a random combination of SNP alleles; for example seven SNPs in principle generating 128 different haplotypes in the absence of association (2^7 where n biallelic SNPs generate 2^n haplotypes) (Fig. 2.2). Across population groups, the observed block structure and haplotypes were very similar, haplotypes being identical in all three populations in half of the cases – and where present in only one population (28% of haplotypes), these were almost all (90%) found in the Yoruban sample. The differences in linkage disequilibrium between populations were felt most likely to be due to differences in demographic history with the observed variation in polymorphism, sites of recombination, and haplotypes all considered consistent with an 'out of Africa' origin for those of European and Asian descent (Box 8.5) (Gabriel et al. 2002).

This study also provided important evidence to support the proposal that it should be possible to genotype a limited number of common SNPs ('**haplotype tagging SNPs**') in order to capture much of the haplotypic diversity in a population (Gabriel et al. 2002). This was shown to be the case, but highlighted the need for a dense set of markers based on a large sample size to do so with confidence. At this time, it was thought that between 300 000 and 1 000 000 haplotype tagging SNPs would be likely to be required for a well powered association study in populations of non-African and African ancestry, respectively (Gabriel et al. 2002).

9.2.4 The International HapMap Project

In October 2002 the International HapMap Project was launched, aiming to produce a route map of genetic diversity to facilitate studies to investigate the genetic basis of disease (TIHMP 2003). The investigators sought to generate a genome-wide SNP map, comprising some 1 million common SNPs (present at greater than 5% frequency in the human population). The HapMap Project would use this map of common genetic diversity to establish the haplotypic structure across different genomic regions in diverse human populations such that in future studies investigating specific disease phenotypes, only a subset of SNPs, the 'haplotype tagging' SNPs, would need to be genotyped in order to capture a similar level of diversity (Fig. 9.1) (TIHMP 2003).

At the time it was estimated that there were 10 million SNPs present across the genome (having a minor allele frequency of greater than 1%), which comprised 90% of the variation seen (Kruglyak and Nickerson 2001): it was anticipated that between 200 000 and 1 million tag SNPs, selected on population haplotypic structure, would be needed (Kruglyak 1999; TIHMP 2003). The HapMap Project thus had a primarily pragmatic goal, to enormously facilitate and enable disease association and linkage studies for candidate gene loci and, it was hoped, genome-wide analysis. By significantly reducing the number of SNPs that would need to be genotyped, it was hoped that such large scale projects might become tractable. The investigators also anticipated that an analysis of diversity at the nucleotide and haplotypic level within and among populations would prove informative across a range of disciplines, as proved to be the case. Indeed the HapMap Project was to develop into a remarkable resource, facilitating research to advance our understanding of genetic diversity and its consequences in ways unanticipated at the outset of the work.

Initially, 270 individuals from four different geographically disparate populations were selected to be genotyped in order to capture as much common variation as possible within the human population and so make the data as potentially broad ranging in their application to future medical studies worldwide as possible (TIHMP 2003). The initial goal was to genotype 1 million SNPs. It was anticipated that genotyping 45 individuals for a given population would provide data on 99% of haplotypes of greater than 5% frequency. The study included individuals of European, African, and Asian ancestry (Box 9.1), both unrelated and family trios. It was hoped that the latter would facilitate establishing accuracy of genotyping, as well as having other advantages in terms of analysis. No data on the phenotype of the individuals recruited were to be recorded and all samples were anonymized.

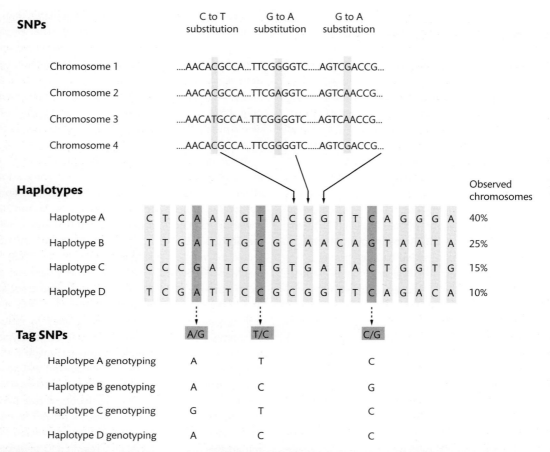

Figure 9.1 Use of tag SNPs to determine common haplotypes. Three SNPs are shown together with flanking sequences for four chromosomes. When a panel of SNPs is genotyped for the population, commonly occurring haplotypes can be defined (here haplotypes A–D are shown but there will be other rarer haplotypes, in this example comprising 10% of observed chromosomes). The three original SNP alleles shown form part of these haplotypes. Selection of informative tag SNPs allows the common haplotypes to be uniquely defined and so significantly reduces the number of SNPs that need to be genotyped to gain the majority of information about SNP diversity at the locus. Redrawn and reprinted by permission from Macmillan Publishers Ltd: Nature (TIHMP 2003), copyright 2003.

The HapMap Project was international in its aims of understanding human diversity among different populations and brought together many different scientific groups into an international collaborative effort. All data were to be publically available through a data coordination centre (www.hapmap.org) and deposition in the dbSNP database (www.ncbi.nlm.nih.gov/SNP). Resources generated as part of the project, such as DNA and immortalized cell lines established from the individuals genotyped in the different populations, were to be available for study from the Coriell Institute for Medical Research (www.coriell.org). Five high throughput genotyping technologies were initially used within the consortium with a randomly selected panel of 1500 SNPs for validating assays; a random selection of SNPs were also regenotyped and internal controls used.

In 2005, Phase I of the International HapMap Project was published, providing publically accessible genotyping data on 1 million SNPs for 269 individuals from four populations (Altshuler *et al.* 2005). Of these SNPs, the

---

Box 9.1 Phase I HapMap populations

For phase I of the International HapMap Project, 270 individuals from four different populations were recruited. Two groups of individuals of Asian ancestry in geographic terms were genotyped: 45 unrelated Han Chinese from Beijing, China, denoted the 'CHB' population, and 45 Japanese from Tokyo, Japan ('JPT'). For the other two populations studied, of African and European ancestry,

each comprised of 30 parent–offspring trios (a total of 90 individuals): one cohort was from Yoruba in Ibadan, Nigeria ('YRI') and the other from Utah, USA ('CEU'), the latter being part of the Centre d'Etude du Polymorphisme Humain (CEPH) collection of reference families established in 1984 and used for constructing genetic maps of the human genome (www.cephb.fr).

---

number that were polymorphic varied between populations, being highest in those of African ancestry (85% polymorphic in YRI) and lowest in those of Asian origin (75% in CHB/JPT); among those of European origin intermediate levels were found (79% in CEU). There were very few fixed differences, with alternate alleles seen only in particular population panels: 11 such differences were reported for example between CEU and YRI populations. The dataset was remarkably accurate (99.7%) and complete (99.3%), with for most genomic regions a common SNP (with a minor allele frequency of greater than 5%) genotyped every 5 kb. Analysis of parent–offspring trios showed that the statistical methods used for the reconstruction of haplotypes was very accurate and high quality, long range haplotypes were established.

The study complimented the genotyping dataset by publishing the results of resequencing ten regions of 500 kb in 48 individuals (16 YRI, 16 CEU, 8 CHB, 8 JPT) to try and capture all sequence diversity (Altshuler *et al.* 2005). These ten regions formed part of the genomic regions analysed in the ENCODE (ENCyclopedia Of DNA Elements) Project, which aimed to define the relationship between DNA sequence and functional regulation (www.genome.gov/10005107) (TEPC 2004). A total of 17 944 single nucleotide variants were identified over 5 Mb of sequence, equating to one every 279 bp: most were rare, with 45% having a minor allele frequency of less than 5%, and 9% of variants were found in only a single individual. However, the dataset confirmed that common SNPs (having a minor allele frequency above 5%) comprised 90% of heterozygous sites and that the strategy of analysing a limited set of informative common SNPs would

capture the majority of diversity. More low frequency variants were seen in the individuals of African ancestry (YRI), consistent with earlier reports and the concept of population (genetic) bottlenecks in the non-African populations (Section 8.5.1). The dataset also allowed estimation of recombination rates and hotspots, with one hotspot identified on average per 57 kb and 80% of all recombination estimated to have occurred in 15% of the sequence.

The resequencing data from the ENCODE regions highlighted the haplotype 'block' structure of linkage disequilibrium (Fig. 9.2) with most sequence in blocks of four or more sequence variants. The average block encompassed many such variants (on average 30–70), with an average of 4.0 (Asian CHB + JPT) to 5.6 (African YRI) common haplotypes per block (Altshuler *et al.* 2005). These data validated the strategic approach and utility of the much sparser genome-wide dataset of common SNPs genotyped in Phase I HapMap. In terms of tag SNP selection, when working with the ENCODE set of variants, SNP density could be reduced by 75–90%. If common SNPs were selected progressively until the tag SNP set encompassed SNPs highly correlated with all common SNPs, it was possible to reduce the density of genotyping to one SNP every 2 kb (in YRI) or one every 5 kb for the other populations studied. For the full Phase I HapMap dataset of common SNPs, the number of tag SNPs required to capture common SNP diversity could be reduced by between one-third (Asian panel) and one-half (African population panel) (Fig. 9.3).

The extent of long range haplotypes varied, and the dataset allowed comparison of unique haplotypes to a

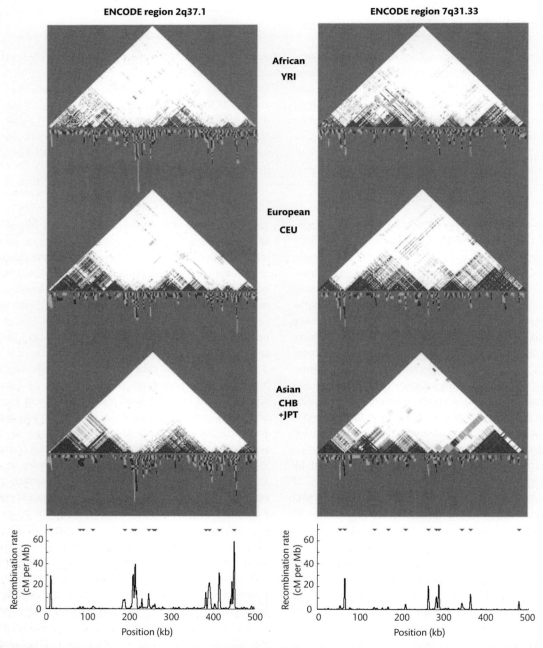

Figure 9.2 Haplotype block structure revealed in the analysis of linkage disequilibrium and recombination for two ENCODE regions. D´ plots are shown for two ENCODE regions according to geographic ancestry of individuals; D´ provides a measure of linkage disequilibrium (Box 2.8) where white is D´ <1 with a lod score of <2, with the darkest shading indicates D´ = 1 with a lod score of >2. Recombination hotspots are indicated by inverted triangles in the lower portion of figure where estimated recombination rates are shown. Reprinted by permission from Macmillan Publishers Ltd: Nature (Altshuler *et al.* 2005), copyright 2005.

Phase I HapMap (1 million SNPs)

r^2 threshold	YRI	CEU	CHB +JPT
$r^2 \geq 0.5$	324 865	178 501	159 029
$r^2 \geq 0.8$	474 409	293 835	259 779
$r^2 = 1.0$	604 886	447 579	434 476

Phase II HapMap (3.1 million SNPs)

r^2 threshold	YRI	CEU	CHB +JPT
$r^2 \geq 0.5$	627 458	290 969	277 831
$r^2 \geq 0.8$	1 093 422	552 853	520 111
$r^2 = 1.0$	1 616 739	1 024 665	1 078 959

Figure 9.3 Tag SNP selection. For the Phase I and Phase II HapMap datasets, the number of tag SNPs required to capture genotyped common SNP diversity is shown for different population panels. r^2 is the square of the correlation coefficient between two SNPs where $r^2 = 1$ indicates complete correlation (Fig. 2.2). Reprinted by permission from Macmillan Publishers Ltd: Nature (Altshuler *et al.* 2005), copyright 2005; Nature (Frazer *et al.* 2007), copyright 2007.

recombination map (Fig. 9.4). For some genomic regions such as around centromeres, the absence of recombination meant haplotypes involving more than 100 SNPs were found across megabase regions of DNA. As anticipated, linkage disequilibrium was found to be higher close to centromeres and low near telomeres; it correlated with chromosome length but also varied depending on gene density and function. Genes involved in immunity, for example, were associated with regions of low linkage disequilibrium, while strong linkage disequilibrium was associated with genes involved in cell cycling and other fundamental cellular processes.

The HapMap dataset was also a rich source of information about the possible action of natural selection on genetic diversity with 'signatures of selection' recognized (Section 10.2). These included SNPs showing extreme variation in frequency between populations, with 926 such SNPs identified. There were 19 regions showing evidence of a '**selective sweep**' with all diversity lost except for a particular allele that had risen to very high levels in the population such that the original mutated allele became 'fixed' and the only allele seen in the population, for example *LCT* encoding lactase which had been previously shown to display such effects (Section 10.4) (Enattah *et al.* 2002; Bersaglieri *et al.* 2004), and long

haplotypes that were candidates for being subject to selection which might be balancing or not sufficient to lead to fixation (Altshuler *et al.* 2005).

In 2007, Phase II HapMap was published in which genotyping data were made available on an additional 2.1 million SNPs for 270 individuals in the four population panels (CEU, YRI, CHB, and JPT; see Box 9.1), taking the total number of genotypes available for each individual to more than 3.1 million (Frazer *et al.* 2007). The coverage of common variants was markedly increased. For example, on reanalysis of the resequenced ENCODE regions, the Phase II SNP set captured common variation to a very high degree (the square of the correlation coefficient, $r^2 = 0.90$ in YRI, and 0.96 in CEU; in phase I $r^2 = 0.67$ for YRI). The number of tag SNPs required increased by two-fold despite the number of genotyped SNPs increasing three-fold (see Fig. 9.3). This remarkable dataset of common SNP variation across the genome was estimated to have a SNP every 1.1 kb and to represent 25–35% of all common SNPs genome-wide based on the assembled genome sequence (Frazer *et al.* 2007).

The SNP set provided increased resolution, notably in the YRI population of African ancestry, with finer scale haplotypic structure resolved (Fig. 9.5). Representation of rarer variants was better than in Phase I, as was

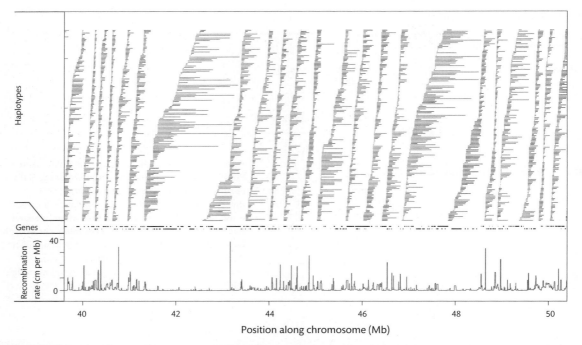

Figure 9.4 Illustration of Phase I HapMap data for chromosome 19q13. Haplotypes with a frequency of at least 5% are shown together with the positions of genes and recombination rate across the region. Reprinted by permission from Macmillan Publishers Ltd: Nature (Altshuler *et al.* 2005), copyright 2005.

definition of recombination hotspots – with nearly 33 000 identified. The HapMap Project continues: in May 2008 a preliminary data release was made for Phase III HapMap in which seven additional populations have been genotyped to increase further the diversity of individuals within the study (www.hapmap.org) (Fig. 9.6). For genome-wide association studies 500 000 SNPs are currently considered to be required for non-African populations and 1 000 000 SNPs for African populations (Kruglyak 2008).

The HapMap Project aimed to deliver a catalogue of patterns of common genetic diversity using SNP markers to inform and enable genome-wide studies. In this aim the work was remarkably successful (Section 9.3). The HapMap Project and other genome-wide SNP studies proved extremely informative about the nature and extent of variation within and between populations allowing significant advances in our understanding of the haplotypic structure of the human genome, and the nature and extent of linkage disequilibrium and recombination

hotspots (Section 9.2.5). These datasets also allowed investigators to identify and advance our understanding of genomic loci subject to positive selection (Section 10.6), of the extent of structural genomic variation (Section 4.2), and into the identification of functionally important genetic variants modulating gene expression (Section 11.4.1). Methodological advances have also been facilitated, notably in terms of **imputation** of SNP genotypes not genotyped on the particular genotyping platform used in the genome-wide association scan but which could be inferred from SNPs previously genotyped using HapMap and other datasets (Marchini *et al.* 2007).

SNPs and other simple nucleotide level diversity are now curated and assembled in a number of online databases (Box 9.2), notably the National Center for Biotechnology Information (NCBI) variation database 'dbSNP'. This is a remarkable catalogue of fine scale sequence diversity including SNPs, deletion/insertion polymorphisms (DIPs), and short tandem repeats (microsatellites). There is no minimum minor allele

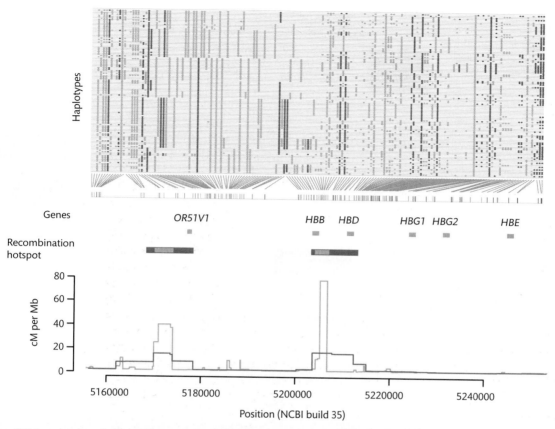

Figure 9.5 Increased resolution of Phase II HapMap illustrated for the *HBB* locus using data for individuals of African ancestry from the YRI population panel. Haplotypic structure shows increased resolution (SNPs genotyped in Phase I are shown in dark grey, additional SNPs in Phase II in light grey). Recombination rates are shown in the line graph (dark line, Phase I, light line, Phase II) with the location of hotspots shown above. Reprinted by permission from Macmillan Publishers Ltd: Nature (Frazer *et al.* 2007), copyright 2007.

frequency for variant inclusion or any functional assumption made, such that both clinically important disease-causing 'mutations' and 'neutral' variants are included. A unique reference identifier is given (the rs number) for a given variant, together with flanking sequences and population allele frequency data, and a range of links (Fig. 1.14).

9.2.5 Large scale SNP mapping and insights into recombination

Allelic recombination between homologous chromosomes at meiosis is a major source of genetic diversity.

Non-allelic homologous recombination also occurs but much more rarely and may be associated with chromosomal rearrangements and genomic disorders (Section 5.2). A number of studies established that recombination rates were highly variable across the genome in human populations. For example, analysis of family pedigrees in the Icelandic population using a dense set of microsatellite markers demonstrated this at a large scale resolution (Fig. 9.7A) (Kong *et al.* 2002). Variation was observed within individual chromosomes and between chromosomes. Recombination rates were much higher in shorter chromosomes, for example chromosomes 21 and 22 exhibited rates twice as high as chromosomes 1 and 2. Finer resolution

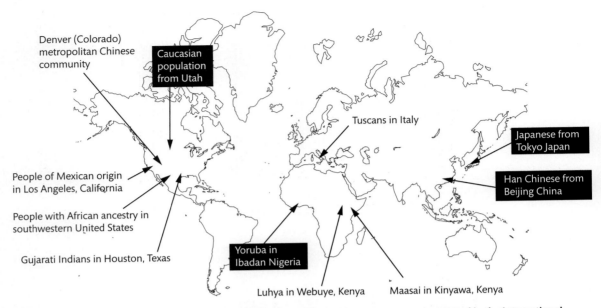

Figure 9.6 HapMap population panels. Individuals of different geographic ancestry were genotyped in the International HapMap Project: the four populations included in Phase I and II are shown in black boxes; those in Phase III in plain black text.

Box 9.2   SNP-related databases

- dbSNP (http://www.ncbi.nlm.nih.gov/SNP/) (Sherry *et al.* 1999, 2001).
- OMIM: Online Mendelian Inheritance in Man (http://www.ncbi.nlm.nih.gov/sites/entrez?db=omim) (Hamosh *et al.* 2005).
- Human Genome Variation Database (http://www.hgvbase.org/) (Fredman *et al.* 2004).
- Human gene mutation database at the Institute of Medical Genetics in Cardiff (http://www.hgmd.cf.ac.uk/ac/index.php) (Stenson *et al.* 2003).
- Population-specific, for example the Finnish disease database: disease mutations are described for 35 monogenic diseases occurring more frequently in Finland than elsewhere in the world due to founder effects and genetic isolation (http://www.findis.org/) (Peltonen *et al.* 1999).
- Locus-specific mutation databases (http://www.hgvs.org/dblist/glsdb .html) (Claustres *et al.* 2002), for example of the *ABCA4* gene encoding a retina-specific ATP-binding cassette (ABC) transporter protein (http://www.retina-international.org/sci-news/abcrmut.htm) curated by Retina International.
- Disease centred central mutation databases (http://www.genomic.unimelb.edu.au/mdi/dblist/disease.html).

---

was achieved using sperm studies of crossover events with specific recombination hotspots resolved including within the MHC on chromosome 6p21 (Jeffreys *et al.* 2001), the β globin locus (Smith *et al.* 1998), or Duchenne muscular dystrophy gene (Grimm *et al.* 1989) (Fig. 9.7B).

The high density genome-wide set of common SNP markers now available as a result of the International HapMap Project (Altshuler *et al.* 2005; Frazer *et al.* 2007) and other large scale surveys of common SNP diversity (Hinds et al. 2005) allowed, for the first time,

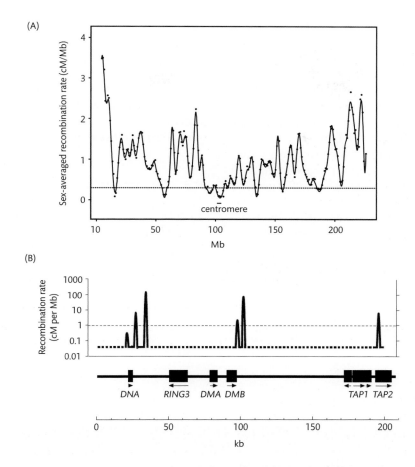

Figure 9.7 Broad and fine scale resolution of recombination rates. (A) Chromosomal level analysis based on microsatellite markers and family pedigree analysis. Data for sex averaged recombination rates are shown for chromosome 3. Reprinted by permission from Macmillan Publishers Ltd: Nature Genetics (Kong *et al.* 2002), copyright 2002. (B) Fine scale mapping in the MHC class II region based on sperm crossover activity. Reprinted by permission from Macmillan Publishers Ltd: Nature Genetics (Jeffreys *et al.* 2001), copyright 2001.

recombination rates to be defined across the genome at a fine scale of resolution. Using a coalescent-based method to estimate variation in rates of recombination, the degree of fine scale variation was found to be very extensive (McVean *et al.* 2004). A genome-wide map of recombination was derived by Myers and colleagues using data on 1.6 million common SNPs for 24 unrelated Americans of European ancestry, 23 African Americans, and 24 American Han Chinese (Hinds *et al.* 2005; Myers *et al.* 2005). This revealed marked local variation, with recombination being focused into narrow 'hotspots' such that more than 80% of recombination was found to occur in 10–20% of genomic sequence (Myers *et al.* 2005) (Fig. 9.8). In total more than 25 000 recombination hotspots were found across the genome (Myers *et al.* 2005).

The denser SNP set and larger population panels within Phase II HapMap meant that this estimate has since risen to nearly 33 000 recombination hotspots (Frazer *et al.* 2007). Recombination hotspots are found on average every 50–100 kb with similar rates of recombination estimated for males and females. Recombination is noted to occur preferentially outside but near to transcribed regions: peaks were noted 5′ to the transcriptional start site and 3′ to the gene, the latter being more pronounced (Fig. 9.9A) (Frazer *et al.* 2007). Striking differences in rates of recombination were noted between gene classes when gene function was considered, being highest in genes involved in defence and immunity (1.9 cM per megabase) and lowest among genes involved in essential cellular

functions such as chaperones (Fig. 9.9B) (Frazer *et al.* 2007). Other characteristics of recombination hotspots include enrichment for GC content and particular mobile DNA elements such as the THE1A/B retrotransposons: when present within recombination hotspots, these elements show marked enrichment for a particular nucleotide sequence CCTCCCT (Fig. 9.9) (Myers *et al.* 2005; Frazer *et al.* 2007).

9.3 Genome-wide association studies

By 2006, genome-wide association scans had become a feasible approach to investigate the role of common sequence diversity in multifactorial diseases and traits. Reaching this position was the result of a number of different factors, notably the elucidation of the nature

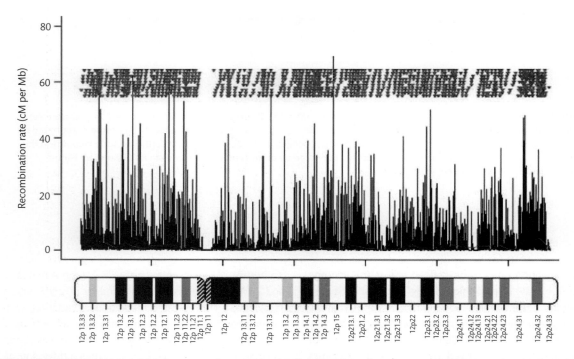

Figure 9.8 Variation in recombination across chromosome 12. The recombination rate is plotted on the *y* axis, with genomic location along the *x* axis indicated by the ideogram of chromosome 12 banding shown below it. From Myers *et al.* (2005), reprinted with permission from AAAS.

and patterns of common nucleotide diversity within and between populations through the International HapMap Project, advances in genotyping technologies to allow accurate and affordable high throughput SNP typing at genome-wide coverage, and establishment of clearly phenotyped large sample collections for analysis (WTCCC 2007).

Genome-wide association studies are concerned with finding associations between common genetic variants and a particular trait by using a high density set of SNP markers to capture a substantial proportion of the common nucleotide diversity in the DNA samples analysed. The number of SNPs analysed is large, for example a

minimum of 100 000 SNPs used in the initial scan is currently the threshold for the inclusion of a study in the catalogue of published genome-wide association studies maintained at the National Human Genome Research Institute (www.genome.gov/GWAstudies) (Hindorff *et al.* 2008). The approach has been very successful over a short space of time, with more than 50 disease susceptibility loci identified across a range of diseases – notably diabetes, inflammatory bowel disease (Section 9.5), and cancer, as well as coronary heart disease and asthma (McCarthy *et al.* 2008). Increasingly, the approach is also yielding major new insights into the genetic variation contributing to continuous traits ranging from height

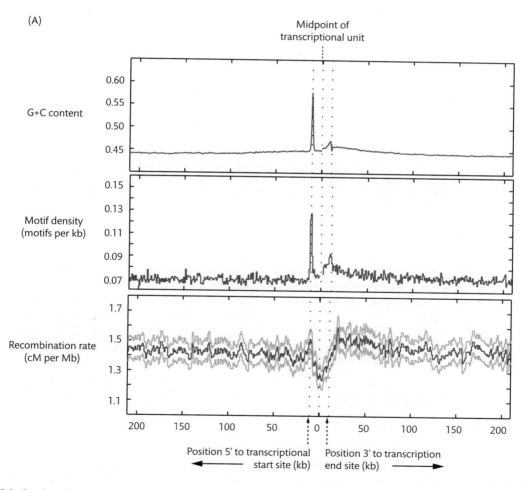

Figure 9.9 *Continued*

(B)

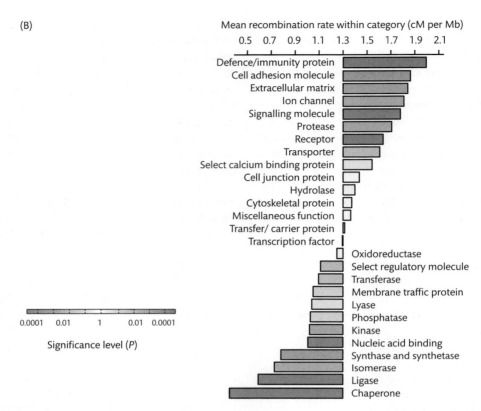

Figure 9.9 Recombination rates vary with GC content, specific motif density, and in relation to start and end sites of transcription. (A) Recombination rates shown in relation to distance from transcriptional start and end sites, GC content, and density of a specific motif CCTCCCTNNCCAC (shown for all motifs varying by 1 bp from this consensus). (B) Recombination rates vary by gene category. Reprinted by permission from Macmillan Publishers Ltd: Nature (Frazer *et al.* 2007), copyright 2007.

and fat mass, to warfarin dose and circulating levels of human immunodeficiency virus (HIV) prior to the onset of acquired immune deficiency syndrome (AIDS) (www.genome.gov/GWAstudies).

9.3.1 Insights into the design, analysis, and interpretation of genome-wide association studies

The design and analysis of genome-wide association studies is a complex subject but common themes are emerging as detailed in a number of reviews (Chanock *et al.* 2007; Kruglyak 2008; Manolio *et al.* 2008; McCarthy *et al.* 2008). A number of principles are outlined here, while in the following sections age-related macular degeneration (Section 9.4) and Crohn's disease (Section 9.5)

are reviewed in detail as examples of two common multifactorial diseases of significant public health importance in which genetic factors are important, and where genome-wide association studies have been particularly fruitful. Review of data related to these two diseases also serves to highlight that while genome-wide association studies have been highly successful for many of the diseases studied, their application comes on a background of several decades of intensive research using other approaches including linkage and candidate gene association studies. Moreover, while genome-wide association studies have dramatically increased the number of disease associated loci which have been identified, the altered disease risk (odds ratios in a case–control design) associated with the possession of specific variants is modest if highly significant (typically, odds ratios of less than 2 and more commonly less than 1.5). Indeed the proportion of the variation in disease risk attributable to genetic factors (for example estimates based on **familial aggregation**) collectively explained by associated SNPs from genome-wide studies remains very small.

Much work remains to be done through increasingly large genome-wide association scans (of the order of thousands of cases) to define the remaining common variants associated with disease with modest odds ratios (of the order of 1.2). To detect such variants, appropriately conducted meta analysis of comparable scans has proved a powerful approach in diabetes (Zeggini *et al.* 2008) and Crohn's disease (Barrett *et al.* 2008). To facilitate this, and to generally promote data sharing, there is a strong consensus that the results of genome-wide association scans should be formally deposited in a transparent fashion. A number of resources have been established to enable this such as the database of Genotype and Phenotype (dbGaP) from the National Center for Biotechnology Information (NCBI; www.ncbi.nlm.nih.gov/sites/entrez?Db=gap) (Mailman *et al.* 2007), the catalogue of published genome-wide association studies at the National Human Genome Research Institute (www.genome.gov/GWAstudies) (Hindorff *et al.* 2008), and the European Genotype Archive maintained by the European Bioinformatics Institute (EMBL-EBI) (www.ebi.ac.uk/ega).

Small effect sizes do not equate to the results of genome-wide association scans being of no biological interest. On the contrary, highly significant replicated associations have highlighted several new biological pathways and processes involved in disease pathogenesis (for example complement factor H in age-related macular degeneration; autophagy and IL23R signalling in Crohn's disease) which may provide new therapeutic targets or opportunities for detecting and monitoring disease with new biomarkers. Moreover, the population attributable risk depends on the risk allele frequency in the population. Given the small magnitude of currently explained risk, however, use of data from genome-wide scans for personalized medicine will require careful prospective evaluation and should be facilitated by ongoing work to fine map observed associations and broadening of the type of genetic variation interrogated.

The Wellcome Trust Case Control Consortium (WTCCC) study published in 2007 (Section 9.3.2) (WTCCC 2007), together with the many other genome-wide association studies published to date, have highlighted some specific issues relating to design and analysis. For example, the use of 'common controls' for a variety of diseases was established for a UK Caucasian population as a valid approach allowing considerable economy of cost and effort, and should provide an important resource as a historical control set for future studies (WTCCC 2007). Large numbers of individuals in control panels help to avoid misclassification bias in which cases of disease may be latently present in the control group. Careful selection of cases for a case–control design to minimize heterogeneity in the phenotype to be studied is also critical, and study of more extreme phenotypes is thought more likely to reveal clear effects. The chances of success will also be much greater if there is robust evidence from familial aggregation or other studies to support a role for genetic factors in contributing to the trait to be studied. Failure to match cases and controls may lead to misleading associations that reflect underlying population stratification (Section 2.5.3), as for any case–control study. Knowledge of population-specific genetic architecture through the HapMap and other studies can help take account of this by allowing the exclusion of specific individuals not corresponding to the recruitment criteria with respect to geographic ancestry.

Accurate genotyping and stringent quality control is extremely important throughout the process of sample collection, handling, data collection, and analysis, while replication is viewed as essential, including both a further set of independent samples and a separate genotyping methodology (McCarthy *et al.* 2008). The number of loci confirmed on replication from genome-wide association scans is high, thought to be significantly higher than from case–control studies based on a candidate gene approach where sample sizes are typically much more modest (Section 2.4).

A notable example of apparent failure to replicate in specific studies was seen with the fat mass and the obesity associated gene *FTO* (Section 9.3.3) due to case heterogeneity (also dubbed 'informative heterogeneity') (McCarthy *et al.* 2008). In a UK population, variants at this gene showed highly significant association with type 2 diabetes on the initial genome-wide scan and independent replication cohorts. However, this related to the high

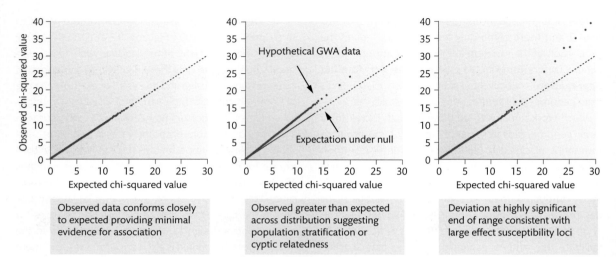

| Observed data conforms closely to expected providing minimal evidence for association | Observed greater than expected across distribution suggesting population stratification or cyptic relatedness | Deviation at highly significant end of range consistent with large effect susceptibility loci |

Figure 9.10 Quantile–quantile plots. The distribution of observed test statistics for a genome-wide association (GWA) scan can be visualized versus expected results using a quantile–quantile plot. Redrawn and reprinted by permission from Macmillan Publishers Ltd: Nature Reviews Genetics (McCarthy *et al.* 2008), copyright 2008.

levels of obesity in the diabetic cases versus controls as when lean diabetic subjects were analysed (Sladek *et al.* 2007) or when cases and controls were matched for body mass index (BMI) (Saxena *et al.* 2007; Scott *et al.* 2007), no association was seen.

The detailed analytical methods used in genome-wide association scans are beyond the scope of this discussion but two important ways of plotting the data are commonly encountered. '**Quantile–quantile plots**' summarize the data in terms of the distribution of observed versus expected test statistics and provide an overview of whether significant results are found (Fig. 9.10) (McCarthy *et al.* 2008). This is illustrated by data from the WTCCC for hypertension and Crohn's disease, which demonstrate very little deviation from the null except at the extreme tail for Crohn's disease indicating that several highly significant associations were present (WTCCC 2007). By contrast, deviation from null across the distribution is likely to indicate unexpected population stratification or cryptic relatedness. 'Genome-wide Manhattan plots' provide a graphic representation of the results of genome-wide scans ordered by genomic location, the genomic skyline interspersed with skyscrapers of association – well illustrated by some of the seven diseases analysed in the WTCCC (Fig. 9.11). The question of how best to correct

for multiple testing remains open to debate, with thresholds for genomic significance proposed such as *P* values below 5×10^{-8} (based on adjusting for 1–2 million independent tests) while Bayesian approaches may prove more informative (McCarthy *et al.* 2008).

To date genome-wide association studies have been carried out predominantly in populations of European origin, but ongoing analysis in other population groups will be highly informative as differences in underlying genetic architecture should provide opportunities to facilitate fine mapping as well as likely yield new associated loci. It is important to remember that genome-wide association scans using common SNP markers identify trait associated loci, not specific causative functional variants (except in rare instances), and much further work is required to fine map associations and establish functional causative variants within trait associated loci.

Indeed for the current set of genome-wide association scans utilizing informative common SNP markers, specific functional variants at or related to associated loci remain to be resolved in almost all cases. How to identify such variants remains a major road block in the field. It is possible to try and define boundaries for the search based on flanking recombination hot spots such that the haplotypic block bearing the associated SNP is scrutinized, with fine mapping

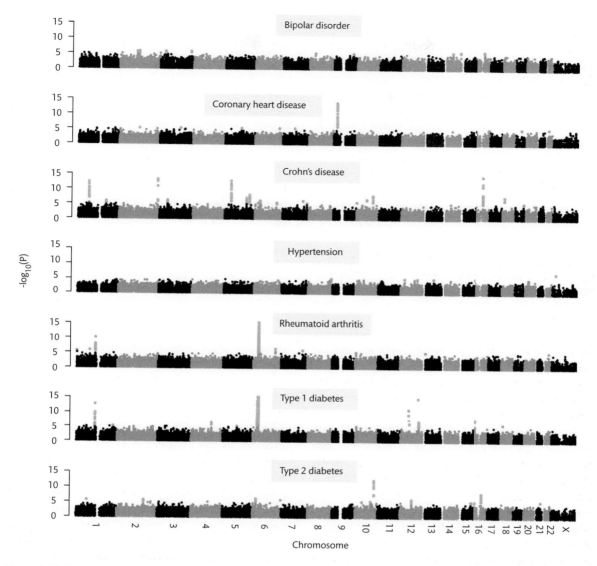

Figure 9.11 Genome-wide association for seven common diseases. Genome-wide Manhattan plots ordered by genomic location are shown for seven common diseases from the Wellcome Trust Case Control Consortium. Reprinted by permission from Macmillan Publishers Ltd: Nature (WTCCC 2007), copyright 2007.

through resequencing to catalogue variation present in the region and search for disease association through additional genotyping. However, even accepting the assumptions inherent in such a strategy, many genes can still be involved. Moreover, there is no evidence to suggest that a majority of causative variants will relate to coding changes in gene sequences, rather noncoding variants modulating gene expression are expected to predominate. Research into gene regulation suggests distant regulatory elements may be highly significant, operating for example through looping mechanisms to bring genomic regions together, in some cases operating as transcriptional factories.

The net may therefore need to be cast at a considerable distance and ongoing efforts to catalogue the functional architecture of the human genome through the ENCODE Project (Birney *et al.* 2007) and other studies should greatly facilitate the search for regulatory variants. Additional wet lab analyses of the impact of genetic diversity on the control of gene expression are required in the relevant cell type and context to the disease of interest. Some important insights can be gained from the accumulating number of studies analysing the genetics of gene expression for lymphoblastoid cell lines derived from HapMap and other samples (Section 11.4.1).

Particularly intriguing are examples of loci where multiple diseases show association, for example several cancers have shown association on genome-wide scans with chromosome 8q24, including prostate (Gudmundsson *et al.* 2007; Yeager *et al.* 2007), breast (Easton *et al.* 2007), and colorectal cancer (Tomlinson *et al.* 2007; Zanke *et al.* 2007). The disease associated locus lies within a 1.18 Mb **gene desert** flanked by the known oncogene *cMYC*; there is evidence that multiple functional variants are likely to occur within this region specific to particular cancer types, most probably operating as regulatory variants modulating gene expression (Ghoussaini *et al.* 2008). Other associated loci have also been found common to different diseases, for example with *PTPN2* at 18p11 with Crohn's disease and type 1 diabetes (WTCCC 2007); and the region containing *CDKN2A* and *CDKN2B* at 9p21 with coronary artery disease (McPherson *et al.* 2007; Samani *et al.* 2007; WTCCC 2007) and type 2 diabetes (Scott *et al.* 2007; Zeggini *et al.* 2007).

Defining variants directly responsible for the altered disease risk at a mechanistic level may in turn demonstrate higher odds ratios. This is illustrated by *NOD2* variants where the SNP marker showing association in genome-wide scans had a modest odds ratio of 1.3 compared to the putative functional variant in linkage disequilibrium with it, an insertion polymorphism that truncates the encoded protein (OR 4.1) (Section 9.5.2). More likely, however, is that a significant proportion of the disease risk attributable to genetic factors remains to be explained, and that responsible variants are not well detected either by genome-wide association scans or linkage approaches.

McCarthy and colleagues categorized disease associated variants based on allele frequency and penetrance, mendelian disease being associated with very rare alleles with high penetrance while common variants found to be associated with common multifactorial diseases have low to modest penetrance (Fig. 9.12) (McCarthy *et al.* 2008). Low frequency variants (with a minor allele frequency of around 1%) with intermediate penetrance are currently poorly characterized as the penetrance is insufficient to support mendelian segregation and hence definition by linkage-based approaches, while the low frequency of the risk allele means they are not well detected by genome-wide association scans using common SNP markers. Examples have been found in this group using a candidate gene approach and it is hypothesized that the associated odds ratios will be higher (of the order of 3.0) than those found for common variants.

Resequencing technologies should facilitate identification and characterization of such variants, together with rare sequence variants which may be highly significant in common diseases and traits but are as yet poorly characterized. The same is true for structural genomic variation, which is not well represented by current SNP genotyping approaches although some information can now be determined. The number of examples where clear associations between structural variation, notably copy number variants, and common disease have been found is increasing and will undoubtedly continue to grow (Section 4.5). If all common SNPs can be identified – estimated by some authors at 11 million SNPs with a minor allele frequency of greater than 1% (Kruglyak and Nickerson 2001) – then direct association studies become possible. Equally, if resequencing technologies continue to develop as predicted in terms of affordable cost and throughput (Box 1.24), the need for genotyping may become obsolete and a full picture of association with common and rare variants will be available, although the technological and analytical challenges of such analyses remain formidable.

9.3.2 The Wellcome Trust Case Control Consortium study of seven common diseases

In 2007 the results of a genome-wide association study into seven common diseases in British individuals were published (WTCCC 2007). This was a pioneering study by a consortium of over 50 UK research groups, which was seen as establishing the utility of the genome-wide

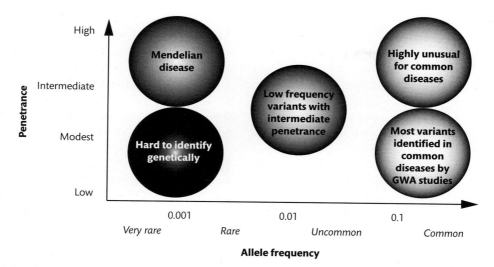

Figure 9.12 Low frequency variants with intermediate penetrance and disease. Schematic diagram illustrating the contrast between very rare variants with high penetrance classically found to be responsible for mendelian disease versus common variants with modest to low penetrance resolved in genome-wide association (GWA) scans of common disease. A third major group may be highly significant in determining disease susceptibility but to date is not detected well by either linkage analysis or genome-wide association scans. Redrawn and reprinted by permission from Macmillan Publishers Ltd: Nature Reviews Genetics (McCarthy *et al.* 2008), copyright 2008.

approach, resolving many design and analytical issues, and providing a very important dataset that clearly demonstrated previously established susceptibility loci as well as numerous novel susceptibility loci in diseases of major public health importance.

A total of 500 568 SNPs were genotyped among 2000 cases for each of the seven diseases and 3000 controls, comprising 1500 individuals from the 1958 British Birth Cohort and 1500 blood donors (WTCCC 2007). Among the methodological and analytical issues related to genome-wide association addressed by this study were quality control, genotype calling, use of imputation to infer genotypes, use of common controls, and statistical power. Novel disease associated loci were shown to be associated with modest effect sizes, namely odds ratios of less than 1.5. This was recognized to have significant consequences for the power of the WTCCC study, having 80% power to detect a ratio of 1.5 but only 43% power to detect a ratio of 1.3, with a *P* value threshold of less than 5×10^{-7}. This was despite this being one of the largest genome-wide association study sample sizes analysed to date: even larger samples sizes would be required to

increase the detection of further novel loci, likely to be associated with odds ratios of 1.2 and below.

The analysis was restricted to individuals of European ancestry; at recruitment almost all individuals self identified as white Europeans. The study demonstrated the utility of the HapMap dataset to look for non-European ancestry based on multidimensional scaling (Fig. 9.13) from which 153 individuals were excluded from analysis. Additional samples were excluded related to contamination, false identity, and relatedness, leaving a total of 16 179 individuals to be analysed for 469 557 SNPs passing quality control assessment. The two control groups were different in terms of sample collection and preparation, and in the age and population groups sampled. However, there were few significant differences between the two control groups in terms of allele frequencies, and they were used as a combined 'shared' control group in subsequent analyses.

A notable result of the study related to the analysis of geographic variation and population structure, performed due to concerns that hidden population structure within the cases and controls from the British population analysed may lead to confounding. Allele frequency

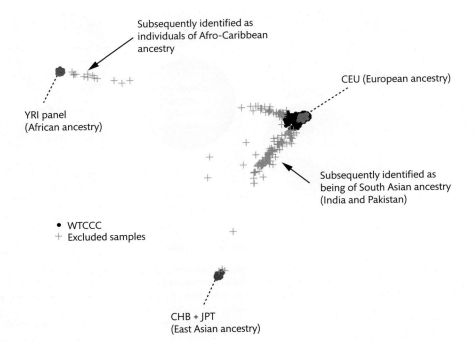

Figure 9.13 Use of data from HapMap to identify individuals with evidence of non-European ancestry. Samples for Wellcome Trust Case Control Consortium (WTCCC) and YRI (African geographic ancestry), CEU (European), CHB (Chinese), and JPT (Japanese) HapMap samples were plotted for the first two principal components obtained by multidimensional scaling. WTCCC samples were then selected for exclusion where they were not clustered with the CEU sample set. Reprinted by permission from Macmillan Publishers Ltd: Nature (WTCCC 2007), copyright 2007.

differences were found between the 12 geographic regions of the UK analysed for 13 genomic regions with variation noted along a northwest–southeast axis, thought to relate to natural selection in ancestral populations. Highly significant associations were found with *LCT* (encoding lactase) at 2q21, HLA at 6p21, and *TLR* genes at 4p14, which had been previously reported, with the remaining associated loci representing new findings that may relate to previous selection by tuberculosis, pellagra, or leprosy (WTCCC 2007). No disease associations were found in the genomic loci showing strong geographic differentiation and it was felt that population structure would not significantly confound the genome-wide disease association mapping study.

In terms of the seven diseases analysed, all had previously been demonstrated to have a significant genetic component in terms of defining disease susceptibility. Classical and Bayesian statistical approaches were used in

the analysis with trend and general genotype tests performed between each case and the pooled set of controls. It was notable that the largest numbers of significant associations were with type 1 diabetes and Crohn's disease, which have the highest sibling relative risks. Among these conditions, and the other five diseases analysed (type 2 diabetes, bipolar disorder, hypertension, coronary artery disease, and rheumatoid arthritis), 13 out of 15 previously reported 'robustly replicated' loci showed association in the WTCCC dataset (WTCCC 2007). Of those that did not, variants at *APOE* on chromosome 19q13 previously associated with coronary artery disease were poorly tagged on the genotyping platform used, while variants at *INS* (encoding insulin) previously associated with type 1 diabetes on 11p15 (Box 7.5) did show strong association but the SNP genotyping narrowly failed quality control. The Consortium used a genome-wide level of significance of 5×10^{-7} to identify 25 strongly associated susceptibility

Disease	Chr	SNP	Trend P value	Genotypic P value	Risk allele	Minor allele	Heterozygous OR (95% CI)	Homozygous OR (95% CI)	Control MAF	Cases MAF
Standard analysis										
BD	16p12	rs420259	2.19×10^{-4}	6.3×10^{-08}	A	G	2.08 (1.60–2.71)	2.07 (1.6–2.69)	0.282	0.248
CAD	9p21	rs1333049	1.79×10^{-14}	1.16×10^{-13}	C	C	1.47 (1.27–1.70)	1.9 (1.61–2.24)	0.474	0.554
CD	1p31	rs11805303	6.45×10^{-13}	5.85×10^{-12}	T	T	1.39 (1.22–1.58)	1.86 (1.54–2.24)	0.317	0.391
CD	2q37	rs10210302	7.1×10^{-14}	5.26×10^{-14}	T	C	1.19 (1.01–1.41)	1.85 (1.56–2.21)	0.481	0.402
CD	3p21	rs9858542	7.71×10^{-07}	3.58×10^{-08}	A	A	1.09 (0.96–1.24)	1.84 (1.49–2.26)	0.282	0.331
CD	5p13	rs17234657	2.13×10^{-13}	1.99×10^{-12}	G	G	1.54 (1.34–1.76)	2.32 (1.59–3.39)	0.125	0.181
CD	5q33	rs1000113	5.1×10^{-08}	3.15×10^{-07}	T	T	1.54 (1.31–1.82)	1.92 (0.92–4.00)	0.067	0.098
CD	10q21	rs10761659	2.68×10^{-07}	1.75×10^{-06}	G	A	1.23 (1.05–1.45)	1.55 (1.3–1.84)	0.461	0.406
CD	10q24	rs10883365	1.41×10^{-08}	5.82×10^{-08}	G	G	1.2 (1.03–1.39)	1.62 (1.37–1.92)	0.477	0.537
CD	16q12	rs17221417	9.36×10^{-12}	3.98×10^{-11}	G	G	1.29 (1.13–1.46)	1.92 (1.58–2.34)	0.287	0.356
CD	18p11	rs2542151	4.56×10^{-08}	2.03×10^{-07}	G	G	1.3 (1.14–1.48)	2.01 (1.46–2.76)	0.163	0.208
RA	1p13	rs6679677	4.9×10^{-26}	5.55×10^{-25}	A	A	1.98 (1.72–2.27)	3.32 (1.93–5.69)	0.096	0.168
RA	6 (MHC)	rs6457617*	3.44×10^{-76}	5.18×10^{-75}	T	T	2.36 (1.97–2.84)	5.21 (4.31–6.30)	0.489	0.685
T1D	1p13	rs6679677	1.17×10^{-26}	5.43×10^{-26}	A	A	1.82 (1.59–2.09)	5.19 (3.15–8.55)	0.096	0.169
T1D	6 (MHC)	rs9272346*	2.42×10^{-134}	5.47×10^{-134}	A	G	5.49 (4.83–6.24)	18.52 (27.03–12.69)	0.387	0.15
T1D	12q13	rs11171739	1.14×10^{-11}	9.71×10^{-11}	C	C	1.34 (1.17–1.54)	1.75 (1.48–2.06)	0.423	0.493
T1D	12q24	rs17696736	2.17×10^{-14}	1.51×10^{-14}	G	G	1.34 (1.16–1.53)	1.94 (1.65–2.29)	0.424	0.506
T1D	16p13	rs12708716	9.24×10^{-07}	4.92×10^{-07}	A	G	1.19 (0.97–1.45)	1.55 (1.27–1.89)	0.35	0.297
T2D	6p22	rs9465871	1.02×10^{-07}	3.34×10^{-07}	C	C	1.18 (1.04–1.34)	2.17 (1.6–2.95)	0.178	0.218
T2D	10q25	rs4506565	5.68×10^{-12}	5.05×10^{-12}	T	T	1.36 (1.2–1.54)	1.88 (1.56–2.27)	0.324	0.395
T2D	16q12	rs9939609	5.24×10^{-07}	1.91×10^{-07}	A	A	1.34 (1.17–1.52)	1.55 (1.3–1.84)	0.398	0.453
Multilocus analysis										
T1D	4q27	rs6534347	4.48×10^{-07}	1.83×10^{-06}	A	A	1.3 (1.1–1.55)	1.49 (1.25–1.78)	0.351	0.402
T1D	12p13	rs3764021	7.19×10^{-05}	5.08×10^{-08}	C	T	1.57 (1.38–1.79)	1.48 (1.25–1.75)	0.467	0.426
Sex differentiated analysis										
RA	7q32	rs11761231	3.91×10^{-07}	1.37×10^{-06}	G	A	1.44 (1.19–1.75)	1.64 (1.35–1.99)	0.375	0.327
Combined cases										
RA1T1D	10p15	rs2104286	5.92×10^{-08}	2.52×10^{-07}	T	C	1.35 (1.11–1.65)	1.62 (1.34–1.97)	0.286	0.245

Figure 9.14 Genomic loci showing the strongest association in the WTCCC study. The regions shown had at least one SNP with a *P* value of less than 5×10^{-7}. Diseases: BD, bipolar disease; CAD, coronary artery disease; CD, Crohn disease; RA, rheumatoid arthritis; T1D, type 1 diabetes; T2D, type 2 diabetes. MAF, minor allele frequency; OR, odds ratio; CI, confidence interval; *, multiple SNPs in MHC region significant, most extreme shown. Reprinted by permission from Macmillan Publishers Ltd: Nature (WTCCC 2007), copyright 2007.

loci (Figs 9.11, 9.14): 12 involved previously identified loci; of the remaining novel loci, nine out of ten had showed significant replication at the time of publication (Frayling *et al.* 2007; Parkes *et al.* 2007; Saxena *et al.* 2007; Todd *et al.* 2007; Zeggini *et al.* 2007). A further 58 loci were of moderate significance (*P* values between 10^{-5} and 10^{-7}) and contained many interesting loci for further investigation.

Among the most strongly associated novel loci was a region that included the genes *CDKN2A* and *CDKN2B* encoding two cyclin-dependent kinase inhibitors at 9p21.3, strongly associated with coronary artery disease. Four novel loci were identified in Crohn's disease as well as five previously recognized (Section 9.5). In rheumatoid arthritis the major associations were with HLA-DRB1

(Box 12.3) and *PTPN22* at 1p13 (Begovich *et al.* 2004). In type 1 diabetes, five existing susceptibility loci were demonstrated, together with five novel regions showing strong evidence of association, of which loci at 12q24, 12q13, 16p13, and 18p11 were subsequently clearly replicated (Todd *et al.* 2007). In type 2 diabetes the previously identified loci involving *PPARG* (Altshuler *et al.* 2000a), *KCNJ11* (Gloyn *et al.* 2003), and *TCF7L2* (Grant *et al.* 2006) were found to be associated, while highly significant associations were found with two novel loci: *FTO* at 16q, which was found to be due to an association with obesity related traits (Section 9.3.3); and chromosome 6p22 where a cluster of SNPs mapped to *CDKAL1* (encoding CDK5 regulatory subunit associated protein 1-like 1), which was subsequently replicated (Zeggini *et al.* 2007).

9.3.3 FTO and common obesity traits

Originally cloned in mice based on the identification of a fused toe phenotype associated with a 1.6 Mb deletion, the human orthologue of the mouse *Fto* gene, *FTO* at chromosome 16q12.2, was renamed by the HUGO Gene Nomenclature Committee as the 'fat mass and obesity associated gene' in view of the striking associations found with common human obesity-related traits in Caucasian populations. Frayling and colleagues described how a cluster of linked SNPs in the first intron of *FTO* showed reproducible association with type 2 diabetes in the WTCCC study, represented by the specific SNP rs9939609 with an odds ratio of 1.27 (1.16–1.37, $P = 5 \times 10^{-8}$) on the initial scan among 1924 cases of type 2 diabetes (Fig. 9.15) and 2938 controls. This was replicated in a further UK cohort of 3257 cases and 5346 controls (Frayling *et al.* 2007). However, with access to data on BMI in the original cases and replication samples, it became clear that the association was actually with raised BMI, as on adjusting for this the association with type 2 diabetes was lost (OR 1.03 (0.96–1.1), $P = 0.44$).

At this point the availability of large population cohorts, including 19 424 adults and 10 172 children, allowed the dramatic association of the A allele of this common *FTO* SNP with BMI, obesity, and risk of being overweight to be clearly demonstrated (Frayling *et al.* 2007). An additive association was found, BMI increasing per copy of the A allele by 0.4 kg/m², which was highly significant among the total of 30 081 participants analysed ($P = 3 \times 10^{-35}$). Homozygotes for the risk allele (AA) are found among 16% of the white population and were found to weigh 3 kg more, have a 1.67 (1.47–1.89) increased risk of obesity, and a 1.38 (1.27–1.52) increased risk of being overweight, compared to those without a copy of the A allele. Significant associations were also found for waist circumference and subcutaneous fat.

The association with genetic variation in the first intron of *FTO* was found in an independent study of obesity-related **quantitative traits** in a genetically isolated population from Sardinia and was replicated in Americans of European and Hispanic ancestry (Scuteri *et al.* 2007), together with a study that identified the *FTO* association while analysing a small set of neutral SNPs to look at population stratification (Dina *et al.* 2007). The association has since been widely replicated among populations of European descent. The association of *FTO* variants with common obesity traits is highly significant and was the first such locus to be reproducibly identified. No association with birth weight was found but association with fat mass is reported by 2 weeks of age (Lopez-Bermejo *et al.* 2008).

Among those of European ancestry, 63% of the population are estimated to carry at least one risk allele, with a population attributable risk for obesity of 20% and for being overweight of 13% (Loos and Bouchard 2008). The public health application of such findings remains to be defined but it is notable that individuals homozygous for the risk allele reported as physically active had the same BMI as those without a copy (Andreasen *et al.* 2008). Genetic factors are important in obesity, estimated as contributing between 40% and 70% of observed variation, yet the *FTO* variants are thought to explain only between 1% and 1.3% of the variance in BMI (Frayling *et al.* 2007; Scuteri *et al.* 2007). Much remains to be done to explain the remainder of this variation, and to functionally characterize the associated variants at *FTO* where both central and peripheral effects are reported, including modulation of appetite with homozygotes for the risk allele showing reduced satiety (Wardle *et al.* 2008).

9.4 Age-related macular degeneration

Investigators studying age-related macular degeneration were among the first to show the power of a genome-wide approach of analysing SNP markers for association with a common multifactorial trait (Klein *et al.* 2005). Age-related macular degeneration is a common and potentially devastating disease, noted to be the leading cause of blindness in the developed world among people over 50 years of age, with a rising prevalence as the population demographics shift to a more elderly population (Box 9.3) (Jager *et al.* 2008). Environmental and genetic risk factors are important, the former including increasing age, white race, smoking, and dietary factors (Jager *et al.* 2008). Evidence to support a genetic predisposition include family studies that show aggregation of cases; among first degree relatives of affected versus unaffected individuals there is an estimated three- to six-fold increased disease risk (Seddon *et al.* 1997; Klaver

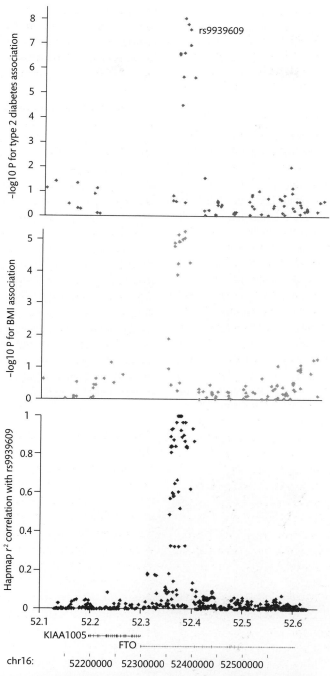

Figure 9.15 SNP associations at the *FTO* gene locus. Data for SNPs from a genome-wide association scan at *FTO* locus are shown for type 2 diabetes (top), body mass index (BMI) in type 2 diabetes patients (middle), and a plot of linkage disequilibrium between rs9939609 and SNPs from the International HapMap Project CEU panel of European geographic ancestry (bottom). From Frayling *et al.* (2007), reprinted with permission from AAAS. Gene plots with genomic location adapted from screenshot of UCSC Genome Browser (Kent *et al.* 2002) (http://genome.ucsc.edu/) (Human May 2004 Assembly).

et al. 1998; Swaroop *et al.* 2007). This was most striking when cases with more severe disease were analysed, for example Klaver and colleagues found an associated lifetime risk of 50% for relatives of patients compared to 12% for relatives of controls (Klaver *et al.* 1998). Twin studies provide further support for a role for inherited factors with comparisons of **monozygotic** and **dizygotic-twins**, concordant and discordant for the disease, allowing estimation of heritability (Hammond *et al.* 2002; Seddon *et al.* 2005). A large study of 840 twins estimated

---

Box 9.3   Age-related macular degeneration (OMIM 603075)

This progressive disease can lead to profound visual loss but spares peripheral vision (Fig. 9.16). This is because the disease affects the macula, a central region of the retina responsible for highest visual acuity through a very dense collection of specialized photoreceptors (cones). Initially extracellular deposits of debris, 'drusen', are seen behind the retina, specifically between the retinal pigment epithelium and a membrane in front of a vascular region called the choroid. Early disease is mild but through damage and chronic inflammatory change, regions of atrophy ('geographic atrophy') can develop which may include the centre of the macula, leading to advanced (sometimes called 'dry') disease (de Jong 2006; Jager *et al.* 2008). The development of new blood vessels (choroidal neovascularization, seen in 'wet' exudative disease) related to vascular endothelial growth factor and other cytokines is responsible for the majority of cases of severe visual loss through complications such as haemorrhage, retinal detachment, or fluid accumulation (de Jong 2006; Jager *et al.* 2008). Age-related macular degeneration is common and rises in incidence with age. In the United States, the estimated prevalence of advanced disease involving geographic atrophy and/or neovascularization is 1.5% among people over the age of 40 years, rising to 15% of white women over 80 years (Friedman *et al.* 2004).

Figure 9.16 Age-related macular degeneration and central vision loss. Image used with permission of the Foundation Fighting Blindness.

heritability at 46% for age-related macular degeneration, rising to 71% for advanced disease (Seddon *et al.* 2005).

To investigate further the genetic contribution to a common multifactorial disease like age-related macular degeneration, some insight may be gained by looking at similar but distinct rare traits inherited in a mendelian fashion. For example, Stargardt's disease (OMIM 248200) leads to severe macular degeneration in childhood with very impaired visual acuity but normal peripheral vision. This recessive disorder was mapped to chromosome 1p22 by linkage analysis (Hoyng *et al.* 1996) and diverse mutations identified involving the causative gene *ABCR*, which encodes an ATP binding transporter expressed in the retina (Allikmets *et al.* 1997). Screening of patients with age-related macular degeneration showed 26 out of 167 had mutations in the same gene with various deletions and nonsynonymous changes found (Allikmets *et al.* 1997). However further work suggested that while these may be rare variants causing age-related macular degeneration, they are not major genetic contributors to

the disease, with a number of studies failing to find this association (Stone *et al.* 1998; Swaroop *et al.* 2007).

Linkage analysis has been used with considerable success in studying genetic factors underlying age-related macular degeneration, despite the apparent difficulties of studying a multifactorial disease that is also late onset. The latter means that generally only a single generation is available to study within a family, as parents may have died and children are too young for the disease to be apparent, precluding the analysis of large multigenerational families. Studies of affected sibling pairs have been extensively and successfully employed with linkage to several loci, notably chromosome 1q31-q32 and 10q26 (Fig. 9.17) (Swaroop *et al.* 2007).

9.4.1 Genome-wide association, linkage, and complement factor H gene

One of the first demonstrations of the power of genome-wide association studies in common disease was the work

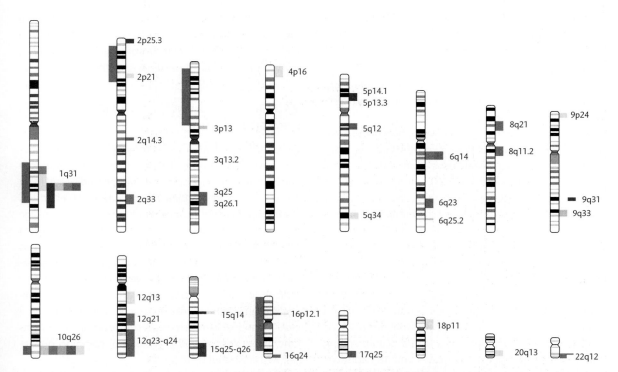

Figure 9.17 Linkage studies in age-related macular degeneration. Summary of linkage studies with loci across 16 autosomes. Each study is represented by a shaded block. Reprinted from Swaroop *et al.* (2007), with permission of Oxford University Press.

of Klein and colleagues, who found a strong association with variants at the *CFH* gene encoding complement factor H, which lies in the chromosome 1q31 region (Klein *et al.* 2005). In the context of more recent genome-wide association studies of other diseases involving several thousand individuals with up to 1 million SNPs, it seems in some ways remarkable that the study should have achieved such success with only 96 cases and 50 controls using 116 204 SNP markers.

A clear and relatively extreme phenotype was studied (the presence of large drusen and sight threatening disease through geographic atrophy or neovascularization) in a white non-Hispanic population. Genome-wide analysis implicated only two linked SNPs, 1.8 kb apart, which were strongly associated with disease (nominal P value $<10^{-7}$) and remained significant even after a conservative **Bonferroni correction** for multiple comparisons. The strongest association was for rs380390, one copy of the C allele was associated with an odds ratio of 4.6 (95% CI 2–11), two copies with an odds ratio of 7.4 (2.9–19). Haplotypic analysis resolved a specific risk haplotype 41 kb in length; resequencing exons and exon/intron boundaries defined 50 SNPs of which three were nonsynonymous. Of chromosomes bearing the risk haplotype, 97% were found to have one of the nonsynonymous SNPs, rs1061170, found in exon 9, which leads to an amino acid substitution of tyrosine for histidine and was within a region of the protein important for interactions (Klein *et al.* 2005).

In the same issue of *Science* in which the work of Klein and colleagues was published, two other papers independently found significant association with the same SNP, rs1061170 (Edwards *et al.* 2005; Haines *et al.* 2005). Six independent linkage studies had previously implicated the region 1q31 in disease susceptibility. Edwards and coworkers focused on this specific region of linkage and looked for disease association with 86 SNPs across 388 kb among 224 cases (many of whom showed familial aggregation or early disease) and 134 controls; as is important in any association study they then studied 14 SNPs in a replication cohort (176 cases and 68 controls) (Edwards *et al.* 2005). Haines and colleagues first resolved the 24 Mb region of linkage to a higher degree among 182 families together with 495 cases and 185 controls, before resequencing 24 cases and 24 controls homozygous for the disease risk haplotype (Haines *et al.* 2005).

All three papers had defined the same nonsynonymous SNP (rs1061170, c.1204T>C, p.Y402H). Possession of one copy of the C allele encoding histidine was associated with a two- to four-fold increased risk of age-related macular degeneration, and two copies with a five- to seven-fold risk. In the groups studied this was thought to account for 20–50% of the risk of disease developing (Daiger 2005).

CFH encodes an important regulator of complement activation making it a very plausible candidate in terms of disease pathogenesis. The association was replicated by a series of independent studies, with a meta analysis of 3697 cases and 2380 controls (from eight studies involving individuals of European origin), that showed possession of each copy of the C allele increased risk 2.5-fold (Thakkinstian *et al.* 2006). However, definitive identification of the causative function variant(s) remains a source of debate with evidence from additional SNP typing showing that multiple alleles at the *CFH* locus may be important (Li *et al.* 2006; Maller *et al.* 2006). A low risk haplotype has also been identified at this locus, present in 8% of cases and 20% of controls, which involves deletion of nearby related genes *CFHR1* and *CFHR3* (Hughes *et al.* 2006). However, other investigators have found the effect of possessing the deletion on disease risk to be modest and suggest that other protective variants may be involved at this locus (Spencer *et al.* 2008).

9.4.2 Disease association at the 10q26 susceptibility locus

Of any genomic locus, the strongest linkage with age-related macular degeneration was found at chromosome 10q26 based on six independent linkage studies and meta analysis (Fisher *et al.* 2005). Intensive research effort has resolved that a specific area within the region of linkage is important, containing the genes *ARMS2* (age-related maculopathy susceptibility 2; previously described as *LOC387715*) and *HTRA1* (high-temperature requirement factor A1) (Fig. 9.18). However it has proved very difficult to fine map associations due to the strong linkage disequilibrium across the region and establishing a clear functional variant remains controversial – problems inherent to studies of the genetics of common disease but well illustrated by this work.

Rivera and colleagues sought to fine map the association within a 22 Mb interval using 93 common

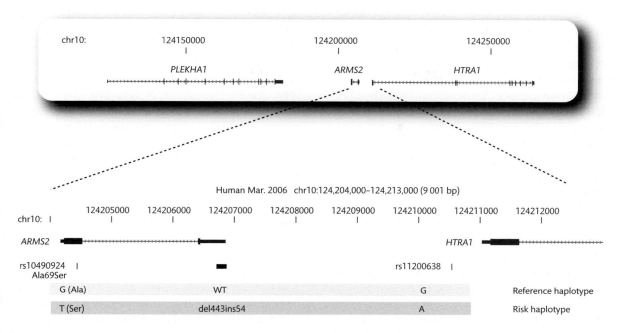

Figure 9.18 The 10q26 locus and age-related macular degeneration. The strongest evidence from linkage studies was with the 10q26 locus within which variation in three candidate genes has been associated with age-related macular degeneration: *PLEKHA1*, *ARMS2* (*LOC387715*), and *HTRA1*. A risk haplotype bearing two SNPs (rs10490924 and rs11200638) and a deletion/ insertion polymorphism has been resolved. Reprinted by permission from Macmillan Publishers Ltd: Nature Genetics (Allikmets and Dean 2008), copyright 2008. Gene plots with genomic location adapted from screenshot of UCSC Genome Browser (Kent *et al.* 2002) (http://genome.ucsc.edu/) (Human March 2006 Assembly).

SNP markers in a total of 1166 cases and 945 controls, identifying the *PLEKHA1* gene region (Rivera *et al.* 2005). Further SNP genotyping identified a 60 kb region including *PLEKHA1* and *ARMS2*, with resequencing highlighting a strong association with age-related macular degeneration for a specific nonsynonymous SNP, rs10490924 (c.205G>T, p.A69S), in exon 1 of *ARMS2* which results in substitution of alanine for serine at amino acid position 69. Individuals homozygous for the SNP had a 7.6-fold increased disease risk (Rivera *et al.* 2005). The same SNP was identified by Jakobsdottir and colleagues (2005) and has been replicated by several independent groups (Conley *et al.* 2006; Maller *et al.* 2006; Schmidt *et al.* 2006). The magnitude of effect was noted to be similar and independent to that seen at *CFH* although Schmidt and coworkers found the risk was higher among smokers (Schmidt *et al.* 2006).

A large case–control study defined five common SNPs as explaining half of the excess risk of age-related macular degeneration seen in siblings of those with the disease: at *CFH* (a noncoding SNP together with the nonsynonymous SNP rs1061170), at *ARMS2* (rs10490924), and at two alleles in the complement factor B/complement 2 genes (*C2-CFB*) (Maller *et al.* 2006). Two subsequent studies highlighted a role for a promoter SNP in the *HTRA1* gene (Fig. 9.18) based on a genome-wide association study in a Chinese population of individuals with neovascular (wet) disease (Dewan *et al.* 2006) and a study of a small number of SNPs flanking rs10490924 in a population of European ancestry (Yang *et al.* 2006). The SNP identified (rs11200638) was associated with increased expression of the heat shock serine protease encoded by *HTRA1* (Dewan *et al.* 2006; Yang *et al.* 2006). However, dissecting away the effects of rs11200638 and rs10490924 is very difficult as the two SNPs are in almost complete linkage disequilibrium and other studies failed to find the reported expression differences (Kanda *et al.* 2007; Fritsche *et al.* 2008).

A definitive answer as to which is (are) the functionally important variants(s) in or near this cluster of genes remains elusive. However the work of Fritsche and colleagues suggests that other classes of genetic diversity are involved, in particular an indel polymorphism of *ARMS2* with dramatic functional consequences for RNA stability (Fritsche *et al.* 2008). Genotyping 28 SNP markers over a 107 kb region in 794 cases and 812 controls identified a specific risk haplotype of 23.3 kb spanning *ARMS2* and the 5′ region of *HTRA1*, which included rs10490924 and rs11200638 (Fig. 9.18) (Fritsche *et al.* 2008). Resequencing of this region among 16 affected individuals resolved 54 variants, including a specific indel (c.372_815del443ins54) in the 3′ untranslated region (UTR) of *ARMS2* with dramatic functional consequences.

The deletion leads to loss of the polyadenylation signal in the transcript while the 54 bp insertion is an AU-rich element controlling mRNA decay: possession of the indel markedly affected expression as the encoded transcript was found to be highly unstable and the encoded protein undetectable (Fritsche *et al.* 2008). Individuals homozygous for the risk haplotype bearing the indel (and SNPs rs10490924 and rs11200638) are found among 2–5% of individuals in European ancestry populations, they fail to express ARMS2 and have an eight-fold increased risk of age-related macular degeneration (Allikmets and Dean 2008; Fritsche *et al.* 2008). The frequency of heterozygotes is 35–40% and these people will only express ARMS2 protein from the wild type allele, and thus have half the normal level of overall expression. Possession of one copy of the indel variant haplotype is associated with a 2.9-fold increased disease risk (Allikmets and Dean 2008; Fritsche *et al.* 2008). A functional role for ARMS2 in mitochondrial homeostasis is postulated.

9.5 Lessons from inflammatory bowel disease

The elucidation of inherited factors determining susceptibility to inflammatory bowel disease, specifically Crohn's disease, has been a major success story in the field of common complex disease genetics. Pioneering research in Crohn's disease demonstrated the successful application of linkage analysis and a positional cloning approach to identify a specific gene and associated variants in a common disease, while more recent genome-wide association studies have been arguably more successful in Crohn's disease than any other multifactorial disease. The wealth of data now accumulated from such genetic studies have provided radical new insights into Crohn's disease pathogenesis and individual susceptibility.

Why has Crohn's disease been such a success story? The answer is unclear, but compared to many common diseases the evidence for a significant role for inherited factors is strong, the disease phenotype is clearly resolved, and the clinical cohorts studied carefully defined and largely restricted to individuals of European descent, among whom the disease is most commonly encountered. The stage was well set and many genomic loci have now been reproducibly identified, but it is important to remember that, as with so many other diseases and associated variants, identification of the causative functional variants remains largely elusive.

9.5.1 A role for inherited factors in inflammatory bowel disease

Crohn's disease and ulcerative colitis are types of inflammatory bowel disease – chronic inflammatory disorders affecting the intestine (Box 9.4). The evidence to support a role for inherited factors is much stronger for Crohn's disease than ulcerative colitis, although clustering of both diseases within families has been recognized since the 1960s (Kirsner and Spencer 1963). A Danish study of 637 patients with inflammatory bowel disease showed that first degree relatives of individuals with either Crohn's disease or ulcerative colitis had a ten-fold increased risk of the same disease (Orholm *et al.* 1991), while a Swedish study showed that for 1048 patients with Crohn's disease, the risk was even greater with a 21-fold higher prevalence among first degree relatives than the general population (Monsen *et al.* 1991). Based on many studies, the risk ratio for siblings relative to the population prevalence, λ_S, is currently estimated at 20–35 for Crohn's disease (Barrett *et al.* 2008).

Other evidence supporting a role for inherited factors has also been found. For families with inflammatory bowel disease, high rates of **concordance** (75% in

---

Box 9.4 Inflammatory bowel disease

Crohn's disease and ulcerative colitis are chronic inflammatory diseases affecting the gastrointestinal tract. Inflammatory bowel disease is common, with a peak age of onset between the second and fourth decade of life, and associated with very significant morbidity and long term complications. The highest incidence and prevalence of inflammatory bowel disease is seen in northern Europe, the United Kingdom, and North America; in the latter the prevalence of Crohn's disease is reported as 26–199 cases per 100 000 population, and ulcerative colitis at 37–246 cases per 100 000 (Loftus 2004). Both types of inflammatory bowel disease are characterized by symptomatic flare-ups manifesting with diarrhoea, abdominal pain, rectal bleeding, and malnutrition. Bloody diarrhoea is less common in Crohn's disease, in which abdominal masses, perianal disease, and malabsorption are other common clinical features.

The pathological changes seen in the two diseases are distinct in a number of ways. Crohn's disease involves inflammatory changes across the bowel wall (transmural), which are often localized with a discontinuous distribution along the intestine (most commonly the ileum and colon are involved); there are often granulomas, cobblestone ulcers, bowel narrowing (strictures), and fistulas. By contrast the inflammatory changes seen in ulcerative colitis always involve the rectum and are continuous in distribution, sometimes extending as far as the caecum; fine ulceration is seen, with superficial inflammatory changes involving the mucosa and submucosa. In both types of inflammatory bowel disease, extraintestinal manifestations may occur such as arthralgia, and other chronic inflammatory diseases are more common such as psoriasis (Box 4.6). Treatment options involve anti-inflammatory and immunosuppressive drugs. Unfortunately for many patients surgical treatment with bowel resection can become necessary. The pathogenesis of inflammatory bowel disease remains unresolved but a combination of environmental factors in a genetically susceptible individual are thought to result in disease, notably involving dysregulation of the normal immune response to commensal bacteria in the intestine (Xavier and Podolsky 2007).

---

child–parent pairs and 81% among affected sib pairs) were observed for disease type, extent, and other clinical features (Satsangi *et al.* 1996). Twin studies were also supportive of a significant role for genetic factors, notably in Crohn's disease. Two large Scandinavian studies demonstrated rates of Crohn's disease among monozygotic twins to be significantly higher than among dizygotic twins (50% versus 0–3.8%), and to a lesser extent in ulcerative colitis (14.3–18.8% versus 0–4.5%) (Orholm *et al.* 2000; Halfvarson *et al.* 2003). It should be noted that intensive study has failed to reveal any simple mendelian model of inheritance for Crohn's disease and that this is demonstrably a multifactorial disease trait with environmental factors being highly significant; a longstanding model is one of an infectious trigger(s) in a genetically susceptible individual.

9.5.2 Crohn's disease and variants of the NOD2 gene

Crohn's disease was mapped to chromosome 16 by Hugot and colleagues in 1996 in the first genome-wide linkage analysis in this disease using affected pedigrees: a total of 25 Caucasian families with at least two affected siblings were analysed using 270 highly polymorphic microsatellite markers across the genome (Hugot *et al.* 1996). A stepwise approach involving additional families and denser sets of markers resolved a pericentromeric region of chromosome 16, dubbed 'inflammatory bowel disease locus 1' (IBD1). The number of alleles 'shared identical by descent' among affected siblings were analysed, in other words those alleles inherited by both siblings that are copies of the same parental alleles with no assumption about the mode of inheritance. Other investigators subsequently replicated and to some extent refined the region of linkage for IBD1, clearly showing that the linkage was with Crohn's disease and not ulcerative colitis (Ohmen *et al.* 1996; Cavanaugh *et al.* 1998; Cavanaugh 2001).

In 2001 evidence was published from different groups resolving a specific gene and sequence variants underlying IBD1 in Crohn's disease (Hampe *et al.* 2001; Hugot *et al.* 2001; Ogura *et al.* 2001a). Hugot and coworkers sought to use a positional cloning strategy, with linkage analysis and linkage disequilibrium mapping: they interrogated the 20 Mb region of IBD1 (a large region thought to contain some 250 genes) with 26 microsatellite markers, analysing two sets of affected families (Hugot *et al.* 2001). Sequencing of cloned DNA associated with disease resolved a number of SNPs and a specific gene, *NOD2*, which encodes 'nucleotide-binding oligomerization domain containing 2' protein, whose role in the inflammatory response to bacteria would suggest a highly plausible model for disease pathogenesis. The exonic sequence of *NOD2* was determined for 50 unrelated patients with further variants identified. Overall, highly significant associations were found for two nonsynonymous SNPs and a single nucleotide insertion polymorphism, with a gene dosage effect such that highest risks were seen in individuals homozygous for a given variant or having two different variants (**compound heterozygotes**) (Fig. 9.19A).

In the same issue of *Nature* that Hugot and colleagues published their results, Ogura and coworkers published a paper in which they identified the same gene and insertion polymorphism as being important in determining Crohn's disease, except that here the approach had been different (Ogura *et al.* 2001a). Ogura and colleagues had sought to identify novel Nod1-like genes and found a sequence with high homology on chromosome 16q12, leading to the identification of *NOD2*, a gene found to be expressed mainly in monocytes, which activated NFkB, and was important to endotoxin responsiveness (Inohara *et al.* 2001; Ogura *et al.* 2001b). Given that *NOD2* was lying under the linkage peak, Ogura and colleagues then proceeded to investigate it as a candidate gene for Crohn's disease by resequencing the exons and flanking intronic sequence. Among 12 patients with Crohn's disease three were found to have a specific insertion polymorphism in exon 11, the same variant identified by Hugot and colleagues. This C nucleotide insertion (rs2066847; also called 3020insC as the insertion is at nucleotide position 3020) alters the encoded amino acid at that codon from leucine to proline (CUU to CCU) and results in a frameshift such that the next codon is a stop codon (UGA) (Fig. 9.20). This truncates the NOD2 protein by 33 amino acids and, as in the other study, the variant was significantly associated with Crohn's disease but not ulcerative colitis, being preferentially transmitted from heterozygous parents to affected children and, in three Caucasian cohorts, being found at a significantly higher frequency among cases of Crohn's disease (8.2%) versus controls (4%) (Ogura *et al.* 2001a).

A further study by Hampe and colleagues, also published in the same year, identified the same insertion polymorphism among British and German affected sib pairs after adopting a candidate gene approach (Hampe *et al.* 2001). In a meta analysis published in 2004, Economou and colleagues found that from 39 studies of Caucasians of non-Jewish descent, the three *NOD2* variants identified by Hugot and others were all significantly associated with Crohn's disease. Possession of one of the high risk alleles was associated with a 2.4 (95% CI 2–2.9) increased odds of disease compared to individuals without the alleles, rising to 17.1 (10.7–27.2) for carriers of at least two of the risk alleles (Fig. 9.19B) (Economou *et al.* 2004). Subsequent genome-wide association studies repeatedly demonstrated association at this locus, showing the *NOD2* locus was one of the most strongly associated genomic regions in Crohn's disease susceptibility (Box 9.5).

Are these *NOD2* alleles functionally important? Intensive research has established that NOD2 is critical to the innate and acquired immune response through its role as an intracellular protein responsible for detecting bacteria and inducing a proinflammatory response (Inohara *et al.* 2005; Kobayashi *et al.* 2005; Watanabe *et al.* 2005; Strober *et al.* 2006). NOD2 is a member of a family of pattern recognition receptors and comprises a central nucleotide binding domain (NBD), N-terminal caspase recruitment domains, and a leucine-rich repeat (LRR) region (see Fig. 9.20). The three genetic variants associated with Crohn's disease discussed to date are found in sequence encoding the LRR, a region responsible for bacterial recognition, specifically recognizing a short motif called muramyl dipeptide (MDP) found in peptoglycans in bacterial cell walls. On recognizing MDP,

(A)

	Number of chromosomes	SNP8	SNP12	SNP13	Total
Unaffected	206	0.04	0.01	0.02	0.07
Ulcerative colitis	318	0.03	0.00	0.01	0.05
Crohn's disease	936	0.11	0.06	0.12	0.29

	No variant	Simple heterozygous	Homozygous	Compound heterozygous
Unaffected	88	15	0	0
Ulcerative colitis	145	13	1	0
Crohn's disease	267	133	28	40
Relative risk for Crohn's disease		3	38	44

(B)

	SNP8	SNP12	SNP13
Nucleotide change	2104C>T	2722G>C	3020insC
rs number	rs2066844	rs2066845	rs2066847
Location	Exon 4	Exon 8	Exon 11
Amino acid change	Arginine (R) to typtophan (W) R702W	Glycine (G) to arginine (R) G908R	Leucine (L) to proline (P) Frameshift and premature stop codon
Odds ratio for Crohn's disease (95% CI)	2.2 (1.8–2.6)	3 (2.4–3.7)	4.1 (3.2–5.2)

Figure 9.19 NOD2 risk alleles. (A) Data from Hugot and colleagues for three major Crohn's disease risk alleles comparing allele frequencies in cases of either Crohn's disease or ulcerative colitis versus controls, and the possession of different combinations of risk alleles (either no variant, a single rare variant 'simple heterozygous', the same variant on both chromosomes 'homozygous', or two different variants 'compound heterozygous'). Reprinted by permission from Macmillan Publishers Ltd: Nature (Hugot *et al.* 2001), copyright 2001. (B) Results of a meta analysis of 39 studies analysing Caucasian individuals of non-Jewish descent. Numbering for the amino acid positions follows Economou *et al.* (2004) and other papers, the original nomenclature with respect to amino acid position for SNP8, according to Hugot and colleagues (2001), is R675W and for SNP 12 is G881R. Quoted odds ratios are based on Economou *et al.* (2004).

(A)

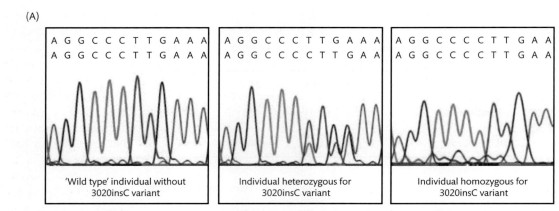

(B)

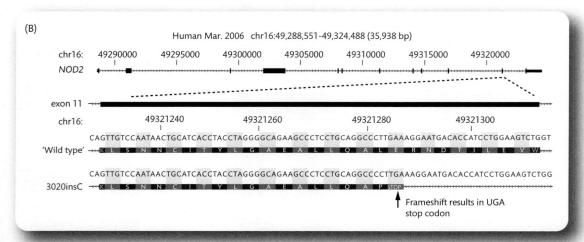

(C)

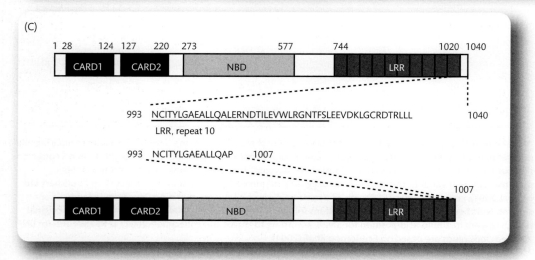

Figure 9.20 *Continued*

a complex series of events leads to NFkB activation, a family of transcription factors involved in regulating many immune and inflammatory genes. Activation of NFkB involves conformational changes in NOD2 on sensing bacterial MDP, interaction with specific kinases, and ubiquitination.

A number of independent *in vitro* studies suggested a loss of function effect for the 3020insC variant, which truncates the LRR of NOD2 with defective NFkB activation on sensing MDP (Ogura *et al.* 2001a; Girardin *et al.* 2003). This contrasted with data from murine studies. When the insertion variant was analysed in mice, the gastrointestinal tract appeared healthy and no difference was found at the RNA level in expression of *Nod2* (Maeda *et al.* 2005). However, mice with the variant had increased NFkB activation in response to MDP and showed increased inflammation and mortality in response to bacterial intestinal challenge, with more extensive and severe inflammation, suggesting this was a gain of function allele (Maeda *et al.* 2005).

Supporting a functional role for Nod2, mice deficient in Nod2 show increased susceptibility to the bacteria *Listeria monocytogenes* when given via the oral route and reduced NFkB activation in response to MDP (Kobayashi *et al.* 2005). Further work is required to more fully define the functional role of the different *NOD2* disease associated variants in the most appropriate cell type in a disease context. *NOD2* is known, for example, to be expressed in the intestinal epithelium in Paneth cells (Lala *et al.* 2003) and effects of *NOD2* disease associated variants may also involve antimicrobial proteins such as alpha defensins (Wehkamp *et al.* 2004). Intriguingly, recent data suggest that *NOD2* transcripts found in the colon are the products of a novel promoter and first exon (King *et al.* 2007).

9.5.3 Linkage studies and other inflammatory bowel disease susceptibility loci

Linkage studies in Crohn's disease proved highly informative and have led to the identification of a number of

---

Box 9.5   NOD2 and genome-wide association studies

As a known disease associated locus, it was reassuring that the *NOD2* region was implicated in genome-wide association studies, but the data serve to highlight some caveats in the interpretation of genome-wide scans based on the specific markers analysed. In the WTCCC study, for example, a strong signal of association was found for *NOD2* but not with the *NOD2* insertion polymorphism or nonsynonymous SNPs, described earlier as showing the strongest disease association, as they were not part of the panel of SNP markers present on the genotyping array used in this study (WTCCC 2007; Barrett *et al.* 2008). Highly significant association was instead found for rs17221417, a SNP in modest linkage with the 3020insC polymorphism rs2066847, with a *P* value of 9.4×10^{-12} and an odds ratio of 1.29 (1.13–1.46) for heterozygotes and 1.92 (1.58–2.34) for homozygotes for the risk allele (WTCCC 2007). When rs2066847 was specifically investigated in replication studies it was, however, very significant ($P = 1.5 \times 10^{-24}$) and showed the largest odds ratio in case–control analysis of any replicated Crohn's disease risk locus (Barrett *et al.* 2008).

Figure 9.20 NOD2 3020insC variant. (A) DNA sequencing traces for part of *NOD2* exon 11 showing an individual without ('wild type'), heterozygous, or homozygous for the C insertion variant (rs2066847). (B) *NOD2* gene exon structure and detail for exon 11 showing the DNA sequence and amino acid sequences for wild type and 3020ins. (C) NOD2 protein domain structure illustrating the position of protein truncation. Redrawn and reprinted by permission from Macmillan Publishers Ltd: Nature (Ogura *et al.* 2001a), copyright 2001. Gene plots with genomic location and sequences adapted from screenshot of UCSC Genome Browser (Kent *et al.* 2002) (http://genome.ucsc.edu/) (Human March 2006 Assembly).

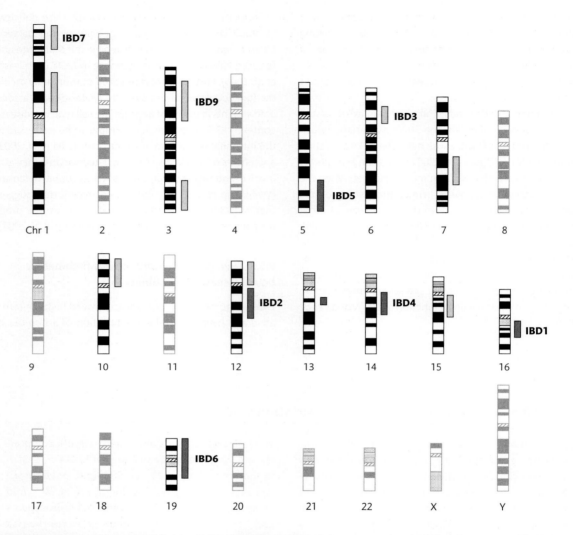

Figure 9.21 Chromosomal regions showing linkage with inflammatory bowel disease from a review of whole genome linkage scans. Ideogram demonstrating regions of significant linkage (dark grey shading) and suggestive linkage (light grey). Inflammatory bowel disease (IBD) loci 1–9 are indicated. Redrawn and reprinted by permission from Macmillan Publishers Ltd: Nature Reviews Genetics (Barrett *et al.* 2008), copyright 2008.

susceptibility loci in addition to IBD1. In 2004, Brant and colleagues reviewed 11 genome-wide linkage screens involving 1200 family pedigrees affected by inflammatory bowel disease analysed using microsatellite markers, with affected sib pairs most commonly analysed (Brant and Shugart 2004). Three loci were noted as showing 'good evidence' of linkage: IBD1 at chromosome 16q, IBD3 at chromosome 6p, and IBD5 at 5q31; 'promising'

loci included IBD2 (chromosome 12q13.2-q24.1), IBD4 (14q11-q12), IBD6 (19p13), IBD7 (1p36), IBD8 (16p), and IBD9 (3p26), together with chromosomes 2q, 4q, 7, 11p, and Xp.

For IBD5 at chromosome 5q31, linkage with Crohn's disease has been reproducibly and convincingly demonstrated and a specific risk haplotype resolved (Rioux *et al.* 2000, 2001). However, despite intensive efforts at

Population Study (year)	Sample size Cases controls	SNPs analysed	Replication Cases controls Trios	Associations observed	
				Chr	Genes or loci
Japanese	94–752	72K	484 cases + 2 European cohorts	16q12	NOD2
Yamazaki *et al.* (2005)				9q32	TNFSF15
North American (European)	946–977	304K (Illumina HumanHap 300)	353–207 530 trios	16q12	NOD2
				1p31	IL23R
Duerr *et al.* (2006) Rioux *et al.* (2007)				2q37	ATG16L1
				10q21	No known genes
				4q13	PHOX2B
				22	NCF4
				16q24	FAM92B
German	735–368	7K non-synonymous	498+1032 380 trios + UK case–control	16q12	NOD2
Hampe *et al.* (2007)				2q37	ATG16L1
				5q31	SLC22A4
Belgian/French	547–928	302K (Illumina HumanHap 300)	1266–559 428 trios	1p31	IL23R
Libioulle *et al.* (2007)				5p13.1	Gene desert
				16q12	NOD2
German	393–399	92K (Affymetrix 100K)	Y	16q12	NOD2
Franke *et al.* (2007)				5q31	
				5p13.1	Gene desert
				11p15	NELL1
British	1748–2938	469K (Affymetrix 500K)	1182–2024	16q12	NOD2
WTCCC (2007) Parkes *et al.* (2007)				1p31	IL23R
				2q37	ATG16L1
				5p13	Gene desert
				10q21	No known genes, flanked by FNF365
				3p21	Includes MST1
				5q33	IRGM
				10q24	NKX2-3
				18p11	PTPN2
Quebec (Founder population, French origin)	382 trios	164K (Perlegen)	2 German cohorts	16q12	NOD2
				5q31	
				1p31	IL23R
Raelson *et al.* (2007)				4p16.1	JAKMIP1
				3p21	Many
				2 regions chr 17	

Figure 9.22 Genome-wide association studies in Crohn disease. Redrawn and reprinted by permission from Macmillan Publishers Ltd: Nature Reviews Genetics (Mathew 2008), copyright 2008.

fine mapping and functional characterization, the identification of causative functional variants remains unresolved. Meta analysis demonstrated that linkage for IBD3
(which includes the MHC region) achieved genome-wide
significance for inflammatory bowel disease (van Heel
et al. 2004).

Markedly fewer loci have been resolved by linkage
analysis in ulcerative colitis than Crohn's disease although
IBD2 is one example where both diseases have shown
linkage, with evidence that the effect may be strongest
for ulcerative colitis (Parkes et al. 2000). An overview of
chromosomal locations of regions showing linkage to
inflammatory bowel disease based on genome-wide
linkage scans was published by Barrett and colleagues,
set in the context of more recent genome-wide association studies (Fig. 9.21) (Barrett et al. 2008).

9.5.4 Genome-wide association studies and inflammatory bowel disease

Perhaps of any common multifactorial disease, the
greatest success seen to date with genome-wide association studies has been with Crohn's disease. From 2005
to 2008, the number of clearly replicated and defined
genomic loci involved in susceptibility to this disease has
increased from a widely quoted figure of two (NOD2 at
16q12 and IBD5 at 5q31, although it could be argued
this is a conservative interpretation of the linkage data)
to more than 30 (Barrett et al. 2008). By 2008, at least
seven genome-wide scans have been performed using
different population groups and SNP marker sets with
extensive replication in independent cohorts (see Fig.
9.22) (Mathew 2008). The concordance between the
studies for the most strongly associated loci was striking, as was the relatively modest magnitude of effect size
observed. This meant that studies were relatively underpowered to define odds ratios of 1.5 or less (1.3 or below
in the case of the largest study, published by the WTCCC).
Meta analysis of three of the studies improved the power
to 74% to detect an odds ratio of 1.2 and provided evidence of an additional 21 loci involved in susceptibility
(Fig. 9.23) (Barrett et al. 2008).

What new genes or genomic loci have genome-wide
association studies of Crohn's disease identified? The first
such study to be reported defined an association with
SNPs in TNFSF15, a gene found at chromosome 9q32

that encodes tumour necrosis factor (TNF) superfamily
member 15 (Yamazaki et al. 2005). TNFSF15 is upregulated in mucosal cells lining the intestine of patients with
Crohn's disease, and in mouse models of colitis, and is an
important modulator of chronic mucosal inflammation
(Takedatsu et al. 2008). Yamazaki and colleagues analysed
72 000 SNPs in 94 affected Japanese individuals, the SNPs
selected as being informative in a Japanese population
(Yamazaki et al. 2005). Then 1888 associated SNPs were
genotyped in 484 additional cases and controls: of the
22 SNPs most strongly associated, seven were at 9q32.
Clustering of disease associated SNPs at a given locus
typically increases confidence that this represents a 'true'
susceptibility locus. A further 143 SNPs were genotyped
based on resequencing at the locus; the strongest association was with a SNP in the third intron of TNFSF15 with
a quoted odds ratio of 2.17 (OR 1.78–2.66, $P = 1.7 \times 10^{-14}$). Association was also found in two European
cohorts investigated by Yamazaki and colleagues.

In the meta analysis by Barrett and colleagues combining three European studies, the association with TNFSF15
was also found, although the odds ratio was more modest (Fig. 9.24) (Yamazaki et al. 2005; Barrett et al. 2008).
As with all the newly identified Crohn susceptibility loci,
further work is required to fine map and resolve functionally important causative variants. However the role
of TNFSF15 in activating NFkB and the inflammatory
response in a cell type relevant to disease makes this
a strong candidate for further study. As described in
the following sections, a series of other loci were to be
implicated in susceptibility to inflammatory bowel disease using the genome-wide association approach.

9.5.5 Genetic diversity in IL23R pathway genes and inflammatory bowel disease

A number of genome-wide association scans have
identified genes in the IL23R signalling pathway as
being associated with inflammatory bowel disease,
providing important new insights into disease pathogenesis and a potential new therapeutic target. IL23R
encodes a subunit of the receptor for IL23. Variants
of IL23R at chromosome 1p31 were identified from
a genome-wide association study of individuals with
Crohn's disease of European ancestry in North America
(Duerr et al. 2006; Rioux et al. 2007) and subsequently

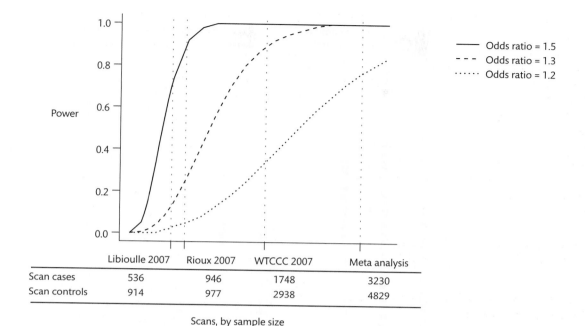

	Libioulle 2007	Rioux 2007	WTCCC 2007	Meta analysis
Scan cases	536	946	1748	3230
Scan controls	914	977	2938	4829

Scans, by sample size

Figure 9.23 Effect of sample size on power to detect common alleles associated with different odds ratios. Three different genome-wide association studies of Crohn disease (Libioulle *et al.* 2007; Rioux *et al.* 2007; WTCCC 2007) and a meta analysis (Barrett *et al.* 2008) were analysed for power to define odds ratios of 1.2, 1.3, and 1.5. Power was defined as probability of $P < 10^{-5}$ in a genome-wide association scan with a multiplicative model and a risk allele frequency of 20%. Reprinted by permission from Macmillan Publishers Ltd: Nature Reviews Genetics (Barrett *et al.* 2008), copyright 2008.

in scans of Belgian and British populations, and a French Quebec founder population (see Fig. 9.22) (Libioulle *et al.* 2007; Parkes *et al.* 2007; Raelson *et al.* 2007; WTCCC 2007).

Duerr and colleagues analysed 304 000 SNPs in patients with ileal Crohn's disease, reporting data for a scan of 547 cases and 548 controls, although this number was increased to a genome-wide association study of 946 cases and 977 controls in the full dataset subsequently published (Duerr *et al.* 2006; Rioux *et al.* 2007). As with many of these studies, a strong association was found for *NOD2* (rs2066843) (see Fig. 9.22) but strikingly, multiple SNPs showed association at *IL23R* (Fig. 9.25). This was confirmed in several replication cohorts (case–control and family-based), including analysis of patients with ulcerative colitis, such that ten SNPs were associated with inflammatory bowel disease with *P* values reported as ranging between 3.5×10^{-9} and 6.6×10^{-19}.

A number of intronic *IL23R* SNPs showed association with susceptibility to Crohn's disease but the most notable association was with a nonsynonymous SNP (rs11209026, c.1142G>A, p.R318Q) that results in an arginine to glutamine substitution at amino acid 381. Possession of the glutamine encoding variant was associated with protection from Crohn's disease, with an odds ratio of 0.26 (0.15–0.43) among individuals of non-Jewish European ancestry (Duerr *et al.* 2006). The reported associations with Crohn's disease were confirmed by several independent groups (note a number of these relate to rs11465804, an intronic SNP in linkage with rs11209026) (see Fig. 9.22). The same nonsynonymous *IL23R* SNP, rs11209026, was also demonstrated to be strongly associated with ulcerative colitis (OR 0.53, $P = 8.9 \times 10^{-8}$) when analysed in 1841 ulcerative colitis cases and 1470 controls; further variants at *IL23R* also showed independent association with disease risk (Fisher *et al.* 2008).

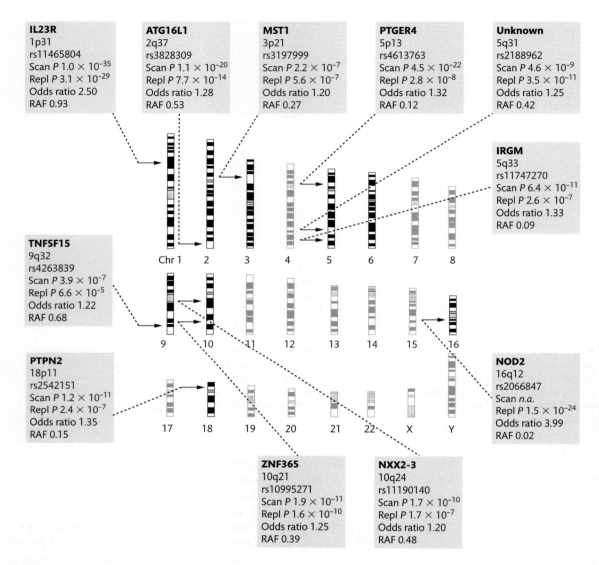

IL23R
1p31
rs11465804
Scan P 1.0 × 10^{-35}
Repl P 3.1 × 10^{-29}
Odds ratio 2.50
RAF 0.93

ATG16L1
2q37
rs3828309
Scan P 1.1 × 10^{-20}
Repl P 7.7 × 10^{-14}
Odds ratio 1.28
RAF 0.53

MST1
3p21
rs3197999
Scan P 2.2 × 10^{-7}
Repl P 5.6 × 10^{-7}
Odds ratio 1.20
RAF 0.27

PTGER4
5p13
rs4613763
Scan P 4.5 × 10^{-22}
Repl P 2.8 × 10^{-8}
Odds ratio 1.32
RAF 0.12

Unknown
5q31
rs2188962
Scan P 4.6 × 10^{-9}
Repl P 3.5 × 10^{-11}
Odds ratio 1.25
RAF 0.42

IRGM
5q33
rs11747270
Scan P 6.4 × 10^{-11}
Repl P 2.6 × 10^{-7}
Odds ratio 1.33
RAF 0.09

TNFSF15
9q32
rs4263839
Scan P 3.9 × 10^{-7}
Repl P 6.6 × 10^{-5}
Odds ratio 1.22
RAF 0.68

PTPN2
18p11
rs2542151
Scan P 1.2 × 10^{-11}
Repl P 2.4 × 10^{-7}
Odds ratio 1.35
RAF 0.15

NOD2
16q12
rs2066847
Scan *n.a.*
Repl P 1.5 × 10^{-24}
Odds ratio 3.99
RAF 0.02

ZNF365
10q21
rs10995271
Scan P 1.9 × 10^{-11}
Repl P 1.6 × 10^{-10}
Odds ratio 1.25
RAF 0.39

NXX2-3
10q24
rs11190140
Scan P 1.7 × 10^{-10}
Repl P 1.7 × 10^{-7}
Odds ratio 1.20
RAF 0.48

Figure 9.24 Established loci in Crohn disease. Eleven loci are shown which had been previously shown to be robustly associated with Crohn disease, achieving genome-wide levels of significance and convincing replication (based on genome-wide association studies and earlier work defining *NOD2* and 5q31 (IBD5) by linkage and other approaches). For each locus, chromosomal location and an rs number of a specific illustrative SNP are shown together with data to highlight the significance of the loci from a meta analysis by Barrett and colleagues (2008) of three genome-wide association scans among individuals of European descent (Libioulle *et al.* 2007; Rioux *et al.* 2007; WTCCC 2007). 'Scan' refers to significance values from the analysis of 3230 cases of Crohn disease and 4829 controls from these three studies; 'Repl' results from a replication study of a further 2325 cases and 1809 controls as well as 1339 family trios; 'RAF' risk allele frequency in controls. Prepared from table 2 and reprinted by permission from Macmillan Publishers Ltd: Nature Reviews Genetics (Barrett *et al.* 2008), copyright 2008.

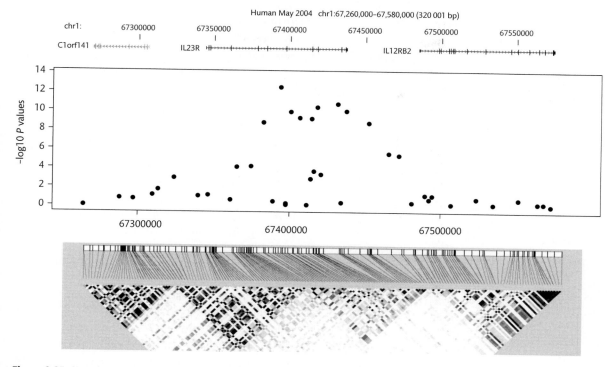

Figure 9.25 SNP markers show association with *IL23R* at 1p31. Data for SNPs from combined case–control cohorts were analysed by Duerr and colleagues. Negative $\log_{10}$ associated *P* values are shown together with a pairwise r^2 plot for individuals of European geographic ancestry from the CEU HapMap panel, demonstrating discrete haplotypic blocks with no association in the block spanning neighbouring gene *IL12RB2*. Adapted from Duerr *et al.* (2006), reprinted with permission from AAAS. Gene plots with genomic location adapted from screenshot of UCSC Genome Browser (Kent *et al.* 2002) (http://genome.ucsc.edu/) (Human May 2004 Assembly).

These genetic studies, added to a range of other data from animal models and studies of affected patients, suggest IL23 may play a critical role in the pathogenesis of inflammatory bowel disease (Neurath 2007). IL23 is an important proinflammatory cytokine made up of two subunits, p19 (encoded by *IL23A*) and p40 (encoded by *IL12B*) (Fig. 9.26). Antibodies to p40 block both IL12 and IL23 signalling, and were shown to suppress inflammation in an animal model of Crohn's disease (Neurath *et al.* 1995) and to significantly improve clinical response and remission in active human disease (Mannon *et al.* 2004). Antibodies to p19 or a knock-out also suppress intestinal inflammation in mouse models and it is thought that IL23 plays an important role in both innate and

T cell mediated inflammation occurring in the intestine, central to inflammatory bowel disease pathogenesis (Neurath 2007). Fundamental to this may be the role of IL23 in activating T helper 17 lymphocytes producing interleukin-17, a key mediator of chronic inflammation found at high levels in the intestinal mucosa of patients with Crohn's disease (Fujino *et al.* 2003); pathogenesis may also involve effects of IL23 on mucosal barrier function (Cho 2008).

Interest in the role of the IL23R pathway in disease pathogenesis is further highlighted by the finding that SNPs in *IL12B* at chromosome 5q33 (encoding the p40 subunit shared by IL23 and IL12), *STAT3* at chromosome 17q21 (encoding signal transducer and activator of

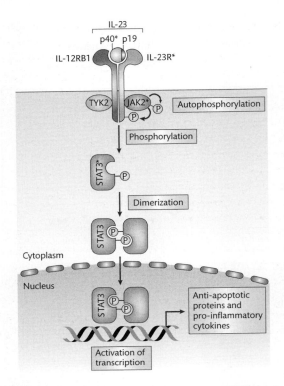

Figure 9.26 Genetic associations with the IL23R pathway in Crohn's disease. Disease associations have been found for SNPs in genes encoding different proteins involved in this signalling pathway (indicated by asterixes) including the p40 subunit of IL23; the IL23R subunit of the IL23 receptor; Janus kinase 2 (JAK2), which is activated on receptor–ligand binding; and signal transducer and activator of transcription (STAT3), which is recruited, phosphorylated, homodimerized, and translocated into the nucleus to activate transcription. Reprinted by permission from Macmillan Publishers Ltd: Nature Reviews Immunology (Cho 2008), copyright 2008.

transcription 3), and *JAK2* at chromosome 9p24 (encoding Janus kinase 2) were associated with Crohn's disease (Fig. 9.26) (Barrett *et al.* 2008). Strikingly, SNPs of *IL12B* and *STAT3* were also associated with ulcerative colitis (Fisher *et al.* 2008; Franke *et al.* 2008). Potential common mechanisms underlying autoimmune disease are also underlined by the finding that SNPs in *IL12B* and *IL23R* are associated with psoriasis (Cargill *et al.* 2007) while an association study of 14 500 nonsynonymous

SNPs showed significant association between variation at *IL23R* and ankylosing spondylitis (Burton *et al.* 2007). This phenomenon is also observed for other Crohn's disease associated loci, for example *PTPN2* was also associated with type 1 diabetes (WTCCC 2007) and 5q31 with psoriasis (Chang *et al.* 2008).

9.5.6 Autophagy and Crohn's disease

An unexpected insight into disease pathogenesis was provided by the genome-wide associations found involving autophagy (Box 9.6), specifically with SNPs at *ATG16L1* on chromosome 2q37 encoding ATG autophagy-related 16-like protein, and *IRGM* at chromosome 5q33 encoding the immunity-related GTPase family M (see Fig. 9.24).

Hampe and colleagues carried out a nonsynonymous SNP scan involving 7159 informative SNPs in 735 Crohn's disease cases and 368 controls with replication of the observed association with *ATG16L1* in independent German and UK cohorts (Hampe *et al.* 2007). Resequencing in affected individuals of exonic, splice site, and promoter sequence of *ATG16L1*, further genotyping, and haplotypic analysis served to confirm the initial observation that a specific nonsynonymous SNP, rs2241880 (c898A>G, p.T300A), was responsible for all the observed increased disease risk at this locus. Possession of the risk G allele resulted in a threonine to alanine substitution at amino acid position 300 in the N-terminal region of ATG16L1 protein. The odds ratio of disease among the German cases analysed was 1.45 (1.21–1.74); no association with susceptibility to ulcerative colitis was found.

This result was replicated by Rioux and colleagues in their large genome-wide association scan of ileal Crohn's disease for North American populations, the same nonsynonymous SNP (rs2241880; $P < 5 \times 10^{-8}$) completely accounting for the association signal at this locus (Rioux *et al.* 2007). *ATG16L1* is expressed in the intestinal epithelium (Hampe *et al.* 2007; Rioux *et al.* 2007) and played a key role in autophagy when investigated by RNA interference experiments to '**knockdown**' gene expression in the setting of *Salmonella typhimurium* infection (Rioux *et al.* 2007). The functional consequence of the specific nonsynonymous *ATG16L1* variant

> ## Box 9.6   Autophagy
>
> Autophagy or 'self devouring' describes the process whereby proteins and other intracellular components can be degraded in lysosomes within a cell. Autophagy plays many critical roles, including removal of intracellular bacteria as part of the innate immune response.

remains unresolved but could promote intracellular bacterial growth.

The WTCCC showed evidence of a novel disease association with a cluster of SNPs around *IRGM* on chromosome 5q33 (WTCCC 2007). The strongest signal was with an intergenic SNP near *IRGM*, rs1000113 ($P = 5.1 \times 10^{-8}$); possession of a copy of the T allele was associated with an odds ratio of 1.54 (1.31–1.82) (WTCCC 2007). The association with *IRGM* was replicated in an independent cohort of 1182 individuals of European ancestry and 2024 population controls, specifically with two SNPs flanking *IRGM*, rs13361189 and rs4958847, with *P* values of 2.1×10^{-10} and 3.8×10^{-9} in the combined panels of 2930 cases and 4962 controls (Parkes *et al.* 2007). Resequencing failed to identify any more significant variants or functional candidates. The association with *IRGM* was, however, highly intriguing; like *ATG16L1*, the encoded protein appears to play a key role in the control of intracellular pathogens with evidence for example that *IRGM* induces autophagy to control intracellular mycobacteria (Singh *et al.* 2006).

Subsequently McCarroll and colleagues found striking evidence that the disease association arises from an insertion/deletion polymorphism in complete linkage disequilibrium with the most strongly associated SNP rs13361189 (McCarroll *et al.* 2008). They resolved that possession of a 20 kb deletion upstream of *IRGM* was associated with increased risk of Crohn's disease and with altered levels of gene expression. Analysis of allele-specific gene expression among cell types and cell lines heterozygous for a linked exonic SNP marker suggests this may be cell type-specific but potentially highly significant given the role of *IRGM* in antibacterial autophagy, which the investigators demonstrated using siRNA and by *IRGM* overexpression experiments.

9.5.7 Insights from the Wellcome Trust Case Control Consortium study

The WTCCC genome-wide association study of Crohn's disease analysed 469 000 SNPs in 2000 cases of Crohn's disease and 3000 shared controls (WTCCC 2007) (Section 9.3.2). A total of nine genomic loci were associated with disease susceptibility at a genome-wide level of significance, including five previously identified regions *NOD2*, *IL23R*, *ATG16L1*, 10q21 (14 kb telomeric to *ZNF365*) (Rioux *et al.* 2007), and a cluster of SNPs in a gene desert on 5p13.1 (Libioulle *et al.* 2007; WTCCC 2007) (see Fig. 9.22). Apart from *IRGM*, three other novel loci were found at genome-wide significance. These include SNPs at 3p21 in a region containing many genes, notably *MST1* (encoding macrophage stimulating 1 protein involved in inducing phagocytosis), SNPs within *NKX2–3* at 10q24.2 (encoding a homeodomain containing transcription factor protein called 'NK2 transcription factor-related locus 3' thought to influence lymphocyte migration and intestinal inflammation), and SNPs upstream of *PTPN2* at 18p11 (encoding protein tyrosine phosphatase non-receptor type 2, a negative regulator of inflammation) (WTCCC 2007). All three loci were replicated in a cohort of European descent (Parkes *et al.* 2007) and later meta analysis (Barrett *et al.* 2008). Association with ulcerative colitis was found for *MST1* at 3p21 and *NKX2–3* but not *PTPN2* (Fisher *et al.* 2008). Many other loci showed some evidence of association in the WTCCC scan but did not achieve genome-wide significance. Twenty-five loci with $P < 10^{-5}$ were followed up in the replication study by Parkes and colleagues (2007). Of these, five loci were significant, including *IL12B*, the IBD5 locus on chromosome 5q31, *FLJ45139* on chromosome 21 (a gene of unknown function), and two gene deserts on chromosome 1q (Parkes *et al.* 2007).

9.5.8 Gene deserts and other loci

The finding of disease associations with SNP markers in regions without any genes (so-called 'gene deserts') is seen for several loci in Crohn's disease as well as other diseases studied by genome-wide association scans. The loci appear highly significant and reproducible in replication studies. For example, Libioulle and colleagues in a genome-wide association scan used 302 000 SNPs to analyse 547 cases and 928 controls in a Belgian population (Libioulle *et al.* 2007). They found three loci to show strong evidence of disease association with P values between 10^{-6} and 10^{-9}: *IL23R* and *NOD2*, together with multiple SNPs in a 250 kb region at 5p13.1, including four at $P < 10^{-7}$. These results were replicated in 1266 additional cases and 559 controls, together with 428 trios and fine mapping performed using 111 SNP markers. The associated markers and haplotypes fell within a 1.25 Mb gene desert flanked by *DAB2* and *PTGER4* (encoding prostaglandin E receptor 4 (subtype EP4)). There is preliminary evidence that the variant(s) may be modulating gene expression, as two disease associated SNPs showed strong association with levels of gene expression of *PTGER4* in lymphoblastoid cell lines (Section 11.4.2) (Dixon *et al.* 2007; Libioulle *et al.* 2007).

Other gene loci have been reported in genome-wide association scans. Franke and colleagues showed association with *NELL1* (nel-like 1 precursor) for Crohn's disease and ulcerative colitis, and postulated a functional link with *PTGER4* as *Nell1*-deficient mice showed downregulation of this gene (Desai *et al.* 2006; Franke *et al.* 2007). In their genome-wide association study, Rioux and colleagues identified a number of other variants including a promoter SNP of *PHOX2B* (encoding paired like homeobox 2B) at chromosome 4p12, *FAM92B* (family with sequence similarity 92, member B) at 16q24, and an intronic SNP of *NCF4* (neutrophil cytosolic factor 4) at 22q13 (Rioux *et al.* 2007). The list of disease associated loci continues to grow, notably with analyses sufficiently powered to detect modest odds ratios of the order of 1.2 – such as the meta analysis by Barrett and colleagues that identified 21 new loci of which 19 showed replication in an independent set of 2325 Crohn's disease cases, 1809 controls, and 1339 trios (Barrett *et al.* 2008). Some have been discussed in the preceding text, others include *CDKAL1* (CDK5 regulatory subunit associated protein 1-like 1) at 6p22, which was also associated with type 2 diabetes, and *PTPN22* (protein tyrosine phosphatase non-receptor type 22) at 1p13, in which the same nonsynonymous SNP was associated with protection from Crohn's disease and susceptibility to type 1 diabetes and rheumatoid arthritis (Barrett *et al.* 2008).

Genetic studies of Crohn's disease have proved a very powerful approach to advancing our understanding of disease susceptibility and pathogenesis, and with time will hopefully lead to advances in disease management and targeted treatment, our ability to diagnose disease, and predict risk. The number of associated genomic loci is now very large but how much of the risk attributable to genetic factors has been explained? Barrett and colleagues estimate that the 32 loci identified explained approximately 20% of the genetic risk and 10% of the overall disease risk variation between individuals (Barrett *et al.* 2008). However, this is likely to be the lower boundary of this estimate as multiple variants are likely to contribute at many loci and the causative variants remain largely to be defined. These are likely to be associated with higher risk, as evidenced by the modest risk associated on genome-wide scans for genotyped markers at *NOD2* versus the 3020insC variant thought to be functionally important.

Meta analysis and further large association scans in diverse populations should help to define further variants with modest effect sizes, while resequencing analyses and other strategies should allow definition of rare variants contributing to disease susceptibility. Structural genomic variation, including copy number variation, is also likely to play an important role as highlighted by the recent report that a 20 kb deletion polymorphism near *IRGM* was responsible for the observed disease association at this locus. Overall, however, fine mapping and definition of functionally important causative variants remain a major challenge to be overcome. Finally, it is worth emphasizing that environmental factors play a critical role in Crohn's disease pathogenesis and it is the synthesis of research in this area with genetic studies that should allow further advances in our understanding of this common and debilitating disease.

9.6 Summary

Crohn's disease provides a remarkable example of a success story in research to define the genetic factors responsible for the observed heritable component in common multifactorial diseases and other traits. The story of research in this disease illustrates well the timeline of different approaches and advances in this field. We have seen how, despite the many caveats attached to linkage studies in common disease, success was achieved using linkage analysis with many loci implicated and in some instances robust definition of specific genes and variants as exemplified by the NOD2 locus. However, among the remaining four inflammatory bowel disease susceptibility loci and other genomic regions identified by linkage, there were significant issues with either replication or implicating specific genes or variants. It has only been with the more recent application of genome-wide association studies that further dramatic insights have been made, with over 30 loci resolved and unexpected new insights into disease pathogenesis achieved. These include genetic variation in genes involved in the IL23R pathway and autophagy, as well as gene deserts or loci containing multiple candidate genes.

The success of the genome-wide association approach in common multifactorial traits should be tempered by realization that the results of such studies still only account for a small proportion of the genetic risk. Moreover, the major road block of fine mapping the observed disease associations and defining the specific causal functional variants remains largely to be overcome. The view that common disease is associated with multiple loci of individually modest magnitude of effect has been upheld by genome-wide association studies, and increasingly large studies are being performed or proposed that will have sufficient statistical power to detect small, if highly significant, effects. The value of meta analysis has also been highlighted in this chapter, with standardised curation and public availability of published studies set to facilitate this process.

The human genome sequencing projects, the SNP Consortium and the International HapMap Project, together with many other studies, have served to highlight the extent and nature of fine scale nucleotide diversity. Dramatic insights have been gained into the nature and diversity of underlying allelic architecture, notably how diversity arises and is inherited with our understanding of haplotypes, recombination, and linkage disequilibrium significantly advanced. Population-specific differences have been highlighted, notably the remarkable diversity at a nucleotide and haplotypic level within populations of African ancestry. Genotyping common, informative, haplotype tagging SNPs has been shown to be a highly effective approach, with the technological advances in ultrahigh throughput genotyping and the availability of large cohorts of carefully phenotyped clinical samples allowing genome-wide association studies to become a reality. Great success has been achieved, as illustrated by specific studies such as the Wellcome Trust Case Control Consortium (WTCCC 2007) and in diseases including age-related macular degeneration and Crohn's disease, as discussed in this chapter.

However, much more remains to be done if the complexities of the genetics of common disease and other traits are to be resolved. As preceding chapters have demonstrated, SNPs and other fine scale nucleotide variations are not the only class of genetic diversity and many types of structural genomic variation are likely to be important. Advances in resequencing and other technologies should enable investigation of the broad range of known diversity among specific phenotypes. Moreover the contribution of intermediate penetrance, low frequency variants remains largely unexplored, falling as they do between rare variants with high penetrance generally well resolved by linkage, and common variants with low penetrance currently being defined by genome-wide association scans (McCarthy et al. 2008). In the following chapters, the implications of the analysis of fine scale diversity for our understanding of the genetics of gene expression and selection are discussed. The tale of SNPs is not yet complete, but again it is worth emphasizing that while SNPs are remarkable markers and tools with which to gain insight into the nature and consequences of genetic diversity, they form only part of the remarkably rich tapestry that is human genetic variation.

9.7 Reviews

Reviews of subjects in this chapter can be found in the following publications:

Topic	References
Genome-wide association studies	Ioannidis 2007; Altshuler *et al.* 2008 Hindorff *et al.* 2008; Kruglyak 2008; McCarthy *et al.* 2008
The International HapMap Project	Manolio *et al.* 2008
Variation at the *FTO* gene and obesity	Loos and Bouchard 2008
Pathophysiology of age-related macular degeneration	de Jong 2006; Jager *et al.* 2008
Genetics of age-related macular degeneration	Swaroop *et al.* 2007; Allikmets and Dean 2008
Epidemiology and pathogenesis of inflammatory bowel disease	Loftus 2004; Xavier and Podolsky 2007; Cho 2008
Genetics of inflammatory bowel disease	Brant and Shugart 2004; Loftus 2004; Duerr 2007; Xavier and Podolsky 2007; Barrett *et al.* 2008; Cho 2008; Mathew 2008

CHAPTER 10

Fine scale sequence diversity and signatures of selection

10.1 Introduction

In the previous chapter the role of fine scale sequence diversity in common disease was discussed, in particular how advances in our understanding of genomic architecture through the establishment of dense maps of single nucleotide polymorphism (SNP) markers has facilitated genome-wide association studies in common multifactorial traits. In this chapter the relationship between genetic diversity and selection is considered at a broader level. Human diseases, notably infectious disease, have been one of a number of major selective pressures operating over our recent evolutionary past for which we can now find evidence through analysis of the nature of genetic diversity within and between populations, and on comparison with other species.

A number of examples of the consequences of selection for observed genetic variation are discussed over the course of this chapter as well as elsewhere in this book, for example when genes involved in susceptibility to major parasitic diseases of man such as malaria are considered (Section 13.2). The publication of draft genome sequences for the common chimpanzee *Pan troglodytes* and rhesus macaque *Macaca mulatta* has dramatically advanced our ability to analyse and understand human genetic diversity, notably in terms of selection. A detailed review of two well characterized examples of genes subject to positive selection, namely variants involving the *LCT* gene and lactase persistence, and the *SLC24A5* gene and skin pigmentation, is presented to illustrate many of the themes inherent to analysis of genetic diversity and defining gene loci subject to positive selection. The role of genome-wide SNP marker sets across populations in resolving evidence of selection is also discussed.

10.2 Genetic diversity and evidence of selection

Patterns of genetic diversity across the human genome reflect in part the selection pressures that have operated to increase or decrease the frequency of particular sequence and structural genomic variants found in the human population. The vast majority of variants are thought to be neutral and vary in frequency within the population, with a balance between new mutations arising and variants being lost from the population by genetic drift (Kimura 1983). If a particular allele is deleterious or possesses a selective advantage it may more rapidly progress to become 'fixed' in the population, either being lost or rising in frequency to 100%, respectively. Deviations in frequency from those expected based on neutral theory may indicate selection, with a variety of statistical tests used such as Tajima's D (Tajima 1989). Here deviation from the null hypothesis of neutrality with an excess of high frequency variation is consistent with balancing selection, as seen within genes encoding human leukocyte antigen (HLA) molecules in the major histocompatibility complex (MHC) (Chapter 12); or with an excess of low frequency variation consistent with newly arising rare mutations after positive selection leading to a selective sweep, as discussed below.

Positive selection leads to a number of characteristic patterns of genetic diversity, dubbed 'signatures of selection', which have been very informative to developing strategies for detecting loci and specific variants that have been subject to positive selection. In a recent review, Sabeti and colleagues described approaches used for detecting evidence of signatures of positive selection and noted that different methods have particular advantages or limitations, for example in terms of the time scale within which selection events can be detected (Fig. 10.1) (Sabeti *et al.* 2006). The basis and nature of different signatures of selection are outlined in the following paragraphs and reflect intensive research across disciplines.

10.2.1 Hitch-hiking and selective sweeps

Genetic variants may be subject to positive selection for a variety of reasons, notable examples being conferring an advantage in terms of susceptibility to infectious diseases such as malaria, or to prevailing environmental factors such as habitat or food sources associated with human migration, climate change, and development of agricultural societies. If a new mutation is subject to strong selective pressure it will rapidly rise in frequency in the population. Associated 'neutral' sequence variants borne by the haplotype on which the new mutation arose will also rise in frequency within the population, a phenomenon known as 'genetic hitch-hiking' (Maynard-Smith and Haigh 1974). The speed with which the beneficial variant increases in frequency will influence the extent to which this occurs, as with time recombination will occur and new mutations will arise, degrading the extent of linkage disequilibrium and haplotypic structure.

However, with variants subject to selection relatively recently, characteristic patterns or signatures of selection may be observed related to genetic hitch-hiking. There may be loss of diversity (heterozygosity) around the selected variant indicative of a 'selective sweep'; in an extreme case with no recombination, all linked variation would be lost and subsequently a relative excess of rare alleles is seen as new mutations arise. The extent of loss of diversity will depend on local levels of recombination, reducing with distance from the positively selected variant, with observed patterns dependent also on time since the hitch-hiking event. A number of statistical approaches have been developed to detect such regions of low heterozygosity from the analysis of sequence diversity (Kim and Stephan 2002; Nielsen *et al.* 2005b) and can allow estimation of the time that has elapsed since a selected allele became fixed in the population (Przeworski 2003). Where the favoured allele has not become fixed in the population, tests of deviation from neutrality for variation

at linked markers have been used based on local linkage and haplotypic variation (Hudson *et al.* 1994; Slatkin and Bertorelle 2001).

10.2.2 Extended haplotypes of high frequency

The haplotype on which the variant has arisen and been subject to selection may be found to be present at a high frequency in the population yet be unexpectedly extensive in length. This contrasts with the usual situation arising under neutral theory in which common haplotypes are typically older and relatively shorter due to degradation by recombination and mutation. A variety of statistical tests have been developed to analyse such long haplotypes, including measuring decay of association over distance for particular haplotypes by the extended haplotype homozygosity (EHH) test (Sabeti *et al.* 2002a). A haplotype showing very high EHH present at high frequency in the population indicates a variant that has been selected and increased in frequency faster than expected based on neutral conditions. Application to loci associated with malarial resistance such as *G6PD*, encoding glucose-6-phosphate dehydrogenase, where low activity alleles have been associated with prevalence and resistance to malaria, has illustrated the utility of the approach for analysing recent events over the last 10 000 years (Fig. 10.2; Section 13.2.5) (Ruwende and Hill 1998; Tishkoff *et al.* 2001; Sabeti *et al.* 2002a).

10.2.3 Differences in allele frequency

Marked differences in allele frequency between populations can also provide evidence that a locus has been subject to selection, with differing selection pressures between the populations (Weir and Cockerham 1984). This has been highly informative for recent events, notably with genes subject to selection by malaria with the allele frequency of particular variants of the *HBB* gene (such as Hb S) and *DARC* (Duffy blood group, chemokine receptor) gene geographically strongly related to malarial endemicity (discussed further in Section 13.2). The utility of the approach requires some degree of reproductive isolation between populations and is therefore more informative for human populations after the proposed migration of modern humans out of Africa over the last 75 000 years (Box 8.5).

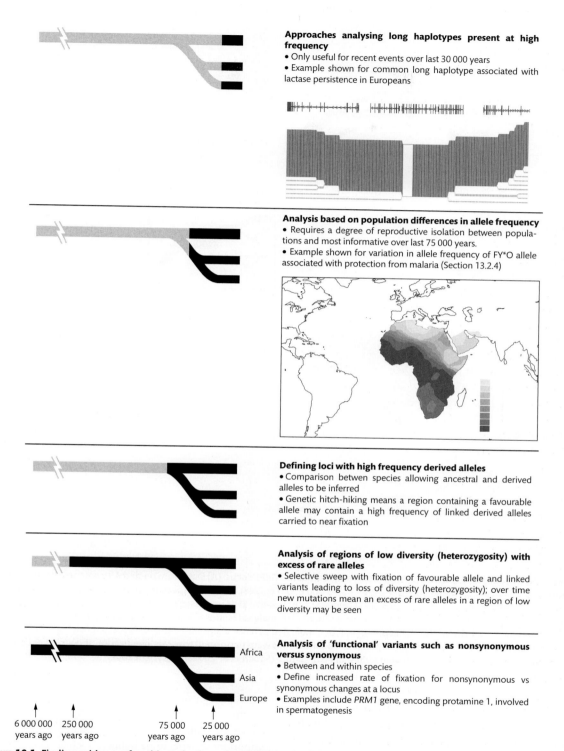

Approaches analysing long haplotypes present at high frequency
- Only useful for recent events over last 30 000 years
- Example shown for common long haplotype associated with lactase persistence in Europeans

Analysis based on population differences in allele frequency
- Requires a degree of reproductive isolation between populations and most informative over last 75 000 years.
- Example shown for variation in allele frequency of FY*O allele associated with protection from malaria (Section 13.2.4)

Defining loci with high frequency derived alleles
- Comparison betwen species allowing ancestral and derived alleles to be inferred
- Genetic hitch-hiking means a region containing a favourable allele may contain a high frequency of linked derived alleles carried to near fixation

Analysis of regions of low diversity (heterozygosity) with excess of rare alleles
- Selective sweep with fixation of favourable allele and linked variants leading to loss of diversity (heterozygosity); over time new mutations mean an excess of rare alleles in a region of low diversity may be seen

Analysis of 'functional' variants such as nonsynonymous versus synonymous
- Between and within species
- Define increased rate of fixation for nonsynonymous vs synonymous changes at a locus
- Examples include *PRM1* gene, encoding protamine 1, involved in spermatogenesis

Africa

Asia

Europe

6 000 000 years ago   250 000 years ago   75 000 years ago   25 000 years ago

Figure 10.1 Finding evidence of positive selection using different approaches. Figures and text adapted from Sabeti *et al.* (2006), reprinted with permission from AAAS.

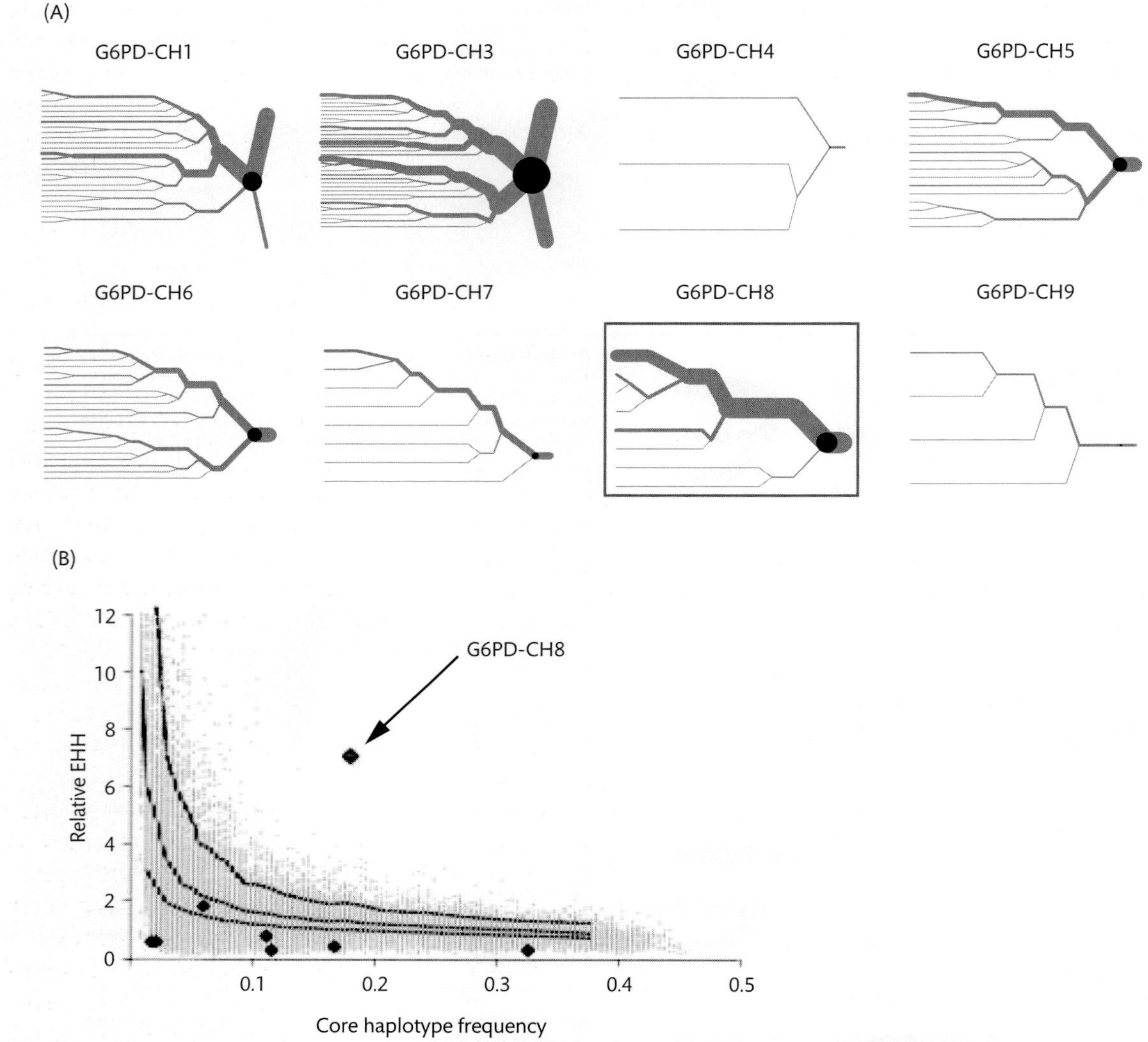

(A)

(B)

Figure 10.2 Examples of haplotype bifurcation diagrams at the *G6PD* locus. (A) Data from pooled African populations illustrates how a specific haplotype (*G6PD* core haplotype 8, G6PD-CH8) has an unexpectedly high degree of extended haplotype homozygosity (EHH) given the allele frequency. (B) Relative EHH was assessed over distance from the core region in which the haplotypes were defined; for the most distant SNP analysed (some 413 kb proximal to *G6PD*) relative EHH versus allele frequency is shown for the eight core haplotypes, indicated as black diamonds. G6PD-CH8 had a much higher EHH than haplotypes of comparable frequency (extensive simulations were performed to assess significance, shown in light grey). Reprinted by permission from Macmillan Publishers Ltd: Nature (Sabeti *et al.* 2002a), copyright 2002.

10.2.4 Comparisons between species

Comparison of sequence diversity between and within species has also been highly informative in studies looking for evidence of positive selection – work greatly facilitated and enabled by publication of the human genome sequence and that of other species, notably the chimpanzee and mouse. Comparing between species allows sites with unusual rates of apparently functionally important variants to be defined, such as loci with high rates of nonsynonymous to synonymous variants (most commonly nonsynonymous changes in coding DNA are deleterious and selected against, but they can rarely be advantageous). The approach has the advantage of being able to look back over distant events up to millions of years ago, however multiple events are needed to distinguish a locus from background 'neutral' rates of change.

For example, an excess of nonsynonymous compared to synonymous or intronic nucleotide substitutions was found among hominids and Old World monkeys in *PRM1*, a gene encoding proteins that bind sperm during spermatogenesis, when compared to ruminants or rodents, suggesting positive selection (Rooney and Zhang 1999). **Comparative genomics** also allows the ancestral state to be defined and, hence, high frequency derived alleles to be resolved that may have arisen through linkage with the selected allele. Finding an excess of derived variants at high frequency is another signature of selection and can be tested for by comparing the excess of high compared to intermediate frequency derived variants (Fay and Wu 2000).

10.3 Evidence for selection at a nucleotide level from sequencing the chimpanzee and macaque genomes

10.3.1 Diversity between the human and chimpanzee genomes

The publication of the draft sequence for the chimpanzee genome (Box 10.1) allowed for significant advances in our ability to interrogate the human genome to find evidence of selection (CSAC 2005). At a single nucleotide level, divergence between the human and chimpanzee genomes occurred at a rate of 1.23%, of which 1.06% or less was fixed between species. The level of divergence varied significantly between and within chromosomes (Fig. 10.3A). Divergence between chromosomes was most marked for the X and Y chromosomes (0.94% and 1.9%, respectively), the higher rate within the Y chromosome felt most likely to reflect the known higher rate of mutation in male germlines associated with higher numbers of cell divisions (Li *et al.* 2002). Within chromosomes, levels of divergence were highest in the terminal 10 Mb, particularly the telomeres, regions associated with increased recombination and G+C content (Fig. 10.3B).

In terms of the specific nucleotide substitutions involved, a marked excess involved CpG dinucleotides (25.2% of all observed substitutions) despite these comprising only a small minority of bases aligned between the two species (2.1%): as noted in Chapter 1, methylation and deamination means these nucleotides have a much higher rate of mutation (Box 1.17).

One major insight into the action of selective pressures, which the availability of the draft sequence for the chimpanzee genome now allowed, was to analyse more completely the rate of nonsynonymous to synonymous nucleotide substitutions within coding sequences for orthologous genes and local intronic substitution rates (the K_A/K_S and K_A/K_I ratios, respectively). Considering divergence within the two hominid genomes, human and chimpanzee, the Chimpanzee Sequencing Analysis Consortium (CSAC) found that both K_A/K_S and K_A/K_I gave ratios of 0.23, consistent with an excess of negative 'purifying' selection leading to loss of amino acid changing variants (CSAC 2005). Intriguingly, the K_A/K_S ratio was not significantly different to that observed for common human alleles within the Phase I HapMap study (Section 9.2.4); the hominid K_A/K_S ratios were, however, much higher than in the mouse and rat genomes ($K_A/K_S = 0.13$).

Only 4.4% of the 13 454 orthologous genes analysed between the human and chimp genomes had a K_A/K_I ratio greater than one, suggesting strong positive selection. These differences included genes involved in our interaction with major human pathogens such as malaria (glycophorin C), intracellular pathogens such as tuberculosis (granylsin), reproduction (protamines, semenogelins), and nociception (mas-related gene family). Perhaps more intriguing were the results of analysis based on local clustering of genes showing similiar evidence of selection as well as analysis based on functional categorization. The former implicated genes involved in host defence and chemosensation while the latter showed genes with evidence of selection-involved immunity, host defence, reproduction, and olfaction. Excess negative (purifying) selection was associated with significantly low K_A/K_S ratios and involved intracellular signalling, metabolism, and brain development and function (CSAC 2005). An intriguing result of analysing divergence between human and chimpanzee lineages was the finding of an

---

Box 10.1 Sequencing the chimpanzee genome

In 2005 the genome sequence for the common chimpanzee was published originating from a single, male, captive-born chimpanzee called Clint, who was a descendant of West African chimpanzees subspecies *Pan troglodytes verus* (CSAC 2005). The sequence was a consensus of each pair of autosomes, the reference sequence based at heterozygous sites on nucleotides arbitrarily selected from the two possible alleles. A whole genome shotgun sequencing approach was used with the draft sequence having 94% genome coverage at more than 98% quality. In the same publication, sequence data were also presented for four other West African chimpanzees and three Central African chimpanzees (*P. t. troglodytes*). This allowed the discovery of 1.66 million SNPs among these chimpanzees, with 1 million heterozygous in Clint. Two-fold higher rates were found among the Central African chimpanzees. Comparison of the human and chimpanzee genomes revealed about 35 million nucleotide substitution events accounting for a 1.23% sequence difference between the two genomes. There were about 5 million insertions/deletions (indels), and in terms of overall sequence this was a much more substantial source of species difference, at 3%, a total of about 90 Mb, of which approximately half was specific to each species. Human–chimpanzee differences at the level of indels are discussed further in Section 6.6.1, insights into structural genomic variation over the course of Chapters 3 to 7, and transposable elements in Chapter 8. Kehrer-Sawatzki and Cooper have also published a detailed review of insights into evolution based on human–chimpanzee comparisons (Kehrer-Sawatzki and Cooper 2007).

---

accumulation of 47% more nonsynonymous variants in genes encoding transcription factor proteins among humans, suggesting selection related to diversity in regulation of gene expression.

The CSAC were also able to look for large regions of reduced genetic diversity between the human and chimpanzee genomes, an important signature of selection as described earlier in this section (CSAC 2005). Six such regions were resolved; these also had a high frequency of derived alleles consistent with strong recent selective sweeps. The regions involved ranged from a gene-dense region of chromosome 22 to a gene desert on chromosome 4. The significance of these findings requires further localization and biological characterization.

Data relating to evidence of selection from genome level analysis of human and chimpanzee sequences by the CSAC differed in some respects from earlier analyses using a more limited dataset based on sequencing chimpanzee DNA using primers flanking human exonic sequences. Clark and colleagues sequenced coding regions for 7645 orthologous genes to allow human–chimpanzee–mouse comparisons and defined 1547 human and 1534 chimpanzee genes as showing evidence of selection – the former including genes involved in olfaction, sensory perception, amino acid catabolism, and hearing (Clark *et al.* 2003a). Subsequently, Nielsen and colleagues analysed 13 731 orthologous genes in the human and chimpanzee genomes and found evidence of selection particularly involving immune defence, sensory perception, and spermatogenesis (Nielsen *et al.* 2005a). Intriguingly, the latter study also highlighted that for 50 genes with the strongest evidence of selection, human populations showed a high frequency of derived nonsynonymous changes consistent with selection. Kehrer-Sawatzki and Cooper note that some of these apparent differences between studies may relate to methodological issues (Kehrer-Sawatzki and Cooper 2007); a later study by Arbiza and colleagues using a different approach, for example, suggested 5% of human genes and 10.2% of chimpanzee genes evolving under positive selection (Arbiza *et al.* 2006).

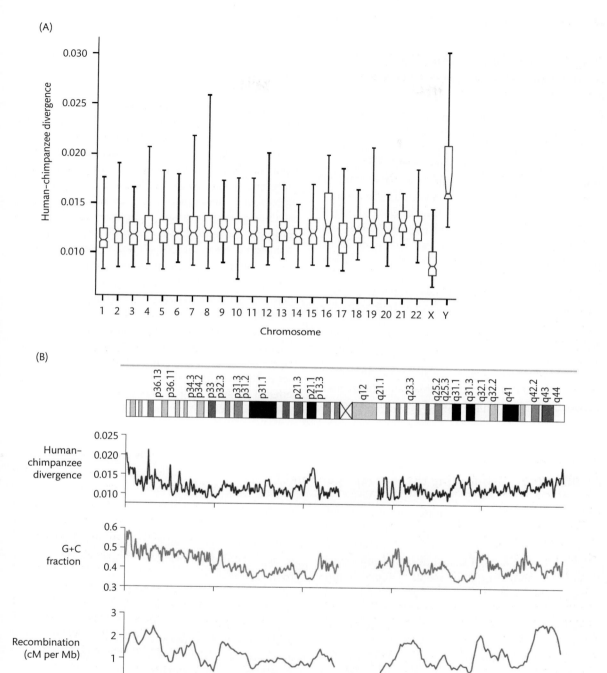

Figure 10.3 Human–chimpanzee nucleotide divergence. (A) Variation in divergence between and within chromosomes is shown as box plots (vertical bars show the range, box edges the quartiles, and notches the standard error of median) (B) For chromosome 1, local variation in divergence is shown together with variation in G+C content and human recombination rates (1 Mb sliding windows). Reprinted by permission from Macmillan Publishers Ltd: Nature (CSAC 2005), copyright 2005.

10.3.2 Sequencing the rhesus macaque provides new insights into genetic diversity and selection

The sequencing of the genome of an Old World monkey the rhesus macaque in 2007 further advanced our ability to look for evidence of positive selection (Box 10.2) (Gibbs *et al.* 2007). Divergence between the macaque lineage and the human–chimp common ancestor is thought to have occurred some 25 million years ago, in contrast to the human and chimpanzee lineages which diverged an estimated 6 million years ago (Fig 6.5) (Kumar and Hedges 1998; Chen and Li 2001; Patterson *et al.* 2006). This increased divergence was reflected in the 93.54% sequence identity observed between the human and macaque genomes, considerably lower than that observed between human and chimpanzee (Gibbs *et al.* 2007). For analysis of putative effects of selection, increased divergence offers more opportunity to detect differences. A further major advance made possible through having an additional primate genome sequence was the ability to define the ancestral state of human and chimpanzee nucleotide diversity using the macaque as an 'outgroup':

a nucleotide difference between human and chimpanzee could now been resolved as a gain or loss depending on its absence or presence in the macaque, respectively.

Comparing orthologous genes between the human and chimpanzee genome showed on average three nonsynonymous and five synonymous substitutions per gene; for human–macaque comparisons this was found to increase to 12 and 22, respectively (Gibbs *et al.* 2007). Gibbs and colleagues analysed 10 376 genes for which orthologous counterparts could be identified among the three genomes, human, chimpanzee, and macaque. Looking at the ratio of nonsynonymous to synonymous changes, the mean rate (0.247) was similar to that previously found on comparing human and chimpanzee (0.23) (CSAC 2005); 2.8% of genes had a ratio greater than one indicating positive selection with enrichment among genes involved in the immune response. Extending the comparison to include the mouse and dog, analysis of 5286 orthologous genes showed ratios were significantly higher among the two primate species (0.169 in humans, 0.175 in chimpanzee) suggesting reduced purifying (negative) selection (rates of 0.124 in macaques, 0.104 in mice, and 0.111 in dogs).

Box 10.2   Sequencing the macaque genome

A draft sequence for the rhesus macaque was published by the Rhesus Macaque Genome Sequencing and Analysis Consortium in 2007 (Gibbs *et al.* 2007). The majority of sequence came from a single female of Indian origin; a whole genome shotgun sequencing approach was used with 5.2-fold coverage for a genome assembly of 2.87 Gb representing 98% of the available genome. A total of 26 479 single base differences were also resolved by sequencing an additional eight Chinese and eight Indian rhesus macaques (equalling 26.2 Mb of sequence). The macaque genome differs from the human one in having only 20 autosomes, and in containing more acrocentric chromosomes. Extensive structural variation was apparent at a microscopic and submicroscopic scale, with 43 microscopically visible breakpoints and more than 820 submicroscopic

rearrangement induced breakpoints seen between the reconstructed human–chimp ancestor and macaque genomes (Gibbs *et al.* 2007). The extent of segmental duplication was considerably lower than for the chimpanzee or human genomes with a lower bound estimate of 2.3%; expansion of gene families could also be resolved with 108 gene families identified among primates (Section 6.2.5). Macaque-specific expansions included HLA-related genes and a further immune-related gene cluster, immunoglobulin lambda-like (IGL): more than half of macaque-specific gene expansions also showed evidence of positive selection based on analysis of coding DNA. As with the chimpanzee genome, sequencing of the macaque genome revealed further insights into the biology and nature of mobile DNA elements (Chapter 8).

Among primates, genes with significantly higher K_A/K_S ratios included those with an involvement in the perception of taste and smell, and transcriptional regulation. Genome-wide analysis among human, chimpanzee, and macaque orthologous genes revealed 178 out of 10 376 genes showed evidence of positive selection using a variety of measures. Enrichment among different functional categories was broadly similar to earlier studies, including defence and immunity, signal transduction, fertilization, and cell adhesion; other gene categories included iron binding and hair shaft formation (Gibbs *et al.* 2007).

10.4   Lactase persistence

A remarkable example of positive selection acting on genetic variation within the human genome is provided by studies of lactase persistence (Box 10.3). This refers to persistence beyond weaning of lactase activity in cells lining the small intestine such that the individual is able to continue to digest lactose – the disaccharide sugar found in milk – into glucose and galactose, which can be easily absorbed. The frequency of this trait varies significantly between populations worldwide, being very high among north Europeans and very low in Asian populations (Swallow 2003; Tishkoff *et al.* 2007). The geographic distribution of lactase persistence mirrors that of dairy farming, being high among European populations as well as pastoralist populations in Africa and the Middle East. It is believed that the significant nutritional benefits associated with consuming dairy products have provided the selective pressure that favoured the genetic factors permitting lactase persistence. The domestication of cattle, sheep, and goats in the Middle East is thought to have occurred some 10 000 years ago (Loftus *et al.* 1994; Zeder and Hesse 2000). At least two major alleles have

---

Box 10.3   Lactase persistence, adult-type hypolactasia, and congenital lactase deficiency

The decline in lactase activity after weaning, 'lactase non-persistence', or 'adult-type hypolactasia' (OMIM 223100), can be viewed as the commonest enzyme deficiency known worldwide but is actually the normal physiological state. Selective pressures associated with nutritional benefits from dairy foods have led to high frequencies of lactase persistence now being found among certain populations worldwide. Lactase persistence is a heritable trait, manifesting in an autosomal dominant manner. Lactase non-persistence is usually manifest in childhood or by early adulthood with failure to digest lactose in milk, the resulting bacterial fermentation of undigested lactose causing unpleasant symptoms including diarrhoea, bloating, flatulence, and abdominal pain if milk is ingested.

Adult-type hypolactasia contrasts with congenital lactase deficiency (OMIM 223000), a rare autosomal recessive disorder affecting infants who have almost total lack of lactase-phlorizin hydrolase activity in the cells lining the intestine leading to failure to digest lactose and severe watery diarrhoea after taking breast milk or lactose-containing feeds. Like a number of autosomal recessive diseases, this disorder is more common in Finland, reaching a reported carrier frequency of one in 35 in certain small towns in central Finland. The molecular basis of this disease was found to be mutations in the coding regions of *LCT*, the gene encoding lactase. Most Finnish cases were due to a nucleotide substitution in exon 9 (c.4170T>A), a nonsense mutation leading to a premature stop codon and nonsense mediated mRNA decay (Kuokkanen *et al.* 2006). Other rare mutations causing congenital lactase deficiency have also been found including a four nucleotide deletion in exon 14 (c.4998–5001-delTGAG) and a two nucleotide deletion in exon 2 (c.653_654delCT) of *LCT* which cause frameshifts and truncation of the encoded protein; nonsynonymous mutations causing amino acid changes are also seen, including histidine to glutamine (c.804G>C) and serine to glycine (c.4084G>A).

now been associated with lactase persistence and are thought to have arisen independently in different populations within this time scale, positive selection through adaptation to milk culture acting to cause their allele frequencies to rapidly rise and provide characteristic genetic signatures of recent strong natural selection (Enattah *et al.* 2002; Tishkoff *et al.* 2007).

10.4.1 Genetic diversity and lactase persistence in European populations

The gene encoding lactase, *LCT* (also known as *LPH*, lactase-phlorizin hydrolase) was localized to chromosome 2 and its organization characterized (Kruse *et al.* 1988; Mantei *et al.* 1988). However, careful examination of sequence differences between lactase persistent and non-persistent individuals within the coding regions, flanking intronic regions, and proximal promoter region of *LCT* failed to find any associated variants (Boll *et al.* 1991). Wang and colleagues showed that among those individuals with lactase persistence, some had allele-specific differences in *LCT* expression, indicating that any genetic variants were likely to be cis-acting (Wang *et al.* 1995). They were able to perform such assays based on exonic sequence differences within the *LCT* gene, as such variants are transcribed and can be used to determine the allelic origin of the transcript (Section 11.5). However, it was analysis of Finnish family pedigrees in which individuals showed lactase non-persistence that was to provide evidence of cis effects with sequence variants acting apparently at a considerable distance from the *LCT* gene (Enattah *et al.* 2002).

Within the Finnish population, lactase non-persistence or hypolactasia occurs only in a minority of people, manifesting between the age of 10 and 20 years. Enattah and colleagues studied nine extended families, initially using seven microsatellite markers in the region of the *LCT* gene on chromosome 2q21, to look for linkage with lactase non-persistence (Enattah *et al.* 2002). Intriguingly the highest lod score (Box 2.4) was seen with a polymorphic microsatellite some distance from the *LCT* gene, telomeric to the *DARS* gene (Fig. 10.4). Fine scale mapping using nine polymorphic markers revolved a 200 kb region; haplotypic analysis showed a strong association with lactase persistence for a conserved haplotype, helping to further refine the critical region to some 47

kb between the dinucleotide repeat markers D2S3013 and D2S3014 – again this was some distance from the *LCT* gene itself and in fact spanned a neighbouring gene *MCM6* (Fig. 10.4).

The 47 kb region was resequenced in three individuals with lactase non-persistence and four with lactase persistence, which revealed 43 SNPs and nine indels (Enattah *et al.* 2002). Of these, two SNPs showed complete segregation with lactase non-persistence among the family pedigrees, all family members with lactase non-persistence being homozygous for C-13910 (rs4988235) and G-22018 (rs182549). The nomenclature refers to the distance of the SNP from the first ATG codon of the *LCT* gene – in other words the two variants lay some 14 and 22 kb upstream of the *LCT* gene, respectively, lying within intron 13 and 9 of the minichromosome maintenance-6 gene, *MCM6* (Fig. 10.4).

The relationship of these linked alleles to lactase persistence/non-persistence was striking. When 196 intestinal biopsy specimens were analyzed, all 59 showing lactase deficiency were homozygous CC for the -13910 SNP, while 53 out of 59 were homozygous GG for -22018; among the 137 biopsy samples showing lactase persistence, 74 were homozygous TT and AA for -13910 and -22018, and 63 were heterozygous for both SNPs (Enattah *et al.* 2002). Possessing a copy of the T-13910 allele was thus a very clear marker for a haplotype associated with lactase persistence, showing 100% association in this Finnish population, with subsequent studies showing between 86% and 98% association with lactase persistence in other European populations (Poulter *et al.* 2003; Coelho *et al.* 2005; Ridefelt and Hakansson 2005). Among European and East Asian individuals, the population allele frequency was also highly correlated with the population levels of lactase persistence, the allele being absent among East Asians where lactase persistence is very rare (Bersaglieri *et al.* 2004).

Experimental evidence to support a functional difference in gene expression or lactase activity associated with the European T-13910 allele was provided by analysis of duodenal biopsy samples from patients of differing genotype (Kuokkanen *et al.* 2003). Such a study allows the consequences of genetic diversity to be studied in an appropriate context, namely the cells in which the gene of interest is normally expressed. Kuokkanen

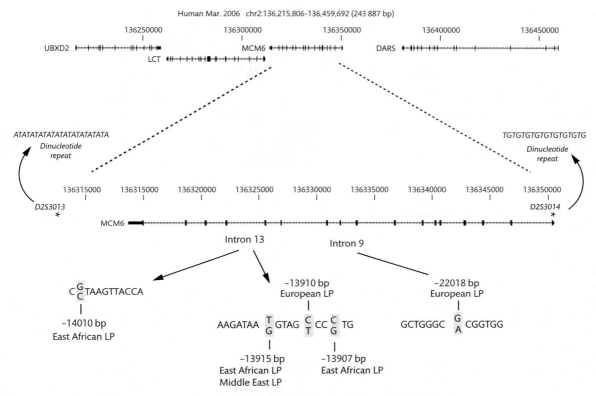

Figure 10.4 Lactase persistence (LP) associated with SNPs in the upstream gene *MCM6* shown in the context of *LCT* and flanking genes. SNPs associated with LP in different populations are shown within intron 13 of *MCM6*. Redrawn and reprinted by permission from Macmillan Publishers Ltd: Nature Genetics (Tishkoff *et al.* 2007), copyright 2007. Genes and genomic location adapted from screenshot of UCSC Genome Browser (Kent *et al.* 2002) (http://genome.ucsc.edu/) (Human March 2006 Assembly).

and colleagues analysed 52 patients and found that lactase activity was strongly associated with genotype, with the highest activity being seen among individuals homozygous for the T-13910 and A-22018 alleles (Fig. 10.5). The investigators were also able to assay the ratio of lactase to sucrase activity and found similar results, segregating according to underlying genotype. Finally, the team analysed **allele-specific gene expression** at the RNA level, comparing relative expression within individuals heterozygous for the variants. They used a transcribed sequence variant to distinguish the allelic origin of the transcript and quantified expression: the results appeared clear cut and consistent with what was seen comparing between individuals, with on average 11.5-fold higher expression seen with the T-13910/A-22018 haplotype.

Is it plausible that a SNP at such a distance from the *LCT* gene, some 14 000 bp away, should be acting to modulate its expression? Reporter gene experiments show that the region containing T-13910 has enhancer activity, increasing gene expression in an orientation-independent manner, a feature classically seen with an enhancer element (Olds and Sibley 2003; Troelsen *et al.* 2003; Lewinsky *et al.* 2005). **DNA footprinting** studies suggest this is a site of multiple protein–DNA interactions, which may act in conjunction with specific transcription factors such as HNF1α at the proximal *LCT* gene promoter. The evidence appears clearest in a colon carcinoma

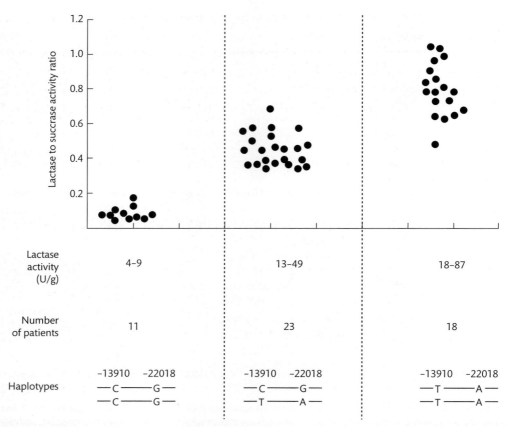

Figure 10.5 Functional analysis of variants associated with lactase persistence. The analysis of lactase activity and lactase to sucrase activity ratio in duodenal mucosal cells for different patients is segregated according to underlying genotype. Redrawn and reproduced from Kuokkanen *et al.* (2003), with permission from BMJ Publishing Group Ltd.

cell line, Caco-2, which expresses lactase; a four-fold greater enhancer activity was seen on reporter gene assays comparing the T versus the C allele for the -13910 SNP (Troelsen *et al.* 2003). Lewinsky and colleagues used DNA affinity purification based on the DNA sequence spanning the SNP to show that the transcription factor Oct-1 was recruited in an allele-specific manner (Lewinsky *et al.* 2005). However, further work is needed including *in vivo* assays such as chromatin immunoprecipitation (ChIP) (Box 11.7) to resolve the functional consequences of this variant. What is not in doubt is the dramatic association of the haplotype bearing T-13910 with lactase persistence in European populations.

The haplotype bearing T-13910 and A-22018 was shown to be both common and remarkably long,

spanning 1 Mb based on initial analysis using 18 SNPs (Poulter *et al.* 2003). This was confirmed in northern European-derived populations by Bersaglieri and colleagues using a more extensive SNP set of 101 markers over 3.2 Mb around the *LCT* gene (Bersaglieri *et al.* 2004). This and other work demonstrated that the region has features of strong recent positive selection within the last 5000–10 000 years. To find such an extended conserved haplotype at such high allele frequency suggests that the favourable allele has been strongly selected for within the population, such that it has risen to a high allele frequency relatively quickly, before there has been time for recombination and mutation to break down the linkage with associated variants. Thus a block of linked variants is seen to also rise in frequency in the population, a further example of genetic hitch-hiking (Section 10.2.1).

10.4.2 Different alleles show association among African pastoralists

Lactase persistence is common among African pastoralist populations yet the T-13910 allele was present in only a few West African populations and at a low allele frequency insufficient to account for the observed lactase persistence (Mulcare *et al.* 2004; Myles *et al.* 2005). Tishkoff and colleagues investigated the genetic basis for lactase persistence among diverse East African populations, which included pastoralist populations showing a high frequency of this phenotype (Tishkoff *et al.* 2007). Field studies were conducted to define the lactase persistence phenotype using a lactose tolerance test to assess the increase in blood sugar after lactose is administered orally (Fig. 10.6). In order to facilitate identification of sequence variants associated with lactase persistence, two sets of individuals were selected from either extreme of the observed phenotype, either lactase persistent and or lactase non-persistent.

Candidate regions of the *MCM6* gene were sequenced, including intron 13 containing the T-13910 allele associated with lactase persistence in Europeans. This showed that three SNPs were associated with lactase persistence in this East African population: two were in very close proximity to T-13910, namely T/G-13915 and C/G-13907, and one was 100 bp upstream, G/C-14010 (see Fig. 10.4). These were further analysed among the whole phenotyped set of 470 individuals to assess the association and haplotypic structure. Tichkoff and colleagues were analysing 43 different ethnic groups, making it difficult to clearly delineate effects for groups with few individuals, however the C-14010 allele was robustly associated with lactase persistence, the strongest of the 123 SNPs genotyped; the other two SNPs were less clear cut with, for example, the T/G-13915 SNP showing association in some Kenyan populations (Fig. 10.6) (Tishkoff *et al.* 2007). The C-14010 allele was associated with increased gene expression when assayed using a reporter gene driven by the *LCT* promoter, but the molecular characterization of this putative functional variant remains to be completed.

Analysis of SNP diversity across a 3 Mb region showed that the C-14010 allele was present at a high frequency with a remarkably long associated stretch of homozygosity (greater than 2 Mb), providing a further striking example of an incomplete selective sweep, the linked sequence variants 'hitch-hiking' with the selected allele and providing a strong signature of positive selection. The investigators estimated that this was a very recent mutational event, occurring some 3000–7000 years ago, in contrast to the European T-13910 mutation: this would be consistent with a relatively late spread of pastoralists into sub-Saharan Africa, arriving in northern Kenya some 4500 years ago (Tishkoff *et al.* 2007).

10.4.3 Lactase persistence in Middle Eastern populations

Consumption of camel milk may have driven selection for an additional major allele associated with lactase persistence among certain Middle Eastern populations in whom the Arabian camel (*Dromedary camelous*) is the main domesticated animal used as a source of milk (Enattah *et al.* 2008). Pastoralist groups in this region, such as among Saudi Arabians, have a high prevalence of lactase persistence, yet the European lactase persistence allele (T-13910) is almost completely absent (Cook and al-Torki 1975; Enattah *et al.* 2008).

Enattah and colleagues first used a haplotype matching approach, resequencing a 47 kb candidate region that included the *LCT* gene and upstream enhancer comparing Saudi samples from lactase persistent individuals, with South Koreans who showed lactase non-persistence, allowing identification of sequence variants that differed between the two groups (Enattah *et al.* 2008). This led to identification of two linked sequence variants that were strongly associated with lactase persistence. Remarkably, one of them, a T to G substitution (G-13915; ss79088033) was located only 5 bp away from the European 'T-13910' variant, lying within the same Oct-1 binding motif; the other was a T to C synonymous variant at -3712 in the *MCM6* gene upstream of *LCT*. These variants were almost completely absent in neighbouring populations such as from Iran and may have a functional role involving HNF1α. It is thought this ancestral allele (bearing C-3712/G-13915) arose some 4000 years ago, independently of the European allele (T-13910), with selection in response to camel milk leading to high allele frequency (Enattah *et al.* 2008).

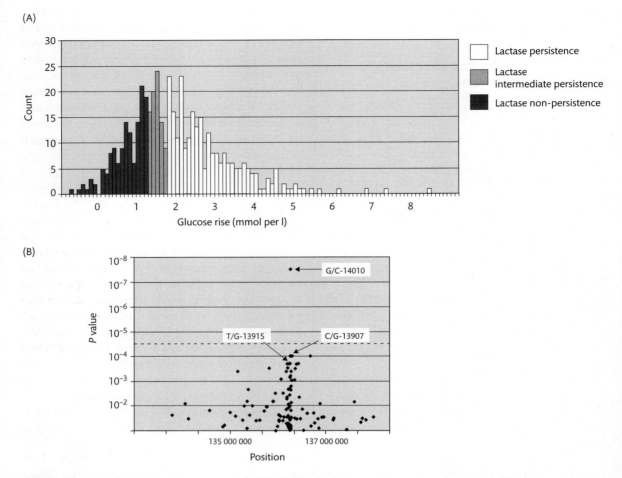

Figure 10.6 Lactase persistence in East Africa. (A) Lactose tolerance among 470 individuals from East Africa comprising 43 different ethnic groups. (B) Meta analysis of 123 SNPs across different groups showing association with lactase persistence, G/C-14010 remaining highly significant after Bonferroni correction. Reprinted by permission from Macmillan Publishers Ltd: Nature Genetics (Tishkoff *et al.* 2007), copyright 2007.

10.4.4 Diversity within an enhancer region modulating LCT expression

A cluster of variants have now been reported from diverse populations within intron 13 of the *MCM6* gene (Fig 10.4). This is a region with enhancer activity of regulatory significance for *LCT* expression that has been the subject of intense and ongoing selection pressure, which has led to some of these alleles rising to high frequency in diverse populations (Enattah *et al.* 2007). Indeed there is evidence from haplotypic analysis of over 1600 individuals in 37 geographically disparate populations that the C/T-13910 mutation has occurred at least twice. The most frequently observed haplotype bearing a mutation is thought to have occurred some 5000–12 000 years ago and is geographically widespread, while a more recent event appears to have introduced the same nucleotide change an estimated 1400–3000 years ago with a much more restricted geographic distribution west of the Urals and north of the Caucasus (Enattah *et al.* 2007).

10.5 Human pigmentation, diversity at *SLC24A5*, and insights from zebrafish

10.5.1 Genetics of pigmentation

Among birds and mammals, melanocytes are specialized cells that produce the pigment melanin within a subcellular organelle called the melanosome. In the eye, melanin is retained in these specialized cells and increases visual acuity through modulation of light scatter, while in the hair and skin the pigment is transported to neighbouring keratinocytes and so defines colour with variation between individuals in the type, number, and size of melanosomes present; in more deeply pigmented skin, for example, melanosomes are larger and more numerous (Sturm 2006).

Progressively lighter skin is seen among human populations with increasing distance from the equator. This is thought in large part to reflect a similar variation in intensity of ultraviolet radiation with latitude from a maximum at the equator, such a that a balance is achieved between protection from harmful effects of ultraviolet radiation (including DNA damage and cancer) associated with darker skin pigmentation, versus lighter skin being able to absorb more of the available ultraviolet radiation for synthesis of vitamin D (Relethford 1997; Jablonski and Chaplin 2000). Other selective pressures related to pigmentation may have involved factors such as heat stress and camouflage.

The genetic basis for pigmentation, whether in animals such as the mouse or in humans, has been the subject of longstanding and intensive research with a complex picture involving many different loci and specific variants. Among human populations diversity in many different genes appears to contribute to pigmentation as manifested in skin, hair, or eye colour. Diversity in these genes contributes in varying degrees as quantitative trait loci (QTLs) to define a continuous range of pigmentation phenotypes, such as seen between and within ethnic groups (Sturm 2006). In mice, numerous alleles at over 130 genetic loci have been resolved that modulate mouse hair, skin, or eye colour (www.espcr.org/micemut/; Bennett and Lamoreux 2003) with notable examples of human homologues defined through such experiments including genes responsible for human albinism. Rare variants at such human genes have been associated with particular syndromes such as oculocutaneous albinism, for example *TYR* encoding tyrosinase, *OCA2* encoding oculocutaneous albinism II protein, *TYRP1* encoding tyrosinase related protein 1, and *MATP* encoding membrane associated transporter protein also known as *SLC45A2* (Sturm 2006).

Common variants have also been associated with particular skin, eye, and hair colour phenotypes, including *MC1R* (encoding a melanocortin receptor) in which specific variants have been associated with red hair, light skin, and a poor tanning response (Valverde *et al.* 1995) and *OCA2* with eye colour (Duffy *et al.* 2007). There is a growing list of genes in which diversity has been associated with pigment variation in Europeans including *OCA2*, *SLC24A5*, *MATP*, *ASIP* (encoding agouti signalling protein), *TYR*, *TPCN2* (two pore pigment channel 2 protein), and *KITLG* (encoding the KIT ligand). Such associations were derived from candidate gene approaches as well as more recent genome-wide association studies which have defined variants at many previously implicated genes as well as new gene loci (Shriver *et al.* 2003; Graf *et al.* 2005; Lamason *et al.* 2005; Norton *et al.* 2007; Sulem *et al.* 2007, 2008).

10.5.2 Golden zebrafish mutants led to the identification of *SLC24A5*

As an illustration of how a gene locus showing strong evidence of positive selection within a particular population was determined, there are few examples more elegant than the work implicating the gene *SLC24A5*. Lamason and colleagues utilized the power of studying a model organism to identify a human gene within which the pattern of nucleotide diversity, and in particular a specific nucleotide substitution, played a major role in determining the light skinned pigmentation phenotype observed among Europeans (Lamason *et al.* 2005).

Among zebrafish, a particular mutation called *golden* had been recognized for many years but the underlying gene remained unidentified: the fish has light 'golden' stripes in contrast to the normal wild-type fish that bear dark stripes (Streisinger *et al.* 1981). Specifically, the *golden* mutants were found to have reduced skin and retinal pigmentation with melanin granules that were produced later and were smaller in size, irregular, and less dense in appearance. Lamason and coworkers were able to identify the gene and specific null mutation responsible for this *golden* phenotype (Lamason *et al.* 2005). A combination of linkage analysis and knockdown experiments allowed cloning of the zebrafish gene *slc24a5*, and a particular nucleotide substitution was identified as causing *golden*: a nonsynonymous C to A substitution led to a change from encoding tyrosine at position 208 in the amino acid sequence to a stop codon, severely truncating the protein and leading to nonsense mediated decay (Lamason *et al.* 2005). As melanosomes are found in the *golden* mutants, the gene product of *slc24a5* is important but not essential for formation of melanosomes.

The golden phenotype could be reproduced on injection of embryos with **morpholino** (the generic name given to a molecule to modify or knockdown gene expression) targeting *slc24a5*, while injection of the wild-type gene transcript 'rescued' embryos homozygous for the *golden* nucleotide substitution leading to pigmentation approaching that seen in the wild-type animals (Lamason *et al.* 2005). *slc24a5* encodes a conserved protein with 69% amino acid identity to human *SLC24A5*, and was identified as a member of the potassium-dependent sodium/calcium exchangers. Consistent with a role in pigmentation, the protein was found to function in intracellular membrane-bound structures. Expression of the mouse homologue was highly enriched in the skin and eye, while the human transcript was found to rescue pigmentation in *golden* zebrafish embryos (Lamason *et al.* 2005).

10.5.3 Variation at *SLC24A5*, skin pigmentation, and evidence of selection

SLC24A5 is located at chromosome 15q21.1. One nonsynonymous SNP (rs1426654) (c.331G>A, p.A111T) had been identified within the gene by the International HapMap Project (Section 9.2.4), a G to A substitution leading to an amino acid substitution of alanine for threonine within the third exon of *SLC24A5* at amino acid 111. G was noted to be the ancestral allele, present with an allele frequency of between 93% and 100% among individuals of African, East Asian, and Indigenous American descent. By contrast the A allele was found to be near or at fixation in European-American individuals (allele frequency of 98–100%). In terms of marked differences in allele frequency between populations, this SNP was among the

most extreme of any HapMap SNP (within the top 0.01%). There were other features in the SNP landscape consistent with recent positive selection. Additional SNPs were present showing striking differences in allele frequency between populations, while the level of heterozygosity was markedly reduced around *SLC24A5* with an average heterozygosity over a 150 kb region of 0.0072 among the European individuals sampled compared to 0.175–0.226 among non-Europeans (Fig. 10.7) (Lamason *et al.* 2005). The authors noted that this was the largest contiguous region of low heterozygosity seen among the autosomal chromosomes in the CEU (European geographic ancestry) HapMap panel, consistent with a major selective sweep in that population (Lamason *et al.* 2005).

When two recently admixed populations were studied, African American and African Caribbean, *SLC24A5* genotype (rs1426654) significantly correlated with skin pigmentation as assayed by reflectometry, with homozygotes for the A allele having lighter skin pigmentation (lower melanin index) compared to heterozygotes or those homozygous for the G allele (Lamason *et al.* 2005). The contribution of *SLC24A5* to the difference in skin pigmentation between Europeans and Africans was estimated at 25–38% but is only part of the story as the same allele is present in both Africans and East Asians despite the marked difference in skin pigmentation between these groups. More recent functional analysis demonstrated both a direct role of *SLC24A5* in melanin synthesis in human melanocytes and a critical functional role for the amino acid residue affected by the nonsynonymous SNP rs1426654 in modulating the intracellular potassium-dependent exchanger activity of the protein – so affecting melanogenesis (Ginger *et al.* 2008). *SLC24A5* is only one of many genetic loci implicated in the genetic determinants of skin pigmentation (Section 10.5.1) but is a highly significant one as highlighted by subsequent genome-wide analyses (Sabeti *et al.* 2006, 2007; Voight *et al.* 2006).

10.6   Genome-wide analyses

By adopting a candidate gene approach some remarkable examples of positive selection have been identified. The availability of the human genome sequence and large scale surveys of fine scale genetic diversity such as the

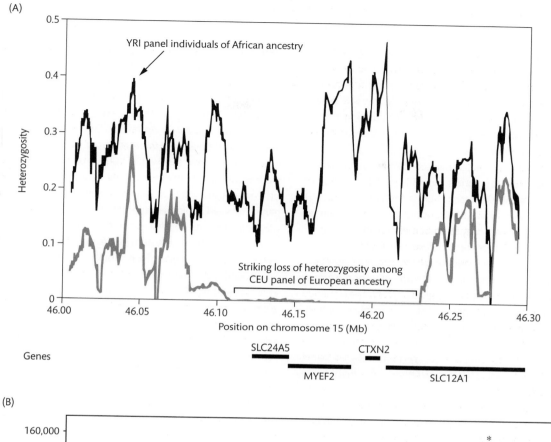

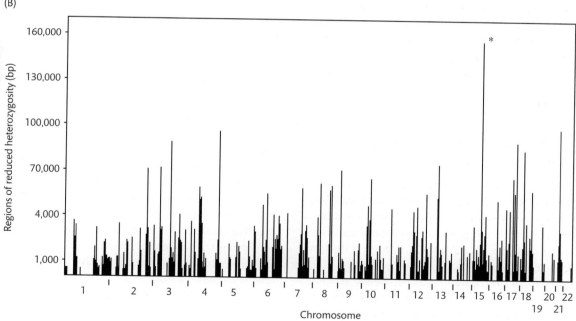

Figure 10.7 Reduced heterozygosity involving *SLC24A5*. (A) Comparison of CEU (European geographic ancestry) and YRI (African) Phase II HapMap SNP marker set heterozygosity for a region of chromosome 15q21, which includes *SLC24A5*. (B) Genome-wide analysis of reduced heterozygosity in the CEU panel; *SLC24A5* is marked by an asterix. Adapted from Lamason *et al.* (2005), reprinted with permission from AAAS.

HapMap Project have more recently enabled genome-wide analyses to define evidence of positive selection. Important caveats to such datasets and their applicability to such analysis remain, however, notably how SNP panels have been selected and underlying population demographics including population size, subdivision, and flow, which may confound analyses.

Voight and colleagues analysed 209 individuals from the Phase I HapMap panels (Box 9.1) to define a map of incomplete selective sweeps across the genome, identifying alleles that showed evidence of selection present at relatively high frequency but not yet at fixation, characterized by a long haplotype of low diversity (Voight et al. 2006). The authors developed an 'integrated haplotype score', building on the extended haplotype homozygosity statistic of Sabeti and colleagues (2002a). The evidence of selection at each common SNP was resolved in turn, looking for clusters of SNPs showing extreme values.

Evidence of selection was seen for known loci such as the region including the LCT gene in the CEU panel of European ancestry (Fig. 10.8). Other genes involved in carbohydrate metabolism showed evidence of selection in the other populations, including MAN2A1 (mannose metabolism) in the YRI panel of African ancestry and East Asians; and SI (sucrose) in East Asians. When analysed across the genome in terms of biological process, many examples were found related to reproduction and fertility; chemosensory perception and olfaction; morphology including skin pigmentation; skeletal development; and metabolism including carbohydrates, lipids, and phosphates (Voight et al. 2006). Specific examples related to fertility and reproduction include sperm protein structure (RSBN1), sperm motility (SPAG4), sperm and egg viability (ACVR1, CPEB2), female immune response to sperm (TGM4), egg fertilization (CRISP), and testis determination (NROB1). In the European panel a number of genes involved in skin pigmentation showed evidence of selection including OCA2, MYO5A, DTNBP1, TYRP1, and SLC24A5 (Voight et al. 2006).

Williamson and colleagues subsequently published work using an analytical approach that the authors felt was more robust to underlying population demographics and recombination rate variation, which explicitly considered SNP ascertainment and defined signatures of selection based on complete selective sweeps with the selected allele having a frequency of around 100%

(Williamson et al. 2007). This contrasted with, but complemented, earlier studies whose approaches mainly defined partial selective sweeps where the favoured allele has not yet reached fixation (Altshuler et al. 2005; Voight et al. 2006; Wang et al. 2006). Data for 1.2 million SNP markers among African American, European American, and Chinese individuals were analysed. Across the genome, 101 genomic loci within 100 kb of a known gene were resolved that had statistically very strong evidence of a recent complete selective sweep. These included genes involved in pigmentation, dystrophin protein complex, olfactory reception, nervous system development and function, immune system genes, and genes involved as molecular chaperones (heat shock).

Availability of data on more than 3 million SNPs from Phase II HapMap (Fig. 9.6) on 210 individuals allowed the application of techniques to define common extended haplotypes, complemented by other methods to detect selected alleles at or near fixation in one population but still polymorphic in another population (Sabeti et al. 2007). This allowed Sabeti and colleagues to define more than 300 candidate loci for positive selection. Top hits included well characterized examples such as LCT and SLC24A5 involved in lactase persistence and skin pigmentation, respectively. Among a number of very intriguing loci, the evidence of selection at genes implicated in the biology of the Lassa fever virus among individuals from Nigeria (YRI panel) was striking (Sabeti et al. 2007). The strongest signal seen with the long range haplotype test across the genome was in the LARGE gene, encoding a protein that modifies the receptor for the virus, a critical event for virus binding (Kunz et al. 2005). Strong evidence for selection was also seen for a second gene, DMD, encoding a protein binding the receptor and modulating its function (Sabeti et al. 2007).

Other investigators have also analysed the Phase II HapMap dataset in terms of population differentiation considering the physical location and functional class of SNPs. Barreiro and colleagues found that low population differentiation (F_{ST}), associated with negative or purifying selection, was more frequent with genic SNPs, in particular nonsynonymous variants. Genes containing such low F_{ST} nonsynonymous SNPs were enriched among those listed in the database of genes and genetic disorders, Online Mendelian Inheritance in Man (OMIM), suggesting as a class that these were likely to be of possible

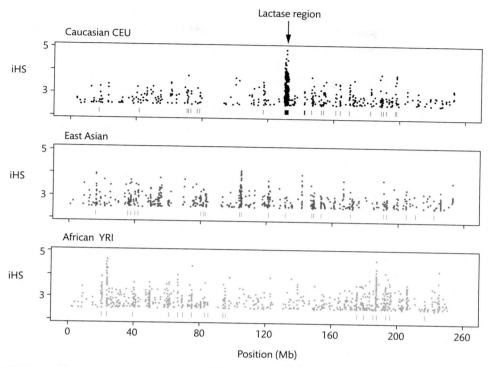

Figure 10.8 Evidence of incomplete selective sweeps on chromosome 2. Common SNPs with extreme integrated haplotype scores (iHS) are shown for the CEU (European geographic ancestry), YRI (African), and combined CHB/JPT (East Asian) Phase I HapMap populations. The short vertical lines below the plots indicate 100 kb windows falling in the top 1% genome-wide analyses. Reprinted with permission from Voight *et al.* (2006).

medical relevance (Barreiro *et al.* 2008). Local evidence of high F_{ST} values, expected to be associated with positive selection, was also enriched among nonsynonymous SNPs and those in the 5′ untranslated region (UTR); a high proportion showed other signatures of positive selection and were implicated in disease much more frequently than expected by chance (Barreiro *et al.* 2008). Such analyses illustrate how approaches to resolve effects of selection are facilitating efforts to uncover gene loci and specific variants which may be involved in disease susceptibility.

10.7 Summary

The genetic diversity observed in human populations provides unique insights into the selective pressures that have shaped our recent evolutionary past. Signatures of positive selection are recognized in which specific patterns of genetic diversity are found, ranging from loss of heterozygosity seen with a recent selective sweep in which there has been genetic hitch-hiking by neutral variants carried on a haplotype with the selected variant(s), to finding unexpectedly long haplotypes of high frequency, or observing marked differences in allele frequency between populations. Insights from comparison between species have been considerably advanced by the publication of the draft sequences of the chimpanzee and rhesus macaque genomes with evidence of selection involving genes in a number of different functional classes including defence and immunity, signal transduction, fertility, iron binding, and hair shaft formation. The review of genetic determinants of lactase persistence and skin pigmentation described in this chapter has highlighted the remarkable progress being made in this area.

Lactase persistence is a trait showing strong evidence of selection whose distribution mirrors that of the geographic distribution of dairy farming. Across wide ranging populations, diversity involving a particular region some 14 kb upstream of the *LCT* gene encoding lactase has been implicated. A mutation associated with lactase persistence in Europeans appears to have arisen some 5000–12 000 years ago and is now found on a specific common extended haplotype. The variant is associated with lactase activity in the duodenum but actually is found within the intron of a neighbouring gene, a genomic region showing enhancer activity for the *LCT* gene. This mutation appears to have occurred independently at least twice and, equally remarkably, further recent mutations within 100 bp have occurred and been independently associated with lactase persistence in East African pastoralists and specific Middle Eastern populations.

The research that led to the discovery of the *SLC24A5* gene illustrates the power of analysing a model organism and how recent mutational events with which a strong selective advantage appears to have been associated can lead to clear signatures of selection, with evidence of marked population differences in allele frequency and of striking reductions in local heterozygosity. The specific allele identified plays a significant role in the lighter skin pigmentation seen in Europeans, but is only part of a complex and incompletely understood story with many different genes and variants involved. Genome-wide analyses based on high density SNP panels are highlighting many new loci showing evidence of selection as candidates for further investigation. These include a variety of gene classes, including fertility and reproduction, immunity, pigmentation, olfaction, metabolism, and the nervous system. The relationship between genetic diversity and selection discussed in this chapter is a theme continued in Chapter 12 when balancing selection involving genetic variation in the major histocompatibility complex is considered, and in Chapter 13 in terms of malaria and other parasitic infections.

10.8 Reviews

Reviews of subjects in this chapter can be found in the following publications:

Topic	References
Genomic signatures of natural selection	Bamshad and Wooding 2003; Harris and Meyer 2006; Sabeti *et al.* 2006
Chimp genome, insights into evolution and selection	CSAC 2005; Kehrer-Sawatzki and Cooper 2007
Macaque genome	Gibbs *et al.* 2007
Genetics of lactase persistence	Swallow 2003; Jarvela 2005
Human pigmentation genetics	Sturm 2006
Genetic determinants of mouse colour	Bennett and Lamoreux 2003

CHAPTER 11

Genetics of gene expression

11.1 Introduction

A recurring theme over the course of this book is the relationship between genetic variation – in all its different guises, from single nucleotide substitutions to large scale structural genomic variation – and the observed phenotype. At the level of the whole organism, the phenotype may be a discrete or continuous trait, which may or may not show mendelian segregation within pedigrees, and in the case of complex traits is the result of multiple genetic and environmental determinants. Over the course of Chapters 2 and 9, approaches to dissecting the genetic factors underlying 'mendelian' and 'complex' traits have been reviewed and the difficulties inherent to such analyses discussed. Even with powerful linkage-based analysis and positional cloning, or more recently the application of genome-wide association analysis to common disease, major roadblocks remain in terms of fine mapping associations and defining specific functional causative variants.

Inherent to the relationship between genetic variation and the observed phenotype is the hypothesis that the underlying genetic diversity has a functional consequence, acting at a molecular or cellular level with potentially diverse pathways and networks involved. Early attention focused on nonsynonymous and other coding variants that had direct consequences for the structure and function of the encoded proteins (Cargill *et al.* 1999). This was exemplified by studies investigating the genetic basis of structural variants of haemoglobin and the thalassaemias (Section 1.3) together with many other examples described over the course of this book. Increasingly sophisticated tools and techniques are available to predict and test the functional consequences of coding variants taking into account the sequence, structure, and pathways involved (Chasman and Adams 2001; Uzun *et al.* 2007). Assigning functionality to such variants remains, however, a complex task and putative regulatory effects of exonic variation, for example involving exonic splicing enhancers and regulation of splicing, need to be considered (Section 11.6.1) (Cartegni *et al.* 2002).

With time it has become apparent, notably with the recent associations from genome-wide analyses in common disease but also in many other contexts including variation in the globin genes, that 'regulatory' variants modulating gene expression play an equally important role in determining phenotypic traits. Regulatory variants are typically located in non-coding DNA sequences although, as noted in the preceding paragraph, the categorization into coding and regulatory variants is in some respects artificial, and structural genomic variation such as copy number variation has clear gene dosage effects on gene expression (Section 4.3). Many examples of regulatory variants are given in this book, ranging from tandem repeats upstream of *INS* encoding insulin modulating gene expression (Section 7.3.3) to single nucleotide polymorphisms (SNPs) modulating the binding sites of specific transcription factors such as *DARC* with dramatic consequences for expression of the encoded receptor and malaria susceptibility (Section 13.2.4), or creating a new promoter as seen in thalassaemia involving the α globin genes (Section 1.3.9).

The challenge of how to define the extent of regulatory variation and identify specific causative variants remains a major roadblock in the field with relatively few tools available. Major advances have been made relatively recently through analysis of the genetics of gene expression. Here the phenotype of interest is far closer, in terms of pathways and networks, to the underlying genetic diversity as it is the **transcriptome**, the transcribed RNA, which is being quantified and analysed in relation to genetic variation. Consideration of gene expression as a quantitative trait in principle removes much of the inherent noise and variability associated with analysing a phenotype at the level of the whole organism, which results from a complex interplay between multiple genetic, environmental, and other factors.

The power of genetic analysis to resolve genetic variation modulating gene expression should in principle be considerably greater than for interrogating phenotypic traits for the whole organism. However, it was not clear at the outset to what extent gene expression varies between individuals and populations, whether such traits are heritable, if multiple genetic loci would be involved, or what the effect sizes would be. Over the course of this chapter advances in genetic analysis of gene expression are discussed and illustrated with work from model organisms and human populations, with significant implications for our ability to resolve susceptibility to common disease and other traits (Dermitzakis 2008).

11.2   Variation in gene expression is common and heritable

Gene expression has been shown to vary within and between populations of a variety of different species, ranging from yeast to humans. Between yeast strains of *Saccharomyces cerevisiae*, for example, 24% of genes analysed showed significant variation (Brem *et al.* 2002). Similar results were found in the fly *Drosophila melanogaster* where 25% of genes showed variation in expression (Jin *et al.* 2001) and telefost fish of the genus *Fundulus* where 18% of genes analysed showed significant variation in expression between individuals in a population (Oleksiak *et al.* 2002). In human populations, significant variation in gene expression was observed between individuals with evidence of familial aggregation and tissue-specific variation (Enard *et al.* 2002; Cheung *et al.* 2003c; Schadt *et al.* 2003; Monks *et al.* 2004; Goring *et al.* 2007; Stranger *et al.* 2007b; Emilsson *et al.* 2008).

In an early study utilizing microarray technology to simultaneously analyse gene expression for many different genes in a high throughput manner, Cheung and colleagues analysed gene expression in 35 lymphoblastoid cell lines established from unrelated individuals of European descent as part of the CEPH collection (Box 11.1) (Cheung *et al.* 2003c). The majority of genes showed greater variation between than within individuals. The top 5% most variable genes in terms of expression were scattered across the genome and included previously characterized genes such as *HLA-DRB1* in the major histocompatibility complex (MHC), known to show significant variation in expression (Section 12.3.4) (Cheung *et al.* 2003c). When the expression of five genes that were highly variable based on the expression array dataset were quantified by real time quantitative reverse transcription polymerase chain reaction (RT-PCR), much greater variation was observed between unrelated individuals than between siblings from the same family, who in turn showed greater variation than monozygotic twins (Fig. 11.1) (Cheung *et al.* 2003c).

Schadt and colleagues also analysed gene expression among lymphoblastoid cell lines from CEPH families and found significant evidence of heritability (Schadt *et al.* 2003; Monks *et al.* 2004). An initial study of 56 individuals was based on four CEPH families who comprised pedigree founders (grandparents), parents, and children. This defined 2726 genes among the 24 479 genes analysed which were differentially expressed in at least half of the founders. Evidence of a significant heritable component was found among 25% of these genes (Schadt *et al.* 2003). In a larger study of 167 individuals among 15 CEPH families, 2430 differentially expressed genes were defined with 31% showing significant heritability at a false discovery rate of 0.05; the median heritability was 0.34 (Monks *et al.* 2004). Interestingly, this group of genes were enriched for genes with immune-related function. Further evidence of significant heritability was seen among lymphoblastoid cell lines established from family trios (parents and child) in the International HapMap Project (Stranger *et al.* 2007b).

Strong evidence for heritability in gene expression came from a more recent large scale analysis of primary blood cells among a cohort of 1240 Mexican Americans, most of whom came from 30 extended families of up to four generations (Goring *et al.* 2007). Gene expression

---

Box 11.1   Lymphoblastoid cell lines

Epstein–Barr virus immortalized lymphoblastoid cell lines have proved a powerful resource for genetic analysis of gene expression, established from individuals and families such as the Centre d'Etude du Polymorphisme Humain (CEPH) reference family panel (Dausset *et al.* 1990) and as part of the International HapMap Project (Box 9.1). Lymphoblastoid cell lines provide an ongoing source of a single cell type in which growth and culture conditions can be standardized to minimize environmental variation. Moreover large numbers of cells can be harvested for analysis and experimental replication achieved. The availability of dense SNP genotyping data for collections of lymphoblastoid cell lines established from diverse human populations such as the HapMap Project made these cells particularly suitable for application of the **genetical genomics** approach (Section 11.3.1). The HapMap Project has greatly advanced our understanding of fine scale genetic variation and genomic architecture with several million common SNPs defined and genotyped across diverse populations with the primary goal of facilitating and enabling

genome-wide association studies of common disease (Section 9.2.4). The panels of lymphoblastoid cell lines generated as part of the HapMap Project to provide an ongoing source of genomic DNA have subsequently been integral to many analyses of the genetics of gene expression. As lymphoblastoid cell lines are established from circulating peripheral blood lymphocytes they are a relevant cell type for many human traits and show detectable expression of a large number of genes. However, there are important caveats to their use, notably the fact that they are immortalized with Epstein–Barr virus which will itself modulate expression of some genes and may show heterogeneity between individual cell lines with viral load and other factors. Further potential issues on prolonged cell culture relate to the establishment of clonality and differential DNA methylation (Pastinen *et al.* 2006). It is also important to remember that functionally important genetic variation is often context-specific, and may be only manifested in specific cell or tissue types under specific conditions of stimulation or other environmental variables.

---

for 18 519 genes was analysed for circulating peripheral blood lymphocytes with a variance components-based heritability analysis performed that included age and sex. This demonstrated that 85% of transcripts show significant heritability with a false discovery rate of 5%. The median heritability of all expressed transcripts was 22.5%. These striking results in a large sample size analysing gene expression in primary blood cells reinforce the substantial genetic component to variation in observed gene expression.

Among primates, when gene expression was analysed at the RNA and protein level for 12 000 genes in blood and liver samples, patterns of gene expression in humans and chimpanzees (*Pan troglodytes*) were found to be more closely related to each other than to that seen in macaques (*Macaca mulatta*) (Enard *et al.* 2002). By contrast, when gene expression in brain

samples was analysed, chimpanzees were more similar to orangutans (*Pongo pygmaeus*) and macaques than to humans, highlighting tissue-specific differences that may be present, and in this study demonstrating accelerated differences in gene expression between human and other primates when brain tissue was analysed (Enard *et al.* 2002).

11.3   Mapping the genetic basis of variation in gene expression

11.3.1 Genetical genomics

In 2001 Jansen and Nap proposed the concept of 'genetical genomics' in which the powerful genetic approaches developed to map quantitative trait loci

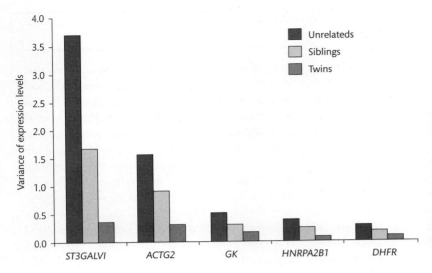

Figure 11.1 Variation in gene expression shows familial aggregation in humans. Gene expression data are shown for five genes among three groups: 49 unrelated individuals, 41 siblings in five CEPH families, and ten pairs of monozygotic twins. Reprinted by permission from Macmillan Publishers Ltd: Nature Genetics (Cheung *et al.* 2003c), copyright 2003.

(QTLs) (Box 11.2) could be applied to gene expression data becoming available through genomic technologies (Jansen and Nap 2001). The rapidly advancing field of **functional genomics**, particularly the quantification of gene expression from the analysis of RNA by microarray, was allowing the expression of multiple genes to be quantified in a high throughput manner. Jansen and Nap described the value of analysing expression of a given gene as a quantitative trait among related individuals in a segregating population, using informative polymorphic genetic markers to map QTLs (Jansen and Nap 2001). They noted that some QTLs would map to the location of the gene whose expression was being analysed, which they described as 'cis-acting', in other words the underlying functionally important genetic variation at the QTL, or in linkage disequilibrium with the genotyped marker(s), was present on the same chromosome and acting locally, while other QTLs would be 'trans-acting', present at other genomic locations including different chromosomes (Section 11.3.2). Jansen and Nap's paper described the application of genetical genomics to plant populations but they noted the approach applied equally to model organisms, animal populations, and humans.

11.3.2 Local and distant regulatory variation

Use of the terms 'cis' and 'trans' acting is however open to some ambiguity if based only on the genomic location

of the mapped genetic marker in relation to the specific gene encoding the transcript whose expression has been quantified. This relationship alone does not define mechanism and indeed there is significant variation between studies in what is considered 'cis-acting' on this basis, whether it is the mapped markers closest to the gene encoding the assayed transcript (Emisson *et al.* 2008), or those within 10 kb (Brem *et al.* 2002), 1 Mb (Stranger at al. 2007b; Schadt *et al.* 2008), or 5 Mb of the gene.

In their review, Rockman and Kruglyak describe how confusion can arise as cis-acting describes a specific relationship between the variant and gene on the same chromosome, which may be local or acting at a distance. Conversely, trans-acting variants can act within a gene by initiating feedback loops or operate at a distance, typically from another chromosome (Rockman and Kruglyak 2006). In the absence of other evidence, Rockman and Kruglyak therefore argue for a classification of mapped markers as defining either 'local' or 'distant' regulatory variation - 'local' mapping close to the physical location of the gene or 'distant' elsewhere in the genome (Rockman and Kruglyak 2006).

Classically, cis-acting variants involve modulation of an upstream regulatory element such as a transcription factor binding site within a promoter, which results in differential transcription of the regulated gene. Regulatory elements can, however, be a considerable distance from the target gene, operating in some cases through DNA

> ## Box 11.2   Quantitative trait loci
>
> Quantitative traits describe phenotypic characteristics that show graded continuous variation in a population, as seen for example with height. Quantitative traits are determined by a combination of genetic variation in more than one gene, and environmental factors. QTLs are genomic regions associated with the particular phenotype being studied. Gene expression can be analysed as a quantitative trait, with resolution of **expression quantitative trait loci (eQTLs)** (Schadt *et al.* 2003).

---

looping to bring a distant enhancer into contact with the gene promoter. Indeed such mechanisms may bring several genes together in 'transcription factories' with coordinate regulation that may involve interchromosomal interactions such that potentially *cis*-acting variation on one chromosome may modulate expression of a gene on a different chromosome (Spilianakis *et al.* 2005; Rockman and Kruglyak 2006). The mechanisms underlying *cis*-acting variation are diverse and include post-transcriptional events such as mRNA stability, processing, and decay as well as alternative splicing and copy number.

While most locally mapped regulatory variants will be *cis*-acting, Rockman and Kruglyak estimate up to a third may act in *trans*, for example through feed back loops (Rockman and Kruglyak 2006). Distant regulatory variants may act in *cis* as described or, more commonly, be *trans*-acting, modulating for example the expression or function of a specific transcription factor and hence the control of expression of distant, often multiple genes. To add to the potential confusion, the mechanism at the transcription factor gene itself may involve a coding variant or *cis*-acting regulatory variant. 'Hotspots' of distant regulatory variants are described, mapping to expression of multiple genes. However the identification of distant regulatory variants through analysis of the genetics of gene expression has proved challenging and, in the absence of further functional data beyond mapping to gene expression, it is most appropriate to describe linked or associated variation as either local or distant to the gene whose expression has been quantified.

11.3.3 Genetical genomics and model organisms

The concept of genetical genomics saw initial application in studies by Brem and colleagues in 2002 using budding yeast to define the role of genetic factors in observed variation in gene expression (Brem *et al.* 2002). *Saccharomyces cerevisiae* has been a remarkably powerful model organism for genetic analysis with an extremely well characterized genome and metabolic pathways, in a species with short generation times that is easily propagated and amenable to population genetic analysis. Genome-wide gene expression of 6215 genes was analysed for six independent cultures of a laboratory strain (BY) and a wild isolate from a vineyard (RM). Many more genes showed differential expression between the strains (1528 genes) than expected by chance (23 genes). The two strains were crossed and the 40 haploid segregants analysed. This showed that the parental differences in gene expression were highly heritable, with 84% of observed variation thought to be genetic. Linkage analysis was then performed using 3312 biallelic markers. When gene expression was analysed, many measured transcripts showed evidence of linkage to at least one locus: with a P value less than 10^{-5}, 570 transcripts showed linkage, while only 53 would be expected by chance; for P less than 10^{-6}, 206 transcripts showed linkage, with less than one transcript expected by chance (Fig. 11.2A) (Brem *et al.* 2002).

This study also allowed estimation of the proportions of markers showing linkage that mapped locally or at a distance from the physical location of the gene encoding the assayed transcript. In 36% of cases, measured transcripts showed linkage to markers within 10 kb of the gene on the same chromosome and constituted locally acting regulatory variation, likely to be *cis*-acting (Brem *et al.* 2002). When distant acting regulatory variation (likely *trans*-acting variants) were considered, a small number of alleles were resolved in linkage with expression of many genes. In fact when the yeast genome was divided into 611 bins each 20 kb in size, 40% of all

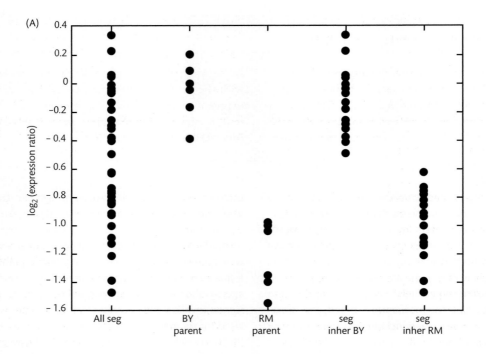

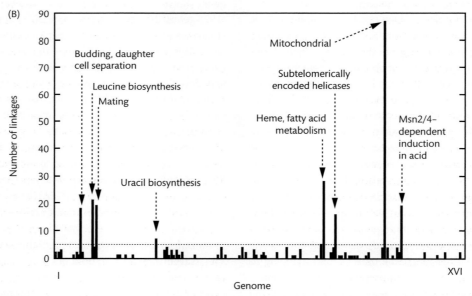

Figure 11.2 Genetic diversity and variation in gene expression in yeast. (A) Data are shown for the gene expression of YLL007C for all 40 segregants (seg) of a cross between two strains BM and MY, for the parental strains, and for segregants inheriting a marker found to be in linkage with YLL007C expression. (B) When the number of linkages was analysed by chromosomal location (based on 20 kb bins) many more linkages were observed at certain bins than expected by chance (no bin would, for example, be expected to contain more than five linkages by chance). Eight of the groups showing more than five linkages are indicated with dashed arrows with the common function of linked expression transcripts indicated. From Brem *et al.* (2002), reprinted with permission from AAAS.

linkages fell into only eight bins (groups) with evidence of common function for linked gene transcripts at a particular group (Fig. 11.2B).

In a subsequent study from the same group, potential *trans*-acting effects of genetic variation were explored in more detail using a cross of the same two yeast strains, BY and RM, analysing 86 segregants (Yvert *et al.* 2003). The strategy here involved hierarchical clustering of genome-wide gene expression data based on similarities between genes in their expression profiles followed by linkage analysis using the mean expression of a cluster as a quantitative phenotype. Many more genes than expected by chance showed significant clustering, with three-quarters involving genes on different chromosomes. Linkage analysis defined significant linkage for 304 clusters containing 1011 genes (ten clusters expected by chance) with the majority of genes and clusters (75% and 80%, respectively) not showing self linkage. This work emphasized the importance of *trans*-acting variation and allowed positional cloning and molecular analysis to be applied with the definition of specific genetic variants responsible for the observed linkage. It also showed that in yeast, at least, genetic variation involving transcription factors did not account for a significant proportion of *trans*-acting loci but rather such variation was broadly dispersed across different genes in the yeast genome (Yvert *et al.* 2003).

Following on from the studies in yeast, Schadt and colleagues demonstrated in mice the utility and application of the genetical genomics approach (Schadt *et al.* 2003). Here two inbred mouse strains (C57BL/6J and DBA/2J) were crossed, producing an F1 generation who were in turn crossed to produce an F2 generation, among whom genetic variation could be used to map particular phenotypes to a specific genomic location. Gene expression profiles were established using a mouse expression microarray. Schadt and colleagues analysed a total of 111 F2 generation mice using 100 microsatellite markers for 7861 gene expression phenotypes that were significantly different between the two parental strains, or more than 10% of the F2 individuals. eQTLs were defined for 2123 genes with a lod score (Box 2.4) greater than 4.3 (*P* <0.00005). As found in yeast, local and distant regulatory variation was resolved, probably *cis*- and *trans*-acting eQTLs, respectively. Here those eQTLs with the highest lod scores were likely *cis*-acting.

An example is provided by the eQTL identified at the C5 gene. DBA/2J mice, in contrast to C57BL/6J mice, are known to be deficient in C5 at both the transcript and protein level as a result of a 2 bp deletion in the C5 gene, which leads to a frameshift and premature stop codon (Wetsel *et al.* 1990; Karp *et al.* 2000). In this case the eQTL is found at the gene itself (with a very high lod score of 27.4) and has a clearly defined functional mechanism with specific genetic variation (the 2 bp deletion) responsible for the difference in gene expression (Schadt *et al.* 2003).

Rodent recombinant inbred lines are a very powerful resource for mapping eQTLs, being a mosaic of the two crossed inbred parental strains after repeated sib matings (Fig. 11.3) (Broman 2005). Such studies in mice and rats allowed the definition of multiple *cis*-acting eQTLs and integration with data available for many other phenotypes on such strains, for example behavioural phenotypes with the eQTLs mapped in mouse brain tissue by Chesler and colleagues (2005). Evidence of tissue-specific effects was also possible, for example between haematopoietic stem cells and brain tissue in the same mice (Bystrykh *et al.* 2005). These datasets and others are available as part of WebQTL at GeneNetwork (www.genenetwork.org) which seeks to integrate networks of genes, transcripts, and traits for different organisms with biological traits and gene expression data (Wang *et al.* 2003b; Chesler *et al.* 2004).

The work of Hubner and colleagues analysing gene expression in recombinant inbred lines from a cross of the spontaneously hypertensive rat and the normotensive Brown Norway rat strains demonstrated more than 1000 eQTLs for gene expression in fat and kidney tissues, with more significant eQTLs predominantly *cis*-acting. Within previously mapped physiological QTLs, specifically hypertension-related loci, 73 significant *cis*-regulated eQTLs with human homologues were defined that were proposed as candidates for gene loci where genetic variation may be important determinants of hypertension (Hubner *et al.* 2005). The genetical genomics approach has also been applied to the fly *Drosophila melanogaster* (Wayne and McIntyre 2002), the worm *Caenorhabditis elegans* (Li *et al.* 2006b), and plants such as maize *Zea mays* (Schadt *et al.* 2003), thale cress *Arabidopsis thaliana* (Keurentjes *et al.* 2007), and *Eucalyptus* (Kirst *et al.* 2004).

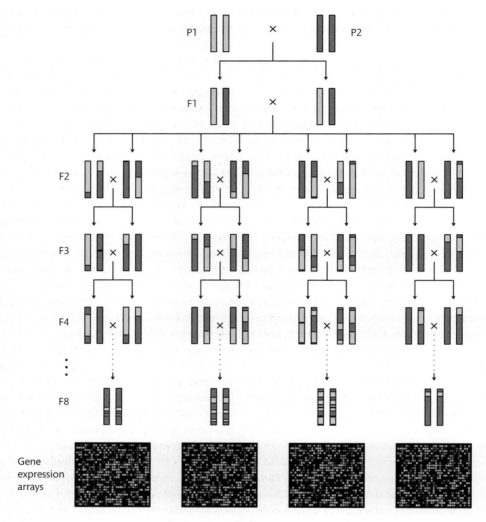

Figure 11.3 Recombinant inbred lines for mapping expression quantitative trait loci. A pair of chromosomes is shown for each of two parental inbred strains from whom recombinant inbred lines are generated by repeated sib mating. This produces mosaics of the original parental genomes and definition of underlying important genetic variation when integrated with gene expression profiling data. Reprinted by permission from Macmillan Publishers Ltd: Nature Genetics (Broman 2005), copyright 2005.

Such studies have been very important in advancing our understanding of this field and the underlying complexities of genetic variation modulating gene expression. The further work of Kruglyak and colleagues in yeast, for example, highlighted the large range of variance explained by genetic variation, even for highly heritable traits, with a median of 27% reported (Brem and Kruglyak 2005). Large effects of eQTLs appear rare with the majority of transcripts mapping to weak eQTLs. Inheritance is complex with only 3% of highly heritable transcripts showing linkage consistent with a single locus; for half of such transcripts, modelling suggests more than five eQTLs are involved (Brem and Kruglyak 2005). It was also striking that interactions between eQTLs are a prominent feature on the analysis of yeast when a two stage approach is taken, analysing primary then secondary eQTLs with such pairs of loci involved in inheritance of 57% of transcripts (Brem *et al.* 2005). Such epistatic effects (Section 13.2.3) are important and care is needed in the choice of analytical approach as the effects seen with the secondary loci were modest, and two-thirds would not be detected without the two stage approach (Brem *et al.* 2005).

11.4 Mapping genetic variation and gene expression in human populations

11.4.1 Insights from lymphoblastoid cell lines

The genetical genomics approach was also successfully applied to human populations and showed the potential great utility of this approach in advancing our understanding of the genetic basis of common complex traits. Initial studies utilized lymphoblastoid cell lines (Box 11.1) and applied linkage-based analysis (Monks *et al.* 2004; Morley *et al.* 2004). For example, Morley and colleagues analysed gene expression for cell lines established from 14 CEPH families comprising grandparents, parents, and children (the latter on average comprising eight siblings per family) (Morley *et al.* 2004). The most variably expressed genes among the cell lines established from the 94 grandparents were selected (3554 of 8500 genes assayed) and genome-wide linkage analysis was performed using 2756 autosomal SNPs. Depending on the level of genome-wide significance chosen, 142 expression phenotypes ($P = 0.001$) or 984 phenotypes ($P = 0.05$) showed significant linkage (Morley *et al.* 2004). Again, the distinction between likely *cis-* and *trans*-acting loci was based on a distance from the target gene, here linked SNPs present less than 5 Mb from the gene were classed as *cis*-acting effects. This classified 19% as *cis*-acting only, and 77.5% as *trans*-acting only. However, the classification refers more specifically to the relative proportion of local and distant regulatory variation and determination of whether they are *cis-* or *trans*-acting remained undefined. Most eQTLs were thus distant and considered to be *trans*-acting, with two hotspots defined at chromosome 14q32 and

chromosome 20q13, regulating expression of seven and six gene expression traits, respectively. Genotyping additional SNPs helped confirm *cis*-acting loci through association and analysis of differential allelic expression, the latter defined for example an eight-fold difference in expression of *PSPHL* between alleles for SNP rs6700 (Morley *et al.* 2004).

Monks and colleagues also studied lymphoblastoid cell lines established from CEPH families, performing linkage analysis for 2430 differentially expressed genes for 167 individuals in 15 families (Monks *et al.* 2004). They defined significant eQTLs for 33 genes with a *P* value of less than 0.000005 or 22 genes at genome-wide significance using a conservative Bonferroni correction. Unlike in yeast, there did not appear to be evidence of hotspots of linkage with particular QTLs controlling expression of multiple genes. The investigators noted their study was powered only to detect major eQTLs and that those identified for the 33 genes explained more than 50% of the observed variance in gene expression (Monks *et al.* 2004).

In 2005 Cheung and colleagues published data showing how genetic association rather than linkage could be successfully employed for such studies, and in particular the power of a genome-wide association approach (Cheung *et al.* 2005). The study was comparatively modest in size, comprising 57 lymphoblastoid cell lines established from individuals in CEPH pedigrees that were included in the CEU panel of European ancestry in the International HapMap Project (Box 9.1). The investigators followed on from their earlier linkage analysis in which they had defined gene expression phenotypes with evidence of *cis*-acting eQTLs (Morley *et al.* 2004) and this allowed the utility of the association approach to now be assessed. Genetic association using dense SNP sets within 50 kb of target genes showed overlap of association with linkage, for 65 out of 374 phenotypes analysed, with a narrower window defined by association compared to linkage peaks (Fig. 11.4) (Cheung *et al.* 2005).

For genome-wide SNP analysis, 27 phenotypes with the strongest evidence of *cis*-acting eQTLs from linkage analysis were analysed using 770 394 SNP markers. This allowed regression analysis of gene expression by marker genotype, and defined evidence of association at a genome-wide level of significance for 14 out of 27 phenotypes ($P < 6.7 \times 10^{-8}$) (Cheung *et al.* 2005). The same region was defined by linkage and genome-wide association for 15 of 27 phenotypes. The study illustrated the power of applying genome-wide association to genetical genomics even with this modest sample size. The investigators also demonstrated for a specific marker SNP showing significant association in the *CHI3L2* gene (encoding chitinase 3-like 2) promoter that the nucleotide substitution modulated reporter gene activity, and allele-specific gene expression based on the haploChIP approach (Section 11.5.3).

The value of the association-based approach was shown in the same year by Stranger and colleagues, who also analysed lymphoblastoid cell lines established from unrelated individuals in the CEU HapMap panel – in this case 60 lymphoblastoid cell lines – for 630 genes from the ENCODE (Section 9.2.4) region using 73 712 common SNPs (present at greater than 5% minor allele frequency) (Stranger *et al.*

2005). Again, even with a modest sample size and gene set, the study was successful in defining up to 40 genes associated with significant local, likely *cis*-acting, eQTLs.

In 2007, Stranger and colleagues published data that built on this work by taking advantage of the denser SNP marker sets now available as part of Phase II of the HapMap Project (Section 9.2.4) (Stranger *et al.* 2007b). They analysed 270 lymphoblastoid cell lines, established from four different populations based on geographic ancestry of populations (Frazer *et al.* 2007). The analysis of cell lines from different human populations provided an important opportunity to assess how reproducible observed associations were between populations. The study was further facilitated by the availability of whole genome expression arrays including some 47 294 probes which, combined with high density, genome-wide marker SNP coverage, allowed for a comprehensive analysis of genome-wide association. Study power was, however, limited by the number of cell lines

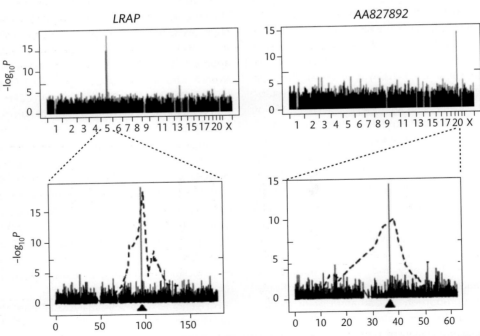

Figure 11.4 Genome-wide association for gene expression phenotypes. Genome-wide association chromosomal plot for *LRAP* and *AA827892* together with higher resolution plots showing superimposed association (bars) and linkage analysis (dashed lines) with target genes indicated by arrow heads. Reprinted by permission from Macmillan Publishers Ltd: Nature (Cheung *et al.* 2005), copyright 2005.

available for interrogation, which remained modest for each population, namely 60 cell lines established from unrelated individuals of European descent (CEU), 60 of African (YRI) ancestry, 45 of Chinese (CHB) ancestry, and 45 of Japanese (JPT) ancestry, giving a total of 210 lymphoblastoid cell lines analysed in this study (Stranger *et al.* 2007b).

For the association study, 2.2 million common SNPs with a minor allele frequency greater than 5% were used to determine eQTLs for 13 643 genes in lymphoblastoid cell lines for each of the four populations (Stranger *et al.* 2007b). Reassuringly, for the 60 CEU individuals there was a significant correlation with the results of their earlier study of a limited number of genes with different RNA samples (Stranger *et al.* 2005). In the current genome-wide analysis, 299 genes in the CEU panel of lymphoblastoid cell lines showed significant local, likely *cis*-acting associations where only 14 would be expected by chance; the definition of *cis*-acting used here was based on SNP markers lying within 1 Mb of the centre of the transcriptional unit whose expression levels were analysed (Stranger *et al.* 2007b). Across all the populations, a total of 1348 genes showed significant local eQTLs of which 37% were found in at least two populations (Stranger *et al.* 2007b). In almost all cases of shared associations between populations, the direction of allelic difference in expression was the same. Local

eQTLs were found to predominantly occur very close to the transcriptional start site and more often than expected within evolutionarily conserved sequences (Stranger *et al.* 2007b).

Population differences in gene expression have been characterized by other investigators using HapMap (Frazer *et al.* 2007) and other sources of lymphoblastoid cell lines (Hinds *et al.* 2005). Spielman and colleagues found that for 4197 genes analysed and expressed in these lines, 25% were significantly different between lines of European ancestry compared to those of Asian descent (Spielman *et al.* 2007). Underlying allele frequency differences between the populations accounted for many of these differences in expression when a small number of the expression phenotypes were characterized more fully. For example, a high producer allele for a SNP of *POMZP3* was much more frequent in Europeans than East Asians, accounting for the observed population differences in expression (Fig. 11.5) (Spielman *et al.* 2007). As a further example, the *UGT2B17* gene showed significant differences in gene expression between European ancestry and East Asian populations consistent with previous data showing population-specific copy number variation for this gene (Wilson *et al.* 2004; McCarroll *et al.* 2006; Xue *et al.* 2008). *UGT2B17* at chromosome 4q13 encodes an enzyme involved in steroid metabolism and is subject to a 117 kb deletion

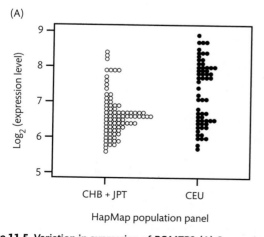
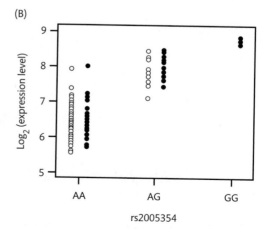

Figure 11.5 Variation in expression of *POMZP3*. (A) Comparison of expression among lymphoblastoid cell lines established from individuals of European (CEU) and East Asian (CHB, JPT) geographic ancestry. (B) Allelic variation for rs2005354. Reprinted by permission from Macmillan Publishers Ltd: Nature (Spielman *et al.* 2007), copyright 2007.

leading to marked differences in observed expression. The gene deletion is rare among European and African populations but common in those of East Asian ancestry (Fig. 11.6) (Xue *et al.* 2008).

These data raise the question of the extent to which fine scale nucleotide diversity and copy number variation underlie the observed eQTLs. Stranger and colleagues analysed copy number variation for association with gene expression in the same HapMap populations as they analysed SNP markers and found significant but distinct associations, which were fewer in number but comparably reproducible across populations (Section 4.3) (Stranger *et al.* 2007a). Many more associations are likely to be found as our ability to resolve and quantify smaller scale copy number variation improves. The study reinforced the need to consider both fine scale nucleotide level diversity and structural genomic variation when analysing genetic determinants of gene expression, and in particular that analysis of either SNPs or copy number variants alone would capture a small minority of the associations due to the untyped class of genetic variation (Stranger *et al.* 2007a).

11.4.2 Insights into genetic susceptibility to asthma

Asthma is a common complex multifactorial disease in which genetic determinants play an important role (Box 11.3). Cookson and colleagues published research in 2007 that elegantly demonstrated the power of complimentary genome-wide association studies of disease susceptibility, and of gene expression (Dixon *et al.* 2007; Moffatt *et al.* 2007). The investigators carried out an initial genome-wide scan in family and case–control panels comprising a total of 994 individuals of British and German descent with childhood onset asthma and 1243 non-asthmatics (Moffatt *et al.* 2007). Analysis of 317 000 SNPs resolved that seven of 12 marker SNPs with a false discovery rate of less than 1% showed association with a 112 kb region at chromosome 17q21 (Fig. 11.7), the most strongly associated SNP being rs7216389 (c.236-1199G>A) ($P = 9 \times 10^{-11}$). The associated markers (with $P < 10^{-6}$) spanned three haplotype blocks over 206 kb and showed low linkage disequilibrium, suggesting one or more causative variants may

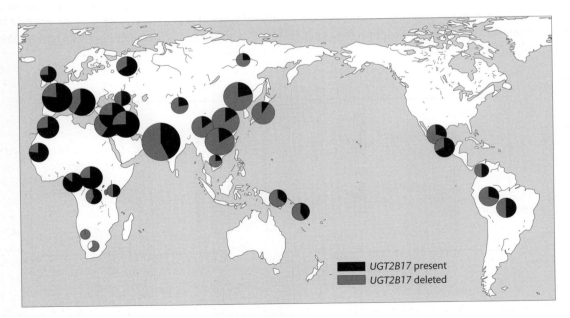

Figure 11.6 Copy number variation in *UGT2B17* for different human populations. Reprinted from *American Journal of Human Genetics* (Xue *et al.* 2008), copyright 2008, with permission from Elsevier .

Box 11.3   Asthma

Asthma is a common, chronic respiratory disease that is associated with significant morbidity and can be fatal. The condition is characterized by chronic airway inflammation and hyper-responsiveness. Symptoms include wheeze, breathlessness, chest tightness, and cough. Airflow obstruction is to a variable degree reversible, with effective bronchodilator medications available. There are many known environmental factors influencing the development of asthma and the precipitation of acute attacks. There is a strong familial component to asthma and the disease has been subject to intensive investigation as a common multifactorial trait without a clear pattern of inheritance. Many linkage and candidate gene association studies have been performed with further evidence more recently from genome-wide association studies.

Linkage analysis and positional cloning has had significant successes, as seen with the identification of *ADAM33* at chromosome 20p13 (Van Eerdewegh *et al.* 2002). Elsewhere genetic variation has been implicated in genes involved in innate immunity and immune regulation such as encoding pattern recognition receptors (*CD14, TLR2*) and cytokines (*IL10, TGFB1*), as well as in antigen presentation (*HLA-DR, -DQ*), Th2 differentiation, and functions integral to allergy and inflammation (*GATA3, IL4,* and *IL13*) together with genes involved in epithelial and mucosal cell biology and airway remodelling (Vercelli 2008). The list of associated genes and variants is extensive, and like many common multifactorial traits there have been difficulties in definitive replication of disease associations and resolution of specific functional variants.

---

be present at this locus; forward stepwise regression revealed three SNPs jointly showed strong association ($P < 10^{-12}$). The association was replicated in further British and German cohorts (Moffatt *et al.* 2007).

In parallel with their genome-wide association study, the investigators established lymphoblastoid cell lines from 400 children of families with a proband who had asthma (Dixon *et al.* 2007). These cell lines were harvested for gene expression studies on initial culture after transformation, minimizing artefacts that may be associated with repeated passaging of such cell lines. Genome-wide SNP genotyping data on a total of 408 273 SNPs were generated and expression profiling for 54 675 transcripts (corresponding to 20 599 genes) was quantified. Of the 19 annotated genes in the disease associated region, gene expression data were available for 14.

Remarkably, the SNPs showing the strongest disease association were found to be highly associated with levels of *ORMDL3* gene expression (correlation 0.67, $P = 0.004$) (Fig. 11.7C) (Dixon *et al.* 2007). *ORMDL3* lies within the disease associated interval although the most strongly associated SNP from the disease association study (rs7216389) is within intron 1 of the neighbouring

gene *GSDML*. This SNP shows a striking association with *ORMDL3* expression ($P < 10^{-22}$), and together the associated SNPs account for 29.5% of the variance observed in expression of *ORMDL3* (Dixon *et al.* 2007). No difference was seen between lymphoblastoid cell lines established from individuals with and without asthma, indeed for all 54 675 transcripts analysed only ten differed significantly between the two groups.

The findings of the genome-wide association studies of disease association and gene expression implicate genetic diversity in a 206 kb interval of chromosome 17q21 within which *ORMDL3* is a candidate gene for genetic susceptibility to childhood asthma. The initial findings have been replicated in a series of independent studies of asthma in a number of different populations, which have defined 17q21 as an asthma susceptibility locus. These have included a study of North American white asthmatics (Sleiman *et al.* 2008), a family-based study of a French Canadian population (Madore *et al.* 2008), a large case–control study from Scotland (Tavendale *et al.* 2008), a family-based study of Mexicans, and a case–control study of African Americans (Galanter *et al.* 2008). In the study by Tavendale and colleagues of 1054 asthmatics and 1465

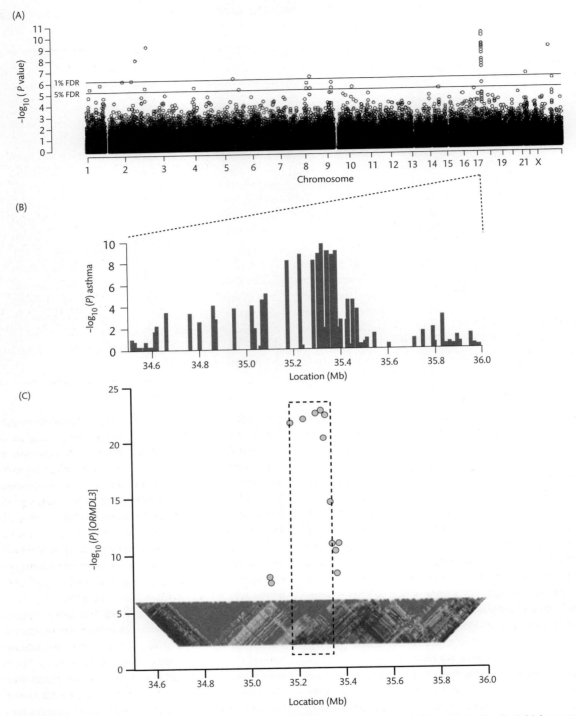

Figure 11.7 Genome-wide association for asthma at 17q21 and expression of *ORMDL3*. (A) Data shown for the initial genome-wide scan of 994 cases and 1243 controls with the strength of association (plotted on *y* axis) by genomic location (plotted on *x* axis) for 317 447 SNPs. FDR, false discovery rate. (B) Detail of disease association to 17q21. (C) Association with *ORMDL3* expression of the same SNP markers. Redrawn and reprinted by permission from Macmillan Publishers Ltd: Nature (Moffatt *et al.* 2007), copyright 2007.

controls, for example, rs7216389 was strongly associated with childhood asthma ($P = 1.7 \times 10^{-12}$) with possession of one copy of the risk T allele associated with an odds ratio of 1.5 (1.2–1.8) and two copies with an odds ratio of 2.1 (1.7–2.6), demonstrating a dose-dependent effect (Tavendale *et al.* 2008).

ORMDL3 encodes a transmembrane protein anchored in the endoplasmic reticulum; it is a member of a novel class of genes and its function is currently unknown. Much remains to be done to define how rs7216389 or the other disease associated SNPs may be acting. It may be that several independent variants are exerting an effect at this locus, and indeed that the specific causative/functional variant(s) have not yet been identified but are in linkage disequilibrium with SNPs resolved by the genome-wide association studies described. What is clear is that Cookson and colleagues demonstrated elegantly the power of the complimentary and synergistic approaches of genome-wide genetic association at the level of disease susceptibility and gene expression. Their expression dataset was established as a searchable database 'mRNA by SNP Browser' (www.sph.umich.edu/csg/liang/asthma/) for use as a public resource (Dixon *et al.* 2007). Application of these data was noted as facilitating definition of a candidate gene in Crohn's disease flanking a gene desert on chromosome 5p13.1 (Section 9.5.4) (Libioulle *et al.* 2007).

It is also worth considering in more detail the genome-wide association scan of gene expression that was performed (Dixon *et al.* 2007). A notable feature of the design was that the 400 children from whom lymphoblastoid cell lines were established comprised 206 families and that this enabled heritability to be assessed for the expression traits that were quantified. In all, 830 offspring and parents were genotyped for the 408 273 SNPs. A total of 14 819 annotated expression traits were identified as showing highly heritable expression levels (H^2 greater than 0.3) indicating genetic regulation was important in determining observed variation in levels of expression. For these traits the peak lod score of association was high, varying from 3.68 to 59.12: when a genome-wide significance threshold of lod score >6 was used, some 1989 transcripts were identified with the peak associated SNP explaining 32.9% of the calculated heritability (Dixon *et al.* 2007).

The most heritable traits were enriched for specific biological processes, notably chaperone and heat shock proteins, as well as genes involved in the cell cycle, RNA processing, DNA repair, immune response, and apoptosis. The data for genes involved in the immune response (Fig. 11.8) highlighted the importance of genetic variation in the MHC on chromosome 6 (Chapter 12), with a number of MHC class II genes showing striking associations between gene expression and genetic variation, which may be highly significant in dissecting the many known associations with infectious and autoimmune disease at this locus. The dataset highlighted the importance of *cis*-acting variation that was responsible for the majority of the strongest observed associations (lod score >9). The authors also emphasized the importance of a sufficiently powered study for such analyses. Their dataset of 206 sibships enabled definition of 16 098 associations for 1989 transcripts; for 100 sibships the number of associations would fall to 4923 associations (736 transcripts) and for 50 sibships to 503 associations (106 transcripts) (Dixon *et al.* 2007).

11.4.3 Genetics of gene expression in primary human cells and tissues

Studies in lymphoblastoid cell lines and animal models demonstrated the utility of mapping gene expression as a quantitative trait to define loci with the potential to modulate gene expression, and specific variation within those loci which may be functionally important. The context of regulatory variants was noted to be highly significant from these and other studies seeking to identify such diversity, with evidence that the effects of particular variants may only be apparent in specific cell types or tissues, in conditions relating to a given physiological or disease state (Cowles *et al.* 2002; Bystrykh *et al.* 2005; Chesler *et al.* 2005; Hubner *et al.* 2005; Petretto *et al.* 2006).

This was shown to be the case in a large study of Icelandic individuals of Caucasian descent when gene expression traits were analysed from blood and adipose samples and correlated with obesity-related traits (Emilsson *et al.* 2008). The individuals recruited were part of three generational pedigrees scored for biometric traits related to obesity, but representative of the general population rather than being recruited on the basis of particular obesity phenotypes. A total of 1002 peripheral blood samples and 673 subcutaneous fat samples were collected and genome-wide expression profiling carried

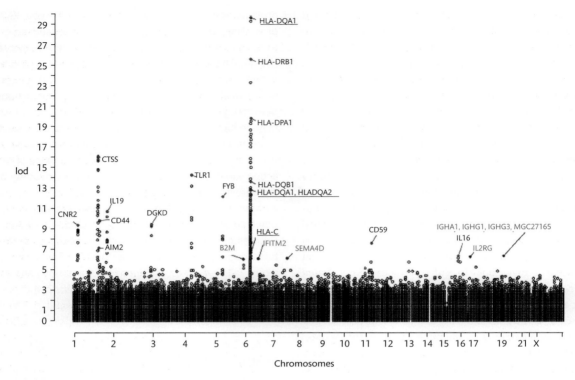

Figure 11.8 Genome-wide association for genes involved in the immune response as defined by Gene Ontology analysis. Genes showing genome-wide significance (lod score >6) are indicated with gene names either in *cis* within 100 kb of the gene, or in *cis* on the same chromosome but more than 100 kb away (gene name underlined), or in *trans* (gene name in grey). Data are plotted by genome position (*x* axis) with the lod score of association on the *y* axis. Reprinted by permission from Macmillan Publishers Ltd: Nature Genetics (Dixon *et al.* 2007), copyright 2007 .

out, quantifying 23 720 transcripts. A significantly higher proportion of gene expression traits correlated with obesity-related biometric traits (such as body mass index, waist–hip ratio) from adipose compared to blood samples (Emilsson *et al.* 2008). These comprised more than 50% of all gene expression traits from adipose compared to less than 10% in blood samples: for body mass index, a 35-fold enrichment of expression traits was seen in adipose compared to blood samples among paired samples (where both blood and subcutaneous fat samples were available for a given individual).

The power of this Icelandic study lay not only in the large number of individuals recruited but the family pedigrees available for analysis, allowing estimation of heritability and linkage analysis to be performed. A high

proportion of transcripts showed evidence of being significantly heritable traits, 59% in blood and 71% in adipose samples, with genetic variance noted to explain on average 30% of observed variation (Emilsson *et al.* 2008). When 1732 microsatellites were used for genome-wide linkage analysis, significant linkage was found for markers closest to the location of the gene encoding the measured transcript (denoted *cis*-acting eQTLs) for 9.4% of traits in blood and 5.8% of traits in adipose samples. When SNP associations were sought using whole genome SNP genotyping (some 317 503 SNPs) for 150 unrelated individuals in whom blood and adipose samples were both available, significant association was found within 2 Mb of a gene for 11.5% of measured expression traits in blood and 14.6% in adipose tissue (Emilsson *et al.* 2008). The results from this large scale analysis of primary human cells and tissues serve to underline both the importance of genetic variation in modulating gene expression, and the relative importance of local *cis*-regulatory variants.

The significance of local regulatory variants was also demonstrated in a large scale analysis of gene expression in peripheral blood lymphocytes from the 1240 Mexican American individuals participating in the San Antonio Family Heart Study (Mitchell *et al.* 1996; Goring *et al.* 2007). Like the Icelandic study by Emilsson and colleagues, this cohort of individuals was extremely powerful for the application of genetic analysis, comprising primarily of 30 extended families over up to four generations with a mean age of 39 years (Goring *et al.* 2007). Genome-wide expression profiling of transcripts from 18 519 genes demonstrated significant heritability in 85% of transcripts (Section 11.3.1), while analysis of 432 highly polymorphic microsatellite markers allowed genome-wide linkage analysis to be performed to define eQTLs. A total of 1345 transcripts (6.8%) showed significant evidence of *cis* regulation with a false discovery rate of 5%. In this study significant *cis*-regulatory effects were defined by calculating the multipoint lod score at the marker nearest the structural gene encoding the specific transcript being analysed (Goring *et al.* 2007). The median effect size of these local, likely *cis*-regulatory eQTLs was 25% and, for more significant loci, genetic variation explained much of the observed variance in expression, with 128 eQTLs having locus-specific heritability greater than 50%. Overall *cis* regulation was estimated to account for 5% of variation in gene expression.

The dataset confirmed many of the significant local *cis* eQTLs demonstrated in lymphoblastoid cells (Morley *et al.* 2004) but distant, likely *trans*-acting loci were much

harder to resolve and may reflect the additional complexity of defining such loci *in vivo*. A striking demonstration of the power of the dataset and approach established by Goring and colleagues was shown for a complex phenotypic trait assayed in the same 1240 individuals, namely plasma concentration of high density lipoprotein (HDL) cholesterol (Box 11.4) (Goring *et al.* 2007).

Goring and colleagues were able to assay plasma HDL cholesterol at the same time as lymphocytes were sampled for gene expression analysis, samples being taken after an overnight fast (Goring *et al.* 2007). Expression levels of all transcripts with significant *cis*-acting eQTLs were correlated with HDL cholesterol, and 67 were found to be significant at a false discovery rate of 5%. Of these transcripts, *VNN1* (encoding vanin 1) at 6q23-q24 was striking with a *cis*-acting eQTL lod score of 11.7 ($P = 1.1 \times 10^{-13}$) and very strong association with HDL cholesterol ($P = 4 \times 10^{-9}$).

Resequencing the putative promoter region of *VNN1* (2 kb upstream of the start site) among 96 individuals resolved 22 SNPs, of which the six most strongly associated with transcript levels were genotyped in the full cohort of 1240 individuals. The results were highly significant, notably for a T to G SNP located 137 nt upstream of the transcriptional start site, which was very strongly associated with *VNN1* expression ($P = 5.7 \times 10^{-83}$) and with HDL cholesterol concentration ($P = 4 \times 10^{-4}$) (Goring *et al.* 2007). Further work is required to define the functional mechanisms and specific variants involved; this SNP disrupts a consensus binding site for the transcription factor Sp1 while vanin 1 has enzyme activity involved in the production of cysteamine, which is important in preventing lipid perioxidation. There is also evidence of association between genetic variation at *VNN1* and hypertension with a nonsynonymous

---

Box 11.4   Genetic determinants of HDL cholesterol

Twin studies show 50% of the variation in levels of HDL cholesterol are genetically determined (Goode *et al.* 2007). Levels of HDL cholesterol have been related to diversity in multiple genes, resolved for example by linkage analysis in hereditary HDL disorders, genetic association studies, and analysis in

mice (Holleboom *et al.* 2008). Clinically, levels of HDL cholesterol are highly significant in terms of cardiovascular disease risk and new insights into its regulation through genetic analysis would be extremely important.

SNP showing significant association (Zhu and Cooper 2007).

The power of analysing the genetics of gene expression in specific tissues and integrating this with data from genome-wide association studies of common disease was seen in the analysis by Schadt and colleagues of liver tissue from a large cohort of 427 individuals, collected either post mortem or from organ donors (Schadt et al. 2008). Here genome-wide gene expression profiles were determined for 39 280 transcripts related to 34 266 known or predicted genes. Genotyping data for 782 476 unique SNPs were established and significant associated 'expression SNPs' (eSNPs) resolved, either using a conservative Bonferroni statistical correction to determine genome-wide significance or a false discovery rate of less than 10%. For local, likely cis-acting effects (SNPs within 1 Mb of the gene), a total of 1350 expression traits were significant based on Bonferroni or 3210 traits based on the false discovery rate cut off; for trans-acting effects the number of significant expression traits were 242 and 491, respectively (Schadt et al. 2008).

The dataset highlighted how significant eSNPs may be a considerable distance from the gene whose expression is being analysed, with 30% of likely cis-acting eSNPs more than 100 kb away from the transcriptional start or stop sites (Schadt et al. 2008). The other dramatic illustration of this study was the power of integration of the data with genome-wide association studies of common disease. In type 1 diabetes, for example, many different susceptibility loci have been defined (Section 9.3), including from the Wellcome Trust Case Control Consortium (WTCCC 2007) and replication studies (Todd et al. 2007). A striking number of the disease associated SNPs showed significant associations with gene expression in the liver samples; for example within the MHC region on chromosome 6p21, an intergenic SNP rs9270986 upstream of *HLA-DRB1* shows very strong association with disease susceptibility (genotypic P value $= 2.3 \times 10^{-122}$) (WTCCC 2007) and with gene expression ($P = 1.14 \times 10^{-36}$) (Schadt et al. 2008). The expression dataset also facilitates resolution of probable susceptibility genes: the disease associated SNP rs3764021 (c.171C>T) located within the *CLEC2D* gene (encoding C-type lectin domain family 2 member D) on chromosome 12p13 showed no association with expression of that gene but did show strong association with expression

of the neighbouring gene *CLECL1* (encoding C-type lectin-like 1) ($P = 5.8 \times 10^{-17}$). A novel disease association with type 1 diabetes had been defined at *ERBB3* at chromosome 12q13 for rs2292239 (c.875-147T>G) in the WTCCC study ($P = 1.5 \times 10^{-9}$) (WTCCC 2007), which was strongly replicated ($P = 1.5 \times 10^{-20}$ for all samples) (Todd et al. 2007). This SNP did not show association with expression of *ERBB3* but did show association with expression of the flanking gene *RPS26*, encoding ribosomal protein S26 ($P = 4 \times 10^{-22}$). This was a striking association with expression: this specific SNP showed the strongest association of any genotyped and explained 40% of the variation measured for *RPS26* (Schadt et al. 2008).

11.5 Allele-specific gene expression

11.5.1 Allele-specific gene expression among autosomal non-imprinted genes

Further important insights into the genetics of gene expression have been provided by the definition of allele-specific gene expression. This is seen in its most extreme form when expression is restricted to a single allele (monoallelic expression), as classically seen among imprinted genes (Box 11.5) where only one of the two copies of an autosomal gene is expressed, or in females where one X chromosome is inactivated (Plath et al. 2002). Among autosomal non-imprinted genes allele-specific expression was found, somewhat surprisingly at the outset of such studies, to occur relatively commonly and show evidence of heritability (Knight 2004). The magnitude of allele-specific expression is variable and often relatively modest, however the definition and analysis of genes showing such differential expression has provided powerful evidence to support the existence of cis-regulatory variation.

By comparing relative allelic expression within a cellular sample, much of the potential heterogeneity inherent to comparison between individual samples can be avoided. Typically, individuals heterozygous for a given transcribed genetic variant are selected for analysis which allows the allelic origin of the transcript to be defined (Fig. 11.9; Box 11.6). In this case the exonic SNP is being used as a genetic marker to define allelic origin; it may be acting as a regulatory variant but more probably the

Box 11.5   Genomic imprinting

Imprinted genes show monoallelic expression in a parent-of-origin-specific manner. Genomic imprinting arises due to epigenetic marks involving methylation of DNA and modifications of **histones**. Imprinting is an example of an epigenetic process – heritable changes in gene expression that do not involve changes in the DNA sequence and are transmitted through cell divisions. Up to 1% of human genes are thought to be imprinted with 90 described to date (Luedi *et al.* 2007; Ideraabdullah *et al.* 2008). Imprinted genes are found across the genome but with a number of clusters identified, characterized by regulation through parent-specific epigenetic modifications of imprinting control regions. These regions have been shown to act, for example, as insulators and promoters for non-coding RNAs (Ideraabdullah *et al.* 2008). Imprinted genes have been shown to be important for normal development with the parental imprint set in the germline; often their expression is tissue- and developmental stage-specific. The consequences of dysregulation of imprinting through fine scale or structural genomic variation can be profound, with at least nine human imprinting syndromes described (Amor and Halliday 2008). These include the Prader–Willi and Angelman syndromes, most often arising through deletions of chromosome 5q11-q13. These lead to loss of function of the paternally imprinted gene *SNRPN* in Prader–Willi syndrome or the maternally imprinted gene *UBE3A* in Angelman syndrome (Section 5.2.6). A number of databases cataloguing imprinted genes are available online including the Imprinted Gene Catalogue (www.otago.ac.nz/IGC; Morison *et al.* 2001) and the genomic imprinting website maintained by the Jirtle lab at Duke University (http://www.geneimprint.com).

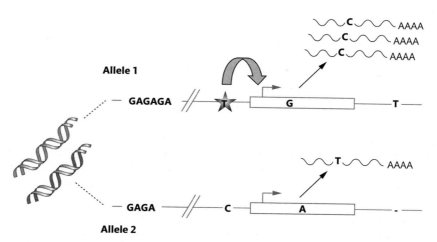

Figure 11.9 Allele-specific gene expression. Quantification of a transcribed SNP, here an exonic G to A nucleotide substitution, allows the allelic origin of the transcript to be defined with increased expression of allele 1 versus allele 2 shown. This allele-specific expression may be the result of a number of different possible regulatory variants, the exonic SNP is serving as a marker to distinguish allelic origin and is not itself necessarily functional. In this figure several genetic variants are shown including a GA dinucleotide repeat and a T insertion polymorphism. However in this example, the allele-specific expression results from a functional *cis*-acting regulatory variant, a T to C single nucleotide substitution which in the presence of the T substitution increases transcription from allele 1. The mechanism for this may involve allele-specific recruitment of a specific transcription factor in the presence of the T substitution. Redrawn and reprinted from *Trends in Genetics* (Knight 2004), copyright 2004, with permission from Elsevier .

Box 11.6   Quantification of allele-specific gene expression

Singer-Sam and colleagues established the utility of allele-specific transcript quantification using transcribed marker SNPs for which the individual was heterozygous. Their methodology involved PCR amplification of a genomic region followed by a primer extension reaction, the primer being designed such that the 3′ end was immediately adjacent to the polymorphic nucleotide. On primer extension, the nucleotide corresponding to the variant nucleotide would be incorporated and detected

(Singer-Sam *et al.* 1992b). The approach was successfully applied to mouse embryos in the analysis of parental imprinting (Singer-Sam *et al.* 1992a) and adapted in later studies with fluorescent-labelled dideoxynucleotides used for primer extension (Yan *et al.* 2002). Quantification of primer extension products by matrix-assisted laser desorption/ionization (MALDI) time-of-flight (TOF) mass spectrometry has also been used (Knight *et al.* 2003; Jurinke *et al.* 2005).

---

functional *cis*-acting variant lies elsewhere on the same chromosome. A clear definition of the underlying genetic variation and haplotypic structure is therefore critical to dissecting the functional basis of observed allele-specific expression. Further resolution can then be achieved through correlation of specific haplotypes and SNPs with allele-specific expression as demonstrated in lymphoblastoid cell lines using dense genotyping data available from the HapMap Project (Pastinen *et al.* 2005; Forton *et al.* 2007).

Given that the transcribed SNP is being used as an allelic marker to analyse individuals heterozygous for that SNP and not necessarily because of a high probability of it being a functional regulatory variant, it is perhaps not surprising that in many cases where allele-specific expression is observed it occurs in only a proportion of individuals analysed for that marker. This was seen in an important early study demonstrating the occurrence of allele-specific expression in human cells by Yan and colleagues (2002). Here lymphoblastoid cell lines established from 96 individuals in the CEPH families were analysed for potential allele-specific expression in 13 different genes, selecting cell lines where individuals were heterozygous for SNPs in those genes.

Allele-specific transcript abundance was quantified using a fluorescence-based dideoxy terminator method (Box 11.6) and found to occur commonly – in six out of 13 genes selected – with between 1.3- and 4.3-fold differences in expression between the alleles resolved. However, allele-specific differences were present in only

3–30% of individuals heterozygous for a given SNP. The availability of family pedigrees allowed the heritability of allele-specific differences in gene expression to be assessed. Clear evidence of coinheritance of allele-specific expression with underlying haplotype as defined by microsatellite markers were seen for *PKD2* (encoding polycystic kidney disease 2) and *CAPN10* (encoding calpain 10) (Fig. 11.10). The potential relevance of such data is apparent from studies showing that genetic variation at *CAPN10* is strongly associated with risk of type 2 diabetes, with allele-specific analysis of gene expression providing a functional approach to defining regulatory variants (Cox *et al.* 2004).

In primary human cells allele-specific differences in gene expression were also seen. Bray and colleagues analysed gene expression using RNA from post mortem brain tissue of 60 different individuals (Bray *et al.* 2003). Following PCR amplification, a primer extension-based technique allowed allele-specific quantification sensitive to detection of as low as a 20% difference in relative allelic expression. For the 15 genes analysed, allele-specific expression greater than this level was seen in seven genes for at least one individual (Bray *et al.* 2003). Like Yan and colleagues, the allelic differences in expression varied between individuals heterozygous for a given SNP marker. Among the genes identified as showing allele-specific expression was *DTNBP1* at chromosome 6p22.3 which encodes a neuronal protein, dystrobrevin binding protein 1. Genetic variation at this locus is significantly associated with schizophrenia from linkage analysis and

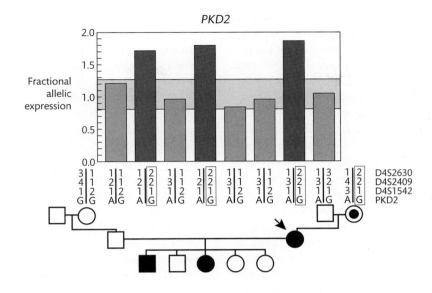

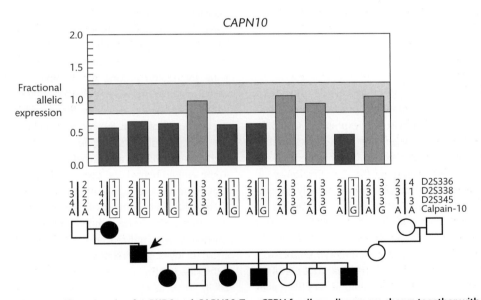

Figure 11.10 Allele-specific expression for *PKD2* and *CAPN10*. Two CEPH family pedigrees are shown together with relative allelic expression for lymphoblastoid cell lines established from individuals heterozygous for a marker SNP. Allele-specific differences were seen for alleles bearing particular haplotypes of *PKD2* (2,2,1,G) and *CAPN10* (1,1,1,G). From Yan *et al.* (2002), reprinted with permission from AAAS.

association studies (Straub *et al.* 1995; Allen *et al.* 2008). The specific functional variants remain elusive but it is notable that expression of *DTNBP1* is reduced in some areas of the brain in patents with schizophrenia (Talbot *et al.* 2004; Weickert *et al.* 2004) and expression of specific haplotypes of *DTNBP1* shows some correlation with risk or protective alleles, being reduced or increased, respectively (Bray *et al.* 2005; Williams *et al.* 2005a).

Further important insights into allele-specific gene expression were provided by the analysis of inbred mouse strains (Cowles *et al.* 2002). Here analysis of gene expression for 69 genes using transcribed SNP markers was carried out in F1 hybrid mice resulting from crossing parental strains homozygous for the markers. Allele-specific gene expression in three different tissues was determined using a quantitative assay based on primer extension with a threshold for reliable detection of allelic imbalance set at a 1.5-fold difference in expression between alleles. Six per cent of genes showed consistent allele-specific differences, confirmed by resequencing to identify additional linked transcribed markers for use in allelic discrimination. The analysis demonstrated that allele-specific effects may be tissue-specific, allelic differences in expression for example being found for *Ccnf* (encoding cyclin F) and *Hmgcr* (encoding HMG coenzyme A reductase) in liver only while for *Uros* (encoding uroporphyrinogen III synthase) differences were seen in liver, spleen, and brain tissue (Cowles *et al.* 2002).

11.5.2 Large scale analysis of allele-specific gene expression

Analysis of allele-specific gene expression provides a complimentary and informative approach to investigate the role of genetic variation in gene expression. Could the approach be applied at a genome-wide level to define the extent of allele-specific expression and specific gene loci, which would be candidates for identification of *cis*-regulatory variants and compliment the rapidly growing datasets of genome-wide association with disease? A number of studies have addressed this question by using array-based technologies to interrogate allele-specific expression in human samples including lymphoblastoid cell lines (Bjornsson *et al.* 2008; Serre *et al.* 2008), fetal tissues (Lo *et al.* 2003), and peripheral blood cells (Pant *et al.* 2006). These studies indicate allele-specific expression is common, involving approximately 20% of genes, with both monoallelic expression consistent with genomic imprinting and more modest differences in expression, which in more than half of cases shows variation among individuals informative for a given marker SNP. This is consistent with the need for further analysis to resolve genetic and epigenetic effects, including the specific *cis*-acting regulatory variant(s) responsible for allele-specific

expression, which may be present on only a proportion of chromosomes bearing the marker SNP.

In reviewing these studies in more detail it is worth considering again yeast as a powerful model organism for studying the genetics of gene expression. Kruglyak and colleagues sought to build on and compliment their earlier linkage analysis of haploid segregants by investigating diploid hybrids of the same two parental yeast strains, RM and BY (Section 11.3.3) (Brem *et al.* 2002; Yvert *et al.* 2003). In particular they utilized an oligonucleotide array to allow simultaneous genotyping and allele-specific expression, which proved informative but had a high false positive rate for the detection of allele-specific expression. A total of 692 unique open reading frames (ORFs) were interrogated with 1049 polymorphic marker probes and significant evidence of allele-specific expression found at 70 ORFs, of which 24 were likely false positives (Ronald *et al.* 2005). The RM-derived allele was underrepresented among these, being preferentially expressed in only 11 of 70 ORFs.

Given the power of genetic analysis in yeast it was possible to investigate whether these allele-specific effects in diploid hybrids corresponded to sites of linkage with gene expression among the haploid segregants from earlier studies. There was indeed significant enrichment, with ORFs showing allele-specific expression also showing evidence of local linkage and higher expression of the predicted allele: the allele associated with relatively higher expression on analysis of allele-specific expression in the diploid hybrid was associated with greater expression in the haploid segregant (Ronald *et al.* 2005).

Early evidence of the extent of allele-specific expression across the human genome was provided by Lo and colleagues who utilized an oligonucleotide genotyping array to investigate relative allelic expression for 1063 transcribed SNPs corresponding to 602 genes (Lo *et al.* 2003). They analysed gene expression for kidney and liver samples from seven different fetuses and were able to detect with confidence allelic differences in expression greater than two-fold in magnitude. A high proportion of genes analysed appeared to show allele-specific expression, 54% at greater than two-fold difference and 28% at greater than four-fold. For known imprinted genes, monoallelic expression was found in five out of six genes analysed. Variation between individuals was noted together with tissue specificity; genes showing allelic differences included known and novel clusters but were generally scattered across the genome. Overall the authors estimated between 20% and 50% of genes showed allele-specific expression.

Other studies suggest that the proportion of genes showing allele-specific expression may be closer to 20%. Pant and colleagues, for example, investigated gene expression for white blood cells from 12 unrelated individuals using oligonucleotide arrays interrogating 8406 exonic SNPs in 4012 genes (Pant *et al.* 2006). Of these genes, 1389 were expressed in white blood cells and 731 showed allele-specific expression greater than 1.5-fold in magnitude between alleles in at least one individual. Further analysis of 60 genes where allele-specific expression was found in at least three individuals revealed that 5% had monoallelic expression, confirmed as restricted to the maternal copy from experiments in lymphoblastoid cell line pedigrees. In 54% of the remaining genes, all heterozygotes showed an allelic difference consistent with the *cis*-regulatory variant being present among one of the pair of alleles assayed. Importantly, independent RNA preparations showed high correlation, there was good concordance between multiple informative SNPs at a given gene, and an independent methodology for quantification of allele-specific expression also showed good concordance. Overall the authors estimated that on average 22% of exonic SNPs assayed showed allele-specific expression for a given individual (Pant *et al.* 2006).

A similar proportion was found in a more recent study by Serre and colleagues who found that approximately 20% of genes showed allele-specific expression in a study of 643 expressed genes among 80 lymphoblastoid cell lines established from individuals of European ancestry (Serre *et al.* 2008). The study utilized an allele-specific expression bead array technology based on primer extension together with quantitative sequencing of the products of RT-PCR reactions. The assay platforms were validated and found to be quantitative; allelic differences greater than 1.5-fold being detected with confidence above experimental noise. The study also showed that the allelic differences in expression were consistent when lymphoblastoid cell lines were repeatedly sampled after different numbers of passages of culture. Comparison with earlier data from Stranger and colleagues based on association of SNP markers with gene expression (Stranger *et al.* 2005) showed six out of 21

genes with strong evidence of *cis*-regulatory effects also had significant evidence of allele-specific expression (Serre *et al.* 2008).

11.5.3 Allele-specific expression based on RNA polymerase loading

Analysis of allele-specific gene expression at a transcript level was dependent on finding a transcribed SNP marker to define allelic origin and hence relative allelic expression. Studying nascent pre-spliced mRNA expanded the number of informative SNPs to include intronic variants (Pastinen *et al.* 2004) although this approach appears limited to more highly expressed genes given the low proportion of unspliced mRNA to spliced transcripts (Serre *et al.* 2008). The number of haplotypes that could potentially be interrogated was increased further by approaches based on quantification of the relative allelic abundance of RNA polymerase II (Knight *et al.* 2003; Maynard *et al.* 2008).

RNA polymerase II binding had been shown to reflect levels of gene expression, notably when specific phosphorylated states of serine residues in the carboxy-terminal domain of the RNA polymerase were assayed (Weeks *et al.* 1993; O'Brien *et al.* 1994). The haplotype-specific chromatin immunoprecipitation (haploChIP) methodology was established to quantify allele-specific recruitment of RNA polymerase II (Box 11.7) (Knight *et al.* 2003). The assay involved immunoprecipitation of RNA polymerase crosslinked to DNA, with the abundance of DNA fragments quantified in an allele-specific manner following reversal of crosslinks. Application of the approach to lymphoblastoid cell lines allowed resolution of a low producer haplotype of *LTA*, encoding lymphotoxin alpha at chromosome 6p21.3. This haplotype was shown to be associated with allele-specific recruitment of the transcriptional repressor protein activated B cell factor 1 (ABF-1) in the presence of the A allele of a specific SNP in the 5′ untranslated region (UTR) of the *LTA* gene (rs2239704) (Knight *et al.* 2004). Possession of this SNP has subsequently been associated with leprosy and malaria (Section 13.2.6).

The analysis of allele-specific RNA polymerase II loading has subsequently been analysed at a genome-wide level to advance our understanding of the nature and extent of allele-specific gene expression (Maynard *et al.* 2008). Microarray platforms have proved a powerful approach

---

Box 11.7 Chromatin immunoprecipitation

The chromatin immunoprecipitation (ChIP) technique allows protein–DNA interactions to be assayed in living cells or tissues. Proteins are crosslinked to DNA, typically by exposure of cells to formaldehyde. Chromatin is then extracted and sonicated, and subjected to immunoprecipitation with specific antibodies, for example to RNA polymerase II or specific histone modifications. The crosslinks between immunoprecipitated protein and DNA are reversed, and relative enrichment of DNA fragments compared to input chromatin quantified. This is done either for specific genomic loci by quantitative PCR; or by amplification, labelling, and hybridization to microarray platforms to allow genome-wide analysis using the 'ChIP-on-chip' approach (Ren *et al.* 2000). Next generation, high throughput sequencing technologies (Box 1.24) will further advance application of ChIP for genome-wide mapping of protein–DNA interactions, dubbed 'ChIP-Seq' analysis (Jothi *et al.* 2008).

---

to analyse the products of ChIP experiments and enable genome-wide analysis of sites of protein–DNA interactions (Box 11.7). Maynard and colleagues extended this to analyse the products of ChIP for RNA polymerase II by hybridization to a SNP genotyping array and define allele-specific binding for heterozygous SNPs. Human lung fibroblasts were analysed using 317 513 SNPs, of which 119 821 were heterozygous in this cell line, and 11 028 showed enrichment relative to input DNA. Of these, 466 SNPs showed significant allele-specific enrichment corresponding to 239 genes including known imprinted loci such as *SNRPN* (Maynard *et al.* 2008). A similar approach based on the analysis of specific histone modifications demonstrated for lymphoblastoid cell lines from CEPH family pedigrees that allele-specific chromatin modifications occur and show familial aggregation (Kadota *et al.* 2007). The relative roles of genetic and epigenetic variation in determining gene expression is complex and incompletely understood, but further dissection of allele-specific effects at specific loci and genome-wide offers an opportunity to advance our understanding of this fundamentally important area.

11.6 Genetic variation and alternative splicing

In this chapter the relationship between gene expression and genetic variation has been explored with strong evidence of heritable differences in expression which can be mapped to specific genomic loci, local or distant to the gene encoding the transcript whose abundance has been quantified. But in terms of the transcript, what has been measured? For the majority of human genes, several different mRNAs will be expressed as a result of alternative splicing (Section 11.6.1). The occurrence of different, alternatively spliced isoforms of varying levels of abundance encoded by a given gene adds a significant additional level of complexity to the analysis of the genetics of gene expression, as clearly such diversity needs to defined and quantified if the consequences of underlying genetic variation are to be resolved. Indeed it is worth considering that almost all the expression array datasets reviewed in this chapter so far used probes targeted at the 3′ ends of genes. Such array designs give an incomplete picture of the true extent of transcript diversity and abundance as they are not designed to resolve and quantify specific splice isoforms. Published studies have been based on the knowledge and technological platforms available at the time and it remains the case that the detection and quantification of alternatively spliced isoforms at a genome-wide level remains a very considerable analytical challenge (Johnson *et al.* 2003; Wang *et al.* 2003a; Xing *et al.* 2006; Anton *et al.* 2008). This has to date largely precluded analysis of the impact of genetic variation on gene expression at the level of the abundance of specific alternatively spliced isoforms. However, the advent of exon-specific microarray platforms (Section 11.6.2) and the future application of 'next generation' sequencing are set to significantly advance this field of research (Salehi-Ashtiani *et al.* 2008).

11.6.1 Alternative splicing in health and disease

The process of splicing involves identifying and joining together coding exonic sequence in pre-mRNA through a complex process involving the splicing machinery (the spliceosome) and 'splicing code', which includes consensus splice site sequences at exon–intron boundaries together with *cis*-regulatory elements to which specific proteins bind (Fig. 11.11) (Wang and Cooper 2007). The latter include enhancer and suppressor elements within intronic and exonic sequences and are important to splice site recognition and regulation, including control of alternative splicing. Current estimates are that between 40% and 70% of human genes show evidence of alternative splicing, with on average four to six alternatively spliced isoforms per gene (Kapranov *et al.* 2002; Johnson *et al.* 2003; Wang and Cooper 2007). These alternatively spliced isoforms can differ in many potential ways including exon skipping, intron retention, alternative splice site usage, and more complex events (Fig. 11.11C) (Kim *et al.* 2008).

The process of alternative splicing is critical to our ability to generate proteomic diversity but also modulates gene expression. The latter occurs through isoform variation affecting control of mRNA stability, translation efficiency, and mRNA localization as well as mRNA degradation resulting from the introduction of premature termination codons (Wang and Cooper 2007). Dysregulation of splicing as a result of genetic variation is a major cause of inherited disease and is increasingly recognized to be involved in common complex traits.

There are many different ways in which genetic variation may modulate splicing. Disease may result directly from disruption of the splicing code by *cis*-acting genetic variants or of the splicing machinery by *trans*-acting variants (Wang and Cooper 2007). These are relatively common events, indeed between 15% and 60% of disease-causing point mutations may act by affecting splicing (Krawczak *et al.* 1992; Lopez-Bigas *et al.* 2005). The complex, often tissue-specific, consequences of dysregulated splicing were highlighted for adult myotonic dystrophy resulting from triplet repeat expansion (Section 7.6.3). As well as causing disease, variants can alter splicing of modifier genes affecting disease severity, as seen for example at the *CFTR* gene in cystic fibrosis (Section 2.3.1) (Niksic *et al.* 1999), and susceptibility to common disease (Wang and Cooper 2007). Examples of the latter include *IRF5* at chromosome 7q32 encoding interferon regulatory factor 5 and systemic lupus erythematosus (Section 11.6.2); *CTLA4* at chromosome 2q33 encoding cytotoxic T lymphocyte-associated protein 4 and autoimmune disease (Ueda *et al.* 2003); *ERBB4* on chromosome 2q34 encoding the neuregulin 1 receptor and schizophrenia (Law *et al.* 2007); and *BTNL2* at chromosome 6p21.3 encoding butyrophilin-like 2 with sarcoidosis (Section 12.8).

11.6.2 Common genetic variation and alternative splicing

A number of studies have highlighted how alternative splicing varies between individuals and shows significant heritability (Hull *et al.* 2007; Kwan *et al.* 2007). Insights into the potential relationship with underlying genetic variation have been gained from databases of alternatively spliced isoforms (Modrek *et al.* 2001) looking for association with transcribed SNPs, which showed that 6–21% of alternatively spliced genes had evidence of complete or relative isoform abundance varying with specific alleles (Nembaware *et al.* 2004). Variation between unrelated individuals for specific splice events has been investigated for simple cassette exon events in a panel of 22 lymphoblastoid cell lines, showing that consistent differences between individuals occur, and could be associated with local SNP diversity (Hull *et al.* 2007). In this study six exons were resolved showing variable skip-inclusion and association with specific SNPs. When further unrelated lymphoblastoid lines were investigated, the SNP genotype accurately predicted the splicing pattern. For five exons, the strongest correlation with splicing pattern was found with the SNP closest to the intron–exon boundary. Intriguingly, four of the six linked SNPs were located within alternative exons.

Further advances have been made through exon targeted arrays such as the Affymetrix GeneChip Human Exon 1.0 ST array which allowed expression to be quantified for more than 1 million known and predicted exons using multiple probes to individual exons (Clark *et al.* 2007). Kwan and colleagues investigated exon level

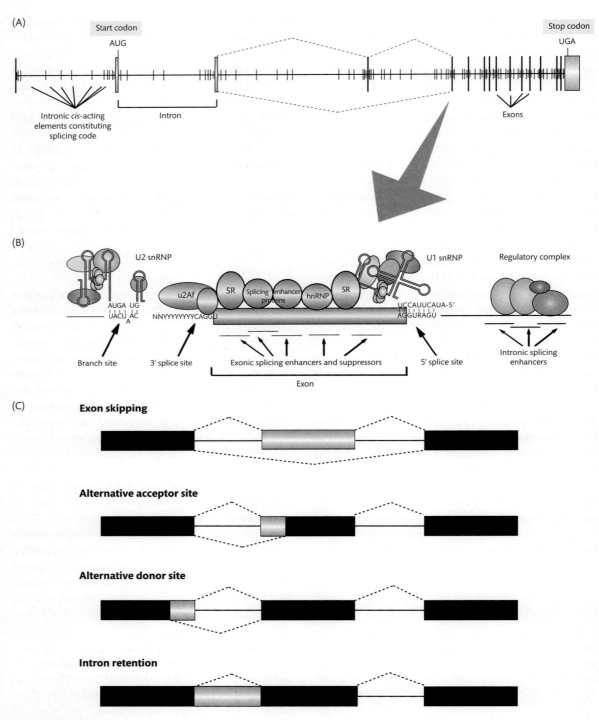

Figure 11.11 *Continued*

variation in gene expression between two unrelated individuals based on a 'splicing index' that divided expression of a given probes set corresponding to one exon by the sum of expression from probes representing the gene (Kwan *et al.* 2007). The splicing index was analysed as a quantitative trait in two lymphoblastoid cell lines with most of the observed variation due to individual differences; up to 2.5% of expressed exons were estimated to show differential expression between the two lines. When a small number of validated alternative splicing events were analysed in a three generation CEPH family, evidence of linkage was found with segregation of splicing pattern and associated haplotype in the pedigree (Kwan *et al.* 2007).

Kwan and colleagues proceeded to a genome-wide analysis relating common SNP diversity to the expression of specific transcript isoforms using the same exon targeted array (Kwan *et al.* 2008). Here gene expression of 57 lymphoblastoid cell lines established from individuals genotyped in the CEU panel of the HapMap Project were analysed, each using three different RNA preparations. Genetic association was sought for expression intensity with SNP markers within a 50 kb region flanking the transcribed region using a linear regression analysis. Based on a 5% false discovery rate and cut-off P value of 9.7×10^{-9}, significant SNP association was found for 324 transcripts. The complexity of potential gene expression differences at all stages of transcript processing was illustrated by the breakdown of these flanking SNP associations: 39% involved whole gene expression changes and 6% were classified as complex; the remaining 55% were at the level of transcript isoforms with 11% involving changes in transcriptional initiation, 26% alternative splicing, and 18% transcription termination changes.

Overall the authors estimated that 50–55% of variation in gene expression is isoform based (Kwan *et al.* 2008). However, this analysis represents only the beginning of the story as significant advances are needed in tools to enable splice isoform reconstruction and accurate quantification (Anton *et al.* 2008).

As the lymphoblastoid cell lines analysed by Kwan and colleagues had been subject to previous detailed array analysis (Cheung *et al.* 2005; Stranger *et al.* 2007a), the additional resolution provided by the exon level probe sets could be compared. For example, a clear isoform-specific effect was resolved at *IRF5* encoding interferon regulatory factor 5 on chromosome 7q32 (Fig. 11.12). This replicated recently published data which showed that the associated SNP (rs10954213) (c.*555G>A) created a functional polyadenylation site in the presence of the A allele, which was correlated with the short isoform (Cunninghame Graham *et al.* 2007). This SNP was part of a haplotype associated with susceptibility to systemic lupus erythematosus (Cunninghame Graham *et al.* 2007). The role of genetic diversity at *IRF5* in disease susceptibility is incompletely understood; the disease association has been robustly demonstrated (Sigurdsson *et al.* 2005; Graham *et al.* 2006) with evidence that complex modulation of splicing may be involved as a further disease associated SNP was shown to create a 5′ donor site in an alternative exon 1 of *IRF5* (Graham *et al.* 2006).

11.7 Genetic variation: from transcriptome to proteome

The relationship between genetic diversity and proteins is less direct than quantification of gene expression

---

Figure 11.11 Splicing events, the splicing code, and spliceosome. (A) Pre-mRNA showing exons and introns with consensus splice site sequences at the boundaries of introns and exons. (B) Detail of a spliceosome and *cis*-regulatory elements. The main splice signals comprise 5′ and 3′ splice sites, a branch site, and a polypyrimidine tract found upstream of the 3′ splice site. Exonic and intronic *cis*-regulatory elements (enhancers and suppressors) are required for the regulation of constitutive and alternative splicing. U1 and U2 small nuclear ribonucleoproteins (snRNPs) bind to the 5′ splice site and branch site. Further details of the splicing code and regulation are reviewed in Wang and Cooper (2007) and Kim *et al.* (2008). (C) The main subgroups of alternative splicing are illustrated: exon skipping (prevalence 38.4%), alternative acceptor site (18.4%), alternative donor site (7.9%), and intron retention (2.8%). More complex events not shown include mutually exclusive events, alternative transcription start sites, and multiple polyadenylation sites (Kim *et al.* 2008). Constitutive exons are shown in black, alternatively spliced regions in light grey, and splicing options by dashed lines. Redrawn and reprinted by permission from Macmillan Publishers Ltd: Nature Reviews Genetics (Wang and Cooper 2007), copyright 2007; and with permission from Kim *et al.* (2008) .

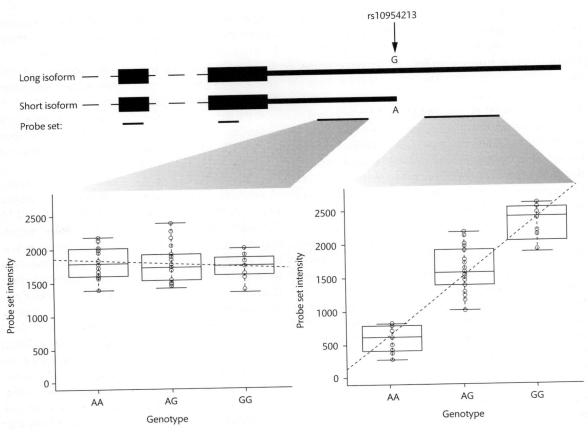

Figure 11.12 Isoform-specific expression of *IRF5*. A schematic is shown for the long and short isoforms of *IRF5* and the corresponding locations of the probe sets. A SNP rs1095413 shows significant association with levels of probe intensity, which is only seen with the more 3′ probe set (right hand panel) ($P = 8.3 \times 10^{-22}$). Data were validated by quantitative RT-PCR. Reprinted by permission from Macmillan Publishers Ltd: Nature Genetics (Kwan *et al.* 2008), copyright 2008.

based on transcript levels but protein levels are of greater relevance to determining a particular cellular and whole organism phenotype. The observed abundance of a given protein will be dependent on many factors following transcription, and indeed changes at the RNA level may not be reflected in protein abundance and vice versa. For example, observed circulating plasma or serum concentration of a particular protein will depend on a complex series of factors ranging from genetic and epigenetic variation to post-translational modifications, rates of secretion and clearance, protein interactions, and multiple environmental variables.

As already described in this chapter, fundamental insights can be gained from model organisms. Kruglyak and colleagues published data in 2007 that continued their pioneering work in yeast by complementing their previous analysis of the genetic basis of variation in the transcriptome with a proteomic analysis (Foss *et al.* 2007). Fundamental to the analysis of protein abundance is an accurate methodology for quantification of many different proteins, which was provided by mass spectrometry and the development of specific analytical tools. A cross between the parental BY and RM yeast strains was analysed, making use of the extensive genotypic and transcript data available (Section 11.3.3) (Yvert *et al.* 2003; Brem and Kruglyak 2005; Brem *et al.* 2005). Peptides were quantified for eight independent cultures of the parents and two for each of the 98 segregants: 6898 peptides were identified with 1873 quantified (Foss *et al.* 2007).

Comparison of the parents showed significant differences in one-third of the **proteome** with abundance amenable to analysis as a quantitative trait. One hundred and fifty-six proteins on which high quality data were available were selected for analysis with an average heritability found of 62%. Linkage analysis with 2951 markers resolved 85 loci with a lod score greater than 3, most were on different chromosomes to the protein encoding gene with 7% within 20 kb of the gene. Four major hotspots were noted affecting expression of between six and 35 different proteins. The data comparing the role of genetic variation on the transcriptome and proteome showed similarities, some loci coinciding precisely but many more being different; overall where both RNA and protein were assayed, linkage was found for 56% of transcripts and 38% of peptides. Hotspots linked to multiple transcripts or proteins varied, some being specific to transcript abundance and vice versa.

Levels of particular proteins have been implicated in a range of different diseases and related to underlying genetic variation predominantly on a candidate-based approach. Application of a genome-wide SNP association approach to define 'protein quantitative trait loci' (pQTLs) has been described for 42 different candidate proteins whose abundance was previously related to a range of infectious, inflammatory, and metabolic diseases (Melzer *et al.* 2008b).

Melzer and colleagues analysed fasting protein levels in a cohort of 1200 individuals from Italy and sought association with 496 032 SNPs (Melzer *et al.* 2008b). They defined eight significant, local, likely *cis*-acting associations, three of which had been previously established. For example, a significant association was found between a SNP at *IL6R* and soluble IL6 receptor levels with mean protein concentrations of 70 (67–73), 101 (97–104), and 138 (130–147) ng/ml for the three possible genotypes CC, CT, and TT, respectively (uncorrected $P = 1.8 \times 10^{-57}$). The associated SNP was correlated ($r^2 = 0.96$) with a previously reported associated SNP (rs8192284) linked with levels of the soluble receptor (Box 11.8). Significant likely *cis*-acting pQTLs were also identified for: *CCL4L1* and levels of macrophage inflammatory protein beta, which may involve copy number variation (Section 14.2.4); gamma glutamyltransferase 1 levels and polymorphism at *GGT1* for which a linked SNP shows association with transcript levels in lymphoblastoid cell lines; C-reactive protein; sex hormone binding globulin; interleukin 18; and interleukin1 receptor antagonist (Melzer *et al.* 2008b). The study demonstrated the utility of a genome-wide approach with large effect sizes shown for the associations found with per allele standard deviation differences in protein levels between 0.20 (95% CI 0.12–0.29) for C-reactive protein and 0.69 (95% CI 0.62–0.77) for soluble IL6 receptor.

Box 11.8   Genetic variation and IL6 receptor levels

Membrane bound and soluble forms of the receptor are found. A nonsynonymous A to C nucleotide substitution in exon 9 of IL6R (rs8192284) was strongly associated with levels of the soluble IL6 receptor protein in a study of healthy volunteers in a Japanese population (Galicia *et al.* 2004). This SNP results in an asparagine to alanine substitution in the proteolytic cleavage site of the protein, critical in determining the relative abundance of membrane bound and soluble forms of the protein. The

association was subsequently found by admixture mapping in African Americans ($P < 10^{-12}$) with replication in European Americans (Reich *et al.* 2007). The SNP is present at higher frequencies among those of European versus African descent (35% versus 4%); levels of IL6 soluble receptor are higher in the former population group. Overall rs8192284 was estimated to explain 33% of the variation in levels of IL6 soluble receptor among African Americans and 49% in European Americans (Reich *et al.* 2007).

11.8 Summary

Genetic variation between individuals plays a significant role in many different phenotypic traits. Considerable advances have been made in our ability to define the relationship between phenotype and genotype, and to map by linkage or association the genetic loci responsible for, or contributing to, a given phenotype. In many cases, however, the causative variant(s) are unknown, notably in common multifactorial traits, where fine mapping genetic associations and assigning causality to a specific functional genetic variant remains the exception rather than the rule.

The analysis of the genetics of gene expression described in this chapter goes an important way along the path to facilitating such work. To treat gene expression as a quantitative trait and apply the powerful genetic analytical approaches based on linkage and association seems in retrospect to be an almost obvious thing to do, but that is true of many notable scientific insights. In contrast to analysis of genetic variation in the context of the whole organism, the relative proximity in mechanistic terms of underlying genetic variation at the DNA level to gene expression within a cell might be expected to reduce the noise inherent to phenotypic analysis and increase the chances of success. We have seen how gene expression varies between and within populations, is a heritable trait, and has been successfully mapped, for transcript originating from a given gene, to specific loci by linkage and association. The high throughput microarray technologies available for gene expression combined with high density SNP genotyping data have made the 'genetical genomics' approach increasingly popular and successfully applied to both the analysis of model organisms such as yeast and mice, as well as to human populations using lymphoblastoid cell lines and primary cells.

The synergy between analysis of the genetics of gene expression and genetic susceptibility to disease was clearly demonstrated in asthma where robust evidence of genome-wide association with disease at chromosome 17q21 was complimented by genome-wide expression profiling and mapping of gene expression by association using dense SNP genotyping (Dixon *et al.* 2007; Moffatt *et al.* 2007). The most strongly disease associated SNPs were also strongly associated with gene expression, in this case of *ORMDL3*. The genome-wide datasets for expression and SNP genotyping established by this and other studies represent important resources for future use by investigators and are publically available in searchable formats. The increasing application to primary human cells and tissues of the genetical genomics approach should further advance our understanding of the relationship between genetic variation and gene expression, which are often highly context-specific. Such studies, involving thousands of individuals often in family pedigrees, are enormously powerful and confirm the high heritability and important contribution of genetic variation to expression traits (Goring *et al.* 2007; Emilsson *et al.* 2008).

Multiple eQTLs are typically found to contribute to a particular gene expression trait, with individually modest effect sizes. Local and distant regulatory variants are implicated with likely *cis*-acting effects most frequently identified among highly significant trait associations. Distant, likely *trans*-acting effects have been more challenging to resolve. Allele-specific gene expression has proved a powerful tool to compliment genetic analysis of gene expression, again with evidence of heritability and being a relatively common occurrence among non-imprinted autosomal genes. Application of genome-wide analysis to the analysis of allele-specific gene expression using either relative transcript abundance or RNA polymerase II loading should prove powerful approaches.

The increasing sophistication of our approaches to the genetics of gene expression, notably technological advances allowing the discrimination of the remarkable diversity that exists at the level of alternatively spliced transcript isoforms, should further advance our ability to define regulatory genetic variants. It is clearly essential to more accurately define and quantify the myriad of splice isoforms that are present across a range of tissues if we are to understand the relationship of gene expression to genetic variation. Similar approaches at a protein level will also prove highly informative and may be more relevant to the whole organism phenotype.

Major challenges remain, however, in ascribing functional mechanisms to specific regulatory variants. The different approaches discussed in this chapter are of great value but determination of a direct effect of the genetic variant requires complimentary testing in experimental systems, whether by manipulation in a model organism or in human cells (Chorley *et al.* 2008). For example, advances in technologies for transfecting DNA into cells allow for reporter gene and other DNA constructs to compare the effects of different variants directly on gene expression and have been widely used in this field (Rockman and Wray 2002). The consequences of sequence variation can also be predicted by bioinformatics with a sophisticated set of tools available for such analyses, for example based on predicted effects of variants on transcription factor binding sites. Direct assays of protein–DNA interactions are also possible both *in vitro* using the electrophoretic mobility shift assay and *in vivo* using ChIP.

Context remains paramount in such studies, as it does in the analysis of the genetics of gene expression. Only by analysis of genetic variation in an appropriate cell type and relevant conditions to the phenotype of interest will functionally important regulatory genetic variation be likely to be found, as exemplified by studies of genetic variation and globin gene expression. The themes of genetic variation and gene expression are continued in the next chapter where the remarkable diversity at the MHC is reviewed – a region highlighted by the genome-wide mapping of gene expression discussed in this chapter, and of genome-wide association studies for a range of autoimmune and infectious diseases (Section 9.3).

11.9   Reviews

Reviews of subjects in this chapter can be found in the following publications:

Topic	References
Genetics of gene expression	de Koning and Haley 2005; Gibson and Weir 2005; Rockman and Kruglyak 2006
Nature and consequences of genetic variation in coding DNA including effects on splicing	Cargill *et al.* 1999; Chasman and Adams 2001; Cartegni *et al.* 2002
Analysis of regulatory variants	Rockman and Wray 2002; Pastinen and Hudson 2004; Pampin and Rodriguez-Rey 2007; Chorley *et al.* 2008; Cobb *et al.* 2008
Imprinting	Amor and Halliday 2008; Ideraabdullah *et al.* 2008
X chromosome inactivation	Plath *et al.* 2002
Allele-specific gene expression	Buckland 2004; Knight 2004; Pastinen and Hudson 2004; Pastinen *et al.* 2006
Alternative splicing in health and disease	Modrek and Lee 2002; Pagani and Baralle 2004; Wang and Cooper 2007; Kim *et al.* 2008
Asthma genetics	Vercelli 2008
Genetic disease associations with schizophrenia	Allen *et al.* 2008
Genetic variation and HDL cholesterol	Holleboom *et al.* 2008

CHAPTER 12

Extreme diversity in the major histocompatibility complex

12.1 Introduction

The rainforest is home to the greatest diversity of animal and plant life on our planet. To those interested in human genetics, a relatively small region of chromosome 6 has proved a similarly remarkable environment. This genomic region, the major histocompatibility complex (MHC), shows extreme levels of genetic polymorphism, it is remarkably gene dense, and genetic variation within it has been associated with susceptibility to more diseases than anywhere else in the genome. The MHC was first discovered in 1936 in the mouse (Gorer 1936) and has subsequently been the subject of intensive research. Genes within the MHC play a fundamental role in our immune response: the first MHC-encoded proteins were discovered on circulating white blood cells and are known as leukocyte antigens – hence the other commonly used term for the region, the 'human leukocyte antigen' (HLA) complex. Early research demonstrated how analysis of MHC molecules could allow successful tissue grafting and organ transplantation ('histocompatibility').

The role of the MHC in immunity is, however, much more extensive, and provides protection against pathogens through presentation of foreign proteins to immune cells such as T cells, as well as encoding inflammatory cytokines and other key mediators of the immune response. Polymorphism in the sequences encoding recognition of foreign proteins is believed to have conferred a selective advantage, providing a driving force in the accumulation of allelic diversity in many MHC genes. Extreme polymorphism at such genes is made more complex by the co-inheritance of such diversity across extended stretches of DNA such that when we look for association with disease susceptibility, dissecting the causal variant(s) has proved very difficult. In this chapter, the biology of the MHC is reviewed, the nature and extent of genetic diversity within it, its co-inheritance and ancestral origins, together with specific examples of MHC genes that exemplify the pivotal role of the region in understanding biology, human diversity, and disease.

12.2 MHC genes, the immune response, and disease

The MHC lies on the short arm of chromosome 6 and encodes a diverse range of proteins, notably antigen presenting molecules by class I and II region genes (Fig. 12.1) (Horton *et al.* 2004). The MHC class I supercluster includes three classical class I genes (*HLA-A*, *-B*, and *-C*) whose products are involved in presentation of antigen to CD8+ T cells; four non-classical class I genes (*HLA-E*, *-F*, *-G*, and *HFE*); and 12 pseudogenes and class I-like genes such as *MICA* and *MICB*. The class II cluster includes the classical genes *HLA-DP*, *-DQ*, and *-DR* that encode α and β chains expressed as heterodimers on the cell surface and are responsible for the presentation of antigens to CD4+ T cells. The class I and II gene clusters span a 3.6 Mb region, with the sequence between the two clusters denoted the class III region. The class III region has the highest gene density of anywhere in the genome with 61 expressed genes within 900 kb (Xie *et al.* 2003).

The MHC is now generally considered to also include flanking regions beyond class I and II such that the extended MHC spans 7.6 Mb of the genome. This demarcation arose through analysis of the patterns of linkage

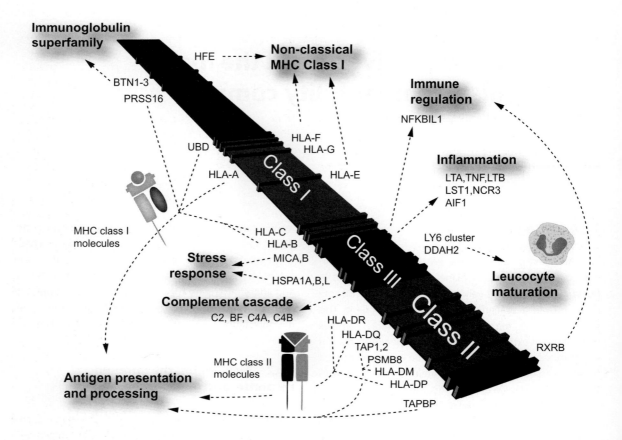

Figure 12.1 Genes encoded within the MHC involved in the immune system. Over the 7.6 Mb of DNA sequence comprising the extended MHC, 28% of genes are estimated to be involved in different immune functions including innate and adaptive immunity. A selection of genes are shown here with arrows to functional groupings. The majority are concerned with antigen presentation and processing. These include classical MHC class I and II molecules, together with antigen processing machinery needed for loading peptides onto class I molecules.

disequilibrium (Box 2.8), comparison between species of gene and sequence conservation, together with the finding of genes of common function to the MHC in flanking DNA such as the *HFE* gene (Section 12.6) and *BTN* family genes (encoding butyrophilin members) (Stephens *et al.* 1999). Allelic variation is greatest within the genes encoding class I and II molecules, with extremes of polymorphism observed in *HLA-B* and the *HLA-DRB1* region (Section 12.3.4) (Bergstrom *et al.* 1998). A further hypervariable region is found in the class III region in the complement locus with copy number variation at the RCCX module (Section 12.7) (Yang *et al.* 1999).

12.2.1 MHC class I and II molecules

MHC class I molecules are present on nearly all nucleated cells and are vital to the cellular immune response (Box 12.1) (Cresswell *et al.* 2005). The transmembrane heavy chain (HC) encoded by MHC class I genes has two polymorphic domains, $\alpha1$ and $\alpha2$, responsible for binding peptides. Bound antigenic peptides are presented via the endogenous pathway to CD8+ cytotoxic T cells (Fig. 12.2). In contrast, MHC class II molecules are responsible for antigen presentation via the exogenous pathway (Box 12.2) (Watts 2004). Bacteria, parasites, and

Box 12.1   Class I molecules and antigen presentation via the endogenous pathway

Foreign (non-self) intracellular proteins such as those originating from viruses and some bacteria present in the cytosol of the cell are degraded by the proteasome into antigenic peptides (Fig 12.2). These are translocated into the endoplasmic reticulum (ER) by the transporter TAP (transporter associated with antigen processing). Here the unfolded MHC class I heavy chain associates with the molecular chaperone calnexin (CNX), initiating folding and disulphide bond formation. Following dissociation, the heavy chain binds β2 microglobulin and is incorporated into the peptide loading complex, which includes TAP, a glycoprotein molecule tapasin, calreticulin, and ERp57. Foreign peptides transported from the cytosol are trimmed to 8–10 amino acids if necessary by the ER associated aminopeptidase, ERAPP. Peptide binding to the HC/β2 microglobulin heterodimer initiates release from the peptide loading complex, allowing the fully assembled molecule bound with antigen to leave the ER and be delivered to the plasma membrane via the Golgi apparatus. Here the antigen is presented to CD8+ cytotoxic T cells via the T cell receptor. Cells presenting non-self peptides are destroyed, usually by cellular apoptosis.

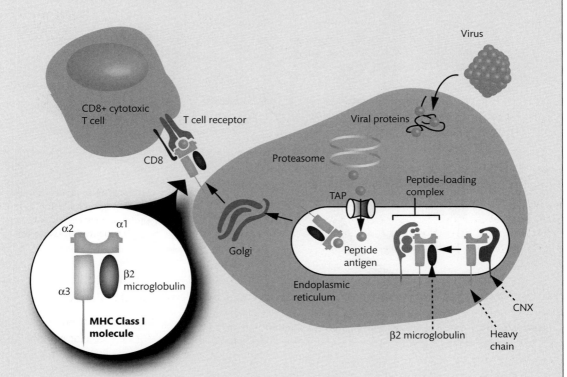

Figure 12.2 MHC class I molecules and antigen presentation.

Box 12.2 Class II molecules and antigen presentation via the exogenous pathway

Exogenous proteins are degraded by denaturation and proteases present in the acidic endosome and lysosomal compartments producing peptide ligands (Fig 12.3). The MHC class II molecules are heterodimers, comprising an α and a β chain. They are assembled in the ER with a molecular chaperone protein, Invariant chain (Ii), the cytoplasmic tail of which targets the complex to the endosomal pathway and prevents premature association of antigenic peptide. Within the endosome, Ii undergoes

stepwise degradation by proteases leaving a small fragment, CLIP (class II invariant chain peptide), lodged in the peptide binding groove. The removal of CLIP is catalysed by DM, a protein encoded by *HLA-DM* within the MHC, which also stabilizes the molecule and assists antigenic peptide selection. Following antigen binding, MHC class II molecules are delivered to the cell surface where they present the antigen to CD4+ helper T cells, triggering an immune response.

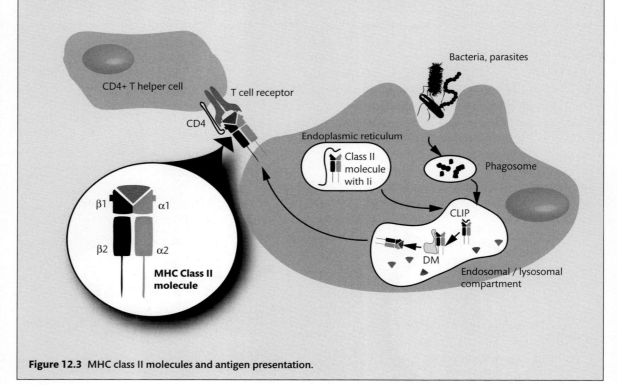

Figure 12.3 MHC class II molecules and antigen presentation.

other extracellular pathogens are endocytosed into antigen presenting cells such as macrophages, B cells, and dendritic cells (Fig. 12.3). Class II molecules present antigen to CD4+ T helper cells and are integral to successful maintenance of self tolerance by the immune system and the adaptive immune response to invading pathogens.

12.2.2 Biological complexity among the many genes found in the human MHC

Currently there are 252 known expressed genes encoded by the MHC with a large proportion involved in the immune response (summarized in Fig. 12.1). In addition to classical class I and II genes, other genes in the MHC are involved in antigen presentation and processing, notably a cluster of genes within the class II region containing *TAP1*, *TAP2*, *PSMB8*, and *PSMB9*. *TAP1* and *TAP2* encode the transporter associated with antigen processing proteins 1 and 2, respectively, encoding the two subunits of TAP that are responsible for binding peptides in the cell cytoplasm and transporting them to the ER (Box 12.1). *PSMB8* (previously known as *LMP2*) and *PSMB9* (*LMP2*) encode protein components of the immuno-proteasome involved in degradation of ubiquitin tagged proteins for presentation by MHC class I molecules. In the extended class II region, *TABP* encodes tapasin – the TAP binding protein involved in the peptide loading complex (Box 12.1).

Elsewhere in the MHC are genes encoding mediators of inflammation and the immune response. These include members of the tumour necrosis factor (TNF) superfamily, immunoglobulins, the complement cascade (Section 12.7), molecular chaperones such as heat shock proteins, and lymphocyte antigen genes. The TNF superfamily is a cluster of genes within the MHC class III region which comprises *TNF* (encoding TNF), *LTA* (lymphotoxin alpha), and *LTB* (lymphotoxin beta), whose protein products have key roles in immunity and inflammation. Genetic variation in the *TNF* locus has been associated with susceptibility to a number of autoimmune and infectious diseases (Section 2.4.4). The class III region also includes five genes (*LY6G5B*, *LY6G5C*, *LY6G6D*, *LY6G6E*, and *LY6G6C*) encoding cell surface proteins which are part of the lymphocyte antigen superfamily and involved in the immune response (Mallya *et al.* 2006). The heat shock cluster of genes *HSPA1A*, *HSPA1B*, and *HSPA1L* is also found in the MHC class III region and encodes members of the heat shock protein 70 family, molecular chaperone proteins involved in the stress response (Milner and Campbell 1990).

Further examples of immune-related genes are found in the MHC, such as the MHC class I-related chain A and B genes (*MICA* and *MICB*) that encode proteins expressed on the cell surface in response to stress. These are ligands for the NKG2D receptor found on natural killer cells, CD8+ cytotoxic T cells, and gamma delta T cells (Bauer *et al.* 1999). Other functional groupings of MHC genes

include large clusters of genes encoding histones and transfer RNA, pheromone and olfactory receptors, and zinc finger proteins such as enzymes and transcription factors (Horton *et al.* 2004).

The reported genetic linkage and association studies implicating genetic variation in the MHC and disease are typically highly significant but have proved hard to localize to specific variants. Almost all autoimmune diseases have been linked to the MHC, many showing their strongest association of any genomic region (Lechler and Warrens 2000). Susceptibility to many infectious diseases has also been linked to the MHC, notably malaria (Section 13.2.6), leprosy, hepatitis B, and human immunodeficiency virus 1 (HIV-1) infection (Section 14.4). In some cases causal relationships have been established between variants and disease, as exemplified by the iron storage disorder haemochromatosis and the *HFE* gene (Section 12.6) (Feder *et al.* 1996). For the majority of diseases, specific haplotypes or individual alleles have been associated with protection or susceptibility to disease. A proportion of the many reported disease associations of the MHC are summarized in Fig. 12.4.

12.2.3 Genetic diversity in the MHC and disease: insights from rheumatoid arthritis

The number of disease associations and the complexity of dissecting their genetic basis within the MHC are illustrated by many different examples through the course of this chapter. One disease that demonstrates the role of diversity in the MHC in disease susceptibility and the difficulties of defining causative variants is rheumatoid arthritis (Newton *et al.* 2004b, Coenen and Gregersen 2009). Rheumatoid arthritis is a chronic systemic disease of unknown cause which primarily affects the joints where it is associated with inflammatory and destructive processes. There is a strong body of evidence from twin and family studies that genetic factors are important in determining susceptibility – estimated at between 40% and 60% of the risk. Polymorphism of the MHC is the most important genomic locus identified, contributing about half of that genetic risk with a number of other non-MHC loci reported (Box 12.3). Within the MHC, polymorphic

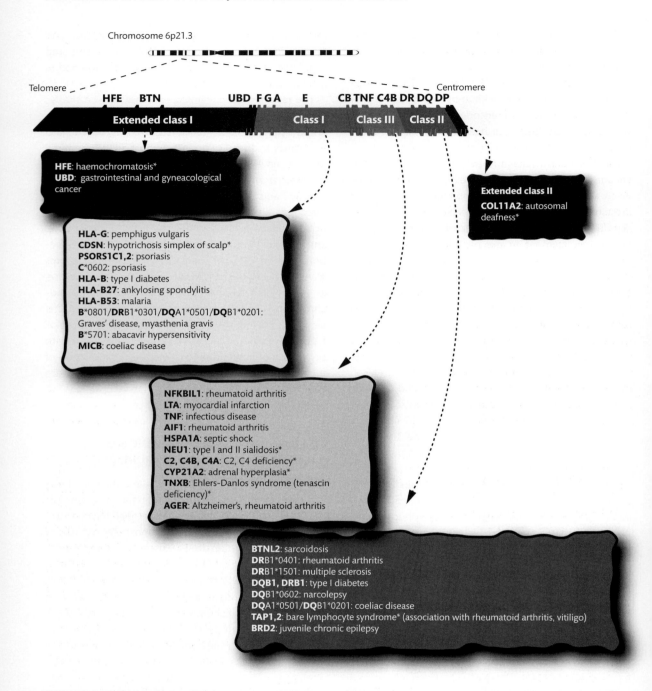

Figure 12.4 MHC associations with disease. Examples of specific genes, alleles, or haplotypes that have been associated with disease susceptibility through linkage analysis or association are indicated. Where a causal relationship has been established an asterix is shown.

HLA-DRB1 alleles were found to be most significant in determining disease susceptibility. These include DRB1*0401, *0404, and *0101 (Stastny 1976). Studies of different disease associated DRB1 alleles across a range of populations implicated a common short sequence of amino acids that was part of the peptide binding groove of the DRβ1 chain, leading to the 'shared epitope' hypothesis (Gregersen et al. 1987).

The molecular mechanism underlying the DRB1 association remains unresolved. It appears that despite a very strong and reproducible genetic association, possessing a shared epitope DRB1 allele is neither necessary nor sufficient for disease to occur (Box 12.3). The situation is complex as the major disease associated alleles vary between ethnic groups and there is a hierarchy of strength of association, with some DRB1 alleles actually associated with protection from disease. The effect has low penetrance as about one-third of the UK population carry DRB1*04. Surprisingly one, rather than two, copies of particular alleles appear associated with more severe disease. Nor can this be the whole story as one-third of patients with rheumatoid arthritis do not carry an allele encoding the shared epitope motif. Intriguingly, there is also evidence that the shared epitope alleles are associated with the development of pathogenic autoantibodies (to anticitrunillated protein)

and only show association with disease in patients with these antibodies (van der Helm-van Mil et al. 2006).

Are there more strongly associated loci or indeed causative variants in linkage disequilibrium with DRB1? Certain haplotypes, for example DRB1*0401-DQB1*0301, have been associated with disease severity but remain unresolved (Cranney et al. 1999). Susceptibility loci in the neighbouring class III region are proposed, for example at NFKBIL1 (nuclear factor of kappa light polypeptide gene enhancer in B-cells inhibitor-like 1) in a Japanese population (Okamoto et al. 2003) and the TNF gene cluster where specific haplotypes have been implicated but causative variants not resolved (Newton et al. 2003). There is some recent evidence that diversity involving a nearby gene, AIF1, encoding allograft inflammatory factor-1 may be involved (Harney et al. 2008). Analyses have not been restricted to disease risk; **pharmacogenomic** studies in rheumatoid arthritis have shown association with the MHC, for example between TNF promoter polymorphisms and disease activity following treatment with anti-TNF monoclonal antibodies (Mugnier et al. 2003).

Why should it prove so hard to localize disease associations? Some reasons are common to studying genetic factors in any multifactorial disease trait such as infectious or autoimmune diseases where genetic, epigenetic,

---

Box 12.3 Non-MHC disease associations with rheumatoid arthritis

Although the MHC is the major genetic risk locus for rheumatoid arthritis, a number of other loci have been implicated (Coenen and Gregersen 2009). These include variants of STAT4 on chromosome 2q32 encoding the transcription factor 'signal transducer and activator of transcription 4' in a region originally highlighted by linkage analysis (Remmers et al. 2007), and of chromosome 1p13 where evidence from linkage analysis and a candidate gene approach resolved a specific nonsynonymous SNP of the intracellular phosphatase gene PTPN22, rs2476601 (c.1858C>T, p.R620W) (Jawaheer et al. 2003; Begovich et al. 2004). Intriguingly this variant also shows association with susceptibility to a number of other autoimmune diseases. PTPN22 was highlighted in

recent genome-wide association studies of rheumatoid arthritis (section 9.3.2) that have also demonstrated significant association with chromosome 9q33-q34, a region which includes tumor necrosis factor receptor-associated factor 1 (TRAF1) and complement component 5 (C5) (Plenge et al. 2007b). A clearer picture is beginning to emerge with other recently reported disease associations including chromosome 6q23 near TNFAIP3 (encoding tumor necrosis factor-alpha-induced protein 3) (Plenge et al. 2007a; Thomson et al. 2007), CD40 (Raychaudhuri et al. 2008) and REL (encoding a member of the NF-kB transcription factor family) (Gregersen et al. 2009) which suggest a key role for the CD40 signalling pathway in the pathogenesis of rheumatoid arthritis.

and environmental factors may be involved. Genetic susceptibility loci in such diseases often involve several independent genetic regions, constitute individually a modest magnitude of effect, and may be confounded by population stratification and other risk factors (Section 2.5). Linkage disequilibrium or coinheritance between genetic polymorphism (Box 2.8) is particularly pronounced in the MHC. This, combined with the level and complexity of polymorphism, the heterogeneity and density of genes in this region, and a paucity of knowledge about the biological function of individual genes and the functional consequences of genetic diversity, have all contributed to a difficult task. In many ways the MHC has provided a paradigm for complex disease genetics: as we will see in the remainder of this chapter there have been success stories but further levels of complexity arise as we look deeper into the story.

12.3 Polymorphism, haplotypes, and disease

The MHC is highly polymorphic including single nucleotide polymorphisms (SNPs), deletion/insertion (indel) polymorphisms, and copy number variation, with strong patterns of coinheritance or linkage disequilibrium between given genetic variants. Such patterns of association between alleles may arise through recombination events, gene conversion, demography, genetic drift, and natural selection. Defining the common combinations of particular alleles, the haplotypic structure, helps us to understand the ancestral origin of genetic diversity. The extensive linkage disequilibrium seen at the MHC facilitated early studies to define the importance of this genomic region in susceptibility to a range of different autoimmune, infectious and other diseases through typing serological or genetic markers. It has also led to considerable challenges in subsequently fine mapping disease associations and resolving specific functional variants. Given the biological and medical interest of genetic variation in the MHC, the patterns of linkage disequilibrium, the hotspots of recombination, and the haplotypic structure of this genomic region have been the subject of intense research for several decades. The extreme levels of polymorphism found in the MHC and the extent of linkage disequilibrium has made defining the haplotypic structure particularly challenging, but a clearer picture is starting to emerge.

12.3.1 Infectious disease, selection, and maintenance of MHC polymorphism

Susceptibility to infectious disease has been postulated as a major driving force in the maintenance of polymorphism at class I and II loci in the MHC. The finding that susceptibility to malaria was associated with the possession of particular HLA alleles was consistent with this hypothesis, a notable study being that by Hill and colleagues who determined that in a large cohort of children with malaria in The Gambia in West Africa, possession of HLA-B53 was associated with protection from severe malaria (Section 13.2.6) (Hill *et al.* 1991). Malaria and other infectious diseases have represented major selective forces during human history, and are significant determinants of the patterns of genetic diversity we observe today (Section 13.2). Individuals who are heterozygous for HLA alleles have been proposed to be at an advantage within a population because the greater repertoire of MHC molecules would allow them to present up to twice as many peptides from pathogens as homozygous individuals, and hence mount a more effective immune response (Doherty and Zinkernagel 1975a).

The much higher than expected rate of nonsynonymous (amino acid changing) nucleotide substitutions within coding regions for peptide binding in MHC molecules compared to other areas of the MHC is consistent with Darwinian selection operating at these gene loci (Hughes and Nei 1988, 1989). Analysis of *HLA-A* and *-B* showed significantly less homozygosity than expected given neutrality (Hedrick and Thomson 1983), with amino acid heterozygosity concentrated in peptide binding regions when diversity across different human populations was studied (Hedrick *et al.* 1991). Finding clear evidence of heterozygote advantage in terms of disease susceptibility has, however, been surprisingly difficult. Examples have been found for a number of viral infections. Maximal heterozygosity at *HLA-A*, *-B*, and *-C* loci has been associated with delayed onset of acquired immuodeficiency syndrome (AIDS) and lower mortality in HIV-1 infection (Section 14.4) (Carrington *et al.* 1999). Heterozygosity of class II alleles has been associated with clearance of hepatitis B (Thursz *et al.* 1997), while heterozygosity at all class I loci was associated with a lower proviral load of human T cell lymphotropic virus type I, consistent with a strong class I restricted cytotoxic T lymphocyte response reducing infection and disease risk (Jeffery *et al.* 2000). Analysis of isolated populations has also proved informative, with substantially less homozygotes than expected found on serological testing for HLA-A and HLA-B among South Amerindian tribes (Black and Hedrick 1997). Black and colleagues studied children of families in 23 tribes from the Amazon and Orinoco basins and found approximately 25% less homozygotes than from mendelian expectations.

Is heterozygote advantage (also known as overdominant selection) sufficient to account for the observed degree of variation at the MHC? The answer remains unclear but other factors such as mate selection and preferential abortion may be important, as well as host–pathogen coevolution. Here, greater evolutionary fitness is associated with individuals possessing new rare MHC alleles with the potential to present peptides from pathogens that have avoided presentation by common MHC molecules (Borghans *et al.* 2004). Such a situation is postulated for HIV infection where the virus adapts to common MHC alleles in a population, giving a selective advantage to those with rare alleles (Trachtenberg *et al.* 2003).

12.3.2 Ancestral haplotypes

Ancestral haplotypes (also known as **conserved extended haplotypes**) are large chromosomal segments that have been conserved *en bloc*, with a fixed constellation of alleles. In the case of the MHC, such haplotypes can span several megabases, for example from *HLA-B* to *HLA-DR* (Degli-Esposti *et al.* 1992). Ancestral haplotypes have been named with reference to the HLA-B allele, followed by a number denoting the order of discovery. The 8.1 haplotype, bearing HLA-A1-B8-Cw7-DR3, is perhaps the most intensively studied of any ancestral haplotype with many significant associations with autoimmune diseases including susceptibility to type 1 diabetes, coeliac disease, systemic lupus erythematosus, myasthenia gravis, dermatitis herpetiformis, common variable immunodeficiency, and IgA deficiency, as well as survival after HIV-1 infection

(Box 12.4) (Price *et al.* 1999). The extent of conservation between *HLA-A* in the class I region and *HLA-DQ* in the class II region is remarkable: recent analysis of 656 SNPs within 4.8 Mb of the MHC among 31 examples of 8.1 haplotypes showed greater than 99% conservation over a 2.9 Mb region (Aly *et al.* 2006).

Why should such a long region be conserved? Natural selection is perhaps the most attractive hypothesis. The high frequency of the haplotype (10% in north Europeans) and extended length are consistent with positive selection favouring the allele(s) such that the haplotype rapidly increases in frequency before it has time to become disrupted by recombination (Section 10.2.2) (Sabeti *et al.* 2002a). Given a recombination rate between *HLA-B* and *HLA-DR* of 1%, the 8.1 haplotype may have arisen 23–40 generations ago (Price *et al.* 1999). Suppression of recombination is another mechanism whereby long range conservation may have arisen. The extreme sequence diversity and other structural differences between homologous MHC chromosomes may prevent crossovers through disruption of pairing and alignment.

12.3.3 Abacavir hypersensitivity

The 57.1 ancestral haplotype, identified by the presence of the alleles HLA-B*5701, C4A6, HLA-DRB1*0701 (DR7) and HLA-DQB1*0303 (DQ3), is of particular interest given its association with hypersensitivity to the anti-retroviral drug abacavir (Mallal *et al.* 2002). Abacavir is a nucleoside analogue which in approximately 5% of white patients with HIV infection will produce a characteristic hypersensitivity reaction including fever, rash, and gastrointestinal symptoms. The reaction shows familial clustering and racial differences in frequency, and can be life threatening on re-exposure to the drug.

In the Western Australian HIV Cohort Study, the 57.1 haplotype showed a remarkable level of association with abacavir hypersensitivity. In a cohort of 200 individuals,

---

Box 12.4 **Functional consequences of the 8.1 haplotype**

This remains controversial. The haplotype has been associated with longevity, with differences in cellular activation and the balance of inflammatory cytokines. Low levels of IL-1 and IFNγ, higher levels of TNF, changes in the autoantibody response, and relative proportions of CD4 and CD8 T cells have all been reported.

the 57.1 haplotype was found among 13 (72%) of hyper-sensitive patients and none of the tolerant patients (P <0·0001) (Mallal *et al.* 2002). Subsequent prospective application as a genetic screening test among HIV patients prior to prescription of the drug in this population showed it performed well with a fall in incidence of drug reaction based on *HLA-B*5701* testing (Rauch *et al.* 2006). The clinical utility of testing for HLA-B*5701 before starting treatment with abacavir has been highlighted by a recent large prospective randomised study (Box 12.5) (Mallal *et al.* 2008). The molecular basis for this association remain unclear but an essential role for HLA-B*5701 and the endogenous pathway for antigen presentation have been defined. The effect is specific to HLA-B*5701 and not closely related allelotypes and involves either the drug itself interacting with the antigen binding groove or causing a modification of it to allow self antigens to bind (Chessman *et al.* 2008).

12.3.4 Extreme polymorphism at *HLA-DRB1*

There are estimated to be 450 different DRB1 alleles in human populations with 13 allelic lineages, a remarkable level of diversity for one genomic locus. Like other MHC class II genes, most variation is found within the anti-gen recognition sites encoded by the second exon of the gene. This sequence, and diversity of it, is critical to the ability of class II molecules to present diverse antigens to T cell receptors. Balancing selection has been proposed as the driving force enhancing sequence diversity within exon 2, with a significantly higher rate of nonsynonymous versus synonymous nucleotide substitutions observed in these sequences (Hughes and Nei 1989). It remains con-troversial whether the different allelic lineages of *HLA-DRB1* involve ancient alleles, predating hominids, or are more recent with generation through mechanisms such as gene conversion-like events.

The work of Bergstrom and colleagues proved highly informative based on analysis of the intronic sequence flanking exon 2 between alleles (Bergstrom *et al.* 1998; Takahata and Satta 1998). They found a small num-ber of characteristic sequence motifs in these regions showing strong linkage disequilibrium, in a pattern con-sistent with allelic lineages based on serological test-ing. The diversity suggested evolutionary isolation over a considerable length of time with no evidence for the

---

Box 12.5 Pharmacogenomics and abacavir hypersensitivity

Pharmacogenomics promises to be one of the first areas of general medical care impacted by the recent rapid advances in our understanding of human gen-etics and its role in disease. The ability to tailor drug therapies to the individual patient so as to maximise clinical benefit while minimizing risk of adverse effects promises a radical advance in how medicine is practiced. A number of examples have already been described in this book including modulation of drug metabolism and dosage as seen with copy num-ber variation of the cytochrome p450 genes such as *CYP2D6* (Section 4.4.2); work relating to warfarin dose is described in chapter 15 (Section 15.5). In terms of risk of adverse effects, abacavir hypersensi-tivity provides one of the clearest examples to date. This was highlighted by PREDICT-1 (Prospective Randomized Evaluation of DNA Screening in a Clinical Trial), a prospective, randomized, multi-center, double-blind study of 1956 adult patients from 19 countries infected with HIV-1 (Mallal *et al.* 2008). The study compared the effect of screen-ing for HLA-B*5701 before antiretroviral ther-apy involving abacavir with a standard of care approach without screening. Overall the prevalence in the study population of HLA-B*5701 was 5.6%. Remarkably, prospective screening completely elim-inated immunologically confirmed abacavir hyper-sensitivity (0% vs 2.7%). The positive predictive value of testing for HLA-B*5701 was 58% (in other words the allele was necessary but not sufficient for hypersensitivity) while most importantly the nega-tive predictive value of the test was 100%. This large prospective trial clearly showed the clinical utility of HLA-B*5701 as a pharmacogenomic biomarker in identifying at risk individuals and testing for HLA-B*5701 before commencing abacavir therapy is now recommended in guidelines from US Department of Health and Human Services.

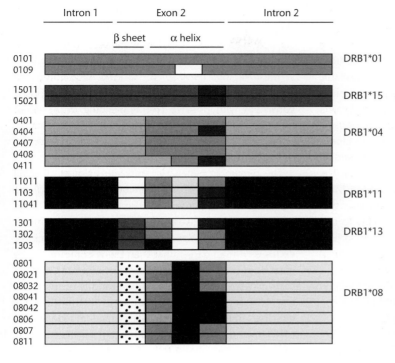

Figure 12.5 Sequence motif polymorphism at *HLA-DRB1* exon 2 and flanking introns. Shading denotes vertically different sequence motifs. Redrawn and reprinted by permission from Macmillan Publishers Ltd: Nature Genetics (Takahata and Satta 1998), copyright 1998.

postulated mutation-driven convergence or conversion events between lineages. This low level of polymorphism extended into the region of exon 2 encoding the β sheet, found at the floor of the antigen binding groove. In contrast, the portion of exon 2 encoding the α helix, found at the ridge of the antigen binding groove, was remarkably polymorphic. Here the pattern was consistent with localized gene conversion at a microscopic level with polymorphism within allelic lineages and sharing of nucleotide substitutions common between lineages (Fig. 12.5). They concluded that recent shuffling of existing alleles by gene conversion generated this highly localized diversity. The low level of intronic diversity was also postulated as being consistent with the single origin hypothesis of human populations with an effective breeding population size of 10 000 over the past million years (estimated based on the very low diversity at non-HLA loci of 0.1%) (Li and Sadler 1991) rather than estimates of 100 000 based on the persistence of HLA allelic lineages over tens of millions of years (Takahata 1993).

More recent analysis based on more extensive sequencing at the DRB1 locus, including full length alleles together with comparisons between primate species

of different lineages, suggest a recent origin for alleles (Bontrop 2006; von Salome *et al.* 2007). The DRB1*03 lineage is found in humans, chimpanzees, bonobos, gorillas, and orangutans, and an ancestral 'proto 03' lineage is postulated. If only the exon 2 sequences are used to construct the phylogenetic tree, all lineages predate the separation of humans and chimpanzees an estimated 5 million years ago. Many lineages are, however, unique to humans (such as *8, *11, *13, *14, *15, and *16) and analysis using the dataset excluding exon 2 is consistent with this. Some of the motifs subject to gene conversion shuffling within the antigen recognition sequences may be very ancient but it is believed that the alleles are being generated relatively rapidly with the average age of within lineage diversity estimated at less than 1 million years (von Salome *et al.* 2007).

12.3.5 The MHC Haplotype Project

The reference DNA sequence for the MHC, like the rest of the human genome, comprised a composite of many haplotypes with DNA fragments derived from different individuals. In 2004 the first results of the MHC Haplotype

Project (http://www.sanger.ac.uk/HGP/Chr6/MHC/) were published (Stewart *et al.* 2004a), providing a resource unique to the MHC whereby a DNA sequence would be available comprising a single contiguous haplotype from one individual. This 'homozygous' DNA was available from consanguineous cell lines each of which comprised a single MHC haplotype. The MHC Haplotype Project aimed to sequence eight different MHC haplotypes using bacterial artificial chromosome (BAC) library clones derived from eight cell lines from the Tenth International Histocompatibility Workshop (Allcock *et al.* 2002). The cell lines were selected as representing common MHC haplotypes found in north European populations that were known to be associated with susceptibility to disease (Box 12.6). The hope was that this work would help identify the specific genetic variants within the extended haplotypes which were responsible for disease susceptibility, as well as broader questions about the nature and frequency of genetic diversity, the ancestral origins and relationships of haplotypes, and their structure.

What has the MHC Haplotype Project told us? We now have a contiguous sequence comprising a single haplotype to use as a reference sequence (the DNA sequence of the PGF cell line). The MHC haplotypes are the first long range single haplotypes sequenced in the human genome. Over 44 000 variants were identified by comparing the three haplotypes (Horton *et al.* 2008). Comparing PGF and QBL, for example, there were 17 695 sequence differences of which 15 345 were single nucleotide substitutions and the remainder insertion/deletion events (indels) (Traherne *et al.* 2006b). The single nucleotide substitutions were most frequent at classical class I and II loci as expected, with the ratio of nonsynonymous to synonymous substitutions 3 : 1 at these regions (*HLA-A, -B, -C, -DR, -DQ*) compared to non-classical loci where the ratio was 1 : 1. This is consistent with positive selection acting on genes in the classical loci resulting in the higher frequency of polymorphism altering the amino acid structure of the encoded proteins.

The remaining genetic variants were almost all small (<96 bp) indels but 34 of these were large (96–5157 bp), including 15 Alu element insertions (Section 8.4). Most of these insertions were 'young', including five members of the Ya5 and Yb8 families which are found almost exclusively in humans rather than being common to the great apes or other species (Carroll *et al.* 2001). Ancient variants were also found, including repeats of the HERV (human endogenous retrovirus) sequences, LINEs (long interspersed nuclear elements), long tandem repeats, and MERs (mammalian interspersed repetitive elements).

The three haplotypes differ for example at the RCCX locus (Section 12.7), with PGF having two copies of the *C4* gene (*C4A* and *C4B*), which were both of the 'long' type as they contained an HERVC4 insertion in intron 9, COX had only a single copy of *C4*, *C4B* of the short type, while QBL had a single *C4A* of the long type. Specific variants were also discovered at disease associated loci – for example at *PSORSICI* (a candidate locus for psoriasis, a skin disorder) in the QBL haplotype a deletion of 1 nt in the poly(C)

Box 12.6   MHC haplotypes associated with disease

The first two cell lines to be sequenced, PGF and COX (Stewart *et al.* 2004a), bore two common MHC haplotypes associated with autoimmune disease susceptibility. HLA-A3-B7-Cw7-DR15, the 7.1 haplotype found in the PGF cell line, is present in 10% of north Europeans and has been associated with protection from type 1 diabetes (relative risk 0.05) and predisposition to multiple sclerosis and systemic lupus erythematosus (relative risks 2–4) (Barcellos *et al.* 2003; Larsen and Alper 2004), while the '8.1' haplotype HLA-A1-B8-Cw7-DR3 was found in the COX cell line. In 2006 the sequence for a third haplotype, found in the QBL cell line (HLA-A26-B18-Cw5-DR3-DQ2; the 18.2 ancestral haplotype), was published (Traherne *et al.* 2006b) associated with susceptibility to Graves' disease and type 1 diabetes (Johansson *et al.* 2003). The MHC Haplotype Project was completed by 2008 with the publication of sequence data and analysis of all eight selected haplotypes (Horton *et al.* 2008).

tract of exon 5 led to a frameshift in the spliced transcript, a premature stop codon, and shortening of the coding sequence by 266 amino acids.

The *HLA-DRB* locus is known to be extremely polymorphic and the highest diversity was observed between PGF and the COX/QBL haplotypes. COX and QBL both have the DR3 haplotype and the detailed analysis provided by the MHC Haplotype Project resolved a 158 kb region with almost no variation, dubbed a 'SNP desert' consistent with a relatively recent ancestry and allowing resolution of the original recombination event (Traherne *et al.* 2006).

12.3.6 A map of diversity across the MHC

In 2006 the MHC Sequencing Consortium published the most extensive map to date of the patterns of linkage disequilibrium across the MHC (de Bakker *et al.* 2006). The group performed extensive typing of HLA genes and of over 7500 polymorphisms (SNPs together with indels) within the 7.5 Mb region of the extended MHC. Four different ethnic groups were studied, African, European, Chinese, and Japanese, with a total of 361 individuals. This allowed the recombination rates and hotspots to be estimated, as well as the inferred haplotypes (Fig. 12.6). Overall the small scale haplotypic block structure in the MHC is comparable to elsewhere in the genome but the extent of linkage disequilibrium between blocks is higher, resulting in the observed extended or ancestral haplotypes that are found to be particularly common among individuals of north European origin. Consistent with studies elsewhere in the genome, the linkage disequilibrium was found to be lower in the African cohort with typically shorter haplotypes observed. The recombination rate overall was lower than for other genomic regions (0.44 cM/Mb versus genome-wide average of 1.2 cM/Mb).

Why is such a map important? For those interested in using genetic and HLA markers to resolve disease susceptibility, a dense map such as this provides the possibility of using specific 'tag SNPs' which potentially significantly reduce the number of markers required. As discussed in Section 9.2, this is because the tag SNPs are selected based on their ability to be informative about the underlying allelic architecture: a small number of SNPs may be sufficient to define the common haplotypes for a given gene or region of DNA. How well do they work? In the paper by de Bakker and colleagues, the selected tag SNPs

showed high sensitivity and specificity as predictors for HLA alleles in two disease cohort studies (de Bakker *et al.* 2006). Knowledge of patterns of linkage disequilibrium and haplotypic structure also provides a route map for future studies of evolutionary dynamics and understanding the ancestral origin of MHC polymorphism as well as fine mapping disease associations.

12.4 Getting in the groove: diversity in MHC class II alleles

The class II region of the MHC contains genes encoding molecules involved in antigen presentation. These molecules are found on the surface of circulating antigen presenting cells as well as epithelial cells of the thymus. Striking genetic associations have been reported for a number of chronic inflammatory disorders (see Fig. 12.4) but defining the causative gene or variant has proved difficult due to linkage disequilibrium and a paucity of knowledge about the underlying disease mechanism. However recent genetic and pathophysiological studies are providing insights into the relationship between possession of particular HLA alleles and individual disease risk. Of particular note are advances in our knowledge of the structural biology of the MHC class II molecules which are highlighting the importance of genetic diversity in defining the specificity of the peptide binding domain (Jones *et al.* 2006).

12.4.1 Narcolepsy

Although the molecular basis for the dramatic HLA association with susceptibility to narcolepsy (Box 12.7) is still incompletely understood, it is striking that two very similar MHC class II alleles, HLA-DQB1*0602 and DQB1*06011, should be strongly associated with risk and protection from the disorder, respectively, and that structural analysis of their respective peptide binding grooves should show specificity for peptides implicated in disease pathogenesis.

The occurrence of narcolepsy in dogs has provided a very important animal model to understand disease pathogenesis. Familial narcolepsy can be bred in Doberman pinschers and Labrador retrievers, comprising an autosomal recessive trait with full penetrance.

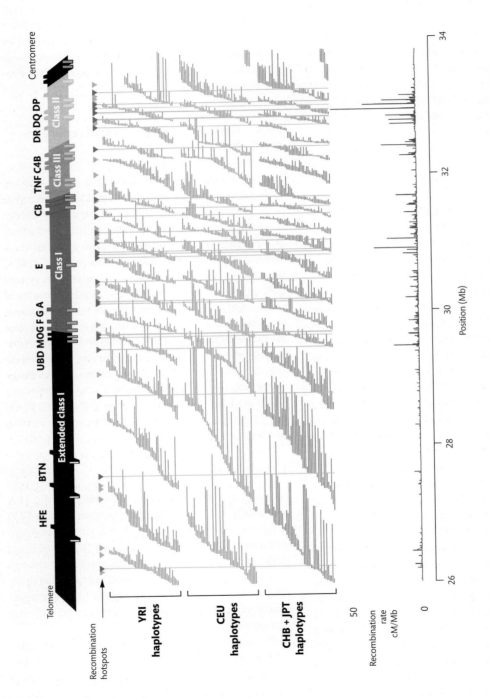

Figure 12.6 Haplotypic structure of the extended MHC region. Analysis of sequence diversity and HLA typing in 361 individuals from four ethnic groups of differing geographic ancestry: YRI (African), CEU (European), CHB (Chinese), and JPT (Japanese) allowed the inferred haplotypic structure across the region to be deduced. Recombination hotspots (triangles) are shown for overall population, with non-redundant haplotypes (lines) shown for each ethnic group. Redrawn and reprinted by permission from Macmillan Publishers Ltd: Nature Genetics (de Bakker *et al.* 2006), copyright 2006.

Box 12.7 Narcolepsy (OMIM 161400)

Narcolepsy is a complex neurological disorder characterized by excessive daytime sleepiness (Overeem et al. 2001). In the majority of cases, cataplexy is also present – a dramatic symptom in which there is sudden bilateral loss of postural muscle tone in association with intense emotions such as laughter.

Other rarer symptoms of narcolepsy include vivid dream-like experiences during the transition between wakefulness and sleep (hypnagogic hallucinations) and sleep paralysis, in which patients are subjectively awake and conscious but unable to move during the onset of sleep or on waking.

Subsequent detailed genetic analysis including linkage analysis and positional cloning led to the identification of specific deletions of the hypocretin receptor 2 gene (Lin et al. 1999). Genetic knock-out studies of the same gene in mice gave a phenotype with some similarity to narcolepsy (Chemelli et al. 1999). There is now substantial evidence to link the 'wake promoting' neuropeptide hypocretin with narcolepsy. Post mortem studies of affected individuals show the destruction of hypothalamic neurones secreting hypocretin (Peyron et al. 2000) while cerebrospinal fluid (CSF) levels of hypocretin are characteristically low or undetectable at the time of making a clinical diagnosis of narcolepsy (Nishino et al. 2000).

How does the hypocretin peptide relate to the very strong genetic association between MHC class II alleles and narcolepsy? The initial association was reported for Japanese patients with HLA-DR2 found in all 40 patients with narcolepsy (Juji et al. 1984). The association was also found among Caucasians, with subsequent studies showing narcolepsy to be most strongly associated with HLA-DQB1*0602 across a broad range of ethnic groups – a feature unusual among HLA disease associations – such that this allele is carried by 88–98% of cases of narcolepsy, but only if cataplexy is present (Mignot et al. 1997). The haplotype is also relatively common among the general population, typically at 10% frequency, indicating incomplete penetrance and the multifactorial nature of the condition. Indeed among monozygotic twins, only about one-third are concordant, indicating a major environmental contribution. Other MHC genes have been implicated such as TNF, while a recent genome-wide association study among Japanese adults highlighted chromosome 21q22.3 where three narcolepsy candidate genes were identified (Kawashima et al. 2006). However it is HLA-DQB1*0602 that shows the strongest association – one of

the strongest identified for any disorder – and the finding that a very closely related allele, HLA-DQB1*06011, protected against narcolepsy (Mignot et al. 2001) prompted structural studies comparing these HLA molecules.

The solving of the crystal structure of HLA-DQB1*0602 complexed with hypocretin has highlighted some potentially functionally important differences with DQB1*06011 (Siebold et al. 2004). The two molecules differ only at nine amino acid residues in the β chain, of which three may modulate T cell receptor recognition while the remaining five are found in the peptide binding groove. Of particular note are polymorphisms altering the P4 pocket of the groove which define the particular peptide side chains that can fit – for DQB1*06011 the pocket becomes closed up such that hypocretin is prevented from binding.

The molecular mechanism relating DQB1*0602 and narcolepsy remains incompletely understood. An autoimmune process with selective destruction of hypocretin-containing neurones in the hypothalamus is an attractive hypothesis but unproven. The environmental trigger that may be required for such an autoimmune reaction is unknown. The finding of functional autoantibodies in serum and CSF from narcolepsy patients with the *0602 haplotype is intriguing. Immunoglobulin from patient serum led to narcolepsy-like behaviour in mice (Smith et al. 2004) while that from CSF was shown to bind to rat hypothalamic proteins. Further work is required, but the HLA association has provided a clinically useful diagnostic tool and important insights into disease pathophysiology.

12.4.2 Coeliac disease

The study of genetic factors underlying coeliac disease has provided important insights into the pathophysiology of this disorder, in particular the role of MHC class II alleles in determining disease susceptibility (reviewed in Kagnoff 2007). Coeliac disease (OMIM 212750) is a common, chronic, inflammatory disease affecting the small bowel, which often presents with symptoms of weight loss, malnutrition, and diarrhoea. The disease is activated in susceptible individuals by eating wheat gluten, together with other similar proteins found in rye and barley; examination of the small intestine in affected people shows damage to the lining intestinal mucosa resulting in malabsorption of nutrients.

Genetic factors have been recognized for many years to be very important in defining susceptible individuals. The disease affects families and shows a very high (70–75%) concordance among monozygotic twins (Greco et al. 2002). The strongest genetic association is with *HLA-DQ*, which is necessary but not sufficient for the disease to occur (Kagnoff 2007). Heterodimers of HLA-DQ2 are present in 90–95% of patients with coeliac disease. Here the β chain of the class II molecule is encoded by DQB1*0201 or *0202 and the α chain by DQA1*05. The remaining 5–15% of patients have DQ8 heterodimers encoded by DQB1*0302 and DQA1*03 (encoding the β and α chain, respectively). Coeliac disease is extremely rare in Japan where the DQ2 susceptibility alleles are very uncommon.

How does possession of particular HLA-DQ2 and -DQ8 alleles relate to susceptibility to coeliac disease? The answer appears to lie once more in the peptide binding groove of the class II molecule (Kim et al. 2004; Bergseng et al. 2005). At first sight the situation appears contradictory. The binding grooves of DQ2 and DQ8 favour binding of negatively charged protein residues at key anchor positions, making them highly unlikely to bind the praline- and glutamine-rich gluten peptides. However, a specific deamidating enzyme, transglutaminase 2 (TGase), was found to be upregulated in the inflamed intestine, which converts neutral glutamine to negatively charged glutamic acid (Dieterich et al. 1997). Moreover structural studies showed that the praline-rich gluten peptide with deaminated glutamines would bind DQ2 without the expected disruption of hydrogen bonds. The DQ2 and DQ8 heterodimers on antigen presenting cells thus bind and present the gluten peptides to CD4-positive T cells in the lining of the intestine, activating them (Fig. 12.7).

The gluten peptides, by the nature of their biochemical make up, are resistant to proteolytic digestion and occur as relatively large peptides in the small intestine. The presence of multiple DQ binding epitopes is thought to explain the greater T cell stimulating ability of larger peptides. Other genetic and immunological factors are likely to be involved in disease pathogenesis as well as concurrent infection, for example with enteric viruses, which is thought to be a key event in triggering disease (Kagnoff 2007). This may involve exposure to gluten peptides at a time of intestinal inflammation with increased permeability due to viral infection associated with interferon production and immunological activation. What has become clear is that a genetic background of HLA-DQ2 or -DQ8 alleles is required for the disease to occur, and that understanding how these specific molecules interact with the peptide antigens of gluten has proved very valuable to our knowledge of disease pathogenesis.

12.4.3 Type 1 diabetes

The MHC has the strongest evidence of linkage and genetic association of any genomic region with susceptibility to, and protection from, the juvenile onset form of diabetes – which is characterized by autoimmune destruction of pancreatic β cells and insulin deficiency. Type 1 diabetes (OMIM 222100; also known as insulin-dependent diabetes) shows strong familial clustering: siblings of affected individuals are at a 15-fold increased risk of the disease (Spielman et al. 1980) with the largest proportion of this increased risk due to inherited factors within the MHC. All reported genome-wide linkage scans have shown strongest linkage to markers in the MHC although non-MHC associations are also found (Box 12.8). It is thought that genetic variation within the MHC accounts for 40% of the observed familial clustering, with relative risks of disease between 3 and 50 depending on the specific genotypes and haplotypes involved (Lambert et al. 2004). Type 1 diabetes is, however, a multifactorial disease with strong environmental influences as demonstrated by the marked rise in incidence and the earlier age of onset over the last 50 years, as well as the incomplete concordance observed in monozygotic twins (Hirschhorn 2003). The role of inherited factors in type 1 diabetes has been of great research interest in the quest for an improved understanding of disease pathogenesis and the possibility of primary prevention and early intervention through ascertainment of genetic risk markers.

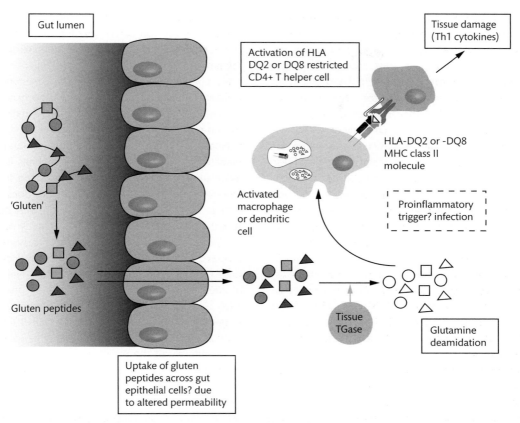

Figure 12.7 Pathogenesis of coeliac disease. Illustration of events following ingestion of gluten in a genetically susceptible person with presentation by HLA-DQ2 and -DQ8 class II molecules to CD4+ T cells of gluten peptides which have undergone glutamine deamidation. Redrawn with permission from Kagnoff (2007).

Box 12.8 Non-MHC associations with type 1 diabetes

Other genomic regions have also been associated with disease susceptibility outside the MHC, notably genetic variation at the *INS* (encoding insulin) gene locus on chromosome 11p15 (Section 7.3.3) (Bell *et al.* 1984), *CTLA4* (encoding cytotoxic T lymphocyte antigen 4 gene) at 2q33 (Ueda *et al.* 2003), *PTPN22* (encoding lymphoid protein tyrosine phosphatase, a suppressor of T cell activation) at 1p13 (Bottini *et al.* 2004), *IL2RA* (encoding interleukin-2 receptor α chain) at 10p15 (Vella *et al.* 2005), and *IFIH1* (innate immunity viral RNA receptor gene region) at 2q24 (Smyth *et al.* 2006). Recent genome-wide association studies have demonstrated a number of other loci notably at 12q13, 12q24, 16p13, 18p11, and 18q22 (Section 9.3) (Hakonarson *et al.* 2007; Todd *et al.* 2007; WTCCC 2007).

The role of the MHC in determining susceptibility to diabetes has been the subject of intense research for over 20 years (Nerup *et al.* 1974; Cudworth and Woodrow 1975). Almost all patients with type 1 diabetes were found to have either HLA-DR3 or -DR4 compared to about half of the unaffected population. Early studies using serological tests for HLA antigens and restriction fragment length polymorphisms (RFLPs) as markers demonstrated that the strongest association was with the linked *HLA-DQ genes*. In 1987 Todd and colleagues sequenced the major expressed polymorphic MHC class II gene products from affected patients and controls (Todd *et al.* 1987). Their work showed the amino acid sequence of the DQ β chain directly correlated with susceptibility to disease, in particular the amino acid found at position 57. The presence of aspartic acid (Asp) at position 57 in one copy of the gene was associated with protection from disease: having two Asp 57 positive DQ β alleles conferred almost complete protection from type 1 diabetes. The non-obese diabetic (NOD) mouse strain, a classic spontaneous autoimmune animal model for human type 1 diabetes, was found to have Ser at this position while all other mouse β chains have Asp. The relationship was, however, found not to be absolute, with particular combinations of DQ and DR conferring susceptibility (Erlich *et al.* 1990).

The situation is thus complex and remains incompletely understood. Specific polymorphisms and haplotypic combinations across the *HLA-DR* and *-DQ* region, mainly involving amino acid substitutions in the α chain of DR, and both α and β chains of DQ, are associated with disease and show a hierarchy of susceptibility (Lambert *et al.* 2004). Thus the HLA-DR2 haplotype, DRB1*1501-DQA1*0102-DQB1*0602, is extremely rare among children with type 1 diabetes (<1%) while present in 20% of the unaffected population (Noble *et al.* 1996). Most Caucasian patients with type 1 diabetes will possess one of the two commonest susceptibility haplotypes (lacking Asp at position 57 of the β chain), of DR3 (DRB1*0301-DQA1*0501-DQB1*0201) and DR4 (DRB1*0401-DQA1*0301-DQB1*0302), with heterozygosity for these haplotypes associated with a 15-fold relative risk and earlier onset of disease.

How do genotypic and haplotypic differences relate to mechanism of action at a molecular level? The answer, or at least part of it, may lie in how these differences change the composition of the peptide binding groove of DR and DQ molecules. Specific features of peptide binding pockets P1, P4, and P9 appear critical. As discussed earlier, residue 57 of the DQ β chain correlates well with disease susceptibility. The substitution of Ala57β for Asp57β at this site within the P9 pocket results in a change in the basic

charge, the accessible volume of the pocket, and disrupts hydrogen bond formation with the peptide backbone (Jones *et al.* 2006). Such changes in structure of the binding groove could have profound consequences for peptide recognition and function. They have been postulated to alter disease susceptibility through mechanisms such as tolerance (presentation of autoantigens in the thymus) as well as peripheral antigen presentation.

Finally, there is also evidence of association of type 1 diabetes with variation in MHC class I genes. The major association has been robustly demonstrated to be with *HLA-DQB1* and *HLA-DRB1*, however this does not fully explain the observed MHC association with disease. A recent study combined a large family study of affected sib pairs together with case–control cohorts, which included individuals genotyped in the Wellcome Trust Case Control Consortium (Section 9.3.2) (Nejentsev *et al.* 2007). Dense genotyping of 254 polymorphic loci and 1475 SNPs within the MHC allowed Nejentsev and colleagues to define associations with *HLA-B* (combined P value $= 2.01 \times 10^{-19}$) and *HLA-A* ($P = 2.35 \times 10^{-13}$) independently of the previously resolved MHC class II associations. The major effect was with HLA-B*39, which showed strong association with disease risk and lower age at diagnosis. Possession of this allele was associated with a relative risk among families of 3.5 (95% CI 2.2–5.7) and among the case–control cohort of 2.4 (95% CI 1.4–3.9) (Nejentsev *et al.* 2007).

12.5 HLA-B27 and susceptibility to ankylosing spondylitis

By the early 1970s there were growing numbers of reports of association between the possession of particular HLA antigens on the surface of white blood cells and susceptibility to disease, notably coeliac disease, multiple myeloma, lymphoma, systemic lupus erythematosus, and psoriasis. During the course of the year in 1972, independently reproducible and highly significant associations were found between the possession of HL-A1 and HL-A8 and coeliac disease (Falchuk *et al.* 1972; Stokes *et al.* 1972), W17 and HL-A13 and psoriasis (Russell *et al.* 1972), and HL-A1 and HL-A8 and chronic active hepatitis (Mackay and Morris 1972). The following year there were dramatic reports of an HLA association with a chronic inflammatory rheumatological condition that characteristically affected the spine and sacroiliac joints called ankylosing spondylitis (OMIM 106300) (Brewerton *et al.* 1973; Caffrey and James 1973; Schlosstein *et al.* 1973). This disease association between HLA-B27 and ankylosing spondylitis withstood repeated genetic investigation over the next 34 years to remain as among the strongest known genetic markers of human disease.

Clinicians and scientists working at the Westminster Hospital in London first reported in a letter to *Nature* in 1973 the dramatic overrepresentation of the HL-A27 antigen on the surface of lymphocytes from patients with ankylosing spondylitis (Caffrey and James 1973). The subsequent paper published in the *Lancet* described their findings from testing a panel of 26 different HLA typing sera. The authors had carefully selected a clear disease phenotype with radiographic evidence of sacroilitis (inflammation of the sacroiliac joint at the base of the spine) and restricted their study to Caucasians. They found 72 out of 75 patients with ankylosing spondylitis possessed HL-A27 while only three out of 75 blood donor controls had that particular antigen (Brewerton *et al.* 1973). The results also made clear that this was not a simple causal relationship: 96% of cases had this particular HLA antigen compared to 52% of first degree relatives and 5% of the general population among Caucasians – and yet the disease prevalence of ankylosing spondylitis was much lower, with only about 0.4% of men and 0.05% of women affected by the disease. Only a minority of those with the HL-A27 antigen had disease.

In the United States, the same association was independently reported for a Caucasian population in April 1973 (Schlosstein *et al.* 1973). In this study a panel of 24 HLA antigens were analysed among patients with ankylosing spondylitis, rheumatoid arthritis, and gout, and the frequencies of particular antigens compared with those among 908 controls. Only among the 40 patients with ankylosing spondylitis was an association found, with 88% of cases having HLA-B27 (denoted W27 in the nomenclature used at the time).

Ankylosing spondylitis shows significant familial clustering with twin studies demonstrating that additive genetic factors explain more than 90% of the observed population variance (Brown *et al.* 1997). A very strong body of evidence from linkage and association studies shows possession of HLA-B27 to be the strongest single genetic factor, accounting for nearly 40% of the total disease risk and being found in over 90% of patients with ankylosing spondylitis of European ancestry. However, only 1–5% of people with HLA-B27 will develop ankylosing spondylitis and there is evidence of a role for modifier genes in disease pathogenesis (Box 12.9) (Brown 2008).

There are thought to be at least 31 molecular subtypes of HLA-B27 (Reveille 2006). Most are defined by polymorphisms in exons 2 and 3 of the gene that encodes the $\alpha 1$ and $\alpha 2$ peptide binding domains of the B27 molecule (Fig. 12.8). HLA-B27*2705 is postulated as the ancestral molecule and is the most common found among European Caucasians. Subtypes vary in frequency among different populations and geographic areas. Only some subtypes are associated with ankylosing spondylitis, these include *2705, *2702, *2704, and *2707. Many subtypes are too rare to allow analysis of association with

---

Box 12.9   Non HLA-B27 associations with ankylosing spondylitis

The higher concordance rate found among monozygotic twin than dizygotic twin pairs who are positive for HLA-B27 implicates other genetic loci in ankylosing spondylitis. Within the MHC, genetic diversity at *HLA-DRB1*, *TNF*, *TAP1*, and *PSMB9* show association with disease while outside the MHC, genome-wide linkage studies identified a number of loci including chromosome 16q (Brown *et al.* 1998; Laval *et al.* 2001), 9q31-q34 (Miceli-Richard *et al.* 2004), 6q and 11q (Zang *et al.* 2004a);

at 2q13 there was evidence of association with the IL-1 complex (Timms *et al.* 2004). A study involving 14,000 non-synonymous SNPs more recently showed significant association with *IL23R* at 1p31 and *ERAP1* (encoding endoplasmic reticulum aminopeptidase 1) at 5q15 (Burton *et al.* 2007); the *IL23R* association is of particular interest given disease associations with psoriasis and Crohn's disease for this locus (Section 9.5.5).

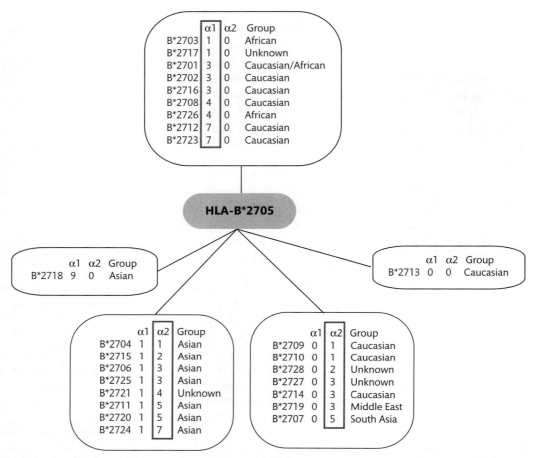

Figure 12.8 Evolutionary relationships of HLA-B27 molecular subtypes in the context of the 'parent' subtype HLA-B*2705. Three major families of HLA-B27 subtypes are shown together with HLA-B*2713 and B*2718, which are assumed to have evolved separately. Subtypes differ from B*2705 by amino acid substitutions in the α1 and α2 domains with the number of differences shown. The predominant ethnic group in which the subtype was described is also listed. Reprinted from Reveille (2006), copyright 2006, with permission from Elsevier.

disease but there are some notable examples of this, such as *2706 found in South East Asia and the *2709 subtype common in Sardinia, which are not associated with disease risk and their structural analysis has led to intriguing insights into disease mechanism (Montserrat *et al.* 2003; Goodall *et al.* 2006).

The particular subtypes result from polymorphism in the amino acid composition of the antigenic cleft of the B27 molecule, changing the potential peptides presented or the interaction with the peptide loading complex molecule tapasin (Fig. 12.9). For example, the only difference between the disease associated *2705 subtype and the

non-associated *2709 subtype is an aspartate to histidine polymorphism in the peptide binding groove at a position important to anchoring the C-terminal peptide residue; this change alters the repertoire of peptides bound by the two molecules.

Given the extent of research effort directed at HLA-B27 it is perhaps surprising that we still do not understand the mechanism underlying the undoubted association between the possession of HLA-B27 and ankylosing spondylitis. A number of models have been proposed but none have been conclusively demonstrated (Box 12.10). Animal studies have given some insight into the

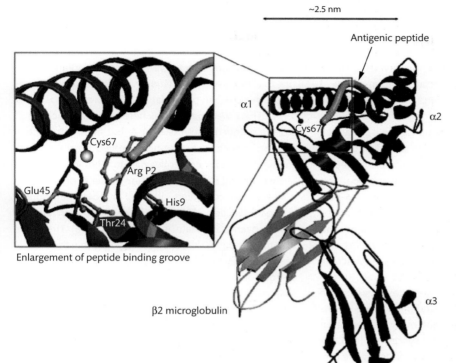

~2.5 nm

Antigenic peptide

α1

Cys67

α2

Enlargement of peptide binding groove

Cys67

Arg P2

Glu45

His9

Thr24

β2 microglobulin

α3

Figure 12.9 Structure of the HLA-B27 molecule. Ribbon diagram showing the HLA heavy chain (α1, α2, and α3 domains) of HLA-B27 molecule, β2 microglobulin (light grey), and the bound peptide. Reprinted from Bowness *et al.* (1999), with permission from Cambridge University Press.

Box 12.10   Models of HLA-B27 and disease mechanism in ankylosing spondylitis

- *Arthritigenic peptide hypothesis.* The antigen presenting ability of HLA-B27 may be crucial to disease pathogenesis. The presentation of particular 'arthritogenic' peptides by HLA-B27 following infection could result in inflammation. Certain disease associated subtypes will, for example, present peptide from *Chlamydia trachomatis* (*2705, *2702, *2704) while others (*2706 and *2709) will not (Ramos *et al.* 2002).

- *Molecular mimicry.* Homology was noted between a particular amino acid sequence of HLA-B27 and *Klebsiella pneumoniae*, with antibody raised to the homologous peptide antigen cross-reacting with HLA-B27 (Schwimmbeck and Oldstone 1988). Cross reactivity by pathological antibodies to foreign antigens may invoke an autoimmune response resulting in disease. If foreign antigens mimicked self peptide constitutively presented by HLA-B27, this could break tolerance and induce chronic inflammation through an autoimmune response (Benjamin and Parham 1990).

- *'Altered self' model of disease.* This model involves the unique property of HLA-B27 molecules to misfold and form homodimers, rather than the usual association with β2 microglobulin. This is promoted by oxidizing conditions such as within activated macrophages and can promote a proinflammatory unfolded protein, intracellular stress response or act as an antigen if presented on the cell surface.

- *Host defence.* HLA-B27 may alter intracellular invasion and killing of pathogens such as *Salmonella*.

molecular mechanism underlying the HLA-B27 association. Informative studies have been predominantly limited to rats where arthritic disease could be experimentally induced. Transgenic rats expressing B27 and human β2 microglobulin showed a phenotype with most of the features of HLA-B27 associated disease seen in humans (Hammer *et al.* 1990). Peripheral and axial arthritis, gut inflammation, and genital and skin lesions were observed. The penetrance of the transgene is, however, affected by the genetic background of the rat studied.

Ankylosing spondylitis is one of a number of rheumatic diseases called seronegative spondyloarthropathies. These are common, chronic inflammatory disorders characterized by inflammation of the spine and sacroiliac joints leading to bone and joint erosions and ankylosis of the spine. Peripheral joint arthritis is also commonly found, with ocular (uveitis), cardiovascular, and pulmonary involvement seen more rarely. In the same year that Brewerton and colleagues reported the association between HLA-B27 and ankylosing spondylitis, they also found significant overrepresentation of HLA-B27 among cases of Reiter syndrome (reactive arthritis) and an inflammatory disorder affecting the eye, anterior uveitis. These results have since been substantiated with significant associations between HLA-B27 and reactive arthritis, juvenile spondyloarthrthritis, psoriatic arthritis and spondylitis, and enteropathic arthropathy (Reveille and Arnett 2005). The underlying pathogenic role of the HLA-B27 molecule in disease pathogenesis remains unresolved but the clinical utility of typing HLA-B27 in the diagnosis of spondyloarthropathies is substantial.

12.6   Genetic variation and haemochromatosis

The classical clinical features of haemochromatosis – diabetes, bronze pigmentation of the skin, and liver cirrhosis – have been recognized as a distinct clinical and pathological entity since the work of Trousseau and Troisier in the 19th century. Haemochromatosis is an iron storage disorder (OMIM 235200) that may be secondary to another disease such as thalassaemia, or constitute a primary illness with no underlying disorder. In 1935 Sheldon proposed that primary haemochromatosis was

an 'inborn error of metabolism' resulting in excess deposits of iron in tissues (Sheldon 1935). The idea that heredity should play an important role in the disease remained controversial but evidence accumulated that not only was an autosomal recessive mode of inheritance likely but that there was association with genetic polymorphism in the MHC, in particular HLA-A3 (Simon *et al.* 1976, 1977). There were data to link hereditary haemochromatosis to within 1–2 cM of *HLA-A*, but finding the actual gene or specific polymorphism proved frustrating until the work of Feder and colleagues in 1996 identified the likely gene and causative mutation (Feder *et al.* 1996).

Feder and coworkers adopted a positional cloning approach (Section 2.3) to identify the gene underlying hereditary haemochromatosis (Feder *et al.* 1996). They were able to define a 250 kb candidate region within the 8 Mb of sequence they took as their starting point by using linkage disequilibrium mapping and haplotypic analysis. Some 45 polymorphic markers were analysed in 101 patients with haemochromatosis and 64 controls. The allele having the highest excess frequency in affected individuals versus controls was defined as 'ancestral'. This allowed reconstruction of the ancestral haplotype on which the authors proposed the causative mutation(s) occurred. A genomic position of maximum linkage disequilibrium was mapped that colocalized to a region they identified by analysis of likely historic recombination events on chromosomes bearing the ancestral haplotype.

This 250 kb candidate region was then carefully analysed to identify genes within it by cDNA selection, exon trapping, and genomic DNA sequencing (Feder *et al.* 1996). This revealed 12 histone genes and three novel genes of unknown function. The products of reverse transcription polymerase chain reaction (RT-PCR) amplified RNA and PCR amplified genomic DNA were then sequenced to try and identify any likely causative mutations in coding DNA of these genes. Two patients with haemochromatosis who were homozygous for the ancestral allele were compared to two control individuals. Eighteen sequence variants were found of which three were nonsynonymous and resulted in an amino acid change in the encoded protein. Two of these lay in the histone H1 gene but one was found in a novel MHC class I-like gene, cDNA 24, now denoted the *HFE* gene.

Feder and colleagues had found the mutation underlying classical hereditary haemochromatosis. In all four patient chromosomes there was a G to A transition at nucleotide 845 of the open reading frame (ORF) which caused a cysteine to tyrosine substitution at amino acid 282 of the encoded protein (rs1800562, c.845G>A, p.C282Y) (Feder *et al.* 1996). In their study population, the p.C282Y variant correlated with the presence of the ancestral haplotype. It was present on 85% of haemochromatosis chromosomes and 3.2% of controls, giving a carrier frequency of 6.4% in their cohort. The nucleotide substitution resulted in a change in a highly conserved residue involved in intramolecular disulphide bridging in other MHC class I proteins and potentially affected the interaction with β2 microglobulin (Fig. 12.10). Of the 178 patients analysed by Feder and colleagues, 148 were homozygous for p.C282Y while nine were heterozygous. They analysed the non-ancestral chromosome present in these heterozygotes and identified a further C to G variant in exon 2 of the gene which results in a histidine to aspartic acid amino acid substitution at position 63 (rs1799945, c.187C>G, p.H63D) (Feder *et al.* 1996).

Subsequent biochemical analysis has shown that the HFE protein does not bind iron but through interaction with transferrin receptor 1 (TfR1) it can facilitate cellular uptake of transferrin-bound iron. HFE is an MHC class I-like protein whose ancestral peptide binding groove is too narrow for antigen presentation. Like other class I molecules, however, HFE still requires interaction with β2 microglobulin for its cell surface expression. The p.C282Y variant, by disrupting the disulphide bond critical for binding to β2 microglobulin, impairs HFE stability, transport, and cell surface expression, thus compromising the interaction of HFE with TfR1 (Waheed *et al.* 1997). Further insights were also gained from studies in mice (Box 12.11).

Primary haemochromatosis is one of the commonest inherited diseases described among people of northern European ancestry. The p.C282Y variant is responsible for the most common form of primary haemochromatosis but the situation is complex, with homozygosity for p.C282Y characterized by low penetrance such that while individuals may be predisposed to the severe phenotype of iron storage disorder it is very difficult to predict whether, and to what extent, the genetic risk will be manifested (Pietrangelo 2004). Mutations of other iron genes have also been identified and found to be involved in 'non-HFE' haemochromatosis such as *TfR2* (encoding transferrin receptor 2), *HAMP* (hepcidin), *HJV* (haemojuvelin), and *FPN* (ferroportin).

The p.C282Y mutation is thought to have arisen relatively recently, perhaps 60–70 generations ago, based on linkage disequilibrium analysis (Ajioka *et al.* 1997). A worldwide study of 2978 people in 1997 showed that the variant was most frequent in northern European populations and did not occur in Africans, Asians, and native Australians (Merryweather-Clarke *et al.* 1997). Further studies suggest that the mutation occurred by chance in a single ancestor of Celtic or Viking descent

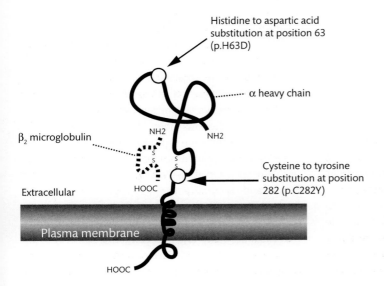

Histidine to aspartic acid substitution at position 63 (p.H63D)

α heavy chain

β₂ microglobulin

NH2   NH2

Cysteine to tyrosine substitution at position 282 (p.C282Y)

Extracellular   HOOC

Plasma membrane

HOOC

Figure 12.10 Schematic of HFE protein. Hypothetical model of HLA-H protein based on similarity to MHC class I proteins as proposed by Feder *et al.* with the positions of the mutations they identified shown. Redrawn and reprinted by permission from Macmillan Publishers Ltd: Nature Genetics (Feder *et al.* 1996), copyright 1996.

Box 12.11   Lessons from murine studies of HFE

The *HFE* gene in mice is structurally identical to that in humans (Riegert *et al.* 1998). This has allowed both knock-out studies and analysis of the effects of specific mutations. *HFE* knock-out mice show impaired iron homeostasis with elevated transferrin saturations and hepatic iron loading (Zhou *et al.* 1998). Iron overloading mimicking human haemochromatosis was seen in knock-in mice homozygous for the C282Y mutation: the effects of the mutation were not, however, to produce a null allele (in other words to give no gene product or product activity) (Levy *et al.* 1999).

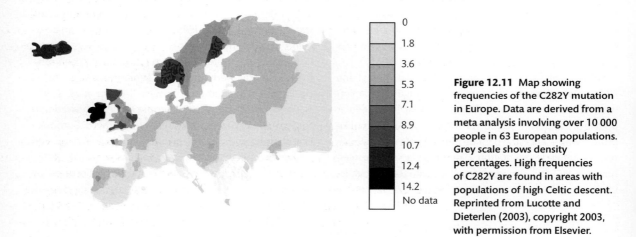

Figure 12.11 Map showing frequencies of the C282Y mutation in Europe. Data are derived from a meta analysis involving over 10 000 people in 63 European populations. Grey scale shows density percentages. High frequencies of C282Y are found in areas with populations of high Celtic descent. Reprinted from Lucotte and Dieterlen (2003), copyright 2003, with permission from Elsevier.

approximately 2000 years ago (Fig. 12.11). The mutation was then passed on and spread by population migration with heterozygote advantage postulated for the individual based on relative protection from iron deficiency and possibly infectious disease. Approximately five people out of every 1000 northern Europeans are now homozygous for p.C282Y. The p.H63D mutation is older, with a higher prevalence and worldwide distribution.

For those individuals homozygous for p.C282Y, a number of biochemical and clinical sequelae may result, culminating in the most severe forms of iron overload (Fig. 12.12) (Pietrangelo 2004). The initial biochemical phase is characterized by slow progressive plasma iron overload, whose effects are rarely manifested before adult-

hood. This can be accelerated by high dietary iron or attenuated by active iron use or loss. The second phase involves progressive iron accumulation in parenchymal tissues before the final stage of organ damage. This is more frequent in men, typically with cirrhotic involvement of the liver, but also potentially causing endocrine problems such as diabetes, cardiac arrhythmias and heart failure, and destructive arthritis of the joints. Screening for HFE haemochromatosis is clinically a very attractive option as therapeutic phlebotomy is safe and effective, reducing morbidity and mortality associated with the disease if instituted early. Further insights into the apparently low penetrance of p.C282Y are, however, required to facilitate this. Why should severe haemochromatosis

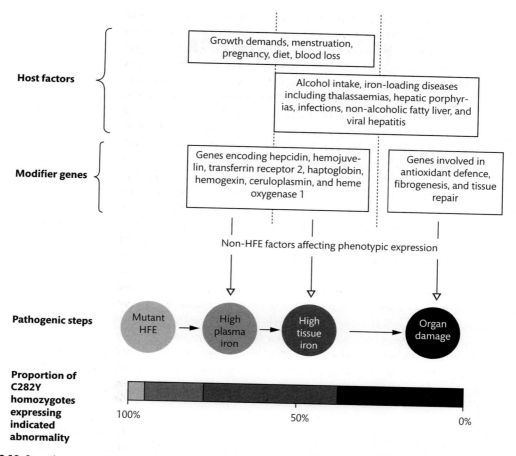

Figure 12.12 Stepwise progression of HFE-related haemochromatosis showing host factors and modifier genes, disease stages, and proportions of p.C282Y homozygotes with resulting abnormalities. Redrawn from Pietrangelo (2004), copyright 2004 Massachusetts Medical Society. All rights reserved.

arise in a small minority of individuals homozygous for p.C282Y while the majority have only a benign biochemical derangement in iron homeostasis? The clinical significance of other mutations of *HFE* such as p.H63D remain unclear, although in a compound heterozygous state with p.C282Y a very small proportion of individuals may be predisposed to disease.

12.7   Many forms of genetic diversity are exhibited by complement C4

Even in the highly polymorphic environment of the human MHC, the genomic landscape of the gene encoding complement component C4 is remarkable. Located within the MHC class III region, the *C4* gene and its flanking sequences vary in gene number, size, and nucleotide composition (Blanchong *et al.* 2001). Such polymorphism is postulated to relate to intense selective pressures to maintain diversity among the C4 proteins to combat a diverse range of pathogens, but heterozygosity has come at a price, with increased risk of gene deletions and inherited disease involving autoimmune and immune complex disorders.

Complement C4 is an essential component of our humoral immune response (Blanchong *et al.* 2001). Following activation, it constitutes a subunit of the C3 and C5 convertases, critical to the classical and lectin activated complement pathways. The presence of activated

C4 on the surface of pathogens aids clearance of immune complexes by immune cells. C4 deficiency is associated with the defective processing of immune complexes, impaired B cell memory, and failure to clear bacterial and viral infections. Telomeric to the *C4* gene are genes encoding other complement proteins constituting the C3 convertases, C2, and factor B.

Polymorphism of the haemolytic activity and electrophoretic mobility of C4 was described in 1969 by Rosenfeld using antibody–antigen crossed electrophoresis (Rosenfeld *et al*. 1969). Two major protein isotypes are found: C4A which is acidic with a lower haemolytic activity, and C4B which is basic and has higher haemolytic activity (Box 12.12) (Awdeh and Alper 1980). The two isotypes also differ in their covalent affinity and serological reactivities (Law *et al*. 1984). Activated C4A will solubilize antibody–antigen aggregates and clear immune complexes through preferentially forming amide bonds with protein antigens such as immune complexes. Following activation, C4B helps propagate activation pathways and formation of the membrane attack complex by reacting to hydroxyl groups to form an ester link with carbohydrate antigens such as bacterial cell walls. Among a Caucasian population, the gene frequency of *C4A* is 55% while *C4B* is 45%. Both may be found in an individual, or either alone.

It is hypothesized that *C4A* evolved from *C4B* after gene duplication based on evolutionary comparisons and analysis (Blanchong *et al*. 2001). Thus while the gorilla has identical *C4A* and *C4B* structure to humans, and other Old World primates have two or more *C4* loci, most mammals have a single *C4* locus that encodes a C4B-like protein.

As well as nucleotide polymorphism, the *C4* gene varies in length and dosage (Yu and Whitacre 2004). The varying length of the *C4* gene (20.6 or 14.2 kb) is due to the integration of the human endogenous retrovirus HERV-K(C4) into intron 9 of the gene (Dangel *et al*. 1994). HERV-K(C4) is typical of endogenous retroviruses, comprising three major genes (gag, pol, and env) flanked by long terminal repeats (LTRs). Among Caucasians, 76% of *C4* genes possess a copy of HERV-K(C4) to give the long form of the *C4* gene while 24% do not. Among African Americans, the proportion having the long form of *C4* is lower with 42% of *C4* genes lacking the insertion. A remarkable level of variation in gene dosage is found for *C4* with just over half of Caucasians possessing four copies (Fig. 12.13); 12–18% of Caucasians have five or six copies of *C4* genes, while among Chinese and Asian Indian people the gene dosage is higher with 24–31% of people having five to seven *C4* genes in a diploid genome. People having higher number of copies of *C4* are associated with higher levels of expression of *C4A* and/or *C4B*. The converse is also true with partial deficiency of *C4A* and *C4B* being the commonest inherited human immune deficiency, present at a combined frequency of 31% in a Caucasian population. Typically these arise due to having only a single *C4A* or *C4B* gene. Partial deficiency may also arise due to the presence of two genes encoding identical C4 isotypes, or by pseudogenes caused by point mutations.

The story of genetic diversity at *C4* then becomes much more complex. The *C4* gene behaves as a modular grouping with three other flanking genes: *RP* (encoding nuclear kinase RP), *CYP21* (cytochrome P450 21 hydroxylase), and *TNX* (tenasein X, an extracellular matrix protein) (reviewed in Blanchong *et al*. 2001). The four contiguous genes form a module called RCCX, such that the four genes are always duplicated together. When the module is duplicated the additional *C4* gene is typically functional, *CYP21* is either a non-functional pseudogene *CYP21A* or

---

Box 12.12   Complement C4A and C4B

The two proteins differ only in four isotype-specific amino acid residues resulting from nucleotide substitutions in exon 26 (PCPVLD in C4A, LSPVIH in C4B) at positions 1101, 1102, 1105, and 1106 (Yu *et al*. 1986). These are critical in determining chemical reactivities: D1106 promotes binding of C4A to immunoglobulin G (IgG) immune aggregates, and H1106 facilitates binding of C4B thioester carbonyl group to hydroxyl group containing antigens. Twenty other polymorphic amino acid residues in C4 protein sequence are known. In total over 41 allotypes for C4A and C4B are proposed (Mauff *et al*. 1998) with diversity in protein levels and functional properties facilitating our intrinsic immune response.

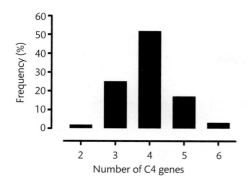

Figure 12.13 Gene dosage for C4 observed in Caucasians. Redrawn from Blanchong *et al.* (2000), copyright Blanchong *et al.*, 2000 (originally published in the *Journal of Experimental Medicine*).

functional *CYP21B* gene, while *RP2* and *TNXA* are fragments and non-functional. Most work has been done on Caucasian populations with between one and three copies of the module found on a given haplotype (Fig. 12.14). A 1–2–3-locus concept has been proposed with over 22 haplotypes of RCCX described comprising different combinations of long and short genes for *C4A*, *C4B*, *CYP21A*, *CYP21B*, *TNXA*, and *TNXB* (Blanchong *et al.* 2000).

The bimodular state is most commonly seen, with most people having a combination of *C4A* and *C4B*. There are, however, ethnic differences (Yu and Whitacre 2004). The bimodular long-long (LL) state is the most common seen in Caucasians and is associated with equal expression of *C4A* and *C4B* but at relatively low levels when C4 protein is measured in the plasma. Among blacks and orientals, long-short (LS) is the most common with higher plasma levels of *C4B* than *C4A*. Monomodular short RCCX with *C4B* alone is found at a relatively high frequency in Caucasians of 10%, but at only 1% in Asians: this variant of the RCCX module is found as part of the 8.1 ancestral haplotype associated with susceptibility to systemic lupus erythematosus (SLE) and type 1 diabetes. In contrast, the bimodular LL RCCX module associated with low expression of *C4A* and *C4B* is found on the 7.2 ancestral haplotype associated with susceptibility to SLE and multiple sclerosis but with protection from type 1 diabetes.

While this diversity has advantages for our immune response, the fact that 69.4% of the population have heterozygous haplotypic combinations in terms of number of modules and gene size promotes misalignment of chromosomes during meiosis, with risk of non-allelic homologous recombination and unequal crossover. This in turn poses a risk of gene deletions and duplications and disease.

The complex possible haplotypic combinations of *C4A* and *C4B* are the major determinants of levels of protein expression. C4 gene dosage, gene size, and body mass index have all been found to be associated with quantitative variation in plasma levels of C4 protein (Yu and Whitacre 2004). Thus people with four copies of the long C4 gene (LL/LL) have 40% lower C4 levels than those having four copies of the gene if they comprise two long and two short versions (LS/LS). This apparent association with the possession of the HERV-K(C4) insertion may be a direct consequence as experiments have shown that sequences from the 3′ LTR of HERV-K(C4) have promoter activity and can direct synthesis of antisense transcripts complementary to human C4 (Mack *et al.* 2004). Serum C4B levels are 40% higher in African Americans than those Americans of European descent (Moulds *et al.* 1991): the former have a high frequency of the short form of C4 lacking the HERV-K(C4) insertion.

There is strong evidence to link deficiencies of C4 proteins with the autoimmune disease SLE. It is striking that in the extremely rare instances where individuals with complete C4A and C4B deficiency have been identified, almost all had SLE or the associated renal or skin symptoms found in lupus (Yang *et al.* 2004b). Among SLE patients in general, low serum C4 protein levels are often seen with two- to five-fold greater rates of homozygous or partial deficiency of C4A than matched control individuals without SLE (Yang *et al.* 2004a). There are, however, ethnic differences reported: among patients with SLE of Spanish, Mexican, or Australian aboriginal origin, a high frequency of C4B deficiency is found. A diverse range of other autoimmune diseases have been associated with polymorphism of C4, including type 1 diabetes, IgA deficiency, common variable immunodeficiency, IgA nephropathy, vitiligo, and pemphigus vulgaris. Elsewhere in the RCCX module, mutations or deletions of CYP21B cause congenital adrenal hyperplasia (Miller and Morel 1989).

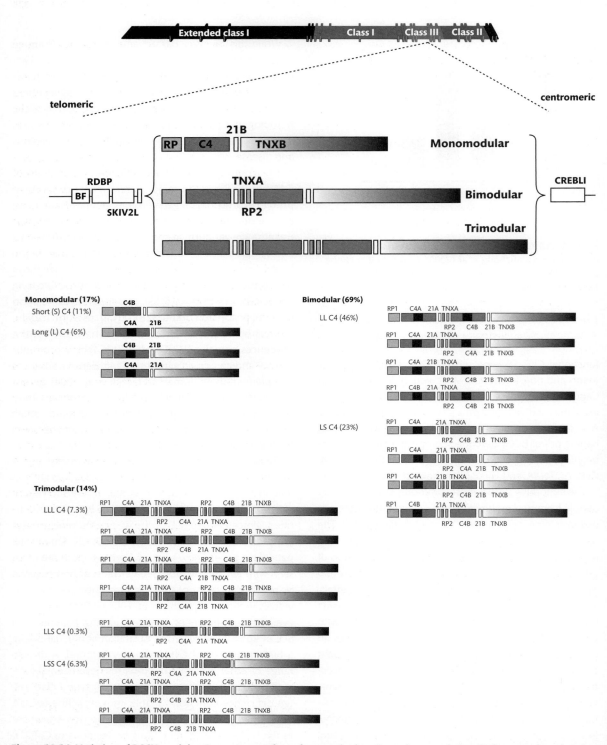

Figure 12.14 Variation of RCCX modules. Percentages refer to frequencies in a Caucasian population. Redrawn and reprinted from Chung *et al.* (2002), copyright 2002; and from Blanchong *et al.* (2001), copyright 2001, with permission from Elsevier.

12.8 A SNP modulating the splicing of the *BTNL2* gene is associated with sarcoidosis

The final example of genetic diversity within the MHC and its functional consequences is the result of many years of investigation and provides an elegant illustration of how regulation of alternative splicing by a SNP can modulate susceptibility to a complex disease (see also Section 11.6.2).

Sarcoidosis is a rare chronic inflammatory disorder (Box 12.13) which has been recognized for many years to show familial clustering, suggesting a genetic basis. For example, a large case–control study of sarcoidosis that included over 10 000 first degree and 17 000 second degree relatives showed that cases with sarcoidosis were five times more likely than controls to have a sibling or parent with a history of the disease (Rybicki *et al.* 2001). A number of candidate genes showed association with susceptibility to sarcoidosis, including within the MHC, but further evidence to implicate genetic diversity in the MHC was to come from genome-wide analysis.

Schreiber and coworkers began by carrying out a genome-wide linkage analysis of families with sarcoidosis (Schurmann *et al.* 2001). They used microsatellite markers to try and localize genomic regions in which genetic variation was linked to disease susceptibility. Their analysis of 63 German families with affected siblings showed significant linkage to chromosome 6p21, a genomic interval that included the MHC. Other chromosomal regions also showed evidence of linkage, indicating multiple loci may be involved. The researchers proceeded to fine map the disease association at the MHC using a denser set of genetic markers and by recruiting affected individuals and families to allow analysis using both case–control and family-based approaches (Fig. 12.15) (Valentonyte *et al.* 2005).

Fine mapping involved progressively narrowing down the association using different panels of SNPs and patients. Initially the 16.3 Mb region corresponding to the 99% confidence region for linkage from the genome scan was analysed using 69 SNPs in 372 patients, with a haplotype-based approach using both transmission disequilibrium testing (TDT) in families and case–control analysis. Consistent associations were seen using the two analysis methods for the *BTNL2* subregion while the *MICB* locus showed an effect only on case–control analysis. Logistic regression using further genetic markers to differentiate primary from secondary effects showed the association to only hold up for the *BTNL2* region.

Further fine mapping narrowed the associated region from 440 to 15 kb through analysis of 620 patients using 48 SNPs from published databases and resequencing of affected individuals. This focused the attention of the investigators to the 3′ end of *BTNL2* where the associated 15 kb region was noted to contain four functional SNPs, three of which lead to amino acid substitutions. However it was rs2076530, a G to A transition, which showed the most significant association, an effect that held up on independent analysis in a replication cohort of 462 cases and 876 controls. A multiplicative effect was seen, strongest in individuals homozygous for the rarer A allele (AA versus GG Odds Ratio, OR = 2.75, 95% CI 1.86–4.05) and present to a lesser extent in the heterozygous state (AG versus GG OR 1.6, 95% CI 1.1–2.32). This highly statistically significant, but relatively modest, individual increased disease risk with the polymorphic allele is substantial at a population level, with an attributable risk of 23% for heterozygotes and homozygotes.

Box 12.13 Sarcoidosis (OMIM 181000)

Sarcoidosis is a disease whose pathogenesis has long remained obscure. The disease can involve many different parts of the body where the characteristic lesion seen is of non-caseating granulomas. The cellular immune response of affected individuals is exaggerated, with the levels of inflammation ranging from sub-clinical to acute (Loefgrens syndrome). About half of cases spontaneously resolve, while others may progress to debilitating lung fibrosis and chronic respiratory failure.

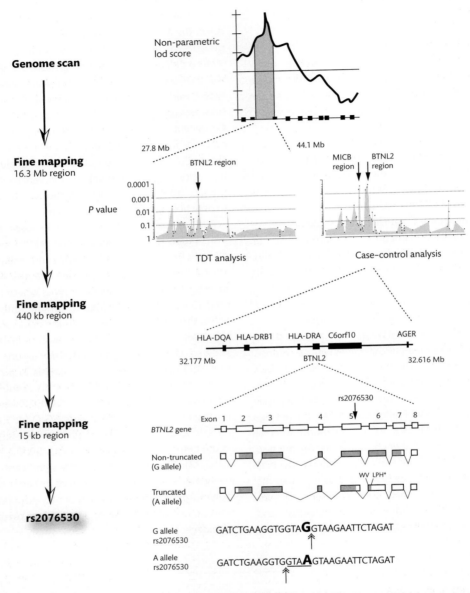

Figure 12.15 Identification of a functional polymorphism of *BTNL2* in sarcoidosis. The stepwise approach from genome-wise linkage analysis to fine mapping is shown. The exonic structure of *BTNL2* is shown with the site of the functional SNP, rs2076530, indicated. Shaded regions denote coding sequence. The four underlined base pairs are those excluded from the spliced mRNA transcript derived from the A allele, resulting in recruitment of an alternative splice site, a frameshift, and premature stop after five codons. Redrawn and reprinted from Traherne *et al.* (2006a), with permission from Oxford University Press; and from Macmillan Publishers Ltd: Nature Genetics (Valentonyte *et al.* 2005), copyright 2005.

That the increased risk associated with the A allele is not greater becomes an important question when the molecular mechanism by which the polymorphism exerts its effects is revealed. rs2076530 lies in exon 5 of the *BTNL2* gene. Schreiber and colleagues demonstrated that the A allele leads to recruitment of an alternative splice site 4 bp upstream of the polymorphism, truncating the A allele (Valentonyte *et al.* 2005). The loss of four bases from the cDNA causes frameshift and premature stop in the downstream exon, leading to loss of the C-terminal IgC domain and transmembrane helix. The functional consequences of this were investigated by cloning the transcripts into a mammalian expression vector: the presence of green fluorescent protein fusions on the transcribed proteins allowed subcellular localization of the wild type and truncated proteins to be visualized in transfected cells. This showed protein derived from the truncated A allele failed to localize to the membrane, being present instead on cytoplasmic vesicular structures.

Why then is there not a greater effect seen in terms of disease risk, given this profound functional effect? The answer may lie in the redundancy of the system within which BTNL2 operates. BTNL2 is a member of the immunoglobulin superfamily, and shows evidence of immunoregulatory activity. Our understanding of the biological role of the molecule encoded by BTNL2 is still preliminary but putative receptors for BTNL2 have been identified on activated T and B cells and the molecule itself has been shown to inhibit T cell proliferation (Nguyen *et al.* 2006). Recent studies in mice have confirmed this effect of the downregulation of T cell activation, with the highest levels of expression seen in the digestive tract (Arnett *et al.* 2007). Other molecules have similar negative costimulatory effects and may to some extent compensate for any loss of function caused by polymorphisms such as rs2076530.

The association has since been replicated in an independent German case–control cohort (Li *et al.* 2006d) and an American study by Rybicki and colleagues (2005), although the situation is complicated. The latter study demonstrated that diversity in a 490 bp region spanning exon 5 of *BTNL2* showed association with risk of sarcoidosis. However, significant ethnic differences were observed, with rs2076530 showing the strongest association in the white case–control cohort studied. Indeed, among African Americans, the association with rs2076530 was not statistically significant and complex interactions with HLA class II alleles were postulated by the authors. This functional polymorphism has also been investigated in other autoimmune diseases as a candidate SNP modulating disease susceptibility. Associations were noted with Graves' disease, type 1 diabetes, rheumatoid arthritis, SLE, and multiple sclerosis but were attributable to linkage disequilibrium with DQB1-DRB1 alleles (Orozco *et al.* 2005; Simmonds *et al.* 2006; Traherne *et al.* 2006a).

12.9   Summary

The MHC has been a focus of intensive research effort for over 50 years and continues to reveal new and unexpected secrets. Study of the MHC has provided novel insights into human origins and ancestry, the nature of human genetic variation, the likely mechanisms operating to generate and maintain it, and the role of such variation in determining disease susceptibility. Extreme polymorphism and linkage disequilibrium have thwarted many efforts to fine map disease association and resolve specific causative variants in the MHC. However, large scale and high density SNP mapping, insights into linkage disequilibrium and haplotypic structure combined with functional studies of molecular mechanisms whereby genetic variants may act, offers the promise of unlocking these secrets.

Structural biology, notably of HLA class II molecules, has provided important insights, for example with narcolepsy and type 1 diabetes associated alleles. Causation has been established for specific gene variants, notably the p.C282Y mutation of the *HFE* gene in hereditary haemochromatosis. Functionality through modulation of alternative splicing at *BTNL2* has been associated with disease risk in sarcoidosis, a mechanism likely to be seen as increasingly common and important in future studies. Elsewhere structural genomic variation in the form of copy number variation, for example at complement C4, has been found to be complex and to provide major determinants of gene expression and autoimmune disease risk. The future for understanding the nature and consequences of genetic diversity at the MHC is looking increasingly bright as new tools and resources become available, not least the remarkable efforts to resequence the MHC and provide maps of genetic diversity for common disease associated haplotypes. Work in the MHC has highlighted the many ways that genetic variation may exert a functional effect, whether at the level of gene expression, control of alternative splicing or alterations in the structure and function of encoded proteins. Further levels of complexity relating to epigenetic and epistatic mechanisms together with the context specificity of prevailing environmental factors are likely to be important with resolution of specific functionally important genetic variants requiring careful characterisation in the relevant cell type and disease context.

12.10   Reviews

Reviews of subjects in this chapter can be found in the following publications:

Topic	References
MHC genes, disease association, and biology	Xie *et al.* 2003; Horton *et al.* 2004; Larsen and Alper 2004; Fernando *et al.* 2008
MHC and antigen presentation	Watts 2004; Cresswell 2005; Cresswell *et al.* 2005
MHC haplotypes	Degli-Esposti *et al.* 1992; Price *et al.* 1999; Allcock *et al.* 2002
Structural biology and understanding of MHC class II disease associations	Jones *et al.* 2006
Rheumatoid arthritis and the MHC	Newton *et al.* 2004, Coenen *et al.* 2009
Narcolepsy	Maret and Tafti 2005
Pathophysiology and genetics of coeliac disease	Kagnoff 2007
Genetics of type 1 diabetes	Hirschhorn 2003; Kim and Polychronakos 2005; Ounissi-Benkalha and Polychronakos 2008
Genetics and pathogenesis of spondyloarthritis including ankylosing spondylitis	Reveille and Arnett 2005; Reveille 2006; Brown 2008
Genetics, clinical features, and pathogenesis of haemochromatosis	Brissot 2003; Pietrangelo 2004; Beutler 2006
Complement gene diversity in biology and disease	Blanchong *et al.* 2001; Yu and Whitacre 2004

CHAPTER 13

Parasite wars

13.1   Introduction

Infectious disease has provided a major selective pressure on genetic diversity in human populations. The idea that genetic factors might significantly influence our susceptibility to infectious diseases seems at first improbable. However, before the awareness of disease-causing microbes towards the end of the 19th century, many people considered diseases such as tuberculosis and leprosy to have an inherited basis (Cooke and Hill 2001). Evidence from twin studies has supported this idea: for tuberculosis, monozygotic twins have a concordance rate of 62% compared to 18% in dizygotic twins; for leprosy concordance rates of 52% versus 22% have been reported (Cooke and Hill 2001). A landmark study of causes of mortality among Danish children born in the 1920s and placed in the care of unrelated adoptive parents highlighted the importance of genetic factors in determining risk of death from infectious disease (Sorensen *et al.* 1988). Early death of a biological parent before the age of 50 years from an infection was associated with a relative risk of death of the adopted child from an infectious disease of 5.8 (95% CI 2.5–13.7) while no effect on mortality risk was seen with the death of an adoptive parent from an infectious cause.

In this chapter the relationship between parasitic disease and human genetic variation is reviewed. As noted in Chapter 10, malaria infection has provided a remarkably powerful selective pressure among human populations with a range of different genetic variants showing complex relationships with disease susceptibility or resistance. Here a number of examples are described, highlighting how particular mendelian traits such as

thalassaemia and structural variants of haemoglobin have been driven to high frequency in populations at risk of malaria (Section 13.2). As with a number of different infectious diseases, insights have also been gained by investigating specific population groups showing a difference in disease susceptibility despite living in the same geographic area and being exposed to the same environmental risk of disease. A number of other parasitic diseases show strong evidence of modulation by host genetic diversity, notably leishmaniasis (Section 13.3) and helminth infections (Section 13.4).

Parasitic diseases have provided some of the clearest examples of genetic association for any infectious disease, often with remarkably strong effects given that these are common complex multifactorial traits. In a review published in 2006, Hill described the 'big six' disease associations in human infectious disease genetics known at that time, all involving common variants (present at greater than 1% allele frequency) with odds ratios greater than 5 (Hill 2006). Three involved malaria: haemoglobin S (Hb S) and *Plasmodium falciparum* (Section 13.2.3), Duffy antigen and *P. vivax* malaria (Section 13.2.4), and South East Asian ovalocytosis. The latter arises due to a deletion in the *SLC4A1* gene on chromosome 17q21-q22 and is common in endemic areas for *P. falciparum* in the western Pacific such as Papua New Guinea, where it is strongly associated with protection from cerebral malaria (Genton *et al.* 1995; Allen *et al.* 1999). The remaining disease associations noted by Hill included homozygosity for the *CCR5* Δ32 deletion and human immunodeficiency virus (HIV) infection (Section 14.2.1); a coding polymorphism of the prion protein gene *PRNP* at codon 129 (c.385A>G) resulting in substitution of valine for methionine, which modulates

341

susceptibility to prion diseases, most strikingly variant Creutzfeld–Jacob disease (Palmer *et al.* 1991; Zeidler *et al.* 1997; Mead 2006); and a nonsense variant of the *FUT2* gene providing resistance to symptomatic gastroenteritis due to Norwalk virus (Lindesmith *et al.* 2003; Thorven *et al.* 2005).

13.2 Malaria, genetic diversity, and selection

13.2.1 Inherited factors and resistance to malaria

Malaria is one of the most important infectious diseases in the world and a leading cause of death, particularly among children. Genetic variation in man is an important determinant of disease susceptibility but has to be considered as part of a remarkably complex multifactorial disorder involving diversity in the mosquito vector, parasites, and environment as well as our own immune response (Box 13.1). The complexity of the relationship between parasite and host, involving multiple points of attack and defence, has left evidence in the human genome of the profound selective pressure malaria has exerted on human populations at risk of the disease living in tropical and subtropical regions of the world. Inherited factors providing a selective advantage are thought to have driven specific human alleles to high frequency in malaria endemic regions providing some of the clearest examples of genomic 'signatures of selection' (Section 10.2) as well as being manifest in the observed global distribution of monogenic diseases such as thalassaemia (Section 13.2.2) and sickle cell disease (Section 13.2.3).

Particular ethnic groups are also seen to have different inherent levels of resistance to malaria in the same geographic location, as seen among the Tharu in Nepal (Section 13.2.2) and Fulani people of West Africa (see Box 13.3). Specific genetic variants associated with malaria resistance are found in some cases to be localized to particular regions. For example there is evidence that a recent mutation of *HBB* resulting in haemoglobin C (Hb C) has risen to relatively high allele frequency in a particular region of central West Africa where, among the Dogon people for example, it is associated with protection from malaria due to *P. falciparum* (Section 13.2.3).

Other variants such as a single nucleotide substitution in the *DARC* gene associated with the Duffy antigen receptor have become more widely established: the variant confers complete resistance to *P. vivax* malaria and is at or near fixation in most West and Central African populations but very rare outside Africa (Section 13.2.4). Malaria has had a relationship with man since prehistory; indeed similar parasite species are found among reptiles and birds as well as other primates. However, malaria is thought to have become a major human pathogen only over the last 10 000 years when a sudden increase in the African malaria parasite population was found to have occurred, concurrent with the development of resident agricultural societies (Kwiatkowski 2005).

The contribution of human genetic diversity to the risk of malarial infection and severe illness is only one of many interrelated factors, which also include the intensity and nature of disease transmission and exposure, genetic variation in the parasite, and the virulence of the particular strain. Other factors include the ecology of the mosquito vector and acquired immunity developed by an individual living in a malaria endemic region, as well as coinfection and multiple environmental factors of which socioeconomic status is very important. A large longitudinal study of mild clinical malarial illness and disease requiring hospital admission among children in the Kilifi District of coastal Kenya has shown that genetic factors account for approximately one-quarter of the variation observed between children in their susceptibility to malaria, equal to the proportion of variability explained by household factors such as spatial distribution in mosquito breeding sites, and use of insecticide and insect repellents (Mackinnon *et al.* 2005). Such a study to estimate the heritability of malarial infection was possible due to careful epidemiological analysis within and between households of individuals of varying genetic relatedness. The estimate of genetic factors accounting for 25% of variation in the incidence of malaria among an African population of children living in a malaria endemic area was broadly similar to that observed among a mainly adult population in Sri Lanka, where disease transmission and incidence was significantly lower, and *P. vivax* as well as *P. falciparum* contributed to the clinical cases (Mackinnon *et al.* 2000).

In the following sections examples of the relationship between genetic variation and malaria are described to

Box 13.1 Malaria

Malaria is the most important parasitic disease of man. The figures are stark. There were an estimated 515 million episodes of clinical malaria due to *Plasmodium falciparum* in 2002 and more than 1 million deaths globally per annum (Snow *et al.* 2005). Some 2.37 billion people were estimated to be at risk of *P. falciparum* transmission in 2007 (Fig. 13.1) (Guerra *et al.* 2008). The major disease burden falls in Africa where 70% of clinical events are thought to occur, with the majority of deaths among children (Snow *et al.* 2005). *P. falciparum* is one of four different *Plasmodium* species known to infect man, the others being *P. vivax*, *P. ovale*, and *P. malariae*. The species differ in their virulence, length of life cycle, and red blood cell preferences. The major mosquito vector of *P. falciparum* in sub-Saharan Africa is *Anopheles gambiae*. Transmission of the sporozoite stage of the parasite to the human host occurs by the bite of an infected mosquito; this is followed by invasion and replication in liver cells and the release of merozoites, which invade red blood cells and multiply with fever and vital organ damage. Gametocytes from infected red cells may then be taken up during a blood meal by a mosquito where they multiply and complete the life cycle of the parasite by developing into sporozoites (Fig. 13.2). The most severe manifestations of malaria are seen with *P. falciparum* where cerebral malaria, severe malarial anaemia, and respiratory distress may result, in some cases with fatal results. It is, however, only a small minority of individuals who develop the life threatening forms of the disease. Major advances have been achieved through malaria control programmes but malaria remains a major global public health problem with ongoing efforts to coordinate and intensify such work, for example through the Roll Back Malaria Partnership launched in 1998 (www.rbm.who.int).

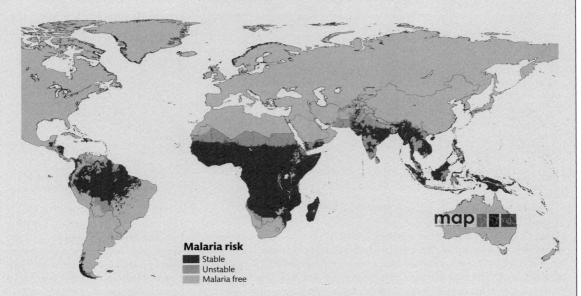

Malaria risk
■ Stable
■ Unstable
■ Malaria free

Figure 13.1 *Plasmodium falciparum* global distribution and malaria risk. Stable transmission refers to a *P. falciparum* annual parasite incidence of 0.1 cases per thousand people per annum; unstable are areas of extremely low transmission (<0.1 per thousand per annum). Reprinted with permission from the Malaria Atlas Project (www.map.ox.ac.uk) (Hay and Snow 2006; Guerra *et al.* 2007, 2008).

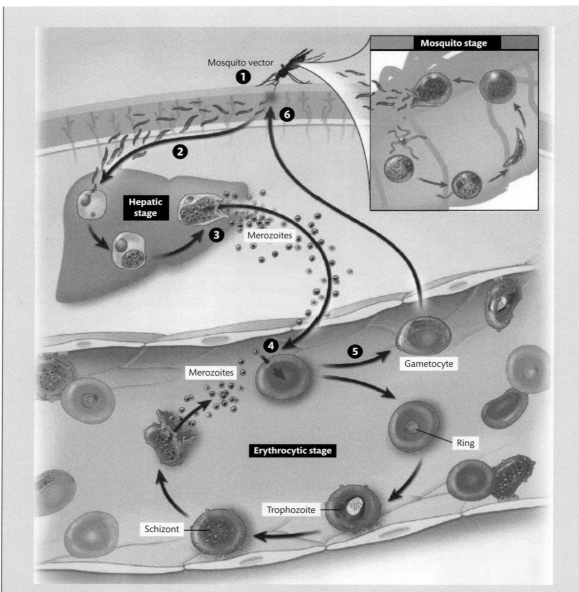

Figure 13.2 Life cycle of human malaria parasites. (1) During the blood meal of the female *Anopheles* mosquito, sporozoites are injected into the human capillaries. (2) Sporozoites can be found in the blood for about 30 minutes, most are destroyed but some invade the cells of the liver. (3) After about a week, infected liver cells containing mature schizonts burst releasing thousands of tiny merozoites into the blood. (4) Merozoites invade red blood cells and develop over 48–72 hours to produce more merozoites. (5) After several generations, some merozoites develop into sexually differentiated forms (gametocytes). (6) Male and female gametocytes are taken up by the mosquito when feeding, gametes fuse in the mosquito gut, and the zygote (ookinete) penetrates the gut wall to form an oocyst. Ten to 14 days later thousands of sporozoites are released from the oocyst and travel to the salivary glands. Redrawn from Rosenthal (2008), copyright 2008 Massachusetts Medical Society. All rights reserved.

highlight how this relationship has been explored and the remarkable insights it has afforded (Fig. 13.3). A large number of variants relate to the red blood cell where malaria has exerted powerful selective pressures, notably involving the α and β globin gene loci with structural variants of haemoglobin such as Hb S, Hb C, and Hb E as well as the thalassaemias providing resistance to infection. Genetic variation involving receptors for parasite entry to red cells and red cell metabolism have also provided substrate on which selection has acted. Diversity in major histocompatibility complex (MHC) molecules involved in the presentation of parasite proteins in liver cells, and polymorphism of immune genes and of receptors involved in the process of cytoadherence of infected red cells, provide further examples of the complexity of the relationship between parasite and host, and the points at which advantage may be gained through the possession of particular variants. A number of excellent reviews address the role of genetic factors in malaria in more detail (Kwiatkowski 2005; Williams 2006; Weatherall 2008).

13.2.2 Thalassaemia, natural selection and malaria

Thalassaemias are inherited disorders causing defective and imbalanced synthesis of globin, first described in the 1920s among children of Mediterranean origin with severe anaemia. This group of disorders are now recognized as the most common single gene diseases found in man, whose consequences range from being asymptomatic to lethal. In Section 1.3 the nature and molecular basis of α and β thalassaemia were reviewed. The name thalassaemia derives from the Greek for the sea ('*thalassa*') and blood ('*aima*'), reflecting the high frequency of thalassaemia in Italian and Greek coastal populations. The occurrence of thalassaemia is however much more extensive, being found in the Middle East, Africa, India, and South East Asia. A correlation between the geographic distribution of thalassaemia and the incidence of malarial infection was noted (Fig. 13.4), with Haldane proposing that heterozygotes may have a selective advantage compared to those without the variant (Haldane 1949). Haldane noted that the red blood cells of carriers of thalassaemia might be more resistant to malarial infection, with the resulting heterozygote advantage leading to increased frequency of thalassaemia gene variants in the population balanced by the selective disadvantage of the homozygous state (Haldane 1949; Weatherall 2008). Evidence to support

the concepts of balanced polymorphism and the malaria hypothesis continue to accumulate with further elegant examples, as will be described for the structural haemoglobin variant, sickle haemoglobin (Section 13.2.3).

Individuals with α^+ thalassemia have reduced production of α chains due to one of the two *HBA* genes in the α globin locus on chromosome 16p13.3 being deleted or inactivated by specific mutations. At a molecular level such individuals are either heterozygotes or homozygotes for deletion events (denoted $-\alpha/\alpha\alpha$ and $-\alpha/-\alpha$ respectively) or particular mutations ($\alpha^T\alpha/\alpha\alpha$ and $\alpha^T\alpha/\alpha^T\alpha$); the haematological phenotype associated with this varies from being clinically silent (heterozygotes) to mild anaemia (homozygotes) although is more severe in non-deletional forms of α^+ thalassaemia. In α^0 thalassaemia there are no α chains produced from a given chromosome as both linked genes are deleted; affected individuals may be either heterozygous $--/\alpha\alpha$ or homozygous $--/--$ for α^0 thalassaemia and this is associated with a severe or fatal phenotype (Box 1.15). (Section 1.3) (Weatherall 2008). In the mountains and swamps of Papua New Guinea, and the islands of the southwest Pacific, a clear relationship was found between the endemicity of malaria and the frequency of deletional forms of α^+ thalassaemia (Flint *et al.* 1986). In Papua New Guinea, rates of malaria were low above 1500 m and the frequency of α^+ thalassaemia dropped ten-fold from northern coastal areas into the highlands. Moreover, a clear progressive reduction in malarial endemicity was observed on progressing south and east from coastal Papua New Guinea until the disease was absent in New Caledonia and Fiji (Fig. 13.5). This was mirrored by the frequency of α^+ thalassaemia, which fell from almost 70% in the northern coastal regions of Papua New Guinea to less than 10% in New Caledonia in the southeast.

In the southern lowlands of the Terasi region of Nepal, rates of malarial infection were so high that it was considered for a long time uninhabitable by most Nepalese. The Tharu people, however, appeared to be naturally immune and had lived there for generations. As malaria control measures improved and other ethnic groups settled in the area, the prevalence of cases of residual malaria were noted to be seven times higher in the non-Tharu population (Terrenato *et al.* 1988). Genetic studies showed almost all Tharu people have α^+ thalassaemia in the heterozygous or homozygous state, with most homozygous $(-\alpha/-\alpha)$, among whom morbidity from malaria was estimated to

Common erythrocyte variants that affect resistance to malaria

Gene	Protein	Function	Reported genetic associations with malaria
FY	Duffy antigen	Chemokine receptor	FY*O allele completely protects against *P. vivax* infection
G6PD	Glucose-6-phosphate dehydogenase	Enzyme that protects against oxidative stress	G6PD deficiency protects against severe malaria
GYPA	Glycophorin A	Sialoglycoprotein	GYPA deficient erythrocytes are resistant to invasion by *P. falciparum*. Also seen with GYPB and GYPC
HBA	α globin	Component of haemoglobin	α⁺ thalassaemia protects against severe malaria but appears to enhance mild malaria episodes in some environments
HBB	β globin	Component of haemoglobin	Hb S and Hb C alleles protect against severe malaria. Hb E allele reduces parasite invasion
HP	Haptoglobin	Haemoglobin binding protein present in plasma (not erythrocyte)	Haptoglobin 1-1 genotype is associated with susceptibility to severe malaria in Sudan and Ghana

Immune gene associations with malaria

Gene	Protein	Function	Reported genetic associations with malaria
FCGR2A	CD32, low affinity receptor for Fc fragment of IgG	Clearance of antigen–antibody complexes	Association with severe malaria in The Gambia
HLA-B	HLA-B, a component of MHC class I	Antigen presentation that leads to cytotoxic T cells	HLA-B53 association with severe malaria in The Gambia
HLA-DR	HLA-DR, a component of MHC class II	Antigen presentation that leads to antibody production	HLA-DRB1 association with severe malaria in The Gambia
IFNAR1	Interferon α receptor component	Cytokine receptor	Association with severe malaria in The Gambia
IFNG	Interferon γ	Cytokine with antiparasitic and proinflammatory properties	Weak associations with severe malaria in The Gambia
IFNGR1	Interferon γ receptor component	Cytokine receptor	Association with severe malaria in Mandinka people of The Gambia
IL1A/IL1B	Interleukin-1a and -1b	Proinflammatory cytokines	Marginal associations with severe malaria in The Gambia
IL10	Interleukin-10	Anti-inflammatory cytokine	Haplotypic association with severe malaria in The Gambia
IL12B	Interleukin-12 β subunit	Promotes development of Th1 cells	Association with severe malaria in Tanzania
IL4	Interleukin-4	Promotes antibody producing B cells	Association with antimalarial antibody levels in Fulani people of Burkina Faso
MBL2	Mannose-binding protein	Activates classic complement	Association with severe malaria in Gabon
NOS2A	Inducible NO synthase	Generates NO, a free radical	Various associations with severe malaria in Gabon, The Gambia, and Tanzania
TNF	Tumour necrosis factor	Cytokine with antiparasitic and proinflammatory properties	Various associations with severe malaria and reinfection risk in the Gambia, Kenya, Gabon, and Sri Lanka
TNFSF5	CD40 ligand	T cell–B cell interactions leading to immunoglobulin class switching	Association with severe malaria in The Gambia

Host molecules mediating cytoadherence by parasitized red blood cells

Gene	Protein	Reported genetic associations with malaria
CD36	CD36 antigen, thrombospondin receptor	CD36 polymorphisms show variable associations with severe malaria in The Gambia, Kenya, and Thailand
CR1	CR1, complement receptor 1	CR1 polymorphisms show variable associations with severe malaria in The Gambia, Thailand, and Papua New Guinea
ICAM1	CD54, intercellular adhesion molecule-1	ICAM1 polymorphisms show variable associations with severe malaria in Kenya, Gabon, and The Gambia
PECAM1	CD31, platelet–endothelial cell–adhesion molecule	PECAM1 polymorphisms show variable associations with severe malaria in Thailand, Kenya, and Papua New Guinea

Figure 13.3 Genetic variation associated with malaria. Reprinted from Kwiatkowski (2005), copyright 2005, with permission from Elsevier.

(A)

(B)

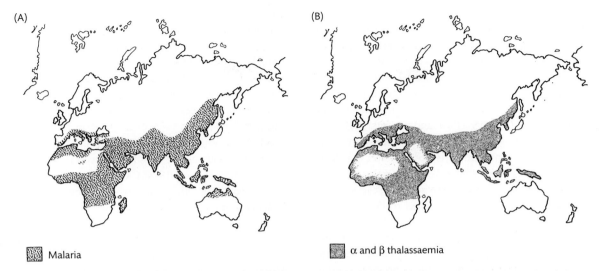

Malaria

α and β thalassaemia

Figure 13.4 Geographic relatedness of malaria and thalassaemia in Europe, Africa, and Asia. **(A)** The distribution of malaria (prior to major control programmes). **(B)** The distribution of α and β thalassaemia. Reprinted with permission from Weatherall (2008) .

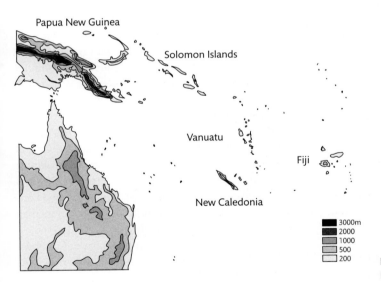

Figure 13.5 Topography of the southwest Pacific region.

be reduced by about ten-fold (Modiano *et al.* 1991). The allele frequency was estimated at 0.8; it is suggested that it has not reached fixation only because of intermarriage with other ethnic groups (Modiano *et al.* 1991).

In the coastal region of northern Papua New Guinea the carrier frequency of α⁺ thalassaemia varies between ethnic groups but in some groups was very high (Madang 0.68, Sepik 0.87) (Allen *et al.* 1997). A prospective case–control study in this region showed that α⁺ thalassemia in either the heterozygous or homozygous form was protective against severe malaria, as well as other infectious diseases (Allen *et al.* 1997). Odds ratios of 0.4 (0.22–0.74) and 0.66 (0.37–1.2) were found for homozygotes and heterozygotes, respectively (Allen *et al.* 1997). Among African populations protection from severe malaria was also observed (Mockenhaupt *et al.* 2004; Williams *et al.* 2005e). In coastal Kenya, for example, a study of 655 cases of severe malaria and 648 controls showed protection associated with heterozygotes for α^+ thalassemia with an odds ratio of 0.73 (0.57–0.94) and for homozygotes of 0.57 (0.40–0.81) versus controls; risk of death with severe malaria was also significantly reduced, by 40% and more than 60%, respectively (Williams *et al.* 2005e).

How α^+ thalassemia acts to protect against severe malaria is unclear, it is not thought to involve effects on parasite growth or invasion but rather to involve differences in the red blood cell membrane and the immune response to infected cells (Weatherall 2008). It remains a complicated story. On the island of Espiritu Sano in Vanuatu in the southwest Pacific a paradox was seen with a higher incidence of uncomplicated malaria and prevalence of splenomegaly among children with α^+ thalassemia, notably among younger children and those with *P. vivax* (Williams *et al.* 1996). Selection was postulated based on a possible cross species selective advantage in terms of immunity gained through infection with *P. vivax* providing protection for the more severe *P. falciparum* infection.

13.2.3 Malaria and structural haemoglobin variants

The single nucleotide substitution responsible for haemoglobin S (Hb S) is a further striking example of a balanced polymorphism in which a relatively high allele frequency is driven and maintained by the selective advantage conferred by possession of one copy of the variant allele (leading to sickle cell trait) in areas where human populations are exposed to the risk of malarial infection, counterbalanced by the deleterious effects of being homozygous for the Hb S allele (resulting in sickle cell disease). Hb S results from a single amino acid substitution, from glutamic acid to valine, due to an A to T transversion (rs334, c.20A>T) in the coding sequence of *HBB* in the β globin locus at chromosome 11p15.5 (Section 1.2). The variant is common in particular populations worldwide, notably in sub-Saharan Africa, the Middle East, and India with carrier frequencies of between 5% and 40% reported, and evidence that the mutation has occurred

at least twice over our recent past (Pagnier *et al.* 1984; Weatherall 2008).

Early observations of a reduced prevalence of parasitaemia on peripheral blood films and of enlarged spleens on clinical examination among individuals with sickle cell trait in Zimbabwe, together with reduced incidence of malaria among mine-workers with the trait, were suggestive of a protective role (Beet 1946; Brain 1952). In 1954, Allison reported that among Ugandan children the incidence of parasitaemia and parasite density was lower in those with sickle cell trait (Allison 1954b). Moreover, on a malaria parasite challenge study of adults who had been away from a malarial area for at least 18 months using two strains of *P. falciparum*, a striking difference was seen with 14 of 15 individuals without sickle trait becoming infected in contrast to only two of 15 with the trait, both with low parasitaemia (Allison 1954b). Allison also noted the geographic variation of the incidence of sickle cell trait within East Africa, being high in areas of hyperendemic malaria such as around Lake Victoria, in contrast to highland and other areas that were either malaria-free or subject to malaria epidemics where low frequencies were found (Allison 1954a).

Many studies have since shown association with protection from malaria for individuals with sickle cell trait (often denoted Hb AS) compared to individuals without a copy of the variant (Hb AA). For example Willcox and colleagues observed that the frequency of Hb AS was 1.8% among 558 patients attending hospital with malaria compared to 7.2% in the local population ($P < 0.001$) (Willcox *et al.* 1983). Similar results were found in a large case–control study of 619 children with severe malaria in The Gambia by Hill and colleagues: the frequency of Hb S carriers was 1.2% compared to 12.9% among 510 mild malaria controls giving a relative risk of 0.08 (0.04–0.16), or more than 90% protection from severe malaria (Hill *et al.* 1991).

More recent studies from East Africa have also shown evidence of protection for individuals with sickle cell trait. A large birth cohort study in Kenya analysing survival between 2 and 16 months after birth showed reduced all-cause mortality associated with Hb AS compared to Hb AA with protection from severe malarial anaemia (Aidoo *et al.* 2002). Further insights were gained from the work of Williams and colleagues using large cohort studies of children in an area endemic for malaria on the coast of Kenya where 15% of the population has Hb AS. They demonstrated that the protective effect was specific to *P. falciparum* infection rather than to other childhood diseases such as respiratory infections or gastroenteritis (Williams *et al.* 2005c). The data confirmed previous studies showing no association with symptomless parasitaemia but rather increasing degrees of protection dependent on the severity of illness: 50% protection for mild clinical malaria, 75% for admission with malaria to hospital, 86% for cerebral malaria, and nearly 90% for severe malarial anaemia. Parasite density was lower in cases of clinical malaria, four-fold lower in those with severe disease admitted to hospital for individuals Hb AS compared to Hb SS (Williams *et al.* 2005c). For mild clinical malaria, a clear age effect was seen with protection rising from 20% in the first 2 years of life, to 56% among children up to 10 years of age, then falling to 30% (Williams *et al.* 2005b).

Perhaps most intriguingly, the significant protective effects associated with Hb AS and with homozygosity for the α^+ thalassaemia allele were lost when they were inherited together (Williams *et al.* 2005d). This appears to be an example of negative **epistasis** whereby the effect of an allele at one genomic locus depends on a genotype coinherited at a second unrelated locus. Such effects may be much more common than we appreciate. Why this should occur is unclear. It may be that the concentration of haemoglobin falls due to differences in binding affinity between the globin chains or that the deleterious effects on the parasite are lost in the presence of both polymorphisms. Looking for malaria susceptibility genes in a population becomes an even harder task as a consequence.

The functional basis for the observed relationship between Hb S and infection with *P. falciparum* has been the subject of much research. Parasitized red blood cells from individuals with sickle cell trait were shown to sickle much more readily than uninfected cells at varying oxygen concentrations, this is presumed to promote clearance of parasitized cells from the circulation by the spleen (Luzzatto *et al.* 1970; Roth *et al.* 1978). Parasite growth rates were reduced in low oxygen concentrations with intracellular death (Friedman 1978; Pasvol *et al.* 1978; Friedman *et al.* 1979). Mechanisms involving acquired immunity have been proposed with greater immune recognition of parasitized red blood cells among children with sickle cell trait (Marsh *et al.* 1989). Cabrera and colleagues, for example, found a highly significant

association between Hb AS and immunoglobulin G (IgG) response to parasite variant surface antigens expressed on infected red blood cells (Cabrera *et al.* 2005).

Other structural variants of haemoglobin have been associated with malarial infection. Haemoglobin C arises due to a G to A single nucleotide substitution in *HBB* (rs33930165, c.19G>A) altering the same encoded amino acid as Hb S but rather than a glutamic acid to valine substitution there is a lysine amino acid substitution (Section 1.3.2). This variant is geographically concentrated among individuals living in central West Africa with, rarely, sporadic cases encountered elsewhere. Early work suggested a role for balancing selection by malaria (Allison 1956) but, in contrast to Hb S, there appear to be no negative consequences of possessing this variant such that it is considered as being subject to unidirectional positive selection in malarial regions. The polymorphism is currently found at relatively low allele frequencies, rarely greater than 20%, and is estimated to have arisen less than 5000 years ago (Wood *et al.* 2005). It may be that with time the allele frequency will rise to fixation and perhaps replace Hb S in particular populations subject to malarial selection (Hedrick 2004).

Studies of the Dogon people of Mali have shown that possessing one or two copies of Hb C is associated with protection from malaria with allele frequencies of 17.4% among uncomplicated malaria controls, 4.5% in severe malaria, and 2.9% among cerebral malaria cases (Agarwal *et al.* 2000). Moreover in a neighbouring West African country, Burkina Faso, a large case–control study of 4348 individuals showed a 29% reduction in risk of clinical malaria associated with heterozygosity, and 93% reduction with homozygosity, for Hb C ($P = 0.008$ and 0.001, respectively) (Modiano *et al.* 2001c). Reduced parasitaemia was noted with Hb C although effects on parasite growth remain controversial (Rihet *et al.* 2004; Williams 2006). A striking result was, however, reported by Fairhurst and colleagues who found reduced expression of the variant surface antigen *Plasmodium falciparum* erythrocyte membrane 1 (PfEMP1) on the surface of infected red blood cells containing Hb C. Reduced adhesion to endothelium and other features were seen with Hb AC or Hb CC genotypes and it may be that the observed association of Hb C with severe malaria relates to reduced parasite sequestration and induction of inflammation in small blood vessels such as the brain

(Fairhurst *et al.* 2005). This process is thought to be critical to the development of cerebral malaria, a relatively rare but dangerous complication of *P. falciparum* infection in which patients rapidly enter a coma and often die.

13.2.4 Duffy antigen and vivax malaria

It is perhaps not surprising that genetic variation in the genes required for producing red blood cells and the haemoglobin they carry should have become so entwined with man's battle with malaria. For the malaria parasite, our red blood cells are an essential staging post in their life cycle as it is there that they differentiate into game-tocytes and are then taken up by a mosquito (Fig 13.2). Merozoites must gain entry into human red blood cells but the way may be shut. *P. vivax* parasites, for example, are unable to enter the red blood cells of people possess-ing a particular variant of a gene encoding a protein nor-mally found on the red cell surface called the Duffy blood group antigen. Those individuals who lack the Duffy pro-tein (Duffy blood group negative individuals) are com-pletely protected from malaria due to this parasite. This explains the innate resistance of West Africans to *P. vivax* malaria (Miller *et al.* 1976). Given that the Duffy blood group is itself a benign genetic trait, it is believed that it was the selective pressure of malaria which drove the genetic variant conferring Duffy negative status to com-plete fixation in West African populations. It is notable however that malaria due to *P. vivax* is currently a much less severe disease than *P. falciparum* infection.

Duffy blood group antigen is a transmembrane glyco-protein encoded by the *DARC* gene (also known as the *FY* gene) on chromosome 1q21-q22. The protein also acts as a chemokine receptor for proinflammatory cytokines such as interleukin 18 (IL-18), hence the formal name of the gene product, 'Duffy blood group, chemokine recep-tor'. A single G to A nucleotide substitution in the cod-ing sequence of the gene results in a structural change in the encoded protein from glycine to asparagine at amino acid position 44 (Tournamille *et al.* 1995b). The two alleles are denoted FY*A and FY*B. The Duffy negative pheno-type was only found with the FY*B allele among black Africans. Among individuals homozygous for the FY*B allele (denoted FY*B^{null} or FY*0), the *DARC* gene was not expressed. The effect is specific to red blood cells as the *DARC* gene continues to be expressed in other cell and tissue types. The molecular basis for this is very elegant and involves a further variant on the same haplotype as the coding variant defining FY*B (Fig. 13.6) (Tournamille *et al.* 1995a). The additional variant, an A to G single nucleotide substitution (rs2814778), occurs in the pro-moter region 46 bases from the transcriptional start site, and disrupts a consensus binding site for the transcrip-tion factor GATA-1. The change from AGATAA to AGGTAA dramatically reduces expression of the *DARC* gene. The beauty of this molecular process is that the specific tran-scription factor involved, GATA-1, is only found in red blood cells. In other cell types, the regulatory variant has no effect and therefore does not disrupt the cellular pro-cesses. It demonstrates how the consequences of genetic variation can be highly context-specific.

The FY*B^{null}/FY*B^{null} genotype, found in Duffy nega-tive individuals, is at or near fixation in most West and Central African populations but is very rare outside Africa (Fig. 13.7). Given that vivax malaria is present at signifi-cant levels in non-African populations, it may be that in evolutionary terms the mutation occurred after the pro-posed major human migrations out of Africa (Box 8.5). Strong subsequent selection pressure could have resulted in non-African populations showing different and greater numbers of polymorphisms at the Duffy locus than in African populations. Greater genomic diversity among non-African populations is unusual, and contrasts with the majority of genomic regions where African popu-lations are more genetically diverse (Hamblin and Di Rienzo 2000).

Remarkably, recent evidence suggests that exactly the same single nucleotide substitution which gives the Duffy negative phenotype has arisen independently in Papua New Guinea, only this time on the FY*A allele (Zimmerman *et al.* 1999). It has only been found in the heterozygous state (FY*A/FY*A^{null}). When the red cells from these people are studied, they have on the cell sur-face only half the amount of Duffy antigen, consistent with a gene dosage effect: only one of the two alleles is prevented from expressing the gene due to loss of bind-ing of the GATA-1 transcription factor. The genetic variant was found in the Abelam-speaking population of low-land Papua New Guinea exposed to all four malaria para-sites. Only a small number of heterozygous individuals were found but the prevalence of vivax malaria was lower among FY*A/FY*A^{null} genotypes compared to FY*A/FY*A.

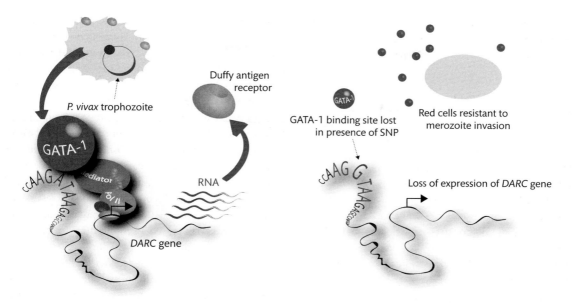

Figure 13.6 Molecular schematic of the functional consequences of an A to G substitution in the promoter of the *DARC* gene. In the homozygous state, GATA-1 binding is lost and the *DARC* gene is not expressed. Red blood cells in these 'Duffy negative' individuals are resistant to invasion by *Plasmodium vivax* as the cells lack the Duffy antigen receptor required by the parasite to gain entry.

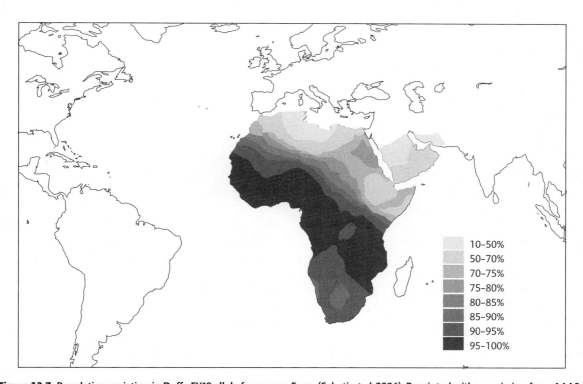

Figure 13.7 Population variation in Duffy FY*0 allele frequency. From (Sabeti *et al.* 2006). Reprinted with permission from AAAS.

13.2.5 Malaria parasites, oxidative stress, and G6PD enzyme deficiency

It may seem remarkable that so many genetic variants are associated with resistance to malaria but this is considered to reflect the remarkably strong selective pressure exerted by malaria over recent human history. The enzyme glucose-6-phosphate dehydrogenase (G6PD) provides protection for the cell against oxidative damage, for example as a result of iron accumulating on the breakdown of haemoglobin by malaria parasites. Inherited deficiency of G6PD is commonly seen (Box 13.2) and a striking correlation between the frequency of deficiency and malarial endemicity was noted when comparing different geographic regions worldwide (Allison 1960; Luzzatto 1979; Ganczakowski *et al.* 1995). Patients with G6PD deficiency have been found to have fewer malaria parasites in their blood due to inhibition of parasite growth (Luzzatto *et al.* 1969; Roth *et al.* 1983).

Box 13.2 Glucose-6-phosphate dehydrogenase deficiency (OMIM 305900)

G6PD deficiency is the commonest enzymopathy known in man, affecting more than 400 million people worldwide, predominantly in tropical and subtropical regions of Africa and Asia (Fig. 13.8). Most people with G6PD deficiency are asymptomatic but the consequences can be severe. These include jaundice in the newborn and haemolytic anaemia, in which the red cells of people with the enzyme deficiency break down in response to the oxidative stress associated with infection, drugs, or other precipitants including eating fava beans (Cappellini and Fiorelli 2008). The *G6PD* gene lies on the long arm of the X chromosome at Xp28 near to the genes responsible for colour blindness and haemophilia A.

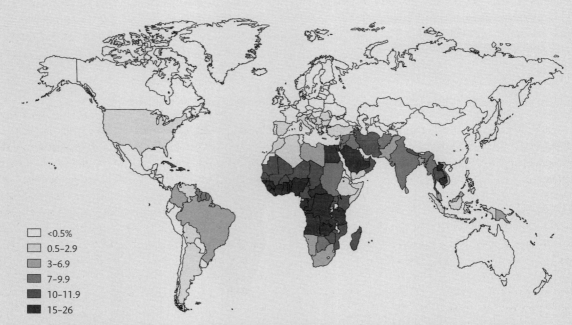

<0.5%
0.5–2.9
3–6.9
7–9.9
10–11.9
15–26

Figure 13.8 Worldwide distribution of G6PD deficiency. Reprinted from Cappellini and Fiorelli (2008), copyright 2008, with permission from Elsevier.

G6PD is remarkably polymorphic: over 400 biochemical variants have been described with more than 140 mutations resulting in variable degrees of enzyme deficiency (Cappellini and Fiorelli 2008). Most are missense mutations, with enzyme stability most commonly affected. A database of mutations involving *G6PD* is available online (www.bioinf.org.uk/g6pd) (Kwok *et al.* 2002). 'A' and 'B' variants are distinguished by an asparagine to aspartic acid substitution at position 126 arising due to an A to G single nucleotide substitution in exon 5; the 'A–' variant carries a further amino acid substitution of valine for methionine at position 68 due to a G to A nucleotide substitution in exon 4. The 'A–' variant accounts for about 90% of G6PD deficiency in tropical Africa (Cappellini and Fiorelli 2008).

There have been several studies investigating G6PD deficiency and susceptibility to malaria. The location of the *G6PD* locus on the X chromosome adds to the complexities of analysis as in men there will only be one copy while in women there will be two, and variable X chromosome inactivation adds significant genetic heterogeneity. Protection from seizures and coma in severe malaria was found among children in Nigeria (Gilles *et al.* 1967). More recently two large studies of over 2000 children found that the common form of G6PD deficiency found in Africa (due to the 'A–' variant) (Box 13.2) was associated with about a 50% reduction in risk of severe malaria in female heterozygotes and male hemizygotes (Ruwende *et al.* 1995). The evidence suggests that, like Hb S, the genetic variability at the *G6PD* gene locus represents an elegant example of a balanced polymorphism. Here, the selective advantage conferred by resistance to malaria appears counterbalanced by a selective disadvantage of blood disorders associated with the enzyme deficiency.

The evolutionary history of genetic variation at the *G6PD* locus has been studied using highly polymorphic microsatellite markers and restriction fragment length polymorphisms (RFLPs) in geographically diverse populations (Tishkoff *et al.* 2001). Analysis of the malaria susceptibility associated 'A–' and 'Med' mutations (the latter is common around the Mediterranean, involving substitution of phenylalanine for serine at position 188) suggests they have arisen independently within the last 3840–11 760 years and 1600–6640 years, respectively. The recent origins of such mutations are consistent with the hypothesis that malaria has had a major impact on human populations only since the introduction of agriculture within the last 10 000 years. Signatures of recent selection are seen involving G6PD, for example on haplotypic analysis with extended haplotypes of high frequency observed (Section 9.2.2).

The mechanism of protection conferred by G6PD deficiency is thought to relate to the susceptibility of parasitized G6PD deficient red blood cells to oxidative stress with earlier phagocytosis (Cappadoro *et al.* 1998). Remarkably, there is evidence that the *P. falciparum* counters this by manufacturing G6PD itself (Usanga and Luzzatto 1985). In G6PD deficient cells, adaptive changes were seen in the parasite over time on culture of *P. falciparum* with induction of a novel G6PD encoded by the parasite genome. The malaria parasite is an active player in its war of attrition with man, having its own genetic arsenal with which to combat variation in the host environment – as further illustrated with the diversity seen in parasite 'var' genes involved in evading the immune response (Section 13.2.6).

13.2.6 Polymorphism of immune genes

Genetic diversity in a number of different immune genes involved in our response to malaria infection have been associated with disease susceptibility, the range of genes involved reflecting the complexities of our immunological battle with the malaria parasite (see Fig. 13.3). Notable among these studies was the early association found with particular MHC class I molecules, glycoproteins that recognize the parasite within infected cells, delivering the foreign protein to the cell surface to be presented to CD8+ cytotoxic T lymphocytes (Box 12.1). The malaria parasite replicates in human liver cells before invading red blood cells. It is in these liver cells that MHC class I molecules can act as they are not expressed in red blood cells.

Possession of a specific human leukocyte antigen HLA-B53 was found to be associated with protection from severe malaria in a large case–control study of children in The Gambia (Hill *et al.* 1991). In this West African population about 1% of children less than 5 years of age die from malaria, and 20% of people carry *P. falciparum* in their blood. Hill and colleagues studied 619 children with severe malaria, initially analysing 45 different class I antigens among half the cases and finding association with the HLA-B53 on comparison with different control groups including children with mild non-malarial illness, mild malaria or severe non-malarial illness, and healthy adult blood donors. Given that the association may have arisen by chance the investigators then analysed the remaining cases only for HLA-B53, this time genotyping the allele directly by a PCR based method. Again significant association was found with a frequency of 16.9% among cases of severe malaria and 25.4% among mild controls with non-malarial illness, and similar frequencies in the other control groups. Taken together the data demonstrated protection from severe malaria associated with possession of HLA-B53 with an odds ratio of 0.59 (0.43–0.81, $P = 0.008$) (Hill *et al.* 1991).

The degree of protection associated with having HLA-B53 was not as marked as for possession of the sickle cell trait but, as the latter is less common in this population, they both appear to contribute a similar level of protection. Each variant has been estimated to prevent about one in ten potential cases of severe malaria. HLA-B53 is particularly common in West Africa with a frequency of 40% in Nigeria and 25% in The Gambia; allele frequencies are lower in other parts of Africa (for example 2% in South Africa) while in Caucasians and South East Asian populations the allele is rare (0–1%) and the allele is absent in Pacific regions (Hill *et al.* 1991). Association was also found with protection from severe malaria for an *HLA-DRB1* variant encoding a specific class II allele, DRB1*1302-DQB1*0501, with the effect specific to protection from severe malarial anaemia (Hill *et al.* 1991). As noted in Chapter 12 and elsewhere, the extensive linkage disequilibrium present in the MHC makes fine mapping such disease associations and assigning causality to specific variants a major challenge.

Insights into the molecular events occurring when HLA-B53 binds to *P. falciparum* were gained through understanding the structure of the molecule using X-ray crystallography studies. These revealed the architecture of a specific pocket on the molecule critical to

recognition of the parasite found on many HLA-B alleles. Within this groove, bound water molecules, acting in concert with the side chains of polymorphic residues, provide functional malleability, which enables high affinity/low specificity binding of multiple peptide epitopes (Smith *et al.* 1996). In the specific case of HLA-B53, the peptide binding groove was observed to be widened (Fig. 13.9). This may be part of the answer to why this particular allele should be associated with protection from severe malaria.

The malaria parasite is not an idle bystander in this process of interaction with HLA-B53. HLA-B53 binds to a specific fragment of *P. falciparum*, the circumsporozoite protein, leading to an attack by cytotoxic T lymphocytes (CTLs) on the parasite. Four variants of the parasite are found in The Gambia that differ in their circumsporozoite protein, of which two, the cp26 and cp29 variants, bind HLA-B53. Together these variants appear to jam the mechanism by which an interaction between HLA-B53 and parasite circumsporozoite protein would normally lead to CTL attack. In experiments, the presence of one variant renders the CTLs unable to kill the parasites carrying the other variant. Moreover, these two variants occur together in infected patients' blood much more often than expected. Remarkably, it appears that each individually suppresses the CTL response to the other. How this occurs is unclear but it seems cooperation between the two parasite strains leads to a mutual survival advantage. This is thought to represent a successful immune evasion strategy by the parasite; selective pressures can clearly act in both directions (Gilbert *et al.* 1998).

Genes encoding other key players in the body's immune response to malarial infection also show evidence of genetic variation associated with disease susceptibility, including innate and adaptive immunity (see Fig. 13.3) (Kwiatkowski 2005). For example, the CD40 ligand is an important component of the immune response, involved in B cell proliferation, activation of antigen presenting cells, and regulation of immunoglobulin class switching. Rare variants of *CD40LG* (also known as *TNFSF5*) encoding the CD40 ligand at chromosome Xq26 are associated with X-linked hyper-IgM syndrome (OMIM 308230), a rare immunodeficiency disorder associated with severe bacterial and life threatening infection. A single nucleotide polymorphism (SNP) in the *CD40LG* promoter has been associated with resistance to severe malaria,

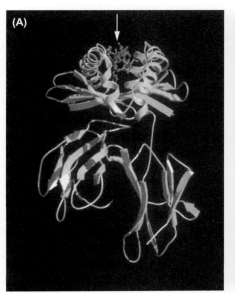 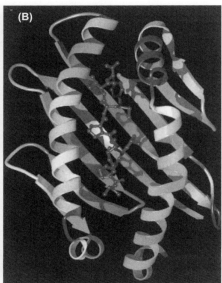

Figure 13.9 Binding by an HLA-B53 molecule to a peptide from *Plasmodium falciparum*. (A) Parasite peptide (indicated by an arrow) in the HLA binding groove of the molecule. (B) Interaction viewed from above. Reprinted by permission from Macmillan Publishers Ltd: Nature Reviews Genetics (Cooke and Hill 2001), copyright 2001.

showing evidence of selection on haplotype analysis (Sabeti *et al.* 2002a, 2002b). Differences in antibody levels against malarial antigens have been found among particular ethnic groups such as the Fulani in West Africa, who show increased resistance to malaria compared to other ethnic groups (Box 13.3).

Genes encoding many different mediators of the inflammatory response have been studied in relation to malaria including pro- and anti-inflammatory cytokines such as tumour necrosis factor (TNF), IL-1, and IL-10, as well as mediators such as IFNγ and the gene encoding inducible nitric oxide (NO) synthase, responsible for the important free radical NO implicated in the pathogenesis of cerebral malaria (see Fig. 13.3) (Rockett *et al.* 1991; Clark *et al.* 2003b; Kwiatkowski 2005). Many of these and other studies of genetic factors determining malaria susceptibility have been based on a candidate gene approach for genetic association and in a number of cases conflicting results have been obtained on replication studies (Section 2.4).

In part this reflects the many potential caveats in such analyses inherent to the study of common disease but also some specific to malaria and the genetically diverse

populations often studied in this disease. African populations for example show high levels of genetic variation with specific and distinct patterns of linkage disequilibrium between populations, which offer greater potential for resolution of specific functional alleles but also more capacity for apparent fail to replicate using a given set of genetic markers. Differences in malarial transmission intensities and prevailing parasite diversity will also vary between locations, which can be highly significant when considering specific disease phenotypes (Kwiatkowski 2005).

As an example to consider in more detail, genetic diversity at or near to *TNF*, the gene encoding TNF, has been associated with susceptibility to severe malaria by a number of studies. *TNF* is one of the earliest genes to be expressed in response to infection and acts to initiate and orchestrate the inflammatory cascade fighting infection, however dysregulation of TNF production can itself cause disease. In septic shock, for example, there is good evidence that TNF has a causative role as it is produced during the septic shock syndrome, it causes the development of the syndrome when given to uninfected animals, and neutralizing TNF in infected septic animals prevents

---

Box 13.3   Resistance to malaria among the Fulani

The Fulani people living in rural savannah areas of Burkina Faso and Mali in West Africa were found to be more resistant to malarial infection with *Plasmodium falciparum* than other sympatric ethnic groups living in the same hyperendemic areas for malaria transmission such as the Mossi and Rimaibe (Modiano *et al.* 1996; Dolo *et al.* 2005). Despite similar exposure rates, markedly lower levels of parasitaemia, fewer episodes of fever or clinical infection, and less severe disease were found among the Fulani. The Fulani are genetically distinct from the other two groups (Modiano *et al.*

2001a) but no known malaria risk alleles such as Hb S, Hb C, thalassaemia, G6PD, or HLA-B53 could be found to account for the observed differences in disease resistance (Modiano *et al.* 2001b). The antibody response to malarial antigens was found to be increased among the Fulani with evidence that genetic variation in the *IL4* gene encoding interleukin-4 was associated with increased antibody levels (Luoni *et al.* 2001). Gene expression profiling suggests that a functional deficit of T regulatory cells may lead to the observed higher resistance to malaria among the Fulani (Torcia *et al.* 2008).

---

the development of shock (Tracey and Cerami 1993). Similarly in cerebral malaria, there is growing evidence that while TNF protects against infection, in excess it can directly cause disease (Kwiatkowski *et al.* 1990). Genetic polymorphism at the *TNF* locus has been associated with susceptibility to many infectious, autoimmune, and inflammatory diseases although many studies appear conflicting and specific functional variants remain hard to localize due to coinheritance of genetic variants across the locus (Section 2.4.4).

There is evidence to implicate *TNF* with disease outcome from malarial infection based on linkage studies in The Gambia and Burkina Faso (Jepson *et al.* 1997; Flori *et al.* 2003) and using a candidate gene approach. Children with severe malaria had a higher allele frequency of a particular SNP in the promoter region of the *TNF* gene, suggesting a role in disease susceptibility (McGuire *et al.* 1994). Individuals homozygous for the 'A' allele of rs1800629 (located 308 bp 5′ of the transcriptional start site and known as 'TNF-308') showed a seven-fold increased risk of cerebral malaria in a large case–control study in The Gambia. Association was also reported with more rapid symptomatic reinfection with *P. falciparum* in Gabon (Meyer *et al.* 2002), with severe malaria and other infectious diseases in Sri Lanka (Wattavidanage *et al.* 1999), and with malaria morbidity in Kenya (Aidoo *et al.* 2001) and some evidence of association with parasitaemia in Burkina Faso (Flori *et al.* 2005). Whether this

specific SNP or linked alleles are functionally important remains controversial (Section 2.4.4).

Other *TNF* promoter SNPs have shown evidence of disease association, for example with severe malarial anaemia (rs361525) (McGuire *et al.* 1999), while a rare variant more distally in the promoter (rs1800750, 'TNF-376') region showed association with severe malaria in The Gambia and Kenya (Knight *et al.* 1999). The latter was identified as a candidate functional variant based on analysis of sites of protein–DNA interaction using the DNA footprinting technique. A single nucleotide substitution was found to modulate the binding of a specific transcription factor, Oct-1, which showed higher affinity to the rs1800750 'A' allele and was associated with increased gene expression in human monocytes (Knight *et al.* 1999). As elsewhere in the MHC there is extensive linkage disequilibrium across the *TNF* locus and association with malaria and other infectious disease have been found at the neighbouring gene *LTA* encoding lymphotoxin alpha. Here a putative functional variant was resolved by analysis of haplotype-specific differences in gene expression with a specific SNP (rs2239704, 'LTA+80'), shown by haplotype-specific chromatin immunoprecipitation (ChIP) (Section 11.5.3) to modulate recruitment of the transcriptional repressor activated B cell factor 1 (Knight *et al.* 2003, 2004). The low producer allele defined by rs2239704 was subsequently associated with reduced parasitaemia in a study in Burkina Faso (Barbier *et al.* 2008) and with early onset leprosy in a study analysing individuals from Vietnam, Brazil, and India (Alcais *et al.* 2007). Whether this SNP explains some of the reported *TNF* associations or indeed is a marker for a further unidentified functional variant remains to be fully resolved, but what is clear is that there is a strong signal of association at the *TNF* locus with malaria and other infectious diseases.

13.2.7 Cytoadhesion and immune evasion: host and parasite diversity

After the parasite invades a red blood cell, a race against time ensues in which the parasite seeks to multiply while avoiding detection and destruction by the human immune system. One parasite strategy involves cytoadherence, parasitized red blood cells sticking to vascular endothelium or other cells to avoid passing through the spleen, which is very efficient at recognizing and removing infected cells from the circulation. This can be achieved by the parasite expressing specific proteins such as *P. falciparum* erythrocyte membrane protein 1 (PfEMP1) which are exported to the cell membrane of the infected red blood cell and interact with host cell receptors found on endothelium, platelets, and other red blood cells. The consequences of such cytoadherence can be severe for the human host as it is thought to be critical to the development of cerebral malaria in which parasites are found sequestered in small blood vessels in the brain, with affected patients at risk of coma and death (Mackintosh *et al.* 2004; van der Heyde *et al.* 2006).

The presence of PfEMP1 on the surface of infected red blood cells provides a target for recognition and attack by the host immune system. The parasite, however, combats this by varying the particular protein present on the infected cell (Roberts *et al.* 1992). PfEMP1 proteins are encoded by a family of 'var' genes found in the malaria parasite genome, with 60 different 'var' genes each encoding a different protein (Su *et al.* 1995; Borst and Genest 2006). By switching the particular protein on the cell surface before the human immune system has had time to mount a full attack, the parasite buys valuable time to multiply and survive. An intricate set of genetic factors in the parasite allow this process to operate (Scherf 2006; Voss *et al.* 2006). Of the 60 genes involved, only one is expressed at a given time but the system can rapidly switch between genes. Molecular

modifications of the chromatin coat covering the parasite DNA act to regulate gene expression, while specific 'silencer' proteins are also recruited to prevent gene expression. The precise location of a given var gene within the parasite nucleus also appears important, with specific expression spots found in which movement of an 'incoming' var gene displaces the currently active gene. The huge selective pressures at play appear to have led to the evolution of an intricate parasite genetic defence system designed to allow the parasites to hide, multiply, and survive.

A number of different host receptors can bind PfEMP1 including intercellular adhesion molecule 1 encoded by *ICAM1* at chromosome 19p13.3-p13.2, a member of the immunoglobulin family (Berendt *et al.* 1989; Smith *et al.* 2000). Sequencing of the N-terminal domain of the protein among 24 Kenyan children revealed a common variant in the coding region of the gene, present at a high allele frequency of about 30% in Kenyan and Gambian populations but rare among Europeans (Fernandez-Reyes *et al.* 1997). The A to T transversion leads to substitution of lysine for methionine at position 56 (rs5491), which among Kenyan children was associated with increased susceptibility to cerebral malaria. Interest in this SNP was heightened by the evidence of differences in adhesion of the variant ICAM1 protein (ICAM1[Kilifi]) to parasitized red blood cells (Adams *et al.* 2000).

However, soon after the Kenyan study showing increased susceptibility, Kun and colleagues found that among children in Gabon in West Africa the variant was associated with protection from severe malaria (Kun *et al.* 1999), and a number of other studies found no association. As described previously, this may reflect the limitations of the candidate gene approach, often with limited statistical power and SNP coverage, or in this case possibly geographic variation in parasite characteristics or timing of infection. Fry and colleagues analysed a denser set of SNPs at the *ICAM1* locus among diverse African populations, including large case–control and family-based cohorts, and found no signal of association with severe malaria (Fry *et al.* 2008).

Other host receptors for PfEMP1 have also been investigated for association with susceptibility to severe malaria (see Fig. 13.3) including mutations leading to deficiency of CD36, again often with apparently conflicting results between studies (Aitman *et al.* 2000; Pain *et al.* 2001).

13.3 Genetic diversity and susceptibility to Leishmaniasis in mouse and man

Genetic variation also appears to play an important role in modulating susceptibility to a second major parasitic disease, leishmaniasis (Box 13.4). Evidence from mouse models highlighted a role for genetic factors in susceptibility to infection (Lipoldova and Demant 2006). Among humans, studies of visceral leishmaniasis in northeastern Brazil provided evidence of familial aggregation and increased risk in siblings of affected sib pairs (Cabello *et al.* 1995; Peacock *et al.* 2001).

Variation between ethnic groups in their susceptibility to leishmaniasis has also been reported with differences in the incidence of clinical disease found between adjacent villages in which people of different ethnic origin lived (Zijlstra *et al.* 1994). For example, the Masalit people living in an endemic region for leishmaniasis in Sudan showed high rates of disease that was fatal unless treated, while neighbouring groups such as the Hawasa tribe did not (Blackwell *et al.* 2004). A further example came from a study in Dinder National Park in northern Sudan, which lies in a highly endemic area for visceral leishmaniasis (Ibrahim *et al.* 1999). Game wardens

---

Box 13.4   Leishmaniasis

Infections with many species of *Leishmania* cause common and often serious diseases in Central and South America, around the Mediterranean, in northern and Central Africa, the Middle East, central and western Asia, and the Indian subcontinent. These intracellular protozoan parasites are transmitted by the bite of infected female phlebotomine sandflies. The parasite predominantly infects macrophages, a type of immune cell found in the blood which normally engulfs and destroys infectious pathogens (Fig. 13.10). The parasites can cause cutaneous disease, which varies from self-healing skin ulcers to highly disfiguring facial mucocutaneous lesions. Disseminated spread is seen in visceral leishmaniasis, known as kala-azar, which results in various clinical manifestations including fever, wasting, an enlarged liver and spleen, lymphadenopathy, and blood pancytopenia. It is usually fatal unless treated and is highly infectious. Devastating epidemics have occurred, as recently as 1984 in Sudan, where death rates of up to 70% in the most affected areas were reported with an estimated 100 000 deaths, among approximately 280 000 people in the epidemic area, thought to be attributable to visceral leishmaniasis (Seaman *et al.* 1996). The particular type of disease that results depends on several factors, notably the individual species of *Leishmania* parasite injected into the blood and

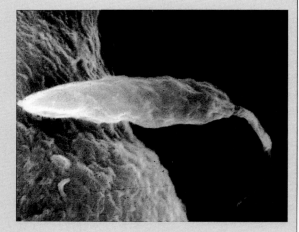

Figure 13.10 *Leishmania major* promastigote attached to a macrophage. The human host is infected with promastigotes by the infected sandfly bite, promastigotes being a particular stage in the *Leishmania* parasite life cycle. The promastigotes invade skin dendritic cells and macrophages. A scanning electron micrograph showing the attachment of a promastigote to a macrophage is shown. Reproduced from Roberts *et al.* (2000), with permission from BMJ Publishing Group Ltd.

the genetic constitution of the host. Other factors include the nutritional status of the host, the sandfly vector, environmental and social factors, and reservoir hosts which may be human or zoonotic (Lipoldova and Demant 2006).

working in the game reserve varied in their susceptibility depending on their ethnic background. Those from the Baria tribe of southern Sudan or the Nuba tribe from western Sudan were more likely to develop visceral leishmaniasis.

Strains of laboratory mice were found to differ in their susceptibility to infection with *Leishamania major* and *L. donovani*, and have proved a very powerful approach to defining genetic susceptibility (Lipoldova and Demant 2006). For example, resistant and susceptible strains were found for *L. donovani* that showed markedly different rates of proliferation of the parasite in the liver. By performing breeding experiments with recombinant inbred strains, it was possible to identify a region of chromosome 1 that conferred resistance to early infection (Bradley 1977). The same genetic locus was found to control other intracellular infections such as *Mycobacterium bovis* (BCG), *M. lepraemurium*, and *Salmonella typhimurium* (Plant and Glynn 1974, 1976; Gros *et al.* 1981; Skamene *et al.* 1982).

Positional cloning resolved a susceptibility gene in the locus called *Nramp1* (also known as *Slc11a1*), encoding 'natural resistance associated macrophage protein' (Vidal *et al.* 1993). A single G to A nucleotide substitution resulted in a glycine to aspartic acid substitution at position 169 in the protein and segregated between the resistant and susceptible mouse strains (Vidal *et al.* 1993). Mice homozygous for this substitution were phenotypically identical to knock-out mice for *Nramp1* (Vidal *et al.* 1995). The Nramp1 protein is found on the membranes of macrophages and neutrophils where it functions as a divalent metal pH-dependent pump, transporting Fe^{2+}, Zn^{2+}, and Mn^{2+} from within cells (Goswami *et al.* 2001). This prevents the parasite from replicating as these metals are needed by the parasite's metabolism as well as resulting in diverse effects promoting macrophage defence (Lipoldova and Demant 2006).

In humans, genetic variation in the *Nramp1* homologue gene *SLC11A1*, encoding solute carrier family 11 (proton-coupled divalent metal ion transporters) member 1, at chromosome 2q35, has been associated with a number of infectious, autoimmune, and inflammatory diseases (O'Brien *et al.* 2008). Studies of affected families among the Masalit tribe in Sudan, a group highly susceptible to visceral leishmaniasis, have shown linkage with

variation in the *SLC11A1* gene (Mohamed *et al.* 2004), a result also found independently in a different ethnic group in Sudan (Bucheton *et al.* 2003b). However, fine mapping and defining specific functional variants has proved challenging. A microsatellite in the promoter region was postulated as functionally important with different (CA)*n* alleles associated with low and high gene expression, and was found to have contrasting associations to infectious and autoimmune disease (O'Brien *et al.* 2008). This microsatellite was one of several variants of *SLC11A1* associated with tuberculosis in a case–control study in The Gambia (Bellamy *et al.* 1998). The disparity between clear cut effects on susceptibility to leishmaniasis seen in mice, and more variable data in humans, reflect the fact that while in mice the *Nramp1* variant dramatically affects the function of the protein, in humans the disease associated polymorphisms identified to date appear to have a much more modest effect (Lipoldova and Demant 2006).

Sequence variation in other genomic regions may also be important. In the village of Barbar El Fugara in eastern Sudan, a devastating outbreak of kala-azar occurred between 1995 and 2002. Antibody studies show nearly everyone was exposed and over a quarter of people developed the disease (Bucheton *et al.* 2002, 2003c). Cases clustered in families and ethnic groups. Analysis showed that Aringa families originally from western Sudan were worst affected (Bucheton *et al.* 2003a). These families were studied to try and resolve any genetic basis for susceptibility. A number of genetic markers were used to carry out a genome-wide linkage study (Bucheton *et al.* 2003a). A total of 63 Sudanese families were studied using 380 markers. This showed evidence of significant linkage for markers on chromosome 22q12. By contrast among the Masalit ethnic group, a genome-wide linkage analysis defined susceptibility loci at 1p22 and 6q27 that were of Y chromosome lineage and were village-specific (Miller *et al.* 2007). Chromosome 6q27 was among a number of loci identified on a genome-wide linkage scan for visceral leishmaniasis carried out in Brazil, other loci were 7q11.22 and 17q11.2 (Jamieson *et al.* 2007). A number of candidate gene studies have also found varying associations with the markers in the MHC, *IL4* and *IFNGR1* (encoding interferon gamma receptor 1) (Lipoldova and Demant 2006).

13.4   Helminth infection

13.4.1   Genetic susceptibility to Ascaris infection

The power of studying genetic variation in an extended family pedigree was highlighted by remarkable studies of roundworm infestation in Nepal by Williams-Blangero and colleagues (Williams-Blangero *et al.* 1999, 2002). *Ascaris lumbricoides* is the most common helminth infection of man, affecting about one-quarter of the world population and causing significant morbidity and even mortality as a result of intestinal blockage and malnutrition. The Jiri region east of Kathmandu in Nepal is home to the Jirels, a genetically isolated population thought to be a hybrid group derived from Sherpa and Sunwar origin between ten and 11 generations ago. Here, as seen elsewhere in the world, ascariasis was found to occur predominantly within family groups and with highest numbers of worms present in a small proportion of those affected.

The genealogical structure of the Jirel community has been very well documented such that a single pedigree containing 1261 people could be studied (Fig. 13.11) (Williams-Blangero *et al.* 1999). A number of different phenotypes were analysed that reflected worm burden including egg and worm counts as well as worm biomass. Heritability was assessed together with the contribution of shared environment. This provided compelling evidence that genetic factors were important in modulating risk. When compared before, and a year after, worm clearing treatment with albendazole, the same individuals were found to be most commonly infected. Environmental factors appeared less significant, with the correlation between spouses for example found to be low. Pedigree studies provide a powerful method to allow the effects of genetic relatedness on worm burdens to be separated from shared household and other environmental effects. Overall, 30–50% of the observed variation between people was attributable to genetic factors, compared to 3–13% due to shared environmental factors (Williams-Blangero *et al.* 1999).

This single pedigree allowed a genome-wide linkage scan to be performed for 444 individuals from the most genetically informative branch of the pedigree (Williams-Blangero *et al.* 2002); 6209 pairs of relationships were informative for genetic analysis with 375

dinucleotide microsatellite markers used. The pedigree structure was postulated by the authors as being 22 times more efficient for determining quantitative trait loci (QTLs) than classical sib pair analysis. Strong evidence of linkage was found at chromosome 13q32-q34 with a lod score (Box 2.4) of 4.43 (genome-wide significance P = 0.0009) for *Ascaris* burden as measured by egg count (Fig. 13.12) (Williams-Blangero *et al.* 2002). The same region was also linked to total IgE levels. The 13q32-q34 region contains at least one strong candidate gene, *TNFSF13B*, encoding a member of the TNF superfamily which acts as a B cell activator. Linkage was also found at chromosome 1p32 with a lod score of 3.01 (P = 0.033) (Williams-Blangero *et al.* 2002).

The family pedigree data were also used to estimate the heritability of levels of immune mediators found in the blood between people (Williams-Blangero *et al.* 2004). This showed that genetic variation between individuals was highly significant in determining levels of secreted cytokine mediators. Overall 40–60% of the variability in levels of secreted cytokines (IFNγ, TNF, IL-2, IL-4, IL-5, and IL-10) between individuals was attributable to genetic factors.

13.4.2   Schistosomiasis and other helminth infections

Schistosomiasis (Box 13.5), together with the gut parasites hookworm, strongyloidiasis, and trichuriasis, are examples of other important diseases caused by helminths infecting man. As with *Ascaris* infection, only a small proportion of people harbouring these parasites are heavily infected so that about 10% of an affected population typically have 70% of the total parasite burden (Anderson and May 1985). Such effects are felt to relate to differences in exposure and to individual susceptibility. Familial aggregation of these different helminth infections is well documented, together with evidence of individual predisposition and ethnic variability in susceptibility – all suggestive evidence of a role of host genetic variation playing a role in modulating disease (Quinnell 2003). It was noted as early as 1925 in the southern United States that a higher prevalence and intensity of hookworm infection was seen in those of European compared to African ancestry (Smillie and Augustine 1925). Overall, up to 44%

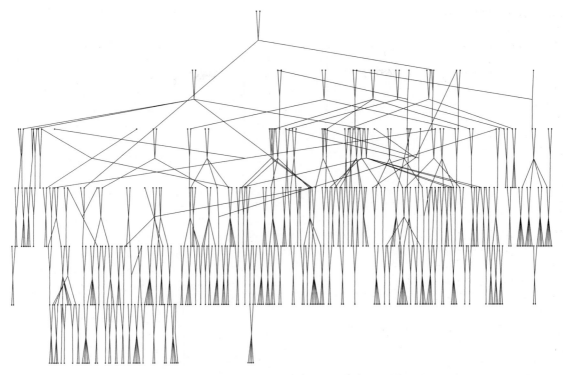

Figure 13.11 Jirel pedigree. The remarkable Jirel population pedigree available for analysis of *Ascaris* infestation in Nepal is shown. Reproduced with permission from Williams-Blangero *et al.* (2002) .

of the variability has been attributed to genetic effects (Quinnell 2003). However the many attempts to localize the specific gene regions or genetic variants within them have proved largely frustrating.

Progress has been made in terms of understanding genetic factors influencing infection with *S. mansoni* in Brazil. Segregation analysis using 20 family pedigrees within an area of Brazil hyperendemic for schistosomiasis indicated a single codominant major gene effect (Abel *et al.* 1991). To try to define the responsible gene, a genome-wide linkage study was performed among 142 individuals from 11 informative Brazilian families (Marquet *et al.* 1996). This showed evidence of linkage to chromosome 5q31-q33, a genomic locus containing many different immune genes including *IL4*, *IL5*, and *IL13* members of the interleukin family (Marquet *et al.* 1996). Further supporting evidence was found from an independent study in an epidemic focus of infection with a shorter history of population exposure in northern Senegal (Muller-Myhsok *et al.* 1997). Linkage was found with the same genomic region but without dominant inheritance (Muller-Myhsok *et al.* 1997).

The 5q31-q33 region has also been associated with levels of IgE production (Marsh *et al.* 1994), and with susceptibility to several other diseases including asthma and inflammatory bowel disease (Yokouchi *et al.* 2000; Rioux *et al.* 2001). Levels of IgE have been shown previously to be an important component of immune resistance to helminth infection (Hagan *et al.* 1991). The specific genetic variants within the 5q31-q33 region responsible for the genetic effects remain to be localized. However, family-based studies in Mali of infection with *S. haematobium*, the cause of urinary schistosomiasis, have provided some intriguing possibilities (Kouriba *et al.* 2005). Here, alleles bearing specific promoter polymorphisms of the *IL13* gene were preferentially inherited by children with the highest 10% levels of infection.

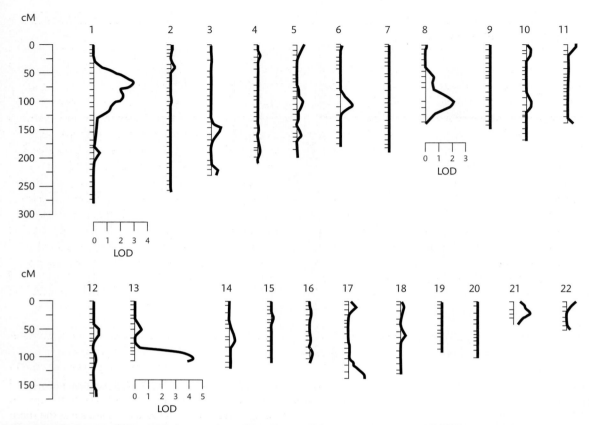

Figure 13.12 Genome linkage scan for *Ascaris lumbricoides* egg counts. Data for 444 individuals from a Jirel pedigree in Nepal. Linkage results from the genome scan are shown plotted for individual chromosomes. LOD, lod score. Reproduced with permission from Williams-Blangero *et al.* (2002) .

A variable but usually small proportion of those infected with *S. mansoni* develop periportal fibrosis of the liver and splenomegaly (Cheever 1968). This is caused by chronic inflammation in the liver, triggered by eggs which embolize from the gut to the liver and release schistosome antigens. Disease was seen to cluster in family pedigrees and to show strong ethnic differences, affecting for example Brazilians of African ancestry less than those of Caucasian origin, despite similar levels of infection intensity (Bina *et al.* 1978).

Lethal disease due to periportal fibrosis also occurs in 2–10% of those infected with *S. mansoni* in Sudan. This was demonstrated to be associated with a major dominant gene effect based on a segregation analysis among 65 pedigrees from a village comprising individuals who had

migrated into an endemic area for schistosomiasis from a non-endemic area 20 years previously (Dessein *et al.* 1999). Analysis of four candidate regions showed linkage to chromosome 6q22-q23, in particular the *IFNGR1* gene, which encodes interferon gamma receptor 1. This was a tantalizing result given the role of IFNγ in regulating production of the extracellular matrix protein found in the periportal spaces. High IFNγ producers have been shown to have a reduced risk of severe periportal fibrosis (Henri *et al.* 2002) and specific SNPs of *IFNGR1* have been associated with risk of disease (Chevillard *et al.* 2003). The molecular basis for these genetic associations remains unclear: the two SNPs lie in the third intron of *IFNGR1* and alter nuclear protein interactions, suggesting that this may involve modulation of gene expression. Linkage

---

> **Box 13.5   Schistosomiasis**
>
> Schistosomiasis is considered second only to malaria in terms of importance as a parasitic disease affecting man – it chronically infects 200 million people in tropical and subtropical regions of Africa, Asia, and South America with the majority in sub-Saharan Africa where some 300 000 deaths per year are estimated to be due to schistosomiasis (van der Werf *et al.* 2003; Chitsulo *et al.* 2004). Three major species infect man, *Schistosoma mansoni*, *S. japoni-cum*, and *S. haematobium*. The life cycle of the parasite involves an intermediate host, aquatic snails, from which infective larvae penetrate human skin in the water and migrate as schistosomulae to the portal circulation and liver where they mature into
>
> male and female worms and mate (Chitsulo *et al.* 2004). Worms of different species show varying predilections for migration, either to mesenteric vessels causing intestinal disease, or into veins draining the urinary system causing genitourinary disease in the case of *S. haematobium*. Severe and potentially fatal disease results from the granulomatous inflammatory reaction to the eggs lodged in small blood vessels in the liver, bladder, urinary tract, or other tissues, leading for example to chronic periportal fibrosis in the liver and portal hypertension, or bladder cancer. Eggs excreted in the faeces or urine complete the life cycle, hatching in water into larvae which infect snails.

was also found in an Egyptian population for *IFNGR1* in a study of 11 candidate genes for hepatic fibrosis due to *S. mansoni* (Blanton *et al.* 2005).

13.5   Summary

Parasitic diseases have exerted major selective pressures on human populations. However, these are common multifactorial diseases in which susceptibility determined by human genetic variation is only part of a much more complex scenario involving multiple environmental factors as well as genetic diversity in the parasite and vector. It is remarkable that inherited factors have been defined at all, given the inherent difficulties of interrogating the genetic basis of common multifactorial disease. Yet the magnitude of effects observed for several genetic variants is substantial with selective advantage driving allele frequencies to high levels, in some populations to fixation as seen for the Duffy antigen receptor in West Africa. The malaria hypothesis advocated by Haldane in which selective advantage of heterozygotes has driven alleles to high frequency despite deleterious effects of the variant allele in the homozygous state has been reviewed for thalassaemia and Hb S. Hb C, found predominantly in central West Africa, appears to be rising in allele frequency with

unidirectional positive selection and no apparent deleterious phenotype in the homozygous form.

The evidence to support a role for genetic variation in determining malaria susceptibility have been highlighted here by the striking examples of particular ethnic groups showing increased resistance to malaria such as the Tharu of Nepal or Fulani of West Africa. The correlation by latitude and altitude of α^+ thalassaemia and malaria endemicity, as well as geographic variation observed for allele frequencies of Hb S, G6PD, Duffy blood group antigen, and other variants is a remarkable testament to the power of selection exerted by malaria. Equally remarkable is the elegance of some of the underlying molecular mechanisms by which functional variants exert their effects which in some instances have been elucidated, notably for the promoter SNP regulating expression of the Duffy blood group antigen which controls entry to red blood cells of the *P. vivax* parasite. We have seen how the parasite is not an idle bystander in this process, varying the surface antigen PfEMP1 to avoid immune detection and producing G6PD, which may be significant given the role of inherited G6PD deficiency in protection from severe malaria. Diversity in the human receptors for the PfEMP1 ligand such as *ICAM1* are important determinants of susceptibility, as is diversity in other immune genes such as HLA-B53 encoding MHC class I molecules.

The range of genes involved in malaria susceptibility, and geographic variation in sites where specific genetic variants are thought to have arisen, or be significant, again reinforces the ongoing selective pressure exerted by this infectious disease. For other parasitic diseases genetic variation has also been shown to be highly significant, as illustrated here for leishmaniasis and helminth infections. Again these are striking examples, in leishmaniasis highlighting the role mouse studies can play in defining susceptibility and resistance genes while in *Ascaris* infections the remarkable power of pedigree analysis for the Jirel population in Nepal was shown. It is notable how linkage analyses in leishmaniasis and helminth infections, including ascariasis and schistosomiasis, have proved powerful approaches despite the inherent difficulties of applying this approach to complex traits (Section 2.3.5). The application of genome-wide association studies to malaria and other parasitic diseases is awaited with interest.

The elucidation of the role of human genetic variation in susceptibility to infectious disease should continue to provide new insights into disease pathogenesis, ranging from the process of red cell invasion and parasite multiplication, to immune recognition and sequestration. New opportunities for intervention, in terms of vaccine and drug development, are provided by such studies while targeting of specific treatments to individual patients is a further longer term objective. The multifaceted interaction between man and parasites continues. Future study of our variable genomes should continue to increase our understanding of susceptibility to parasitic disease, hopefully leading to better treatments for millions of the world's poorest people still plagued by such diseases.

13.6 Reviews

Reviews of subjects in this chapter can be found in the following publications:

Topic	References
Genetic susceptibility to infectious disease	Cooke and Hill 2001; Hill 2001, 2006; Burgner *et al.* 2006
Malaria and human genetic variation	Kwiatkowski 2005; Williams 2006; Weatherall 2008
Severe malaria	Mackintosh *et al.* 2004; van der Heyde *et al.* 2006
G6PD deficiency	Cappellini and Fiorelli 2008
Leishmaniasis	Roberts *et al.* 2000
Genetic susceptibility to leishmaniasis	Blackwell *et al.* 2004; Lipoldova and Demant 2006
Genetic of susceptibility to human helminth infection	Quinnell 2003

CHAPTER 14

Human genetic diversity and HIV

14.1 Introduction

The current devastating pandemic of the acquired immunodeficiency syndrome (AIDS) due to the human immunodeficiency virus (HIV) is the latest in a series of retroviral infections to which we have been subjected over our evolutionary history. A record of past infections in the form of integrated retroviral sequences can be found within the human genome, and that of other primate species, and a substantial body of research now highlights the dramatic selection pressures that have driven genetic diversity in both host and virus. The 2006 report on the global AIDS epidemic by the joint United Nations Programme on HIV/AIDS highlighted that worldwide an estimated 65 million people have been infected with HIV, of whom over 25 million have died of AIDS and nearly 40 million are currently infected with HIV (Joint United Nations Programme on HIV/AIDS & NetLibrary Inc. 2006) (Box 14.1).

Box 14.1 HIV and AIDS

The first cases of AIDS were reported in 1981, an immunodeficiency syndrome characterized by depletion of a specific type of T cell, opportunistic infections (such as pneumonia due to the fungus *Pneumocystis carinii*, and infection with cytomegalovirus and mycobacteria), and specific cancers (such as Kaposi's sarcoma and B cell lymphoma). How was HIV identified as the agent responsible for AIDS? Gallo and Monagnier (2003) describe how a specific subgroup of T cells carrying the CD4 surface antigen were noted to be significantly reduced in patients with AIDS, suggesting an agent specifically targeting CD4+ T cells such as the recently identified human T cell leukaemia virus (HTLV). The modes of transmission of HTLV through blood and sexual contact, and between mother and infant, mirrored what was known about the epidemiology of AIDS, and an AIDS-like wasting syndrome was seen in animal models of lymphotropic retroviruses, leading to a search for HTLV-like viruses in AIDS patients. This culminated in the isolation of HIV, and establishment of a causal relationship with AIDS that was broadly accepted by the scientific community by 1984. The epidemiological data that followed the development and application of a specific blood test for HIV, and the clinical efficacy of drugs inhibiting HIV enzymes, supported this view. In fact there are at least two types of HIV: HIV-1 which is more virulent and responsible for most cases of AIDS worldwide, and HIV-2 which is more geographically restricted and endemic in West Africa.

The origin of HIV lies in simian immunodeficiency viruses (SIV) found in primates (Heeney

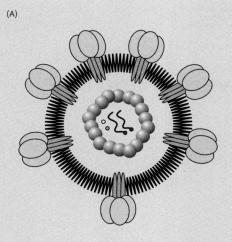

(A)

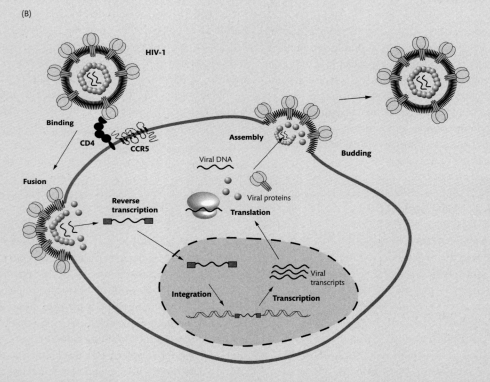

(B)

Figure 14.1 Overview of HIV-1 entry and replication in host cells. (A) The virion of HIV-1 contains integrase (O), protease (O) and reverse transcriptase (●) enzymes together with two copies of the RNA genome. (B) Viral glycoproteins in the viral envelope interact with host cell surface receptor and coreceptor proteins, the envelope then fuses with the cellular membrane, and viral genetic material enters the cytoplasm as a nucleoprotein core. The virus uses the host cell's cellular proteins and machinery, together with some specific viral proteins present in the nucleoprotein core, to replicate. The two copies of the viral RNA genome are reverse transcribed into DNA, transported into the nucleus, and become integrated into host DNA. The integrated proviral DNA is then transcribed and the viral mRNAs are processed and exported from the nucleus. Following translation, viral proteins such as Tat and Rev further amplify transcription of proviral DNA and transport out of the nucleus while late proteins such as Gag, Pol, and Env are assembled with viral RNA into virions, which escape the host cell at budding sites.

et al. 2006). HIV-1 is believed to have evolved on at least three separate occasions from a particular SIV strain, SIVcpz, found in one subspecies of chimpanzee (*Pan troglodytes troglodytes*) whose natural habit includes southern Cameroon (Keele *et al.* 2006). HIV-2 appears to have crossed between species on many occasions and has its origin in SIVsm found in sooty mangabeys (*Cercocebus atys*) (Santiago *et al.* 2005). HIV is a an RNA virus that encodes reverse transcriptase, an enzyme which enables the RNA genome to be copied into DNA within the infected cell and become integrated into the genomic DNA of the host cell (Fig. 14.1). Within the retrovirus family, HIV is a member of the lentivirus group and related to a number of animal lentiviruses which also infect immune cells and result in immunodeficiency and slow progressive disorders.

However, not all people exposed to HIV become infected, and those who do, progress to AIDS at significantly different rates (Box 14.2). Studies of highly exposed yet persistently seronegative individuals together with prospective cohort studies of groups at risk of HIV exposure have proved highly informative in resolving genetic determinants of disease susceptibility and progression (Kulkarni *et al.* 2003; O'Brien and Nelson 2004). Genetic variation in genes encoding proteins involved in HIV entry to cells, barriers to retroviral infection within cells, cytokines, and cell mediated and innate immunity have all been shown to be involved in HIV infection and development of AIDS (Fig. 14.2) (O'Brien and Nelson 2004; Heeney *et al.* 2006). In this chapter some of this remarkable work is reviewed, highlighting the complex relationship between human genetic diversity and viral infection, and the insights such analysis gives into the current AIDS pandemic and our evolutionary history.

14.2 Genetic variation in coreceptors and coreceptor ligands

14.2.1 Polymorphism of CCR5 and HIV-1 infection

In 1996 a series of landmark studies were published which established the role of specific chemokine coreceptor proteins that enable viral invasion of T cells expressing CD4. For the main viral strains of HIV-1 transmitted through sexual activity, denoted R5 strains (Box 14.3), the CC chemokine receptor CCR5 (CC-CKR-5) was identified as the major host coreceptor protein (Alkhatib *et al.* 1996; Choe *et al.* 1996; Deng *et al.* 1996; Doranz *et al.* 1996; Dragic *et al.* 1996). CCR5 is necessary for fusion of the viral envelope protein with the host cell membrane, allowing entry of the viral core into the cytoplasm of the cell (Fig. 14.4). In the same year, a mutation in the gene encoding CCR5 was found to be responsible for the resistance of a small minority of individuals of European descent to infection by HIV-1.

A number of research groups had documented the apparent resistance of a minority of individuals to infection by HIV-1 despite repeated high risk exposures to the virus (Rowland-Jones *et al.* 1995). Study of this extreme phenotype was to prove rewarding. Paxton and colleagues analysed in detail 25 people with a history of multiple high risk sexual exposures who remained free of HIV-1 infection based on available testing at the time using an enzyme-linked immunosorbent assay (ELISA) and diagnostic polymerase chain reaction (PCR) (Paxton *et al.* 1996). The T lymphocytes expressing CD4 in these individuals were significantly less susceptible to infection by different primary isolates of the virus than non-exposed controls. This was most evident for cells derived from two men, designated EU2 and EU3, that required a 1000-fold more virus to establish infection *in vitro*. These individuals had reported sex with multiple HIV-1 infected partners yet had remained uninfected. The genetic basis for this resistance was found to relate to CCR5 (Liu *et al.* 1996). *CCR5* cDNA was amplified and cloned into an expression vector then cotransfected with a *CD4* expression vector. This showed that the *CCR5* constructs prepared using cDNA derived from EU2 and EU3 were inactive.

Gene products	Allele(s)	Effect
Barriers to retroviral infection		
TRIM5α	SPRY species specific	Infection resistance, capsid specific
ABOBEC3G	Polymorphisms	Infection resistance, hypermutation
Influence on HIV-1 infection		
Coreceptor/ligand		
· CCR5	Δ32 homozygous	↓ Infection
· CCL2, CCL 7, CCL11		↑ Infection
Cytokine		
· IL-10	5' A dominant	↓ Infection
Influence on development of AIDS		
Coreceptor/ligand		
· CCR5	Δ32 heterozygous	↓ Disease progression
· CCR2	164 dominant	↓ Disease progression
· CCL5 (RANTES)	In1.1c dominant	↑ Disease progression
· CCL3L1 (MIP1α)	Copy number	↓ Disease progression
· DC-SIGN	Promoter variant	↑ Parenteral infection
Cytokine		
· IL-10	5' A dominant	↑ Disease progression
· IFN-Y	179T dominant	↑ Disease progression
Innate		
· KIR3DS1 (with HLA-Bw4)	3DS1 epistatic	↓ Disease progression
Adaptive		
· HLA-A, HLA-B, HLA-C	Homozygous	↑ Disease progression
· HLA-B*5802, HLA-B*18	Codominant	↑ Disease progression
· HLA-B*35-Px	Codominant	↑ Disease progression
· HLA-B*27	Codominant	↓ Disease progression
· HLA-B*57, HLA-B*5801	Codominant	↓ Disease progression

Figure 14.2 Human genetic variation associated with HIV infection and disease. From Heeney *et al.* (2006), reprinted with permission from AAAS.

Liu and coworkers proceeded to sequence the coding region of the *CCR5* gene using cDNA from EU2, EU3, and a normal donor (Liu *et al.* 1996). This demonstrated a 32 bp deletion in each of the EU2 and EU3 cDNAs that corresponded to the second extracellular loop of the CCR5 receptor. The effect was dramatic: a frameshift occurs, prematurely terminating translation which truncates the CCR5 protein from 352 amino acids to 215

Box 14.2   HIV infection

Primary infection with HIV results in a 'flu-like illness in most cases, an acute viraemia, reduction in circulating CD4+ T cells, and activation of CD8+ cells which kill infected cells. Following the appearance of antibodies to HIV (seroconversion), an asymptomatic period is seen of variable duration during which there is persistence of the virus and gradual reduction in number and function of CD4+ T cells leading ultimately to the appearance of opportunistic infections and the development of AIDS (Fig. 14.3). The development of highly active antiretroviral therapy (HAART) and effective prophylaxis for opportunistic infections has dramatically improved the life expectancy and clinical course for people infected with HIV-1 when such therapies are available. Long term cohort studies such as CASCADE (Concerted Action on Seroconversion to AIDS and Death in Europe; Smit *et al.* 2006) have analysed survival

following HAART and highlighted cofactors such as age which are important in determining HIV-1 disease progression, with older people progressing faster to AIDS. The particular determinants of disease risk and progression for the individual are complex and incompletely understood. What is recognized is that this is highly variable, with a proportion of the population who appear resistant to infection with HIV-1 despite high risk exposure; while others demonstrate long term survival following seroconversion and remain symptom-free with no antiretroviral therapy for over 10 years, and in some cases 25 years. Specific characteristics of the virus may influence replication and evasion of the host immune response, while host cellular factors can modulate the ability of the virus to gain entry to cells and successfully replicate (reviewed in Lama and Planelles 2007).

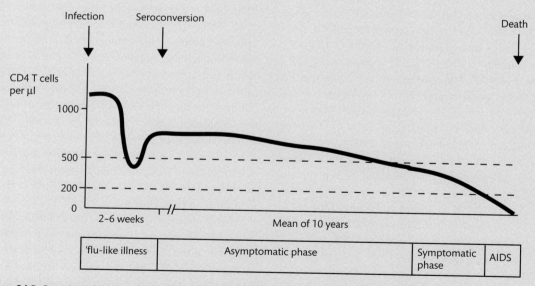

Figure 14.3 Overview of disease progression in untreated HIV infection. Circulating levels of CD4+ cells are shown during the initial acute viraemic phase, which is usually associated with an acute influenza-like illness, followed by an asymptomatic phase of variable duration at the end of which opportunistic infections become more common as the CD4 T cell count falls (<500 cells/μl). AIDS typically develops when the CD4 count is <200 cells/μl. Reprinted with permission from Janeway *et al.* (2005).

Box 14.3 Variants of HIV, coreceptor specificity, and disease progression

Variants of HIV have been described specific to particular cell types and coreceptors, which correlate with disease progression (Connor *et al.* 1997). The terminology can be confusing and related initially to the ability of the virus to infect particular cells *in vitro*. The main variants of HIV *in vivo* infect T cells, dendritic cells, and macrophages using the CCR5 coreceptor and are found in newly infected people: they were initially described as macrophage tropic (M-tropic) as *in vitro* they infected macrophages

but not T cells, but are now usually described as R5 based on their coreceptor use. Later in the course of disease, lymphocyte tropic variants are much more common; *in vivo* they are restricted to infecting CD4+ T cells using the CXCR4 receptor and are known as X4 viruses (previously as T-tropic based on growth *in vitro* in T cell lines). This phenotypic switch between R5 and X4 is associated with faster disease progression.

---

amino acids, and causes it to fail to be expressed at the cell surface or be detectable in the cytoplasm (Fig. 14.5). This deletion, which has since been dubbed the CCR5 Δ32 allele, was confirmed on sequencing genomic DNA of EU2 and EU3. When the investigators looked in more detail at the group of 25 individuals, one further individual was noted to be highly resistant to viral infection and that person was also, like EU2 and EU3, homozygous for CCR5 Δ32. The polymorphism was found to be relatively common among Europeans with 24 of 122 individuals heterozygous for CCR5 Δ32, giving a minor allele frequency of 0.098. The resistance to HIV-1 was specific to R5 viruses found in primary infective isolates. These account for more than 95% of incident HIV-1 infections: the cells were, however, fully infectable by X4 strains of HIV-1 that use a different coreceptor, CXCR4.

The role of CCR5 Δ32 in resistance to HIV-1 infection was independently demonstrated and published by Samson and colleagues (1996) in the journal *Nature* and Dean *et al.* (1996) in *Science* at the same time as Liu *et al.* (1996) published their findings in the journal *Cell*. Samson and coworkers had found the CCR5 Δ32 deletion by sequencing *CCR5* in three HIV-1 infected patients in whom disease progression was slow. They postulated that a 10 bp repeat region flanking the deletion may have promoted the recombination event resulting in the Δ32 deletion. They noted a similar minor allele frequency in Europeans as had Liu and coworkers, at 0.09, but strikingly the deletion was not found in a cohort of individuals

from West and Central Africa or from Japan. The allele frequency in a Caucasian cohort of 723 seropositive patients from Belgian and Parisian hospitals was significantly lower than among the general population, at 0.054, but no individuals homozygous for the CCR5 Δ32 deletion were present among this cohort infected with HIV-1.

Dean *et al.* had mapped the genetic locus to chromosome 3p21 and then identified the Δ32 deletion (Dean *et al.* 1996). They reported that homozygosity for the deletion occurred only among HIV-1 antibody negative individuals and not among HIV-1 infected subjects. Similar results were found by Huang *et al.* with significant overrepresentation of CCR5 Δ32 deletion in at-risk but uninfected people (3.6% of 446 Caucasian subjects) and significant underrepresentation among infected people (Huang *et al.* 1996). In fact among the 2741 HIV-1 infected individuals reported by the three independent cohort studies in 1996, no subjects were found who were homozygous for the CCR5 Δ32 deletion (Dean *et al.* 1996; Samson *et al.* 1996; Huang *et al.* 1998).

What of those individuals who were heterozygous for the CCR5 Δ32 deletion, possessing one 'normal' allele and one bearing the deletion? Analysis of lymphocytes heterozygous for CCR5 Δ32 showed very reduced expression of CCR5 (Wu *et al.* 1997) with subsequent analysis demonstrating this was due to a gene dosage effect rather than receptor sequestration (Venkatesan *et al.* 2002). The frequency of individuals heterozygous for the CCR5 Δ32 deletion was reported to be 35% lower in a cohort

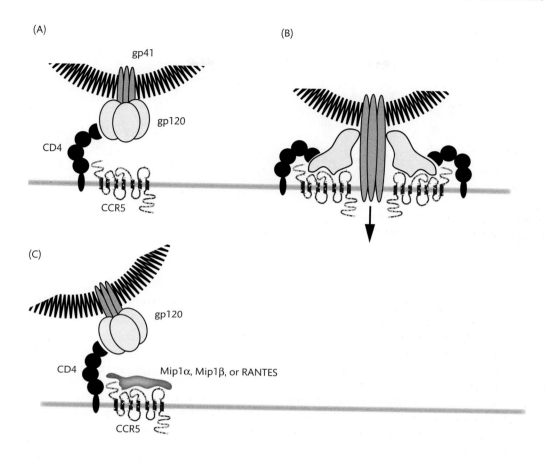

Figure 14.4 HIV virion, CD4 receptor, and CCR5 coreceptor. (A) The HIV virion includes an outer lipoprotein membrane made up of 72 glycoprotein complexes. Individual complexes comprise an external glycoprotein gp120 and a transmembrane protein gp41. The CD4 molecule binds to gp120, 'docking' the HIV-1 virion surface envelope protein. (B) Subsequently, interaction between gp120 and the coreceptor protein CCR5 leads to a change in conformation of the R5 viral gp41 protein which penetrates the cell membrane. The viral envelope then fuses with the host cell membrane allowing entry of the viral core into the cell cytoplasm. (C) The chemokine ligands Mip1α (macrophage inflammatory protein-1 alpha), Mip1β, or RANTES (regulated upon activation T cell expressed and secreted), which normally bind to CCR5, block the infection of cells by R5 HIV-1. During the course of HIV-1 infection in most Caucasian populations where HIV clade B is predominant, a change in coreceptor preference occurs from CCR5 to CXCR4 due to a mutational shift in the gene encoding the envelope glycoproteins. For CXCR4, the normal ligand is SDF-1 (stromal cell-derived factor 1) and this protein will inhibit entry of X4 HIV-1. Redrawn by permission from Macmillan Publishers Ltd: Nature Genetics, copyright 2004, and from Dr Robert Doms (University of Pennsylvania).

of Caucasian individuals infected with HIV-1 than in an uninfected cohort (Samson *et al.* 1996). Protection from infection was also reported among high risk seronegative groups (Marmor *et al.* 2001). In contrast, other studies found no clear evidence of protection from infection but rather suggested or clearly demonstrated a role in disease progression (Dean *et al.* 1996; Huang *et al.* 1996; Zimmerman *et al.* 1997). Heterozygosity has been associated with a slower rate of progression, being present in some cohorts at twice the frequency in those surviving

(A)   (B)

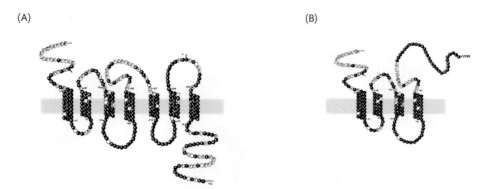

Figure 14.5 Structure of the CCR5 coreceptor. (A) The CCR5 chemokine coreceptor showing its seven transmembrane domain G protein-coupled structure. (B) Non-functional CCR5 Δ32 protein lacks the final three transmembrane segments and the regions involved in G coupling. Redrawn from McNicholl *et al.* (1997).

greater than 10 years with infection compared to rapid progressors (Dean *et al.* 1996). There is evidence that the consequences of heterozygosity for the CCR5 Δ32 deletion depends on the nature of the other allele, with protective or deleterious effects dependent on the particular CCR5 haplotype present on that partner allele (Gonzalez *et al.* 1999; Mangano *et al.* 2001; Hladik 2005) (Section 14.2.2). It is also worth noting that the CCR5 Δ32 deletion, in the homozygous or heterozygous state, explains only a small proportion of cases of highly exposed individuals who are persistently seronegative for HIV-1.

Among populations from Asia or Africa, CCR5 Δ32 is extremely rare (Su *et al.* 2000). Resequencing of the coding region of *CCR5* in different ethnic groups identified a number of rare nonsynonymous polymorphisms including the CCR5–893 variant, which was restricted to Japanese and Chinese individuals (Ansari-Lari *et al.* 1997). This polymorphism was associated with substantially reduced coreceptor activity and cell surface expression as it prematurely terminates translation of the CCR5 protein through a frameshift at codon 299, causing the receptor to lack a cytoplasmic tail. The polymorphism is rare and the relationship to HIV-1 infection unclear; in a small case–control study of HIV-1 infected versus uninfected individuals in Japan, no significant effect was found (Shioda *et al.* 2001).

A variety of other mechanisms have been proposed to explain why highly exposed individuals should be persistently seronegative. These include the cellular and humoral immune response acting at a systemic and mucosal level,

production of soluble suppressive factors, and coreceptor mutations (Kulkarni *et al.* 2003).

14.2.2 Haplotypic structure of the CCR5 locus: evolutionary insights, variation between ethnic groups, and relationship to HIV-1 disease susceptibility

The *CCR5* gene and flanking regions is highly polymorphic. Attention was focused initially on the coding gene sequences but it became evident that understanding the extent of variation was important in resolving underlying associations with disease. This is made more difficult by the finding of significant linkage disequilibrium across the locus, including the neighbouring gene *CCR2* – a site itself associated with delayed progression to AIDS (Box 14.4). Could there be functional diversity elsewhere in the region of *CCR5* responsible for such an observation, perhaps modulating *CCR5* gene expression?

The promoter region of *CCR5* was analysed among AIDS patients by Martin and colleagues who demonstrated four common allelic variants with at least ten SNPs. Strikingly, one *CCR5* promoter haplotype, denoted CCR5 P1, which was present in over 12% of Caucasians, was significantly associated with faster progression to AIDS (Martin *et al.* 1998). In the same year, 1998, McDermott and coworkers reported a screen of allelic variation in a region of the *CCR5* promoter among blood donors which showed a specific SNP was associated with rate of progression to AIDS (McDermott *et al.* 1998). Further work

---

Box 14.4   CCR2 polymorphism and disease progression in HIV-1 infection

The chemokine receptor gene *CCR2* lies within 10 kb of the *CCR5* gene on chromosome 3. On certain cell types, *CCR2* acts as a minor coreceptor for HIV-1. Screening for polymorphism in *CCR2* identified a G to A single nucleotide polymorphism (SNP) that resulted in a substitution of valine for isoleucine at position 64 (rs1799864, c.190G>A, p.V64I; the rarer allele of the SNP was dubbed CCR2–64I) resulting in a conservative change in the first transmembrane domain of the protein (Smith *et al.* 1997). Possession of one or two copies of CCR2–64I did not appear to modulate susceptibility to infection with HIV-1 but was associated with a delay in disease progression. The effect

appeared independent of CCR5 Δ32. The disease association was confirmed in an independent study but this study found evidence of linkage between CCR2–64I and a SNP in the *CCR5* regulatory region (Kostrikis *et al.* 1998). The effects of CCR2–64I appeared most significant early in infection (Mulherin *et al.* 2003) before the emergence of X4 strains (van Rij *et al.* 1998b). The mechanism whereby CCR2–64I may exert a functional effect remains unclear; it may be serving as a marker for a coinherited variant, for example in *CCR5*, or be exerting a direct effect as there is some evidence that it may act to downregulate surface expression of CCR5 (Nakayama *et al.* 2004).

---

was needed to delineate the haplotypic structure of the locus, understand the regulation of gene expression, and determine the functional significance of noncoding DNA sequence polymorphism. Regulation of *CCR5* gene expression was found to be complex, with at least two promoters implicated and multiple alternatively spliced isoforms (Mummidi *et al.* 1997, 2000).

Sequencing of the *CCR5* open reading frame (ORF) and *cis*-regulatory region for 60 human alleles defined 27 haplotypes while sequencing of 43 non-human primates (and genotyping of an additional 40) allowed Mummidi and colleagues to define the ancestral haplotype and to resolve an evolutionary framework from which a rooted phylogenetic network could be constructed (Mummidi *et al.* 2000). This allowed seven distinct clusters of haplotypes to be defined, denoted as haplogroups HHA to HHG (Fig. 14.6). HHA included the ancestral haplotype, while the CCR5 Δ32 deletion was found in haplogroup HHG. The previously reported allele CCR5 P1 was found to be a mixture of HHE, HHF*1 and HHG*1, illustrating the value of using an evolutionary framework to organize and classify haplotypes.

This classification is important in many ways. It allows the variability between frequencies of haplotypes among human populations to be assessed, and indeed significant variability between populations is found (Fig. 14.7)

(Gonzalez *et al.* 1999, 2001). The ancestral haplotype HHA was found to be most common in people of African descent with the highest frequency among Mbuti and Biaka pygmies. Haplogroups HHB and HHD were noted to be African-specific, while HHC haplotypes had a higher frequency in Caucasians than African Americans.

Does consideration of haplogroups provide a rational basis for analysis of any association between genetic variation at *CCR5* and disease? The picture remains unclear but certain haplogroups or allelic pairs of haplogroups are reported to show significant associations with disease progression. The pairing of HHA and HHF*2 was the most common found in Africans, and was associated with slower disease progression in an African American cohort, while HHF*1 was associated with accelerated progression to AIDS across populations groups (Gonzalez *et al.* 1999, 2001). It has also been shown that individuals homozygous for HHE have increased risk of acquisition and disease progression (Gonzalez et al. 1999; Mangano *et al.* 2001). Consideration of haplogroups also facilitates the definition of putative regulatory variants in terms of gene regulation. For example, the HHA haplotype showed the lowest level of transcriptional activity in reporter gene constructs and there was evidence of haplotype-specific protein–DNA binding by transcription factors such as NFkB (Mummidi *et al.* 2000).

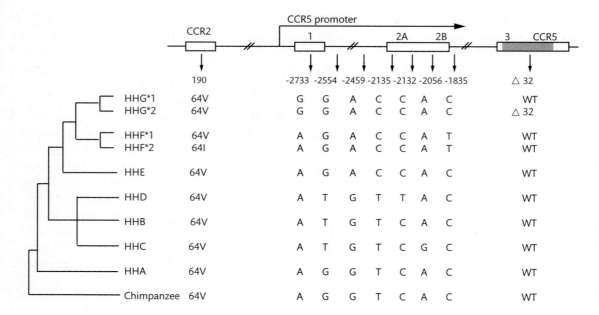

Figure 14.6 *CCR5* haplotypes. *CCR5* haplogroups are shown in the context of the genomic location of the *CCR5* and *CCR2* genes. Numbering 1–3 refers to *CCR5* exons with an open reading frame in exon 3. Alleles for seven SNPs in the *CCR5* cis-regulatory region are shown for each haplogroup, together with the CCR5 Δ32 deletion and CCR2–64V/I. Redrawn with permission from Mummidi *et al.* (2000).

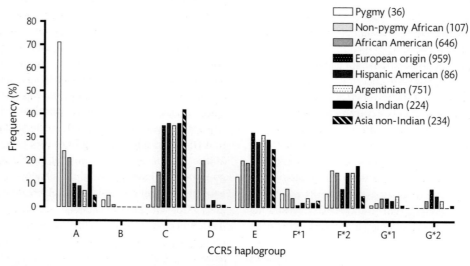

Figure 14.7 *CCR5* haplotype frequencies vary between populations. The frequencies for haplogroups in different populations are shown as percentages, with the total number of individuals given in brackets. Data are plotted from tabulated data in Gonzalez *et al.* (2001).

Specific SNPs in the *cis*-regulatory region of *CCR5* have also been associated with disease progression, notably an A to G SNP found at −2459 nt (rs1799864) a SNP in linkage disequilibrium with the CCR5 P1 allele and found on the HHE, HHF, and HHG haplogroups (Martin *et al.* 1998; McDermott *et al.* 1998; Clegg *et al.* 2000; Knudsen *et al.* 2001). For example, HIV-1 infected homosexual men homozygous for the G allele were reported to show significantly slower progression to AIDS than those homozygous for the A allele, on average 3.8 years more slowly (McDermott *et al.* 1998). This study also found that the G allele was associated with 45% lower reporter gene expression. Subsequently, the G allele was found to be associated with an ability of macrophage-tropic HIV-1 to propagate *in vitro* in peripheral blood mononuclear cells from healthy Caucasian donors (Salkowitz *et al.* 2003). Low, medium, and high viral propagation was seen in cells from individuals with GG, GA, and AA genotypes, respectively. Moreover, flow cytometry of CD14+ donor monocytes showed a lower density of CCR5 on cells from people homozygous GG, or heterozygous GA, compared with those homozygous AA for this SNP. Consistent results were reported in Langerhans cells, which show significant variation in their ability to be infected when cells from a panel of healthy volunteers were studied (Kawamura *et al.* 2003). Those volunteers heterozygous for both the CCR5 Δ32 deletion and the −2459 SNP (a haplotype carried by about 5% of Caucasians) were nearly four-fold less susceptible to macrophage tropic HIV-1 infection than volunteers homozygous for the common alleles. Among a small cohort of 93 exposed seronegative subjects, this combination of being heterozygous for the CCR5 Δ32 deletion and the −2459 SNP was significantly more common than among low risk controls (Hladik *et al.* 2005). Flow cytometry of the compound heterozygotes showed a lower density of CCR5 on CD4+ T cells and monocytes. Data from both population genetic studies and functional assays illustrate the value of haplotype-based analysis at the *CCR5* locus, and show how observed effects are dependent on the particular combination of haplotypes possessed by a given individual.

14.2.3   Coreceptor ligands and HIV-1

HIV-1 suppressing factors produced by CD8+ T cells were identified in 1995 as the chemokines RANTES (regulated on activation normal T cell expressed and secreted), MIP1α (macrophage inflammatory protein-1 alpha), and MIP1β (Cocchi *et al.* 1995). These chemokines were found to be natural ligands for the CCR5 coreceptor necessary for macrophage tropic viral entry, and are associated with HIV-1 transmission and disease progression (see Fig. 14.4) (Gallo *et al.* 1999). RANTES inhibits *in vitro* replication of R5 HIV-1 strains that use CCR5, and high circulating levels of RANTES have been associated with resistance to infection and with delayed onset of AIDS. RANTES is encoded by the *CCL5* gene and two promoter SNPs (-403G/A and -28C/G) (rs2107538 and rs2280788) were found with an ancestral haplotype AC present at highest frequencies among individuals of African descent (Gonzalez *et al.* 2001). Both SNPs have been associated with differential gene expression and with risk of disease acquisition and progression (Liu *et al.* 1999; McDermott *et al.* 2000; Nickel *et al.* 2000; Gonzalez *et al.* 2001). However, effects appear specific to particular ethnic groups. Homozygosity for the AC haplotype for example was associated with increased risk of acquiring infection and worse outcome among European Americans but no effect was seen in African Americans (Gonzalez *et al.* 2001); the AG haplotype was associated with delayed disease progression but appears restricted to Asian populations (Liu *et al.* 1999). Other studies show particular haplotypic combinations may be associated with both risk of increased risk of acquisition and slower disease progression (McDermott *et al.* 2000). Additional SNP discovery then suggested the major effect may lie with an intronic SNP (rs2280789, c.76+231C>T, dubbed 'In1.1') present in a regulatory element, of which the C allele was found to modulate nuclear protein binding and be associated with reduced transcription and faster progression to AIDS (An *et al.* 2002).

The CXC chemokine ligand 12 (CXCL12), also known as stromal cell-derived factor 1 (SDF-1), inhibits X4 viruses usually found in the late stage of disease by downregulation of the HIV-1 coreceptor CXC chemokine receptor 4 (CXCR4) (Bleul *et al.* 1996; Oberlin *et al.* 1996). SDF-1 is the only known natural ligand of CXCR4 and two splice variants of the *SDF-1* gene have been identified. A common G to A transition in a conserved area of the 3′ untranslated region (UTR) of one of the two transcripts was found (rs1801157, c.*519G>A); among individuals homozygous for the SNP (SDF1-3′A) an initial study found an association with delayed disease progression to AIDS, particularly in the later stages of the disease (Winkler *et al.* 1998). However this has proved controversial, with other studies supporting this association, showing no effect, or indeed accelerated progression to AIDS (Mummidi *et al.* 1998; van Rij *et al.* 1998a; Brambilla *et al.* 2000). Further haplotypic analysis including nine SNPs spanning *SDF-1* has shown a modest association with infection and disease progression (Modi *et al.* 2005). A study of the role of chemokine and chemokine–receptor gene polymorphisms in response to HAART did highlight an association between homozygotes for SDF1-3′A and undetectable plasma HIV-1 RNA after one year of treatment (Puissant *et al.* 2006).

14.2.4 Copies count: copy number variation in CCL3L1, a natural ligand of CCR5, and HIV disease

A landmark study by Ahuja and colleagues was published in the journal *Science* in 2005 showing that the number of copies of a gene possessed by an individual could be a major determinant of susceptibility to an infectious disease (Section 4.5) (Gonzalez *et al.* 2005). Copy number variation in the segmental duplication that contains the chemokine gene, *CCL3L1*, was shown to be strongly associated with susceptibility to HIV-1 and the rate of disease progression. Copy number varied significantly between populations and the effect was only apparent when the copy number for an individual was considered with respect to the median for the ancestral population of that individual.

CCL3L1 (also known as MIP1αP and LD78β) is found on the long arm of chromosome 17, together with *CCL4L1*, and these genes are thought to represent segmental duplications of neighbouring genes *CCL3* and *CCL4*. Unlike *CCL3* and *CCL4*, for which only two copies of each are found in the diploid genome, *CCL3L1* and *CC4L1* are present at variable numbers in a population, being found at frequencies of one to six copies per diploid genome in Caucasians and in some individuals are absent altogether (Townson *et al.* 2002).

CCL3L1 was found to have a high affinity for chemokine receptors, notably CCR5, being significantly greater than CCL3 or RANTES such that it is thought to be the most potent ligand of CCR5, and the ligand with the most potent HIV suppressive effects (Nibbs *et al.* 1999). CCL3L1 significantly inhibited infection of peripheral blood mononuclear cells by HIV-1 strains which used CCR5 (Menten *et al.* 1999) and had potent antiviral activity in macrophages (Aquaro *et al.* 2001). The truncated form of CCL3L1, naturally produced by the lymphocyte surface glycoprotein and protease CD26, had even higher affinity for CCR1 and CCR5 receptors (Proost *et al.* 2000). CCL3L1 is thus a highly active natural HIV entry inhibitor.

Given the reported variation in copy number of *CCL3L1* and its potent HIV suppressing activity, Ahuja and colleagues investigated the relationship between copy number and disease susceptibility (Gonzalez *et al.* 2005). They first studied how copy number varied within and between populations of different geographic and ethnic origin. It was a large study group, 1064 people from 57 populations, as well as 83 chimpanzees. This showed clear variation between populations, notably between African and non-Africans with an average of four copies of *CCL3L1* in Africans and two in non-Africans. The investigators proceeded to study large cohorts of more than 5000 HIV infected and uninfected people and clearly showed that *CCL3L1* copy number was a major determinant of susceptibility to HIV-1 for an individual when compared to the median for their ancestral population. This was true for individuals infected by mother-to-child transmission and adult-to-adult transmission: with a copy number lower than the median, risk was higher; at copy numbers higher than the median, the risk fell in a dose-dependent stepwise fashion for the cohort of children exposed perinatally to HIV-1. For each copy of the *CCL3L1* gene, the risk of acquiring HIV-1 reduced by 4.5–10.5% depending on the specific population studied. Overall, HIV positive persons had lower *CCL3L1* copy numbers than uninfected people. Moreover the rate of disease progression was also associated with copy number: at a lower gene dosage than the median for the cohort or population there was a dose-dependent increased risk of faster progression to AIDS or death.

When *CCL3L1* copy number variants were analysed together with the known disease associated variants of CCR5, the authors estimated that variation in these factors accounted for up to 42% of the burden of HIV/AIDS and 30% of the accelerated rate of progression to AIDS (Gonzalez *et al.* 2005). It is thus postulated that in the presence of increased copies of *CCL3L1* there is greater binding of the encoded chemokine to the CCR5 receptor, reducing the amount of CCR5 accessible to HIV-1 on the surface of leukocytes.

14.3   Barriers to retroviral infection

14.3.1 Genetic diversity in TRIM5α gives insights into the impact of retroviruses during primate evolution

Retroviruses have been colonizing vertebrate hosts for hundreds of millions of years, leaving a calling card of integrated retroviral sequence transmitted between generations (Gifford and Tristem 2003). An early discovery in research into the current HIV-1 pandemic was the finding that this particular retrovirus had a limited host range, being restricted to humans and apes. The barrier to infection in other primate cells was identified in 2004 as being mediated by a protein named TRIM5α, which acted to block HIV-1 replication soon after viral entry into cells and before reverse transcription (Fig. 14.8) (Stremlau et al. 2004).

The discovery of TRIM5α was made after a screen of cDNA clones prepared from HIV-1 resistant rhesus macaque cells and tested for their ability to protect an HIV-1 susceptible human cell line (Stremlau et al. 2004). Subsequent small interfering RNA knockdown experiments to endogenous TRIM5α in rhesus cells removed the block to HIV-1 replication naturally occurring in these cells. This blocking protein was, however, specific to HIV-1 as almost no protection was found to the related lentivirus, SIVmac, when the cDNA for rhesus TRIM5α was introduced into human cells. Indeed the pattern of restriction of viral replication varies significantly between primate species and between viruses, with human TRIM5α poorly restricting HIV-1 but efficiently restricting N-MLV, a murine γ retrovirus (Perron et al. 2004).

TRIM5α was found to act by binding to the retroviral capsid core, leading to accelerated uncoating and degradation of the viral capsid (Chatterji et al. 2006; Stremlau et al. 2006). A particular domain on TRIM5α was found to be involved in recognizing the viral capsids entering the cell and to be a critical determinant of species-specific restriction of primate lentiviruses. This pattern recognition molecule is denoted the B30.2/SPRY domain (Fig. 14.9). Between primate species, there is evidence of marked sequence diversity in TRIM5α including deletions, insertions, and nucleotide substitutions and duplications (Newman and Johnson 2007). The greatest amino acid variation is seen in the B30.2/SPRY domain. The domain comprises a conserved central hydrophobic core with a binding surface comprised of protruding loops of variable length containing non-conserved amino acids (Fig. 14.9). It appears no coincidence that the key region of this domain determining retroviral specificity (dubbed the SPRY 'patch') should also have the most evidence of positive selection, with variation in the protein found in different primate lineages thought to reflect past infections with pathogenic retroviruses (Box 14.5) (Sawyer et al. 2005; Stremlau et al. 2005; Yap et al. 2005).

Different primate species encode different TRIM5α proteins, each with a different antiretroviral restriction activity (Song et al. 2005). There is marked sequence variation between Old and New World monkeys with specific regions and individual amino acids altering TRIM5α activity. For example, substitution of arginine (found in humans and chimpanzees) to proline (found in rhesus monkeys) in position 332 of human TRIM5α within the SPRY patch restricts HIV-1 (Yap et al. 2005; Li et al. 2006c).

Moreover there is evidence that amino acid variation at this site will cause retrovirus-specific restriction. An ancient endogenous retrovirus reconstructed from the chimp and gorilla genomes (PtERV1) was found to be restricted by human TRIM5α but this restriction was lost on substituting the arginine (R) for a glutamine (Q), the hominid ancestral residue (see Fig. 14.9). In contrast, this change in amino acid improved the ability of the TRIM5α to restrict HIV-1 (Kaiser et al. 2007). This remarkable piece of scientific work, resurrecting an extinct retrovirus found at more than 100 copies in chimpanzee and gorilla genomes but absent from the human genome, suggests that TRIM5α became fixed in our early human ancestors based on its ability to confer resistance to a retrovirus such as PtERV1 some 3 to 4 million years ago and that this may, in part, be responsible for our current low resistance to HIV-1 (Kaiser et al. 2007). A large body of work has now clearly demonstrated that evolution of diversity in TRIM5α has arisen due to species-specific ancestral retroviral challenges.

Resequencing TRIM5 from indigenous human subjects from geographically distinct regions (Sawyer et al. 2006) as well as HIV infected and exposed but seronegative individuals (Speelmon et al. 2006) has revealed many SNPs, including five common nonsynonymous polymorphisms. Analysis of the distribution of SNPs has shown a marked

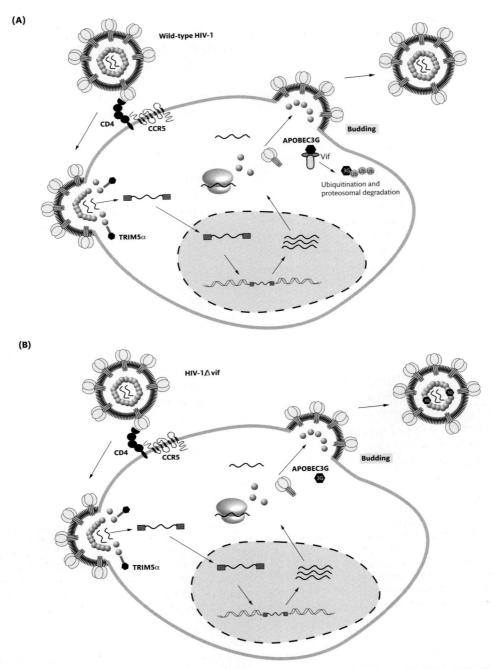

Figure 14.8 Site of action of TRIM5α and APOBEC3G during HIV-1 infection. TRIM5α acts to restrict HIV-1 after viral entry into the cell by binding to the retroviral capsid core. APOBEC3G acts later, but this is critically dependent on whether viral Vif protein is present. (A) In wild-type HIV-1, the integrated HIV-1 provirus is transcribed and translated. The Vif protein acts to bind human APOBEC3G, targeting it for ubiquitination and proteosomal degradation. (B) In contrast with HIV-1/Δvif, no Vif protein is synthesized and APOBEC3G is therefore not degraded but rather can become incorporated into the nascent virion and restrict further infection. Redrawn and reprinted from Holmes *et al.* (2007), copyright 2007, with permission from Elsevier.

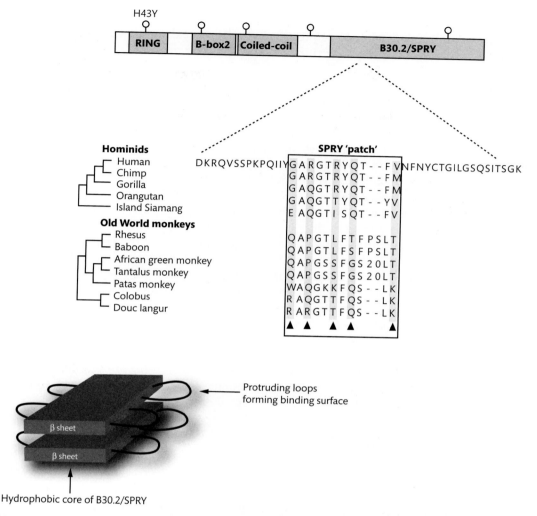

Figure 14.9 Structure and polymorphism of TRIM5α. TRIM5α is a member of the tripartite motif (TRIM) family of proteins and comprises RING finger, B-box, and coiled coil domains; an additional SPRY domain is specific to this α isoform of the TRIM5 proteins (a domain found also in the immunoglobulin family). TRIM5α is the longest of four splice variants of the *TRIM5* gene found on chromosome 11. 'Lollipops' shown above the domains are at the locations of reported common human nonsynonymous polymorphisms (Newman and Johnson 2007). The SPRY patch is shown with black triangles highlighting the five codons in the whole protein that have evidence of evolving under positive selection based on analysis of hominids and Old World monkeys. Redrawn with permission from Newman and Johnson (2007) and from Sawyer *et al.* (2005) .

skew for the SPRY domain consistent with a recent selective sweep in the population removing variation in this region of the protein (Sawyer *et al.* 2006). Only a limited analysis of within species variation has been carried out among other primate species. Among Asian macaques and African monkeys, sequence diversity was clustered in the coiled coil domain and variable portions of B20.2/SPRY – the latter in marked contrast to human subjects.

HIV-1 disease association studies have found no consistent effect of the *TRIM5* genotype on disease progression and variable, but modest, effects on disease susceptibility (Javanbakht *et al.* 2006; Speelmon *et al.* 2006). No

---

Box 14.5 Analysis of primate sequence diversity reveals ancient positive selection in TRIM5α and defines a key functional element of the protein

Sawyer and colleagues sequenced the coding region of the TRIM5α gene for 17 primate genomes and analysed rates of synonymous and nonsynonymous changes between species (Sawyer *et al.* 2005). Such analysis allows estimation of selective pressures that have acted on a gene: in the majority of protein coding genes, rates of synonymous changes are greater than nonsynonymous as amino acid changes are usually deleterious to protein function and subject to purifying selection, resulting in their loss from the population (Section 10.3). For TRIM5α the reverse was true with ratios of nonsynonymous to synonymous changes greater than one for many branches of the primate phylogeny analysed: the range of hominids and Old and New World monkeys sequenced revealed that positive selection had been acting on this protein for more than 33 million years (Sawyer *et al.* 2005). Different branches of the phylogeny showed different ratios with marked evidence of positive selection for the hominid clade when sequences from the whole gene were analysed.

Strikingly, the region of TRIM5α that was found to have undergone the most intense positive selection was the SPRY domain. When hominids and Old World monkeys were analysed, only five residues in the whole protein were identified with a high level of confidence as having undergone positive selection, and all were found to lie in a very short region of the SPRY domain, dubbed the SPRY patch. Such a tight clustering of positive selection was felt by Sawyer *et al.* to be likely to predict a 'point of physical contact between two proteins locked in genetic conflict' and to be reminiscent of the situation in the major histocompatibility complex (MHC) where colocalization of positive selection to regions of the protein involved in antigen recognition has been found (Section 12.4) (Sawyer *et al.* 2005). Functional analysis demonstrated this was indeed the case: substituting the patch region between human and rhesus TRIM5α using chimeric proteins showed it contained the key amino acids conferring species-specific restriction for HIV-1.

---

individual effect of *TRIM5* SNPs was seen for *in vitro* HIV-1 susceptibility of CD4+ T cells (Speelmon *et al.* 2006). In contrast, a SNP (rs3740996, c.127C>T, p.H43Y) in the loop region of the RING domain of TRIM5α (see Fig. 14.9) was found to be associated with impaired replication of murine N-MLV retrovirus (Sawyer *et al.* 2006). Based on frequencies in different populations, the estimated age of the H43Y haplotype was 500 000 years, similar to the estimate of 600 000 years for an average neutral human polymorphism (Kreitman and Di Rienzo 2004). While rare in Africans, the allele was found to be very common in Central and South Americans suggesting that there has not been a detrimental selective pressure to reduce the allele frequency in these populations from the time of the small founder populations on human migration to the Americas some 15 000–30 000 years ago (Sawyer *et al.* 2006). While not having any significant effect on HIV-1

restriction in human cells, the high frequency of this variant in particular populations may predispose them to greater or reduced risk in future retroviral infections.

The analysis of TRIM5α in humans and primates has provided dramatic insights into our evolutionary history with exposure to ancient and more recent retroviruses likely to determine our current and future susceptibilities.

14.3.2 APOBEC3G: an innate host defence mechanism against retroviral infection

Another part of our defence arsenal against retroviral infection after viral entry into cells was identified in humans in 2002 by Malim and colleagues as apolipoprotein B-editing catalytic polypeptide 3G (APOBEC3G) (Sheehy *et al.* 2002). It had been recognized since 1987 that the virion infectivity factor (Vif) protein of HIV-1 was essential for viral infectivity (Fisher *et al.* 1987; Strebel *et al.* 1987) and that this protein was required to overcome a restriction factor encoded by the host cell (Madani and Kabat 1998; Simon *et al.* 1998). This 'restriction factor' was found to be APOBEC3G. In the absence of Vif, APOBEC3G is incorporated into the nascent budding virion from the producer cell and, on infection of the next target host cell, is delivered into the cell with the viral genome where APOBEC3G acts to block infection (Harris *et al.* 2003). This block relates to the cytidine deaminase activity of APOBEC3G that results in hypermutation of nascent retroviral transcripts and degradation (see Fig. 14.8). Unfortunately most lentiviruses encode the Vif protein, which targets APOBEC3G for ubiquitination and proteasome-dependent degradation, preventing incorporation into the nascent virion (Madani and Kabat 1998; Marin *et al.* 2003; Sheehy *et al.* 2003; Yu *et al.* 2003). A state of evolutionary conflict has therefore been proposed between APOBEC3G and Vif with positive selection leading to rapid fixation of mutations that alter amino acids (Sawyer *et al.* 2004).

APOBEC3 encoding genes have been found to be restricted to mammals, while the flanking genes are highly conserved, being found for example in puffer fish. In the mouse and rat, only one *APOBEC3* gene has been identified, while in primates there are many more with eight found in humans: six lie in a cluster on chromosome 22 (*APOBEC3A, B, C, D, F* and *G*), with a further gene found 14 kb downstream (*APOBEC3H*) and one pseudogene is present on chromosome 12. This rapid expansion in the number of genes may have been driven and selected for by past viral interactions. It is postulated that the diversity in primates at the *APOBEC3* locus on chromosome 22 has arisen through unequal crossover and recombination facilitated by retroviral element relics which make up 19% of the locus in humans (Conticello *et al.* 2005).

Sequencing analysis of *APOBEC3G* in different primate species (hominids, Old World and New World monkeys) revealed that, like *TRIM5α*, most branches in the phylogeny show evidence of positive selection and that this predates the presence of lentiviruses (Sawyer *et al.* 2004). In contrast to *TRIM5α*, however, the evidence for positive selection (ratio of nonsynonymous to synonymous changes) was not localized to a particular domain, with all domains appearing to have undergone multiple

distinct episodes of selection (Sawyer *et al.* 2004). The current interaction with Vif appears to have played only a very small role and much remains to be told of the ancestral story of APOBEC3G.

14.4 Genetic diversity in HLA, KIR, and HIV-1: strategies for survival

Viral co-evolution with molecules of the immune system is thought to have been a major force in shaping the genetic diversity observed today in genes involved in our immune defence. The consequences of this are seen in the human leukocyte antigen (HLA) genes where particular alleles have been found to be associated with differences in disease progression after infection with HIV-1. Research in this field has provided evidence to support the hypothesis of 'heterozygous advantage' (also called 'overdominant selection') – the concept that individuals heterozygous for HLA loci were at an immunological advantage compared to homozygotes as they were able to present a greater variety of antigenic peptides (Section 12.3.1) (Doherty and Zinkernagel 1975b). Carrington and coworkers studied patients who became HIV antibody positive after enrolling in at-risk AIDS cohort studies, and in whom rates of progression to AIDS endpoints could be carefully defined (Carrington *et al.* 1999). They found that among 498 seroconverters, those individuals maximally heterozygous at MHC class I loci had significantly delayed onset of AIDS compared to individuals homozygous at one or more loci (Fig. 14.10) – class I loci encoding proteins responsible for presentation of viral antigen to cytotoxic T lymphocytes (Section 12.2.1).

Are specific class I alleles associated with disease progression? Carrington and colleagues found that six of 63 class I alleles tested were associated with disease progression, notably HLA-B*35 and Cw*04 among Caucasians (Carrington *et al.* 1999). These were large effects: people homozygous for HLA-B*35 progressed to AIDS in half the median time of those without HLA-B*35. A more detailed examination of the role of specific HLA-B*35 subtypes provided clues about the underlying mechanisms involved (Gao *et al.* 2001). Presentation of viral peptides by the B*35 encoded class I molecules on CD4+ T lymphocytes to cytotoxic T cells showed highly specific peptide recognition. Among the five different HLA-B*35

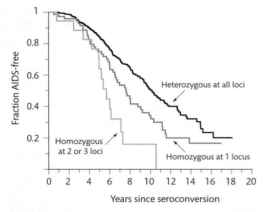

Figure 14.10 Homozygosity at HLA class I and progression to AIDS. A Kaplan–Meier survival curve for 498 seroconverters showing the influence of HLA class I genotype on progression to AIDS (defined as AIDS-1987, namely HIV-1 infection plus an AIDS-defining illness or death). For individuals with one locus homozygous, the relative hazard (RH) was 1.8 ($P = 0.0005$; $n = 74$), for two or three loci homozygous the RH was 4.1 ($P = 1 \times 10^{-6}$; $n = 18$). P values are calculated by the Cox proportional hazards model and analyses are stratified by age and race. From Carrington *et al.* (1999), reprinted with permission from the AAAS.

subtypes, two groups could be resolved with different peptide binding specificities. The HLA-B*35-PY group (mainly comprising B*3501, the commonest subtype of B*35) and the HLA-B*35-Px group differ in the ability to recognize particular amino acids at position 9 in the HIV-1 peptide (the PY group are specific to tyrosine at position 9, the Px group are specific to different amino acids but not tyrosine). All the effect found for HLA-B*35 was due to the HLA-B*35-Px group: a change in preference of a single amino acid in a class I molecule sufficient to dramatically alter disease progression (Gao *et al.* 2001). The observed association with Cw*04 was felt to be attributable to linkage disequilibrium with HLA-B*35-Px alleles.

The strong selective pressure exerted by HLA on the genetic diversity of HIV-1 has also been documented. There is a selective advantage for variant viral epitopes that may escape the cytotoxic T lymphocyte (CTL) response by failing to bind to class I molecules or interact with T cell receptors. Moore and colleagues analysed the reverse transcriptase sequences from 473 people enrolled in the Western Australian HIV Cohort Study

(Moore *et al.* 2002). They chose to look at the reverse transcriptase protein given the high level of expression in virions and immunogenicity in the early host response to HIV-1. Immune escape was found to be common and restricted by a range of different class I molecules. An excess of mutations was found in regions with motifs recognized by HLA alleles associated with rapid progression. Is there evidence that HLA types specific to regions of viral proteins important to viral survival (and therefore harder to mutate without deleteriously affecting function) are advantageous? HLA-B*27 and B*57 encoded molecules are specific to epitopes in conserved parts of p24 viral capsid protein for which at least two mutations are required for immune escape; both HLA types are associated with a survival advantage (Migueles *et al.* 2000; Gao *et al.* 2001; Kelleher *et al.* 2001).

The lysis of HIV-1 infected cells by CD8+ T cells through recognition of viral peptides presented by class I molecules on the infected cell surface is critical to the control of HIV-1 infection. The virus is able to combat this by downregulating expression of class I molecules on the infected cell surface and so make cells less likely to be recognized and lysed (Collins *et al.* 1998). There is a potential problem with such a strategy: downregulation of class I molecules carries the risk of destruction by natural killer (NK) cells. NK cells play a critical role in combating viral infection through killing infected cells, secreting anti-inflammatory cytokines such as interferon gamma (IFNγ), and coordinating, with dendritic cells, the adaptive immune response. NK cells express activating and inhibitory receptors for HLA class I molecules which are critical to determining the NK response between the tolerance of healthy cells and destruction of infected ones (Box 14.6).

The answer for the virus appears to have been selective downregulation of class I molecules by the HIV-1 accessory protein Nef. HLA-A and HLA-B molecules were found to be removed but not HLA-C or HLA-E (Cohen *et al.* 1999). A systematic study of 221 cell lines created to stably express CD4 and specific class I proteins were analysed after infection with HIV-1. HLA-C and HLA-E encoded proteins were protected from viral downregulation by single specific residues in the cytoplasmic tails of the proteins differing between alleles. Retention of expression of HLA-C was sufficient to make the HIV-1 infected cells 'invisible' to most NK cells.

Box 14.6   Polymorphism of KIRs in health, evolution, and disease

NK cell receptors for MHC class I molecules include lectin-like receptors encoded by the NK complex gene cluster on chromosome 12p12-p13 and immunoglobulin-like receptors encoded by genes in the leukocyte receptor complex on 19q13.4, which includes the killer immunoglobulin-like receptors (KIRs). KIRs directly recognize the polymorphic determinants of the class I molecules and are themselves remarkably polymorphic and rapidly evolving (Vilches and Parham 2002). Extreme diversity in gene number and composition is seen between KIR haplotypes, a degree of diversity comparable to that found for HLA class I alleles. Among KIR haplotypes the distribution of Alu repetitive elements (Section 8.4) indicates rapid expansion of the gene family (Martin *et al.* 2000). The KIR genes are highly homologous, as are the highly conserved 2 kb intergenic sequences between them; this is thought to have facilitated non-reciprocal recombination leading to deletion, duplication, and recombination of genes (Parham 2005). One of the intergenic sequences is unique: it is 14 kb, lies in the centre of the cluster (between *KIR3DP1* and *KIR2DL4*), and is the preferred site of reciprocal recombination of the centromeric and telomeric motifs leading to further haplotypic diversity (Martin *et al.* 2003). HLA-A molecules such as A3 and A11 are recognized by receptors encoded by *KIR3DL2* and *KIR2DS1* at the telomeric end of the KIR locus, HLA-B Bw4 is a ligand for KIR3DL1, and in the centromeric part receptors for HLA-C molecules are encoded. HLA-C1 allotypes with Asn at position 80 are recognized by receptors encoded by *KIR2DL3* and *KIR2DL2*; *KIR2DL1* encodes KIR receptors with a methionine rather than lysine at position 44 and these recognize HLA-C2 allotypes with lysine at position 80. As well as the disease associations with HIV-1 infection, genetic diversity at the KIR and HLA class I is important in other infections, autoimmunity, reproduction, and cancer. For example, maternal KIR genotype and fetal HLA-C type are important in pre-eclampsia (Hiby *et al.* 2004); homozygosity of *KIR2DL3* and of HLA-C1 allotypes modulate hepatitis C infection (Khakoo *et al.* 2004); and *KIR2DS1* and Cw*06 are associated with psoriasis (Suzuki *et al.* 2004).

A recent study has highlighted how a combination of specific KIR subtypes and their complementary HLA-B alleles can modulate HIV-1 disease progression and viral load (Martin *et al.* 2007). Martin and colleagues sought to investigate the role of the inhibitory KIR receptor allotypes KIR3DL1 and the class I molecules to which they bind, HLA-B molecules containing the Bw4 motif with isoleucine at position 80 (Bw4-80I). HLA-B Bw4 alleles include HLA-B*27 and B*57, and are known to be highly protective in terms of HIV disease progression. Specific KIR3DL1 allotypes vary in their surface expression on NK cells and thus in their ability to bind class I molecules. Analysis of over 1500 HIV-1 infected individuals revealed that those people with HLA-B alleles Bw4-80I and high expressing KIR3DL1 allotypes had significantly slower disease progression than those individuals with either the low expression allotype or lacking the Bw4 allele (Martin *et al.* 2007). Resequencing KIR3DL1 among diverse human populations has demonstrated remarkable levels of variation with at least 38 alleles and evidence of positive selection notably in the extracellular immunoglobulin domains (Norman *et al.* 2007).

Genome-wide association studies are providing dramatic insights into many common diseases (Section 9.3). For HIV-1 infection, such a study was recently performed to investigate genetic determinants of viral set point, the amount of circulating virus found in the plasma during the asymptomatic phase prior to progression to AIDS (Fellay *et al.* 2007). Over 500 000 SNPs were successfully genotyped for 486 accurately phenotyped patients. The analysis found that genetic variation in the MHC was most significant, in particular that polymorphisms near HLA-B and HLA-C explained 9.6% and 6.5% of the total variation in HIV-1 set point, respectively. Further replication and fine mapping is required but the most strongly associated SNP was in *HCP5*, 100 kb centromeric of HLA-B and in strong linkage disequilibrium with HLA-B*5701, an allele strongly associated with protection against disease progression (Migueles *et al.* 2000) and control of HIV-1 viraemia (Altfeld *et al.* 2003).

14.5   Summary

The intense and sustained research effort to understand the biological and genetic basis of individual variability in susceptibility to HIV/AIDS over the last 25 years has been driven by the devastating scale and consequences of a worldwide pandemic, which has already left an estimated 25 million people dead. The success of this research has been remarkable. There is now evidence of how successful viral entry may be determined by diversity in the specific viral coreceptor or receptor ligand, secreted antiviral peptides such as defensins, or specific proteins involved in viral transmission such as DC-SIGN. The host innate and adaptive immune response to infection is highly significant: interleukins, interferons, and toll-like receptors have been shown to modulate infection, while a complex interrelationship between the virus, MHC class I molecules, and KIRs are critical determinants of disease progression. Cellular factors can influence events early after viral entry into cells to act as barriers to infection (through factors such as TRIM5α, cyclophilin A, and APOBEC) while later acting factors can modulate viral replication, assembly, and release (for example TSG101 and HIV-1 budding, HP68, and assembly of capsid proteins).

The discovery that genetic variation in the gene encoding the major coreceptor for HIV-1, CCR5, had highly significant effects on susceptibility to infection contributed to CCR5 being a drug target for small molecules to act as CCR5 entry inhibitors. This has led to the development of drugs such as maraviroc (Fatkenheuer *et al.* 2005), which have recently been approved by the US Food and Drug Administration (Groeschen 2007). The most dramatic disease association, with the CCR5 Δ32 deletion, has a major effect but the polymorphic variant itself is rare. In contrast, more common polymorphisms of the *CCR5* promoter, and haplotypes spanning the gene region, have been identified with modest effects on disease progression but a greater attributable risk due to their higher allele frequency – the specific variants and underlying functional mechanisms, however, remain controversial. The quest to define host genetic determinants of disease progression in HIV infection has had ramifications throughout the field of human genetics, as illustrated by work on copy number variation in *CCL3L1* which was one of the first studies to demonstrate this very important mechanism determining susceptibility to common disease (Gonzalez *et al.* 2005).

Analysis of genetic diversity at many disease susceptibility loci among different species, notably primates, has also highlighted how positive selection has been operating at a number of genes, allowing insights into molecular function (as seen for the SPRY domain of TRIM5α), species-specific retroviral restriction, and human evolution. The complexity of the task of unravelling the relationship between genetic diversity in the virus and host is daunting, as exemplified by the role of MHC class I molecules. Research into HIV-1 has provided one of the clearest examples of a selective advantage conferred by heterozygosity at HLA of any infectious disease (Carrington *et al.* 1999). However, the extraordinary diversity at this locus in humans, the significant and complex interrelationships with HIV sequence diversity and with the human KIR genes makes for a tough assignment – particularly when you realize that diversity in the KIR genes rivals that of HLA. The recent findings of the clear and dramatic epistatic relationship of *KIR3DL1* and HLA-B Bw4 in HIV-1 infection (Martin *et al.* 2007) are likely to be only the beginning of the story.

Are there more genetic modulators of HIV/AIDS to be found? The answer appears to be 'yes', with only a minority of the genetic risk yet explained. A recent genome-wide association study of determinants of viral set point indicated that the major genetic loci showing association, *HLA-B* and *HLA-C*, explained about 14% of the total variation (Fellay *et al.* 2007). In 2004, O'Brien and Nelson suggested that about 9% of the variation in AIDS disease progression was explained by the sum of the known AIDS restriction genes (O'Brien and Nelson 2004). Other authors have given higher estimates, with Gonzalez and colleagues suggesting in 2005 that variation in *CCR5* and *CCL3L1* accounted for 42% of the disease burden of HIV/AIDS and 30% of accelerated disease progression

(Gonzalez *et al.* 2005). Much remains to be understood and the implications of research on human genetic diversity done in this field are likely to continue to have ramifications far beyond HIV infection.

14.6 Reviews

Reviews of subjects in this chapter can be found in the following publications:

Topic	References
Human genes influencing disease susceptibility and progression in HIV-1/AIDS	Theodorou *et al.* 2003; O'Brien and Nelson 2004; Lama and Planelles 2007
Resistance to HIV-1 among highly exposed persistently seronegative individuals	Kulkarni *et al.* 2003
Origin of HIV and evolution of resistance to AIDS	Heeney *et al.* 2006
TRIM5α and retroviral infection	Newman and Johnson 2007
APOBEC proteins and retroviral restriction	Harris and Liddament 2004
KIRs in innate and adaptive immunity	Vilches and Parham 2002
Genetic diversity of KIR and MHC class I molecules	Parham 2005

CHAPTER 15

Concluding remarks and future directions

15.1 Introduction

The rapid advances in our understanding of human genetic diversity over the last 20 years have been a remarkable feat of scientific endeavour. Over the course of this book, the nature of major classes of genetic variation have been described ranging from large scale microscopic and submicroscopic structural genomic variation to fine scale sequence level variation. As our ability to detect and screen for smaller scale structural variation has improved, through approaches such as microarray-based comparative genome hybridization (array CGH), the extent and importance of such variation, notably copy number variants, has become apparent both among healthy individuals and those with disease. Meanwhile at the sequence level, high throughput DNA sequencing and genotyping have been essential in the quest to determine the sequence of the human genome, understand the nature of variation in human allelic architecture, and perform large scale association studies of disease susceptibility.

In this final short chapter some particular topics will be highlighted that illustrate how the field is advancing and possible future directions. Like elsewhere in the book, the aim is not to be exhaustive in coverage but rather to illustrate important principles and convey some of the excitement felt within this area of research at the rapid pace of change. There is enormous potential for advancing not only our understanding of human genetics and a range of other disciplines from evolutionary biology to forensic medicine, but also to dramatically impact on human health through insights into the pathophysiology of disease and personalized medicine.

15.2 Cataloguing human genetic variation

From early cytogenetic studies revealing major chromosomal abnormalities to more recent analysis of smaller scale structural and fine scale sequence level variation, remarkable diversity in our individual genetic makeup has been demonstrated. The establishment of clear terminology for use in reporting such variation is essential to the continued advancement of the field, and has been highlighted by many investigators. The classification based on size advocated by Scherer and Lee (Fig. 1.29) provides a very useful framework for considering genetic variation (Scherer *et al.* 2007). There is a need for consistency in the specific nomenclature used to describe individual variants following, for example, the recommendations of the Human Genome Variation Society (Box 1.13). The growth in online databases cataloguing human genetic variation such as dbSNP (www.ncbi. nlm.nih.gov/projects/SNP/) for fine scale sequence variation also provides greater opportunities for standardization in reporting through the use of unique identifiers such as rs numbers for specific single nucleotide polymorphisms (SNPs). The establishment of standard reference samples of DNA for comparative study is also very important, notably for the analysis of structural variation.

The completion of the sequencing of the human genome (Section 1.4.2) was a pivotal event in human genetic research, providing the necessary route map and resources with which remarkable progress is now being made in defining genetic diversity. The genome sequence arising from the Human Genome Project and from Celera Genomics was not specific to one diploid individual but rather a composite from a diverse array of individuals, which allowed for the identification of considerable fine scale variation. More recent advances in 'next generation' high throughput sequencing have allowed resequencing of individual diploid genomes (Section 1.4.5), and ongoing work such as a project to sequence at least 1000 individuals of differing geographic ancestry (Hayden 2008; Siva 2008) will further advance our understanding of the nature of human genetic variation, in particular rare variants of which our knowledge is currently very incomplete.

The extent of copy number variation has been highlighted by recent surveys among phenotypically normal individuals (Section 4.2) with evidence that such variation contributes to greater diversity between individuals than that observed at a nucleotide level (Sebat 2007). Copy number variation has important consequences for understanding genomic disorders, as well as common diseases such as Alzheimer's disease, human immunodeficiency virus 1 (HIV-1) infection and psoriasis; as well as individual differences in drug metabolism and gene expression. Ongoing and future studies to accurately quantify and map smaller scale copy number variation, and to establish high throughout approaches and methods of analysis for large scale screening and use in disease association studies should further reinforce the critical role played by this class of genetic variation in human health and disease.

The International HapMap Project (Section 9.2.4) and other large scale surveys of SNP diversity have been of fundamental importance in advancing our understanding of the nature of fine scale genomic architecture and how this can be captured using informative SNPs for genotyping using high throughput technologies. This has enabled recent genome-wide association studies but has also significantly facilitated our understanding of the signatures of selection present in the human genome (Section 10.2), the nature of linkage disequilibrium and recombination (Section 9.2), and the genetics of gene expression (Section 11.4). Technological and scientific advances have allowed SNPs to be used as highly informative genetic markers at a genome-wide level for association studies but it is important to remember the

utility of other variants, notably polymorphic tandem repeats (Section 7.5), and the remarkable work which has taken place in the past to establish genetic linkage maps and to use linkage to identify specific genes and variants, notably in the context of mendelian diseases (Section 2.2).

The analysis of human genetic diversity cannot be considered in isolation and the critical scientific advances achieved through sequencing other species, notably model organisms, has provided essential resources with which to understand human evolutionary origins and the nature of observed human genetic variation. Comparative genomics provides powerful tools to understand the nature of the human genome, how genetic diversity has arisen, and the underlying molecular mechanisms responsible for observed variation. The remarkable extent of mobile DNA elements within the human genome provides unique opportunities to understand past transposition events and how eukaryotic genomes have evolved (Chapter 8). Polymorphism within more recent Alu and other mobile elements has also been very important in understanding human population genetics and evolutionary history.

As our understanding of human genetic variation has advanced, the amounts of data arising from DNA sequencing and other methods of defining genetic variation have increased to a phenomenal extent, a situation which will only increase as resequencing projects at a genomic level progress. This has led to a growing and essential role for bioinformatics as a speciality within this field of research. Statistical genetic approaches applied at a genome-wide level have been fundamental to advancing our understanding of human genetic variation and its impact on disease, as well as on gene expression where mapping of quantitative traits is proving a highly productive approach (Section 11.3).

The multidisciplinary nature of current analyses in human genetic research is highlighted by the role of bioinformatics, but this also serves to emphasize that for the future there is an increasing need for a more collaborative approach, bringing together diverse disciplines. The common themes underpinning human genetic diversity mean that greater communication and interaction is needed between researchers interested in human and population genetics, molecular biologists, evolutionary biologists, biological anthropologists, bioinformaticians, statisticians, and individuals working in the health sciences and clinical medicine.

15.3   Genetics of disease

The power of genome-wide association studies in dissecting the genetic contribution to multifactorial traits has been emphasized by studies such as those delineating common variants underlying age-related macular degeneration (Section 9.4) and the Wellcome Trust Case Control Consortium analysis of seven common diseases (Section 9.3.2). There has been an explosion of studies published over the period 2007–2009 utilizing the genome-wide association study approach with considerable success, notably in Crohn's disease and diabetes but also in other traits such as obesity and drug response (Section 9.3). Such studies have been made possible by the carefully phenotyped and curated sample collections that have been established, and by the progress which has been achieved in understanding the genomic architecture of allelic variation, the establishment of panels of informative SNP markers, and the high throughput technologies needed to genotype them. Advances in statistical analysis have been essential and future work both utilizing existing studies through meta analysis and establishing larger sample collections have been advocated.

A word of caution is, however, appropriate as the proportion of the genetic contribution to a given trait that has been explained by such genome-wide association studies remains very small, typically of the order of 1–5%. The effect sizes attributable to associated variants have been found to be almost always modest, and our knowledge of specific functional variants remains very incomplete. There is a clear argument for increasing the power of studies to detect small effect sizes through increased sample sizes but of perhaps equal importance is the need for clear insight into informative phenotypes to choose to study, and approaches which capture both rarer variants at the sequence level and structural variation.

The role of rarer variants, and in particular rarer variants with incomplete penetrance, remains unclear but may be substantial and is currently not well captured by genome-wide association studies (McCarthy *et al.* 2008). Rare variants with high penetrance have been shown to play a critical role in mendelian diseases, and ongoing efforts to define rare variants through resequencing studies should further advance our understanding of this area although it represents a formidable challenge at both a technological and analytical level.

There was, of course, a world before genome-wide association studies and remarkable insights have been gained into multifactorial diseases using linkage-based approaches and candidate gene association studies, as well as data from diverse approaches including microbial challenge studies. A number of robust examples have been illustrated over the course of this book, ranging from the apoε4 allele in Alzheimer's disease (Section 2.5.2), factor V Leiden and venous thrombosis (Section 2.6.1), copy number variation at *CCL3L1* and HIV-1 (Section 14.2.4), tandem repeats upstream of *INS* and type 1 diabetes (Box 7.5), and a coding variant of *PRNP* and prion disease (Section 13.1), to haemoglobin S and thalassaemia in malaria (Section 13.2). Large effect sizes proved the exception rather than the norm, and many early studies were underpowered and did not appreciate or interrogate the underlying genomic architecture, which led to many failures to replicate results. A clear place remains, however, for a targeted candidate approach based on knowledge of underlying disease biology or a clear functional role for specific variants. Appreciation of the nature and extent of genetic variation at a locus, and the potential for linkage disequilibrium and confounding, are essential.

15.4   Functional consequences of genetic variation

The relationship between genotype and phenotype has, at its core, a premise that associated genetic variation is exerting a functional consequence. In mendelian traits, high penetrance, typically rare, alleles have been identified that result in often dramatic effects for the encoded protein or regulation of gene expression. This relationship is not necessarily clear cut and while major effects are seen involving a particular gene, often variation in modifier genes may play a role in the observed phenotypic diversity, as seen for example in the iron storage disorder haemochromatosis. The situation becomes even harder to dissect in common multifactorial traits where, typically, fine mapping is difficult and defining specific causative functional variants is even harder.

The functional consequences of genetic variation are diverse and complex, as illustrated in Chapter 1 for the globin locus and elsewhere in this volume. Effects involving modulation of the structure or function of the encoded protein range from amino acid substitutions and truncation of the length of the polypeptide chain, to diverse and complex effects on the regulation of splicing. The control of gene expression has also been shown to be regulated to varying extents by genetic variation including gene dosage effects associated with copy number variation as well as effects for example on local and distant regulatory elements associated with sequence level diversity, modulating recruitment of specific transcription factors and even resulting in the creation of a novel gene promoter (Section 1.3.9).

How to define functionally important genetic variation remains a major challenge and significant roadblock in the field. Advances in structural biology and knowledge of the underlying biology of transcription and translation are significantly advancing our ability to predict and test the consequences of coding and other variants for protein structure and function. The nature and complexity of the biology underlying alternative splicing is also being increasingly appreciated and the sophistication of approaches to understand the consequences of genetic diversity for splicing are rapidly increasing. Such knowledge is also critical to the analysis of gene expression with recent advances in mapping genetic determinants of expression quantitative traits needing to include the diversity in alternative spliced transcript isoforms in any analysis (Section 11.6). Mapping quantitative trait loci (QTLs) using the 'genetical genomics' approach looks set to be a very powerful approach. Exciting data are now available from model organisms, as well as human cell lines and increasingly from primary human cells and tissues (Section 11.3).

In trying to define the functional consequences of genetic diversity for gene expression, ongoing work to understand the control of this complex process has been essential and will continue to be so, with fundamental insights from international collaborative studies such as the ENCODE Project (Birney *et al.* 2007). Transcriptional regulation is a multilayered and intricate process which genetic variation may modulate at many different points, ranging from promoter function to disruption of enhancer elements. Through using bioinformatic approaches and wet lab experiments at the bench, our ability to investigate and test such effects continues to develop. However, many current approaches remain technically demanding and not amenable to high throughput analysis making further methodological advances in this area a continued priority.

15.5 Medical applications and pharmacogenomics

A major goal of current genetic research in the medical arena is to understand how medical care can be tailored to the patient based on their individual genetic makeup. The development of 'personalized medicine' promises significant potential benefits but remains some way from being fully realized. This may be most readily achieved in terms of specific therapeutics, aiming to maximize the benefit for the patient while minimizing the risk of adverse effects. Other applications relate to targeted use of screening and monitoring, for example if an individual is at high risk of colonic or breast cancer, or of hypercholesterolaemia, as a result of possession of a particular underlying genetic variant. There are also less direct but potentially equally important applications of current research in terms of our knowledge of underlying disease pathogenesis and potential novel targets for therapeutic interventions. Recent genome-wide association studies in Crohn's disease for example have highlighted novel genes and gene pathways in disease susceptibility, such as involving the IL23R pathway and autophagy, which were not previously considered in models of disease pathogenesis (Section 9.5.5).

The field of pharmacogenomics looks set to be very important in future health care with a number of striking examples already existing of how knowledge of underlying genetic variation can be used in tailoring treatments to the individual patient. There is robust evidence, for example, that genetic variation is a significant determinant of variation in dosing for the blood thinning drug warfarin as well as risk of adverse effects. Such knowledge is important as it is currently very difficult to predict the long term dose of warfarin that a patient will require, and intensive monitoring with blood tests is required in the early stages of treatment to optimize warfarin dose and to avoid potentially fatal over-anticoagulation. The underlying genetic variants robustly identified to date involve the *CYP2C9* gene on chromosome 10q24, which encodes cytochrome P450 2C9, an important enzyme involved in warfarin metabolism (Higashi *et al.* 2002); and the vitamin K epoxide reductase gene (*VKORC1*) encoding an enzyme which is a pharmacological target of warfarin involved in activation of vitamin K (Rieder *et al.* 2005).

Overall, associated variants in *CYP2C9* and *VKORC1* are thought to account for 30–35% of variability in warfarin dosing, and to be significant predictors of risk of adverse events (Lesko 2008). Further clinical trials are proposed in a range of population groups investigating optimal initial dosing and cost effectiveness. Routine application to clinical care is close to being realized with algorithms available combining genetic and clinical data (www.warfarindosing.org), and the US Food and Drug Administration in 2007 adding reported genetic effects to drug labelling and approving clinical tests for specific variants (Lesko 2008; Schwarz *et al.* 2008; Vladutiu 2008).

The highly polymorphic cytochrome P450 enzymes provide many further examples of the consequences of genetic variation for drug metabolism and response (Section 4.4.2). 'Poor' and 'ultrarapid' drug metabolizers can be defined within populations who are at risk of drug toxicity and lack of therapeutic response, respectively. Underlying genetic variation involving point mutations, deletions, insertions, and copy number variation have been defined. For example 63 different alleles at *CYP2D6* have been defined with significant consequences for many different drugs including antidepressants, analgesics, anti-emetics, anti-arrhythmics, and neuroleptics (Ingelman-Sundberg *et al.* 2007). However, to enter routine clinical practice, such pharmacogenetic approaches will require significantly more prospective data demonstrating improved clinical outcome and cost effectiveness; interpretation and application of results is also likely to be complex and require specialist input (Swen *et al.* 2008).

A further example of the development of pharmacogenomics and applications to avoid serious drug side effects are seen with the reverse transcriptase inhibitor abacavir. Abacavir hypersensitivity was found to be associated with the HLA-B*5701 haplotype (Section 12.3.3) (Mallal *et al.* 2002). Prospective application of testing for this haplotype as a screening test prior to starting drug treatment significantly reduced the incidence of adverse drug reaction among patients with HIV-1, both in initial studies in Australia (Rauch *et al.* 2006) and in a more recent international double-blind, prospective, randomized study (Mallal *et al.* 2008). Such genetic testing is thought to be cost effective and likely to become part of standard clinical practice as routine testing becomes available (Hughes *et al.* 2008).

Much work remains to be done to establish the ethical implications of genetic testing and how as a society we would want to make use of contemporary genetic research. Already commercial genetic testing services are being directly marketed despite a lack of firm evidence in many cases regarding clinical relevance and effectiveness (Melzer *et al.* 2008a; Pearson 2008). A series of questions are being posed with very few answers to date. What is the role of genetic counselling in this situation? How should the consumer interpret and make use of genetic information, particularly in the context of a multifactorial trait where specific variants may be contributing a very small risk? Who has a right to access to such data and what are the implications for insurance and employment? For health care providers and institutions, including those at a policy-making level, is the evidence base sufficiently robust to proceed? What regulatory mechanisms need to be put in place?

This debate needs to take place with the involvement of all relevant stake holders and the wider public. Critical to this is greater input from ethicists and from public health specialists, such that the current 'genetic revolution' can be effectively translated into improvements in human health. Finally, it is essential that we consider this in a truly international context as the potential implications of contemporary genetic research are as relevant to people living in developing areas of the world as to any other region.

15.6   Lessons from the past, looking to the future

This book began with a review of the remarkable progress made in delineating the nature and consequences of genetic variation involving the globin genes. Our subsequent journey across the genome has considered the many different types of genetic variation and their diverse consequences. It has necessarily been a subjective account and not an exhaustive review, with many areas remaining to be explored, not least the fields of mitochondrial genetics and epigenetics. At the end of this journey it is perhaps appropriate to return to the α and β globin gene clusters which have provided so many fundamental insights into human genetic variation.

Why should study of the globin genes have been so productive? In part this relates to the distinct and diverse phenotypes that may result from variation at these loci, in part it reflects the decision of scientists to remain focused on particular genomic regions and specific clinical phenotypes. Their work serves to remind us that in this current era of genome-wide analysis, detailed characterization of a locus can be extremely rewarding. It also emphasizes that a full appreciation of the genomic and regulatory context within which genetic variation is occurring has set the stage for many novel insights into the consequences of genetic variation, and vice versa. The ongoing projects aiming to derive genome-wide mapping of regulatory elements such as the ENCODE Project should in time provide a similar context for analysis of genetic variation in other genomic regions. Context is key, both in terms of understanding the regulatory landscape but also in terms of being able to do appropriate experiments to investigate the consequences of genetic variation.

In general the effects of regulatory variants for gene expression are highly specific, in terms of cell or tissue type, developmental stage, particular environmental exposure, or other prevailing condition such that the true effect of any variant may be missed unless investigated in the appropriate context. This principle is being reinforced by a diverse array of studies investigating different classes of genetic variation. For future studies seeking to identify causative functional variants, this remains a very important consideration.

Such is the pace of change in the field of human genetic variation that work done only a few years ago can appear redundant and irrelevant. In some senses this is true and there is no doubt that very substantial advances have been achieved, radically altering our views of the nature and consequences of genetic diversity. Rapidly developing technologies are allowing genome-wide analyses to become a reality with vast amounts of data generated, providing in turn bottlenecks in data analysis and interpretation. We should not forget, however, the many elegant and pioneering studies that have gone before, and the lessons that such work can continue to teach us. Human genetic variation is complex and can impact on phenotype at many different levels. It is at once inspiring and overwhelming, a potential treasure trove for biomedical science but one which continues to keep many of its secrets.

Glossary

allele alternative (variant) forms of a gene at a given chromosomal location (locus)

allele-specific gene expression differential gene expression dependent on the allele from which the transcript originates; a transcribed SNP marker for which an individual is heterozygous can be used to define the allelic origin of a transcript

alternative splicing variation in splicing, the process whereby coding RNA sequences are identified and joined together, such that multiple gene products (alternatively spliced isoforms) arise from a single gene

Alu elements family of short interspersed mobile DNA elements comprising a characteristic noncoding DNA sequence approximately 300 bp long and containing a recognition site specific to the AluI restriction enzyme

ancestral haplotypes (conserved extended haplotypes) large chromosomal segments conserved *en bloc* with a fixed constellation of alleles

aneuploidy possession of an abnormal number of chromosomes, in contrast to euploidy

autosome a non-sex chromosome

bacterial artificial chromosome (BAC) synthetic DNA molecule used in cloning DNA segments up to 300 kb in size which contains the sequences needed for propagation in bacteria

biomarkers a measurable marker of a biological state

Bonferroni correction a statistical method of correcting for multiple testing when interpreting the significance of a result

carriers individuals possessing one copy of a recessive allele

case-control design/study an epidemiological method of comparing exposures (for example possession of a genotype) among people affected by a disease with those unaffected by the disease

centimorgan (cM) measure of genetic distance between two markers based on the chance that the two will become separated at crossing over in a single generation, a 1% chance being equal to 1 cM

centromere a region visible as a constriction during metaphase typically in the middle of chromosomes and separating the long and short arms of the chromosome; includes site of attachment of mitotic or meiotic spindle fibres

CEPH Centre d'Etude du Polymorphisme Humain (CEPH) collection of reference families

chromatids the two exact copies of a replicated chromosome during the process of cell division (mitosis or meiosis), specifically at the time they are joined at the centromere and before separation; on separation (anaphase) the two are said to be 'daughter chromosomes'

chromatin a complex of proteins and nuclear DNA that serves to package DNA into nucleosomes, condensed chromatin fibres (arrays of nucleosomes), and higher order packaging such as condensed chromosomal appearance seen during metaphase; chromatin structure plays an important role in regulating gene expression and replication; the major class of chromatin proteins are histones

chromatin immunoprecipitation (ChIP) a laboratory method to assay protein–DNA interactions in living cells or tissues based on crosslinking proteins to DNA, typically by exposure to formaldehyde, followed by extraction of chromatin, sonication, and immunoprecipitation

chromosome a very long molecule of duplex DNA with associated proteins found in the nucleus of the cell, which is visible morphologically in a highly coiled and condensed state during metaphase with a constriction (centromere) typically separating a short (p) and a long (q) arm: where the arms are of equal length the chromosome is said to be metacentric; where unequal, submetacentric; when the p arm is extremely short chromosomes are said to be acrocentric

cis-acting the *cis*-acting regulatory element and the regulated gene are present on the same chromosome

codon non-overlapping sets of three nucleotides each defining a particular amino acid or stop signal; a nucleotide insertion or deletion will disrupt the 'reading frame' of the codons causing a 'frameshift'

cohort a group with a common characteristic or experience within a defined time period; a birth cohort for example may comprise individuals born in a particular year

comparative genome hybridization (CGH) a method to detect gain or loss of chromosomal material

comparative genomics analysis of genome structure and function among different species

complementary DNA (cDNA) single-stranded DNA synthesized by reverse transcriptase from RNA such that the DNA is complementary to the RNA

compound heterozygotes two different mutations, present on each of two alleles at a given locus

concerted evolution evolution of two genes such that they are more closely related to each other within a gene family in one species compared to members of the same gene family in other species

concordance in twin studies, refers to the probability of a twin having a particular trait if that trait is present in the other twin

confidence intervals (CIs) range of values within which the true population value is statistically likely to be found, based on an estimate from a sample; for example the 95% CI

contig a contiguous stretch of DNA sequence without gaps assembled using direct sequencing information

copy number polymorphism copy number variation present in more than 1% of the population

copy number variation structural variation involving DNA segments larger than 1 kb in which there are relative gains (by insertion or duplication) or losses (by deletions or null genotypes) of genomic copy number relative to a designated reference genome sequence

CpG island regions of DNA enriched for unmethylated CpG dinucleotides (cytosine followed by guanine separated by a linking phosphate), typically 0.3–3 kb in size and commonly found in gene promoter regions or first exons

dbSNP online catalogue (www.ncbi.nlm.nih.gov/projects/SNP/) of fine scale sequence diversity including single nucleotide substitutions, short deletion/insertion polymorphisms and short tandem repeats (microsatellites)

deme single, geographically localized, interbreeding population

diploid possession of two copies of each chromosome; in most human somatic cells there are 22 pairs of autosomes (numbered 1 to 22 from largest to smallest in size) and one pair of sex chromosomes, giving 23 chromosome pairs and a diploid number of 46

dizygotic ('non-identical') twins twins arising when two eggs, each fertilized by different sperm, implant successfully in the uterine wall at the same time

DNA deoxyribonucleic acid, the double-stranded molecule found within cell nuclei that carries genetic information

DNA fingerprinting a technique that analyses highly variable repeats found at certain minisatellites in a population to define a unique pattern for a particular individual; was used for legal and forensic purposes

DNA footprinting a laboratory method of mapping protein–DNA binding and determining sequence-specific interactions

DNA methylation addition of methyl groups to a nucleotide, for example conversion of cytosine to 5-methyl cytosine at a CpG dinucleotide; an epigenetic modification with important consequences for gene regulation and development

DNase hypersensitive sites regions of DNA susceptible to cleavage by the enzyme DNase I due to a more open chromatin conformation, for example associated with the presence of regulatory elements

dominant a phenotypic trait manifest when an individual possesses one copy of a particular allele (heterozygous state)

duplicon a duplication traceable to an ancestral or donor location

ENCODE ENCyclopedia Of DNA Elements (ENCODE), a project which aims to define the relationship between DNA sequence and functional regulation

endonuclease an enzyme that cleaves phosphodiester bonds within polynucleotide chains (including single- or double-stranded DNA, RNA); includes restriction endo-nucleases (restriction enzymes)

enhancer *cis*-acting DNA element that increases transcriptional activity of a gene, often acting at a distance and classically orientation and position independent

epigenetic heritable changes in gene expression that do not involve changes in the DNA sequence and can be transmitted through cell divisions, for example DNA methylation

epistasis effect of an allele at one genomic locus depends on a genotype coinherited at a second unrelated locus

euchromatin light staining, less tightly compacted regions of chromosomes which includes transcriptionally active or potentially active genes; contrasts with dark staining heterochromatin, which accounts for remainder of genome

exon segments of a gene found in mature RNA; includes protein coding DNA sequence

expression quantitative trait loci (eQTLs) analysis of gene expression as a quantitative trait to determine associated genomic regions

familial aggregation observed clustering of disease in families

fluorescence in situ hybridization (FISH) cytogenetic technique using fluorescently labelled probes to bind (hybridize) specific regions of DNA or RNA

fosmid an f-factor cosmid able to hold large pieces of DNA

founder effect change in allele frequency arising in a population established by a small number of founder individuals

frameshift mutations insertion or deletion event that is not in multiples of three base pairs, so disrupting the normal 'reading frame' of the messenger RNA sequence

functional genomics large scale analysis of gene function, for example genome-wide gene expression profiling

gain of function mutation mutation leading to new activity, usually dominant

GC content percentage of bases (A, G, C, or T) that are C or G in a specific region or whole genome

gene discrete unit of heredity passed from parents to offspring

gene conversion non-reciprocal transfer of sequence information between an acceptor and a donor sequence such that one strand becomes identical (complimentary) to another

gene desert genomic regions without any known genes

gene dosage number of copies of a particular gene in an individual genome

genetic anticipation disease severity in affected individuals increases with an earlier age of onset across successive generations, as seen for example in myotonic dystrophy

genetic drift change in gene frequency in a population arising by chance, notably in small breeding populations

genetical genomics analysis of gene expression as a quantitative trait among related individuals in a segregating population using informative polymorphic genetic markers

genetic hitch-hiking process whereby 'neutral' sequence variants on a particular haplotype rapidly increase in frequency in a population when a newly arising mutation on that haplotype is subject to strong selective pressure

genome complete genetic complement of a cell or organism, which includes the nuclear chromosomes and mitochondrial DNA

genome-wide association studies studies investigating associations between the possession of common variants and particular traits by using a high density set of SNP markers to capture a substantial proportion of the common nucleotide diversity across the genome

genomic disorders diverse group of genetic diseases in which genomic rearrangements result in gain, loss, or disruption of dosage sensitive genes

genomics study of genes and their function

genotype hereditary or genetic constitution of an individual, either as a whole or for a specific locus within the genome

haploid half the diploid number of chromosomes, found for example in gametes (single copies of the 22 autosomes plus one of the sex chromosomes)

haploinsufficiency for diploid organisms, this may arise following a mutation at a specific locus if possessing a single normal gene copy is insufficient for a normal phenotype

haplotype combination of genetic markers or alleles found in a specific region of a single chromosome of a given individual; commonly occurring haplotypes can be defined for a population identifying genetic markers that tend to be coinherited together

haplotype block structure resolved for the human genome with regions of strong allelic association or linkage disequilibrium, interspersed with sites where association broke down which correlated with hotspots of meiotic recombination

haplotype tagging SNPs common SNPs selected to be genotyped in order to capture much of the haplotypic diversity in a population

hemizygous having one copy of a gene in diploid cells

heritability variation in a trait attributable to genetic causes

heterochromatin a tightly packaged condensed state of typically transcriptionally suppressed DNA seen as darkly staining chromosomal regions, which is found in regions rich in repetitive satellite DNA such as centromeres, pericentromeres, and telomeres

heterozygote individual with two different alleles for a given locus

histone DNA binding protein; principal component of chromatin

homologous chromosomes chromosomes in a diploid cell, one inherited from each parent, which pair during meiosis; in humans there are 22 pairs of homologous autosomes

homologous genes genes with very similar sequences in different species derived from a common ancestor (orthologous genes) and genes within a species arising by duplication (paralogous genes)

homologous recombination (crossing over) process of breakage of non-sister chromatids of a pair of homologues and the rejoining of the fragments to generate new recombinant strands such that there are equal exchanges between allelic sequences at the same positions within the alleles; the process allows shuffling of genetic material and potential for increased genetic variation during the formation of eggs and sperm

homozygote individual with identical alleles for a given locus

hotspot genomic location with high frequency of mutation or recombination

Human Genome Project international project to map all human genes and completely sequence the human genome

identity by descent in a family pedigree, two siblings may share a gene (allele) identical by descent from a demonstrable common ancestor

imprinting monoallelic expression of a gene in a parent-of-origin-specific manner, arising due to epigenetic marks involving methylation of DNA and histone modifications

imputation a statistical method used to infer SNP genotypes

incidence number of new cases of a disease in a defined time period

indels genetic variation involving gain or loss (insertion or deletion) of one or more contiguous nucleotides in genomic sequence

insulator sequence element preventing inappropriate interactions between domains, for example blocking the action of an enhancer on a promoter

International HapMap Project international collaborative research project to catalogue SNP diversity across the genome for a number of human populations, capturing patterns of common human sequence variation and so informing and enabling future genetic studies of common human disease

interstitial deletion or duplication exchange within a chromosomal arm such that the original telomere is retained

intron noncoding sequence in the primary RNA transcript separating exons, which is removed by splicing

inversion chromosomal rearrangement which may involve two breaks on different arms (pericentric inversion) or the same arm (paracentric inversion)

isochromosome result of duplication of one chromosome arm leading to an abnormal symmetrical chromosome with two identical arms

isoform different forms of a protein

karyotype summary of chromosome complement (constitution) for a cell or person: for example 46,XX in a female and 46,XY in a male

kilobase (kb) measure of DNA sequence length; 1 kb is equal to 1000 base pairs (bp)

knockdown experiments targeted repression of gene expression, for example using RNA interference (RNAi) against a specific gene

linkage cosegregation of characters such as phenotypes and marker alleles in a pedigree which occurs because the underlying determinants are present close together on a chromosome

linkage disequilibrium non-random association of alleles at two or more loci; measures of linkage disequilibrium describe the observed degree of association or correlation between any two genetic variants (alleles) in a population

linkage map genetic map based on recombination frequency, relative positions of markers are defined by how often loci are inherited together

locus/loci unique chromosomal location of a gene or sequence

lod score logarithm of the odds of linkage, a measure of likelihood of linkage between loci, with a lod score of greater than +3 providing evidence of linkage for mendelian characters

low copy repeats (LCRs) sequences of 10 kb or more in size, with at least 95% sequence identity, which are separated by 50 kb to 10 Mb of intervening sequence

lymphoblastoid cell lines lymphocytes immortalized using Epstein–Barr virus

major histocompatibility complex (MHC) genomic locus on chromosome 6p containing genes with a fundamental role in the immune response, including successful tissue grafting and organ transplantation ('histocompatibility'); the first MHC encoded proteins were discovered on circulating white blood cells and known as leukocyte antigens – hence the other commonly used term for the region, the human leukocyte antigen (HLA) complex

mariner mobile element family of transposable mobile DNA elements

marker chromosomes small structurally abnormal chromosomes that are found in addition to the normal complement of 46 chromosomes

megabase (Mb) measure of DNA sequence length; 1000 kb are equal to 1 Mb

meiosis cell division resulting in haploid state occurring in reproductive tissues

meta analysis statistical analysis combining data from several studies

metaphase the point in cell division when the condensed chromosomes align with each other

microarray high throughput approach used for gene expression analysis and for genotyping; utilizes short DNA sequences bound to a chip or beads as probes to which a cDNA or DNA sample are hybridized and quantified

microarray-based comparative genomic hybridization (array CGH) a microarray-based technique to determine structural genomic variation, of particular value in studying copy number variation

microsatellite DNA/microsatellites arrays of tandemly repeated DNA sequence less than 100 bp in length made up of simple repeats up to 6 bp in length (also known as short tandem repeats)

minisatellite DNA/minisatellites arrays of tandemly repeated DNA sequence typically between 100 bp and 20 kb in size with repeat units between 7 and 100 bp in length

missense mutation mutation resulting in a change in the encoded amino acid; missense mutations may be conservative (the amino acid change has minimal consequence for protein function) or non-conservative (a significant structural change)

mitosis process of cell division such that daughter cells contain the same amount of genetic material as the parent cell

mobile DNA elements segments of DNA that can transport or duplicate themselves (transpose) to other genomic regions; in the human genome only a very small minority remain competent for transposition

monogenic involving a single gene

monomorphic presence of a single allele at a locus

monosomy loss of a single chromosome, for example one X chromosome in Turner syndrome

monozygotic ('identical') twins twins arising from a single fertilized egg, from which a single zygote leads to two separate embryos

morpholino a molecule used to modify or knockdown gene expression

mosaicism presence of two or more genetically distinct cell lines in a single conceptus

mutation a permanent structural change in the DNA

NFkB a family of transcription factors involved in regulating many immune and inflammatory genes

non-allelic homologous recombination recombination or crossover between non-allelic sequences which can lead to chromosomal rearrangements and genomic disorders

non-homologous chromosomes set of chromosomes in an organism; a diploid human cell contains two sets of 23 chromosomes each derived from one parent

nonsense mediated mRNA decay pathway degrading RNA with a premature termination codon

nonsense mutation mutation resulting in a codon change from encoding a particular amino acid to being a stop codon; such events typically have dramatic consequences for gene function and are rare

nonsynonymous variant/substitution nucleotide change altering the encoded protein; these include missense and nonsense changes

nucleosides comprise a sugar residue (deoxyribose) that is covalently linked to a nitrogenous base; in DNA there are four types of nitrogenous base, adenine (A), guanine (G), cytosine (C), and thymine (T)

nucleosome repeating structural unit of chromatin comprising a DNA loop around a histone octamer

nucleotides basic repeating unit of DNA, comprise nucleosides with a phosphate group attached

null alleles complete loss of gene product (function) resulting from a mutant allele

nullisomy loss of a chromosome pair

odds ratio (OR) used in a case–control design, describes the ratio of the odds of having a disease among those with an exposure (for example a particular genotype) versus the odds of disease among those not having the exposure

oligonucleotide short DNA or RNA sequence used as probe or primer based on ability to hybridize specifically to a sequence

OMIM Mendelian Inheritance in Man (MIM) is a definitive reference for human genes and genetic disorders available as Online Mendelian Inheritance in Man (OMIM) (www.ncbi.nlm.nih.gov/omim/)

open reading frame (ORF) sequence of bases that could potentially encode a protein

palindrome sequence reading the same on each complimentary strand of DNA when read 5′ to 3′

pedigree shows family relationships and pattern of inheritance of a trait

penetrance proportion of individuals with a particular genotype who manifest a given phenotype

pharmacogenomics study of how genetic variation may modulate individual response to drugs

phenotype an observable characteristic, which may range from appearance to a structural, biochemical, physiological, or behavioural character; may not be genetic in origin and is often the product of underlying genotype and environmental factors

phylogenetic describes evolutionary relatedness between organisms

phylogenetic (evolutionary) tree tree denoting evolutionary relatedness of different organisms (species) with a common ancestor and inferred lines of descent based on different types of evidence, including for example palaeontological and molecular sources

polyadenylation poly(A) tail added at the 3′ cleavage site at the end of the transcript, about 15–30 nucleotides

downstream of the AAUAAA polyadenylation signal sequence

polygenic involving several genes

polymerase chain reaction (PCR) a laboratory method that allows exponential amplification of DNA

polymorphism in genetics, describes DNA sequence variants (alleles, structural variants) present at a frequency of 1% or more in a population without reference to a particular phenotypic effect

polyploid possession of more than the normal complement of two chromosome sets (diploidy, 2n); cells may for example be triploid (3n) or tetraploid (4n)

population (genetic) bottleneck evolutionary event involving transient reduction in population size causing reduced genetic variation

population stratification (structure) differences in allele frequencies between population subgroups that may confound genetic association studies

positional cloning cloning or identification of a gene based on its chromosomal location rather than knowledge of the encoded protein

power statistical term describing the probability of a study of a given size finding a true effect, such as an association between a particular genotype and disease susceptibility; power = 1β, where β is the type II error

prevalence number of people affected by a disease in a population at a particular point in time

primer short oligonucleotide used in PCR to initiate synthesis of the complimentary strand after annealing to a specific genomic location

promoter sequence of DNA typically found upstream of a gene integral to initiation of transcription; specific DNA sequence elements in the promoter bind different transcription factors, recruiting the transcriptional machinery and RNA polymerase enzyme responsible for transcription

proteome the full complement of proteins encoded by the genome of an organism

pseudogene DNA sequence showing a high degree of homology to an active non-allelic functional gene; the pseudogene is non-functional, usually due to a lack of protein coding ability (for example due to a nonsense mutation, frameshift mutation, or partial nucleotide deletion)

purines the nitrogenous bases adenine (A) and guanine (G), which comprise two interlocked heterocyclic rings of carbon and nitrogen atoms

pyrimidines the nitrogenous bases cytosine (C) and thymine (T), which comprise one heterocyclic ring of carbon and nitrogen atoms

quantile–quantile plot graphical plot of observed test statistics versus expected, used for example in the analysis of a genome-wide association scan

quantitative trait loci (QTLs) genomic regions associated with particular quantitative traits

quantitative traits phenotypic characteristics that show graded continuous variation in a population, as seen for example with height, and are determined by a combination of genetic variation in more than one gene and environmental factors

recessive phenotype not seen among individuals heterozygous for a particular allele, only in the homozygous state

recombination hotspots genomic regions where recombination events tend to be more concentrated

regulatory variants genetic variation known or postulated to modulate gene expression

reporter gene assays a laboratory method to investigate the ability of particular DNA sequences to modulate gene expression

resequencing sequencing of a gene, genomic region, or genome of an individual in order to define genetic variation on comparison with a reference sequence or other individuals

restriction enzyme enzyme that cleaves double-stranded DNA at specific sites dependent on recognition of particular nucleotide sequences by the enzyme

restriction fragment length polymorphisms (RFLPs) fine scale sequence variation resulting in gain or loss of specific sequence recognized by a restriction enzyme; manifests as variation in the length of observed products of restriction enzyme digestion dependent on the presence or absence of the restriction cleavage site

retrotransposons (class I mobile DNA elements) transpose via an RNA copy which is then reverse transcribed into DNA and inserted into a new genomic location; includes long interspersed elements (LINEs) and short interspersed elements (SINEs)

reverse transcriptase enzyme used to synthesize DNA from RNA

risk factor character, attribute, exposure, or other factor increasing risk of disease or condition; may or may not be causal

RNA ribonucleic acid

RNA polymerase enzyme synthesizing RNA from DNA

Robertsonian translocations the most common recurrent type of translocation, specifically involves the acrocentric chromosomes (chromosomes 13, 14, 15, 21, 22) in which breaks in the very short p arm lead to fusion of the remaining long arms

rs number unique reference identifier for a sequence variant listed at the dbSNP database

satellite DNA/satellites very large arrays of tandemly repeated DNA sequence spanning hundreds to thousands of kilobases of DNA found at centromeres as well as pericentromeric and telomeric regions

segmental duplications continuous portions of DNA greater than 1 kb in size that occur in two or more copies per haploid genome with the copies sharing more than 90% sequence identity; segmental duplications map to more than one location in the genome and may be found either arranged in tandem or at interspersed locations

selective sweep result of positive selection for a given variant leading to local loss of diversity (heterozygosity) due to genetic hitch-hiking

shotgun sequencing sequencing of individual, randomly generated DNA fragments that are reassembled based on overlaps into a complete sequence; includes whole genome shotgun sequencing

signatures of selection characteristic patterns of genetic diversity indicating positive selection

single nucleotide polymorphism (SNP) single nucleotide substitution present at greater than 1% frequency in the population; SNPs are typically biallelic and of great utility as genetic markers as they occur commonly across the genome and can be genotyped using ultrahigh throughput technologies

single nucleotide substitutions variation in DNA sequence in which one nucleotide is replaced with another

splicing identifying and joining together coding exonic sequence in pre-mRNA through a complex process involving the splicing machinery (the spliceosome) and 'splicing code', which includes consensus splice site sequences at exon–intron boundaries together with *cis*-regulatory elements to which specific proteins bind

structural genomic variation variation involving segments of DNA more than 1 kb in size (may be microscopic to submicroscopic) with either a change in DNA dosage (copy number variants) or a positional change (for example balanced translocations) or an alteration in orientation, as seen with inversions

subtelomeres transition zones ranging in size from 10 to 300 kb which are found near the tips of chromosomes, between chromosome-specific sequence and the arrays of telomeric repeats that cap each chromosome

synonymous variant/substitution nucleotide change in DNA sequence resulting in a change in the RNA codon to another codon that encodes the same amino acid

tandem repeats tandemly repeated DNA sequences in a head-to-tail configuration without sequences between repeating units, typically polymorphic in nature due to expansion or contraction of the number of repeating units; classified on the basis of the size of the repeat array into satellites, minisatellites, and microsatellites

telomeres long tracts of tandemly repeated DNA located at the ends of individual chromosomes, made up of simple repeats of the sequence (TTAGGG)n and typically 3–20 kb in length

tetrasomy gain of a chromosome pair

transcription process of synthesizing RNA from a DNA template

transcription factor required for recruitment of RNA polymerase and initiation of transcription, typically DNA binding proteins

transcriptome complete set of RNAs found in a cell, tissue, or organism

transition substitution of a pyrimidine for a pyrimidine or a purine for a purine

translation synthesis of proteins using an mRNA template

translocation chromosomal rearrangements involving transfer of chromosomal regions between two non-homologous chromosomes as a result of breakage and reattachment; in a balanced translocation there is no net gain or loss of genetic material, in unbalanced translocations there is a net gain or loss

transposon a mobile DNA element that when competent can move to a new part of the genome (transpose) by a 'cut and paste' process involving a transposase enzyme

transversions substitutions of pyrimidines for purines, or purines for pyrimidines

trisomy having three rather than two copies of a particular chromosome

type I error ('false positive') rejection of a true null hypothesis, for example believing a genetic association with disease is true when there is no association (a false positive conclusion); probability of making a type I error ('calculated probability or P value) that is acceptable is referred to as 'alpha' and is often set at less than one in 20 ($P <0.05$)

type II error ('false negative') failure to reject a false null hypothesis (a false negative conclusion); probability of making a type II error is 'beta' and often set at 20%

variable number tandem repeats (VNTRs) usually describes polymorphic minisatellites

X chromosome inactivation (lyonization) process leading to inactivation of one of the two copies of the X chromosome in female mammals

Resources

Topic	References
Glossaries of genetic terms	'Genetics Home Reference', US National Library of Medicine, http://ghr.nlm.nih.gov/glossary
	'Talking Glossary of Genetic Terms', National Human Genome Research Institute (NHGRI), www.genome.gov/glossary.cfm
	Genetics Education Center, University of Kansas Medical Center, www.kumc.edu/gec/glossnew.html
	Links to other glossaries and sources of information, www.kumc.edu/gec/glossary.html
	US Department of Energy Human Genome Program, www.ornl.gov/sci/techresources/Human_Genome/glossary/
Oxford English Dictionary	www.oed.com/
Wikipedia	http://en.wikipedia.org/wiki/Main_Page
Human Molecular Genetics	www.ncbi.nlm.nih.gov/books/bv.fcgi?rid=hmg.glossary.3037; Strachan and Read 2004
Genes VIII	Lewin 2004
Specific articles	Kaushansky 1996; Helmuth 2001; Malats and Calafell 2003a, 2003b

References

Aaltonen, L.A., Peltomaki, P., Leach, F.S. *et al.* (1993). Clues to the pathogenesis of familial colorectal cancer. *Science*, **260**, 812–816.

Aartsma-Rus, A., Van Deutekom, J.C., Fokkema, I.F., Van Ommen, G.J. and Den Dunnen, J.T. (2006). Entries in the Leiden Duchenne muscular dystrophy mutation database: an overview of mutation types and paradoxical cases that confirm the reading-frame rule. *Muscle Nerve*, **34**, 135–144.

Abel, L., Demenais, F., Prata, A., Souza, A.E. and Dessein, A. (1991). Evidence for the segregation of a major gene in human susceptibility/resistance to infection by Schistosoma mansoni. *Am J Hum Genet*, **48**, 959–970.

Adams, M.D., Celniker, S.E., Holt, R.A. et al. (2000). The genome sequence of Drosophila melanogaster. *Science*, **287**, 2185–2195.

Adams, S., Turner, G.D., Nash, G.B., Micklem, K., Newbold, C.I. and Craig, A.G. (2000). Differential binding of clonal variants of Plasmodium falciparum to allelic forms of intracellular adhesion molecule 1 determined by flow adhesion assay. *Infect Immun*, **68**, 264–269.

Agarwal, A., Guindo, A., Cissoko, Y. et al. (2000). Hemoglobin C associated with protection from severe malaria in the Dogon of Mali, a West African population with a low prevalence of hemoglobin S. *Blood*, **96**, 2358–2363.

Aguileta, G., Bielawski, J.P. and Yang, Z. (2004). Gene conversion and functional divergence in the beta-globin gene family. *J Mol Evol*, **59**, 177–189.

AHCMN (1996). Update on nomenclature for human gene mutations. Ad Hoc Committee on Mutation Nomenclature. *Hum Mutat*, **8**, 197–202.

Aidoo, M., McElroy, P.D., Kolczak, M.S. *et al.* (2001). Tumor necrosis factor-alpha promoter variant 2 (TNF2) is associated with pre-term delivery, infant mortality, and malaria morbidity in western Kenya: Asembo Bay Cohort Project IX. *Genet Epidemiol*, **21**, 201–211.

Aidoo, M., Terlouw, D.J., Kolczak, M.S. *et al.* (2002). Protective effects of the sickle cell gene against malaria morbidity and mortality. *Lancet*, **359**, 1311–1312.

Aitman, T.J., Cooper, L.D., Norsworthy, P.J. et al. (2000). Malaria susceptibility and CD36 mutation. *Nature*, **405**, 1015–1016.

Aitman, T.J., Dong, R., Vyse, T.J. et al. (2006). Copy number polymorphism in Fcgr3 predisposes to glomerulonephritis in rats and humans. *Nature*, **439**, 851–855.

Ajioka, R.S., Jorde, L.B., Gruen, J.R. et al. (1997). Haplotype analysis of hemochromatosis: evaluation of different linkage-disequilibrium approaches and evolution of disease chromosomes. *Am J Hum Genet*, **60**, 1439–1447.

Aklillu, E., Persson, I., Bertilsson, L., Johansson, I., Rodrigues, F. and Ingelman-Sundberg, M. (1996). Frequent distribution of ultrarapid metabolizers of debrisoquine in an Ethiopian population carrying duplicated and multiduplicated functional CYP2D6 alleles. *J Pharmacol Exp Ther*, **278**, 441–446.

Albanese, V., Biguet, N.F., Kiefer, H., Bayard, E., Mallet, J. and Meloni, R. (2001). Quantitative effects on gene silencing by allelic variation at a tetranucleotide microsatellite. *Hum Mol Genet*, **10**, 1785–1792.

Albrecht, A. and Mundlos, S. (2005). The other trinucleotide repeat: polyalanine expansion disorders. *Curr Opin Genet Dev*, **15**, 285–293.

Albrecht, U., Sutcliffe, J.S., Cattanach, B.M. *et al.* (1997). Imprinted expression of the murine Angelman syndrome gene, Ube3a, in hippocampal and Purkinje neurons. *Nat Genet*, **17**, 75–78.

Alcais, A., Alter, A., Antoni, G. *et al.* (2007). Stepwise replication identifies a low-producing lymphotoxin-alpha allele as a major risk factor for early-onset leprosy. *Nat Genet*, **39**, 517–522.

Alkhatib, G., Combadiere, C., Broder, C.C. *et al.* (1996). CC CKR5: a RANTES, MIP-1alpha, MIP-1beta receptor as a fusion cofactor for macrophage-tropic HIV-1. *Science*, **272**, 1955–1958.

Allcock, R.J., Atrazhev, A.M., Beck, S. *et al.* (2002). The MHC haplotype project: a resource for HLA-linked association studies. *Tissue Antigens*, **59**, 520–521.

Allen, N.C., Bagade, S., McQueen, M.B. *et al.* (2008). Systematic meta-analyses and field synopsis of genetic association studies in schizophrenia: the SzGene database. *Nat Genet*, **40**, 827–834.

Allen, S.J., O'Donnell, A., Alexander, N.D. *et al.* (1997). alpha+-Thalassemia protects children against disease caused by other infections as well as malaria. *Proc Natl Acad Sci U S A*, **94**, 14736–14741.

Allen, S.J., O'Donnell, A., Alexander, N.D. *et al.* (1999). Prevention of cerebral malaria in children in Papua New Guinea by southeast Asian ovalocytosis band 3. *Am J Trop Med Hyg*, **60**, 1056–1060.

Allikmets, R. and Dean, M. (2008). Bringing age-related macular degeneration into focus. *Nat Genet*, **40**, 820–821.

Allikmets, R., Singh, N., Sun, H. *et al.* (1997). A photoreceptor cell-specific ATP-binding transporter gene (ABCR) is mutated in recessive Stargardt macular dystrophy. *Nat Genet*, **15**, 236–246.

Allison, A.C. (1954a). The distribution of the sickle-cell trait in East Africa and elsewhere, and its apparent relationship to the incidence of subtertian malaria. *Trans R Soc Trop Med Hyg*, **48**, 312–318.

Allison, A.C. (1954b). Protection afforded by sickle-cell trait against subtertian malarial infection. *Br Med J*, **1**, 290–294.

Allison, A.C. (1956). The sickle-cell and haemoglobin C genes in some African populations. *Ann Hum Genet*, **21**, 67–89.

Allison, A.C. (1960). Glucose-6-phosphate dehydrogenase deficiency in red blood cells of East Africans. *Nature*, **186**, 531–532.

Allison, A.C. (1964). Polymorphism and natural selection in human populations. *Cold Spring Harb Symp Quant Biol*, **29**, 137–149.

Altfeld, M., Addo, M.M., Rosenberg, E.S. *et al.* (2003). Influence of HLA-B57 on clinical presentation and viral control during acute HIV-1 infection. *Aids*, **17**, 2581–2591.

Altshuler, D., Brooks, L.D., Chakravarti, A., Collins, F.S., Daly, M.J. and Donnelly, P. (2005). A haplotype map of the human genome. *Nature*, **437**, 1299–1320.

Altshuler, D., Daly, M.J. and Lander, E.S. (2008). Genetic mapping in human disease. *Science*, **322**, 881–888.

Altshuler, D., Hirschhorn, J.N., Klannemark, M. *et al.* (2000a). The common PPARgamma Pro12Ala polymorphism is associated with decreased risk of type 2 diabetes. *Nat Genet*, **26**, 76–80.

Altshuler, D., Pollara, V.J., Cowles, C.R. *et al.* (2000b). An SNP map of the human genome generated by reduced representation shotgun sequencing. *Nature*, **407**, 513–516.

Aly, T.A., Eller, E., Ide, A. *et al.* (2006). Multi-SNP analysis of MHC region: remarkable conservation of HLA-A1-B8-DR3 haplotype. *Diabetes*, **55**, 1265–1269.

Amador, M.L., Oppenheimer, D., Perea, S. *et al.* (2004). An epidermal growth factor receptor intron 1 polymorphism mediates response to epidermal growth factor receptor inhibitors. *Cancer Res*, **64**, 9139–9143.

Amano, K., Sago, H., Uchikawa, C. *et al.* (2004). Dosage-dependent over-expression of genes in the trisomic region of Ts1Cje mouse model for Down syndrome. *Hum Mol Genet*, **13**, 1333–1340.

Amir, R.E., Van den Veyver, I.B., Wan, M., Tran, C.Q., Francke, U. and Zoghbi, H.Y. (1999). Rett syndrome is caused by mutations in X-linked MECP2, encoding methyl-CpG-binding protein 2. *Nat Genet*, **23**, 185–188.

Amor, D.J. and Halliday, J. (2008). A review of known imprinting syndromes and their association with assisted reproduction technologies. *Hum Reprod*, **23**, 2826–2834.

Amos-Landgraf, J.M., Ji, Y., Gottlieb, W. *et al.* (1999). Chromosome breakage in the Prader–Willi and Angelman syndromes involves recombination between large, transcribed repeats at proximal and distal breakpoints. *Am J Hum Genet*, **65**, 370–386.

An, P., Nelson, G.W., Wang, L. *et al.* (2002). Modulating influence on HIV/AIDS by interacting RANTES gene variants. *Proc Natl Acad Sci U S A*, **99**, 10002–10007.

Anderson, R.M. and May, R.M. (1985). Helminth infections of humans: mathematical models, population dynamics, and control. *Adv Parasitol*, **24**, 1–101.

Andrade, M.A. and Bork, P. (1995). HEAT repeats in the Huntington's disease protein. *Nat Genet*, **11**, 115–116.

Andreasen, C.H., Stender-Petersen, K.L., Mogensen, M.S. *et al.* (2008). Low physical activity accentuates the effect of the FTO rs9939609 polymorphism on body fat accumulation. *Diabetes*, **57**, 95–101.

Andrew, S.E., Goldberg, Y.P., Kremer, B. *et al.* (1993). The relationship between trinucleotide (CAG) repeat length and clinical features of Huntington's disease. *Nat Genet*, **4**, 398–403.

Anonymous (1994). International Huntington Association and the World Federation of Neurology Research Group on Huntington's Chorea. Guidelines for the molecular genetics predictive test in Huntington's disease. *J Med Genet*, **31**, 555–559.

Ansari-Lari, M.A., Liu, X.M., Metzker, M.L., Rut, A.R. and Gibbs, R.A. (1997). The extent of genetic variation in the CCR5 gene. *Nat Genet*, **16**, 221–222.

Anton, M.A., Gorostiaga, D., Guruceaga, E. *et al.* (2008). SPACE: an algorithm to predict and quantify alternatively spliced isoforms using microarrays. *Genome Biol*, **9**, R46.

Antonarakis, S.E. (1998). Recommendations for a nomenclature system for human gene mutations. Nomenclature Working Group. *Hum Mutat*, **11**, 1–3.

Antonarakis, S.E. and Epstein, C.J. (2006). The challenge of Down syndrome. *Trends Mol Med*, **12**, 473–479.

Antonarakis, S.E., Lyle, R., Dermitzakis, E.T., Reymond, A. and Deutsch, S. (2004). Chromosome 21 and Down syndrome: from genomics to pathophysiology. *Nat Rev Genet*, **5**, 725–738.

Aparicio, C. and Dahlback, B. (1996). Molecular mechanisms of activated protein C resistance. Properties of factor V isolated from an individual with homozygosity for the Arg506 to Gln mutation in the factor V gene. *Biochem J*, **313**, 467–472.

Aquaro, S., Menten, P., Struyf, S. *et al.* (2001). The LD78beta isoform of MIP-1alpha is the most potent CC-chemokine in inhibiting CCR5-dependent human immunodeficiency virus type 1 replication in human macrophages. *J Virol*, **75**, 4402–4406.

Arbiza, L., Dopazo, J. and Dopazo, H. (2006). Positive selection, relaxation, and acceleration in the evolution of the human and chimp genome. *PLoS Comput Biol*, **2**, e38.

Archer, H.L., Whatley, S.D., Evans, J.C. *et al.* (2006). Gross rearrangements of the MECP2 gene are found in both classical and atypical Rett syndrome patients. *J Med Genet*, **43**, 451–456.

Armour, J.A., Anttinen, T., May, C.A. *et al.* (1996). Minisatellite diversity supports a recent African origin for modern humans. *Nat Genet*, **13**, 154–160.

Arnett, H.A., Escobar, S.S., Gonzalez-Suarez, E. *et al.* (2007). BTNL2, a butyrophilin/B7-like molecule, is a negative costimulatory molecule modulated in intestinal inflammation. *J Immunol*, **178**, 1523–1533.

Arron, J.R., Winslow, M.M., Polleri, A. *et al.* (2006). NFAT dysregulation by increased dosage of DSCR1 and DYRK1A on chromosome 21. *Nature*, **441**, 595–600.

Ashley, C.T., Jr., Wilkinson, K.D., Reines, D. and Warren, S.T. (1993). FMR1 protein: conserved RNP family domains and selective RNA binding. *Science*, **262**, 563–566.

Ashley-Koch, A., Yang, Q. and Olney, R.S. (2000). Sickle hemoglobin (HbS) allele and sickle cell disease: a HuGE review. *Am J Epidemiol*, **151**, 839–845.

Atsuta, N., Watanabe, H., Ito, M. *et al.* (2006). Natural history of spinal and bulbar muscular atrophy (SBMA): a study of 223 Japanese patients. *Brain*, **129**, 1446–1455.

Attems, J., Konig, C., Huber, M., Lintner, F. and Jellinger, K.A. (2005). Cause of death in demented and non-demented elderly inpatients; an autopsy study of 308 cases. *J Alzheimers Dis*, **8**, 57–62.

Avent, N.D., Madgett, T.E., Lee, Z.E., Head, D.J., Maddocks, D.G. and Skinner, L.H. (2006). Molecular biology of Rh proteins and relevance to molecular medicine. *Expert Rev Mol Med*, **8**, 1–20.

Avery, O.T., MacLeod, C.M. and McCarty, M. (1944). Studies of the chemical nature of the substance inducing transformation of pneumococcal types. Induction of transformation by a desoxyribonucleic acid fraction isolated from Pneumococcus Type III. *J Exp Med*, **79**, 137–158.

Awdeh, Z.L. and Alper, C.A. (1980). Inherited structural polymorphism of the fourth component of human complement. *Proc Natl Acad Sci U S A*, **77**, 3576–3580.

Badge, R.M., Alisch, R.S. and Moran, J.V. (2003). ATLAS: a system to selectively identify human-specific L1 insertions. *Am J Hum Genet*, **72**, 823–838.

Baglioni, C. (1962). The fusion of two peptide chains in hemoglobin Lepore and its interpretation as a genetic deletion. *Proc Natl Acad Sci U S A*, **48**, 1880–1886.

Bagnall, R.D., Waseem, N., Green, P.M. and Giannelli, F. (2002). Recurrent inversion breaking intron 1 of the factor VIII gene is a frequent cause of severe hemophilia A. *Blood*, **99**, 168–174.

Bailey, J.A. and Eichler, E.E. (2006). Primate segmental duplications: crucibles of evolution, diversity and disease. *Nat Rev Genet*, **7**, 552–564.

Bailey, J.A., Church, D.M., Ventura, M., Rocchi, M. and Eichler, E.E. (2004). Analysis of segmental duplications and genome assembly in the mouse. *Genome Res*, **14**, 789–801.

Bailey, J.A., Gu, Z., Clark, R.A. *et al.* (2002). Recent segmental duplications in the human genome. *Science*, **297**, 1003–1007.

Bailey, J.A., Liu, G. and Eichler, E.E. (2003). An Alu transposition model for the origin and expansion of human segmental duplications. *Am J Hum Genet*, **73**, 823–834.

Bailey, J.A., Yavor, A.M., Massa, H.F., Trask, B.J. and Eichler, E.E. (2001). Segmental duplications: organization and impact within the current human genome project assembly. *Genome Res*, **11**, 1005–1017.

Baird, D.M., Rowson, J., Wynford-Thomas, D. and Kipling, D. (2003). Extensive allelic variation and ultrashort telomeres in senescent human cells. *Nat Genet*, **33**, 203–207.

Baker, M., Litvan, I., Houlden, H. *et al.* (1999). Association of an extended haplotype in the tau gene with progressive supranuclear palsy. *Hum Mol Genet*, **8**, 711–715.

Baker, P., Piven, J., Schwartz, S. and Patil, S. (1994). Brief report: duplication of chromosome 15q11–13 in two individuals with autistic disorder. *J Autism Dev Disord*, **24**, 529–535.

Ballabio, A. (1993). The rise and fall of positional cloning? *Nat Genet*, **3**, 277–279.

Ballif, B.C., Sulpizio, S.G., Lloyd, R.M. *et al.* (2007). The clinical utility of enhanced subtelomeric coverage in array CGH. *Am J Med Genet*, **143**, 1850–1857.

Baltimore, D. (2001). Our genome unveiled. *Nature*, **409**, 814–816.

Bamshad, M. and Wooding, S.P. (2003). Signatures of natural selection in the human genome. *Nat Rev Genet*, **4**, 99–111.

Bandyopadhyay, R., Heller, A., Knox-DuBois, C. *et al.* (2002). Parental origin and timing of de novo Robertsonian translocation formation. *Am J Hum Genet*, **71**, 1456–1462.

Banfi, S., Servadio, A., Chung, M.Y. *et al.* (1994). Identification and characterization of the gene causing type 1 spinocerebellar ataxia. *Nat Genet*, **7**, 513–520.

Bansal, V., Bashir, A. and Bafna, V. (2007). Evidence for large inversion polymorphisms in the human genome from HapMap data. *Genome Res*, **17**, 219–230.

Barbier, M., Delahaye, N.F., Fumoux, F. and Rihet, P. (2008). Family-based association of a low producing lymphotoxin-alpha allele with reduced Plasmodium falciparum parasitemia. *Microbes Infect*, **10**, 673–679.

Barbour, V.M., Tufarelli, C., Sharpe, J.A. *et al.* (2000). Alpha-thalassemia resulting from a negative chromosomal position effect. *Blood*, **96**, 800–807.

Barcellos, L.F., Oksenberg, J.R., Begovich, A.B. *et al.* (2003). HLA-DR2 dose effect on susceptibility to multiple sclerosis and influence on disease course. *Am J Hum Genet*, **72**, 710–716.

Barlow, G.M., Chen, X.N., Shi, Z.Y. *et al.* (2001). Down syndrome congenital heart disease: a narrowed region and a candidate gene. *Genet Med*, **3**, 91–101.

Barnabas, J. and Muller, C.J. (1962). Haemoglobin Lepore (Hollandia). *Nature*, **194**, 931–932.

Barratt, B.J., Payne, F., Lowe, C.E. *et al.* (2004). Remapping the insulin gene/IDDM2 locus in type 1 diabetes. *Diabetes*, **53**, 1884–1889.

Barreiro, L.B., Laval, G., Quach, H., Patin, E. and Quintana-Murci, L. (2008). Natural selection has driven population differentiation in modern humans. *Nat Genet*, **40**, 340–345.

Barrett, J.C., Hansoul, S., Nicolae, D.L. *et al.* (2008). Genome-wide association defines more than 30 distinct susceptibility loci for Crohn's disease. *Nat Genet*, **40**, 955–962.

Batzer, M.A. and Deininger, P.L. (2002). Alu repeats and human genomic diversity. *Nat Rev Genet*, **3**, 370–379.

Batzer, M.A., Stoneking, M., Alegria-Hartman, M. *et al.* (1994). African origin of human-specific polymorphic Alu insertions. *Proc Natl Acad Sci U S A*, **91**, 12288–12292.

Bauer, S., Groh, V., Wu, J. *et al.* (1999). Activation of NK cells and T cells by NKG2D, a receptor for stress-inducible MICA. *Science*, **285**, 727–729.

Baumer, A., Dutly, F., Balmer, D. *et al.* (1998). High level of unequal meiotic crossovers at the origin of the 22q11. 2 and 7q11.23 deletions. *Hum Mol Genet*, **7**, 887–894.

Bayley, J.P., Ottenhoff, T.H. and Verweij, C.L. (2004). Is there a future for TNF promoter polymorphisms? *Genes Immun*, **5**, 315–329.

Bear, M.F., Huber, K.M. and Warren, S.T. (2004). The mGluR theory of fragile X mental retardation. *Trends Neurosci*, **27**, 370–377.

Beaudet, A.L. and Tsui, L.C. (1993). A suggested nomenclature for designating mutations. *Hum Mutat*, **2**, 245–248.

Beaudet, A.L., Bowcock, A., Buchwald, M. *et al.* (1986). Linkage of cystic fibrosis to two tightly linked DNA markers: joint report from a collaborative study. *Am J Hum Genet*, **39**, 681–693.

Beckman, J.S. and Weber, J.L. (1992). Survey of human and rat microsatellites. *Genomics*, **12**, 627–631.

Beet, E. (1946). Sickle cell disease in the Balovale district of northern Rhodesia. *East Afr Med J*, **23**, 75–86.

Begovich, A.B., Carlton, V.E., Honigberg, L.A. *et al.* (2004). A missense single-nucleotide polymorphism in a gene encoding a protein tyrosine phosphatase (PTPN22) is associated with rheumatoid arthritis. *Am J Hum Genet*, **75**, 330–337.

Bejerano, G., Lowe, C.B., Ahituv, N. *et al.* (2006). A distal enhancer and an ultraconserved exon are derived from a novel retroposon. *Nature*, **441**, 87–90.

Bell, G.I., Horita, S. and Karam, J.H. (1984). A polymorphic locus near the human insulin gene is associated with insulin-dependent diabetes mellitus. *Diabetes*, **33**, 176–183.

Bell, G.I., Selby, M.J. and Rutter, W.J. (1982). The highly polymorphic region near the human insulin gene is composed of simple tandemly repeating sequences. *Nature*, **295**, 31–35.

Bellamy, R., Ruwende, C., Corrah, T., McAdam, K.P., Whittle, H.C. and Hill, A.V. (1998). Variations in the NRAMP1 gene and susceptibility to tuberculosis in West Africans. *N Engl J Med*, **338**, 640–644.

Benjamin, R. and Parham, P. (1990). Guilt by association: HLA-B27 and ankylosing spondylitis. *Immunol Today*, **11**, 137–142.

Bennett, D.C. and Lamoreux, M.L. (2003). The color loci of mice – a genetic century. *Pigment Cell Res*, **16**, 333–344.

Bennett, S.T., Lucassen, A.M., Gough, S.C. *et al.* (1995). Susceptibility to human type 1 diabetes at IDDM2 is determined by tandem repeat variation at the insulin gene minisatellite locus. *Nat Genet*, **9**, 284–292.

Bennett, S.T., Wilson, A.J., Cucca, F. *et al.* (1996). IDDM2-VNTR-encoded susceptibility to type 1 diabetes: dominant protection and parental transmission of alleles of the insulin gene-linked minisatellite locus. *J Autoimmun*, **9**, 415–421.

Bentley, D.R. (2006). Whole-genome re-sequencing. *Curr Opin Genet Dev*, **16**, 545–552.

Berendt, A.R., Simmons, D.L., Tansey, J., Newbold, C.I. and Marsh, K. (1989). Intercellular adhesion molecule-1 is an endothelial cell adhesion receptor for Plasmodium falciparum. *Nature*, **341**, 57–59.

Berg, J.S., Brunetti-Pierri, N., Peters, S.U. *et al.* (2007). Speech delay and autism spectrum behaviors are frequently associated with duplication of the 7q11.23 Williams–Beuren syndrome region. *Genet Med*, **9**, 427–441.

Berger, J., Suzuki, T., Senti, K.A., Stubbs, J., Schaffner, G. and Dickson, B.J. (2001). Genetic mapping with SNP markers in Drosophila. *Nat Genet*, **29**, 475–481.

Bergseng, E., Xia, J., Kim, C.Y., Khosla, C. and Sollid, L.M. (2005). Main chain hydrogen bond interactions in the binding of proline-rich gluten peptides to the celiac disease-associated HLA-DQ2 molecule. *J Biol Chem*, **280**, 21791–21796.

Bergstrom, T.F., Josefsson, A., Erlich, H.A. and Gyllensten, U. (1998). Recent origin of HLA-DRB1 alleles and implications for human evolution. *Nat Genet*, **18**, 237–242.

Bersaglieri, T., Sabeti, P.C., Patterson, N. *et al.* (2004). Genetic signatures of strong recent positive selection at the lactase gene. *Am J Hum Genet*, **74**, 1111–1120.

Bertina, R.M., Koeleman, B.P., Koster, T. *et al.* (1994). Mutation in blood coagulation factor V associated with resistance to activated protein C. *Nature*, **369**, 64–67.

Bertram, L., McQueen, M., Mullin, K., Blacker, D. and Tanzi, R. (2007a). The AlzGene Database. Alzheimer Research Forum. Available at: http://www.alzgene.org. Accessed 18 Aug 2008.

Bertram, L., McQueen, M.B., Mullin, K., Blacker, D. and Tanzi, R.E. (2007b). Systematic meta-analyses of Alzheimer disease genetic association studies: the AlzGene Database. *Nat Genet*, **39**, 17–23.

Beutler, E. (1993). The designation of mutations. *Am J Hum Genet*, **53**, 783–785.

Beutler, E. (2006). Hemochromatosis: genetics and pathophysiology. *Annu Rev Med*, **57**, 331–347.

Beutler, E., McKusick, V.A., Motulsky, A.G., Scriver, C.R. and Hutchinson, F. (1996). Mutation nomenclature: nicknames, systematic names, and unique identifiers. *Hum Mutat*, **8**, 203–206.

Bhangale, T.R., Rieder, M.J., Livingston, R.J. and Nickerson, D.A. (2005). Comprehensive identification and characterization of diallelic insertion-deletion polymorphisms in 330 human candidate genes. *Hum Mol Genet*, **14**, 59–69.

Bhangale, T.R., Stephens, M. and Nickerson, D.A. (2006). Automating resequencing-based detection of insertion-deletion polymorphisms. *Nat Genet*, **38**, 1457–1462.

Bidichandani, S.I., Ashizawa, T. and Patel, P.I. (1998). The GAA triplet-repeat expansion in Friedreich ataxia interferes with transcription and may be associated with an unusual DNA structure. *Am J Hum Genet*, **62**, 111–121.

Biggar, W.D., Klamut, H.J., Demacio, P.C., Stevens, D.J. and Ray, P.N. (2002). Duchenne muscular dystrophy: current knowledge, treatment, and future prospects. *Clin Orthop Relat Res*, **401**, 88–106.

Bina, J.C., Tavares-Neto, J., Prata, A. and Azevedo, E.S. (1978). Greater resistance to development of severe schistosomiasis in Brazilian Negroes. *Hum Biol*, **50**, 41–49.

Bird, A.P. (1986). CpG-rich islands and the function of DNA methylation. *Nature*, **321**, 209–213.

Bird, T.D. (2008). Genetic aspects of Alzheimer disease. *Genet Med*, **10**, 231–239.

Bird, T.D., Lampe, T.H., Nemens, E.J., Miner, G.W., Sumi, S.M. and Schellenberg, G.D. (1988). Familial Alzheimer's disease in American descendants of the Volga Germans: probable genetic founder effect. *Ann Neurol*, **23**, 25–31.

Bird, T.D., Levy-Lahad, E., Poorkaj, P. *et al.* (1996). Wide range in age of onset for chromosome 1-related familial Alzheimer's disease. *Ann Neurol*, **40**, 932–936.

Birney, E., Stamatoyannopoulos, J.A., Dutta, A. *et al.* (2007). Identification and analysis of functional elements in 1% of the human genome by the ENCODE pilot project. *Nature*, **447**, 799–816.

Bjornsson, H.T., Albert, T.J., Ladd-Acosta, C.M. *et al.* (2008). SNP-specific array-based allele-specific expression analysis. *Genome Res*, **18**, 771–779.

Black, F.L. and Hedrick, P.W. (1997). Strong balancing selection at HLA loci: evidence from segregation in South Amerindian families. *Proc Natl Acad Sci U S A*, **94**, 12452–12456.

Blackburn, E.H. (1991). Structure and function of telomeres. *Nature*, **350**, 569–573.

Blacker, D., Haines, J.L., Rodes, L. *et al.* (1997). ApoE-4 and age at onset of Alzheimer's disease: the NIMH genetics initiative. *Neurology*, **48**, 139–147.

Blackwell, J.M., Mohamed, H.S. and Ibrahim, M.E. (2004). Genetics and visceral leishmaniasis in the Sudan: seeking a link. *Trends Parasitol*, **20**, 268–274.

Blanchong, C.A., Chung, E.K., Rupert, K.L. *et al.* (2001). Genetic, structural and functional diversities of human complement components C4A and C4B and their mouse homologues, Slp and C4. *Int Immunopharmacol*, **1**, 365–392.

Blanchong, C.A., Zhou, B., Rupert, K.L. *et al.* (2000). Deficiencies of human complement component C4A and C4B and heterozygosity in length variants of RP-C4-CYP21-TNX (RCCX) modules in caucasians. The load of RCCX genetic diversity on major histocompatibility complex-associated disease. *J Exp Med*, **191**, 2183–2196.

Blanton, R.E., Salam, E.A., Ehsan, A., King, C.H. and Goddard, K.A. (2005). Schistosomal hepatic fibrosis and the interferon gamma receptor: a linkage analysis using single-nucleotide polymorphic markers. *Eur J Hum Genet*, **13**, 660–668.

Blennow, E., Bui, T.H., Kristoffersson, U. *et al.* (1994). Swedish survey on extra structurally abnormal chromosomes in 39 105 consecutive prenatal diagnoses: prevalence and characterization by fluorescence in situ hybridization. *Prenat Diagn*, **14**, 1019–1028.

Bleul, C.C., Farzan, M., Choe, H. *et al.* (1996). The lymphocyte chemoattractant SDF-1 is a ligand for LESTR/fusin and blocks HIV-1 entry. *Nature*, **382**, 829–833.

Board, P., Coggan, M., Johnston, P., Ross, V., Suzuki, T. and Webb, G. (1990). Genetic heterogeneity of the human glutathione transferases: a complex of gene families. *Pharmacol Ther*, **48**, 357–369.

Bobadilla, J.L., Macek, M., Jr., Fine, J.P. and Farrell, P.M. (2002). Cystic fibrosis: a worldwide analysis of CFTR mutations – correlation with incidence data and application to screening. *Hum Mutat*, **19**, 575–606.

Boissinot, S., Chevret, P. and Furano, A.V. (2000). L1 (LINE-1) retrotransposon evolution and amplification in recent human history. *Mol Biol Evol*, **17**, 915–928.

Boll, W., Wagner, P. and Mantei, N. (1991). Structure of the chromosomal gene and cDNAs coding for lactase-phlorizin hydrolase in humans with adult-type hypolactasia or persistence of lactase. *Am J Hum Genet*, **48**, 889–902.

Bondeson, M.L., Dahl, N., Malmgren, H. *et al.* (1995). Inversion of the IDS gene resulting from recombination with IDS-related sequences is a common cause of the Hunter syndrome. *Hum Mol Genet*, **4**, 615–621.

Bontrop, R.E. (2006). Comparative genetics of MHC polymorphisms in different primate species: duplications and deletions. *Hum Immunol*, **67**, 388–397.

Borghans, J.A., Beltman, J.B. and De Boer, R.J. (2004). MHC polymorphism under host–pathogen coevolution. *Immunogenetics*, **55**, 732–739.

Borst, P. and Genest, P.-A. (2006). Switching like for like. *Nature*, **439**, 926–927.

Borstnik, B. and Pumpernik, D. (2002). Tandem repeats in protein coding regions of primate genes. *Genome Res*, **12**, 909–915.

Botstein, D. and Risch, N. (2003). Discovering genotypes underlying human phenotypes: past successes for mendelian disease, future approaches for complex disease. *Nat Genet*, **33**, 228–237.

Botstein, D., White, R.L., Skolnick, M. and Davis, R.W. (1980). Construction of a genetic linkage map in man using restriction fragment length polymorphisms. *Am J Hum Genet*, **32**, 314–331.

Bottini, N., Musumeci, L., Alonso, A. *et al.* (2004). A functional variant of lymphoid tyrosine phosphatase is associated with type I diabetes. *Nat Genet*, **36**, 337–338.

Bowcock, A.M., Kidd, J.R., Mountain, J.L. *et al.* (1991). Drift, admixture, and selection in human evolution: a study with DNA polymorphisms. *Proc Natl Acad Sci U S A*, **88**, 839–843.

Bowcock, A.M., Ruiz-Linares, A., Tomfohrde, J., Minch, E., Kidd, J.R. and Cavalli-Sforza, L.L. (1994). High resolution of human evolutionary trees with polymorphic microsatellites. *Nature*, **368**, 455–457.

Bowness, P., Zaccai, N., Bird, L. and Jones, E.Y. (1999). HLA-B27 and disease pathogenesis: new structural and functional insights. *Expert Rev Mol Med*, **1999**, 1–10.

Bradley, D.J. (1977). Regulation of Leishmania populations within the host. II. Genetic control of acute susceptibility of mice to Leishmania donovani infection. *Clin Exp Immunol*, **30**, 130–140.

Brain, P. (1952). The sickle cell trait; its clinical significance. *S Afr Med J*, **26**, 925–928.

Brambilla, A., Villa, C., Rizzardi, G. *et al.* (2000). Shorter survival of SDF1-3′A/3′A homozygotes linked to CD4+ T cell decrease in advanced human immunodeficiency virus type 1 infection. *J Infect Dis*, **182**, 311–315.

Brant, S.R. and Shugart, Y.Y. (2004). Inflammatory bowel disease gene hunting by linkage analysis: rationale, methodology, and present status of the field. *Inflamm Bowel Dis*, **10**, 300–311.

Bray, N.J., Buckland, P.R., Owen, M.J. and O'Donovan, M.C. (2003). Cis-acting variation in the expression of a high proportion of genes in human brain. *Hum Genet*, **113**, 149–153.

Bray, N.J., Preece, A., Williams, N.M. *et al.* (2005). Haplotypes at the dystrobrevin binding protein 1 (DTNBP1) gene locus mediate risk for schizophrenia through reduced DTNBP1 expression. *Hum Mol Genet*, **14**, 1947–1954.

Breitner, J.C. (1996). APOE genotyping and Alzheimer's disease. *Lancet*, **347**, 1184–1185.

Brem, R.B. and Kruglyak, L. (2005). The landscape of genetic complexity across 5,700 gene expression traits in yeast. *Proc Natl Acad Sci U S A*, **102**, 1572–1577.

Brem, R.B., Storey, J.D., Whittle, J. and Kruglyak, L. (2005). Genetic interactions between polymorphisms that affect gene expression in yeast. *Nature*, **436**, 701–703.

Brem, R.B., Yvert, G., Clinton, R. and Kruglyak, L. (2002). Genetic dissection of transcriptional regulation in budding yeast. *Science*, **296**, 752–755.

Brewer, C., Holloway, S., Zawalnyski, P., Schinzel, A. and FitzPatrick, D. (1998). A chromosomal deletion map of human malformations. *Am J Hum Genet*, **63**, 1153–1159.

Brewer, C., Holloway, S., Zawalnyski, P., Schinzel, A. and FitzPatrick, D. (1999). A chromosomal duplication map of malformations: regions of suspected haplo- and triplolethality – and tolerance of segmental aneuploidy – in humans. *Am J Hum Genet*, **64**, 1702–1708.

Brewerton, D.A., Hart, F.D., Nicholls, A., Caffrey, M., James, D.C. and Sturrock, R.D. (1973). Ankylosing spondylitis and HL-A 27. *Lancet*, **1**, 904–907.

Brinkmann, B., Klintschar, M., Neuhuber, F., Huhne, J. and Rolf, B. (1998). Mutation rate in human microsatellites: influence of the structure and length of the tandem repeat. *Am J Hum Genet*, **62**, 1408–1415.

Brissot, P. (2003). The discovery of the new haemochromatosis gene. 1996. *J Hepatol*, **38**, 704–709.

Broman, K.W. (2005). Mapping expression in randomized rodent genomes. *Nat Genet*, **37**, 209–210.

Broman, K.W., Murray, J.C., Sheffield, V.C., White, R.L. and Weber, J.L. (1998). Comprehensive human genetic maps: individual and sex-specific variation in recombination. *Am J Hum Genet*, **63**, 861–869.

Brook, J.D., McCurrach, M.E., Harley, H.G. *et al.* (1992). Molecular basis of myotonic dystrophy: expansion of a trinucleotide (CTG) repeat at the 3′ end of a transcript encoding a protein kinase family member. *Cell*, **68**, 799–808.

Brookfield, J.F. (2005). The ecology of the genome – mobile DNA elements and their hosts. *Nat Rev Genet*, **6**, 128–136.

Brouha, B., Schustak, J., Badge, R.M. *et al.* (2003). Hot L1s account for the bulk of retrotransposition in the human population. *Proc Natl Acad Sci U S A*, **100**, 5280–5285.

Brown, M.A. (2008). Breakthroughs in genetic studies of ankylosing spondylitis. *Rheumatology (Oxford)*, **47**, 132–137.

Brown, M.A., Kennedy, L.G., MacGregor, A.J. *et al.* (1997). Susceptibility to ankylosing spondylitis in twins: the role of genes, HLA, and the environment. *Arthritis Rheum*, **40**, 1823–1828.

Brown, M.A., Pile, K.D., Kennedy, L.G. *et al.* (1998). A genome-wide screen for susceptibility loci in ankylosing spondylitis. *Arthritis Rheum*, **41**, 588–595.

Bruford, E.A., Lush, M.J., Wright, M.W., Sneddon, T.P., Povey, S. and Birney, E. (2008). The HGNC Database in 2008: a resource for the human genome. *Nucleic Acids Res*, **36**, D445–8.

Bruford, M.W. and Wayne, R.K. (1993). Microsatellites and their application to population genetic studies. *Curr Opin Genet Dev*, **3**, 939–943.

Bucheton, B., Abel, L., El-Safi, S. *et al.* (2003a). A major susceptibility locus on chromosome 22q12 plays a critical role in the control of kala-azar. *Am J Hum Genet*, **73**, 1052–1060.

Bucheton, B., Abel, L., Kheir, M.M. *et al.* (2003b). Genetic control of visceral leishmaniasis in a Sudanese population: candidate gene testing indicates a linkage to the NRAMP1 region. *Genes Immun*, **4**, 104–109.

Bucheton, B., Kheir, M.M., El-Safi, S.H. *et al.* (2002). The interplay between environmental and host factors during an outbreak of visceral leishmaniasis in eastern Sudan. *Microbes Infect*, **4**, 1449–1457.

Bucheton, B., El-Safi, S.H., Hammad, A. *et al.* (2003c). Antileishmanial antibodies in an outbreak of visceral leishmaniasis in eastern Sudan: high antibody responses occur in resistant subjects and are not predictive of disease. *Trans R Soc Trop Med Hyg*, **97**, 463–468.

Buchwald, M., Tsui, L.C. and Riordan, J.R. (1989). The search for the cystic fibrosis gene. *Am J Physiol*, **257**, L47–52.

Buckland, P.R. (2004). Allele-specific gene expression differences in humans. *Hum Mol Genet*, **13**, R255–260.

Buckle, V.J., Higgs, D.R., Wilkie, A.O., Super, M. and Weatherall, D.J. (1988). Localisation of human alpha globin to 16p13.3----pter. *J Med Genet*, **25**, 847–849.

Buiting, K., Saitoh, S., Gross, S. *et al.* (1995). Inherited microdeletions in the Angelman and Prader–Willi syndromes define an imprinting centre on human chromosome 15. *Nat Genet*, **9**, 395–400.

Bulayeva, K., Jorde, L.B., Ostler, C., Watkins, S., Bulayev, O. and Harpending, H. (2003). Genetics and population history of Caucasus populations. *Hum Biol*, **75**, 837–853.

Burgner, D., Jamieson, S.E. and Blackwell, J.M. (2006). Genetic susceptibility to infectious diseases: big is beautiful, but will bigger be even better? *Lancet Infect Dis*, **6**, 653–663.

Burns, J.C., Shimizu, C., Gonzalez, E. *et al.* (2005). Genetic variations in the receptor-ligand pair CCR5 and CCL3L1 are important determinants of susceptibility to Kawasaki disease. *J Infect Dis*, **192**, 344–349.

Burton, P.R., Clayton, D.G., Cardon, L.R. *et al.* (2007). Association scan of 14,500 nonsynonymous SNPs in four diseases identifies autoimmunity variants. *Nat Genet*, **39**, 1329–1337.

Buschiazzo, E. and Gemmell, N.J. (2006). The rise, fall and renaissance of microsatellites in eukaryotic genomes. *Bioessays*, **28**, 1040–1050.

Butler, J.M. (2006). Genetics and genomics of core short tandem repeat loci used in human identity testing. *J Forensic Sci*, **51**, 253–265.

Butler, J.M. (2007). Short tandem repeat typing technologies used in human identity testing. *Biotechniques*, **43**, ii–v.

Buxton, J., Shelbourne, P., Davies, J. *et al.* (1992). Detection of an unstable fragment of DNA specific to individuals with myotonic dystrophy. *Nature*, **355**, 547–548.

Bystrykh, L., Weersing, E., Dontje, B. *et al.* (2005). Uncovering regulatory pathways that affect hematopoietic stem cell function using 'genetical genomics'. *Nat Genet*, **37**, 225–232.

Cabello, P.H., Lima, A.M., Azevedo, E.S. and Krieger, H. (1995). Familial aggregation of Leishmania chagasi infection in northeastern Brazil. *Am J Trop Med Hyg*, **52**, 364–365.

Cabrera, G., Cot, M., Migot-Nabias, F., Kremsner, P.G., Deloron, P. and Luty, A.J. (2005). The sickle cell trait is associated with enhanced immunoglobulin G antibody responses to Plasmodium falciparum variant surface antigens. *J Infect Dis*, **191**, 1631–1638.

Caffrey, M.F. and James, D.C. (1973). Human lymphocyte antigen association in ankylosing spondylitis. *Nature*, **242**, 121.

Callinan, P.A., Wang, J., Herke, S.W., Garber, R.K., Liang, P. and Batzer, M.A. (2005). Alu retrotransposition-mediated deletion. *J Mol Biol*, **348**, 791–800.

Campion, D., Brice, A., Hannequin, D. *et al.* (1995). A large pedigree with early-onset Alzheimer's disease: clinical, neuropathologic, and genetic characterization. *Neurology*, **45**, 80–85.

Campion, D., Dumanchin, C., Hannequin, D. *et al.* (1999). Early-onset autosomal dominant Alzheimer disease: prevalence, genetic heterogeneity, and mutation spectrum. *Am J Hum Genet*, **65**, 664–670.

Campuzano, V., Montermini, L., Molto, M.D. *et al.* (1996). Friedreich's ataxia: autosomal recessive disease caused by an intronic GAA triplet repeat expansion. *Science*, **271**, 1423–1427.

Cann, R.L., Stoneking, M. and Wilson, A.C. (1987). Mitochondrial DNA and human evolution. *Nature*, **325**, 31–36.

Cappadoro, M., Giribaldi, G., O'Brien, E. *et al.* (1998). Early phagocytosis of glucose-6-phosphate dehydrogenase (G6PD)-deficient erythrocytes parasitized by Plasmodium falciparum may explain malaria protection in G6PD deficiency. *Blood*, **92**, 2527–2534.

Cappellini, M.D. and Fiorelli, G. (2008). Glucose-6-phosphate dehydrogenase deficiency. *Lancet*, **371**, 64–74.

Cardon, L.R. and Bell, J.I. (2001). Association study designs for complex diseases. *Nat Rev Genet*, **2**, 91–99.

Cargill, M., Altshuler, D., Ireland, J. *et al.* (1999). Characterization of single-nucleotide polymorphisms in coding regions of human genes. *Nat Genet*, **22**, 231–238.

Cargill, M., Schrodi, S.J., Chang, M. *et al.* (2007). A large-scale genetic association study confirms IL12B and leads to the identification of IL23R as psoriasis-risk genes. *Am J Hum Genet*, **80**, 273–290.

Carlson, C.S., Eberle, M.A., Kruglyak, L. and Nickerson, D.A. (2004). Mapping complex disease loci in whole-genome association studies. *Nature*, **429**, 446–452.

Carrano, A.V., Gray, J.W., Langlois, R.G., Burkhart-Schultz, K.J. and Van Dilla, M.A. (1979). Measurement and purification of human chromosomes by flow cytometry and sorting. *Proc Natl Acad Sci U S A*, **76**, 1382–1384.

Carrel, L. and Willard, H.F. (2005). X-inactivation profile reveals extensive variability in X-linked gene expression in females. *Nature*, **434**, 400–404.

Carrell, R.W., Lehmann, H. and Hutchison, H.E. (1966). Haemoglobin Koln (beta-98 valine–methionine): an unstable protein causing inclusion-body anaemia. *Nature*, **210**, 915–916.

Carrington, M., Nelson, G.W., Martin, M.P. *et al.* (1999). HLA and HIV-1: heterozygote advantage and B*35-Cw*04 disadvantage. *Science*, **283**, 1748–1752.

Carroll, M.L., Roy-Engel, A.M., Nguyen, S.V. *et al.* (2001). Large-scale analysis of the Alu Ya5 and Yb8 subfamilies and their contribution to human genomic diversity. *J Mol Biol*, **311**, 17–40.

Cartegni, L., Chew, S.L. and Krainer, A.R. (2002). Listening to silence and understanding nonsense: exonic mutations that affect splicing. *Nat Rev Genet*, **3**, 285–298.

Carter, A.B., Salem, A.H., Hedges, D.J. *et al.* (2004). Genome-wide analysis of the human Alu Yb-lineage. *Hum Genomics*, **1**, 167–178.

Carter, N.P. (2007). Methods and strategies for analyzing copy number variation using DNA microarrays. *Nat Genet*, **39**, S16–21.

Caspersson, T., Farber, S., Foley, G.E. *et al.* (1968). Chemical differentiation along metaphase chromosomes. *Exp Cell Res*, **49**, 219–222.

Cassidy, S.B., Lai, L.W., Erickson, R.P. *et al.* (1992). Trisomy 15 with loss of the paternal 15 as a cause of Prader–Willi syndrome due to maternal disomy. *Am J Hum Genet*, **51**, 701–708.

Castaldo, G., D'Argenio, V., Nardiello, P. *et al.* (2007). Haemophilia A: molecular insights. *Clin Chem Lab Med*, **45**, 450–461.

Cavalli-Sforza, L.L., Menozzi, P. and Piazza, A. (1994). *The History and Geography of Human Genes*. Princeton University Press, Princeton, NJ.

Cavalli-Sforza, L.L., Piazza, A., Menozzi, P. and Mountain, J. (1988). Reconstruction of human evolution: bringing together genetic, archaeological, and linguistic data. *Proc Natl Acad Sci U S A*, **85**, 6002–6006.

Cavanaugh, J. (2001). International collaboration provides convincing linkage replication in complex disease through analysis of a large pooled data set: Crohn disease and chromosome 16. *Am J Hum Genet*, **68**, 1165–1171.

Cavanaugh, J.A., Callen, D.F., Wilson, S.R. *et al.* (1998). Analysis of Australian Crohn's disease pedigrees refines the localization for susceptibility to inflammatory bowel disease on chromosome 16. *Ann Hum Genet*, **62**, 291–298.

Cerruti Mainardi, P. (2006). Cri du chat syndrome. *Orphanet J Rare Dis*, **1**, 33.

CESC (C. elegans Sequencing Consortium) (1998). Genome sequence of the nematode C. elegans: a platform for investigating biology. *Science*, **282**, 2012–2018.

Chakraborty, R., Kimmel, M., Stivers, D.N., Davison, L.J. and Deka, R. (1997). Relative mutation rates at di-, tri-, and tetranucleotide microsatellite loci. *Proc Natl Acad Sci U S A*, **94**, 1041–1046.

Chance, P.F., Alderson, M.K., Leppig, K.A. *et al.* (1993). DNA deletion associated with hereditary neuropathy with liability to pressure palsies. *Cell*, **72**, 143–151.

Chang, J.C. and Kan, Y.W. (1979). Beta 0 thalassemia, a nonsense mutation in man. *Proc Natl Acad Sci U S A*, **76**, 2886–2889.

Chang, J.C. and Kan, Y.W. (1982). A sensitive test for prenatal diagnosis of sickle cell anemia: direct analysis of amniocyte DNA with MstII. *Trans Assoc Am Physicians*, **95**, 71–78.

Chang, M., Li, Y., Yan, C. *et al.* (2008). Variants in the 5q31 cytokine gene cluster are associated with psoriasis. *Genes Immun*, **9**, 176–181.

Chanock, S.J., Manolio, T., Boehnke, M. *et al.* (2007). Replicating genotype–phenotype associations. *Nature*, **447**, 655–660.

Charache, S., Terrin, M.L., Moore, R.D. *et al.* (1995). Effect of hydroxyurea on the frequency of painful crises in sickle cell anemia. Investigators of the Multicenter Study of Hydroxyurea in Sickle Cell Anemia. *N Engl J Med*, **332**, 1317–1322.

Charcot, J.-M. and Marie, P. (1886). Sur une forme particuliere d'atrophie musculaire progressive souvent familiale debutant par les pieds et les jambes et atteignant plus tard les mains. *Rev Med*, **6**, 97–138.

Charlesworth, B., Sniegowski, P. and Stephan, W. (1994). The evolutionary dynamics of repetitive DNA in eukaryotes. *Nature*, **371**, 215–220.

Charlet, B.N., Savkur, R.S., Singh, G., Philips, A.V., Grice, E.A. and Cooper, T.A. (2002). Loss of the muscle-specific chloride channel in type 1 myotonic dystrophy due to misregulated alternative splicing. *Mol Cell*, **10**, 45–53.

Chartier-Harlin, M.C., Kachergus, J., Roumier, C. *et al.* (2004). Alpha-synuclein locus duplication as a cause of familial Parkinson's disease. *Lancet*, **364**, 1167–1169.

Chasman, D. and Adams, R.M. (2001). Predicting the functional consequences of non-synonymous single nucleotide polymorphisms: structure-based assessment of amino acid variation. *J Mol Biol*, **307**, 683–706.

Chatterji, U., Bobardt, M.D., Gaskill, P., Sheeter, D., Fox, H. and Gallay, P.A. (2006). Trim5alpha accelerates degradation of cytosolic capsid associated with productive HIV-1 entry. *J Biol Chem*, **281**, 37025–37033.

Cheever, A.W. (1968). A quantitative post-mortem study of Schistosomiasis mansoni in man. *Am J Trop Med Hyg*, **17**, 38–64.

Chemelli, R.M., Willie, J.T., Sinton, C.M. *et al.* (1999). Narcolepsy in orexin knockout mice: molecular genetics of sleep regulation. *Cell*, **98**, 437–451.

Chen, F.C. and Li, W.H. (2001). Genomic divergences between humans and other hominoids and the effective population size of the common ancestor of humans and chimpanzees. *Am J Hum Genet*, **68**, 444–456.

Chen, F.C., Chen, C.J., Li, W.H. and Chuang, T.J. (2007). Human-specific insertions and deletions inferred from mammalian genome sequences. *Genome Res*, **17**, 16–22.

Chen, J.M., Ferec, C. and Cooper, D.N. (2006). LINE-1 endonuclease-dependent retrotranspositional events causing human genetic disease: mutation detection bias and multiple mechanisms of target gene disruption. *J Biomed Biotechnol*, **2006**, 56182.

Chen, J.M., Stenson, P.D., Cooper, D.N. and Ferec, C. (2005). A systematic analysis of LINE-1 endonuclease-dependent retrotranspositional events causing human genetic disease. *Hum Genet*, **117**, 411–427.

Cheng, Z., Ventura, M., She, X. *et al.* (2005). A genome-wide comparison of recent chimpanzee and human segmental duplications. *Nature*, **437**, 88–93.

Cherif-Zahar, B., Mattei, M.G., Le Van Kim, C., Bailly, P., Cartron, J.P. and Colin, Y. (1991). Localization of the human Rh blood group gene structure to chromosome region 1p34.3–1p36.1 by in situ hybridization. *Hum Genet*, **86**, 398–400.

Chesler, E.J., Lu, L., Shou, S. *et al.* (2005). Complex trait analysis of gene expression uncovers polygenic and pleiotropic networks that modulate nervous system function. *Nat Genet*, **37**, 233–242.

Chesler, E.J., Lu, L., Wang, J., Williams, R.W. and Manly, K.F. (2004). WebQTL: rapid exploratory analysis of gene expression and genetic networks for brain and behavior. *Nat Neurosci*, **7**, 485–486.

Chessman, D., Kostenko, L., Lethborg, T. et al. (2008). Human leukocyte antigen class I-restricted activation of CD8+ T cells provides the immunogenetic basis of a systemic drug hypersensitivity. *Immunity*, **28**, 822–832.

Cheung, J., Estivill, X., Khaja, R. *et al.* (2003a). Genome-wide detection of segmental duplications and potential assembly errors in the human genome sequence. *Genome Biol*, **4**, R25.

Cheung, J., Wilson, M.D., Zhang, J. *et al.* (2003b). Recent segmental and gene duplications in the mouse genome. *Genome Biol*, **4**, R47.

Cheung, K.H., Shineman, D., Muller, M. *et al.* (2008). Mechanism of Ca2+ disruption in Alzheimer's disease by presenilin regulation of InsP3 receptor channel gating. *Neuron*, **58**, 871–883.

Cheung, V.G., Conlin, L.K., Weber, T.M. *et al.* (2003c). Natural variation in human gene expression assessed in lymphoblastoid cells. *Nat Genet*, **33**, 422–425.

Cheung, V.G., Nowak, N., Jang, W. *et al.* (2001). Integration of cytogenetic landmarks into the draft sequence of the human genome. *Nature*, **409**, 953–958.

Cheung, V.G., Spielman, R.S., Ewens, K.G., Weber, T.M., Morley, M. and Burdick, J.T. (2005). Mapping determinants

of human gene expression by regional and genome-wide association. *Nature*, **437**, 1365–1369.

Chevillard, C., Moukoko, C.E., Elwali, N.E. *et al.* (2003). IFN-gamma polymorphisms (IFN-gamma +2109 and IFN-gamma +3810) are associated with severe hepatic fibrosis in human hepatic schistosomiasis (Schistosoma mansoni). *J Immunol*, **171**, 5596–5601.

Chiba-Falek, O., Kowalak, J.A., Smulson, M.E. and Nussbaum, R.L. (2005). Regulation of alpha-synuclein expression by poly (ADP ribose) polymerase-1 (PARP-1) binding to the NACP-Rep1 polymorphic site upstream of the SNCA gene. *Am J Hum Genet*, **76**, 478–492.

Chitsulo, L., Loverde, P. and Engels, D. (2004). Schistosomiasis. *Nat Rev Microbiol*, **2**, 12–13.

Chiurazzi, P., Tabolacci, E. and Neri, G. (2004). X-linked mental retardation (XLMR): from clinical conditions to cloned genes. *Crit Rev Clin Lab Sci*, **41**, 117–158.

Cho, H.J., Meira-Lima, I., Cordeiro, Q. *et al.* (2005). Population-based and family-based studies on the serotonin transporter gene polymorphisms and bipolar disorder: a systematic review and meta-analysis. *Mol Psychiatry*, **10**, 771–781.

Cho, J.H. (2008). The genetics and immunopathogenesis of inflammatory bowel disease. *Nat Rev Immunol*, **8**, 458–466.

Choe, H., Farzan, M., Sun, Y. *et al.* (1996). The beta-chemokine receptors CCR3 and CCR5 facilitate infection by primary HIV-1 isolates. *Cell*, **85**, 1135–1148.

Chorley, B.N., Wang, X., Campbell, M.R., Pittman, G.S., Noureddine, M.A. and Bell, D.A. (2008). Discovery and verification of functional single nucleotide polymorphisms in regulatory genomic regions: current and developing technologies. *Mutat Res*, **659**, 147–157.

Chotai, J. (1984). On the lod score method in linkage analysis. *Ann Hum Genet*, **48**, 359–378.

Chung, E.K., Yang, Y., Rennebohm, R.M. *et al.* (2002). Genetic sophistication of human complement components C4A and C4B and RP-C4-CYP21-TNX (RCCX) modules in the major histocompatibility complex. *Am J Hum Genet*, **71**, 823–837.

Clark, A.G., Glanowski, S., Nielsen, R. *et al.* (2003a). Inferring nonneutral evolution from human–chimp–mouse orthologous gene trios. *Science*, **302**, 1960–1963.

Clark, I.A., Rockett, K.A. and Burgner, D. (2003b). Genes, nitric oxide and malaria in African children. *Trends Parasitol*, **19**, 335–337.

Clark, T.A., Schweitzer, A.C., Chen, T.X. *et al.* (2007). Discovery of tissue-specific exons using comprehensive human exon microarrays. *Genome Biol*, **8**, R64.

Clarke, C.A. (1959). Correlations of ABO blood groups with peptic ulcer, cancer, and other diseases. *Am J Hum Genet*, **11**, 400–404.

Claustres, M., Horaitis, O., Vanevski, M. and Cotton, R.G. (2002). Time for a unified system of mutation description and reporting: a review of locus-specific mutation databases. *Genome Res*, **12**, 680–688.

Clayton-Smith, J., Webb, T., Cheng, X.J., Pembrey, M.E. and Malcolm, S. (1993). Duplication of chromosome 15 in the region 15q11-13 in a patient with developmental delay and ataxia with similarities to Angelman syndrome. *J Med Genet*, **30**, 529–531.

Clegg, A.O., Ashton, L.J., Biti, R.A. *et al.* (2000). CCR5 promoter polymorphisms, CCR5 59029A and CCR5 59353C, are under represented in HIV-1-infected long-term non-progressors. The Australian Long-Term Non-Progressor Study Group. *Aids*, **14**, 103–108.

Clendenin, T.M. and Benirschke, K. (1963). Chromosome studies on spontaneous abortions. *Lab Invest*, **12**, 1281–1292.

Cobb, J., Busst, C., Petrou, S., Harrap, S. and Ellis, J. (2008). Searching for functional genetic variants in non-coding DNA. *Clin Exp Pharmacol Physiol*, **35**, 372–375.

Cocchi, F., DeVico, A.L., Garzino-Demo, A., Arya, S.K., Gallo, R.C. and Lusso, P. (1995). Identification of RANTES, MIP-1 alpha, and MIP-1 beta as the major HIV-suppressive factors produced by CD8+ T cells. *Science*, **270**, 1811–1815.

Cocozza, S., Riccardi, G., Monticelli, A. *et al.* (1988). Polymorphism at the 5′ end flanking region of the insulin gene is associated with reduced insulin secretion in healthy individuals. *Eur J Clin Invest*, **18**, 582–586.

Coelho, M., Luiselli, D., Bertorelle, G. *et al.* (2005). Microsatellite variation and evolution of human lactase persistence. *Hum Genet*, **117**, 329–339.

Coenen, M.J. and Gregersen, P.K. (2009). Rheumatoid arthritis: a view of the current genetic landscape. *Genes Immun*, **10**, 101-111.

Cohen, G.B., Gandhi, R.T., Davis, D.M. *et al.* (1999). The selective downregulation of class I major histocompatibility complex proteins by HIV-1 protects HIV-infected cells from NK cells. *Immunity*, **10**, 661–671.

Cohen, S.N., Chang, A.C., Boyer, H.W. and Helling, R.B. (1973). Construction of biologically functional bacterial plasmids in vitro. *Proc Natl Acad Sci U S A*, **70**, 3240–3244.

Cole, C.G., McCann, O.T., Collins, J.E. *et al.* (2008). Finishing the finished human chromosome 22 sequence. *Genome Biol*, **9**, R78.

Colin, Y., Cherif-Zahar, B., Le Van Kim, C., Raynal, V., Van Huffel, V. and Cartron, J.P. (1991). Genetic basis of the RhD-positive and RhD-negative blood group poly-morphism as determined by Southern analysis. *Blood*, **78**, 2747–2752.

Collins, F.S. (1991). Of needles and haystacks: finding human disease genes by positional cloning. *Clin Res*, **39**, 615–623.

Collins, F.S. (1992). Positional cloning: let's not call it reverse anymore. *Nat Genet*, **1**, 3–6.

Collins, F.S. (1995). Positional cloning moves from perdition to traditional. *Nat Genet*, **9**, 347–350.

Collins, F.S., Guyer, M.S. and Charkravarti, A. (1997). Variations on a theme: cataloging human DNA sequence variation. *Science*, **278**, 1580–1581.

Collins, K.L., Chen, B.K., Kalams, S.A., Walker, B.D. and Baltimore, D. (1998). HIV-1 Nef protein protects infected primary cells against killing by cytotoxic T lymphocytes. *Nature*, **391**, 397–401.

Collins, P.J., Hennessy, L.K., Leibelt, C.S., Roby, R.K., Reeder, D.J. and Foxall, P.A. (2004). Developmental validation of a single-tube amplification of the 13 CODIS STR loci, D2S1338, D19S433, and amelogenin: the AmpFlSTR Identifiler PCR Amplification Kit. *J Forensic Sci*, **49**, 1265–1277.

Condit, C.M., Achter, P.J., Lauer, I. and Sefcovic, E. (2002). The changing meanings of "mutation:" a contextualized study of public discourse. *Hum Mutat*, **19**, 69–75.

Conley, Y.P., Jakobsdottir, J., Mah, T. *et al.* (2006). CFH, ELOVL4, PLEKHA1 and LOC387715 genes and susceptibility to age-related maculopathy: AREDS and CHS cohorts and meta-analyses. *Hum Mol Genet*, **15**, 3206–3218.

Connor, R.I., Sheridan, K.E., Ceradini, D., Choe, S. and Landau, N.R. (1997). Change in coreceptor use correlates with disease progression in HIV-1-infected individuals. *J Exp Med*, **185**, 621–628.

Conrad, B. and Antonarakis, S.E. (2007). Gene duplication: a drive for phenotypic diversity and cause of human disease. *Annu Rev Genomics Hum Genet*, **8**, 17–35.

Conrad, D.F., Andrews, T.D., Carter, N.P., Hurles, M.E. and Pritchard, J.K. (2006). A high-resolution survey of deletion polymorphism in the human genome. *Nat Genet*, **38**, 75–81.

Contente, A., Dittmer, A., Koch, M.C., Roth, J. and Dobbelstein, M. (2002). A polymorphic microsatellite that mediates induction of PIG3 by p53. *Nat Genet*, **30**, 315–320.

Contente, A., Zischler, H., Einspanier, A. and Dobbelstein, M. (2003). A promoter that acquired p53 responsiveness during primate evolution. *Cancer Res*, **63**, 1756–1758.

Conticello, S.G., Thomas, C.J., Petersen-Mahrt, S.K. and Neuberger, M.S. (2005). Evolution of the AID/APOBEC family of polynucleotide (deoxy)cytidine deaminases. *Mol Biol Evol*, **22**, 367–377.

Cook, G.C. and al-Torki, M.T. (1975). High intestinal lactase concentrations in adult Arabs in Saudi Arabia. *Br Med J*, **3**, 135–136.

Cooke, G.S. and Hill, A.V. (2001). Genetics of susceptibility to human infectious disease. *Nat Rev Genet*, **2**, 967–977.

Coon, K.D., Myers, A.J., Craig, D.W. *et al.* (2007). A high-density whole-genome association study reveals that APOE is the major susceptibility gene for sporadic late-onset Alzheimer's disease. *J Clin Psychiatry*, **68**, 613–618.

Cooper, D.N. and Krawczak, M. (1990). The mutational spectrum of single base-pair substitutions causing human genetic disease: patterns and predictions. *Hum Genet*, **85**, 55–74.

Cooper, G.M., Nickerson, D.A. and Eichler, E.E. (2007). Mutational and selective effects on copy-number variants in the human genome. *Nat Genet*, **39**, S22–29.

Cordaux, R., Udit, S., Batzer, M.A. and Feschotte, C. (2006). Birth of a chimeric primate gene by capture of the transposase gene from a mobile element. *Proc Natl Acad Sci U S A*, **103**, 8101–8106.

Corder, E.H., Saunders, A.M., Risch, N.J. *et al.* (1994). Protective effect of apolipoprotein E type 2 allele for late onset Alzheimer disease. *Nat Genet*, **7**, 180–184.

Corder, E.H., Saunders, A.M., Strittmatter, W.J. *et al.* (1993). Gene dose of apolipoprotein E type 4 allele and the risk of Alzheimer's disease in late onset families. *Science*, **261**, 921–923.

Corneo, G., Ginelli, E. and Polli, E. (1968). Isolation of the complementary strands of a human satellite DNA. *J Mol Biol*, **33**, 331–335.

Cossee, M., Schmitt, M., Campuzano, V. *et al.* (1997). Evolution of the Friedreich's ataxia trinucleotide repeat expansion: founder effect and premutations. *Proc Natl Acad Sci U S A*, **94**, 7452–7457.

Cotton, R.G. and Scriver, C.R. (1998). Proof of "disease causing" mutation. *Hum Mutat*, **12**, 1–3.

Cowles, C.R., Joel, N.H., Altshuler, D. and Lander, E.S. (2002). Detection of regulatory variation in mouse genes. *Nat Genet*, **32**, 432–437.

Cox, N.J., Hayes, M.G., Roe, C.A., Tsuchiya, T. and Bell, G.I. (2004). Linkage of calpain 10 to type 2 diabetes: the biological rationale. *Diabetes*, **53**, S19–25.

Cranney, A., Goldstein, R., Pham, B., Newkirk, M.M. and Karsh, J. (1999). A measure of limited joint motion and deformity correlates with HLA-DRB1 and DQB1 alleles in patients with rheumatoid arthritis. *Ann Rheum Dis*, **58**, 703–708.

Creighton, S., Almqvist, E.W., MacGregor, D. *et al.* (2003). Predictive, pre-natal and diagnostic genetic testing for Huntington's disease: the experience in Canada from 1987 to 2000. *Clin Genet*, **63**, 462–475.

Cresswell, P. (2005). Antigen processing and presentation. *Immunol Rev*, **207**, 5–7.

Cresswell, P., Ackerman, A.L., Giodini, A., Peaper, D.R. and Wearsch, P.A. (2005). Mechanisms of MHC class I-restricted antigen processing and cross-presentation. *Immunol Rev*, **207**, 145–157.

Croager, E.J., Gout, A.M. and Abraham, L.J. (2000). Involvement of Sp1 and microsatellite repressor sequences in the transcriptional control of the human CD30 gene. *Am J Pathol*, **156**, 1723–1731.

Cruts, M. and Van Broeckhoven, C. (1998a). Molecular genetics of Alzheimer's disease. *Ann Med*, **30**, 560–565.

Cruts, M. and Van Broeckhoven, C. (1998b). Presenilin mutations in Alzheimer's disease. *Hum Mutat*, **11**, 183–190.

CSAC (Chimpanzee Sequencing and Analysis Consortium) (2005). Initial sequence of the chimpanzee genome and comparison with the human genome. *Nature*, **437**, 69–87.

Csink, A.K. and Henikoff, S. (1998). Something from nothing: the evolution and utility of satellite repeats. *Trends Genet*, **14**, 200–204.

Cudworth, A.G. and Woodrow, J.C. (1975). Evidence for HL-A-linked genes in "juvenile" diabetes mellitus. *Br Med J*, **3**, 133–135.

Cunninghame Graham, D.S., Manku, H., Wagner, S. *et al.* (2007). Association of IRF5 in UK SLE families identifies a variant involved in polyadenylation. *Hum Mol Genet*, **16**, 579–591.

Cushman, M. (2007). Epidemiology and risk factors for venous thrombosis. *Semin Hematol*, **44**, 62–69.

Czelusniak, J., Goodman, M., Hewett-Emmett, D., Weiss, M.L., Venta, P.J. and Tashian, R.E. (1982). Phylogenetic origins and adaptive evolution of avian and mammalian haemoglobin genes. *Nature*, **298**, 297–300.

Dahlback, B. (2008). Advances in understanding pathogenic mechanisms of thrombophilic disorders. *Blood*, **112**, 19–27.

Dahlback, B. and Hildebrand, B. (1994). Inherited resistance to activated protein C is corrected by anticoagulant cofactor activity found to be a property of factor V. *Proc Natl Acad Sci U S A*, **91**, 1396–1400.

Dahlback, B., Carlsson, M. and Svensson, P.J. (1993). Familial thrombophilia due to a previously unrecognized mechanism characterized by poor anticoagulant response to activated protein C: prediction of a cofactor to activated protein C. *Proc Natl Acad Sci U S A*, **90**, 1004–1008.

Daiger, S.P. (2005). Genetics. Was the Human Genome Project worth the effort? *Science*, **308**, 362–364.

Daly, M.J., Rioux, J.D., Schaffner, S.F., Hudson, T.J. and Lander, E.S. (2001). High-resolution haplotype structure in the human genome. *Nat Genet*, **29**, 229–232.

Dangel, A.W., Mendoza, A.R., Baker, B.J. *et al.* (1994). The dichotomous size variation of human complement C4 genes is mediated by a novel family of endogenous retroviruses, which also establishes species-specific genomic patterns among Old World primates. *Immunogenetics*, **40**, 425–436.

Daniels, G. (1995). *Human Blood Groups*. Blackwell Science, Oxford, UK.

Daniels, R.J., Peden, J.F., Lloyd, C. *et al.* (2001). Sequence, structure and pathology of the fully annotated terminal 2 Mb of the short arm of human chromosome 16. *Hum Mol Genet*, **10**, 339–352.

Dausset, J., Cann, H., Cohen, D., Lathrop, M., Lalouel, J.M. and White, R. (1990). Centre d'Etude du Polymorphisme Humain (CEPH): collaborative genetic mapping of the human genome. *Genomics*, **6**, 575–577.

Davies, J.L., Kawaguchi, Y., Bennett, S.T. *et al.* (1994). A genome-wide search for human type 1 diabetes susceptibility genes. *Nature*, **371**, 130–136.

Davies, K.E., Smith, T.J., Bundey, S. *et al.* (1988). Mild and severe muscular dystrophy associated with deletions in Xp21 of the human X chromosome. *J Med Genet*, **25**, 9–13.

Davies, S.W., Turmaine, M., Cozens, B.A. *et al.* (1997). Formation of neuronal intranuclear inclusions underlies the neurological dysfunction in mice transgenic for the HD mutation. *Cell*, **90**, 537–548.

Davis, B.M., McCurrach, M.E., Taneja, K.L., Singer, R.H. and Housman, D.E. (1997). Expansion of a CUG trinucleotide repeat in the 3′ untranslated region of myotonic dystrophy protein kinase transcripts results in nuclear retention of transcripts. *Proc Natl Acad Sci U S A*, **94**, 7388–7393.

Dawn Teare, M. and Barrett, J.H. (2005). Genetic linkage studies. *Lancet*, **366**, 1036–1044.

Dawson, E., Chen, Y., Hunt, S. *et al.* (2001). A SNP resource for human chromosome 22: extracting dense clusters of SNPs from the genomic sequence. *Genome Res*, **11**, 170–178.

de Bakker, P.I., McVean, G., Sabeti, P.C. *et al.* (2006). A high-resolution HLA and SNP haplotype map for disease association studies in the extended human MHC. *Nat Genet*, **38**, 1166–1172.

de Biase, I., Rasmussen, A., Endres, D. *et al.* (2007a). Progressive GAA expansions in dorsal root ganglia of Friedreich's ataxia patients. *Ann Neurol*, **61**, 55–60.

de Biase, I., Rasmussen, A., Monticelli, A. *et al.* (2007b). Somatic instability of the expanded GAA triplet-repeat sequence in Friedreich ataxia progresses throughout life. *Genomics*, **90**, 1–5.

de Gobbi, M., Viprakasit, V., Hughes, J.R. *et al.* (2006). A regulatory SNP causes a human genetic disease by creating a new transcriptional promoter. *Science*, **312**, 1215–1217.

de Jong, P.T. (2006). Age-related macular degeneration. *N Engl J Med*, **355**, 1474–1485.

de Jong, W.W., Went, L.N. and Bernini, L.F. (1968). Haemoglobin Leiden: deletion of beta-6 or 7 glutamic acid. *Nature*, **220**, 788–790.

de Jongh, G.J., Zeeuwen, P.L., Kucharekova, M. *et al.* (2005). High expression levels of keratinocyte antimicrobial proteins in psoriasis compared with atopic dermatitis. *J Invest Dermatol*, **125**, 1163–1173.

de Koning, D.J. and Haley, C.S. (2005). Genetical genomics in humans and model organisms. *Trends Genet*, **21**, 377–381.

de la Chapelle, A., Herva, R., Koivisto, M. and Aula, P. (1981). A deletion in chromosome 22 can cause DiGeorge syndrome. *Hum Genet*, **57**, 253–256.

de la Chaux, N., Messer, P.W. and Arndt, P.F. (2007). DNA indels in coding regions reveal selective constraints on protein evolution in the human lineage. *BMC Evol Biol*, **7**, 191.

de Lange, T. (2005). Shelterin: the protein complex that shapes and safeguards human telomeres. *Genes Dev*, **19**, 2100–2110.

de Sanctis, L., Corrias, A., Romagnolo, D. *et al.* (2004). Familial PAX8 small deletion (c.989_992delACCC) associated with extreme phenotype variability. *J Clin Endocrinol Metab*, **89**, 5669–5674.

de Strooper, B., Saftig, P., Craessaerts, K. *et al.* (1998). Deficiency of presenilin-1 inhibits the normal cleavage of amyloid precursor protein. *Nature*, **391**, 387–390.

de Vries, B.B., Pfundt, R., Leisink, M. *et al.* (2005). Diagnostic genome profiling in mental retardation. *Am J Hum Genet*, **77**, 606–616.

Dean, M., Carrington, M., Winkler, C. *et al.* (1996). Genetic restriction of HIV-1 infection and progression to AIDS by a deletion allele of the CKR5 structural gene. Hemophilia Growth and Development Study, Multicenter AIDS Cohort Study, Multicenter Hemophilia Cohort Study, San Francisco City Cohort, ALIVE Study. *Science*, **273**, 1856–1862.

Deconinck, N. and Dan, B. (2007). Pathophysiology of duchenne muscular dystrophy: current hypotheses. *Pediatr Neurol*, **36**, 1–7.

Deeb, S.S. (2005). The molecular basis of variation in human color vision. *Clin Genet*, **67**, 369–377.

Deeb, S.S., Fajas, L., Nemoto, M. *et al.* (1998). A Pro12Ala substitution in PPARgamma2 associated with decreased receptor activity, lower body mass index and improved insulin sensitivity. *Nat Genet*, **20**, 284–287.

Degli-Esposti, M.A., Leaver, A.L., Christiansen, F.T., Witt, C.S., Abraham, L.J. and Dawkins, R.L. (1992). Ancestral haplotypes: conserved population MHC haplotypes. *Hum Immunol*, **34**, 242–252.

Dehal, P. and Boore, J.L. (2005). Two rounds of whole genome duplication in the ancestral vertebrate. *PLoS Biol*, **3**, e314.

Deininger, P.L. and Batzer, M.A. (1999). Alu repeats and human disease. *Mol Genet Metab*, **67**, 183–193.

Deininger, P.L. and Batzer, M.A. (2002). Mammalian retroelements. *Genome Res*, **12**, 1455–1465.

del Gaudio, D., Fang, P., Scaglia, F. *et al.* (2006). Increased MECP2 gene copy number as the result of genomic duplication in neurodevelopmentally delayed males. *Genet Med*, **8**, 784–792.

Demuth, J.P., De Bie, T., Stajich, J.E., Cristianini, N. and Hahn, M.W. (2006). The evolution of mammalian gene families. *PLoS ONE*, **1**, e85.

den Dunnen, J.T. and Antonarakis, S.E. (2000). Mutation nomenclature extensions and suggestions to describe complex mutations: a discussion. *Hum Mutat*, **15**, 7–12.

Deng, H., Liu, R., Ellmeier, W. *et al.* (1996). Identification of a major co-receptor for primary isolates of HIV-1. *Nature*, **381**, 661–666.

Dermitzakis, E.T. (2008). From gene expression to disease risk. *Nat Genet*, **40**, 492–493.

Desai, J., Shannon, M.E., Johnson, M.D. *et al.* (2006). Nell1-deficient mice have reduced expression of extracellular matrix proteins causing cranial and vertebral defects. *Hum Mol Genet*, **15**, 1329–1341.

Dessein, A.J., Hillaire, D., Elwali, N.E. *et al.* (1999). Severe hepatic fibrosis in Schistosoma mansoni infection is controlled by a major locus that is closely linked to the interferon-gamma receptor gene. *Am J Hum Genet*, **65**, 709–721.

Dewan, A., Liu, M., Hartman, S. *et al.* (2006). HTRA1 promoter polymorphism in wet age-related macular degeneration. *Science*, **314**, 989–992.

Dewannieux, M., Esnault, C. and Heidmann, T. (2003). LINE-mediated retrotransposition of marked Alu sequences. *Nat Genet*, **35**, 41–48.

Dib, C., Faure, S., Fizames, C. *et al.* (1996). A comprehensive genetic map of the human genome based on 5,264 microsatellites. *Nature*, **380**, 152–154.

Dieterich, W., Ehnis, T., Bauer, M. *et al.* (1997). Identification of tissue transglutaminase as the autoantigen of celiac disease. *Nat Med*, **3**, 797–801.

DiFiglia, M., Sapp, E., Chase, K.O. *et al.* (1997). Aggregation of huntingtin in neuronal intranuclear inclusions and dystrophic neurites in brain. *Science*, **277**, 1990–1993.

DiGeorge, A.M. (1968). Congenital absence of the thymus and its immunologic consequences: concurrence with congenital hypoparathyroidism. *Birth Defects Orig Art Ser*, **IV**, 116–121.

Dina, C., Meyre, D., Gallina, S. *et al.* (2007). Variation in FTO contributes to childhood obesity and severe adult obesity. *Nat Genet*, **39**, 724–726.

Dixon, A.L., Liang, L., Moffatt, M.F. *et al.* (2007). A genome-wide association study of global gene expression. *Nat Genet*, **39**, 1202–1207.

Dixon, J., Gladwin, A.J., Loftus, S.K. *et al.* (1994). A YAC contig encompassing the Treacher Collins syndrome critical region at 5q31.3–32. *Am J Hum Genet*, **55**, 372–378.

Dixon, M.J., Dixon, J., Houseal, T. *et al.* (1993). Narrowing the position of the Treacher Collins syndrome locus to a small interval between three new microsatellite markers at 5q32–33.1. *Am J Hum Genet*, **52**, 907–914.

Dixon, M.J., Haan, E., Baker, E. *et al.* (1991). Association of Treacher Collins syndrome and translocation 6p21.31/16p13.11: exclusion of the locus from these candidate regions. *Am J Hum Genet*, **48**, 274–280.

Dodge, J.A., Lewis, P.A., Stanton, M. and Wilsher, J. (2007). Cystic fibrosis mortality and survival in the UK: 1947–2003. *Eur Respir J*, **29**, 522–526.

Doherty, P.C. and Zinkernagel, R.M. (1975a). A biological role for the major histocompatibility antigens. *Lancet*, **1**, 1406–1409.

Doherty, P.C. and Zinkernagel, R.M. (1975b). Enhanced immunological surveillance in mice heterozygous at the H-2 gene complex. *Nature*, **256**, 50–52.

Dolo, A., Modiano, D., Maiga, B. *et al.* (2005). Difference in susceptibility to malaria between two sympatric ethnic groups in Mali. *Am J Trop Med Hyg*, **72**, 243–248.

Donis-Keller, H., Green, P., Helms, C. *et al.* (1987). A genetic linkage map of the human genome. *Cell*, **51**, 319–337.

Donner, H., Rau, H., Walfish, P.G. *et al.* (1997). CTLA4 alanine-17 confers genetic susceptibility to Graves' disease and to type 1 diabetes mellitus. *J Clin Endocrinol Metab*, **82**, 143–146.

Doranz, B.J., Rucker, J., Yi, Y. *et al.* (1996). A dual-tropic primary HIV-1 isolate that uses fusin and the beta-chemokine receptors CKR-5, CKR-3, and CKR-2b as fusion cofactors. *Cell*, **85**, 1149–1158.

Dorschner, M.O., Sybert, V.P., Weaver, M., Pletcher, B.A. and Stephens, K. (2000). NF1 microdeletion breakpoints are clustered at flanking repetitive sequences. *Hum Mol Genet*, **9**, 35–46.

Dragatsis, I., Levine, M.S. and Zeitlin, S. (2000). Inactivation of Hdh in the brain and testis results in progressive neurodegeneration and sterility in mice. *Nat Genet*, **26**, 300–306.

Dragic, T., Litwin, V., Allaway, G.P. *et al.* (1996). HIV-1 entry into CD4+ cells is mediated by the chemokine receptor CC-CKR-5. *Nature*, **381**, 667–673.

Driscoll, D.J., Waters, M.F., Williams, C.A. *et al.* (1992). A DNA methylation imprint, determined by the sex of the parent, distinguishes the Angelman and Prader–Willi syndromes. *Genomics*, **13**, 917–924.

Driscoll, M.C., Dobkin, C.S. and Alter, B.P. (1989). Gamma delta beta-thalassemia due to a de novo mutation deleting the 5′ beta-globin gene activation-region hypersensitive sites. *Proc Natl Acad Sci U S A*, **86**, 7470–7474.

Dubrova, Y.E., Jeffreys, A.J. and Malashenko, A.M. (1993). Mouse minisatellite mutations induced by ionizing radiation. *Nat Genet*, **5**, 92–94.

Dubrova, Y.E., Nesterov, V.N., Krouchinsky, N.G. *et al.* (1996). Human minisatellite mutation rate after the Chernobyl accident. *Nature*, **380**, 683–686.

Dubrova, Y.E., Nesterov, V.N., Krouchinsky, N.G. *et al.* (1997). Further evidence for elevated human minisatellite mutation rate in Belarus eight years after the Chernobyl accident. *Mutat Res*, **381**, 267–278.

Duerr, R.H. (2007). Genome-wide association studies herald a new era of rapid discoveries in inflammatory bowel disease research. *Gastroenterology*, **132**, 2045–2049.

Duerr, R.H., Taylor, K.D., Brant, S.R. *et al.* (2006). A genome-wide association study identifies IL23R as an inflammatory bowel disease gene. *Science*, **314**, 1461–1463.

Duffy, D.L., Montgomery, G.W., Chen, W. *et al.* (2007). A three-single-nucleotide polymorphism haplotype in intron 1 of OCA2 explains most human eye-color variation. *Am J Hum Genet*, **80**, 241–252.

Dunham, I., Shimizu, N., Roe, B.A. *et al.* (1999). The DNA sequence of human chromosome 22. *Nature*, **402**, 489–495.

Duyao, M., Ambrose, C., Myers, R. *et al.* (1993). Trinucleotide repeat length instability and age of onset in Huntington's disease. *Nat Genet*, **4**, 387–392.

Duyao, M.P., Auerbach, A.B., Ryan, A. *et al.* (1995). Inactivation of the mouse Huntington's disease gene homolog Hdh. *Science*, **269**, 407–410.

Early, P., Huang, H., Davis, M., Calame, K. and Hood, L. (1980). An immunoglobulin heavy chain variable region gene is generated from three segments of DNA: VH, D and JH. *Cell*, **19**, 981–992.

Easton, D.F., Pooley, K.A., Dunning, A.M. *et al.* (2007). Genome-wide association study identifies novel breast cancer susceptibility loci. *Nature*, **447**, 1087–1093.

Economou, M., Trikalinos, T.A., Loizou, K.T., Tsianos, E.V. and Ioannidis, J.P. (2004). Differential effects of NOD2 variants on Crohn's disease risk and phenotype in diverse populations: a metaanalysis. *Am J Gastroenterol*, **99**, 2393–2404.

Edelmann, L., Pandita, R.K. and Morrow, B.E. (1999a). Low-copy repeats mediate the common 3-Mb deletion in patients with velo-cardio-facial syndrome. *Am J Hum Genet*, **64**, 1076–1086.

Edelmann, L., Pandita, R.K., Spiteri, E. *et al.* (1999b). A common molecular basis for rearrangement disorders on chromosome 22q11. *Hum Mol Genet*, **8**, 1157–1167.

Edelmann, L., Spiteri, E., Koren, K. *et al.* (2001). AT-rich palindromes mediate the constitutional t(11;22) translocation. *Am J Hum Genet*, **68**, 1–13.

Edwards, A.O., Ritter, R., 3rd, Abel, K.J., Manning, A., Panhuysen, C. and Farrer, L.A. (2005). Complement factor H polymorphism and age-related macular degeneration. *Science*, **308**, 421–424.

Edwards, J.H., Harnden, D.G., Cameron, A.H., Crosse, V.M. and Wolff, O.H. (1960). A new trisomic syndrome. *Lancet*, **1**, 787–790.

Efstratiadis, A., Kafatos, F.C. and Maniatis, T. (1977). The primary structure of rabbit beta-globin mRNA as determined from cloned DNA. *Cell*, **10**, 571–585.

Egeberg, O. (1965). Inherited antithrombin deficiency causing thrombophilia. *Thromb Diath Haemorrh*, **13**, 516–530.

Eiben, B., Bartels, I., Bahr-Porsch, S. *et al.* (1990). Cytogenetic analysis of 750 spontaneous abortions with the direct-preparation method of chorionic villi and its implications for studying genetic causes of pregnancy wastage. *Am J Hum Genet*, **47**, 656–663.

Eichelbaum, M., Ingelman-Sundberg, M. and Evans, W.E. (2006). Pharmacogenomics and individualized drug therapy. *Annu Rev Med*, **57**, 119–137.

Eichler, E.E. (2006). Widening the spectrum of human genetic variation. *Nat Genet*, **38**, 9–11.

Eichler, E.E., Budarf, M.L., Rocchi, M. *et al.* (1997). Interchromosomal duplications of the adrenoleukodystrophy locus: a phenomenon of pericentromeric plasticity. *Hum Mol Genet*, **6**, 991–1002.

Eichler, E.E., Holden, J.J., Popovich, B.W. *et al.* (1994). Length of uninterrupted CGG repeats determines instability in the FMR1 gene. *Nat Genet*, **8**, 88–94.

Eichler, E.E., Nickerson, D.A., Altshuler, D. *et al.* (2007). Completing the map of human genetic variation. *Nature*, **447**, 161–165.

Ellegren, H. (1991). DNA typing of museum birds. *Nature*, **354**, 113.

Ellegren, H. (2004). Microsatellites: simple sequences with complex evolution. *Nat Rev Genet*, **5**, 435–445.

Emamian, E.S., Kaytor, M.D., Duvick, L.A. *et al.* (2003). Serine 776 of ataxin-1 is critical for polyglutamine-induced disease in SCA1 transgenic mice. *Neuron*, **38**, 375–387.

Embury, S.H., Lebo, R.V., Dozy, A.M. and Kan, Y.W. (1979). Organization of the alpha-globin genes in the Chinese alpha-thalassemia syndromes. *J Clin Invest*, **63**, 1307–1310.

Emilsson, V., Thorleifsson, G., Zhang, B. *et al.* (2008). Genetics of gene expression and its effect on disease. *Nature*, **452**, 423–428.

Emmerich, J., Rosendaal, F.R., Cattaneo, M. *et al.* (2001). Combined effect of factor V Leiden and prothrombin 20210A on the risk of venous thromboembolism – pooled analysis of 8 case–control studies including 2310 cases and 3204 controls. Study Group for Pooled-Analysis in Venous Thromboembolism. *Thromb Haemost*, **86**, 809–816.

Enard, W., Khaitovich, P., Klose, J. *et al.* (2002). Intra- and interspecific variation in primate gene expression patterns. *Science*, **296**, 340–343.

Enattah, N.S., Jensen, T.G., Nielsen, M. *et al.* (2008). Independent introduction of two lactase-persistence alleles into human populations reflects different history of adaptation to milk culture. *Am J Hum Genet*, **82**, 57–72.

Enattah, N.S., Sahi, T., Savilahti, E., Terwilliger, J.D., Peltonen, L. and Jarvela, I. (2002). Identification of a variant associated with adult-type hypolactasia. *Nat Genet*, **30**, 233–237.

Enattah, N.S., Trudeau, A., Pimenoff, V. *et al.* (2007). Evidence of still-ongoing convergence evolution of the lactase persistence T-13910 alleles in humans. *Am J Hum Genet*, **81**, 615–625.

Erlich, H.A., Bugawan, T.L., Scharf, S., Nepom, G.T., Tait, B. and Griffith, R.L. (1990). HLA-DQ beta sequence polymorphism and genetic susceptibility to IDDM. *Diabetes*, **39**, 96–103.

Ewart, A.K., Jin, W., Atkinson, D., Morris, C.A. and Keating, M.T. (1994). Supravalvular aortic stenosis associated with a deletion disrupting the elastin gene. *J Clin Invest*, **93**, 1071–1077.

Ewart, A.K., Morris, C.A., Atkinson, D. *et al.* (1993). Hemizygosity at the elastin locus in a developmental disorder, Williams syndrome. *Nat Genet*, **5**, 11–16.

Ewens, W.J. and Spielman, R.S. (1995). The transmission/disequilibrium test: history, subdivision, and admixture. *Am J Hum Genet*, **57**, 455–464.

Fahn, S. (2003). Description of Parkinson's disease as a clinical syndrome. *Ann N Y Acad Sci*, **991**, 1–14.

Fairhurst, R.M., Baruch, D.I., Brittain, N.J. *et al.* (2005). Abnormal display of PfEMP-1 on erythrocytes carrying haemoglobin C may protect against malaria. *Nature*, **435**, 1117–1121.

Falchuk, Z.M., Rogentine, G.N. and Strober, W. (1972). Predominance of histocompatibility antigen HL-A8 in patients with gluten-sensitive enteropathy. *J Clin Invest*, **51**, 1602–1605.

Fan, H. and Chu, J.Y. (2007). A brief review of short tandem repeat mutation. *Genomics Proteomics Bioinformatics*, **5**, 7–14.

Fanciulli, M., Norsworthy, P.J., Petretto, E. *et al.* (2007). FCGR3B copy number variation is associated with susceptibility to systemic, but not organ-specific, autoimmunity. *Nat Genet*, **39**, 721–723.

Fang, P., Lev-Lehman, E., Tsai, T.F. *et al.* (1999). The spectrum of mutations in UBE3A causing Angelman syndrome. *Hum Mol Genet*, **8**, 129–135.

Farrer, L.A., Cupples, L.A., Haines, J.L. *et al.* (1997). Effects of age, sex, and ethnicity on the association between apolipoprotein E genotype and Alzheimer disease. A meta-analysis. APOE and Alzheimer Disease Meta Analysis Consortium. *JAMA*, **278**, 1349–1356.

Farrer, M. (2006). Genetics of Parkinson disease: paradigm shifts and future prospects. *Nat Rev Genet*, **7**, 306–318.

Farrer, M., Kachergus, J., Forno, L. *et al.* (2004). Comparison of kindreds with parkinsonism and alpha-synuclein genomic multiplications. *Ann Neurol*, **55**, 174–179.

Farrer, M., Maraganore, D.M., Lockhart, P. *et al.* (2001). alpha-Synuclein gene haplotypes are associated with Parkinson's disease. *Hum Mol Genet*, **10**, 1847–1851.

Fatkenheuer, G., Pozniak, A.L., Johnson, M.A. *et al.* (2005). Efficacy of short-term monotherapy with maraviroc, a new CCR5 antagonist, in patients infected with HIV-1. *Nat Med*, **11**, 1170–1172.

Fay, J.C. and Wu, C.I. (2000). Hitchhiking under positive Darwinian selection. *Genetics*, **155**, 1405–1413.

Fazen, L.E., Elmore, J. and Nadler, H.L. (1967). Mandibulofacial dysostosis (Treacher-Collins syndrome). *Am J Dis Child*, **113**, 405–410.

Feder, J.N., Gnirke, A., Thomas, W. *et al.* (1996). A novel MHC class I-like gene is mutated in patients with hereditary haemochromatosis. *Nat Genet*, **13**, 399–408.

Felber, B.K., Orkin, S.H. and Hamer, D.H. (1982). Abnormal RNA splicing causes one form of alpha thalassemia. *Cell*, **29**, 895–902.

Fellay, J., Shianna, K.V., Ge, D. *et al.* (2007). A whole-genome association study of major determinants for host control of HIV-1. *Science*, **317**, 944–947.

Fellermann, K., Stange, D.E., Schaeffeler, E. *et al.* (2006). A chromosome 8 gene-cluster polymorphism with low human beta-defensin 2 gene copy number predisposes to Crohn disease of the colon. *Am J Hum Genet*, **79**, 439–448.

Ferguson-Smith, M.A. (1966). X-Y chromosomal interchange in the aetiology of true hermaphroditism and of XX Klinefelter's syndrome. *Lancet*, **2**, 475–476.

Fernandez-Funez, P., Nino-Rosales, M.L., de Gouyon, B. *et al.* (2000). Identification of genes that modify ataxin-1-induced neurodegeneration. *Nature*, **408**, 101–106.

Fernandez-Reyes, D., Craig, A.G., Kyes, S.A. *et al.* (1997). A high frequency African coding polymorphism in the N-terminal domain of ICAM-1 predisposing to cerebral malaria in Kenya. *Hum Mol Genet*, **6**, 1357–1360.

Fernando, M.M., Stevens, C.R., Walsh, E.C. *et al.* (2008). Defining the role of the MHC in autoimmunity: a review and pooled analysis. *PLoS Genet*, **4**, e1000024.

Feschotte, C. (2008). Transposable elements and the evolution of regulatory networks. *Nat Rev Genet*, **9**, 397–405.

Feuk, L., Carson, A.R. and Scherer, S.W. (2006a). Structural variation in the human genome. *Nat Rev Genet*, **7**, 85–97.

Feuk, L., MacDonald, J.R., Tang, T. *et al.* (2005). Discovery of human inversion polymorphisms by comparative analysis of human and chimpanzee DNA sequence assemblies. *PLoS Genet*, **1**, e56.

Feuk, L., Marshall, C.R., Wintle, R.F. and Scherer, S.W. (2006b). Structural variants: changing the landscape of chromosomes and design of disease studies. *Hum Mol Genet*, **15**, R57–66.

Filla, A., De Michele, G., Cavalcanti, F. *et al.* (1996). The relationship between trinucleotide (GAA) repeat length and clinical features in Friedreich ataxia. *Am J Hum Genet*, **59**, 554–560.

Fischbeck, K.H. (1997). Kennedy disease. *J Inherit Metab Dis*, **20**, 152–158.

Fischbeck, K.H., Ionasescu, V., Ritter, A.W. *et al.* (1986). Localization of the gene for X-linked spinal muscular atrophy. *Neurology*, **36**, 1595–1598.

Fishel, R., Lescoe, M.K., Rao, M.R. *et al.* (1993). The human mutator gene homolog MSH2 and its association with hereditary nonpolyposis colon cancer. *Cell*, **75**, 1027–1038.

Fisher, A.G., Ensoli, B., Ivanoff, L. *et al.* (1987). The sor gene of HIV-1 is required for efficient virus transmission in vitro. *Science*, **237**, 888–893.

Fisher, S.A., Abecasis, G.R., Yashar, B.M. *et al.* (2005). Meta-analysis of genome scans of age-related macular degeneration. *Hum Mol Genet*, **14**, 2257–2264.

Fisher, S.A., Tremelling, M., Anderson, C.A. *et al.* (2008). Genetic determinants of ulcerative colitis include the ECM1 locus and five loci implicated in Crohn's disease. *Nat Genet*, **40**, 710–712.

Fiskerstrand, C.E., Lovejoy, E.A. and Quinn, J.P. (1999). An intronic polymorphic domain often associated with susceptibility to affective disorders has allele dependent differential enhancer activity in embryonic stem cells. *FEBS Lett*, **458**, 171–174.

Fiston-Lavier, A.S., Anxolabehere, D. and Quesneville, H. (2007). A model of segmental duplication formation in Drosophila melanogaster. *Genome Res*, **17**, 1458–1470.

Flatz, G., Kinderlerer, J.L., Kilmartin, J.V. and Lehmann, H. (1971). Haemoglobin Tak: a variant with additional residues at the end of the beta-chains. *Lancet*, **1**, 732–733.

Flavell, R.A., Kooter, J.M., De Boer, E., Little, P.F. and Williamson, R. (1978). Analysis of the beta-delta-globin gene loci in normal and Hb Lepore DNA: direct determination of gene linkage and intergene distance. *Cell*, **15**, 25–41.

Fleischer, B. (1918). Ober myotonische Dystrophie mit Katarakt: eine hereditire, familifire Degeneration. *Arch Ophthalmol (Berlin)*, **96**, 91–133.

Flint, J. and Wilkie, A.O. (1996). The genetics of mental retardation. *Br Med Bull*, **52**, 453–464.

Flint, J., Craddock, C.F., Villegas, A. *et al.* (1994). Healing of broken human chromosomes by the addition of telomeric repeats. *Am J Hum Genet*, **55**, 505–512.

Flint, J., Hill, A.V., Bowden, D.K. *et al.* (1986). High frequencies of alpha-thalassaemia are the result of natural selection by malaria. *Nature*, **321**, 744–750.

Flint, J., Rochette, J., Craddock, C.F. *et al.* (1996). Chromosomal stabilisation by a subtelomeric rearrangement involving two closely related Alu elements. *Hum Mol Genet*, **5**, 1163–1169.

Flint, J., Wilkie, A.O., Buckle, V.J., Winter, R.M., Holland, A.J. and McDermid, H.E. (1995). The detection of subtelomeric chromosomal rearrangements in idiopathic mental retardation. *Nat Genet*, **9**, 132–140.

Flori, L., Delahaye, N.F., Iraqi, F.A., Hernandez-Valladares, M., Fumoux, F. and Rihet, P. (2005). TNF as a malaria candidate gene: polymorphism-screening and family-based association analysis of mild malaria attack and parasitemia in Burkina Faso. *Genes Immun*, **6**, 472–480.

Flori, L., Sawadogo, S., Esnault, C., Delahaye, N.F., Fumoux, F. and Rihet, P. (2003). Linkage of mild malaria to the major histocompatibility complex in families living in Burkina Faso. *Hum Mol Genet*, **12**, 375–378.

Force, A., Lynch, M., Pickett, F.B., Amores, A., Yan, Y.L. and Postlethwait, J. (1999). Preservation of duplicate genes by complementary, degenerative mutations. *Genetics*, **151**, 1531–1545.

Ford, C.E., Jones, K.W., Polani, P.E., De Almeida, J.C. and Briggs, J.H. (1959). A sex-chromosome anomaly in a case of gonadal dysgenesis (Turner's syndrome). *Lancet*, **1**, 711–713.

Forton, J.T., Udalova, I.A., Campino, S., Rockett, K.A., Hull, J. and Kwiatkowski, D.P. (2007). Localization of a long-range cis-regulatory element of IL13 by allelic transcript ratio mapping. *Genome Res*, **17**, 82–87.

Foss, E.J., Radulovic, D., Shaffer, S.A. *et al.* (2007). Genetic basis of proteome variation in yeast. *Nat Genet*, **39**, 1369–1375.

Franchina, M., Woo, A.J., Dods, J. *et al.* (2008). The CD30 gene promoter microsatellite binds transcription factor Yin Yang 1 (YY1) and shows genetic instability in anaplastic large cell lymphoma. *J Pathol*, **214**, 65–74.

Frank, B., Rigas, S.H., Bermejo, J.L. *et al.* (2007). The CASP8 -652 6N del promoter polymorphism and breast cancer risk: a multicenter study. *Breast Cancer Res Treat*, **111**, 139–144.

Franke, A., Balschun, T., Karlsen, T.H. *et al.* (2008). Replication of signals from recent studies of Crohn's disease identifies previously unknown disease loci for ulcerative colitis. *Nat Genet*, **40**, 713–715.

Franke, A., Hampe, J., Rosenstiel, P. *et al.* (2007). Systematic association mapping identifies NELL1 as a novel IBD disease gene. *PLoS ONE*, **2**, e691.

Frayling, T.M., Timpson, N.J., Weedon, M.N. *et al.* (2007). A common variant in the FTO gene is associated with body mass index and predisposes to childhood and adult obesity. *Science*, **316**, 889–894.

Frazer, K.A., Ballinger, D.G., Cox, D.R. *et al.* (2007). A second generation human haplotype map of over 3.1 million SNPs. *Nature*, **449**, 851–861.

Fredman, D., Munns, G., Rios, D. *et al.* (2004). HGVbase: a curated resource describing human DNA variation and phenotype relationships. *Nucleic Acids Res*, **32**, D516–519.

Freeman, J.L., Perry, G.H., Feuk, L. *et al.* (2006). Copy number variation: new insights in genome diversity. *Genome Res*, **16**, 949–961.

Freese, E. (1959). The difference between spontaneous and base-analogue induced mutations of phage T4. *Proc Natl Acad Sci U S A*, **45**, 622–633.

Frenette, P.S. and Atweh, G.F. (2007). Sickle cell disease: old discoveries, new concepts, and future promise. *J Clin Invest*, **117**, 850–858.

Friedman, D.S., O'Colmain, B.J., Munoz, B. *et al.* (2004). Prevalence of age-related macular degeneration in the United States. *Arch Ophthalmol*, **122**, 564–572.

Friedman, J.M., Baross, A., Delaney, A.D. *et al.* (2006). Oligonucleotide microarray analysis of genomic imbalance in children with mental retardation. *Am J Hum Genet*, **79**, 500–513.

Friedman, M.J. (1978). Erythrocytic mechanism of sickle cell resistance to malaria. *Proc Natl Acad Sci U S A*, **75**, 1994–1997.

Friedman, M.J., Roth, E.F., Nagel, R.L. and Trager, W. (1979). Plasmodium falciparum: physiological interactions with the human sickle cell. *Exp Parasitol*, **47**, 73–80.

Fritsche, L.G., Loenhardt, T., Janssen, A. *et al.* (2008). Age-related macular degeneration is associated with an unstable ARMS2 (LOC387715) mRNA. *Nat Genet*, **40**, 892–896.

Fry, A.E., Auburn, S., Diakite, M. *et al.* (2008). Variation in the ICAM1 gene is not associated with severe malaria phenotypes. *Genes Immun*, **9**, 462–469.

Fu, Y.H., Pizzuti, A., Fenwick, R.G., Jr. *et al.* (1992). An unstable triplet repeat in a gene related to myotonic muscular dystrophy. *Science*, **255**, 1256–1258.

Fujino, S., Andoh, A., Bamba, S. *et al.* (2003). Increased expression of interleukin 17 in inflammatory bowel disease. *Gut*, **52**, 65–70.

Gabriel, S.B., Schaffner, S.F., Nguyen, H. *et al.* (2002). The structure of haplotype blocks in the human genome. *Science*, **296**, 2225–2229.

Gaedigk, A., Ndjountche, L., Divakaran, K. *et al.* (2007). Cytochrome P4502D6 (CYP2D6) gene locus heterogeneity: characterization of gene duplication events. *Clin Pharmacol Ther*, **81**, 242–251.

Gajecka, M., Mackay, K.L. and Shaffer, L.G. (2007). Monosomy 1p36 deletion syndrome. *Am J Med Genet C Semin Med Genet*, **145**, 346–356.

Galanter, J., Choudhry, S., Eng, C. *et al.* (2008). ORMDL3 gene is associated with asthma in three ethnically diverse populations. *Am J Respir Crit Care Med*, **177**, 1194–1200.

Galicia, J.C., Tai, H., Komatsu, Y., Shimada, Y., Akazawa, K. and Yoshie, H. (2004). Polymorphisms in the IL-6 receptor (IL-6R) gene: strong evidence that serum levels of soluble IL-6R are genetically influenced. *Genes Immun*, **5**, 513–516.

Gallo, R.C. and Montagnier, L. (2003). The discovery of HIV as the cause of AIDS. *N Engl J Med*, **349**, 2283–2285.

Gallo, R.C., Garzino-Demo, A. and DeVico, A.L. (1999). HIV infection and pathogenesis: what about chemokines? *J Clin Immunol*, **19**, 293–299.

Ganczakowski, M., Town, M., Bowden, D.K. *et al.* (1995). Multiple glucose 6-phosphate dehydrogenase-deficient variants correlate with malaria endemicity in the Vanuatu archipelago (southwestern Pacific). *Am J Hum Genet*, **56**, 294–301.

Gao, X., Nelson, G.W., Karacki, P. *et al.* (2001). Effect of a single amino acid change in MHC class I molecules on the rate of progression to AIDS. *N Engl J Med*, **344**, 1668–1675.

Garcia-Barcelo, M., Chow, L.Y., Lam, K.L., Chiu, H.F., Wing, Y.K. and Waye, M.M. (2000). Occurrence of CYP2D6 gene duplication in Hong Kong Chinese. *Clin Chem*, **46**, 1411–1413.

Gardner R.J.M. and Sutherland G.R. (2004) *Chromosome Abnormalities and Genetic Counseling*. Oxford University Press, Oxford, UK.

Garrick, D., Fiering, S., Martin, D.I. and Whitelaw, E. (1998). Repeat-induced gene silencing in mammals. *Nat Genet*, **18**, 56–59.

Garrigan, D. and Hammer, M.F. (2006). Reconstructing human origins in the genomic era. *Nat Rev Genet*, **7**, 669–680.

Gaston, M. and Rosse, W.F. (1982). The cooperative study of sickle cell disease: review of study design and objectives. *Am J Pediatr Hematol Oncol*, **4**, 197–201.

Gaston, M.H., Verter, J.I., Woods, G. *et al.* (1986). Prophylaxis with oral penicillin in children with sickle cell anemia. A randomized trial. *N Engl J Med*, **314**, 1593–1599.

Gatchel, J.R. and Zoghbi, H.Y. (2005). Diseases of unstable repeat expansion: mechanisms and common principles. *Nat Rev Genet*, **6**, 743–755.

Gatz, M., Reynolds, C.A., Fratiglioni, L. *et al.* (2006). Role of genes and environments for explaining Alzheimer disease. *Arch Gen Psychiatry*, **63**, 168–174.

Gebhardt, F., Zanker, K.S. and Brandt, B. (1999). Modulation of epidermal growth factor receptor gene transcription by a polymorphic dinucleotide repeat in intron 1. *J Biol Chem*, **274**, 13176–13180.

Geever, R.F., Wilson, L.B., Nallaseth, F.S., Milner, P.F., Bittner, M. and Wilson, J.T. (1981). Direct identification of sickle cell anemia by blot hybridization. *Proc Natl Acad Sci U S A*, **78**, 5081–5085.

Genton, B., al-Yaman, F., Mgone, C.S. *et al.* (1995). Ovalocytosis and cerebral malaria. *Nature*, **378**, 564–565.

Gerald, P.S. and Diamond, L.K. (1958). A new hereditary hemoglobinopathy (the Lepore trait) and its interaction with thalassemia trait. *Blood*, **13**, 835–844.

Gerald, P.S., Cook, C.D. and Diamond, L.K. (1957). Hemoglobin M. *Science*, **126**, 300–301.

Gerstein, M.B., Bruce, C., Rozowsky, J.S. *et al.* (2007). What is a gene, post-ENCODE? History and updated definition. *Genome Res*, **17**, 669–681.

Ghilardi, G., Biondi, M.L., Mangoni, J. *et al.* (2001). Matrix metalloproteinase-1 promoter polymorphism 1G/2G is correlated with colorectal cancer invasiveness. *Clin Cancer Res*, **7**, 2344–2346.

Ghoussaini, M., Song, H., Koessler, T. *et al.* (2008). Multiple loci with different cancer specificities within the 8q24 gene desert. *J Natl Cancer Inst*, **100**, 962–966.

Giardine, B., van Baal, S., Kaimakis, P. *et al.* (2007). HbVar database of human hemoglobin variants and thalassemia mutations: 2007 update. *Hum Mutat*, **28**, 206.

Gibbs, R.A., Rogers, J., Katze, M.G. *et al.* (2007). Evolutionary and biomedical insights from the rhesus macaque genome. *Science*, **316**, 222–234.

Gibson, G. and Weir, B. (2005). The quantitative genetics of transcription. *Trends Genet*, **21**, 616–623.

Gifford, R. and Tristem, M. (2003). The evolution, distribution and diversity of endogenous retroviruses. *Virus Genes*, **26**, 291–315.

Giglio, S., Broman, K.W., Matsumoto, N. *et al.* (2001). Olfactory receptor-gene clusters, genomic-inversion polymorphisms, and common chromosome rearrangements. *Am J Hum Genet*, **68**, 874–883.

Giglio, S., Calvari, V., Gregato, G. *et al.* (2002). Heterozygous submicroscopic inversions involving olfactory receptor-gene clusters mediate the recurrent t(4;8)(p16;p23) translocation. *Am J Hum Genet*, **71**, 276–285.

Gilbert, S.C., Plebanski, M., Gupta, S. *et al.* (1998). Association of malaria parasite population structure, HLA, and immunological antagonism. *Science*, **279**, 1173–1177.

Gill, P. (2002). Role of short tandem repeat DNA in forensic casework in the UK – past, present, and future perspectives. *Biotechniques*, **32**, 366–368.

Gill, P. and Werrett, D.J. (1987). Exclusion of a man charged with murder by DNA fingerprinting. *Forensic Sci Int*, **35**, 145–148.

Gilles, H.M., Fletcher, K.A., Hendrickse, R.G., Lindner, R., Reddy, S. and Allan, N. (1967). Glucose-6-phosphate-dehydrogenase deficiency, sickling, and malaria in African children in South Western Nigeria. *Lancet*, **1**, 138–140.

Gilman, J.G., Huisman, T.H. and Abels, J. (1984). Dutch beta β-thalassaemia: a 10 kilobase DNA deletion associated with significant gamma-chain production. *Br J Haematol*, **56**, 339–348.

Gilson, E. and Londono-Vallejo, A. (2007). Telomere length profiles in humans: all ends are not equal. *Cell Cycle*, **6**, 2486–2494.

Ginger, R.S., Askew, S.E., Ogborne, R.M. *et al.* (2008). SLC24A5 encodes a trans-Golgi network protein with potassium-dependent sodium-calcium exchange activity that regulates human epidermal melanogenesis. *J Biol Chem*, **283**, 5486–5495.

Girardin, S.E., Boneca, I.G., Viala, J. *et al.* (2003). Nod2 is a general sensor of peptidoglycan through muramyl dipeptide (MDP) detection. *J Biol Chem*, **278**, 8869–8872.

Glenner, G.G. and Wong, C.W. (1984a). Alzheimer's disease and Down's syndrome: sharing of a unique cerebrovascular amyloid fibril protein. *Biochem Biophys Res Commun*, **122**, 1131–1135.

Glenner, G.G. and Wong, C.W. (1984b). Alzheimer's disease: initial report of the purification and characterization of a novel cerebrovascular amyloid protein. *Biochem Biophys Res Commun*, **120**, 885–890.

Gloyn, A.L., Weedon, M.N., Owen, K.R. *et al.* (2003). Large-scale association studies of variants in genes encoding the pancreatic beta-cell KATP channel subunits Kir6.2 (KCNJ11) and SUR1 (ABCC8) confirm that the KCNJ11 E23K variant is associated with type 2 diabetes. *Diabetes*, **52**, 568–572.

Glusman, G., Yanai, I., Rubin, I. and Lancet, D. (2001). The complete human olfactory subgenome. *Genome Res*, **11**, 685–702.

Goate, A., Chartier-Harlin, M.C., Mullan, M. *et al.* (1991). Segregation of a missense mutation in the amyloid precursor protein gene with familial Alzheimer's disease. *Nature*, **349**, 704–706.

Goldgaber, D., Lerman, M.I., McBride, O.W., Saffiotti, U. and Gajdusek, D.C. (1987). Characterization and chromosomal localization of a cDNA encoding brain amyloid of Alzheimer's disease. *Science*, **235**, 877–880.

Gonzalez, E., Bamshad, M., Sato, N. *et al.* (1999). Race-specific HIV-1 disease-modifying effects associated with CCR5 haplotypes. *Proc Natl Acad Sci U S A*, **96**, 12004–12009.

Gonzalez, E., Dhanda, R., Bamshad, M. *et al.* (2001). Global survey of genetic variation in CCR5, RANTES, and MIP-1alpha: impact on the epidemiology of the HIV-1 pandemic. *Proc Natl Acad Sci U S A*, **98**, 5199–5204.

Gonzalez, E., Kulkarni, H., Bolivar, H. *et al.* (2005). The influence of CCL3L1 gene-containing segmental duplications on HIV-1/AIDS susceptibility. *Science*, **307**, 1434–1440.

Gonzalez, I.L. and Sylvester, J.E. (2001). Human rDNA: evolutionary patterns within the genes and tandem arrays derived from multiple chromosomes. *Genomics*, **73**, 255–263.

Goodall, J.C., Ellis, L. and Hill Gaston, J.S. (2006). Spondylarthritis-associated and non-spondylarthritis-associated B27 subtypes differ in their dependence upon tapasin for surface expression and their incorporation into the peptide loading complex. *Arthritis Rheum*, **54**, 138–147.

Goode, E.L., Cherny, S.S., Christian, J.C., Jarvik, G.P. and de Andrade, M. (2007). Heritability of longitudinal measures of body mass index and lipid and lipoprotein levels in aging twins. *Twin Res Hum Genet*, **10**, 703–711.

Goodman, M., Miyamoto, M.M. and Czelusniak, J. (1987). Pattern and process in vertebrate phylogeny revealed by coevolution of molecules and phylogenies. In: *Molecules and Morphology in Evolution: Conflict or Compromise?* ed. C. Patterson. Cambridge University Press, Cambridge, UK: 140–176.

Goossens, M., Dozy, A.M., Embury, S.H. *et al.* (1980). Triplicated alpha-globin loci in humans. *Proc Natl Acad Sci U S A*, **77**, 518–521.

Gorer, P. (1936). The detection of a hereditary antigenic difference in the blood of mice by means of human group A serum. *J Genet*, **32**, 17–31.

Gorgoulis, V.G., Liloglou, T., Sigala, F. *et al.* (2004). Absence of association with cancer risk and low frequency of alterations at a p53 responsive PIG3 gene polymorphism in breast and lung carcinomas. *Mutat Res*, **556**, 143–150.

Goring, H.H., Curran, J.E., Johnson, M.P. *et al.* (2007). Discovery of expression QTLs using large-scale transcriptional profiling in human lymphocytes. *Nat Genet*, **39**, 1208–1216.

Goswami, T., Bhattacharjee, A., Babal, P. *et al.* (2001). Natural-resistance-associated macrophage protein 1 is an H+/bivalent cation antiporter. *Biochem J*, **354**, 511–519.

Graf, J., Hodgson, R. and van Daal, A. (2005). Single nucleotide polymorphisms in the MATP gene are associated with normal human pigmentation variation. *Hum Mutat*, **25**, 278–284.

Graff-Radford, N.R., Green, R.C., Go, R.C. *et al.* (2002). Association between apolipoprotein E genotype and Alzheimer disease in African American subjects. *Arch Neurol*, **59**, 594–600.

Graham, R.R., Kozyrev, S.V., Baechler, E.C. *et al.* (2006). A common haplotype of interferon regulatory factor 5 (IRF5) regulates splicing and expression and is associated with increased risk of systemic lupus erythematosus. *Nat Genet*, **38**, 550–555.

Grant, S.F., Thorleifsson, G., Reynisdottir, I. *et al.* (2006). Variant of transcription factor 7-like 2 (TCF7L2) gene confers risk of type 2 diabetes. *Nat Genet*, **38**, 320–323.

Greco, L., Romino, R., Coto, I. *et al.* (2002). The first large population based twin study of coeliac disease. *Gut*, **50**, 624–628.

Greenfield, J. (1911). Notes on a family of "myotonia atrophica" and early cataract, with a report of an additional case of "myotonia atrophica". *Rev Neuro Psychiatry*, **9**, 169–181.

Greenstein, R.M., Reardon, M.P. and Chan, T.S. (1977). An X-autosome translocation in a girl with Duchenne muscular dystrophy (DMD): evidence for DMD gene localization. *Pediat Res*, **11**, 457.

Gregersen, P.K., Silver, J. and Winchester, R.J. (1987). The shared epitope hypothesis. An approach to understanding the molecular genetics of susceptibility to rheumatoid arthritis. *Arthritis Rheum*, **30**, 1205–1213.

Gregersen, P.K., Amos, C.I., Lee, A.T. *et al.* (2009). REL, encoding a member of the NF-kappaB family of transcription factors, is a newly defined risk locus for rheumatoid arthritis. *Nat Genet*, **41**, 820–823.

Grewal, S.I. and Elgin, S.C. (2007). Transcription and RNA interference in the formation of heterochromatin. *Nature*, **447**, 399–406.

Griese, E.U., Asante-Poku, S., Ofori-Adjei, D., Mikus, G. and Eichelbaum, M. (1999). Analysis of the CYP2D6 gene mutations and their consequences for enzyme function in a West African population. *Pharmacogenetics*, **9**, 715–723.

Griese, E.U., Ilett, K.F., Kitteringham, N.R. *et al.* (2001). Allele and genotype frequencies of polymorphic cytochromes P4502D6, 2C19 and 2E1 in aborigines from Western Australia. *Pharmacogenetics*, **11**, 69–76.

Griffin, J.H., Evatt, B., Wideman, C. and Fernandez, J.A. (1993). Anticoagulant protein C pathway defective in majority of thrombophilic patients. *Blood*, **82**, 1989–1993.

Griffin, J.H., Evatt, B., Zimmerman, T.S., Kleiss, A.J. and Wideman, C. (1981). Deficiency of protein C in congenital thrombotic disease. *J Clin Invest*, **68**, 1370–1373.

Griffiths, C.E. and Barker, J.N. (2007). Pathogenesis and clinical features of psoriasis. *Lancet*, **370**, 263–271.

Grimm, T., Muller, B., Dreier, M. *et al.* (1989). Hot spot of recombination within DXS164 in the Duchenne muscular dystrophy gene. *Am J Hum Genet*, **45**, 368–372.

Groeschen, H.M. (2007). Novel HIV treatment approved. *Am J Health Syst Pharm*, **64**, 1886.

Groot, P.C., Bleeker, M.J., Pronk, J.C. *et al.* (1989). The human alpha-amylase multigene family consists of haplotypes with variable numbers of genes. *Genomics*, **5**, 29–42.

Gros, P., Skamene, E. and Forget, A. (1981). Genetic control of natural resistance to Mycobacterium bovis (BCG) in mice. *J Immunol*, **127**, 2417–2421.

Gu, W., Zhang, F. and Lupski, J.R. (2008). Mechanisms for human genomic rearrangements. *Pathogenetics*, **1**, 4.

Gu, X., Wang, Y. and Gu, J. (2002). Age distribution of human gene families shows significant roles of both large- and small-scale duplications in vertebrate evolution. *Nat Genet*, **31**, 205–209.

Gu, Y.C., Landman, H. and Huisman, T.H. (1987). Two different quadruplicated alpha globin gene arrangements. *Br J Haematol*, **66**, 245–250.

Gu, Z., Steinmetz, L.M., Gu, X., Scharfe, C., Davis, R.W. and Li, W.H. (2003). Role of duplicate genes in genetic robustness against null mutations. *Nature*, **421**, 63–66.

Gudmundsson, J., Sulem, P., Manolescu, A. *et al.* (2007). Genome-wide association study identifies a second prostate cancer susceptibility variant at 8q24. *Nat Genet*, **39**, 631–637.

Guerra, C.A., Gikandi, P.W., Tatem, A.J. *et al.* (2008). The limits and intensity of Plasmodium falciparum transmission: implications for malaria control and elimination worldwide. *PLoS Med*, **5**, e38.

Guerra, C.A., Hay, S.I., Lucioparedes, L.S. *et al.* (2007). Assembling a global database of malaria parasite prevalence for the Malaria Atlas Project. *Malar J*, **6**, 17.

Gusella, J.F., Wexler, N.S., Conneally, P.M. *et al.* (1983). A polymorphic DNA marker genetically linked to Huntington's disease. *Nature*, **306**, 234–238.

Hagan, P., Blumenthal, U.J., Dunn, D., Simpson, A.J. and Wilkins, H.A. (1991). Human IgE, IgG4 and resistance to reinfection with Schistosoma haematobium. *Nature*, **349**, 243–245.

Hagelberg, E., Gray, I.C. and Jeffreys, A.J. (1991). Identification of the skeletal remains of a murder victim by DNA analysis. *Nature*, **352**, 427–429.

Hagerman, R.J., Leehey, M., Heinrichs, W. *et al.* (2001). Intention tremor, parkinsonism, and generalized brain atrophy in male carriers of fragile X. *Neurology*, **57**, 127–130.

Haiman, C.A., Garcia, R.R., Kolonel, L.N., Henderson, B.E., Wu, A.H. and Le Marchand, L. (2008). A promoter polymorphism in the CASP8 gene is not associated with cancer risk. *Nat Genet*, **40**, 259–260.

Haines, J.L., Hauser, M.A., Schmidt, S. *et al.* (2005). Complement factor H variant increases the risk of age-related macular degeneration. *Science*, **308**, 419–421.

Hakonarson, H., Grant, S.F., Bradfield, J.P. *et al.* (2007). A genome-wide association study identifies KIAA0350 as a type 1 diabetes gene. *Nature*, **448**, 591–594.

Haldane, J.B.S. (1949). The rate of mutation of human genes. Proceedings of the Eight International Congress of Genetics and Heredity. *Hereditas*, **35**, 267–273.

Halfvarson, J., Bodin, L., Tysk, C., Lindberg, E. and Jarnerot, G. (2003). Inflammatory bowel disease in a Swedish twin cohort: a long-term follow-up of concordance and clinical characteristics. *Gastroenterology*, **124**, 1767–1773.

Hamada, H. and Kakunaga, T. (1982). Potential Z-DNA forming sequences are highly dispersed in the human genome. *Nature*, **298**, 396–398.

Hamada, H., Petrino, M.G. and Kakunaga, T. (1982). A novel repeated element with Z-DNA-forming potential is widely found in evolutionarily diverse eukaryotic genomes. *Proc Natl Acad Sci U S A*, **79**, 6465–6469.

Hamada, H., Seidman, M., Howard, B.H. and Gorman, C.M. (1984). Enhanced gene expression by the poly(dT-dG). poly(dC-dA) sequence. *Mol Cell Biol*, **4**, 2622–2630.

Hamblin, M.T. and Di Rienzo, A. (2000). Detection of the signature of natural selection in humans: evidence from the Duffy blood group locus. *Am J Hum Genet*, **66**, 1669–1679.

Hamerton, J.L., Canning, N., Ray, M. and Smith, S. (1975). A cytogenetic survey of 14,069 newborn infants. I. Incidence of chromosome abnormalities. *Clin Genet*, **8**, 223–243.

Hammer, R.E., Maika, S.D., Richardson, J.A., Tang, J.P. and Taurog, J.D. (1990). Spontaneous inflammatory disease in transgenic rats expressing HLA-B27 and human beta 2m: an animal model of HLA-B27-associated human disorders. *Cell*, **63**, 1099–1112.

Hammond, C.J., Webster, A.R., Snieder, H., Bird, A.C., Gilbert, C.E. and Spector, T.D. (2002). Genetic influence on early age-related maculopathy: a twin study. *Ophthalmology*, **109**, 730–736.

Hamosh, A., Scott, A.F., Amberger, J.S., Bocchini, C.A. and McKusick, V.A. (2005). Online Mendelian Inheritance in Man (OMIM), a knowledgebase of human genes and genetic disorders. *Nucleic Acids Res*, **33**, D514–517.

Hampe, J., Cuthbert, A., Croucher, P.J. *et al.* (2001). Association between insertion mutation in NOD2 gene and Crohn's disease in German and British populations. *Lancet*, **357**, 1925–1928.

Hampe, J., Franke, A., Rosenstiel, P. *et al.* (2007). A genome-wide association scan of nonsynonymous SNPs identifies a susceptibility variant for Crohn disease in ATG16L1. *Nat Genet*, **39**, 207–211.

Han, K., Sen, S.K., Wang, J. *et al.* (2005). Genomic rearrangements by LINE-1 insertion-mediated deletion in the human and chimpanzee lineages. *Nucleic Acids Res*, **33**, 4040–4052.

Han, S.W., Jeon, Y.K., Lee, K.H. *et al.* (2007). Intron 1 CA dinucleotide repeat polymorphism and mutations of epidermal growth factor receptor and gefitinib responsiveness in non-small-cell lung cancer. *Pharmacogenet Genomics*, **17**, 313–319.

Hanh, E.V. and Gillespie, E.B. (1927). Sickle cell anaemia. *Arch Internal Med*, **39**, 233.

Hannes, F.D., Sharp, A.J., Mefford, H.C. *et al.* (2009). Recurrent reciprocal deletions and duplications of 16p13.11: the deletion is a risk factor for MR/MCA while the duplication may be a rare benign variant. *J Med Genet*, 46, 223–232.

Harada, N., Visser, R., Dawson, A. *et al.* (2004). A 1-Mb critical region in six patients with 9q34.3 terminal deletion syndrome. *J Hum Genet*, **49**, 440–444.

Harding, R.M., Fullerton, S.M., Griffiths, R.C. *et al.* (1997). Archaic African and Asian lineages in the genetic ancestry of modern humans. *Am J Hum Genet*, **60**, 772–789.

Hardy, J. (2006). Amyloid double trouble. *Nat Genet*, **38**, 11–12.

Hardy, J. and Selkoe, D.J. (2002). The amyloid hypothesis of Alzheimer's disease: progress and problems on the road to therapeutics. *Science*, **297**, 353–356.

Harjes, P. and Wanker, E.E. (2003). The hunt for huntingtin function: interaction partners tell many different stories. *Trends Biochem Sci*, **28**, 425–433.

Harley, C.B., Futcher, A.B. and Greider, C.W. (1990). Telomeres shorten during ageing of human fibroblasts. *Nature*, **345**, 458–460.

Harney, S.M., Vilarino-Guell, C., Adamopoulos, I.E. *et al.* (2008). Fine mapping of the MHC Class III region demonstrates association of AIF1 and rheumatoid arthritis. *Rheumatology (Oxford)*, **47**, 1761–1767.

Harper, P.S. (2006). The discovery of the human chromosome number in Lund, 1955–1956. *Hum Genet*, **119**, 226–232.

Harris, E.E. and Hey, J. (1999). X chromosome evidence for ancient human histories. *Proc Natl Acad Sci U S A*, **96**, 3320–3324.

Harris, E.E. and Meyer, D. (2006). The molecular signature of selection underlying human adaptations. *Am J Phys Anthropol*, **43**, 89–130.

Harris, R.S. and Liddament, M.T. (2004). Retroviral restriction by APOBEC proteins. *Nat Rev Immunol*, **4**, 868–877.

Harris, R.S., Bishop, K.N., Sheehy, A.M. *et al.* (2003). DNA deamination mediates innate immunity to retroviral infection. *Cell*, **113**, 803–809.

Harrison, C.J., Jack, E.M., Allen, T.D. and Harris, R. (1983). The fragile X: a scanning electron microscope study. *J Med Genet*, **20**, 280–285.

Hassold, T. and Hunt, P. (2001). To err (meiotically) is human: the genesis of human aneuploidy. *Nat Rev Genet*, **2**, 280–291.

Hassold, T., Chen, N., Funkhouser, J. *et al.* (1980). A cytogenetic study of 1000 spontaneous abortions. *Ann Hum Genet*, **44**, 151–178.

Hastbacka, J., de la Chapelle, A., Kaitila, I., Sistonen, P., Weaver, A. and Lander, E. (1992). Linkage disequilibrium mapping in isolated founder populations: diastrophic dysplasia in Finland. *Nat Genet*, **2**, 204–211.

Hastbacka, J., de la Chapelle, A., Mahtani, M.M. *et al.* (1994). The diastrophic dysplasia gene encodes a novel sulfate transporter: positional cloning by fine-structure linkage disequilibrium mapping. *Cell*, **78**, 1073–1087.

Hastbacka, J., Kaitila, I., Sistonen, P. and de la Chapelle, A. (1990). Diastrophic dysplasia gene maps to the distal long arm of chromosome 5. *Proc Natl Acad Sci U S A*, **87**, 8056–8059.

Hastbacka, J., Kerrebrock, A., Mokkala, K. *et al.* (1999). Identification of the Finnish founder mutation for diastrophic dysplasia (DTD). *Eur J Hum Genet*, **7**, 664–670.

Hastbacka, J., Sistonen, P., Kaitila, I., Weiffenbach, B., Kidd, K.K. and de la Chapelle, A. (1991). A linkage map spanning the locus for diastrophic dysplasia (DTD). *Genomics*, **11**, 968–973.

Hatch, F.T. and Mazrimas, J.A. (1974). Fractionation and characterization of satellite DNAs of the kangaroo rat (Dipodomys ordii). *Nucleic Acids Res*, **1**, 559–575.

Hatton, C.S., Wilkie, A.O., Drysdale, H.C. *et al.* (1990). Alpha-thalassemia caused by a large (62 kb) deletion upstream of the human alpha globin gene cluster. *Blood*, **76**, 221–227.

Hattori, M., Fujiyama, A., Taylor, T.D. *et al.* (2000). The DNA sequence of human chromosome 21. *Nature*, **405**, 311–319.

Hay, S.I. and Snow, R.W. (2006). The malaria Atlas Project: developing global maps of malaria risk. *PLoS Med*, **3**, e473.

Hayden, E.C. (2008). International genome project launched. *Nature*, **451**, 378–379.

Hayes, J.D. and Strange, R.C. (2000). Glutathione S-transferase polymorphisms and their biological consequences. *Pharmacology*, **61**, 154–166.

HDCRG (Huntington's Disease Collaborative Research Group) (1993). A novel gene containing a trinucleotide repeat that is expanded and unstable on Huntington's disease chromosomes. The Huntington's Disease Collaborative Research Group. *Cell*, **72**, 971–983.

Healy, D.G. (2006). Case-control studies in the genomic era: a clinician's guide. *Lancet Neurol*, **5**, 701–707.

Hebert, L.E., Scherr, P.A., Bienias, J.L., Bennett, D.A. and Evans, D.A. (2003). Alzheimer disease in the US population: prevalence estimates using the 2000 census. *Arch Neurol*, **60**, 1119–1122.

Hedges, S.B. (2000). Human evolution. A start for population genomics. *Nature*, **408**, 652–653.

Hedrick, P. (2004). Estimation of relative fitnesses from relative risk data and the predicted future of haemoglobin alleles S and C. *J Evol Biol*, **17**, 221–224.

Hedrick, P.W. and Thomson, G. (1983). Evidence for balancing selection at HLA. *Genetics*, **104**, 449–456.

Hedrick, P.W., Whittam, T.S. and Parham, P. (1991). Heterozygosity at individual amino acid sites: extremely high levels for HLA-A and -B genes. *Proc Natl Acad Sci U S A*, **88**, 5897–5901.

Heeney, J.L., Dalgleish, A.G. and Weiss, R.A. (2006). Origins of HIV and the evolution of resistance to AIDS. *Science*, **313**, 462–466.

Helmuth, L. (2001). A genome glossary. *Science* **291**: 1197.

Henri, S., Chevillard, C., Mergani, A. *et al.* (2002). Cytokine regulation of periportal fibrosis in humans infected with Schistosoma mansoni: IFN-gamma is associated with protection against fibrosis and TNF-alpha with aggravation of disease. *J Immunol*, **169**, 929–936.

Herbig, U., Jobling, W.A., Chen, B.P., Chen, D.J. and Sedivy, J.M. (2004). Telomere shortening triggers senescence of human cells through a pathway involving ATM, p53, and p21(CIP1), but not p16(INK4a). *Mol Cell*, **14**, 501–513.

Herrick, J.B. (1910). Peculiar elongated and sickle-shaped red blood corpuscules in a case of severe anemia. *Arch Internal Med*, **6**, 517–521.

Heston, L.L., Mastri, A.R., Anderson, V.E. and White, J. (1981). Dementia of the Alzheimer type. Clinical genetics, natural history, and associated conditions. *Arch Gen Psychiatry*, **38**, 1085–1090.

Hiby, S.E., Walker, J.J., O'Shaughnessy K.M. *et al.* (2004). Combinations of maternal KIR and fetal HLA-C genes influence the risk of preeclampsia and reproductive success. *J Exp Med*, **200**, 957–965.

Higashi, M.K., Veenstra, D.L., Kondo, L.M. *et al.* (2002). Association between CYP2C9 genetic variants and anticoagulation-related outcomes during warfarin therapy. *JAMA*, **287**, 1690–1698.

Higgs, D.R. (2004). Gene regulation in hematopoiesis: new lessons from thalassemia. *Hematology (Am Soc Hematol Educ Program)*, 1–13.

Higgs, D.R. and Wood, W.G. (2008a). Genetic complexity in sickle cell disease. *Proc Natl Acad Sci U S A*, **105**, 11595–11596.

Higgs, D.R. and Wood, W.G. (2008b). Long-range regulation of alpha globin gene expression during erythropoiesis. *Curr Opin Hematol*, **15**, 176–183.

Higgs, D.R., Garrick, D., Anguita, E. *et al.* (2005). Understanding {alpha}-globin gene regulation: aiming to improve the management of thalassemia. *Ann N Y Acad Sci*, **1054**, 92–102.

Higgs, D.R., Goodbourn, S.E., Lamb, J., Clegg, J.B., Weatherall, D.J. and Proudfoot, N.J. (1983). Alpha-thalassaemia caused by a polyadenylation signal mutation. *Nature*, **306**, 398–400.

Higgs, D.R., Goodbourn, S.E., Wainscoat, J.S., Clegg, J.B. and Weatherall, D.J. (1981). Highly variable regions of DNA flank the human alpha globin genes. *Nucleic Acids Res*, **9**, 4213–4224.

Higgs, D.R., Old, J.M., Pressley, L., Clegg, J.B. and Weatherall, D.J. (1980). A novel alpha-globin gene arrangement in man. *Nature*, **284**, 632–635.

Higgs, D.R., Vickers, M.A., Wilkie, A.O., Pretorius, I.M., Jarman, A.P. and Weatherall, D.J. (1989). A review of the molecular genetics of the human alpha-globin gene cluster. *Blood*, **73**, 1081–1104.

Higgs, D.R., Wainscoat, J.S., Flint, J. *et al.* (1986). Analysis of the human alpha-globin gene cluster reveals a highly informative genetic locus. *Proc Natl Acad Sci U S A*, **83**, 5165–5169.

Hill, A.V. (2001). The genomics and genetics of human infectious disease susceptibility. *Annu Rev Genomics Hum Genet*, **2**, 373–400.

Hill, A.V. (2006). Aspects of genetic susceptibility to human infectious diseases. *Annu Rev Genet*, **40**, 469–486.

Hill, A.V., Allsopp, C.E., Kwiatkowski, D. *et al.* (1991). Common west African HLA antigens are associated with protection from severe malaria. *Nature*, **352**, 595–600.

Hindorff, L., Junkins, H. and Manolio, T. (2008). A catalog of published genome-wide association studies. Available at: www.genome.gov/26525384. Accessed 8 May 2008.

Hinds, D.A., Kloek, A.P., Jen, M., Chen, X. and Frazer, K.A. (2006). Common deletions and SNPs are in linkage disequilibrium in the human genome. *Nat Genet*, **38**, 82–85.

Hinds, D.A., Stuve, L.L., Nilsen, G.B. *et al.* (2005). Whole-genome patterns of common DNA variation in three human populations. *Science*, **307**, 1072–1079.

Hinoda, Y., Okayama, N., Takano, N. *et al.* (2002). Association of functional polymorphisms of matrix metalloproteinase (MMP)-1 and MMP-3 genes with colorectal cancer. *Int J Cancer*, **102**, 526–529.

Hirschhorn, J.N. (2003). Genetic epidemiology of type 1 diabetes. *Pediatr Diabetes*, **4**, 87–100.

Hirschhorn, J.N., Lohmueller, K., Byrne, E. and Hirschhorn, K. (2002). A comprehensive review of genetic association studies. *Genet Med*, **4**, 45–61.

Hitman, G.A., Tarn, A.C., Winter, R.M. *et al.* (1985). Type 1 (insulin-dependent) diabetes and a highly variable locus close to the insulin gene on chromosome 11. *Diabetologia*, **28**, 218–222.

Hixson, J.E. and Vernier, D.T. (1990). Restriction isotyping of human apolipoprotein E by gene amplification and cleavage with HhaI. *J Lipid Res*, **31**, 545–548.

Hladik, F., Liu, H., Speelmon, E. *et al.* (2005). Combined effect of CCR5-delta32 heterozygosity and the CCR5 promoter polymorphism -2459 A/G on CCR5 expression and resistance to human immunodeficiency virus type 1 transmission. *J Virol*, **79**, 11677–11684.

Ho, H.J., Ray, D.A., Salem, A.H., Myers, J.S. and Batzer, M.A. (2005a). Straightening out the LINEs: LINE-1 orthologous loci. *Genomics*, **85**, 201–207.

Ho, T.H., Bundman, D., Armstrong, D.L. and Cooper, T.A. (2005b). Transgenic mice expressing CUG-BP1 reproduce splicing mis-regulation observed in myotonic dystrophy. *Hum Mol Genet*, **14**, 1539–1547.

Ho, W.K., Hankey, G.J., Quinlan, D.J. and Eikelboom, J.W. (2006). Risk of recurrent venous thromboembolism in patients with common thrombophilia: a systematic review. *Arch Internal Med*, **166**, 729–736.

Hoffman, E.P., Brown, R.H., Jr. and Kunkel, L.M. (1987). Dystrophin: the protein product of the Duchenne muscular dystrophy locus. *Cell*, **51**, 919–928.

Holleboom, A.G., Vergeer, M., Hovingh, G.K., Kastelein, J.J. and Kuivenhoven, J.A. (2008). The value of HDL genetics. *Curr Opin Lipidol*, **19**, 385–394.

Hollegaard, M.V. and Bidwell, J.L. (2006). Cytokine gene polymorphism in human disease: on-line databases, Supplement 3. *Genes Immun*, **7**, 269–276.

Hollox, E.J., Armour, J.A. and Barber, J.C. (2003). Extensive normal copy number variation of a beta-defensin antimicrobial-gene cluster. *Am J Hum Genet*, **73**, 591–600.

Hollox, E.J., Huffmeier, U., Zeeuwen, P.L. *et al.* (2008). Psoriasis is associated with increased beta-defensin genomic copy number. *Nat Genet*, **40**, 23–25.

Holmes, R.K., Malim, M.H. and Bishop, K.N. (2007). APOBEC-mediated viral restriction: not simply editing? *Trends Biochem Sci*, **32**, 118–128.

Horton, R., Gibson, R., Coggill, P. *et al.* (2008). Variation analysis and gene annotation of eight MHC haplotypes: the MHC Haplotype Project. *Immunogenetics*, **60**, 1–18.

Horton, R., Wilming, L., Rand, V. *et al.* (2004). Gene map of the extended human MHC. *Nat Rev Genet*, **5**, 889–899.

Houck, C.M., Rinehart, F.P. and Schmid, C.W. (1979). A ubiquitous family of repeated DNA sequences in the human genome. *J Mol Biol*, **132**, 289–306.

Housman, D., Forget, B.G., Skoultchi, A. and Benz, E.J., Jr. (1973). Quantitative deficiency of chain-specific globin messenger ribonucleic acids in the thalassemia syndromes. *Proc Natl Acad Sci U S A*, **70**, 1809–1813.

Houwen, R.H., Baharloo, S., Blankenship, K. *et al.* (1994). Genome screening by searching for shared segments: mapping a gene for benign recurrent intrahepatic cholestasis. *Nat Genet*, **8**, 380–386.

Howeler, C.J., Busch, H.F., Geraedts, J.P., Niermeijer, M.F. and Staal, A. (1989). Anticipation in myotonic dystrophy: fact or fiction? *Brain*, **112**, 779–797.

Hoyng, C.B., Poppelaars, F., van de Pol, T.J. *et al.* (1996). Genetic fine mapping of the gene for recessive Stargardt disease. *Hum Genet*, **98**, 500–504.

Hsu, L.Y., Benn, P.A., Tannenbaum, H.L., Perlis, T.E. and Carlson, A.D. (1987). Chromosomal polymorphisms of 1, 9, 16, and Y in 4 major ethnic groups: a large prenatal study. *Am J Med Genet*, **26**, 95–101.

Huang, D., Xia, S., Zhou, Y., Pirskanen, R., Liu, L. and Lefvert, A.K. (1998). No evidence for interleukin-4 gene conferring susceptibility to myasthenia gravis. *J Neuroimmunol*, **92**, 208–211.

Huang, T.S., Lee, C.C., Chang, A.C. *et al.* (2003). Shortening of microsatellite deoxy(CA) repeats involved in GL331-induced down-regulation of matrix metalloproteinase-9 gene expression. *Biochem Biophys Res Commun*, **300**, 901–907.

Huang, Y., Paxton, W.A., Wolinsky, S.M. *et al.* (1996). The role of a mutant CCR5 allele in HIV-1 transmission and disease progression. *Nat Med*, **2**, 1240–1243.

Hubner, N., Wallace, C.A., Zimdahl, H. *et al.* (2005). Integrated transcriptional profiling and linkage analysis for identification of genes underlying disease. *Nat Genet*, **37**, 243–253.

Huddart, J. (1777). An account of persons who could not distinguish colours. *Phil Trans R Soc*, **67**, 260.

Hudson, R.R., Bailey, K., Skarecky, D., Kwiatowski, J. and Ayala, F.J. (1994). Evidence for positive selection in the superoxide dismutase (Sod) region of Drosophila melanogaster. *Genetics*, **136**, 1329–1340.

Hughes, A.E., Orr, N., Esfandiary, H., Diaz-Torres, M., Goodship, T. and Chakravarthy, U. (2006). A common CFH haplotype, with deletion of CFHR1 and CFHR3, is associated with lower risk of age-related macular degeneration. *Nat Genet*, **38**, 1173–1177.

Hughes, A.L. and Nei, M. (1988). Pattern of nucleotide substitution at major histocompatibility complex class I loci reveals overdominant selection. *Nature*, **335**, 167–170.

Hughes, A.L. and Nei, M. (1989). Nucleotide substitution at major histocompatibility complex class II loci: evidence for overdominant selection. *Proc Natl Acad Sci U S A*, **86**, 958–962.

Hughes, A.R., Mosteller, M., Bansal, A.T. *et al.* (2004). Association of genetic variations in HLA-B region with hypersensitivity to abacavir in some, but not all, populations. *Pharmacogenomics*, **5**, 203–211.

Hughes, C.A., Foisy, M.M., Dewhurst, N. *et al.* (2008). Abacavir hypersensitivity reaction: an update. *Ann Pharmacother*, **42**, 387–396.

Hugot, J.P., Chamaillard, M., Zouali, H. *et al.* (2001). Association of NOD2 leucine-rich repeat variants with susceptibility to Crohn's disease. *Nature*, **411**, 599–603.

Hugot, J.P., Laurent-Puig, P., Gower-Rousseau, C. *et al.* (1996). Mapping of a susceptibility locus for Crohn's disease on chromosome 16. *Nature*, **379**, 821–823.

Huisman, T.H.J., Carver, M.F.H. and Efremov, G.D. (1996). *A Syllabus of Human Hemoglobin Variants*. The Sickle Cell Anemia Foundation, Augusta, GA.

Hull, J., Campino, S., Rowlands, K. *et al.* (2007). Identification of common genetic variation that modulates alternative splicing. *PLoS Genet*, **3**, e99.

Huntington, G. (1872). On chorea. *Med Surg Reporter*, **26**, 320–321.

Hutchison, C.A., 3rd (2007). DNA sequencing: bench to bedside and beyond. *Nucleic Acids Res*, **35**, 6227–6237.

Hutter, B., Helms, V. and Paulsen, M. (2006). Tandem repeats in the CpG islands of imprinted genes. *Genomics*, **88**, 323–332.

Iafrate, A.J., Feuk, L., Rivera, M.N. *et al.* (2004). Detection of large-scale variation in the human genome. *Nat Genet*, **36**, 949–951.

Ibanez, P., Bonnet, A.M., Debarges, B. *et al.* (2004). Causal relation between alpha-synuclein gene duplication and familial Parkinson's disease. *Lancet*, **364**, 1169–1171.

Ibrahim, M.E., Lambson, B., Yousif, A.O. *et al.* (1999). Kala-azar in a high transmission focus: an ethnic and geographic dimension. *Am J Trop Med Hyg*, **61**, 941–944.

Ideraabdullah, F.Y., Vigneau, S. and Bartolomei, M.S. (2008). Genomic imprinting mechanisms in mammals. *Mutat Res*, **647**, 77–85.

IHGSC (International Human Genome Sequencing Consortium) (2004). Finishing the euchromatic sequence of the human genome. *Nature*, **431**, 931–945.

Inagaki, H., Ohye, T., Kogo, H. *et al.* (2009). Chromosomal instability mediated by non-B DNA: cruciform conformation and not DNA sequence is responsible for recurrent translocation in humans. *Genome Res*, 19, 191–198.

Ingelman-Sundberg, M. (2004). Pharmacogenetics of cytochrome P450 and its applications in drug therapy: the past, present and future. *Trends Pharmacol Sci*, **25**, 193–200.

Ingelman-Sundberg, M. (2005). Genetic polymorphisms of cytochrome P450 2D6 (CYP2D6): clinical consequences, evolutionary aspects and functional diversity. *Pharmacogenomics J*, **5**, 6–13.

Ingelman-Sundberg, M., Sim, S.C., Gomez, A. and Rodriguez-Antona, C. (2007). Influence of cytochrome P450 polymorphisms on drug therapies: pharmacogenetic, pharmacoepigenetic and clinical aspects. *Pharmacol Ther*, **116**, 496–526.

Ingram, V.M. (1957). Gene mutations in human haemoglobin: the chemical difference between normal and sickle cell haemoglobin. *Nature*, **180**, 326–328.

Ingram, V.M. (1958). Abnormal human haemoglobins. I. The comparison of normal human and sickle-cell haemoglobins by fingerprinting. *Biochim Biophys Acta*, **28**, 539–545.

Ingram, V.M. (1959). Abnormal human haemoglobins. III. The chemical difference between normal and sickle cell haemoglobins. *Biochim Biophys Acta*, **36**, 402–411.

Ingram, V.M. (1961). Gene evolution and the haemoglobins. *Nature*, **189**, 704–708.

Ingram, V.M. (2004). Sickle-cell anemia hemoglobin: the molecular biology of the first "molecular disease" – the crucial importance of serendipity. *Genetics*, **167**, 1–7.

Innan, H. (2003). A two-locus gene conversion model with selection and its application to the human RHCE and RHD genes. *Proc Natl Acad Sci U S A*, **100**, 8793–8798.

Inohara, N., Chamaillard, M., McDonald, C. and Nunez, G. (2005). NOD-LRR proteins: role in host–microbial interactions and inflammatory disease. *Annu Rev Biochem*, **74**, 355–383.

Inohara, N., Ogura, Y., Chen, F.F., Muto, A. and Nunez, G. (2001). Human Nod1 confers responsiveness to bacterial lipopolysaccharides. *J Biol Chem*, **276**, 2551–2554.

Inoue, K. and Lupski, J.R. (2002). Molecular mechanisms for genomic disorders. *Annu Rev Genomics Hum Genet*, **3**, 199–242.

Inoue, K., Osaka, H., Imaizumi, K. *et al.* (1999). Proteolipid protein gene duplications causing Pelizaeus–Merzbacher disease: molecular mechanism and phenotypic manifestations. *Ann Neurol*, **45**, 624–632.

Ioannidis, J.P. (2007). Non-replication and inconsistency in the genome-wide association setting. *Hum Hered*, **64**, 203–213.

Ioannidis, J.P., Ntzani, E.E., Trikalinos, T.A. and Contopoulos-Ioannidis, D.G. (2001). Replication validity of genetic association studies. *Nat Genet*, **29**, 306–309.

Ionov, Y., Peinado, M.A., Malkhosyan, S., Shibata, D. and Perucho, M. (1993). Ubiquitous somatic mutations in simple repeated sequences reveal a new mechanism for colonic carcinogenesis. *Nature*, **363**, 558–561.

Itano, H.A. and Neel, J.V. (1950). A new inherited abnormality of human hemoglobin. *Proc Natl Acad Sci U S A*, **36**, 613–617.

Ito, M., Nishiyama, H., Watanabe, J. *et al.* (2006). Association of the PIG3 promoter polymorphism with invasive bladder cancer in a Japanese population. *Jpn J Clin Oncol*, **36**, 116–120.

Jablonski, N.G. and Chaplin, G. (2000). The evolution of human skin coloration. *J Hum Evol*, **39**, 57–106.

Jabs, E.W., Li, X., Coss, C.A., Taylor, E.W., Meyers, D.A. and Weber, J.L. (1991). Mapping the Treacher Collins syndrome locus to 5q31.3-q33.3. *Genomics*, **11**, 193–198.

Jacobs, G.H. (1996). Primate photopigments and primate color vision. *Proc Natl Acad Sci U S A*, **93**, 577–581.

Jacobs, P.A. (1992). The chromosome complement of human gametes. *Oxf Rev Reprod Biol*, **14**, 47–72.

Jacobs, P.A. and Strong, J.A. (1959). A case of human intersexuality having a possible XXY sex-determining mechanism. *Nature*, **183**, 302–303.

Jacobs, P.A., Dalton, P., James, R. *et al.* (1997). Turner syndrome: a cytogenetic and molecular study. *Ann Hum Genet*, **61**, 471–483.

Jacobson, J.W., Medhora, M.M. and Hartl, D.L. (1986). Molecular structure of a somatically unstable transposable element in Drosophila. *Proc Natl Acad Sci U S A*, **83**, 8684–8688.

Jacq, C., Miller, J.R. and Brownlee, G.G. (1977). A pseudogene structure in 5S DNA of Xenopus laevis. *Cell*, **12**, 109–120.

Jacquemont, S., Hagerman, R.J., Leehey, M. *et al.* (2003). Fragile X premutation tremor/ataxia syndrome: molecular, clinical, and neuroimaging correlates. *Am J Hum Genet*, **72**, 869–878.

Jager, R.D., Mieler, W.F. and Miller, J.W. (2008). Age-related macular degeneration. *N Engl J Med*, **358**, 2606–2617.

Jakobsdottir, J., Conley, Y.P., Weeks, D.E., Mah, T.S., Ferrell, R.E. and Gorin, M.B. (2005). Susceptibility genes for age-related maculopathy on chromosome 10q26. *Am J Hum Genet*, **77**, 389–407.

James, R.S., Dalton, P., Gustashaw, K. *et al.* (1997). Molecular characterization of isochromosomes of Xq. *Ann Hum Genet*, **61**, 485–490.

Jameson, K.A., Highnote, S.M. and Wasserman, L.M. (2001). Richer color experience in observers with multiple photopigment opsin genes. *Psychon Bull Rev*, **8**, 244–261.

Jamieson, M.E., Coutts, J.R. and Connor, J.M. (1994). The chromosome constitution of human preimplantation embryos fertilized in vitro. *Hum Reprod*, **9**, 709–715.

Jamieson, S.E., Miller, E.N., Peacock, C.S. *et al.* (2007). Genome-wide scan for visceral leishmaniasis susceptibility genes in Brazil. *Genes Immun*, **8**, 84–90.

Janeway, C.A., Travers, P., Walport, M. and Sholomchik, M.J. (2005). *Immunobiology: the Immune System in Health and Disease*. Garland Science, New York.

Jansen, R.C. and Nap, J.P. (2001). Genetical genomics: the added value from segregation. *Trends Genet*, **17**, 388–391.

Jarman, A.P. and Higgs, D.R. (1988). A new hypervariable marker for the human alpha-globin gene cluster. *Am J Hum Genet*, **43**, 249–256.

Jarman, A.P., Nicholls, R.D., Weatherall, D.J., Clegg, J.B. and Higgs, D.R. (1986). Molecular characterisation of a hypervariable region downstream of the human alpha-globin gene cluster. *EMBO J*, **5**, 1857–1863.

Jarvela, I.E. (2005). Molecular genetics of adult-type hypolactasia. *Ann Med*, **37**, 179–185.

Javanbakht, H., An, P., Gold, B. *et al.* (2006). Effects of human TRIM5alpha polymorphisms on antiretroviral function and susceptibility to human immunodeficiency virus infection. *Virology*, **354**, 15–27.

Jawaheer, D., Seldin, M.F., Amos, C.I. *et al.* (2003). Screening the genome for rheumatoid arthritis susceptibility genes: a replication study and combined analysis of 512 multicase families. *Arthritis Rheum*, **48**, 906–916.

Jeffery, K.J., Siddiqui, A.A., Bunce, M. *et al.* (2000). The influence of HLA class I alleles and heterozygosity on the outcome of human T cell lymphotropic virus type I infection. *J Immunol*, **165**, 7278–7284.

Jeffreys, A.J., Brookfield, J.F. and Semeonoff, R. (1985a). Positive identification of an immigration test-case using human DNA fingerprints. *Nature*, **317**, 818–819.

Jeffreys, A.J., Kauppi, L. and Neumann, R. (2001). Intensely punctate meiotic recombination in the class II region of the major histocompatibility complex. *Nat Genet*, **29**, 217–222.

Jeffreys, A.J., Turner, M. and Debenham, P. (1991). The efficiency of multilocus DNA fingerprint probes for individualization and establishment of family relationships, determined from extensive casework. *Am J Hum Genet*, **48**, 824–840.

Jeffreys, A.J., Wilson, V. and Thein, S.L. (1985b). Hypervariable 'minisatellite' regions in human DNA. *Nature*, **314**, 67–73.

Jeffreys, A.J., Wilson, V. and Thein, S.L. (1985c). Individual-specific 'fingerprints' of human DNA. *Nature*, **316**, 76–79.

Jepson, A., Sisay-Joof, F., Banya, W. *et al.* (1997). Genetic linkage of mild malaria to the major histocompatibility complex in Gambian children: study of affected sibling pairs. *Br Med J*, **315**, 96–97.

Jin, P., Zarnescu, D.C., Ceman, S. *et al.* (2004). Biochemical and genetic interaction between the fragile X mental retardation protein and the microRNA pathway. *Nat Neurosci*, **7**, 113–117.

Jin, W., Riley, R.M., Wolfinger, R.D., White, K.P., Passador-Gurgel, G. and Gibson, G. (2001). The contributions of sex, genotype and age to transcriptional variance in Drosophila melanogaster. *Nat Genet*, **29**, 389–395.

Jobling, M.A., Hurles, M. and Tyler-Smith, C. (2004). *Human Evolutionary Genetics: Origins, Peoples and disease*. Garland, New York/London.

Johansson, I., Lundqvist, E., Bertilsson, L., Dahl, M.L., Sjoqvist, F. and Ingelman-Sundberg, M. (1993). Inherited amplification of an active gene in the cytochrome P450 CYP2D locus as a cause of ultrarapid metabolism of debrisoquine. *Proc Natl Acad Sci U S A*, **90**, 11825–11829.

Johansson, S., Lie, B.A., Todd, J.A. *et al.* (2003). Evidence of at least two type 1 diabetes susceptibility genes in the HLA complex distinct from HLA-DQB1, -DQA1 and -DRB1. *Genes Immun*, **4**, 46–53.

John, B. and Miklos, G.L. (1979). Functional aspects of satellite DNA and heterochromatin. *Int Rev Cytol*, **58**, 1–114.

Johnson, J.M., Castle, J., Garrett-Engele, P. *et al.* (2003). Genome-wide survey of human alternative pre-mRNA splicing with exon junction microarrays. *Science*, **302**, 2141–2144.

Johnson, M.E., Viggiano, L., Bailey, J.A. *et al.* (2001). Positive selection of a gene family during the emergence of humans and African apes. *Nature*, **413**, 514–519.

Joint United Nations Programme on HIV/AIDS & NetLibrary Inc. (2006). *2006 Report on the Global AIDS Epidemic*. UNAIDS, Geneva.

Jones, E.Y., Fugger, L., Strominger, J.L. and Siebold, C. (2006). MHC class II proteins and disease: a structural perspective. *Nat Rev Immunol*, **6**, 271–282.

Jordan, G. and Mollon, J.D. (1993). A study of women heterozygous for colour deficiencies. *Vision Res*, **33**, 1495–1508.

Jordan, I.K., Rogozin, I.B., Glazko, G.V. and Koonin, E.V. (2003). Origin of a substantial fraction of human regulatory sequences from transposable elements. *Trends Genet*, **19**, 68–72.

Jothi, R., Cuddapah, S., Barski, A., Cui, K. and Zhao, K. (2008). Genome-wide identification of in vivo protein–DNA binding sites from ChIP-Seq data. *Nucleic Acids Res*, **36**, 5221–5231.

Juji, T., Satake, M., Honda, Y. and Doi, Y. (1984). HLA antigens in Japanese patients with narcolepsy. All the patients were DR2 positive. *Tissue Antigens*, **24**, 316–319.

Julier, C., Hyer, R.N., Davies, J. *et al.* (1991). Insulin-IGF2 region on chromosome 11p encodes a gene implicated in HLA-DR4-dependent diabetes susceptibility. *Nature*, **354**, 155–159.

Jurinke, C., Denissenko, M.F., Oeth, P., Ehrich, M., van den Boom, D. and Cantor, C.R. (2005). A single nucleotide polymorphism based approach for the identification and characterization of gene expression modulation using MassARRAY. *Mutat Res*, **573**, 83–95.

Jurka, J. and Pethiyagoda, C. (1995). Simple repetitive DNA sequences from primates: compilation and analysis. *J Mol Evol*, **40**, 120–126.

Jurka, J., Kapitonov, V.V., Pavlicek, A., Klonowski, P., Kohany, O. and Walichiewicz, J. (2005). Repbase Update, a database of eukaryotic repetitive elements. *Cytogenet Genome Res*, **110**, 462–467.

Kadota, M., Yang, H.H., Hu, N. *et al.* (2007). Allele-specific chromatin immunoprecipitation studies show genetic influence on chromatin state in human genome. *PLoS Genet*, **3**, e81.

Kagnoff, M.F. (2007). Celiac disease: pathogenesis of a model immunogenetic disease. *J Clin Invest*, **117**, 41–49.

Kaiser, S.M., Malik, H.S. and Emerman, M. (2007). Restriction of an extinct retrovirus by the human TRIM5alpha antiviral protein. *Science*, **316**, 1756–1758.

Kajikawa, M. and Okada, N. (2002). LINEs mobilize SINEs in the eel through a shared 3′ sequence. *Cell*, **111**, 433–444.

Kallioniemi, A., Kallioniemi, O.P., Sudar, D. *et al.* (1992). Comparative genomic hybridization for molecular cytogenetic analysis of solid tumors. *Science*, **258**, 818–821.

Kan, Y.W. and Dozy, A.M. (1978). Polymorphism of DNA sequence adjacent to human beta-globin structural gene: relationship to sickle mutation. *Proc Natl Acad Sci U S A*, **75**, 5631–5635.

Kanadia, R.N., Johnstone, K.A., Mankodi, A. *et al.* (2003). A muscleblind knockout model for myotonic dystrophy. *Science*, **302**, 1978–1980.

Kanamori, Y., Matsushima, M., Minaguchi, T. *et al.* (1999). Correlation between expression of the matrix metalloproteinase-1 gene in ovarian cancers and an insertion/deletion polymorphism in its promoter region. *Cancer Res*, **59**, 4225–4227.

Kanda, A., Chen, W., Othman, M. *et al.* (2007). A variant of mitochondrial protein LOC387715/ARMS2, not HTRA1, is strongly associated with age-related macular degeneration. *Proc Natl Acad Sci U S A*, **104**, 16227–16232.

Kang, J., Lemaire, H.G., Unterbeck, A. *et al.* (1987). The precursor of Alzheimer's disease amyloid A4 protein resembles a cell-surface receptor. *Nature*, **325**, 733–736.

Kapranov, P., Cawley, S.E., Drenkow, J. *et al.* (2002). Large-scale transcriptional activity in chromosomes 21 and 22. *Science*, **296**, 916–919.

Karban, A.S., Okazaki, T., Panhuysen, C.I. *et al.* (2004). Functional annotation of a novel NFKB1 promoter polymorphism that increases risk for ulcerative colitis. *Hum Mol Genet*, **13**, 35–45.

Karp, C.L., Grupe, A., Schadt, E. *et al.* (2000). Identification of complement factor 5 as a susceptibility locus for experimental allergic asthma. *Nat Immunol*, **1**, 221–226.

Kato, T., Inagaki, H., Yamada, K. *et al.* (2006). Genetic variation affects de novo translocation frequency. *Science*, **311**, 971.

Katsuno, M., Adachi, H., Kume, A. *et al.* (2002). Testosterone reduction prevents phenotypic expression in a transgenic mouse model of spinal and bulbar muscular atrophy. *Neuron*, **35**, 843–854.

Kaushansky, K. (1996). Glossary of molecular biology terminology. *Biologicals* **24**, 157–175.

Kawamura, T., Gulden, F.O., Sugaya, M. *et al.* (2003). R5 HIV productively infects Langerhans cells, and infection levels are regulated by compound CCR5 polymorphisms. *Proc Natl Acad Sci U S A*, **100**, 8401–8406.

Kawasaki, K., Minoshima, S., Nakato, E. et al. (2001). Evolutionary dynamics of the human immunoglobulin kappa locus and the germline repertoire of the Vkappa genes. *Eur J Immunol*, **31**, 1017–1028.

Kawashima, M., Tamiya, G., Oka, A. *et al.* (2006). Genomewide association analysis of human narcolepsy and a new resistance gene. *Am J Hum Genet*, **79**, 252–263.

Kayser, M., Roewer, L., Hedman, M. *et al.* (2000). Characteristics and frequency of germline mutations at microsatellite loci from the human Y chromosome, as revealed by direct observation in father/son pairs. *Am J Hum Genet*, **66**, 1580–1588.

Kazazian, H.H., Jr. (2004). Mobile elements: drivers of genome evolution. *Science*, **303**, 1626–1632.

Kazazian, H.H., Jr., Orkin, S.H., Boehm, C.D., Sexton, J.P. and Antonarakis, S.E. (1983). Beta-thalassemia due to a deletion of the nucleotide which is substituted in the beta S-globin gene. *Am J Hum Genet*, **35**, 1028–1033.

Kazazian, H.H., Jr., Wong, C., Youssoufian, H., Scott, A.F., Phillips, D.G. and Antonarakis, S.E. (1988). Haemophilia A resulting from de novo insertion of L1 sequences represents a novel mechanism for mutation in man. *Nature*, **332**, 164–166.

Keele, B.F., Van Heuverswyn, F., Li, Y. *et al.* (2006). Chimpanzee reservoirs of pandemic and nonpandemic HIV-1. *Science*, **313**, 523–526.

Kehrer-Sawatzki, H. and Cooper, D.N. (2007). Understanding the recent evolution of the human genome: insights from human–chimpanzee genome comparisons. *Hum Mutat*, **28**, 99–130.

Kelleher, A.D., Long, C., Holmes, E.C. *et al.* (2001). Clustered mutations in HIV-1 gag are consistently required for escape from HLA-B27-restricted cytotoxic T lymphocyte responses. *J Exp Med*, **193**, 375–386.

Kennedy, L., Evans, E., Chen, C.M. *et al.* (2003). Dramatic tissue-specific mutation length increases are an early molecular event in Huntington disease pathogenesis. *Hum Mol Genet*, **12**, 3359–3367.

Kennedy, W.R., Alter, M. and Sung, J.H. (1968). Progressive proximal spinal and bulbar muscular atrophy of late onset. A sex-linked recessive trait. *Neurology*, **18**, 671–680.

Kent, W.J., Sugnet, C.W., Furey, T.S. *et al.* (2002). The human genome browser at UCSC. *Genome Res*, **12**, 996–1006.

Kerem, B., Rommens, J.M., Buchanan, J.A. *et al.* (1989). Identification of the cystic fibrosis gene: genetic analysis. *Science*, **245**, 1073–1080.

Keurentjes, J.J., Fu, J., Terpstra, I.R. *et al.* (2007). Regulatory network construction in Arabidopsis by using genome-wide gene expression quantitative trait loci. *Proc Natl Acad Sci U S A*, **104**, 1708–1713.

Khachaturian, A.S., Corcoran, C.D., Mayer, L.S., Zandi, P.P. and Breitner, J.C. (2004). Apolipoprotein E epsilon4 count affects age at onset of Alzheimer disease, but not lifetime susceptibility: the Cache County Study. *Arch Gen Psychiatry*, **61**, 518–524.

Khaja, R., Zhang, J., MacDonald, J.R. *et al.* (2006). Genome assembly comparison identifies structural variants in the human genome. *Nat Genet*, **38**, 1413–1418.

Khakoo, S.I., Thio, C.L., Martin, M.P. *et al.* (2004). HLA and NK cell inhibitory receptor genes in resolving hepatitis C virus infection. *Science*, **305**, 872–874.

Kim, C.Y., Quarsten, H., Bergseng, E., Khosla, C. and Sollid, L.M. (2004). Structural basis for HLA-DQ2-mediated presentation of gluten epitopes in celiac disease. *Proc Natl Acad Sci U S A*, **101**, 4175–4179.

Kim, E., Goren, A. and Ast, G. (2008). Alternative splicing: current perspectives. *Bioessays*, **30**, 38–47.

Kim, M.S. and Polychronakos, C. (2005). Immunogenetics of type 1 diabetes. *Horm Res*, **64**, 180–188.

Kim, Y. and Stephan, W. (2002). Detecting a local signature of genetic hitchhiking along a recombining chromosome. *Genetics*, **160**, 765–777.

Kimura, A., Ohta, Y., Fukumaki, Y. and Takagi, Y. (1984). A fusion gene in man: DNA sequence analysis of the abnormal globin gene of hemoglobin Miyada. *Biochem Biophys Res Commun*, **119**, 968–974.

Kimura, M. (1983). *The Neutral Theory of Molecular Evolution*. Cambridge University Press, Cambridge, UK.

King, K., Bagnall, R., Fisher, S.A. *et al.* (2007). Identification, evolution, and association study of a novel promoter and first exon of the human NOD2 (CARD15) gene. *Genomics*, **90**, 493–501.

Kioussis, D., Vanin, E., deLange, T., Flavell, R.A. and Grosveld, F.G. (1983). Beta-globin gene inactivation by DNA translocation in gamma beta-thalassaemia. *Nature*, **306**, 662–666.

Kirsner, J.B. and Spencer, J.A. (1963). Family occurrences of ulcerative colitis, regional enteritis, and ileocolitis. *Ann Internal Med*, **59**, 133–144.

Kirst, M., Myburg, A.A., De Leon, J.P., Kirst, M.E., Scott, J. and Sederoff, R. (2004). Coordinated genetic regulation of growth and lignin revealed by quantitative trait locus analysis of cDNA microarray data in an interspecific backcross of eucalyptus. *Plant Physiol*, **135**, 2368–2378.

Klauck, S.M. (2006). Genetics of autism spectrum disorder. *Eur J Hum Genet*, **14**, 714–720.

Klaver, C.C., Wolfs, R.C., Assink, J.J., van Duijn, C.M., Hofman, A. and de Jong, P.T. (1998). Genetic risk of age-related maculopathy. Population-based familial aggregation study. *Arch Ophthalmol*, **116**, 1646–1651.

Klein, R.J., Zeiss, C., Chew, E.Y. *et al.* (2005). Complement factor H polymorphism in age-related macular degeneration. *Science*, **308**, 385–389.

Kleinjan, D.A. and van Heyningen, V. (2005). Long-range control of gene expression: emerging mechanisms and disruption in disease. *Am J Hum Genet*, **76**, 8–32.

Klenova, E., Scott, A.C., Roberts, J. *et al.* (2004). YB-1 and CTCF differentially regulate the 5-HTT polymorphic intron 2 enhancer which predisposes to a variety of neurological disorders. *J Neurosci*, **24**, 5966–5973.

Klinefelter, H., Reifenstein, E. and Albright, F. (1942). Syndrome characterized by gynecomastia aspermatogenes without A-Leydigism and increased excretion of follicle stimulating hormone. *J Clin Endocrinol Metab*, **2**, 615–627.

Knight, J.C. (2004). Allele-specific gene expression uncovered. *Trends Genet*, **20**, 113–116.

Knight, J.C. (2005). Regulatory polymorphisms underlying complex disease traits. *J Mol Med*, **83**, 97–109.

Knight, J.C. and Kwiatkowski, D. (1999). Inherited variability of tumor necrosis factor production and susceptibility to infectious disease. *Proc Assoc Am Physicians*, **111**, 290–298.

Knight, J.C., Keating, B.J. and Kwiatkowski, D.P. (2004). Allele-specific repression of lymphotoxin-alpha by activated B cell factor-1. *Nat Genet*, **36**, 394–399.

Knight, J.C., Keating, B.J., Rockett, K.A. and Kwiatkowski, D.P. (2003). In vivo characterization of regulatory polymorphisms by allele-specific quantification of RNA polymerase loading. *Nat Genet*, **33**, 469–475.

Knight, J.C., Udalova, I., Hill, A.V. *et al.* (1999). A polymorphism that affects OCT-1 binding to the TNF promoter region is associated with severe malaria. *Nat Genet*, **22**, 145–150.

Knight, S.J. and Flint, J. (2004). The use of subtelomeric probes to study mental retardation. *Methods Cell Biol*, **75**, 799–831.

Knight, S.J. and Regan, R. (2006). Idiopathic learning disability and genome imbalance. *Cytogenet Genome Res*, **115**, 215–224.

Knoll, J.H., Nicholls, R.D., Magenis, R.E., Graham, J.M., Jr., Lalande, M. and Latt, S.A. (1989). Angelman and Prader–Willi syndromes share a common chromosome 15 deletion but differ in parental origin of the deletion. *Am J Med Genet*, **32**, 285–290.

Knowler, W.C., Williams, R.C., Pettitt, D.J. and Steinberg, A.G. (1988). Gm3;5,13,14 and type 2 diabetes mellitus: an association in American Indians with genetic admixture. *Am J Hum Genet*, **43**, 520–526.

Knudsen, T.B., Kristiansen, T.B., Katzenstein, T.L. and Eugen-Olsen, J. (2001). Adverse effect of the CCR5 promoter -2459A allele on HIV-1 disease progression. *J Med Virol*, **65**, 441–444.

Kobayashi, K.S., Chamaillard, M., Ogura, Y. *et al.* (2005). Nod2-dependent regulation of innate and adaptive immunity in the intestinal tract. *Science*, **307**, 731–734.

Koch, M.C., Grimm, T., Harley, H.G. and Harper, P.S. (1991). Genetic risks for children of women with myotonic dystrophy. *Am J Hum Genet*, **48**, 1084–1091.

Koenig, M., Hoffman, E.P., Bertelson, C.J., Monaco, A.P., Feener, C. and Kunkel, L.M. (1987). Complete cloning of the Duchenne muscular dystrophy (DMD) cDNA and preliminary genomic organization of the DMD gene in normal and affected individuals. *Cell*, **50**, 509–517.

Kominato, Y., Tsuchiya, T., Hata, N., Takizawa, H. and Yamamoto, F. (1997). Transcription of human ABO histo-blood group genes is dependent upon binding of transcription factor CBF/NF-Y to minisatellite sequence. *J Biol Chem*, **272**, 25890–25898.

Kong, A., Gudbjartsson, D.F., Sainz, J. *et al.* (2002). A high-resolution recombination map of the human genome. *Nat Genet*, **31**, 241–247.

Koolen, D.A., Sharp, A.J., Hurst, J.A. *et al.* (2008). Clinical and molecular delineation of the 17q21.31 microdeletion syndrome. *J Med Genet*, **45**, 710–720.

Koolen, D.A., Vissers, L.E., Pfundt, R. *et al.* (2006). A new chromosome 17q21.31 microdeletion syndrome asso-ciated with a common inversion polymorphism. *Nat Genet*, **38**, 999–1001.

Koopmans, R.T., van der Sterren, K.J. and van der Steen, J.T. (2007). The 'natural' endpoint of dementia: death from cachexia or dehydration following palliative care? *Int J Geriatr Psychiatry*, **22**, 350–355.

Kopelman, N.M., Lancet, D. and Yanai, I. (2005). Alternative splicing and gene duplication are inversely correlated evolutionary mechanisms. *Nat Genet*, **37**, 588–589.

Korbel, J.O., Urban, A.E., Affourtit, J.P. *et al.* (2007). Paired-end mapping reveals extensive structural variation in the human genome. *Science*, **318**, 420–426.

Koren, G., Cairns, J., Chitayat, D., Gaedigk, A. and Leeder, S.J. (2006). Pharmacogenetics of morphine poisoning in a breastfed neonate of a codeine-prescribed mother. *Lancet*, **368**, 704.

Koster, T., Rosendaal, F.R., de Ronde, H., Briet, E., Vandenbroucke, J.P. and Bertina, R.M. (1993). Venous thrombosis due to poor anticoagulant response to activated protein C: Leiden Thrombophilia Study. *Lancet*, **342**, 1503–1506.

Kostrikis, L.G., Huang, Y., Moore, J.P. *et al.* (1998). A chemokine receptor CCR2 allele delays HIV-1 disease progression and is associated with a CCR5 promoter mutation. *Nat Med*, **4**, 350–353.

Kouriba, B., Chevillard, C., Bream, J.H. *et al.* (2005). Analysis of the 5q31-q33 locus shows an association between IL13-1055C/T IL-13-591A/G polymorphisms and Schistosoma haematobium infections. *J Immunol*, **174**, 6274–6281.

Kozman, H.M., Keith, T.P., Donis-Keller, H. *et al.* (1995). The CEPH consortium linkage map of human chromosome 16. *Genomics*, **25**, 44–58.

Krawczak, M., Reiss, J. and Cooper, D.N. (1992). The mutational spectrum of single base-pair substitutions in mRNA splice junctions of human genes: causes and consequences. *Hum Genet*, **90**, 41–54.

Kreitman, M. and Di Rienzo, A. (2004). Balancing claims for balancing selection. *Trends Genet*, **20**, 300–304.

Kruglyak, L. (1999). Prospects for whole-genome linkage disequilibrium mapping of common disease genes. *Nat Genet*, **22**, 139–144.

Kruglyak, L. (2008). The road to genome-wide association studies. *Nat Rev Genet*, **9**, 314–318.

Kruglyak, L. and Nickerson, D.A. (2001). Variation is the spice of life. *Nat Genet*, **27**, 234–236.

Kruglyak, S., Durrett, R.T., Schug, M.D. and Aquadro, C.F. (1998). Equilibrium distributions of microsatellite repeat length resulting from a balance between slippage events and point mutations. *Proc Natl Acad Sci U S A*, **95**, 10774–10778.

Kruse, T.A., Bolund, L., Grzeschik, K.H. *et al.* (1988). The human lactase-phlorizin hydrolase gene is located on chromosome 2. *FEBS Lett*, **240**, 123–126.

Krynetski, E.Y., Schuetz, J.D., Galpin, A.J., Pui, C.H., Relling, M.V. and Evans, W.E. (1995). A single point mutation leading to loss of catalytic activity in human thiopurine S-methyltransferase. *Proc Natl Acad Sci U S A*, **92**, 949–953.

Kukull, W.A., Higdon, R., Bowen, J.D. *et al.* (2002). Dementia and Alzheimer disease incidence: a prospective cohort study. *Arch Neurol*, **59**, 1737–1746.

Kulkarni, P.S., Butera, S.T. and Duerr, A.C. (2003). Resistance to HIV-1 infection: lessons learned from studies of highly exposed persistently seronegative (HEPS) individuals. *AIDS Rev*, **5**, 87–103.

Kulozik, A.E., Bellan-Koch, A., Bail, S., Kohne, E. and Kleihauer, E. (1991). Thalassemia intermedia: moderate reduction of beta globin gene transcriptional activity by a novel mutation of the proximal CACCC promoter element. *Blood*, **77**, 2054–2058.

Kumar, S. and Hedges, S.B. (1998). A molecular timescale for vertebrate evolution. *Nature*, **392**, 917–920.

Kun, J.F., Klabunde, J., Lell, B. *et al.* (1999). Association of the ICAM-1 Kilifi mutation with protection against severe malaria in Lambarene, Gabon. *Am J Trop Med Hyg*, **61**, 776–779.

Kunz, S., Rojek, J.M., Kanagawa, M. *et al.* (2005). Posttranslational modification of alpha-dystroglycan, the cellular receptor for arenaviruses, by the glycosyltransferase LARGE is critical for virus binding. *J Virol*, **79**, 14282–14296.

Kuokkanen, M., Enattah, N.S., Oksanen, A., Savilahti, E., Orpana, A. and Jarvela, I. (2003). Transcriptional regulation of the lactase-phlorizin hydrolase gene by polymorphisms associated with adult-type hypolactasia. *Gut*, **52**, 647–652.

Kuokkanen, M., Kokkonen, J., Enattah, N.S. *et al.* (2006). Mutations in the translated region of the lactase gene (LCT) underlie congenital lactase deficiency. *Am J Hum Genet*, **78**, 339–344.

Kurahashi, H. and Emanuel, B.S. (2001). Long AT-rich palindromes and the constitutional t(11;22) breakpoint. *Hum Mol Genet*, **10**, 2605–2617.

Kurahashi, H., Inagaki, H., Hosoba, E. *et al.* (2007). Molecular cloning of a translocation breakpoint hotspot in 22q11. *Genome Res*, **17**, 461–469.

Kurahashi, H., Inagaki, H., Yamada, K. *et al.* (2004). Cruciform DNA structure underlies the etiology for palindrome-mediated human chromosomal translocations. *J Biol Chem*, **279**, 35377–35383.

Kwan, T., Benovoy, D., Dias, C. *et al.* (2007). Heritability of alternative splicing in the human genome. *Genome Res*, **17**, 1210–1218.

Kwan, T., Benovoy, D., Dias, C. *et al.* (2008). Genome-wide analysis of transcript isoform variation in humans. *Nat Genet*, **40**, 225–231.

Kwiatkowski, D. (2005). How malaria has affected the human genome and what human genetics can teach us about malaria. *Am J Hum Genet*, **77**, 171–192.

Kwiatkowski, D., Hill, A.V.S., Sambou, I. *et al.* (1990). TNF concentration in fatal cerebral, non-fatal cerebral, and uncomplicated *Plasmodium falciparum* malaria. *Lancet*, **336**, 1201–1204.

Kwok, C.J., Martin, A.C., Au, S.W. and Lam, V.M. (2002). G6PDdb, an integrated database of glucose-6-phosphate dehydrogenase (G6PD) mutations. *Hum Mutat*, **19**, 217–224.

La Spada, A.R., Wilson, E.M., Lubahn, D.B., Harding, A.E. and Fischbeck, K.H. (1991). Androgen receptor gene mutations in X-linked spinal and bulbar muscular atrophy. *Nature*, **352**, 77–79.

Labie, D., Schroeder, W.A. and Huisman, T.H. (1966). The amino acid sequence of the delta-beta chains of hemoglobin Lepore Augusta = Lepore Washington. *Biochim Biophys Acta*, **127**, 428–437.

Lafreniere, R.G., Rochefort, D.L., Chretien, N. *et al.* (1997). Unstable insertion in the 5′ flanking region of the cystatin B gene is the most common mutation in progressive myoclonus epilepsy type 1, EPM1. *Nat Genet*, **15**, 298–302.

Lai, Y. and Sun, F. (2003). The relationship between microsatellite slippage mutation rate and the number of repeat units. *Mol Biol Evol*, **20**, 2123–2131.

Lakich, D., Kazazian, H.H., Jr., Antonarakis, S.E. and Gitschier, J. (1993). Inversions disrupting the factor VIII gene are a common cause of severe haemophilia A. *Nat Genet*, **5**, 236–241.

Lala, S., Ogura, Y., Osborne, C. *et al.* (2003). Crohn's disease and the NOD2 gene: a role for paneth cells. *Gastroenterology*, **125**, 47–57.

Lalioti, M.D., Scott, H.S. and Antonarakis, S.E. (1999). Altered spacing of promoter elements due to the dodecamer repeat expansion contributes to reduced expression of the cystatin B gene in EPM1. *Hum Mol Genet*, **8**, 1791–1798.

Lalioti, M.D., Scott, H.S., Buresi, C. *et al.* (1997). Dodecamer repeat expansion in cystatin B gene in progressive myoclonus epilepsy. *Nature*, **386**, 847–851.

Lam, Y.C., Bowman, A.B., Jafar-Nejad, P. *et al.* (2006). ATAXIN-1 interacts with the repressor Capicua in its native complex to cause SCA1 neuropathology. *Cell*, **127**, 1335–1347.

Lama, J. and Planelles, V. (2007). Host factors influencing susceptibility to HIV infection and AIDS progression. *Retrovirology*, **4**, 52.

Lamason, R.L., Mohideen, M.A., Mest, J.R. *et al.* (2005). SLC24A5, a putative cation exchanger, affects pigmentation in zebrafish and humans. *Science*, **310**, 1782–1786.

Lamb, J., Harris, P.C., Wilkie, A.O., Wood, W.G., Dauwerse, J.G. and Higgs, D.R. (1993). De novo truncation of chromosome 16p and healing with (TTAGGG)n in the alpha-thalassemia/mental retardation syndrome (ATR-16). *Am J Hum Genet*, **52**, 668–676.

Lamb, J., Wilkie, A.O., Harris, P.C. *et al.* (1989). Detection of breakpoints in submicroscopic chromosomal translocation, illustrating an important mechanism for genetic disease. *Lancet*, **2**, 819–824.

Lambert, A.P., Gillespie, K.M., Thomson, G. *et al.* (2004). Absolute risk of childhood-onset type 1 diabetes defined by human leukocyte antigen class II genotype: a population-based study in the United Kingdom. *J Clin Endocrinol Metab*, **89**, 4037–4043.

Lamy, M. and Maroteaux, P. (1960). [Diastrophic nanism.] *Presse Med*, **68**, 1977–1980.

Landegent, J.E., Jansen in de Wal, N., van Ommen, G.J. *et al.* (1985). Chromosomal localization of a unique gene by non-autoradiographic in situ hybridization. *Nature*, **317**, 175–177.

Lander, E.S. (1996). The new genomics: global views of biology. *Science*, **274**, 536–539.

Lander, E.S. and Botstein, D. (1986). Mapping complex genetic traits in humans: new methods using a complete RFLP linkage map. *Cold Spring Harb Symp Quant Biol*, **51**, 49–62.

Lander, E.S., Linton, L.M., Birren, B. *et al.* (2001). Initial sequencing and analysis of the human genome. *Nature*, **409**, 860–921.

Landsteiner, K. and Wiener, A. (1940). An agglutinable factor in human blood recognised by immune sera for Rhesus blood. *Proc Soc Exp Biol N Y*, **43**, 223–224.

Larsen, C.E. and Alper, C.A. (2004). The genetics of HLA-associated disease. *Curr Opin Immunol*, **16**, 660–667.

Laval, S.H., Timms, A., Edwards, S. *et al.* (2001). Whole-genome screening in ankylosing spondylitis: evidence of non-MHC genetic-susceptibility loci. *Am J Hum Genet*, **68**, 918–926.

Law, A.J., Kleinman, J.E., Weinberger, D.R. and Weickert, C.S. (2007). Disease-associated intronic variants in the ErbB4 gene are related to altered ErbB4 splice-variant expression in the brain in schizophrenia. *Hum Mol Genet*, **16**, 129–141.

Law, S.K., Dodds, A.W. and Porter, R.R. (1984). A comparison of the properties of two classes, C4A and C4B, of the human complement component C4. *EMBO J*, **3**, 1819–1823.

Lawn, R.M., Efstratiadis, A., O'Connell, C. and Maniatis, T. (1980). The nucleotide sequence of the human beta-globin gene. *Cell*, **21**, 647–651.

Le Marechal, C., Masson, E., Chen, J.M. *et al.* (2006). Hereditary pancreatitis caused by triplication of the trypsinogen locus. *Nat Genet*, **38**, 1372–1374.

Lechler, R. and Warrens, A., eds. (2000). *HLA in Health and Disease*. Academic Press, San Diego, CA.

Ledbetter, D.H., Riccardi, V.M., Airhart, S.D., Strobel, R.J., Keenan, B.S. and Crawford, J.D. (1981). Deletions of chromosome 15 as a cause of the Prader–Willi syndrome. *N Engl J Med*, **304**, 325–329.

Lee, J.A., Carvalho, C.M. and Lupski, J.R. (2007). A DNA replication mechanism for generating nonrecurrent rearrangements associated with genomic disorders. *Cell*, **131**, 1235–1247.

Lee, J.A., Madrid, R.E., Sperle, K. *et al.* (2006). Spastic paraplegia type 2 associated with axonal neuropathy and apparent PLP1 position effect. *Ann Neurol*, **59**, 398–403.

Lehesjoki, A.E., Koskiniemi, M., Norio, R. *et al.* (1993). Localization of the EPM1 gene for progressive myoclonus epilepsy on chromosome 21: linkage disequilibrium allows high resolution mapping. *Hum Mol Genet*, **2**, 1229–1234.

Lehesjoki, A.E., Koskiniemi, M., Sistonen, P. *et al.* (1991). Localization of a gene for progressive myoclonus epilepsy to chromosome 21q22. *Proc Natl Acad Sci U S A*, **88**, 3696–3699.

Lejeune, J., Gautier, M. and Turpin, R. (1959). [Study of somatic chromosomes from 9 mongoloid children.] *C R Hebd Seances Acad Sci*, **248**, 1721–1722.

Lejeune, J., Lafourcade, J., Berger, R. *et al.* (1963). Trois ca de deletion partielle du bras court d'un chromosome 5. *C R Acad Sci (Paris)*, **257**, 3098.

Lemmers, R.J., Wohlgemuth, M., van der Gaag, K.J. *et al.* (2007). Specific sequence variations within the 4q35 region are associated with facioscapulohumeral muscular dystrophy. *Am J Hum Genet*, **81**, 884–894.

Lengauer, C., Kinzler, K.W. and Vogelstein, B. (1998). Genetic instabilities in human cancers. *Nature*, **396**, 643–649.

Lesko, L.J. (2008). The critical path of warfarin dosing: finding an optimal dosing strategy using pharmacogenetics. *Clin Pharmacol Ther*, **84**, 301–303.

Letvin, N.L., Linch, D.C., Beardsley, G.P., McIntyre, K.W. and Nathan, D.G. (1984). Augmentation of fetal-hemoglobin production in anemic monkeys by hydroxyurea. *N Engl J Med*, **310**, 869–873.

Levine, P. and Stetson, R. (1939). An unusual case of intra-group agglutination. *JAMA*, **113**, 126–127.

Levinson, G. and Gutman, G.A. (1987). High frequencies of short frameshifts in poly-CA/TG tandem repeats borne by bacteriophage M13 in Escherichia coli K-12. *Nucleic Acids Res*, **15**, 5323–5338.

Levy, E., Carman, M.D., Fernandez-Madrid, I.J. *et al.* (1990). Mutation of the Alzheimer's disease amyloid gene in hereditary cerebral hemorrhage, Dutch type. *Science*, **248**, 1124–1126.

Levy, J.E., Montross, L.K., Cohen, D.E., Fleming, M.D. and Andrews, N.C. (1999). The C282Y mutation causing hereditary hemochromatosis does not produce a null allele. *Blood*, **94**, 9–11.

Levy, S., Sutton, G., Ng, P.C. *et al.* (2007). The diploid genome sequence of an individual human. *PLoS Biol*, **5**, e254.

Levy-Lahad, E., Wijsman, E.M., Nemens, E. *et al.* (1995). A familial Alzheimer's disease locus on chromosome 1. *Science*, **269**, 970–973.

Lewander, A., Butchi, A.K., Gao, J. *et al.* (2007). Polymorphism in the promoter region of the NFKB1 gene increases the risk of sporadic colorectal cancer in Swedish but not in Chinese populations. *Scand J Gastroenterol*, **42**, 1332–1338.

Lewin, B. (2004). *Genes VIII*. Pearson Prentice Hall, Upper Saddle River, NJ.

Lewinsky, R.H., Jensen, T.G., Moller, J., Stensballe, A., Olsen, J. and Troelsen, J.T. (2005). T-13910 DNA variant associated with lactase persistence interacts with Oct-1 and stimulates lactase promoter activity in vitro. *Hum Mol Genet*, **14**, 3945–3953.

Lewontin, R.C. and Kojima, K. (1960). The evolutionary dynamics of complex polymorphisms. *Evolution*, **14**, 458–472.

Li, D.Y., Toland, A.E., Boak, B.B. *et al.* (1997). Elastin point mutations cause an obstructive vascular disease, supravalvular aortic stenosis. *Hum Mol Genet*, **6**, 1021–1028.

Li, M., Atmaca-Sonmez, P., Othman, M. *et al.* (2006a). CFH haplotypes without the Y402H coding variant show strong association with susceptibility to age-related macular degeneration. *Nat Genet*, **38**, 1049–1054.

Li, W.H. and Sadler, L.A. (1991). Low nucleotide diversity in man. *Genetics*, **129**, 513–523.

Li, W.H., Yi, S. and Makova, K. (2002). Male-driven evolution. *Curr Opin Genet Dev*, **12**, 650–656.

Li, Y., Alvarez, O.A., Gutteling, E.W. *et al.* (2006b). Mapping determinants of gene expression plasticity by genetical genomics in C. elegans. *PLoS Genet*, **2**, e222.

Li, Y., Li, X., Stremlau, M., Lee, M. and Sodroski, J. (2006c). Removal of arginine 332 allows human TRIM5alpha to bind human immunodeficiency virus capsids and to restrict infection. *J Virol*, **80**, 6738–6744.

Li, Y., Wollnik, B., Pabst, S. *et al.* (2006d). BTNL2 gene variant and sarcoidosis. *Thorax*, **61**, 273–274.

Libioulle, C., Louis, E., Hansoul, S. *et al.* (2007). Novel Crohn disease locus identified by genome-wide association maps to a gene desert on 5p13.1 and modulates expression of PTGER4. *PLoS Genet*, **3**, e58.

Liehr, T. and Weise, A. (2007). Frequency of small supernumerary marker chromosomes in prenatal, newborn, developmentally retarded and infertility diagnostics. *Int J Mol Med*, **19**, 719–731.

Lim, J., Crespo-Barreto, J., Jafar-Nejad, P. *et al.* (2008). Opposing effects of polyglutamine expansion on native protein complexes contribute to SCA1. *Nature*, **452**, 713–718.

Lin, L., Faraco, J., Li, R. *et al.* (1999). The sleep disorder canine narcolepsy is caused by a mutation in the hypocretin (orexin) receptor 2 gene. *Cell*, **98**, 365–376.

Linardopoulou, E.V., Williams, E.M., Fan, Y., Friedman, C., Young, J.M. and Trask, B.J. (2005). Human subtelomeres are hot spots of interchromosomal recombination and segmental duplication. *Nature*, **437**, 94–100.

Lindesmith, L., Moe, C., Marionneau, S. *et al.* (2003). Human susceptibility and resistance to Norwalk virus infection. *Nat Med*, **9**, 548–553.

Lindqvist, P.G., Svensson, P.J., Marsaal, K., Grennert, L., Luterkort, M. and Dahlback, B. (1999). Activated protein C resistance (FV:Q506) and pregnancy. *Thromb Haemost*, **81**, 532–537.

Lindsay, E.A., Vitelli, F., Su, H. *et al.* (2001). Tbx1 haploinsufficieny in the DiGeorge syndrome region causes aortic arch defects in mice. *Nature*, **410**, 97–101.

Lipoldova, M. and Demant, P. (2006). Genetic susceptibility to infectious disease: lessons from mouse models of leishmaniasis. *Nat Rev Genet*, **7**, 294–305.

Liquori, C.L., Ricker, K., Moseley, M.L. *et al.* (2001). Myotonic dystrophy type 2 caused by a CCTG expansion in intron 1 of ZNF9. *Science*, **293**, 864–867.

Litt, M. and Luty, J.A. (1989). A hypervariable microsatellite revealed by in vitro amplification of a dinucleotide repeat within the cardiac muscle actin gene. *Am J Hum Genet*, **44**, 397–401.

Liu, D., Bischerour, J., Siddique, A., Buisine, N., Bigot, Y. and Chalmers, R. (2007). The human SETMAR protein preserves most of the activities of the ancestral Hsmar1 transposase. *Mol Cell Biol*, **27**, 1125–1132.

Liu, H., Chao, D., Nakayama, E.E. *et al.* (1999). Polymorphism in RANTES chemokine promoter affects HIV-1 disease progression. *Proc Natl Acad Sci U S A*, **96**, 4581–4585.

Liu, J.S., Molchanova, T.P., Gu, L.H. *et al.* (1992). Hb Graz or alpha 2 beta 2(2)(NA2)His–>Leu; a new beta chain variant observed in four families from southern Austria. *Hemoglobin*, **16**, 493–501.

Liu, R., Paxton, W.A., Choe, S. *et al.* (1996). Homozygous defect in HIV-1 coreceptor accounts for resistance of some multiply-exposed individuals to HIV-1 infection. *Cell*, **86**, 367–377.

Lo, H.S., Wang, Z., Hu, Y. *et al.* (2003). Allelic variation in gene expression is common in the human genome. *Genome Res*, **13**, 1855–1862.

Lo, Y.M., Hjelm, N.M., Fidler, C. *et al.* (1998). Prenatal diagnosis of fetal RhD status by molecular analysis of maternal plasma. *N Engl J Med*, **339**, 1734–1738.

Locke, D.P., Segraves, R., Nicholls, R.D. *et al.* (2004). BAC microarray analysis of 15q11-q13 rearrangements and the impact of segmental duplications. *J Med Genet*, **41**, 175–182.

Locke, D.P., Sharp, A.J., McCarroll, S.A. *et al.* (2006). Linkage disequilibrium and heritability of copy-number polymorphisms within duplicated regions of the human genome. *Am J Hum Genet*, **79**, 275–290.

Loftus, B.J., Kim, U.J., Sneddon, V.P. *et al.* (1999). Genome duplications and other features in 12 Mb of DNA sequence from human chromosome 16p and 16q. *Genomics*, **60**, 295–308.

Loftus, E.V., Jr. (2004). Clinical epidemiology of inflammatory bowel disease: incidence, prevalence, and environmental influences. *Gastroenterology*, **126**, 1504–1517.

Loftus, R.T., MacHugh, D.E., Bradley, D.G., Sharp, P.M. and Cunningham, P. (1994). Evidence for two independent domestications of cattle. *Proc Natl Acad Sci U S A*, **91**, 2757–2761.

Loftus, S.K., Dixon, J., Koprivnikar, K., Dixon, M.J. and Wasmuth, J.J. (1996). Transcriptional map of the Treacher Collins candidate gene region. *Genome Res*, **6**, 26–34.

Loftus, S.K., Edwards, S.J., Scherpbier-Heddema, T., Buetow, K.H., Wasmuth, J.J. and Dixon, M.J. (1993). A combined genetic and radiation hybrid map surrounding the Treacher Collins syndrome locus on chromosome 5q. *Hum Mol Genet*, **2**, 1785–1792.

Lohmueller, K.E., Pearce, C.L., Pike, M., Lander, E.S. and Hirschhorn, J.N. (2003). Meta-analysis of genetic association studies supports a contribution of common variants to susceptibility to common disease. *Nat Genet*, **33**, 177–182.

Loirat, F., Hazout, S. and Lucotte, G. (1997). G542X as a probable Phoenician cystic fibrosis mutation. *Hum Biol*, **69**, 419–425.

Loos, R.J. and Bouchard, C. (2008). FTO: the first gene contributing to common forms of human obesity. *Obes Rev*, **9**, 246–250.

Lopez-Bermejo, A., Petry, C.J., Diaz, M. *et al.* (2008). The association between the FTO gene and fat mass in humans develops by the postnatal age of two weeks. *J Clin Endocrinol Metab*, **93**, 1501–1505.

Lopez-Bigas, N., Audit, B., Ouzounis, C., Parra, G. and Guigo, R. (2005). Are splicing mutations the most frequent cause of hereditary disease? *FEBS Lett*, **579**, 1900–1903.

Luan, D.D., Korman, M.H., Jakubczak, J.L. and Eickbush, T.H. (1993). Reverse transcription of R2Bm RNA is primed by a nick at the chromosomal target site: a mechanism for non-LTR retrotransposition. *Cell*, **72**, 595–605.

Lubs, H.A. (1969). A marker X chromosome. *Am J Hum Genet*, **21**, 231–244.

Lucassen, A.M., Julier, C., Beressi, J.P. *et al.* (1993). Susceptibility to insulin dependent diabetes mellitus maps to a 4.1 kb segment of DNA spanning the insulin gene and associated VNTR. *Nat Genet*, **4**, 305–310.

Lucassen, A.M., Screaton, G.R., Julier, C., Elliott, T.J., Lathrop, M. and Bell, J.I. (1995). Regulation of insulin gene expression by the IDDM associated, insulin locus haplotype. *Hum Mol Genet*, **4**, 501–506.

Lucotte, G. and Dieterlen, F. (2003). A European allele map of the C282Y mutation of hemochromatosis: Celtic versus Viking origin of the mutation? *Blood Cells Mol Dis*, **31**, 262–267.

Luedi, P.P., Dietrich, F.S., Weidman, J.R., Bosko, J.M., Jirtle, R.L. and Hartemink, A.J. (2007). Computational and experimental identification of novel human imprinted genes. *Genome Res*, **17**, 1723–1730.

Luoni, G., Verra, F., Arca, B. *et al.* (2001). Antimalarial antibody levels and IL4 polymorphism in the Fulani of West Africa. *Genes Immun*, **2**, 411–414.

Lupski, J.R. (1998). Genomic disorders: structural features of the genome can lead to DNA rearrangements and human disease traits. *Trends Genet*, **14**, 417–422.

Lupski, J.R. (2006). Genome structural variation and sporadic disease traits. *Nat Genet*, **38**, 974–976.

Lupski, J.R. and Stankiewicz, P. (2005). Genomic disorders: molecular mechanisms for rearrangements and conveyed phenotypes. *PLoS Genet*, **1**, e49.

Lupski, J.R., de Oca-Luna, R.M., Slaugenhaupt, S. *et al.* (1991). DNA duplication associated with Charcot–Marie–Tooth disease type 1A. *Cell*, **66**, 219–232.

Lupski, J.R., Wise, C.A., Kuwano, A. *et al.* (1992). Gene dosage is a mechanism for Charcot–Marie–Tooth disease type 1A. *Nat Genet*, **1**, 29–33.

Luzzatto, L. (1979). Genetics of red cells and susceptibility to malaria. *Blood*, **54**, 961–976.

Luzzatto, L., Nwachuku-Jarrett, E.S. and Reddy, S. (1970). Increased sickling of parasitised erythrocytes as mechanism of resistance against malaria in the sickle-cell trait. *Lancet*, **1**, 319–321.

Luzzatto, L., Usanga, F.A. and Reddy, S. (1969). Glucose-6-phosphate dehydrogenase deficient red cells: resistance to infection by malarial parasites. *Science*, **164**, 839–842.

Lynch, M. and Conery, J.S. (2000). The evolutionary fate and consequences of duplicate genes. *Science*, **290**, 1151–1155.

Lynch, M. and Force, A. (2000). The probability of duplicate gene preservation by subfunctionalization. *Genetics*, **154**, 459–473.

Machuca-Tzili, L., Brook, D. and Hilton-Jones, D. (2005). Clinical and molecular aspects of the myotonic dystrophies: a review. *Muscle Nerve*, **32**, 1–18.

Mack, M., Bender, K. and Schneider, P.M. (2004). Detection of retroviral antisense transcripts and promoter activity of the HERV-K(C4) insertion in the MHC class III region. *Immunogenetics*, **56**, 321–332.

Mackay, I.R. and Morris, P.J. (1972). Association of autoimmune active chronic hepatitis with HL-A1,8. *Lancet*, **2**, 793–795.

MacKenzie, A. and Quinn, J. (1999). A serotonin transporter gene intron 2 polymorphic region, correlated with affective disorders, has allele-dependent differential enhancer-like properties in the mouse embryo. *Proc Natl Acad Sci U S A*, **96**, 15251–15255.

Mackinnon, M.J., Gunawardena, D.M., Rajakaruna, J., Weerasingha, S., Mendis, K.N. and Carter, R. (2000). Quantifying genetic and nongenetic contributions to malarial infection in a Sri Lankan population. *Proc Natl Acad Sci U S A*, **97**, 12661–12666.

Mackinnon, M.J., Mwangi, T.W., Snow, R.W., Marsh, K. and Williams, T.N. (2005). Heritability of malaria in Africa. *PLoS Med*, **2**, e340.

Mackintosh, C.L., Beeson, J.G. and Marsh, K. (2004). Clinical features and pathogenesis of severe malaria. *Trends Parasitol*, **20**, 597–603.

Madani, N. and Kabat, D. (1998). An endogenous inhibitor of human immunodeficiency virus in human lymphocytes is overcome by the viral Vif protein. *J Virol*, **72**, 10251–10255.

Madore, A.M., Tremblay, K., Hudson, T.J. and Laprise, C. (2008). Replication of an association between 17q21 SNPs and asthma in a French-Canadian familial collection. *Hum Genet*, **123**, 93–95.

Maeda, S., Hsu, L.C., Liu, H. *et al.* (2005). Nod2 mutation in Crohn's disease potentiates NF-kappaB activity and IL-1beta processing. *Science*, **307**, 734–738.

Maere, S., De Bodt, S., Raes, J. *et al.* (2005). Modeling gene and genome duplications in eukaryotes. *Proc Natl Acad Sci U S A*, **102**, 5454–5459.

Mahadevan, M., Tsilfidis, C., Sabourin, L. *et al.* (1992). Myotonic dystrophy mutation: an unstable CTG repeat in the 3′ untranslated region of the gene. *Science*, **255**, 1253–1255.

Mahley, R.W. (1988). Apolipoprotein E: cholesterol transport protein with expanding role in cell biology. *Science*, **240**, 622–630.

Mahtani, M.M. and Willard, H.F. (1990). Pulsed-field gel analysis of alpha-satellite DNA at the human X chromosome centromere: high-frequency polymorphisms and array size estimate. *Genomics*, **7**, 607–613.

Mailman, M.D., Feolo, M., Jin, Y. *et al.* (2007). The NCBI dbGaP database of genotypes and phenotypes. *Nat Genet*, **39**, 1181–1186.

Malats, N. and F. Calafell (2003a). Advanced glossary on genetic epidemiology. *J Epidemiol Community Health* **57**, 562–564.

Malats, N. and F. Calafell (2003b). Basic glossary on genetic epidemiology. *J Epidemiol Community Health* **57**, 480–482.

Malcolm, S., Clayton-Smith, J., Nichols, M. *et al.* (1991). Uniparental paternal disomy in Angelman's syndrome. *Lancet*, **337**, 694–697.

Mallal, S., Nolan, D., Witt, C. *et al.* (2002). Association between presence of HLA-B*5701, HLA-DR7, and HLA-DQ3 and hypersensitivity to HIV-1 reverse-transcriptase inhibitor abacavir. *Lancet*, **359**, 727–732.

Mallal, S., Phillips, E., Carosi, G. *et al.* (2008). HLA-B*5701 screening for hypersensitivity to abacavir. *N Engl J Med*, **358**, 568–579.

Maller, J., George, S., Purcell, S. *et al.* (2006). Common variation in three genes, including a noncoding variant in CFH, strongly influences risk of age-related macular degeneration. *Nat Genet*, **38**, 1055–1059.

Mallya, M., Campbell, R.D. and Aguado, B. (2006). Characterization of the five novel Ly-6 superfamily members encoded in the MHC, and detection of cells expressing their potential ligands. *Protein Sci*, **15**, 2244–2256.

Malter, H.E., Iber, J.C., Willemsen, R. *et al.* (1997). Characterization of the full fragile X syndrome mutation in fetal gametes. *Nat Genet*, **15**, 165–169.

Mamtani, M., Rovin, B., Brey, R. *et al.* (2007). CCL3L1 gene-containing segmental duplications and polymorphisms in CCR5 affect risk of systemic lupus erythematosus. *Ann Rheum Dis*, **67**, 1076–1083.

Mangano, A., Gonzalez, E., Dhanda, R. *et al.* (2001). Concordance between the CC chemokine receptor 5 genetic determinants that alter risks of transmission and disease progression in children exposed perinatally to human immunodeficiency virus. *J Infect Dis*, **183**, 1574–1585.

Mangiarini, L., Sathasivam, K., Seller, M. *et al.* (1996). Exon 1 of the HD gene with an expanded CAG repeat is sufficient to cause a progressive neurological phenotype in transgenic mice. *Cell*, **87**, 493–506.

Maniatis, T. and Reed, R. (2002). An extensive network of coupling among gene expression machines. *Nature*, **416**, 499–506.

Mankodi, A., Logigian, E., Callahan, L. *et al.* (2000). Myotonic dystrophy in transgenic mice expressing an expanded CUG repeat. *Science*, **289**, 1769–1773.

Mankodi, A., Takahashi, M.P., Jiang, H. *et al.* (2002). Expanded CUG repeats trigger aberrant splicing of ClC-1 chloride channel pre-mRNA and hyperexcitability of skeletal muscle in myotonic dystrophy. *Mol Cell*, **10**, 35–44.

Mannon, P.J., Fuss, I.J., Mayer, L. *et al.* (2004). Anti-interleukin-12 antibody for active Crohn's disease. *N Engl J Med*, **351**, 2069–2079.

Manolio, T.A., Brooks, L.D. and Collins, F.S. (2008). A HapMap harvest of insights into the genetics of common disease. *J Clin Invest*, **118**, 1590–1605.

Mantei, N., Villa, M., Enzler, T. *et al.* (1988). Complete primary structure of human and rabbit lactase-phlorizin hydrolase: implications for biosynthesis, membrane anchoring and evolution of the enzyme. *EMBO J*, **7**, 2705–2713.

Maraganore, D.M., de Andrade, M., Elbaz, A. *et al.* (2006). Collaborative analysis of alpha-synuclein gene promoter variability and Parkinson disease. *JAMA*, **296**, 661–670.

Marchini, J., Howie, B., Myers, S., McVean, G. and Donnelly, P. (2007). A new multipoint method for genome-wide association studies by imputation of genotypes. *Nat Genet*, **39**, 906–913.

Marchiori, A., Mosena, L., Prins, M.H. and Prandoni, P. (2007). The risk of recurrent venous thromboembolism among heterozygous carriers of factor V Leiden or prothrombin G20210A mutation. A systematic review of prospective studies. *Haematologica*, **92**, 1107–1114.

Mardis, E.R. (2006). Anticipating the 1,000 dollar genome. *Genome Biol*, **7**, 112.

Maret, S. and Tafti, M. (2005). Genetics of narcolepsy and other major sleep disorders. *Swiss Med Wkly*, **135**, 662–665.

Margulies, M., Egholm, M., Altman, W.E. *et al.* (2005). Genome sequencing in microfabricated high-density picolitre reactors. *Nature*, **437**, 376–380.

Marin, M., Rose, K.M., Kozak, S.L. and Kabat, D. (2003). HIV-1 Vif protein binds the editing enzyme APOBEC3G and induces its degradation. *Nat Med*, **9**, 1398–1403.

Marmor, M., Sheppard, H.W., Donnell, D. *et al.* (2001). Homozygous and heterozygous CCR5-Delta32 genotypes are associated with resistance to HIV infection. *J Acquir Immune Defic Syndr*, **27**, 472–481.

Marques-Bonet, T., Kidd, J.M., Ventura, M. et al. (2009). A burst of segmental duplications in the genome of the African great ape ancestor. *Nature*, **457**, 877-881.

Marquet, S., Abel, L., Hillaire, D. *et al.* (1996). Genetic localization of a locus controlling the intensity of infection by Schistosoma mansoni on chromosome 5q31-q33. *Nat Genet*, **14**, 181–184.

Marsh, D.G., Neely, J.D., Breazeale, D.R. *et al.* (1994). Linkage analysis of IL4 and other chromosome 5q31.1 markers and total serum immunoglobulin E concentrations. *Science*, **264**, 1152–1156.

Marsh, K., Otoo, L., Hayes, R.J., Carson, D.C. and Greenwood, B.M. (1989). Antibodies to blood stage antigens of Plasmodium falciparum in rural Gambians and their relation to protection against infection. *Trans R Soc Trop Med Hyg*, **83**, 293–303.

Marshall, J.H. (2002). On the changing meanings of "mutation". *Hum Mutat*, **19**, 76–78.

Martens, U.M., Zijlmans, J.M., Poon, S.S. *et al.* (1998). Short telomeres on human chromosome 17p. *Nat Genet*, **18**, 76–80.

Martin, A.M., Freitas, E.M., Witt, C.S. and Christiansen, F.T. (2000). The genomic organization and evolution of the natural killer immunoglobulin-like receptor (KIR) gene cluster. *Immunogenetics*, **51**, 268–280.

Martin, A.M., Nolan, D., Gaudieri, S. *et al.* (2004a). Predisposition to abacavir hypersensitivity conferred by HLA-B*5701 and a haplotypic Hsp70-Hom variant. *Proc Natl Acad Sci U S A*, **101**, 4180–4185.

Martin, J., Han, C., Gordon, L.A. *et al.* (2004b). The sequence and analysis of duplication-rich human chromosome 16. *Nature*, **432**, 988–994.

Martin, M.P., Bashirova, A., Traherne, J., Trowsdale, J. and Carrington, M. (2003). Cutting edge: expansion of the KIR locus by unequal crossing over. *J Immunol*, **171**, 2192–2195.

Martin, M.P., Dean, M., Smith, M.W. *et al.* (1998). Genetic acceleration of AIDS progression by a promoter variant of CCR5. *Science*, **282**, 1907–1911.

Martin, M.P., Qi, Y., Gao, X. *et al.* (2007). Innate partnership of HLA-B and KIR3DL1 subtypes against HIV-1. *Nat Genet*, **39**, 733–740.

Masters, C.L., Simms, G., Weinman, N.A., Multhaup, G., McDonald, B.L. and Beyreuther, K. (1985). Amyloid plaque core protein in Alzheimer disease and Down syndrome. *Proc Natl Acad Sci U S A*, **82**, 4245–4249.

Matassi, G., Cherif-Zahar, B., Pesole, G., Raynal, V. and Cartron, J.P. (1999). The members of the RH gene family (RH50 and RH30) followed different evolutionary pathways. *J Mol Evol*, **48**, 151–159.

Mathew, C.G. (2008). New links to the pathogenesis of Crohn disease provided by genome-wide association scans. *Nat Rev Genet*, **9**, 9–14.

Mathias, S.L., Scott, A.F., Kazazian, H.H., Jr., Boeke, J.D. and Gabriel, A. (1991). Reverse transcriptase encoded by a human transposable element. *Science*, **254**, 1808–1810.

Matilla, A., Roberson, E.D., Banfi, S. *et al.* (1998). Mice lacking ataxin-1 display learning deficits and decreased hippocampal paired-pulse facilitation. *J Neurosci*, **18**, 5508–5516.

Matsunami, N., Smith, B., Ballard, L. *et al.* (1992). Peripheral myelin protein-22 gene maps in the duplication in chromosome 17p11.2 associated with Charcot–Marie–Tooth 1A. *Nat Genet*, **1**, 176–179.

Matsuura, T., Yamagata, T., Burgess, D.L. *et al.* (2000). Large expansion of the ATTCT pentanucleotide repeat in spinocerebellar ataxia type 10. *Nat Genet*, **26**, 191–194.

Mauff, G., Luther, B., Schneider, P.M. *et al.* (1998). Reference typing report for complement component C4. *Exp Clin Immunogenet*, **15**, 249–260.

Maxam, A.M. and Gilbert, W. (1977). A new method for sequencing DNA. *Proc Natl Acad Sci U S A*, **74**, 560–564.

Maynard, N.D., Chen, J., Stuart, R.K., Fan, J.B. and Ren, B. (2008). Genome-wide mapping of allele-specific protein-DNA interactions in human cells. *Nat Methods*, **5**, 307–309.

Maynard-Smith, J. and Haigh, J. (1974). The hitch-hiking effect of a favourable gene. *Genet Res*, **23**, 23–25.

McCarroll, S.A., Hadnott, T.N., Perry, G.H. *et al.* (2006). Common deletion polymorphisms in the human genome. *Nat Genet*, **38**, 86–92.

McCarroll, S.A., Huett, A., Kuballa, P. *et al.* (2008). Deletion polymorphism upstream of IRGM associated with altered IRGM expression and Crohn's disease. *Nat Genet*, **40**, 1107–1112.

McCarthy, M.I., Abecasis, G.R., Cardon, L.R. *et al.* (2008). Genome-wide association studies for complex traits: consensus, uncertainty and challenges. *Nat Rev Genet*, **9**, 356–369.

McDermid, H.E. and Morrow, B.E. (2002). Genomic disorders on 22q11. *Am J Hum Genet*, **70**, 1077–1088.

McDermott, D.H., Beecroft, M.J., Kleeberger, C.A. *et al.* (2000). Chemokine RANTES promoter polymorphism affects risk of both HIV infection and disease progression in the Multicenter AIDS Cohort Study. *AIDS*, **14**, 2671–2678.

McDermott, D.H., Zimmerman, P.A., Guignard, F., Kleeberger, C.A., Leitman, S.F. and Murphy, P.M. (1998). CCR5 promoter polymorphism and HIV-1 disease progression. Multicenter AIDS Cohort Study (MACS). *Lancet*, **352**, 866–870.

McDougall, I., Brown, F.H. and Fleagle, J.G. (2005). Stratigraphic placement and age of modern humans from Kibish, Ethiopia. *Nature*, **433**, 733–736.

McGuire, W., Hill, A.V., Allsopp, C.E., Greenwood, B.M. and Kwiatkowski, D. (1994). Variation in the TNF-alpha promoter region associated with susceptibility to cerebral malaria. *Nature*, **371**, 508–510.

McGuire, W., Knight, J.C., Hill, A.V., Allsopp, C.E., Greenwood, B.M. and Kwiatkowski, D. (1999). Severe malarial anemia and cerebral malaria are associated with different tumor necrosis factor promoter alleles. *J Infect Dis*, **179**, 287–290.

McKinney, C., Merriman, M.E., Chapman, P.T. *et al.* (2007). Evidence for an influence of chemokine ligand 3-like 1 (CCL3L1) gene copy number on susceptibility to rheumatoid arthritis. *Ann Rheum Dis*, **67**, 409–413.

McKusick, V.A. and Antonarakis, S.E. (1998). *Mendelian Inheritance in Man: a Catalog of Human Genes and Genetic Disorders*. Johns Hopkins University Press, Baltimore, MD/London.

McLellan, R.A., Oscarson, M., Alexandrie, A.K. *et al.* (1997a). Characterization of a human glutathione S-transferase mu cluster containing a duplicated GSTM1 gene that causes ultrarapid enzyme activity. *Mol Pharmacol*, **52**, 958–965.

McLellan, R.A., Oscarson, M., Seidegard, J., Evans, D.A. and Ingelman-Sundberg, M. (1997b). Frequent occurrence of CYP2D6 gene duplication in Saudi Arabians. *Pharmacogenetics*, **7**, 187–191.

McNicholl, J.M., Smith, D.K., Qari, S.H. and Hodge, T. (1997). Host genes and HIV: the role of the chemokine receptor gene CCR5 and its allele. *Emerg Infect Dis*, **3**, 261–271.

McPherson, R., Pertsemlidis, A., Kavaslar, N. *et al.* (2007). A common allele on chromosome 9 associated with coronary heart disease. *Science*, **316**, 1488–1491.

McVean, G.A., Myers, S.R., Hunt, S., Deloukas, P., Bentley, D.R. and Donnelly, P. (2004). The fine-scale structure of recombination rate variation in the human genome. *Science*, **304**, 581–584.

Mead, S. (2006). Prion disease genetics. *Eur J Hum Genet*, **14**, 273–281.

Mefford, H.C. and Trask, B.J. (2002). The complex structure and dynamic evolution of human subtelomeres. *Nat Rev Genet*, **3**, 91–102.

Meissen, G.J., Myers, R.H., Mastromauro, C.A. *et al.* (1988). Predictive testing for Huntington's disease with use of a linked DNA marker. *N Engl J Med*, **318**, 535–542.

Meloni, R., Albanese, V., Ravassard, P., Treilhou, F. and Mallet, J. (1998). A tetranucleotide polymorphic microsatellite, located in the first intron of the tyrosine hydroxylase gene, acts as a transcription regulatory element in vitro. *Hum Mol Genet*, **7**, 423–428.

Meloni, R., Laurent, C., Campion, D. *et al.* (1995a). A rare allele of a microsatellite located in the tyrosine hydroxylase gene found in schizophrenic patients. *C R Acad Sci III*, **318**, 803–809.

Meloni, R., Leboyer, M., Bellivier, F. *et al.* (1995b). Association of manic-depressive illness with tyrosine hydroxylase microsatellite marker. *Lancet*, **345**, 932.

Melzer, D., Hogarth, S., Liddell, K., Ling, T., Sanderson, S. and Zimmern, R.L. (2008a). Genetic tests for common diseases: new insights, old concerns. *Br Med J*, **336**, 590–593.

Melzer, D., Perry, J.R., Hernandez, D. *et al.* (2008b). A genome-wide association study identifies protein quantitative trait loci (pQTLs). *PLoS Genet*, **4**, e1000072.

Menten, P., Struyf, S., Schutyser, E. *et al.* (1999). The LD78beta isoform of MIP-1alpha is the most potent CCR5 agonist and HIV-1-inhibiting chemokine. *J Clin Invest*, **104**, R1–5.

Merbs, S.L. and Nathans, J. (1992). Absorption spectra of human cone pigments. *Nature*, **356**, 433–435.

Merryweather-Clarke, A.T., Pointon, J.J., Shearman, J.D. and Robson, K.J. (1997). Global prevalence of putative haemochromatosis mutations. *J Med Genet*, **34**, 275–278.

Merscher, S., Funke, B., Epstein, J.A. *et al.* (2001). TBX1 is responsible for cardiovascular defects in velo-cardio-facial/DiGeorge syndrome. *Cell*, **104**, 619–629.

Messer, P.W. and Arndt, P.F. (2007). The majority of recent short DNA insertions in the human genome are tandem duplications. *Mol Biol Evol*, **24**, 1190–1197.

Messier, W., Li, S.H. and Stewart, C.B. (1996). The birth of microsatellites. *Nature*, **381**, 483.

Messing, J., Gronenborn, B., Muller-Hill, B. and Hans Hopschneider, P. (1977). Filamentous coliphage M13 as a cloning vehicle: insertion of a HindII fragment of the lac regulatory region in M13 replicative form in vitro. *Proc Natl Acad Sci U S A*, **74**, 3642–3646.

Metzgar, D., Bytof, J. and Wills, C. (2000). Selection against frameshift mutations limits microsatellite expansion in coding DNA. *Genome Res*, **10**, 72–80.

Meyer, C.G., May, J., Luty, A.J., Lell, B. and Kremsner, P.G. (2002). TNFalpha-308A associated with shorter intervals of Plasmodium falciparum reinfections. *Tissue Antigens*, **59**, 287–292.

Meyer, M.R., Tschanz, J.T., Norton, M.C. *et al.* (1998). APOE genotype predicts when – not whether – one is predisposed to develop Alzheimer disease. *Nat Genet*, **19**, 321–322.

Micale, L., Fusco, C., Augello, B. *et al.* (2008). Williams–Beuren syndrome TRIM50 encodes an E3 ubiquitin ligase. *Eur J Hum Genet*, **16**, 1038–1049.

Miceli-Richard, C., Zouali, H., Said-Nahal, R. *et al.* (2004). Significant linkage to spondyloarthropathy on 9q31–34. *Hum Mol Genet*, **13**, 1641–1648.

Miesfeld, R., Krystal, M. and Arnheim, N. (1981). A member of a new repeated sequence family which is conserved throughout eucaryotic evolution is found between the human delta and beta globin genes. *Nucleic Acids Res*, **9**, 5931–5947.

Mignot, E., Hayduk, R., Black, J., Grumet, F.C. and Guilleminault, C. (1997). HLA DQB1*0602 is associated with cataplexy in 509 narcoleptic patients. *Sleep*, **20**, 1012–1020.

Mignot, E., Lin, L., Rogers, W. *et al.* (2001). Complex HLA-DR and -DQ interactions confer risk of narcolepsy-cataplexy in three ethnic groups. *Am J Hum Genet*, **68**, 686–699.

Mignot, E., Lin, X., Arrigoni, J. *et al.* (1994). DQB1*0602 and DQA1*0102 (DQ1) are better markers than DR2 for narcolepsy in Caucasian and black Americans. *Sleep*, **17**, S60–67.

Migueles, S.A., Sabbaghian, M.S., Shupert, W.L. *et al.* (2000). HLA B*5701 is highly associated with restriction of virus replication in a subgroup of HIV-infected long term nonprogressors. *Proc Natl Acad Sci U S A*, **97**, 2709–2714.

Miki, Y., Katagiri, T., Kasumi, F., Yoshimoto, T. and Nakamura, Y. (1996). Mutation analysis in the BRCA2 gene in primary breast cancers. *Nat Genet*, **13**, 245–247.

Miki, Y., Nishisho, I., Horii, A. *et al.* (1992). Disruption of the APC gene by a retrotransposal insertion of L1 sequence in a colon cancer. *Cancer Res*, **52**, 643–645.

Milet, J., Dehais, V., Bourgain, C. *et al.* (2007). Common variants in the BMP2, BMP4, and HJV genes of the hepcidin regulation pathway modulate HFE hemochromatosis penetrance. *Am J Hum Genet*, **81**, 799–807.

Miller, E.N., Fadl, M., Mohamed, H.S. *et al.* (2007). Y chromosome lineage- and village-specific genes on chromosomes 1p22 and 6q27 control visceral leishmaniasis in Sudan. *PLoS Genet*, **3**, e71.

Miller, J.W., Urbinati, C.R., Teng-Umnuay, P. *et al.* (2000). Recruitment of human muscleblind proteins to (CUG)(n) expansions associated with myotonic dystrophy. *EMBO J*, **19**, 4439–4448.

Miller, L.H., Mason, S.J., Clyde, D.F. and McGinniss, M.H. (1976). The resistance factor to Plasmodium vivax in blacks. The Duffy-blood-group genotype, FyFy. *N Engl J Med*, **295**, 302–304.

Miller, W.L. and Morel, Y. (1989). The molecular genetics of 21-hydroxylase deficiency. *Annu Rev Genet*, **23**, 371–393.

Mills, R.E., Luttig, C.T., Larkins, C.E. *et al.* (2006). An initial map of insertion and deletion (INDEL) variation in the human genome. *Genome Res*, **16**, 1182–1190.

Milner, C.M. and Campbell, R.D. (1990). Structure and expression of the three MHC-linked HSP70 genes. *Immunogenetics*, **32**, 242–251.

Mirza, M.M., Fisher, S.A., Onnie, C. *et al.* (2005). No association of the NFKB1 promoter polymorphism with ulcerative colitis in a British case control cohort. *Gut*, **54**, 1205–1206.

Mitchell, B.D., Kammerer, C.M., Blangero, J. *et al.* (1996). Genetic and environmental contributions to cardiovascular risk factors in Mexican Americans. The San Antonio Family Heart Study. *Circulation*, **94**, 2159–2170.

Mizunuma, M., Fujimori, S., Ogino, H., Ueno, T., Inoue, H. and Kamatani, N. (2001). A recurrent large Alu-mediated deletion in the hypoxanthine phosphoribosyltransferase (HPRT1) gene associated with Lesch–Nyhan syndrome. *Hum Mutat*, **18**, 435–443.

Mockenhaupt, F.P., Ehrhardt, S., Gellert, S. *et al.* (2004). Alpha(+)-thalassemia protects African children from severe malaria. *Blood*, **104**, 2003–2006.

Modi, W.S., Scott, K., Goedert, J.J. *et al.* (2005). Haplotype analysis of the SDF-1 (CXCL12) gene in a longitudinal HIV-1/AIDS cohort study. *Genes Immun*, **6**, 691–698.

Modiano, D., Luoni, G., Petrarca, V. *et al.* (2001a). HLA class I in three West African ethnic groups: genetic distances from sub-Saharan and Caucasoid populations. *Tissue Antigens*, **57**, 128–137.

Modiano, D., Luoni, G., Sirima, B.S. *et al.* (2001b). The lower susceptibility to Plasmodium falciparum malaria of Fulani of Burkina Faso (west Africa) is associated with low frequencies of classic malaria-resistance genes. *Trans R Soc Trop Med Hyg*, **95**, 149–152.

Modiano, D., Luoni, G., Sirima, B.S. *et al.* (2001c). Haemoglobin C protects against clinical Plasmodium falciparum malaria. *Nature*, **414**, 305–308.

Modiano, D., Petrarca, V., Sirima, B.S. *et al.* (1996). Different response to Plasmodium falciparum malaria in west African sympatric ethnic groups. *Proc Natl Acad Sci U S A*, **93**, 13206–13211.

Modiano, G., Morpurgo, G., Terrenato, L. *et al.* (1991). Protection against malaria morbidity: near-fixation of the alpha-thalassemia gene in a Nepalese population. *Am J Hum Genet*, **48**, 390–397.

Modrek, B. and Lee, C. (2002). A genomic view of alternative splicing. *Nat Genet*, **30**, 13–19.

Modrek, B., Resch, A., Grasso, C. and Lee, C. (2001). Genome-wide detection of alternative splicing in expressed sequences of human genes. *Nucleic Acids Res*, **29**, 2850–2859.

Moffatt, M.F., Kabesch, M., Liang, L. *et al.* (2007). Genetic variants regulating ORMDL3 expression contribute to the risk of childhood asthma. *Nature*, **448**, 470–473.

Mohamed, H.S., Ibrahim, M.E., Miller, E.N. *et al.* (2004). SLC11A1 (formerly NRAMP1) and susceptibility to visceral leishmaniasis in the Sudan. *Eur J Hum Genet*, **12**, 66–74.

Moi, P., Cash, F.E., Liebhaber, S.A., Cao, A. and Pirastu, M. (1987). An initiation codon mutation (AUG----GUG) of the human alpha 1-globin gene. Structural characterization and evidence for a mild thalassemic phenotype. *J Clin Invest*, **80**, 1416–1421.

Monks, S.A., Leonardson, A., Zhu, H. *et al.* (2004). Genetic inheritance of gene expression in human cell lines. *Am J Hum Genet*, **75**, 1094–1105.

Monsen, U., Bernell, O., Johansson, C. and Hellers, G. (1991). Prevalence of inflammatory bowel disease among relatives of patients with Crohn's disease. *Scand J Gastroenterol*, **26**, 302–306.

Montserrat, V., Marti, M. and Lopez de Castro, J.A. (2003). Allospecific T cell epitope sharing reveals extensive conservation of the antigenic features of peptide ligands among HLA-B27 subtypes differentially associated with spondyloarthritis. *J Immunol*, **170**, 5778–5785.

Moore, C.B., John, M., James, I.R., Christiansen, F.T., Witt, C.S. and Mallal, S.A. (2002). Evidence of HIV-1 adaptation to HLA-restricted immune responses at a population level. *Science*, **296**, 1439–1443.

Moore, S., Woodrow, C.F. and McClelland, D.B. (1982). Isolation of membrane components associated with human red cell antigens Rh(D), (c), (E) and Fy. *Nature*, **295**, 529–531.

Morgan, G.T. (1995). Identification in the human genome of mobile elements spread by DNA-mediated transposition. *J Mol Biol*, **254**, 1–5.

Morison, I.M., Paton, C.J. and Cleverley, S.D. (2001). The imprinted gene and parent-of-origin effect database. *Nucleic Acids Res*, **29**, 275–276.

Morley, M., Molony, C.M., Weber, T.M. *et al.* (2004). Genetic analysis of genome-wide variation in human gene expression. *Nature*, **430**, 743–747.

Morral, N., Bertranpetit, J., Estivill, X. *et al.* (1994). The origin of the major cystic fibrosis mutation (delta F508) in European populations. *Nat Genet*, **7**, 169–175.

Morshauser, R.C., Hu, W., Wang, H., Pang, Y., Flynn, G.C. and Zuiderweg, E.R. (1999). High-resolution solution structure of the 18 kDa substrate-binding domain of the mammalian chaperone protein Hsc70. *J Mol Biol*, **289**, 1387–1403.

Moulds, J.M., Warner, N.B. and Arnett, F.C. (1991). Quantitative and antigenic differences in complement component C4 between American blacks and whites. *Complement Inflamm*, **8**, 281–287.

Mounsey, A., Bauer, P. and Hope, I.A. (2002). Evidence suggesting that a fifth of annotated Caenorhabditis elegans genes may be pseudogenes. *Genome Res*, **12**, 770–775.

Moyzis, R.K., Buckingham, J.M., Cram, L.S. *et al.* (1988). A highly conserved repetitive DNA sequence, (TTAGGG)n, present at the telomeres of human chromosomes. *Proc Natl Acad Sci U S A*, **85**, 6622–6626.

Mugnier, B., Balandraud, N., Darque, A., Roudier, C., Roudier, J. and Reviron, D. (2003). Polymorphism at position -308 of the tumor necrosis factor alpha gene influences outcome of infliximab therapy in rheumatoid arthritis. *Arthritis Rheum*, **48**, 1849–1852.

Mulcare, C.A., Weale, M.E., Jones, A.L. *et al.* (2004). The T allele of a single-nucleotide polymorphism 13.9 kb upstream of the lactase gene (LCT) (C-13.9kbT) does not predict or cause the lactase-persistence phenotype in Africans. *Am J Hum Genet*, **74**, 1102–1110.

Mulherin, S.A., O'Brien, T.R., Ioannidis, J.P. *et al.* (2003). Effects of CCR5-Delta32 and CCR2-64I alleles on HIV-1 disease progression: the protection varies with duration of infection. *Aids*, **17**, 377–387.

Muller-Myhsok, B., Stelma, F.F., Guisse-Sow, F. *et al.* (1997). Further evidence suggesting the presence of a locus, on human chromosome 5q31-q33, influencing the intensity of infection with Schistosoma mansoni. *Am J Hum Genet*, **61**, 452–454.

Mullikin, J.C., Hunt, S.E., Cole, C.G. *et al.* (2000). An SNP map of human chromosome 22. *Nature*, **407**, 516–520.

Mullis, K.B. and Faloona, F.A. (1987). Specific synthesis of DNA in vitro via a polymerase-catalyzed chain reaction. *Methods Enzymol*, **155**, 335–350.

Mummidi, S., Ahuja, S.S., Gonzalez, E. *et al.* (1998). Genealogy of the CCR5 locus and chemokine system gene variants associated with altered rates of HIV-1 disease progression. *Nat Med*, **4**, 786–793.

Mummidi, S., Ahuja, S.S., McDaniel, B.L. and Ahuja, S.K. (1997). The human CC chemokine receptor 5 (CCR5) gene. Multiple transcripts with 5′-end heterogeneity, dual promoter usage, and evidence for polymorphisms within the regulatory regions and noncoding exons. *J Biol Chem*, **272**, 30662–30671.

Mummidi, S., Bamshad, M., Ahuja, S.S. *et al.* (2000). Evolution of human and non-human primate CC chemokine receptor 5 gene and mRNA. Potential roles for

haplotype and mRNA diversity, differential haplotype-specific transcriptional activity, and altered transcription factor binding to polymorphic nucleotides in the pathogenesis of HIV-1 and simian immunodeficiency virus. *J Biol Chem*, **275**, 18946–18961.

Murakami, T., Garcia, C.A., Reiter, L.T. and Lupski, J.R. (1996). Charcot–Marie–Tooth disease and related inherited neuropathies. *Medicine (Baltimore)*, **75**, 233–250.

Myers, J.S., Vincent, B.J., Udall, H. *et al.* (2002). A comprehensive analysis of recently integrated human Ta L1 elements. *Am J Hum Genet*, **71**, 312–326.

Myers, S., Bottolo, L., Freeman, C., McVean, G. and Donnelly, P. (2005). A fine-scale map of recombination rates and hotspots across the human genome. *Science*, **310**, 321–324.

Myles, S., Bouzekri, N., Haverfield, E., Cherkaoui, M., Dugoujon, J.M. and Ward, R. (2005). Genetic evidence in support of a shared Eurasian–North African dairying origin. *Hum Genet*, **117**, 34–42.

Nadeau, J.H. and Lee, C. (2006). Genetics: copies count. *Nature*, **439**, 798–799.

Nadir, E., Margalit, H., Gallily, T. and Ben-Sasson, S.A. (1996). Microsatellite spreading in the human genome: evolutionary mechanisms and structural implications. *Proc Natl Acad Sci U S A*, **93**, 6470–6475.

Nakamura, Y., Koyama, K. and Matsushima, M. (1998). VNTR (variable number of tandem repeat) sequences as transcriptional, translational, or functional regulators. *J Hum Genet*, **43**, 149–152.

Nakamura, Y., Leppert, M., O'Connell, P. *et al.* (1987). Variable number of tandem repeat (VNTR) markers for human gene mapping. *Science*, **235**, 1616–1622.

Nakayama, E.E., Tanaka, Y., Nagai, Y., Iwamoto, A. and Shioda, T. (2004). A CCR2-V64I polymorphism affects stability of CCR2A isoform. *Aids*, **18**, 729–738.

Narita, N., Nishio, H., Kitoh, Y. *et al.* (1993). Insertion of a 5′ truncated L1 element into the 3′ end of exon 44 of the dystrophin gene resulted in skipping of the exon during splicing in a case of Duchenne muscular dystrophy. *J Clin Invest*, **91**, 1862–1867.

Naslund, K., Saetre, P., von Salome, J., Bergstrom, T.F., Jareborg, N. and Jazin, E. (2005). Genome-wide prediction of human VNTRs. *Genomics*, **85**, 24–35.

Nathans, J., Davenport, C.M., Maumenee, I.H. *et al.* (1989). Molecular genetics of human blue cone monochromacy. *Science*, **245**, 831–838.

Nathans, J., Piantanida, T.P., Eddy, R.L., Shows, T.B. and Hogness, D.S. (1986). Molecular genetics of inherited variation in human color vision. *Science*, **232**, 203–210.

Naylor, L.H. and Clark, E.M. (1990). d(TG)n.d(CA)n sequences upstream of the rat prolactin gene form Z-DNA and inhibit gene transcription. *Nucleic Acids Res*, **18**, 1595–1601.

Neel, J.V. (1949). The inheritance of sickle cell anemia. *Science*, **110**, 64–66.

Nei, M. and Hughes, A. (1992). Balanced polymorphism and evolution by the birth-and-death process in the MHC loci. In: *11th Histocompatibility Workshop and Conference*. ed. K. Tsuji, M. Aizawa and T. Sasazuki. Oxford University Press, Oxford, UK: 27–38.

Nei, M. and Rooney, A.P. (2005). Concerted and birth-and-death evolution of multigene families. *Annu Rev Genet*, **39**, 121–152.

Nei, M. and Roychoudhury, A.K. (1993). Evolutionary relationships of human populations on a global scale. *Mol Biol Evol*, **10**, 927–943.

Nejentsev, S., Howson, J.M., Walker, N.M. *et al.* (2007). Localization of type 1 diabetes susceptibility to the MHC class I genes HLA-B and HLA-A. *Nature*, **450**, 887–892.

Nembaware, V., Wolfe, K.H., Bettoni, F., Kelso, J. and Seoighe, C. (2004). Allele-specific transcript isoforms in human. *FEBS Lett*, **577**, 233–238.

Nerup, J., Platz, P., Andersen, O.O. *et al.* (1974). HL-A antigens and diabetes mellitus. *Lancet*, **2**, 864–866.

Neumann, B., Kubicka, P. and Barlow, D.P. (1995). Characteristics of imprinted genes. *Nat Genet*, **9**, 12–13.

Neurath, M.F. (2007). IL-23: a master regulator in Crohn disease. *Nat Med*, **13**, 26–28.

Neurath, M.F., Fuss, I., Kelsall, B.L., Stuber, E. and Strober, W. (1995). Antibodies to interleukin 12 abrogate established experimental colitis in mice. *J Exp Med*, **182**, 1281–1290.

Nevin, N.C., Hughes, A.E., Calwell, M. and Lim, J.H. (1986). Duchenne muscular dystrophy in a female with a translocation involving Xp21. *J Med Genet*, **23**, 171–173.

Newman, R.M. and Johnson, W.E. (2007). A brief history of TRIM5alpha. *AIDS Rev*, **9**, 114–125.

Newman, T.L., Tuzun, E., Morrison, V.A. *et al.* (2005). A genome-wide survey of structural variation between human and chimpanzee. *Genome Res*, **15**, 1344–1356.

Newton, J., Brown, M.A., Milicic, A. *et al.* (2003). The effect of HLA-DR on susceptibility to rheumatoid arthritis is influenced by the associated lymphotoxin alpha-tumor necrosis factor haplotype. *Arthritis Rheum*, **48**, 90–96.

Newton, J.L., Harney, S.M., Timms, A.E. *et al.* (2004a). Dissection of class III major histocompatibility complex haplotypes associated with rheumatoid arthritis. *Arthritis Rheum*, **50**, 2122–2129.

Newton, J.L., Harney, S.M., Wordsworth, B.P. and Brown, M.A. (2004b). A review of the MHC genetics of rheumatoid arthritis. *Genes Immun*, **5**, 151–157.

Nguyen, T., Liu, X.K., Zhang, Y. and Dong, C. (2006). BTNL2, a butyrophilin-like molecule that functions to inhibit T cell activation. *J Immunol*, **176**, 7354–7360.

Nibbs, R.J., Yang, J., Landau, N.R., Mao, J.H. and Graham, G.J. (1999). LD78beta, a non-allelic variant of human MIP-1alpha (LD78alpha), has enhanced receptor interactions and potent HIV suppressive activity. *J Biol Chem*, **274**, 17478–17483.

Nicholls, R.D., Fischel-Ghodsian, N. and Higgs, D.R. (1987). Recombination at the human alpha-globin gene cluster: sequence features and topological constraints. *Cell*, **49**, 369–378.

Nickel, R.G., Casolaro, V., Wahn, U. *et al.* (2000). Atopic dermatitis is associated with a functional mutation in the promoter of the C-C chemokine RANTES. *J Immunol*, **164**, 1612–1616.

Niebuhr, E. (1978). The cri du chat syndrome: epidemiology, cytogenetics, and clinical features. *Hum Genet*, **44**, 227–275.

Nielsen, R., Bustamante, C., Clark, A.G. *et al.* (2005a). A scan for positively selected genes in the genomes of humans and chimpanzees. *PLoS Biol*, **3**, e170.

Nielsen, R., Williamson, S., Kim, Y., Hubisz, M.J., Clark, A.G. and Bustamante, C. (2005b). Genomic scans for selective sweeps using SNP data. *Genome Res*, **15**, 1566–1575.

NIH/CEPH (1992). A comprehensive genetic linkage map of the human genome. NIH/CEPH Collaborative Mapping Group. *Science*, **258**, 67–86.

Niimura, Y. and Nei, M. (2005). Comparative evolutionary analysis of olfactory receptor gene clusters between humans and mice. *Gene*, **346**, 13–21.

Niksic, M., Romano, M., Buratti, E., Pagani, F. and Baralle, F.E. (1999). Functional analysis of cis-acting elements regulating the alternative splicing of human CFTR exon 9. *Hum Mol Genet*, **8**, 2339–2349.

Nirenberg, M.W. (1963). The genetic code. II. *Sci Am*, **208**, 80–94.

Nishida, Y., Fukuda, T., Yamamoto, I. and Azuma, J. (2000). CYP2D6 genotypes in a Japanese population: low frequencies of CYP2D6 gene duplication but high frequency of CYP2D6*10. *Pharmacogenetics*, **10**, 567–570.

Nishino, S., Ripley, B., Overeem, S., Lammers, G.J. and Mignot, E. (2000). Hypocretin (orexin) deficiency in human narcolepsy. *Lancet*, **355**, 39–40.

Noble, J.A., Valdes, A.M., Cook, M., Klitz, W., Thomson, G. and Erlich, H.A. (1996). The role of HLA class II genes in insulin-dependent diabetes mellitus: molecular analysis of 180 Caucasian, multiplex families. *Am J Hum Genet*, **59**, 1134–1148.

Nolin, S.L., Brown, W.T., Glicksman, A. *et al.* (2003). Expansion of the fragile X CGG repeat in females with premutation or intermediate alleles. *Am J Hum Genet*, **72**, 454–464.

Norman, P.J., Abi-Rached, L., Gendzekhadze, K. *et al.* (2007). Unusual selection on the KIR3DL1/S1 natural killer cell receptor in Africans. *Nat Genet*, **39**, 1092–1099.

Norton, H.L., Kittles, R.A., Parra, E. *et al.* (2007). Genetic evidence for the convergent evolution of light skin in Europeans and East Asians. *Mol Biol Evol*, **24**, 710–722.

Nowell, P. and Hungerford, D. (1960). A minute chromosome in human chronic granulocytic leukaemia. *Science*, **1332**, 1497–1501.

Oberlin, E., Amara, A., Bachelerie, F. *et al.* (1996). The CXC chemokine SDF-1 is the ligand for LESTR/fusin and prevents infection by T-cell-line-adapted HIV-1. *Nature*, **382**, 833–835.

O'Brien, B.A., Archer, N.S., Simpson, A.M., Torpy, F.R. and Nassif, N.T. (2008). Association of SLC11A1 promoter polymorphisms with the incidence of autoimmune and inflammatory diseases: a meta-analysis. *J Autoimmun*, **31**, 42–51.

O'Brien, S.J. and Nelson, G.W. (2004). Human genes that limit AIDS. *Nat Genet*, **36**, 565–574.

O'Brien, T., Hardin, S., Greenleaf, A. and Lis, J.T. (1994). Phosphorylation of RNA polymerase II C-terminal domain and transcriptional elongation. *Nature*, **370**, 75–77.

Ogata, T., Matsuo, N. and Nishimura, G. (2001). SHOX haploinsufficiency and overdosage: impact of gonadal function status. *J Med Genet*, **38**, 1–6.

Ogura, Y., Bonen, D.K., Inohara, N. *et al.* (2001a). A frameshift mutation in NOD2 associated with susceptibility to Crohn's disease. *Nature*, **411**, 603–606.

Ogura, Y., Inohara, N., Benito, A., Chen, F.F., Yamaoka, S. and Nunez, G. (2001b). Nod2, a Nod1/Apaf-1 family member that is restricted to monocytes and activates NF-kappaB. *J Biol Chem*, **276**, 4812–4818.

Ohene-Frempong, K., Weiner, S.J., Sleeper, L.A. *et al.* (1998). Cerebrovascular accidents in sickle cell disease: rates and risk factors. *Blood*, **91**, 288–294.

Ohmen, J.D., Yang, H.Y., Yamamoto, K.K. *et al.* (1996). Susceptibility locus for inflammatory bowel disease on chromosome 16 has a role in Crohn's disease, but not in ulcerative colitis. *Hum Mol Genet*, **5**, 1679–1683.

Ohno, S. (1970). *Evolution by Gene Duplication*. Springer-Verlag, Berlin.

Ohno, S. (1972). So much "junk" DNA in our genome. *Brookhaven Symp Biol*, **23**, 366–370.

Ohshima, K., Montermini, L., Wells, R.D. and Pandolfo, M. (1998). Inhibitory effects of expanded GAA.TTC triplet repeats from intron I of the Friedreich ataxia gene on transcription and replication in vivo. *J Biol Chem*, **273**, 14588–14595.

Ohta, T. (2000). Evolution of gene families. *Gene*, **259**, 45–52.

Okada, N. and Ohshima, K. (1993). A model for the mechanism of initial generation of short interspersed elements (SINEs). *J Mol Evol*, **37**, 167–170.

Okada, T., Ohzeki, J., Nakano, M. *et al.* (2007). CENP-B controls centromere formation depending on the chromatin context. *Cell*, **131**, 1287–1300.

Okamoto, K., Makino, S., Yoshikawa, Y. *et al.* (2003). Identification of I kappa BL as the second major histocompatibility complex-linked susceptibility locus for rheumatoid arthritis. *Am J Hum Genet*, **72**, 303–312.

Old, J.M. (2006). Globin genes: polymorphic variants and mutations. In: *Encyclopedia of Life Sciences*. John Wiley & Sons Ltd, Chichester, UK: 1–8.

Old, J.M. (2007). Screening and genetic diagnosis of haemoglobinopathies. *Scand J Clin Lab Invest*, **67**, 71–86.

Oldridge, M., Zackai, E.H., McDonald-McGinn, D.M. *et al.* (1999). De novo alu-element insertions in FGFR2 identify a distinct pathological basis for Apert syndrome. *Am J Hum Genet*, **64**, 446–461.

Olds, L.C. and Sibley, E. (2003). Lactase persistence DNA variant enhances lactase promoter activity in vitro: functional role as a cis regulatory element. *Hum Mol Genet*, **12**, 2333–2340.

Oleksiak, M.F., Churchill, G.A. and Crawford, D.L. (2002). Variation in gene expression within and among natural populations. *Nat Genet*, **32**, 261–266.

Oliver, J., Gomez-Garcia, M., Paco, L. *et al.* (2005). A functional polymorphism of the NFKB1 promoter is not associated with ulcerative colitis in a Spanish population. *Inflamm Bowel Dis*, **11**, 576–579.

Oliver, S.G., van der Aart, Q.J., Agostoni-Carbone, M.L. *et al.* (1992). The complete DNA sequence of yeast chromosome III. *Nature*, **357**, 38–46.

Oosumi, T., Belknap, W.R. and Garlick, B. (1995). Mariner transposons in humans. *Nature*, **378**, 672.

Orholm, M., Binder, V., Sorensen, T.I., Rasmussen, L.P. and Kyvik, K.O. (2000). Concordance of inflammatory bowel disease among Danish twins. Results of a nationwide study. *Scand J Gastroenterol*, **35**, 1075–1081.

Orholm, M., Munkholm, P., Langholz, E., Nielsen, O.H., Sorensen, T.I. and Binder, V. (1991). Familial occurrence of inflammatory bowel disease. *N Engl J Med*, **324**, 84–88.

Orkin, S.H., Antonarakis, S.E. and Kazazian, H.H., Jr. (1984). Base substitution at position -88 in a beta-thalassemic globin gene. Further evidence for the role of distal promoter element ACACCC. *J Biol Chem*, **259**, 8679–8681.

Orkin, S.H., Goff, S.C. and Hechtman, R.L. (1981). Mutation in an intervening sequence splice junction in man. *Proc Natl Acad Sci U S A*, **78**, 5041–5045.

Orkin, S.H., Kazazian, H.H., Jr., Antonarakis, S.E. *et al.* (1982a). Linkage of beta-thalassaemia mutations and beta-globin gene polymorphisms with DNA polymorphisms in human beta-globin gene cluster. *Nature*, **296**, 627–631.

Orkin, S.H., Kazazian, H.H., Jr., Antonarakis, S.E., Ostrer, H., Goff, S.C. and Sexton, J.P. (1982b). Abnormal RNA processing due to the exon mutation of beta E-globin gene. *Nature*, **300**, 768–769.

Orkin, S.H., Old, J.M., Weatherall, D.J. and Nathan, D.G. (1979). Partial deletion of beta-globin gene DNA in certain patients with beta β-thalassemia. *Proc Natl Acad Sci U S A*, **76**, 2400–2404.

Orozco, G., Eerligh, P., Sanchez, E. *et al.* (2005). Analysis of a functional BTNL2 polymorphism in type 1 diabetes, rheumatoid arthritis, and systemic lupus erythematosus. *Hum Immunol*, **66**, 1235–1241.

Orr, H.T. and Zoghbi, H.Y. (2007). Trinucleotide repeat disorders. *Annu Rev Neurosci*, **30**, 575–621.

Orr, H.T., Chung, M.Y., Banfi, S. *et al.* (1993). Expansion of an unstable trinucleotide CAG repeat in spinocerebellar ataxia type 1. *Nat Genet*, **4**, 221–226.

Osborne, L.R. and Mervis, C.B. (2007). Rearrangements of the Williams–Beuren syndrome locus: molecular basis and implications for speech and language development. *Expert Rev Mol Med*, **9**, 1–16.

Osborne, L.R., Li, M., Pober, B. *et al.* (2001). A 1.5 million-base pair inversion polymorphism in families with Williams–Beuren syndrome. *Nat Genet*, **29**, 321–325.

Ostertag, E.M. and Kazazian, H.H., Jr. (2001). Biology of mammalian L1 retrotransposons. *Annu Rev Genet*, **35**, 501–538.

Ostertag, W. and Smith, E.W. (1969). Hemoglobin-Lepore-Baltimore, a third type of a delta, beta crossover (delta 50, beta 86). *Eur J Biochem*, **10**, 371–376.

Ota, T. and Nei, M. (1994). Divergent evolution and evolution by the birth-and-death process in the immunoglobulin VH gene family. *Mol Biol Evol*, **11**, 469–482.

Othman, M., Notley, C., Lavender, F.L. *et al.* (2005). Identification and functional characterization of a novel 27-bp deletion in the macroglycopeptide-coding region of the GPIBA gene resulting in platelet-type von Willebrand disease. *Blood*, **105**, 4330–4336.

Otterness, D.M., Szumlanski, C.L., Wood, T.C. and Weinshilboum, R.M. (1998). Human thiopurine methyltransferase pharmacogenetics. Kindred with a terminal exon splice junction mutation that results in loss of activity. *J Clin Invest*, **101**, 1036–1044.

Ottolenghi, S., Lanyon, W.G., Paul, J. *et al.* (1974). The severe form of alpha thalassaemia is caused by a haemoglobin gene deletion. *Nature*, **251**, 389–392.

Ounissi-Benkalha, H. and Polychronakos, C. (2008). The molecular genetics of type 1 diabetes: new genes and emerging mechanisms. *Trends Mol Med*, **14**, 268–275.

Overeem, S., Mignot, E., van Dijk, J.G. and Lammers, G.J. (2001). Narcolepsy: clinical features, new pathophysiologic insights, and future perspectives. *J Clin Neurophysiol*, **18**, 78–105.

Overhauser, J., Huang, X., Gersh, M. *et al.* (1994). Molecular and phenotypic mapping of the short arm of chromosome 5: sublocalization of the critical region for the cri-du-chat syndrome. *Hum Mol Genet*, **3**, 247–252.

Ozaki, K., Ohnishi, Y., Iida, A. *et al.* (2002). Functional SNPs in the lymphotoxin-alpha gene that are associated with susceptibility to myocardial infarction. *Nat Genet*, **32**, 650–654.

Ozcelik, T., Leff, S., Robinson, W. *et al.* (1992). Small nuclear ribonucleoprotein polypeptide N (SNRPN), an expressed gene in the Prader–Willi syndrome critical region. *Nat Genet*, **2**, 265–269.

Pace, J.K., 2nd and Feschotte, C. (2007). The evolutionary history of human DNA transposons: evidence for intense activity in the primate lineage. *Genome Res*, **17**, 422–432.

Padanilam, B.J., Felice, A.E. and Huisman, T.H. (1984). Partial deletion of the 5′ beta-globin gene region causes beta zero-thalassemia in members of an American black family. *Blood*, **64**, 941–944.

Pagani, F. and Baralle, F.E. (2004). Genomic variants in exons and introns: identifying the splicing spoilers. *Nat Rev Genet*, **5**, 389–396.

Page, S.L., Shin, J.C., Han, J.Y., Choo, K.H. and Shaffer, L.G. (1996). Breakpoint diversity illustrates distinct mechanisms for Robertsonian translocation formation. *Hum Mol Genet*, **5**, 1279–1288.

Pagnier, J., Mears, J.G., Dunda-Belkhodja, O. *et al.* (1984). Evidence for the multicentric origin of the sickle cell hemoglobin gene in Africa. *Proc Natl Acad Sci U S A*, **81**, 1771–1773.

Pain, A., Urban, B.C., Kai, O. *et al.* (2001). A non-sense mutation in Cd36 gene is associated with protection from severe malaria. *Lancet*, **357**, 1502–1503.

Palmer, L.J. (2007). UK Biobank: bank on it. *Lancet*, **369**, 1980–1982.

Palmer, M.S., Dryden, A.J., Hughes, J.T. and Collinge, J. (1991). Homozygous prion protein genotype predisposes to sporadic Creutzfeldt–Jakob disease. *Nature*, **352**, 340–342.

Pampin, S. and Rodriguez-Rey, J.C. (2007). Functional analysis of regulatory single-nucleotide polymorphisms. *Curr Opin Lipidol*, **18**, 194–198.

Pant, P.V., Tao, H., Beilharz, E.J., Ballinger, D.G., Cox, D.R. and Frazer, K.A. (2006). Analysis of allelic differential expression in human white blood cells. *Genome Res*, **16**, 331–339.

Parham, P. (2005). MHC class I molecules and KIRs in human history, health and survival. *Nat Rev Immunol*, **5**, 201–214.

Park, J.P., Wojiski, S.A., Spellman, R.A., Rhodes, C.H. and Mohandas, T.K. (1998). Human chromosome 9 pericentric homologies: implications for chromosome 9 heteromorphisms. *Cytogenet Cell Genet*, **82**, 192–194.

Parkes, M., Barmada, M.M., Satsangi, J., Weeks, D.E., Jewell, D.P. and Duerr, R.H. (2000). The IBD2 locus shows linkage heterogeneity between ulcerative colitis and Crohn disease. *Am J Hum Genet*, **67**, 1605–1610.

Parkes, M., Barrett, J.C., Prescott, N.J. *et al.* (2007). Sequence variants in the autophagy gene IRGM and multiple other replicating loci contribute to Crohn's disease susceptibility. *Nat Genet*, **39**, 830–832.

Pastinen, T. and Hudson, T.J. (2004). Cis-acting regulatory variation in the human genome. *Science*, **306**, 647–650.

Pastinen, T., Ge, B., Gurd, S. *et al.* (2005). Mapping common regulatory variants to human haplotypes. *Hum Mol Genet*, **14**, 3963–3971.

Pastinen, T., Ge, B. and Hudson, T.J. (2006). Influence of human genome polymorphism on gene expression. *Hum Mol Genet*, **15**, R9–16.

Pastinen, T., Sladek, R., Gurd, S. *et al.* (2004). A survey of genetic and epigenetic variation affecting human gene expression. *Physiol Genomics*, **16**, 184–193.

Pasvol, G., Weatherall, D.J. and Wilson, R.J. (1978). Cellular mechanism for the protective effect of haemoglobin S against P. falciparum malaria. *Nature*, **274**, 701–703.

Patau, K., Smith, D.W., Therman, E., Inhorn, S.L. and Wagner, H.P. (1960). Multiple congenital anomaly caused by an extra autosome. *Lancet*, **1**, 790–793.

Patel, P.I., Roa, B.B., Welcher, A.A. *et al.* (1992). The gene for the peripheral myelin protein PMP-22 is a candidate for Charcot–Marie–Tooth disease type 1A. *Nat Genet*, **1**, 159–165.

Patrinos, G.P., Giardine, B., Riemer, C. *et al.* (2004). Improvements in the HbVar database of human hemoglobin variants and thalassemia mutations for population and sequence variation studies. *Nucleic Acids Res*, **32**, D537–41.

Patterson, N., Richter, D.J., Gnerre, S., Lander, E.S. and Reich, D. (2006). Genetic evidence for complex speciation of humans and chimpanzees. *Nature*, **441**, 1103–1108.

Paulding, C.A., Ruvolo, M. and Haber, D.A. (2003). The Tre2 (USP6) oncogene is a hominoid-specific gene. *Proc Natl Acad Sci U S A*, **100**, 2507–2511.

Pauling, L., Itano, H.A., Singer, S. J. *et al.* (1949). Sickle cell anemia a molecular disease. *Science*, **110**, 543–548.

Paxton, W.A., Martin, S.R., Tse, D. *et al.* (1996). Relative resistance to HIV-1 infection of CD4 lymphocytes from persons who remain uninfected despite multiple high-risk sexual exposure. *Nat Med*, **2**, 412–417.

Peacock, C.S., Collins, A., Shaw, M.A. *et al.* (2001). Genetic epidemiology of visceral leishmaniasis in northeastern Brazil. *Genet Epidemiol*, **20**, 383–396.

Pearson, C.E., Nichol Edamura, K. and Cleary, J.D. (2005). Repeat instability: mechanisms of dynamic mutations. *Nat Rev Genet*, **6**, 729–742.

Pearson, H. (2008). Genetic testing for everyone. *Nature*, **453**, 570–571.

Pearson, P.L. (2006). Historical development of analysing large-scale changes in the human genome. *Cytogenet Genome Res*, **115**, 198–204.

Pelletier, V., Jambou, M., Delphin, N. *et al.* (2007). Comprehensive survey of mutations in RP2 and RPGR in patients affected with distinct retinal dystrophies: genotype-phenotype correlations and impact on genetic counseling. *Hum Mutat*, **28**, 81–91.

Peltonen, L., Jalanko, A. and Varilo, T. (1999). Molecular genetics of the Finnish disease heritage. *Hum Mol Genet*, **8**, 1913–1923.

Pemble, S., Schroeder, K.R., Spencer, S.R. *et al.* (1994). Human glutathione S-transferase theta (GSTT1): cDNA cloning and the characterization of a genetic polymorphism. *Biochem J*, **300** (Pt 1), 271–276.

Penagarikano, O., Mulle, J.G. and Warren, S.T. (2007). The pathophysiology of fragile X syndrome. *Annu Rev Genomics Hum Genet*, **8**, 109–129.

Pennacchio, L.A., Lehesjoki, A.E., Stone, N.E. *et al.* (1996). Mutations in the gene encoding cystatin B in progressive myoclonus epilepsy (EPM1). *Science*, **271**, 1731–1734.

Penrose, L. (1933). The relative effects of paternal and maternal age in mongolism. *J Genet*, **27**, 219–224.

Perez Jurado, L.A., Peoples, R., Kaplan, P., Hamel, B.C. and Francke, U. (1996). Molecular definition of the chromosome 7 deletion in Williams syndrome and parent-of-origin effects on growth. *Am J Hum Genet*, **59**, 781–792.

Pericak-Vance, M.A., Bebout, J.L., Gaskell, P.C., Jr. *et al.* (1991). Linkage studies in familial Alzheimer disease: evidence for chromosome 19 linkage. *Am J Hum Genet*, **48**, 1034–1050.

Perron, M.J., Stremlau, M., Song, B., Ulm, W., Mulligan, R.C. and Sodroski, J. (2004). TRIM5alpha mediates the postentry block to N-tropic murine leukemia viruses in human cells. *Proc Natl Acad Sci U S A*, **101**, 11827–11832.

Perry, G.H., Dominy, N.J., Claw, K.G. *et al.* (2007). Diet and the evolution of human amylase gene copy number variation. *Nat Genet*, **39**, 1256–1260.

Peter, H., Deutschmann, S., Reichel, C. and Hallier, E. (1989). Metabolism of methyl chloride by human erythrocytes. *Arch Toxicol*, **63**, 351–355.

Petretto, E., Mangion, J., Dickens, N.J. *et al.* (2006). Heritability and tissue specificity of expression quantitative trait loci. *PLoS Genet*, **2**, e172.

Peyron, C., Faraco, J., Rogers, W. *et al.* (2000). A mutation in a case of early onset narcolepsy and a generalized absence of hypocretin peptides in human narcoleptic brains. *Nat Med*, **6**, 991–997.

Philips, A.V., Timchenko, L.T. and Cooper, T.A. (1998). Disruption of splicing regulated by a CUG-binding protein in myotonic dystrophy. *Science*, **280**, 737–741.

Pieretti, M., Zhang, F.P., Fu, Y.H. *et al.* (1991). Absence of expression of the FMR-1 gene in fragile X syndrome. *Cell*, **66**, 817–822.

Pietrangelo, A. (2004). Hereditary hemochromatosis – a new look at an old disease. *N Engl J Med*, **350**, 2383–2397.

Pinkel, D., Segraves, R., Sudar, D. *et al.* (1998). High resolution analysis of DNA copy number variation using comparative genomic hybridization to microarrays. *Nat Genet*, **20**, 207–211.

Pinol, V., Castells, A., Andreu, M. *et al.* (2005). Accuracy of revised Bethesda guidelines, microsatellite instability, and immunohistochemistry for the identification of patients with hereditary nonpolyposis colorectal cancer. *JAMA*, **293**, 1986–1994.

Pirastu, M., Saglio, G., Chang, J.C., Cao, A. and Kan, Y.W. (1984). Initiation codon mutation as a cause of alpha thalassemia. *J Biol Chem*, **259**, 12315–12317.

Plant, J. and Glynn, A.A. (1974). Natural resistance to Salmonella infection, delayed hypersensitivity and Ir genes in different strains of mice. *Nature*, **248**, 345–347.

Plant, J. and Glynn, A.A. (1976). Genetics of resistance to infection with Salmonella typhimurium in mice. *J Infect Dis*, **133**, 72–78.

Plath, K., Mlynarczyk-Evans, S., Nusinow, D.A. and Panning, B. (2002). Xist RNA and the mechanism of X chromosome inactivation. *Annu Rev Genet*, **36**, 233–278.

Platt, O.S., Brambilla, D.J., Rosse, W.F. *et al.* (1994). Mortality in sickle cell disease. Life expectancy and risk factors for early death. *N Engl J Med*, **330**, 1639–1644.

Plenge, R.M., Cotsapas, C., Davies, L. *et al.* (2007a). Two independent alleles at 6q23 associated with risk of rheumatoid arthritis. *Nat Genet*, **39**, 1477–1482.

Plenge, R.M., Seielstad, M., Padyukov, L. *et al.* (2007b). TRAF1-C5 as a risk locus for rheumatoid arthritis--a genomewide study. *N Engl J Med*, **357**, 1199–1209.

Plohl, M., Luchetti, A., Mestrovic, N. and Mantovani, B. (2008). Satellite DNAs between selfishness and functionality: structure, genomics and evolution of tandem repeats in centromeric (hetero)chromatin. *Gene*, **409**, 72–82.

Pober, B.R., Johnson, M. and Urban, Z. (2008). Mechanisms and treatment of cardiovascular disease in Williams-Beuren syndrome. *J Clin Invest*, **118**, 1606–1615.

Polymeropoulos, M.H., Lavedan, C., Leroy, E. *et al.* (1997). Mutation in the alpha-synuclein gene identified in families with Parkinson's disease. *Science*, **276**, 2045–2047.

Poncz, M., Schwartz, E., Ballantine, M. and Surrey, S. (1983). Nucleotide sequence analysis of the delta beta-globin gene region in humans. *J Biol Chem*, **258**, 11599–11609.

Poort, S.R., Rosendaal, F.R., Reitsma, P.H. and Bertina, R.M. (1996). A common genetic variation in the 3´-untranslated region of the prothrombin gene is associated with elevated plasma prothrombin levels and an increase in venous thrombosis. *Blood*, **88**, 3698–3703.

Popovich, B.W., Rosenblatt, D.S., Kendall, A.G. and Nishioka, Y. (1986). Molecular characterization of an atypical beta-thalassemia caused by a large deletion in the 5´ beta-globin gene region. *Am J Hum Genet*, **39**, 797–810.

Potocki, L., Chen, K.S., Park, S.S. *et al.* (2000). Molecular mechanism for duplication 17p11.2 – the homologous recombination reciprocal of the Smith-Magenis microdeletion. *Nat Genet*, **24**, 84–87.

Poulter, M., Hollox, E., Harvey, C.B. *et al.* (2003). The causal element for the lactase persistence/non-persistence polymorphism is located in a 1 Mb region of linkage disequilibrium in Europeans. *Ann Hum Genet*, **67**, 298–311.

Price, P., Witt, C., Allcock, R. *et al.* (1999). The genetic basis for the association of the 8.1 ancestral haplotype (A1, B8, DR3) with multiple immunopathological diseases. *Immunol Rev*, **167**, 257–274.

Pritchard, J.K. and Rosenberg, N.A. (1999). Use of unlinked genetic markers to detect population stratification in association studies. *Am J Hum Genet*, **65**, 220–228.

Proost, P., Menten, P., Struyf, S., Schutyser, E., De Meester, I. and Van Damme, J. (2000). Cleavage by CD26/dipeptidyl peptidase IV converts the chemokine LD78beta into a most efficient monocyte attractant and CCR1 agonist. *Blood*, **96**, 1674–1680.

Prosser, J., Reisner, A.H., Bradley, M.L., Ho, K. and Vincent, P.C. (1981). Buoyant density and hybridization analysis of human DNA sequences, including three satellite DNAs. *Biochim Biophys Acta*, **656**, 93–102.

Proudfoot, N.J. (1977). Complete 3′ noncoding region sequences of rabbit and human beta-globin messenger RNAs. *Cell*, **10**, 559–570.

Pruitt, K.D., Tatusova, T. and Maglott, D.R. (2007). NCBI reference sequences (RefSeq): a curated non-redundant sequence database of genomes, transcripts and proteins. *Nucleic Acids Res*, **35**, D61–65.

Przeworski, M. (2003). Estimating the time since the fixation of a beneficial allele. *Genetics*, **164**, 1667–1676.

Puffenberger, E.G., Kauffman, E.R., Bolk, S. *et al.* (1994). Identity-by-descent and association mapping of a recessive gene for Hirschsprung disease on human chromosome 13q22. *Hum Mol Genet*, **3**, 1217–1225.

Pugliese, A., Zeller, M., Fernandez, A., Jr. *et al.* (1997). The insulin gene is transcribed in the human thymus and transcription levels correlated with allelic variation at the INS VNTR-IDDM2 susceptibility locus for type 1 diabetes. *Nat Genet*, **15**, 293–297.

Puissant, B., Roubinet, F., Massip, P. *et al.* (2006). Analysis of CCR5, CCR2, CX3CR1, and SDF1 polymorphisms in HIV-positive treated patients: impact on response to HAART and on peripheral T lymphocyte counts. *AIDS Res Hum Retroviruses*, **22**, 153–162.

Pulsinelli, P.D., Perutz, M.F. and Nagel, R.L. (1973). Structure of hemoglobin M Boston, a variant with a five-coordinated ferric heme. *Proc Natl Acad Sci U S A*, **70**, 3870–3874.

Qin, W. and Jia, J. (2008). Down-regulation of insulin-degrading enzyme by presenilin 1 V97L mutant potentially underlies increased levels of amyloid beta 42. *Eur J Neurosci*, **27**, 2425–2432.

Quinnell, R.J. (2003). Genetics of susceptibility to human helminth infection. *Int J Parasitol*, **33**, 1219–1231.

Raelson, J.V., Little, R.D., Ruether, A. *et al.* (2007). Genome-wide association study for Crohn's disease in the Quebec Founder Population identifies multiple validated disease loci. *Proc Natl Acad Sci U S A*, **104**, 14747–14752.

Ramos, M., Alvarez, I., Sesma, L., Logean, A., Rognan, D. and Lopez de Castro, J.A. (2002). Molecular mimicry of an HLA-B27-derived ligand of arthritis-linked subtypes with chlamydial proteins. *J Biol Chem*, **277**, 37573–37581.

Rao, E., Weiss, B., Fukami, M. *et al.* (1997). Pseudoautosomal deletions encompassing a novel homeobox gene cause growth failure in idiopathic short stature and Turner syndrome. *Nat Genet*, **16**, 54–63.

Rao, F., Zhang, L., Wessel, J. *et al.* (2007). Tyrosine hydroxylase, the rate-limiting enzyme in catecholamine biosynthesis: discovery of common human genetic variants governing transcription, autonomic activity, and blood pressure in vivo. *Circulation*, **116**, 993–1006.

Rauch, A., Nolan, D., Martin, A., McKinnon, E., Almeida, C. and Mallal, S. (2006). Prospective genetic screening decreases the incidence of abacavir hypersensitivity reactions in the Western Australian HIV cohort study. *Clin Infect Dis*, **43**, 99–102.

Ravid, K., Lu, J., Zimmet, J.M. and Jones, M.R. (2002). Roads to polyploidy: the megakaryocyte example. *J Cell Physiol*, **190**, 7–20.

Ray, P.N., Belfall, B., Duff, C. *et al.* (1985). Cloning of the breakpoint of an X;21 translocation associated with Duchenne muscular dystrophy. *Nature*, **318**, 672–675.

Raychaudhuri, S., Remmers, E.F., Lee, A.T. *et al.* (2008). Common variants at CD40 and other loci confer risk of rheumatoid arthritis. *Nat Genet*, **40**, 1216–1223.

Redon, R., Ishikawa, S., Fitch, K.R. et al. (2006). Global variation in copy number in the human genome. *Nature*, **444**, 444–454.

Rees, D.C., Cox, M. and Clegg, J.B. (1995). World distribution of factor V Leiden. *Lancet*, **346**, 1133–1134.

Reich, D.E. and Goldstein, D.B. (2001). Detecting association in a case–control study while correcting for population stratification. *Genet Epidemiol*, **20**, 4–16.

Reich, D.E., Patterson, N., Ramesh, V. *et al.* (2007). Admixture mapping of an allele affecting interleukin 6 soluble receptor and interleukin 6 levels. *Am J Hum Genet*, **80**, 716–726.

Reiman, E.M., Webster, J.A., Myers, A.J. *et al.* (2007). GAB2 alleles modify Alzheimer's risk in APOE epsilon4 carriers. *Neuron*, **54**, 713–720.

Reissmann, K.R., Ruth, W.E. and Nomura, T. (1961). A human hemoglobin with lowered oxygen affinity and impaired heme-heme interactions. *J Clin Invest*, **40**, 1826–1833.

Reiter, L.T., Murakami, T., Koeuth, T. *et al.* (1996). A recombination hotspot responsible for two inherited peripheral neuropathies is located near a mariner transposon-like element. *Nat Genet*, **12**, 288–297.

Relethford, J.H. (1997). Hemispheric difference in human skin color. *Am J Phys Anthropol*, **104**, 449–457.

Remmers, E.F., Plenge, R.M., Lee, A.T. *et al.* (2007). STAT4 and the risk of rheumatoid arthritis and systemic lupus erythematosus. *N Engl J Med*, **357**, 977–986.

Ren, B., Robert, F., Wyrick, J.J. *et al.* (2000). Genome-wide location and function of DNA binding proteins. *Science*, **290**, 2306–2309.

Repping, S., van Daalen, S.K., Brown, L.G. *et al.* (2006). High mutation rates have driven extensive structural polymorphism among human Y chromosomes. *Nat Genet*, **38**, 463–467.

Reveille, J.D. (2006). Major histocompatibility genes and ankylosing spondylitis. *Best Pract Res Clin Rheumatol*, **20**, 601–609.

Reveille, J.D. and Arnett, F.C. (2005). Spondyloarthritis: update on pathogenesis and management. *Am J Med*, **118**, 592–603.

Ridefelt, P. and Hakansson, L.D. (2005). Lactose intolerance: lactose tolerance test versus genotyping. *Scand J Gastroenterol*, **40**, 822–826.

Ridley, R.M., Frith, C.D., Crow, T.J. and Conneally, P.M. (1988). Anticipation in Huntington's disease is inherited through the male line but may originate in the female. *J Med Genet*, **25**, 589–595.

Rieder, M.J., Reiner, A.P., Gage, B.F. *et al.* (2005). Effect of VKORC1 haplotypes on transcriptional regulation and warfarin dose. *N Engl J Med*, **352**, 2285–2293.

Riegert, P., Gilfillan, S., Nanda, I., Schmid, M. and Bahram, S. (1998). The mouse HFE gene. *Immunogenetics*, **47**, 174–177.

Rihet, P., Flori, L., Tall, F., Traore, A.S. and Fumoux, F. (2004). Hemoglobin C is associated with reduced Plasmodium falciparum parasitemia and low risk of mild malaria attack. *Hum Mol Genet*, **13**, 1–6.

Riordan, J.R., Rommens, J.M., Kerem, B. *et al.* (1989). Identification of the cystic fibrosis gene: cloning and characterization of complementary DNA. *Science*, **245**, 1066–1073.

Rioux, J.D., Daly, M.J., Silverberg, M.S. *et al.* (2001). Genetic variation in the 5q31 cytokine gene cluster confers susceptibility to Crohn disease. *Nat Genet*, **29**, 223–228.

Rioux, J.D., Silverberg, M.S., Daly, M.J. *et al.* (2000). Genomewide search in Canadian families with inflammatory bowel disease reveals two novel susceptibility loci. *Am J Hum Genet*, **66**, 1863–1870.

Rioux, J.D., Xavier, R.J., Taylor, K.D. *et al.* (2007). Genome-wide association study identifies new susceptibility loci for Crohn disease and implicates autophagy in disease pathogenesis. *Nat Genet*, **39**, 596–604.

Risch, N.J. (2000). Searching for genetic determinants in the new millennium. *Nature*, **405**, 847–856.

Risch, N.J. and Merikangas, K. (1996). The future of genetic studies of complex human diseases. *Science*, **273**, 1516–1517.

Rivera, A., Fisher, S.A., Fritsche, L.G. *et al.* (2005). Hypothetical LOC387715 is a second major susceptibility gene for age-related macular degeneration, contributing independently of complement factor H to disease risk. *Hum Mol Genet*, **14**, 3227–3236.

Robakis, N.K., Ramakrishna, N., Wolfe, G. and Wisniewski, H.M. (1987). Molecular cloning and characterization of a cDNA encoding the cerebrovascular and the neuritic plaque amyloid peptides. *Proc Natl Acad Sci U S A*, **84**, 4190–4194.

Roberts, D.J., Craig, A.G., Berendt, A.R. *et al.* (1992). Rapid switching to multiple antigenic and adhesive phenotypes in malaria. *Nature*, **357**, 689–692.

Roberts, J., Scott, A.C., Howard, M.R. *et al.* (2007). Differential regulation of the serotonin transporter gene by lithium is mediated by transcription factors, CCCTC binding protein and Y-box binding protein 1, through the polymorphic intron 2 variable number tandem repeat. *J Neurosci*, **27**, 2793–2801.

Roberts, L.J., Handman, E. and Foote, S.J. (2000). Science, medicine, and the future: Leishmaniasis. *Br Med J*, **321**, 801–804.

Rockett, K.A., Awburn, M.M., Cowden, W.B. and Clark, I.A. (1991). Killing of Plasmodium falciparum in vitro by nitric oxide derivatives. *Infect Immun*, **59**, 3280–3283.

Rockman, M.V. and Kruglyak, L. (2006). Genetics of global gene expression. *Nat Rev Genet*, **7**, 862–872.

Rockman, M.V. and Wray, G.A. (2002). Abundant raw material for cis-regulatory evolution in humans. *Mol Biol Evol*, **19**, 1991–2004.

Roeleveld, N., Zielhuis, G.A. and Gabreels, F. (1997). The prevalence of mental retardation: a critical review of recent literature. *Dev Med Child Neurol*, **39**, 125–132.

Rogaev, E.I., Sherrington, R., Rogaeva, E.A. *et al.* (1995). Familial Alzheimer's disease in kindreds with missense mutations in a gene on chromosome 1 related to the Alzheimer's disease type 3 gene. *Nature*, **376**, 775–778.

Romas, S.N., Santana, V., Williamson, J. *et al.* (2002). Familial Alzheimer disease among Caribbean Hispanics: a reexamination of its association with APOE. *Arch Neurol*, **59**, 87–91.

Rommens, J.M., Iannuzzi, M.C., Kerem, B. *et al.* (1989). Identification of the cystic fibrosis gene: chromosome walking and jumping. *Science*, **245**, 1059–1065.

Ronald, J., Akey, J.M., Whittle, J., Smith, E.N., Yvert, G. and Kruglyak, L. (2005). Simultaneous genotyping, gene-expression measurement, and detection of allele-specific expression with oligonucleotide arrays. *Genome Res*, **15**, 284–291.

Rooney, A.P. and Zhang, J. (1999). Rapid evolution of a primate sperm protein: relaxation of functional constraint or positive Darwinian selection? *Mol Biol Evol*, **16**, 706–710.

Rosendaal, F.R. (1999). Venous thrombosis: a multicausal disease. *Lancet*, **353**, 1167–1173.

Rosenfeld, S.I., Ruddy, S. and Austen, K.F. (1969). Structural polymorphism of the fourth component of human complement. *J Clin Invest*, **48**, 2283–2292.

Rosenthal, P.J. (2008). Artesunate for the treatment of severe falciparum malaria. *N Engl J Med*, **358**, 1829–1836.

Roth, E.F., Jr., Friedman, M., Ueda, Y., Tellez, I., Trager, W. and Nagel, R.L. (1978). Sickling rates of human AS red cells infected in vitro with Plasmodium falciparum malaria. *Science*, **202**, 650–652.

Roth, E.F., Jr., Raventos-Suarez, C., Rinaldi, A. and Nagel, R.L. (1983). Glucose-6-phosphate dehydrogenase deficiency inhibits in vitro growth of Plasmodium falciparum. *Proc Natl Acad Sci U S A*, **80**, 298–299.

Rovelet-Lecrux, A., Hannequin, D., Raux, G. *et al.* (2006). APP locus duplication causes autosomal dominant early-onset Alzheimer disease with cerebral amyloid angiopathy. *Nat Genet*, **38**, 24–26.

Rovin, S., Dachi, S.F., Borenstein, D.B. and Cotter, W.B. (1964). Mandibulofacial dysostosis, a familial study of five generations. *J Pediatr*, **65**, 215–221.

Rowland-Jones, S., Sutton, J., Ariyoshi, K. *et al.* (1995). HIV-specific cytotoxic T-cells in HIV-exposed but uninfected Gambian women. *Nat Med*, **1**, 59–64.

Rowley, J.D. (1973). Letter: A new consistent chromosomal abnormality in chronic myelogenous leukaemia identified by quinacrine fluorescence and Giemsa staining. *Nature*, **243**, 290–293.

Roy-Engel, A.M., Carroll, M.L., Vogel, E. *et al.* (2001). Alu insertion polymorphisms for the study of human genomic diversity. *Genetics*, **159**, 279–290.

Rubin, G.M., Yandell, M.D., Wortman, J.R. *et al.* (2000). Comparative genomics of the eukaryotes. *Science*, **287**, 2204–2215.

Rubinsztein, D.C., Leggo, J., Coles, R. *et al.* (1996). Phenotypic characterization of individuals with 30–40 CAG repeats in the Huntington disease (HD) gene reveals HD cases with 36 repeats and apparently normal elderly individuals with 36–39 repeats. *Am J Hum Genet*, **59**, 16–22.

Ruddle, F.H., Chapman, V.M., Ricciuti, F., Murnane, M., Klebe, R. and Meera Khan, P. (1971). Linkage relationships of seventeen human gene loci as determined by man–mouse somatic cell hybrids. *Nat New Biol*, **232**, 69–73.

Russell, T.J., Schultes, L.M. and Kuban, D.J. (1972). Histocompatibility (HL-A) antigens associated with psoriasis. *N Engl J Med*, **287**, 738–740.

Rustgi, A.K. (2007). The genetics of hereditary colon cancer. *Genes Dev*, **21**, 2525–2538.

Rutter, J.L., Mitchell, T.I., Buttice, G. *et al.* (1998). A single nucleotide polymorphism in the matrix metalloproteinase-1 promoter creates an Ets binding site and augments transcription. *Cancer Res*, **58**, 5321–5325.

Ruwende, C. and Hill, A. (1998). Glucose-6-phosphate dehydrogenase deficiency and malaria. *J Mol Med*, **76**, 581–588.

Ruwende, C., Khoo, S.C., Snow, R.W. *et al.* (1995). Natural selection of hemi- and heterozygotes for G6PD deficiency in Africa by resistance to severe malaria. *Nature*, **376**, 246–249.

Rybicki, B.A., Iannuzzi, M.C., Frederick, M.M. *et al.* (2001). Familial aggregation of sarcoidosis. A case–control etiologic study of sarcoidosis (ACCESS). *Am J Respir Crit Care Med*, **164**, 2085–2091.

Rybicki, B.A., Walewski, J.L., Maliarik, M.J., Kian, H. and Iannuzzi, M.C. (2005). The BTNL2 gene and sarcoidosis susceptibility in African Americans and whites. *Am J Hum Genet*, **77**, 491–499.

Sabeti, P.C., Reich, D.E., Higgins, J.M. *et al.* (2002a). Detecting recent positive selection in the human genome from haplotype structure. *Nature*, **419**, 832–837.

Sabeti, P.C., Schaffner, S.F., Fry, B. *et al.* (2006). Positive natural selection in the human lineage. *Science*, **312**, 1614–1620.

Sabeti, P.C., Usen, S., Farhadian, S. *et al.* (2002b). CD40L association with protection from severe malaria. *Genes Immun*, **3**, 286–291.

Sabeti, P.C., Varilly, P., Fry, B. *et al.* (2007). Genome-wide detection and characterization of positive selection in human populations. *Nature*, **449**, 913–918.

Sachidanandam, R., Weissman, D., Schmidt, S.C. *et al.* (2001). A map of human genome sequence variation containing 1.42 million single nucleotide polymorphisms. *Nature*, **409**, 928–933.

Sachse, C., Brockmoller, J., Bauer, S. and Roots, I. (1997). Cytochrome P450 2D6 variants in a Caucasian population: allele frequencies and phenotypic consequences. *Am J Hum Genet*, **60**, 284–295.

Saiki, R.K., Bugawan, T.L., Horn, G.T., Mullis, K.B. and Erlich, H.A. (1986). Analysis of enzymatically amplified beta-globin and HLA-DQ alpha DNA with allele-specific oligonucleotide probes. *Nature*, **324**, 163–166.

Saiki, R.K., Scharf, S., Faloona, F. *et al.* (1985). Enzymatic amplification of beta-globin genomic sequences and restriction site analysis for diagnosis of sickle cell anemia. *Science*, **230**, 1350–1354.

Sakamoto, N., Chastain, P.D., Parniewski, P. *et al.* (1999). Sticky DNA: self-association properties of long GAA. TTC repeats in R.R.Y triplex structures from Friedreich's ataxia. *Mol Cell*, **3**, 465–475.

Saleh, M., Vaillancourt, J.P., Graham, R.K. *et al.* (2004). Differential modulation of endotoxin responsiveness by human caspase-12 polymorphisms. *Nature*, **429**, 75–79.

Salehi-Ashtiani, K., Yang, X., Derti, A. *et al.* (2008). Isoform discovery by targeted cloning, 'deep-well' pooling and parallel sequencing. *Nat Methods*, **5**, 597–600.

Salem, A.H., Kilroy, G.E., Watkins, W.S., Jorde, L.B. and Batzer, M.A. (2003a). Recently integrated Alu elements and human genomic diversity. *Mol Biol Evol*, **20**, 1349–1361.

Salem, A.H., Myers, J.S., Otieno, A.C., Watkins, W.S., Jorde, L.B. and Batzer, M.A. (2003b). LINE-1 preTa elements in the human genome. *J Mol Biol*, **326**, 1127–1146.

Salem, A.H., Ray, D.A., Xing, J. *et al.* (2003c). Alu elements and hominid phylogenetics. *Proc Natl Acad Sci U S A*, **100**, 12787–12791.

Salkowitz, J.R., Bruse, S.E., Meyerson, H. *et al.* (2003). CCR5 promoter polymorphism determines macrophage CCR5 density and magnitude of HIV-1 propagation in vitro. *Clin Immunol*, **108**, 234–240.

Samani, N.J., Erdmann, J., Hall, A.S. *et al.* (2007). Genomewide association analysis of coronary artery disease. *N Engl J Med*, **357**, 443–453.

Samonte, R.V. and Eichler, E.E. (2002). Segmental duplications and the evolution of the primate genome. *Nat Rev Genet*, **3**, 65–72.

Samson, M., Libert, F., Doranz, B.J. *et al.* (1996). Resistance to HIV-1 infection in caucasian individuals bearing mutant alleles of the CCR-5 chemokine receptor gene. *Nature*, **382**, 722–725.

Sanders-Haigh, L., Anderson, W.F. and Francke, U. (1980). The beta-globin gene is on the short arm of human chromosome 11. *Nature*, **283**, 683–686.

Sanger, F. and Coulson, A.R. (1975). A rapid method for determining sequences in DNA by primed synthesis with DNA polymerase. *J Mol Biol*, **94**, 441–448.

Sanger, F., Nicklen, S. and Coulson, A.R. (1977). DNA sequencing with chain-terminating inhibitors. *Proc Natl Acad Sci U S A*, **74**, 5463–5467.

Sanguansermsri, T., Pape, M., Laig, M., Hundrieser, J. and Flatz, G. (1990). Beta zero-thalassemia in a Thai family is caused by a 3.4 kb deletion including the entire beta-globin gene. *Hemoglobin*, **14**, 157–168.

Santangelo, A.M., de Souza, F.S., Franchini, L.F., Bumaschny, V.F., Low, M.J. and Rubinstein, M. (2007). Ancient exaptation of a CORE-SINE retroposon into a highly conserved mammalian neuronal enhancer of the proopiomelanocortin gene. *PLoS Genet*, **3**, 1813–1826.

Santiago, M.L., Range, F., Keele, B.F. *et al.* (2005). Simian immunodeficiency virus infection in free-ranging sooty mangabeys (Cercocebus atys atys) from the Tai Forest, Cote d'Ivoire: implications for the origin of epidemic human immunodeficiency virus type 2. *J Virol*, **79**, 12515–12527.

Sassaman, D.M., Dombroski, B.A., Moran, J.V. *et al.* (1997). Many human L1 elements are capable of retrotransposition. *Nat Genet*, **16**, 37–43.

Satsangi, J., Grootscholten, C., Holt, H. and Jewell, D.P. (1996). Clinical patterns of familial inflammatory bowel disease. *Gut*, **38**, 738–741.

Saunders, A.M., Strittmatter, W.J., Schmechel, D. *et al.* (1993). Association of apolipoprotein E allele epsilon 4 with late-onset familial and sporadic Alzheimer's disease. *Neurology*, **43**, 1467–1472.

Savkur, R.S., Philips, A.V. and Cooper, T.A. (2001). Aberrant regulation of insulin receptor alternative splicing is associated with insulin resistance in myotonic dystrophy. *Nat Genet*, **29**, 40–47.

Sawyer, S.L., Emerman, M. and Malik, H.S. (2004). Ancient adaptive evolution of the primate antiviral DNA-editing enzyme APOBEC3G. *PLoS Biol*, **2**, e275.

Sawyer, S.L., Wu, L.I., Akey, J.M., Emerman, M. and Malik, H.S. (2006). High-frequency persistence of an impaired allele of the retroviral defense gene TRIM5alpha in humans. *Curr Biol*, **16**, 95–100.

Sawyer, S.L., Wu, L.I., Emerman, M. and Malik, H.S. (2005). Positive selection of primate TRIM5alpha identifies a critical species-specific retroviral restriction domain. *Proc Natl Acad Sci U S A*, **102**, 2832–2837.

Saxena, R., Voight, B.F., Lyssenko, V. *et al.* (2007). Genome-wide association analysis identifies loci for type 2 diabetes and triglyceride levels. *Science*, **316**, 1331–1336.

Schachenmann, G., Schmid, W., Fraccaro, M. *et al.* (1965). Chromosomes in coloboma and anal atresia. *Lancet*, **2**, 290.

Schadt, E.E., Molony, C., Chudin, E. *et al.* (2008). Mapping the genetic architecture of gene expression in human liver. *PLoS Biol*, **6**, e107.

Schadt, E.E., Monks, S.A., Drake, T.A. *et al.* (2003). Genetics of gene expression surveyed in maize, mouse and man. *Nature*, **422**, 297–302.

Schellenberg, G.D., Bird, T.D., Wijsman, E.M. *et al.* (1992). Genetic linkage evidence for a familial Alzheimer's disease locus on chromosome 14. *Science*, **258**, 668–671.

Scherer, S.W., Lee, C., Birney, E. *et al.* (2007). Challenges and standards in integrating surveys of structural variation. *Nat Genet*, **39**, S7–15.

Scherf, A. (2006). A greedy promoter controls malarial variations. *Cell*, **124**, 251–253.

Scheuner, D., Eckman, C., Jensen, M. *et al.* (1996). Secreted amyloid beta-protein similar to that in the senile plaques of Alzheimer's disease is increased in vivo by the presenilin 1 and 2 and APP mutations linked to familial Alzheimer's disease. *Nat Med*, **2**, 864–870.

Schlosstein, L., Terasaki, P.I., Bluestone, R. and Pearson, C.M. (1973). High association of an HL-A antigen, W27, with ankylosing spondylitis. *N Engl J Med*, **288**, 704–706.

Schlotterer, C. and Tautz, D. (1992). Slippage synthesis of simple sequence DNA. *Nucleic Acids Res*, **20**, 211–215.

Schlotterer, C., Amos, B. and Tautz, D. (1991). Conservation of polymorphic simple sequence loci in cetacean species. *Nature*, **354**, 63–65.

Schmechel, D.E., Saunders, A.M., Strittmatter, W.J. *et al.* (1993). Increased amyloid beta-peptide deposition in cerebral cortex as a consequence of apolipoprotein E genotype in late-onset Alzheimer disease. *Proc Natl Acad Sci U S A*, **90**, 9649–9653.

Schmidt, S., Hauser, M.A., Scott, W.K. *et al.* (2006). Cigarette smoking strongly modifies the association of LOC387715 and age-related macular degeneration. *Am J Hum Genet*, **78**, 852–864.

Schueler, M.G., Higgins, A.W., Rudd, M.K., Gustashaw, K. and Willard, H.F. (2001). Genomic and genetic definition of a functional human centromere. *Science*, **294**, 109–115.

Schurmann, M., Reichel, P., Muller-Myhsok, B., Schlaak, M., Muller-Quernheim, J. and Schwinger, E. (2001). Results from a genome-wide search for predisposing genes in sarcoidosis. *Am J Respir Crit Care Med*, **164**, 840–846.

Schwahn, U., Lenzner, S., Dong, J. *et al.* (1998). Positional cloning of the gene for X-linked retinitis pigmentosa 2. *Nat Genet*, **19**, 327–332.

Schwarz, H.P., Fischer, M., Hopmeier, P., Batard, M.A. and Griffin, J.H. (1984). Plasma protein S deficiency in familial thrombotic disease. *Blood*, **64**, 1297–1300.

Schwarz, U.I., Ritchie, M.D., Bradford, Y. *et al.* (2008). Genetic determinants of response to warfarin during initial anticoagulation. *N Engl J Med*, **358**, 999–1008.

Schwimmbeck, P.L. and Oldstone, M.B. (1988). Molecular mimicry between human leukocyte antigen B27 and Klebsiella. Consequences for spondyloarthropathies. *Am J Med*, **85**, 51–53.

Scott, L.J., Mohlke, K.L., Bonnycastle, L.L. *et al.* (2007). A genome-wide association study of type 2 diabetes in Finns detects multiple susceptibility variants. *Science*, **316**, 1341–1345.

Scriver, J.R. and Waugh, T.R. (1930). Studies on a case of sickle cell anemia. *Canadian Medical Association*, **23**, 375–380.

Scuteri, A., Sanna, S., Chen, W.M. *et al.* (2007). Genome-wide association scan shows genetic variants in the FTO gene are associated with obesity-related traits. *PLoS Genet*, **3**, e115.

Seaman, J., Mercer, A.J. and Sondorp, E. (1996). The epidemic of visceral leishmaniasis in western Upper Nile, southern Sudan: course and impact from 1984 to 1994. *Int J Epidemiol*, **25**, 862–871.

Sebat, J. (2007). Major changes in our DNA lead to major changes in our thinking. *Nat Genet*, **39**, S3–5.

Sebat, J., Lakshmi, B., Malhotra, D. *et al.* (2007). Strong association of de novo copy number mutations with autism. *Science*, **316**, 445–449.

Sebat, J., Lakshmi, B., Troge, J. *et al.* (2004). Large-scale copy number polymorphism in the human genome. *Science*, **305**, 525–528.

Seddon, J.M., Ajani, U.A. and Mitchell, B.D. (1997). Familial aggregation of age-related maculopathy. *Am J Ophthalmol*, **123**, 199–206.

Seddon, J.M., Cote, J., Page, W.F., Aggen, S.H. and Neale, M.C. (2005). The US twin study of age-related macular degeneration: relative roles of genetic and environmental influences. *Arch Ophthalmol*, **123**, 321–327.

Segal, D.J. and McCoy, E.E. (1974). Studies on Down's syndrome in tissue culture. I. Growth rates and protein contents of fibroblast cultures. *J Cell Physiol*, **83**, 85–90.

Seidegard, J., Vorachek, W.R., Pero, R.W. and Pearson, W.R. (1988). Hereditary differences in the expression of the human glutathione transferase active on trans-stilbene oxide are due to a gene deletion. *Proc Natl Acad Sci U S A*, **85**, 7293–7297.

Sen, S.K., Han, K., Wang, J. *et al.* (2006). Human genomic deletions mediated by recombination between Alu elements. *Am J Hum Genet*, **79**, 41–53.

Serjeant, G.R., Ashcroft, M.T. and Serjeant, B.E. (1973). The clinical features of haemoglobin SC disease in Jamaica. *Br J Haematol*, **24**, 491–501.

Serjeant, G.R., Sommereux, A.M., Stevenson, M., Mason, K. and Serjeant, B.E. (1979). Comparison of sickle cell-beta0 thalassaemia with homozygous sickle cell disease. *Br J Haematol*, **41**, 83–93.

Serre, D., Gurd, S., Ge, B. *et al.* (2008). Differential allelic expression in the human genome: a robust approach to identify genetic and epigenetic cis-acting mechanisms regulating gene expression. *PLoS Genet*, **4**, e1000006.

Seznec, H., Agbulut, O., Sergeant, N. *et al.* (2001). Mice transgenic for the human myotonic dystrophy region with expanded CTG repeats display muscular and brain abnormalities. *Hum Mol Genet*, **10**, 2717–2726.

Shaffer, L.G. and Lupski, J.R. (2000). Molecular mechanisms for constitutional chromosomal rearrangements in humans. *Annu Rev Genet*, **34**, 297–329.

Shaffer, L.G. and Tommerup, N., eds (2005). *ISCN 2005: An International System for Human Cytogenetic Nomenclature*. S. Karger AG, Basel, Switzerland.

Shaffer, L.G., Kashork, C.D., Saleki, R. *et al.* (2006). Targeted genomic microarray analysis for identification of chromosome abnormalities in 1500 consecutive clinical cases. *J Pediatr*, **149**, 98–102.

Shaikh, T.H. (2007). Oligonucleotide arrays for high-resolution analysis of copy number alteration in mental retardation/multiple congenital anomalies. *Genet Med*, **9**, 617–625.

Shaikh, T.H., Budarf, M.L., Celle, L., Zackai, E.H. and Emanuel, B.S. (1999). Clustered 11q23 and 22q11 breakpoints and 3:1 meiotic malsegregation in multiple unrelated t(11;22) families. *Am J Hum Genet*, **65**, 1595–1607.

Shaikh, T.H., Kurahashi, H., Saitta, S.C. *et al.* (2000). Chromosome 22-specific low copy repeats and the 22q11.2 deletion syndrome: genomic organization and deletion endpoint analysis. *Hum Mol Genet*, **9**, 489–501.

Shapira, S.K., McCaskill, C., Northrup, H. *et al.* (1997). Chromosome 1p36 deletions: the clinical phenotype and molecular characterization of a common newly delineated syndrome. *Am J Hum Genet*, **61**, 642–650.

Sharma, P., Hingorani, A., Jia, H. *et al.* (1998). Positive association of tyrosine hydroxylase microsatellite marker to essential hypertension. *Hypertension*, **32**, 676–682.

Sharp, A.J., Hansen, S., Selzer, R.R. *et al.* (2006). Discovery of previously unidentified genomic disorders from the duplication architecture of the human genome. *Nat Genet*, **38**, 1038–1042.

Sharp, A.J., Locke, D.P., McGrath, S.D. *et al.* (2005). Segmental duplications and copy-number variation in the human genome. *Am J Hum Genet*, **77**, 78–88.

Shaw, C.J. and Lupski, J.R. (2004). Implications of human genome architecture for rearrangement-based disorders: the genomic basis of disease. *Hum Mol Genet*, **13**, R57–64.

She, X., Horvath, J.E., Jiang, Z. *et al.* (2004). The structure and evolution of centromeric transition regions within the human genome. *Nature*, **430**, 857–864.

She, X., Liu, G., Ventura, M. *et al.* (2006). A preliminary comparative analysis of primate segmental duplications shows elevated substitution rates and a great-ape expansion of intrachromosomal duplications. *Genome Res*, **16**, 576–583.

Sheehy, A.M., Gaddis, N.C., Choi, J.D. and Malim, M.H. (2002). Isolation of a human gene that inhibits HIV-1 infection and is suppressed by the viral Vif protein. *Nature*, **418**, 646–650.

Sheehy, A.M., Gaddis, N.C. and Malim, M.H. (2003). The antiretroviral enzyme APOBEC3G is degraded by the proteasome in response to HIV-1 Vif. *Nat Med*, **9**, 1404–1407.

Sheldon, J. (1935). *Haemochromatosis*. Oxford University Press, London.

Shen, M.R., Batzer, M.A. and Deininger, P.L. (1991). Evolution of the master Alu gene(s). *J Mol Evol*, **33**, 311–320.

Shen, Y., Holman, K., Doggett, N.A., Callen, D.F., Sutherland, G.R. and Richards, R.I. (1993). Three dinucleotide repeat polymorphisms on human chromosome 16p13.11–p13.3. *Hum Mol Genet*, **2**, 1506.

Sherard, J., Bean, C., Bove, B. *et al.* (1986). Long survival in a 69,XXY triploid male. *Am J Med Genet*, **25**, 307–312.

Sherrington, R., Rogaev, E.I., Liang, Y. *et al.* (1995). Cloning of a gene bearing missense mutations in early-onset familial Alzheimer's disease. *Nature*, **375**, 754–760.

Sherry, S.T., Ward, M. and Sirotkin, K. (1999). dbSNP-database for single nucleotide polymorphisms and other classes of minor genetic variation. *Genome Res*, **9**, 677–679.

Sherry, S.T., Ward, M.H., Kholodov, M. *et al.* (2001). dbSNP: the NCBI database of genetic variation. *Nucleic Acids Res*, **29**, 308–311.

Shimajiri, S., Arima, N., Tanimoto, A. *et al.* (1999). Shortened microsatellite d(CA)21 sequence down-regulates promoter activity of matrix metalloproteinase 9 gene. *FEBS Lett*, **455**, 70–74.

Shioda, T., Nakayama, E.E., Tanaka, Y. *et al.* (2001). Naturally occurring deletional mutation in the C-terminal cytoplasmic tail of CCR5 affects surface trafficking of CCR5. *J Virol*, **75**, 3462–3468.

Shizuya, H., Birren, B., Kim, U.J. *et al.* (1992). Cloning and stable maintenance of 300-kilobase-pair fragments of human DNA in Escherichia coli using an F-factor-based vector. *Proc Natl Acad Sci U S A*, **89**, 8794–8797.

Shoja, V. and Zhang, L. (2006). A roadmap of tandemly arrayed genes in the genomes of human, mouse, and rat. *Mol Biol Evol*, **23**, 2134–2141.

Shprintzen, R.J., Goldberg, R.B., Young, D. and Wolford, L. (1981). The velo-cardio-facial syndrome: a clinical and genetic analysis. *Pediatrics*, **67**, 167–172.

Shriver, M.D., Parra, E.J., Dios, S. *et al.* (2003). Skin pigmentation, biogeographical ancestry and admixture mapping. *Hum Genet*, **112**, 387–399.

Shyue, S.K., Hewett-Emmett, D., Sperling, H.G. *et al.* (1995). Adaptive evolution of color vision genes in higher primates. *Science*, **269**, 1265–1267.

Siebold, C., Hansen, B.E., Wyer, J.R. *et al.* (2004). Crystal structure of HLA-DQ0602 that protects against type 1 diabetes and confers strong susceptibility to narcolepsy. *Proc Natl Acad Sci U S A*, **101**, 1999–2004.

Sigurdsson, S., Nordmark, G., Goring, H.H. *et al.* (2005). Polymorphisms in the tyrosine kinase 2 and interferon regulatory factor 5 genes are associated with systemic lupus erythematosus. *Am J Hum Genet*, **76**, 528–537.

Simmonds, M.J., Heward, J.M., Barrett, J.C., Franklyn, J.A. and Gough, S.C. (2006). Association of the BTNL2 rs2076530 single nucleotide polymorphism with Graves' disease appears to be secondary to DRB1 exon 2 position beta74. *Clin Endocrinol (Oxford)*, **65**, 429–432.

Simon, D., Seznec, H., Gansmuller, A. *et al.* (2004). Friedreich ataxia mouse models with progressive cerebellar and sensory ataxia reveal autophagic neurodegeneration in dorsal root ganglia. *J Neurosci*, **24**, 1987–1995.

Simon, J.H., Gaddis, N.C., Fouchier, R.A. and Malim, M.H. (1998). Evidence for a newly discovered cellular anti-HIV-1 phenotype. *Nat Med*, **4**, 1397–1400.

Simon, M., Bourel, M., Fauchet, R. and Genetet, B. (1976). Association of HLA-A3 and HLA-B14 antigens with idiopathic haemochromatosis. *Gut*, **17**, 332–334.

Simon, M., Bourel, M., Genetet, B. and Fauchet, R. (1977). Idiopathic hemochromatosis. Demonstration of recessive transmission and early detection by family HLA typing. *N Engl J Med*, **297**, 1017–1021.

Simon-Sanchez, J., Scholz, S., Del Mar Matarin, M. *et al.* (2007). Genomewide SNP assay reveals mutations underlying Parkinson disease. *Hum Mutat*, **29**, 315–322.

Singer-Sam, J., Chapman, V., LeBon, J.M. and Riggs, A.D. (1992a). Parental imprinting studied by allele-specific primer extension after PCR: paternal X chromosome-linked genes are transcribed prior to preferential paternal X chromosome inactivation. *Proc Natl Acad Sci U S A*, **89**, 10469–10473.

Singer-Sam, J., LeBon, J.M., Dai, A. and Riggs, A.D. (1992b). A sensitive, quantitative assay for measurement of allele-specific transcripts differing by a single nucleotide. *PCR Methods Appl*, **1**, 160–163.

Singh, S.B., Davis, A.S., Taylor, G.A. and Deretic, V. (2006). Human IRGM induces autophagy to eliminate intracellular mycobacteria. *Science*, **313**, 1438–1441.

Singleton, A.B., Farrer, M., Johnson, J. *et al.* (2003). Alpha-synuclein locus triplication causes Parkinson's disease. *Science*, **302**, 841.

Singleton, B.K., Green, C.A., Avent, N.D. *et al.* (2000). The presence of an RHD pseudogene containing a 37 base pair duplication and a nonsense mutation in africans with the Rh D-negative blood group phenotype. *Blood*, **95**, 12–18.

Sistermans, E.A., de Coo, R.F., De Wijs, I.J. and Van Oost, B.A. (1998). Duplication of the proteolipid protein gene is the major cause of Pelizaeus–Merzbacher disease. *Neurology*, **50**, 1749–1754.

Sitnikova, T. and Nei, M. (1998). Evolution of immunoglobulin kappa chain variable region genes in vertebrates. *Mol Biol Evol*, **15**, 50–60.

Siva, N. (2008). 1000 Genomes Project. *Nat Biotechnol*, **26**, 256.

Skamene, E., Gros, P., Forget, A., Kongshavn, P.A., St Charles, C. and Taylor, B.A. (1982). Genetic regulation of resistance to intracellular pathogens. *Nature*, **297**, 506–509.

Skipper, L., Wilkes, K., Toft, M. *et al.* (2004). Linkage disequilibrium and association of MAPT H1 in Parkinson disease. *Am J Hum Genet*, **75**, 669–677.

Skre, H. (1974). Genetic and clinical aspects of Charcot-Marie-Tooth's disease. *Clin Genet*, **6**, 98–118.

Sladek, R., Rocheleau, G., Rung, J. *et al.* (2007). A genome-wide association study identifies novel risk loci for type 2 diabetes. *Nature*, **445**, 881–885.

Slagboom, P.E., Droog, S. and Boomsma, D.I. (1994). Genetic determination of telomere size in humans: a twin study of three age groups. *Am J Hum Genet*, **55**, 876–882.

Slatkin, M. (1991). Inbreeding coefficients and coalescence times. *Genet Res*, **58**, 167–175.

Slatkin, M. (2008). Linkage disequilibrium – understanding the evolutionary past and mapping the medical future. *Nat Rev Genet*, **9**, 477–485.

Slatkin, M. and Bertorelle, G. (2001). The use of intraallelic variability for testing neutrality and estimating population growth rate. *Genetics*, **158**, 865–874.

Sleiman, P.M., Annaiah, K., Imielinski, M. *et al.* (2008). ORMDL3 variants associated with asthma susceptibility in North Americans of European ancestry. *J Allergy Clin Immunol*, **122**, 1225–1227.

Small, K., Iber, J. and Warren, S.T. (1997). Emerin deletion reveals a common X-chromosome inversion mediated by inverted repeats. *Nat Genet*, **16**, 96–99.

Smillie, W.G. and Augustine, D.L. (1925). Intensity of hookworm infestation in Alabama. Its relationship to residence, occupation, age, sex and race. *JAMA* **85**, 1958–1963.

Smit, A.F. and Riggs, A.D. (1996). Tiggers and DNA transposon fossils in the human genome. *Proc Natl Acad Sci U S A*, **93**, 1443–1448.

Smit, A.F., Toth, G., Riggs, A.D. and Jurka, J. (1995). Ancestral, mammalian-wide subfamilies of LINE-1 repetitive sequences. *J Mol Biol*, **246**, 401–417.

Smit, C., Geskus, R., Walker, S. *et al.* (2006). Effective therapy has altered the spectrum of cause-specific mortality following HIV seroconversion. *Aids*, **20**, 741–749.

Smith, A.J., Jackson, M.W., Neufing, P., McEvoy, R.D. and Gordon, T.P. (2004). A functional autoantibody in narcolepsy. *Lancet*, **364**, 2122–2124.

Smith, J.D., Craig, A.G., Kriek, N. *et al.* (2000). Identification of a Plasmodium falciparum intercellular adhesion molecule-1 binding domain: a parasite adhesion trait implicated in cerebral malaria. *Proc Natl Acad Sci U S A*, **97**, 1766–1771.

Smith, K.J., Reid, S.W., Harlos, K. *et al.* (1996). Bound water structure and polymorphic amino acids act together to allow the binding of different peptides to MHC class I HLA-B53. *Immunity*, **4**, 215–228.

Smith, L.M., Sanders, J.Z., Kaiser, R.J. *et al.* (1986). Fluorescence detection in automated DNA sequence analysis. *Nature*, **321**, 674–679.

Smith, M.W., Dean, M., Carrington, M. *et al.* (1997). Contrasting genetic influence of CCR2 and CCR5 variants on HIV-1 infection and disease progression. Hemophilia Growth and Development Study (HGDS), Multicenter AIDS Cohort Study (MACS), Multicenter Hemophilia Cohort Study (MHCS), San Francisco City Cohort (SFCC), ALIVE Study. *Science*, **277**, 959–965.

Smith, R.A., Ho, P.J., Clegg, J.B., Kidd, J.R. and Thein, S.L. (1998). Recombination breakpoints in the human beta-globin gene cluster. *Blood*, **92**, 4415–4421.

Smyth, D.J., Cooper, J.D., Bailey, R. *et al.* (2006). A genome-wide association study of nonsynonymous SNPs identifies a type 1 diabetes locus in the interferon-induced helicase (IFIH1) region. *Nat Genet*, **38**, 617–619.

Snow, R.W., Guerra, C.A., Noor, A.M., Myint, H.Y. and Hay, S.I. (2005). The global distribution of clinical episodes of Plasmodium falciparum malaria. *Nature*, **434**, 214–217.

Sobue, G., Hashizume, Y., Mukai, E., Hirayama, M., Mitsuma, T. and Takahashi, A. (1989). X-linked recessive bulbospinal neuronopathy. A clinicopathological study. *Brain*, **112**, 209–232.

Somerville, M.J., Mervis, C.B., Young, E.J. *et al.* (2005). Severe expressive-language delay related to duplication of the Williams–Beuren locus. *N Engl J Med*, **353**, 1694–1701.

Song, B., Javanbakht, H., Perron, M., Park, D.H., Stremlau, M. and Sodroski, J. (2005). Retrovirus restriction by TRIM5alpha variants from Old World and New World primates. *J Virol*, **79**, 3930–3937.

Sorensen, T.I., Nielsen, G.G., Andersen, P.K. and Teasdale, T.W. (1988). Genetic and environmental influences on premature death in adult adoptees. *N Engl J Med*, **318**, 727–732.

Southern, E.M. (1975). Detection of specific sequences among DNA fragments separated by gel electrophoresis. *J Mol Biol*, **98**, 503–517.

Speelmon, E.C., Livingston-Rosanoff, D., Li, S.S. *et al.* (2006). Genetic association of the antiviral restriction factor TRIM5alpha with human immunodeficiency virus type 1 infection. *J Virol*, **80**, 2463–2471.

Spencer, K.L., Hauser, M.A., Olson, L.M. *et al.* (2008). Deletion of CFHR3 and CFHR1 genes in age-related macular degeneration. *Hum Mol Genet*, **17**, 971–977.

Spielman, R.S., Baker, L. and Zmijewski, C.M. (1980). Gene dosage and suceptibility to insulin-dependent diabetes. *Ann Hum Genet*, **44**, 135–150.

Spielman, R.S., Bastone, L.A., Burdick, J.T., Morley, M., Ewens, W.J. and Cheung, V.G. (2007). Common genetic variants account for differences in gene expression among ethnic groups. *Nat Genet*, **39**, 226–231.

Spielman, R.S., McGinnis, R.E. and Ewens, W.J. (1993). Transmission test for linkage disequilibrium: the insulin gene region and insulin-dependent diabetes mellitus (IDDM). *Am J Hum Genet*, **52**, 506–516.

Spilianakis, C.G., Lalioti, M.D., Town, T., Lee, G.R. and Flavell, R.A. (2005). Interchromosomal associations between alternatively expressed loci. *Nature*, **435**, 637–645.

Spillantini, M.G., Schmidt, M.L., Lee, V.M., Trojanowski, J.Q., Jakes, R. and Goedert, M. (1997). Alpha-synuclein in Lewy bodies. *Nature*, **388**, 839–840.

Spire-Vayron de la Moureyre, C., Debuysere, H., Fazio, F. *et al.* (1999). Characterization of a variable number tandem repeat region in the thiopurine S-methyltransferase gene promoter. *Pharmacogenetics*, **9**, 189–198.

Spire-Vayron de la Moureyre, C., Debuysere, H., Mastain, B. *et al.* (1998). Genotypic and phenotypic analysis of the polymorphic thiopurine S-methyltransferase gene (TPMT) in a European population. *Br J Pharmacol*, **125**, 879–887.

Splendore, A., Fanganiello, R.D., Masotti, C., Morganti, L.S. and Passos-Bueno, M.R. (2005). TCOF1 mutation database: novel mutation in the alternatively spliced exon 6A and update in mutation nomenclature. *Hum Mutat*, **25**, 429–434.

Splendore, A., Jabs, E.W. and Passos-Bueno, M.R. (2002). Screening of TCOF1 in patients from different populations: confirmation of mutational hot spots and identification of a novel missense mutation that suggests an important functional domain in the protein treacle. *J Med Genet*, **39**, 493–495.

Spritz, R.A. (1981). Duplication/deletion polymorphism 5′-to the human beta globin gene. *Nucleic Acids Res*, **9**, 5037–5047.

St George-Hyslop, P.H., Haines, J.L., Farrer, L.A. *et al.* (1990). Genetic linkage studies suggest that Alzheimer's disease is not a single homogeneous disorder. FAD Collaborative Study Group. *Nature*, **347**, 194–197.

Stankiewicz, P. and Lupski, J.R. (2002). Genome architecture, rearrangements and genomic disorders. *Trends Genet*, **18**, 74–82.

Stastny, P. (1976). Mixed lymphocyte cultures in rheumatoid arthritis. *J Clin Invest*, **57**, 1148–1157.

Stefansson, H., Helgason, A., Thorleifsson, G. *et al.* (2005). A common inversion under selection in Europeans. *Nat Genet*, **37**, 129–137.

Steinberg, M.H., Barton, F., Castro, O. *et al.* (2003). Effect of hydroxyurea on mortality and morbidity in adult sickle cell anemia: risks and benefits up to 9 years of treatment. *JAMA*, **289**, 1645–1651.

Stenson, P.D., Ball, E.V., Mort, M. *et al.* (2003). Human Gene Mutation Database (HGMD): 2003 update. *Hum Mutat*, **21**, 577–581.

Stephens, R., Horton, R., Humphray, S., Rowen, L., Trowsdale, J. and Beck, S. (1999). Gene organisation, sequence variation and isochore structure at the centromeric boundary of the human MHC. *J Mol Biol*, **291**, 789–799.

Stewart, C.A., Horton, R., Allcock, R.J. *et al.* (2004a). Complete MHC haplotype sequencing for common disease gene mapping. *Genome Res*, **14**, 1176–1187.

Stewart, D.R., Huang, A., Faravelli, F. *et al.* (2004b). Subtelomeric deletions of chromosome 9q: a novel microdeletion syndrome. *Am J Med Genet A*, **128A**, 340–351.

Stokes, P.L., Asquith, P., Holmes, G.K., Mackintosh, P. and Cooke, W.T. (1972). Histocompatibility antigens associated with adult coeliac disease. *Lancet*, **2**, 162–164.

Stone, E.M., Webster, A.R., Vandenburgh, K. *et al.* (1998). Allelic variation in ABCR associated with Stargardt disease but not age-related macular degeneration. *Nat Genet*, **20**, 328–329.

Strachan, T. and Read, A.P. (2004). *Human Molecular Genetics 3*. Garland Science, London.

Strand, M., Prolla, T.A., Liskay, R.M. and Petes, T.D. (1993). Destabilization of tracts of simple repetitive DNA in yeast by mutations affecting DNA mismatch repair. *Nature*, **365**, 274–276.

Stranger, B.E., Forrest, M.S., Clark, A.G. *et al.* (2005). Genome-wide associations of gene expression variation in humans. *PLoS Genet*, **1**, e78.

Stranger, B.E., Forrest, M.S., Dunning, M. *et al.* (2007a). Relative impact of nucleotide and copy number variation on gene expression phenotypes. *Science*, **315**, 848–853.

Stranger, B.E., Nica, A.C., Forrest, M.S. *et al.* (2007b). Population genomics of human gene expression. *Nat Genet*, **39**, 1217–1224.

Straub, R.E., MacLean, C.J., O'Neill, F.A. *et al.* (1995). A potential vulnerability locus for schizophrenia on chromosome 6p24–22: evidence for genetic heterogeneity. *Nat Genet*, **11**, 287–293.

Strebel, K., Daugherty, D., Clouse, K., Cohen, D., Folks, T. and Martin, M.A. (1987). The HIV 'A' (sor) gene product is essential for virus infectivity. *Nature*, **328**, 728–730.

Streisinger, G., Walker, C., Dower, N., Knauber, D. and Singer, F. (1981). Production of clones of homozygous diploid zebra fish (Brachydanio rerio). *Nature*, **291**, 293–296.

Stremlau, M., Owens, C.M., Perron, M.J., Kiessling, M., Autissier, P. and Sodroski, J. (2004). The cytoplasmic body component TRIM5alpha restricts HIV-1 infection in Old World monkeys. *Nature*, **427**, 848–853.

Stremlau, M., Perron, M., Lee, M. *et al.* (2006). Specific recognition and accelerated uncoating of retroviral capsids by the TRIM5alpha restriction factor. *Proc Natl Acad Sci U S A*, **103**, 5514–5519.

Stremlau, M., Perron, M., Welikala, S. and Sodroski, J. (2005). Species-specific variation in the B30.2(SPRY) domain of TRIM5alpha determines the potency of human immunodeficiency virus restriction. *J Virol*, **79**, 3139–3145.

Stringer, C.B. and Andrews, P. (1988). Genetic and fossil evidence for the origin of modern humans. *Science*, **239**, 1263–1268.

Strittmatter, W.J. and Roses, A.D. (1996). Apolipoprotein E and Alzheimer's disease. *Annu Rev Neurosci*, **19**, 53–77.

Strittmatter, W.J., Saunders, A.M., Schmechel, D. *et al.* (1993). Apolipoprotein E: high-avidity binding to beta-amyloid and increased frequency of type 4 allele in late-onset familial Alzheimer disease. *Proc Natl Acad Sci U S A*, **90**, 1977–1981.

Strober, W., Murray, P.J., Kitani, A. and Watanabe, T. (2006). Signalling pathways and molecular interactions of NOD1 and NOD2. *Nat Rev Immunol*, **6**, 9–20.

Stromme, P., Bjornstad, P.G. and Ramstad, K. (2002). Prevalence estimation of Williams syndrome. *J Child Neurol*, **17**, 269–271.

Strong, T.V., Tagle, D.A., Valdes, J.M. *et al.* (1993). Widespread expression of the human and rat Huntington's disease gene in brain and nonneural tissues. *Nat Genet*, **5**, 259–265.

Sturm, R.A. (2006). A golden age of human pigmentation genetics. *Trends Genet*, **22**, 464–468.

Su, B., Sun, G., Lu, D. *et al.* (2000). Distribution of three HIV-1 resistance-conferring polymorphisms (SDF1-3′A, CCR2-641, and CCR5-delta32) in global populations. *Eur J Hum Genet*, **8**, 975–979.

Su, X.Z., Heatwole, V.M., Wertheimer, S.P. *et al.* (1995). The large diverse gene family var encodes proteins involved in cytoadherence and antigenic variation of Plasmodium falciparum-infected erythrocytes. *Cell*, **82**, 89–100.

Su, Z., Wang, J., Yu, J., Huang, X. and Gu, X. (2006). Evolution of alternative splicing after gene duplication. *Genome Res*, **16**, 182–189.

Subramanian, S., Mishra, R.K. and Singh, L. (2003). Genome-wide analysis of microsatellite repeats in humans: their abundance and density in specific genomic regions. *Genome Biol*, **4**, R13.

Sulem, P., Gudbjartsson, D.F., Stacey, S.N. *et al.* (2007). Genetic determinants of hair, eye and skin pigmentation in Europeans. *Nat Genet*, **39**, 1443–1452.

Sulem, P., Gudbjartsson, D.F., Stacey, S.N. *et al.* (2008). Two newly identified genetic determinants of pigmentation in Europeans. *Nat Genet*, **40**, 835–837.

Sun, T., Gao, Y., Tan, W. *et al.* (2007). A six-nucleotide insertion-deletion polymorphism in the CASP8 promoter is associated with susceptibility to multiple cancers. *Nat Genet*, **39**, 605–613.

Suter, U., Welcher, A.A., Ozcelik, T. *et al.* (1992). Trembler mouse carries a point mutation in a myelin gene. *Nature*, **356**, 241–244.

Suzuki, Y., Hamamoto, Y., Ogasawara, Y. *et al.* (2004). Genetic polymorphisms of killer cell immunoglobulin-like receptors are associated with susceptibility to psoriasis vulgaris. *J Invest Dermatol*, **122**, 1133–1136.

Svensson, P.J. and Dahlback, B. (1994). Resistance to activated protein C as a basis for venous thrombosis. *N Engl J Med*, **330**, 517–522.

Swallow, D.M. (2003). Genetics of lactase persistence and lactose intolerance. *Annu Rev Genet*, **37**, 197–219.

Swaroop, A., Branham, K.E., Chen, W. and Abecasis, G. (2007). Genetic susceptibility to age-related macular degeneration: a paradigm for dissecting complex disease traits. *Hum Mol Genet*, **16**, R174–182.

Swen, J.J., Wilting, I., de Goede, A.L. *et al.* (2008). Pharmacogenetics: from bench to byte. *Clin Pharmacol Ther*, **83**, 781–787.

Szamalek, J.M., Cooper, D.N., Schempp, W. *et al.* (2006). Polymorphic micro-inversions contribute to the genomic variability of humans and chimpanzees. *Hum Genet*, **119**, 103–112.

Tae, H.J., Luo, X. and Kim, K.H. (1994). Roles of CCAAT/enhancer-binding protein and its binding site on repression and derepression of acetyl-CoA carboxylase gene. *J Biol Chem*, **269**, 10475–10484.

Tajima, F. (1989). Statistical method for testing the neutral mutation hypothesis by DNA polymorphism. *Genetics*, **123**, 585–595.

Takahata, N. (1993). Allelic genealogy and human evolution. *Mol Biol Evol*, **10**, 2–22.

Takahata, N. and Satta, Y. (1998). Improbable truth in human MHC diversity? *Nat Genet*, **18**, 204–206.

Takedatsu, H., Michelsen, K.S., Wei, B. *et al.* (2008). TL1A (TNFSF15) regulates the development of chronic colitis by modulating both T-helper 1 and T-helper 17 activation. *Gastroenterology*, **135**, 552–567.

Talbot, K., Eidem, W.L., Tinsley, C.L. *et al.* (2004). Dysbindin-1 is reduced in intrinsic, glutamatergic terminals of the hippocampal formation in schizophrenia. *J Clin Invest*, **113**, 1353–1363.

Tamaki, K. and Jeffreys, A.J. (2005). Human tandem repeat sequences in forensic DNA typing. *Leg Med (Tokyo)*, **7**, 244–250.

Tanzi, R.E. and Bertram, L. (2005). Twenty years of the Alzheimer's disease amyloid hypothesis: a genetic perspective. *Cell*, **120**, 545–555.

Tanzi, R.E., Gusella, J.F., Watkins, P.C. *et al.* (1987). Amyloid beta protein gene: cDNA, mRNA distribution, and genetic linkage near the Alzheimer locus. *Science*, **235**, 880–884.

Tautz, D. (1989). Hypervariability of simple sequences as a general source for polymorphic DNA markers. *Nucleic Acids Res*, **17**, 6463–6471.

Tautz, D., Trick, M. and Dover, G.A. (1986). Cryptic simplicity in DNA is a major source of genetic variation. *Nature*, **322**, 652–656.

Tavendale, R., Macgregor, D.F., Mukhopadhyay, S. and Palmer, C.N. (2008). A polymorphism controlling ORMDL3 expression is associated with asthma that is poorly controlled by current medications. *J Allergy Clin Immunol*, **121**, 860–863.

Tawil, R. and Van Der Maarel, S.M. (2006). Facioscapulohumeral muscular dystrophy. *Muscle Nerve*, **34**, 1–15.

Taylor, J.M., Dozy, A., Kan, Y.W. *et al.* (1974). Genetic lesion in homozygous alpha thalassaemia (hydrops fetalis). *Nature*, **251**, 392–393.

Taylor, J.S. and Raes, J. (2004). Duplication and divergence: the evolution of new genes and old ideas. *Annu Rev Genet*, **38**, 615–643.

Taylor, J.S., Durkin, J.M. and Breden, F. (1999). The death of a microsatellite: a phylogenetic perspective on microsatellite interruptions. *Mol Biol Evol*, **16**, 567–572.

TCSCG (Treacher Collins Syndrome Collaborative Group) (1996). Positional cloning of a gene involved in the pathogenesis of Treacher Collins syndrome. The Treacher Collins Syndrome Collaborative Group. *Nat Genet*, **12**, 130–136.

Telenius, H., Kremer, B., Goldberg, Y.P. *et al.* (1994). Somatic and gonadal mosaicism of the Huntington disease gene CAG repeat in brain and sperm. *Nat Genet*, **6**, 409–414.

Telenius, H., Kremer, H.P., Theilmann, J. *et al.* (1993). Molecular analysis of juvenile Huntington disease: the major influence on (CAG)n repeat length is the sex of the affected parent. *Hum Mol Genet*, **2**, 1535–1540.

TEPC (The ENCODE Project Consortium) (2004). The ENCODE (ENCyclopedia Of DNA Elements) Project. *Science*, **306**, 636–640.

Terrenato, L., Shrestha, S., Dixit, K.A. *et al.* (1988). Decreased malaria morbidity in the Tharu people compared to sympatric populations in Nepal. *Ann Trop Med Parasitol*, **82**, 1–11.

Thakkinstian, A., Han, P., McEvoy, M. *et al.* (2006). Systematic review and meta-analysis of the association between complement factor H Y402H polymorphisms and age-related macular degeneration. *Hum Mol Genet*, **15**, 2784–2790.

Theodorou, I., Capoulade, C., Combadiere, C. and Debre, P. (2003). Genetic control of HIV disease. *Trends Microbiol*, **11**, 392–397.

Therman, E., Susman, B. and Denniston, C. (1989). The non-random participation of human acrocentric chromosomes in Robertsonian translocations. *Ann Hum Genet*, **53**, 49–65.

Thomson, W., Barton, A., Ke, X. *et al.* (2007). Rheumatoid arthritis association at 6q23. *Nat Genet*, **39**, 1431–1433.

Thorven, M., Grahn, A., Hedlund, K.O. *et al.* (2005). A homozygous nonsense mutation (428G-->A) in the human secretor (FUT2) gene provides resistance to symptomatic norovirus (GGII) infections. *J Virol*, **79**, 15351–15355.

Thursz, M.R., Thomas, H.C., Greenwood, B.M. and Hill, A.V. (1997). Heterozygote advantage for HLA class-II type in hepatitis B virus infection. *Nat Genet*, **17**, 11–12.

Tibben, A. (2007). Predictive testing for Huntington's disease. *Brain Res Bull*, **72**, 165–171.

TIHMP (The International HapMap Project Consortium) (2003). The International HapMap Project. *Nature*, **426**, 789–796.

Timchenko, L.T., Miller, J.W., Timchenko, N.A. *et al.* (1996). Identification of a (CUG)n triplet repeat RNA-binding protein and its expression in myotonic dystrophy. *Nucleic Acids Res*, **24**, 4407–4414.

Timmerman, V., Nelis, E., Van Hul, W. *et al.* (1992). The peripheral myelin protein gene PMP-22 is contained within the Charcot–Marie–Tooth disease type 1A duplication. *Nat Genet*, **1**, 171–175.

Timms, A.E., Crane, A.M., Sims, A.M. *et al.* (2004). The interleukin 1 gene cluster contains a major susceptibility locus for ankylosing spondylitis. *Am J Hum Genet*, **75**, 587–595.

Tishkoff, S.A., Reed, F.A., Ranciaro, A. *et al.* (2007). Convergent adaptation of human lactase persistence in Africa and Europe. *Nat Genet*, **39**, 31–40.

Tishkoff, S.A., Varkonyi, R., Cahinhinan, N. *et al.* (2001). Haplotype diversity and linkage disequilibrium at human G6PD: recent origin of alleles that confer malarial resistance. *Science*, **293**, 455–462.

Tjio, H. and Levan, A. (1956). The chromosome numbers of man. *Hereditas*, **42**, 1–6.

Todd, J.A., Bell, J.I. and McDevitt, H.O. (1987). HLA-DQ beta gene contributes to susceptibility and resistance to insulin-dependent diabetes mellitus. *Nature*, **329**, 599–604.

Todd, J.A., Walker, N.M., Cooper, J.D. *et al.* (2007). Robust associations of four new chromosome regions from genome-wide analyses of type 1 diabetes. *Nat Genet*, **39**, 857–864.

Tomlinson, I., Webb, E., Carvajal-Carmona, L. *et al.* (2007). A genome-wide association scan of tag SNPs identifies a susceptibility variant for colorectal cancer at 8q24.21. *Nat Genet*, **39**, 984–988.

Tonegawa, S. (1983). Somatic generation of antibody diversity. *Nature*, **302**, 575–581.

Tooth, H. (1886). *The Peroneal Type of Progressive Muscular Atrophy*. Lewis, London.

Torcia, M.G., Santarlasci, V., Cosmi, L. *et al.* (2008). Functional deficit of T regulatory cells in Fulani, an ethnic group with low susceptibility to Plasmodium falciparum malaria. *Proc Natl Acad Sci U S A*, **105**, 646–651.

Torres, E.M., Sokolsky, T., Tucker, C.M. *et al.* (2007). Effects of aneuploidy on cellular physiology and cell division in haploid yeast. *Science*, **317**, 916–924.

Toth, G., Gaspari, Z. and Jurka, J. (2000). Microsatellites in different eukaryotic genomes: survey and analysis. *Genome Res*, **10**, 967–981.

Tournamille, C., Colin, Y., Cartron, J.P. and Le Van Kim, C. (1995a). Disruption of a GATA motif in the Duffy gene promoter abolishes erythroid gene expression in Duffy-negative individuals. *Nat Genet*, **10**, 224–228.

Tournamille, C., Le Van Kim, C., Gane, P., Cartron, J.P. and Colin, Y. (1995b). Molecular basis and PCR-DNA typing of the Fya/fyb blood group polymorphism. *Hum Genet*, **95**, 407–410.

Townson, J.R., Barcellos, L.F. and Nibbs, R.J. (2002). Gene copy number regulates the production of the human chemokine CCL3-L1. *Eur J Immunol*, **32**, 3016–3026.

Tracey, K.J. and Cerami, A. (1993). Tumor necrosis factor, other cytokines and disease. *Annu Rev Cell Biol*, **9**, 317–343.

Trachtenberg, E., Korber, B., Sollars, C. *et al.* (2003). Advantage of rare HLA supertype in HIV disease progression. *Nat Med*, **9**, 928–935.

Traeger, J., Wood, W.G., Clegg, J.B. and Weatherall, D.J. (1980). Defective synthesis of HbE is due to reduced levels of beta E mRNA. *Nature*, **288**, 497–499.

Traherne, J.A., Barcellos, L.F., Sawcer, S.J. *et al.* (2006a). Association of the truncating splice site mutation in BTNL2 with multiple sclerosis is secondary to HLA-DRB1*15. *Hum Mol Genet*, **15**, 155–161.

Traherne, J.A., Horton, R., Roberts, A.N. *et al.* (2006b). Genetic analysis of completely sequenced disease-associated MHC haplotypes identifies shuffling of segments in recent human history. *PLoS Genet*, **2**, e9.

Trask, B.J. (2002). Human cytogenetics: 46 chromosomes, 46 years and counting. *Nat Rev Genet*, **3**, 769–778.

Treacher Collins, E. (1900). Cases with symmetrical congenital notches in the outer part of each lid and defective development of the malar bones. *Trans Opthalmol Soc UK*, **20**, 190–192.

Treisman, R., Orkin, S.H. and Maniatis, T. (1983). Specific transcription and RNA splicing defects in five cloned beta-thalassaemia genes. *Nature*, **302**, 591–596.

Treisman, R., Proudfoot, N.J., Shander, M. and Maniatis, T. (1982). A single-base change at a splice site in a beta β-thalassemic gene causes abnormal RNA splicing. *Cell*, **29**, 903–911.

Trembath, R.C., Clough, R.L., Rosbotham, J.L. *et al.* (1997). Identification of a major susceptibility locus on chromosome 6p and evidence for further disease loci revealed by a two stage genome-wide search in psoriasis. *Hum Mol Genet*, **6**, 813–820.

Troelsen, J.T., Olsen, J., Moller, J. and Sjostrom, H. (2003). An upstream polymorphism associated with lactase persistence has increased enhancer activity. *Gastroenterology*, **125**, 1686–1694.

Turner, G., Webb, T., Wake, S. and Robinson, H. (1996). Prevalence of fragile X syndrome. *Am J Med Genet*, **64**, 196–197.

Turner, H. (1938). A syndrome of infantilism, congenital webbed neck, and cubitus valgus. *Endocrinology*, **23**, 566–574.

Turri, M.G., Cuin, K.A. and Porter, A.C. (1995). Characterisation of a novel minisatellite that provides multiple splice donor sites in an interferon-induced transcript. *Nucleic Acids Res*, **23**, 1854–1861.

Tuzun, E., Bailey, J.A. and Eichler, E.E. (2004). Recent segmental duplications in the working draft assembly of the brown Norway rat. *Genome Res*, **14**, 493–506.

Tuzun, E., Sharp, A.J., Bailey, J.A. *et al.* (2005). Fine-scale structural variation of the human genome. *Nat Genet*, **37**, 727–732.

Ueda, H., Howson, J.M., Esposito, L. *et al.* (2003). Association of the T-cell regulatory gene CTLA4 with susceptibility to autoimmune disease. *Nature*, **423**, 506–511.

Ullu, E. and Tschudi, C. (1984). Alu sequences are processed 7SL RNA genes. *Nature*, **312**, 171–172.

Ullu, E. and Weiner, A.M. (1985). Upstream sequences modulate the internal promoter of the human 7SL RNA gene. *Nature*, **318**, 371–374.

Usanga, E.A. and Luzzatto, L. (1985). Adaptation of Plasmodium falciparum to glucose 6-phosphate dehydrogenase-deficient host red cells by production of parasite-encoded enzyme. *Nature*, **313**, 793–795.

Uzun, A., Leslin, C.M., Abyzov, A. and Ilyin, V. (2007). Structure SNP (StSNP): a web server for mapping and modeling nsSNPs on protein structures with linkage to metabolic pathways. *Nucleic Acids Res*, **35**, W384–392.

Vafiadis, P., Bennett, S.T., Colle, E., Grabs, R., Goodyer, C.G. and Polychronakos, C. (1996). Imprinted and genotype-specific expression of genes at the IDDM2 locus in pancreas and leucocytes. *J Autoimmun*, **9**, 397–403.

Vafiadis, P., Bennett, S.T., Todd, J.A. *et al.* (1997). Insulin expression in human thymus is modulated by INS VNTR alleles at the IDDM2 locus. *Nat Genet*, **15**, 289–292.

Valentijn, L.J., Baas, F., Wolterman, R.A. *et al.* (1992a). Identical point mutations of PMP-22 in Trembler-J mouse and Charcot–Marie–Tooth disease type 1A. *Nat Genet*, **2**, 288–291.

Valentijn, L.J., Bolhuis, P.A., Zorn, I. *et al.* (1992b). The peripheral myelin gene PMP-22/GAS-3 is duplicated in Charcot–Marie–Tooth disease type 1A. *Nat Genet*, **1**, 166–170.

Valentonyte, R., Hampe, J., Huse, K. *et al.* (2005). Sarcoidosis is associated with a truncating splice site mutation in BTNL2. *Nat Genet*, **37**, 357–364.

Valverde, P., Healy, E., Jackson, I., Rees, J.L. and Thody, A.J. (1995). Variants of the melanocyte-stimulating hormone receptor gene are associated with red hair and fair skin in humans. *Nat Genet*, **11**, 328–330.

van Bon, B.W., Koolen, D.A., Borgatti, R. *et al.* (2008). Clinical and molecular characteristics of 1qter microdeletion syndrome: delineating a critical region for corpus callosum agenesis/hypogenesis. *J Med Genet*, **45**, 346–354.

van Broeckhoven, C., Backhovens, H., Cruts, M. *et al.* (1992). Mapping of a gene predisposing to early-onset Alzheimer's disease to chromosome 14q24.3. *Nat Genet*, **2**, 335–339.

van der Helm-van Mil, A.H., Verpoort, K.N., Breedveld, F.C., Huizinga, T.W., Toes, R.E. and de Vries, R.R. (2006). The HLA-DRB1 shared epitope alleles are primarily a risk factor for anti-cyclic citrullinated peptide antibodies and are not an independent risk factor for development of rheumatoid arthritis. *Arthritis Rheum*, **54**, 1117–1121.

van der Heyde, H.C., Nolan, J., Combes, V., Gramaglia, I. and Grau, G.E. (2006). A unified hypothesis for the genesis of cerebral malaria: sequestration, inflammation and hemostasis leading to microcirculatory dysfunction. *Trends Parasitol*, **22**, 503–508.

van der Werf, M.J., de Vlas, S.J., Brooker, S. *et al.* (2003). Quantification of clinical morbidity associated with schistosome infection in sub-Saharan Africa. *Acta Trop*, **86**, 125–139.

van Deutekom, J.C., Bakker, E., Lemmers, R.J. *et al.* (1996). Evidence for subtelomeric exchange of 3.3 kb tandemly repeated units between chromosomes 4q35 and 10q26: implications for genetic counselling and etiology of FSHD1. *Hum Mol Genet*, **5**, 1997–2003.

van Dyke, D.L., Weiss, L., Roberson, J.R. and Babu, V.R. (1983). The frequency and mutation rate of balanced autosomal rearrangements in man estimated from prenatal genetic studies for advanced maternal age. *Am J Hum Genet*, **35**, 301–308.

van Eerdewegh, P., Little, R.D., Dupuis, J. *et al.* (2002). Association of the ADAM33 gene with asthma and bronchial hyperresponsiveness. *Nature*, **418**, 426–430.

van Esch, H., Bauters, M., Ignatius, J. *et al.* (2005). Duplication of the MECP2 region is a frequent cause of severe mental retardation and progressive neurological symptoms in males. *Am J Hum Genet*, **77**, 442–453.

van Heel, D.A., Fisher, S.A., Kirby, A., Daly, M.J., Rioux, J.D. and Lewis, C.M. (2004). Inflammatory bowel disease susceptibility loci defined by genome scan meta-analysis of 1952 affected relative pairs. *Hum Mol Genet*, **13**, 763–770.

van Kim, C.L., Colin, Y. and Cartron, J.P. (2006). Rh proteins: key structural and functional components of the red cell membrane. *Blood Rev*, **20**, 93–110.

van Luenen, H.G., Colloms, S.D. and Plasterk, R.H. (1994). The mechanism of transposition of Tc3 in C. elegans. *Cell*, **79**, 293–301.

van Overveld, P.G., Enthoven, L., Ricci, E. *et al.* (2005). Variable hypomethylation of D4Z4 in facioscapulohumeral muscular dystrophy. *Ann Neurol*, **58**, 569–576.

van Rij, R.P., Broersen, S., Goudsmit, J., Coutinho, R.A. and Schuitemaker, H. (1998a). The role of a stromal cell-derived factor-1 chemokine gene variant in the clinical course of HIV-1 infection. *Aids*, **12**, F85–90.

van Rij, R.P., de Roda Husman, A.M., Brouwer, M., Goudsmit, J., Coutinho, R.A. and Schuitemaker, H. (1998b). Role of CCR2 genotype in the clinical course of syncytium-inducing (SI) or non-SI human immunodeficiency virus type 1 infection and in the time to conversion to SI virus variants. *J Infect Dis*, **178**, 1806–1811.

Vance, J.M., Nicholson, G.A., Yamaoka, L.H. *et al.* (1989). Linkage of Charcot–Marie–Tooth neuropathy type 1a to chromosome 17. *Exp Neurol*, **104**, 186–189.

Vekilov, P.G. (2007). Sickle-cell haemoglobin polymerization: is it the primary pathogenic event of sickle-cell anaemia? *Br J Haematol*, **139**, 173–184.

Vella, A., Cooper, J.D., Lowe, C.E. *et al.* (2005). Localization of a type 1 diabetes locus in the IL2RA/CD25 region by use of tag single-nucleotide polymorphisms. *Am J Hum Genet*, **76**, 773–779.

Venkatesan, S., Petrovic, A., Van Ryk, D.I., Locati, M., Weissman, D. and Murphy, P.M. (2002). Reduced cell surface expression of CCR5 in CCR5Delta 32 heterozygotes is mediated by gene dosage, rather than by receptor sequestration. *J Biol Chem*, **277**, 2287–2301.

Venter, J.C., Adams, M.D., Myers, E.W. *et al.* (2001). The sequence of the human genome. *Science*, **291**, 1304–1351.

Vercelli, D. (2008). Discovering susceptibility genes for asthma and allergy. *Nat Rev Immunol*, **8**, 169–182.

Vergnaud, G., Mariat, D., Apiou, F., Aurias, A., Lathrop, M. and Lauthier, V. (1991). The use of synthetic tandem repeats to isolate new VNTR loci: cloning of a human hypermutable sequence. *Genomics*, **11**, 135–144.

Verkerk, A.J., Pieretti, M., Sutcliffe, J.S. *et al.* (1991). Identification of a gene (FMR-1) containing a CGG repeat coincident with a breakpoint cluster region exhibiting length variation in fragile X syndrome. *Cell*, **65**, 905–914.

Verrelli, B.C. and Tishkoff, S.A. (2004). Signatures of selection and gene conversion associated with human color vision variation. *Am J Hum Genet*, **75**, 363–375.

Vidal, S.M., Malo, D., Vogan, K., Skamene, E. and Gros, P. (1993). Natural resistance to infection with intracellular parasites: isolation of a candidate for Bcg. *Cell*, **73**, 469–485.

Vidal, S.M., Tremblay, M.L., Govoni, G. *et al.* (1995). The Ity/Lsh/Bcg locus: natural resistance to infection with intracellular parasites is abrogated by disruption of the Nramp1 gene. *J Exp Med*, **182**, 655–666.

Vigilant, L., Stoneking, M., Harpending, H., Hawkes, K. and Wilson, A.C. (1991). African populations and the evolution of human mitochondrial DNA. *Science*, **253**, 1503–1507.

Vilches, C. and Parham, P. (2002). KIR: diverse, rapidly evolving receptors of innate and adaptive immunity. *Annu Rev Immunol*, **20**, 217–251.

Virtaneva, K., D'Amato, E., Miao, J. *et al.* (1997). Unstable minisatellite expansion causing recessively inherited myoclonus epilepsy, EPM1. *Nat Genet*, **15**, 393–396.

Visootsak, J. and Graham, J.M., Jr. (2006). Klinefelter syndrome and other sex chromosomal aneuploidies. *Orphanet J Rare Dis*, **1**, 42.

Vladutiu, G.D. (2008). The FDA announces new drug labeling for pharmacogenetic testing: is personalized medicine becoming a reality? *Mol Genet Metab*, **93**, 1–4.

Voight, B.F., Kudaravalli, S., Wen, X. and Pritchard, J.K. (2006). A map of recent positive selection in the human genome. *PLoS Biol*, **4**, e72.

Volarcik, K., Sheean, L., Goldfarb, J., Woods, L., Abdul-Karim, F.W. and Hunt, P. (1998). The meiotic competence of in-vitro matured human oocytes is influenced by donor age: evidence that folliculogenesis is compromised in the reproductively aged ovary. *Hum Reprod*, **13**, 154–160.

Vollrath, D., Nathans, J. and Davis, R.W. (1988). Tandem array of human visual pigment genes at Xq28. *Science*, **240**, 1669–1672.

von Salome, J., Gyllensten, U. and Bergstrom, T.F. (2007). Full-length sequence analysis of the HLA-DRB1 locus suggests a recent origin of alleles. *Immunogenetics*, **59**, 261–271.

Vonsattel, J.P., Myers, R.H., Stevens, T.J., Ferrante, R.J., Bird, E.D. and Richardson, E.P., Jr. (1985). Neuropathological classification of Huntington's disease. *J Neuropathol Exp Neurol*, **44**, 559–577.

Vorstman, J.A., Staal, W.G., van Daalen, E., van Engeland, H., Hochstenbach, P.F. and Franke, L. (2006). Identification of novel autism candidate regions through analysis of reported cytogenetic abnormalities associated with autism. *Mol Psychiatry*, **11**, 1, 18–28.

Voss, T.S., Healer, J., Marty, A.J. *et al.* (2006). A var gene promoter controls allelic exclusion of virulence genes in Plasmodium falciparum malaria. *Nature*, **439**, 1004–1008.

Wagner, F.F. and Flegel, W.A. (2000). RHD gene deletion occurred in the Rhesus box. *Blood*, **95**, 3662–3668.

Wagner, F.F. and Flegel, W.A. (2002). RHCE represents the ancestral RH position, while RHD is the duplicated gene. *Blood*, **99**, 2272–2273.

Wagner, F.F. and Flegel, W.A. (2004). Review: the molecular basis of the Rh blood group phenotypes. *Immunohematol*, **20**, 23–36.

Wagner, F.F., Frohmajer, A. and Flegel, W.A. (2001). RHD positive haplotypes in D negative Europeans. *BMC Genet*, **2**, 10.

Waheed, A., Parkkila, S., Zhou, X.Y. *et al.* (1997). Hereditary hemochromatosis: effects of C282Y and H63D mutations on association with beta2-microglobulin, intracellular processing, and cell surface expression of the HFE protein in COS-7 cells. *Proc Natl Acad Sci U S A*, **94**, 12384–12389.

Wahls, W.P., Wallace, L.J. and Moore, P.D. (1990a). Hypervariable minisatellite DNA is a hotspot for homologous recombination in human cells. *Cell*, **60**, 95–103.

Wahls, W.P., Wallace, L.J. and Moore, P.D. (1990b). The Z-DNA motif d(TG)30 promotes reception of information during gene conversion events while stimulating homologous recombination in human cells in culture. *Mol Cell Biol*, **10**, 785–793.

Wainscoat, J.S., Hill, A.V., Boyce, A.L. *et al.* (1986). Evolutionary relationships of human populations from an analysis of nuclear DNA polymorphisms. *Nature*, **319**, 491–493.

Wainscoat, J.S., Kanavakis, E., Weatherall, D.J. *et al.* (1981). Regional localisation of the human alpha-globin genes. *Lancet*, **2**, 301–302.

Waldman, A.S. and Liskay, R.M. (1988). Dependence of intrachromosomal recombination in mammalian cells on uninterrupted homology. *Mol Cell Biol*, **8**, 5350–5357.

Walker, B.A., Scott, C.I., Hall, J.G., Murdoch, J.L. and McKusick, V.A. (1972). Diastrophic dwarfism. *Medicine (Baltimore)*, **51**, 41–59.

Wallace, M.R., Andersen, L.B., Saulino, A.M., Gregory, P.E., Glover, T.W. and Collins, F.S. (1991). A de novo Alu insertion results in neurofibromatosis type 1. *Nature*, **353**, 864–866.

Wang, D.G., Fan, J.B., Siao, C.J. *et al.* (1998). Large-scale identification, mapping, and genotyping of single-nucleotide polymorphisms in the human genome. *Science*, **280**, 1077–1082.

Wang, E.T., Kodama, G., Baldi, P. and Moyzis, R.K. (2006). Global landscape of recent inferred Darwinian selection for Homo sapiens. *Proc Natl Acad Sci U S A*, **103**, 135–140.

Wang, G.S. and Cooper, T.A. (2007). Splicing in disease: disruption of the splicing code and the decoding machinery. *Nat Rev Genet*, **8**, 749–761.

Wang, H., Hubbell, E., Hu, J.S. *et al.* (2003a). Gene structure-based splice variant deconvolution using a microarray platform. *Bioinformatics*, **19**, i315–322.

Wang, J., Williams, R.W. and Manly, K.F. (2003b). WebQTL: web-based complex trait analysis. *Neuroinformatics*, **1**, 299–308.

Wang, T., Zeng, J., Lowe, C.B. *et al.* (2007). Species-specific endogenous retroviruses shape the transcriptional network of the human tumor suppressor protein p53. *Proc Natl Acad Sci U S A*, **104**, 18613–18618.

Wang, Y., Harvey, C.B., Pratt, W.S. *et al.* (1995). The lactase persistence/non-persistence polymorphism is controlled by a cis-acting element. *Hum Mol Genet*, **4**, 657–662.

Wang, Y., Macke, J.P., Merbs, S.L. *et al.* (1992). A locus control region adjacent to the human red and green visual pigment genes. *Neuron*, **9**, 429–440.

Wang, Y.H. and Griffith, J. (1995). Expanded CTG triplet blocks from the myotonic dystrophy gene create the strongest known natural nucleosome positioning elements. *Genomics*, **25**, 570–573.

Wang, Y.H., Amirhaeri, S., Kang, S., Wells, R.D. and Griffith, J.D. (1994). Preferential nucleosome assembly at DNA triplet repeats from the myotonic dystrophy gene. *Science*, **265**, 669–671.

Wardle, J., Carnell, S., Haworth, C.M., Farooqi, I.S., O'Rahilly, S. and Plomin, R. (2008). Obesity-associated genetic variation in FTO is associated with diminished satiety. *J Clin Endocrinol Metab*, **93**, 3640–3643.

Waring, S.C. and Rosenberg, R.N. (2008). Genome-wide association studies in Alzheimer disease. *Arch Neurol*, **65**, 329–334.

Warpeha, K.M., Xu, W., Liu, L. *et al.* (1999). Genotyping and functional analysis of a polymorphic (CCTTT)(n) repeat of NOS2A in diabetic retinopathy. *FASEB J*, **13**, 1825–1832.

Warren, S.T. (1997). Polyalanine expansion in synpolydactyly might result from unequal crossing-over of HOXD13. *Science*, **275**, 408–409.

Watanabe, H., Fujiyama, A., Hattori, M. *et al.* (2004). DNA sequence and comparative analysis of chimpanzee chromosome 22. *Nature*, **429**, 382–388.

Watanabe, T., Kitani, A. and Strober, W. (2005). NOD2 regulation of Toll-like receptor responses and the pathogenesis of Crohn's disease. *Gut*, **54**, 1515–1518.

Watase, K., Weeber, E.J., Xu, B. *et al.* (2002). A long CAG repeat in the mouse Sca1 locus replicates SCA1 features and reveals the impact of protein solubility on selective neurodegeneration. *Neuron*, **34**, 905–919.

Waterston, R.H., Lindblad-Toh, K., Birney, E. *et al.* (2002). Initial sequencing and comparative analysis of the mouse genome. *Nature*, **420**, 520–562.

Watkins, W.S., Ricker, C.E., Bamshad, M.J. *et al.* (2001). Patterns of ancestral human diversity: an analysis of Alu-insertion and restriction-site polymorphisms. *Am J Hum Genet*, **68**, 738–752.

Watkins, W.S., Rogers, A.R., Ostler, C.T. *et al.* (2003). Genetic variation among world populations: inferences from 100 Alu insertion polymorphisms. *Genome Res*, **13**, 1607–1618.

Watson, J.D. and Crick, F.H. (1953). Molecular structure of nucleic acids; a structure for deoxyribose nucleic acid. *Nature*, **171**, 737–738.

Wattavidanage, J., Carter, R., Perera, K.L. *et al.* (1999). TNFalpha*2 marks high risk of severe disease during Plasmodium falciparum malaria and other infections in Sri Lankans. *Clin Exp Immunol*, **115**, 350–355.

Watts, C. (2004). The exogenous pathway for antigen presentation on major histocompatibility complex class II and CD1 molecules. *Nat Immunol*, **5**, 685–692.

Waye, J.S. and Willard, H.F. (1986). Structure, organization, and sequence of alpha satellite DNA from human chromosome 17: evidence for evolution by unequal crossing-over and an ancestral pentamer repeat shared with the human X chromosome. *Mol Cell Biol*, **6**, 3156–3165.

Wayne, M.L. and McIntyre, L.M. (2002). Combining mapping and arraying: an approach to candidate gene identification. *Proc Natl Acad Sci U S A*, **99**, 14903–14906.

Weatherall, D.J. (2000). Single gene disorders or complex traits: lessons from the thalassaemias and other monogenic diseases. *Br Med J*, **321**, 1117–1120.

Weatherall, D.J. (2001). Phenotype-genotype relationships in monogenic disease: lessons from the thalassaemias. *Nat Rev Genet*, **2**, 245–255.

Weatherall, D.J. (2004a). 2003 William Allan Award address. The thalassemias: the role of molecular genetics in an evolving global health problem. *Am J Hum Genet*, **74**, 385–392.

Weatherall, D.J. (2004b). Thalassaemia: the long road from bedside to genome. *Nat Rev Genet*, **5**, 625–631.

Weatherall, D.J. (2008). Genetic variation and susceptibility to infection: the red cell and malaria. *Br J Haematol*, **141**, 276–286.

Weatherall, D.J. and Clegg, J.B. (2001). *The Thalassaemia Syndromes*. Blackwell Science, Oxford, UK.

Weatherall, D.J., Clegg, J.B. and Boon, W.H. (1970). The haemoglobin constitution of infants with the haemoglobin Bart's hydrops foetalis syndrome. *Br J Haematol*, **18**, 357–367.

Weatherall, D.J., Higgs, D.R., Bunch, C. *et al.* (1981). Hemoglobin H disease and mental retardation: a new syndrome or a remarkable coincidence? *N Engl J Med*, **305**, 607–612.

Weber, J.L. and May, P.E. (1989). Abundant class of human DNA polymorphisms which can be typed using the polymerase chain reaction. *Am J Hum Genet*, **44**, 388–396.

Weber, J.L. and Wong, C. (1993). Mutation of human short tandem repeats. *Hum Mol Genet*, **2**, 1123–1128.

Weeks, J.R., Hardin, S.E., Shen, J., Lee, J.M. and Greenleaf, A.L. (1993). Locus-specific variation in phosphorylation state of RNA polymerase II in vivo: correlations with gene activity and transcript processing. *Genes Dev*, **7**, 2329–2344.

Wehkamp, J., Harder, J., Weichenthal, M. *et al.* (2004). NOD2 (CARD15) mutations in Crohn's disease are associated with diminished mucosal alpha-defensin expression. *Gut*, **53**, 1658–1664.

Wei, C.L., Wu, Q., Vega, V.B. *et al.* (2006). A global map of p53 transcription-factor binding sites in the human genome. *Cell*, **124**, 207–219.

Weickert, C.S., Straub, R.E., McClintock, B.W. *et al.* (2004). Human dysbindin (DTNBP1) gene expression in normal brain and in schizophrenic prefrontal cortex and midbrain. *Arch Gen Psychiatry*, **61**, 544–555.

Weigert, M., Gatmaitan, L., Loh, E., Schilling, J. and Hood, L. (1978). Rearrangement of genetic information may produce immunoglobulin diversity. *Nature*, **276**, 785–790.

Weigert, M.G., Cesari, I.M., Yonkovich, S.J. and Cohn, M. (1970). Variability in the lambda light chain sequences of mouse antibody. *Nature*, **228**, 1045–1047.

Weiler, I.J., Irwin, S.A., Klintsova, A.Y. *et al.* (1997). Fragile X mental retardation protein is translated near synapses in response to neurotransmitter activation. *Proc Natl Acad Sci U S A*, **94**, 5395–5400.

Weir, B.S. and Cockerham, C.C. (1984). Estimating F-statistics for the analysis of population structure. *Evolution*, **38**, 1358–1370.

Weiss, M.C. and Green, H. (1967). Human–mouse hybrid cell lines containing partial complements of human chromosomes and functioning human genes. *Proc Natl Acad Sci U S A*, **58**, 1104–1111.

Weissenbach, J., Gyapay, G., Dib, C. *et al.* (1992). A second-generation linkage map of the human genome. *Nature*, **359**, 794–801.

Wells, R.D. (2008). DNA triplexes and Friedreich ataxia. *FASEB J*, **22**, 1625–1634.

Welsh, M.J. and Smith, A.E. (1993). Molecular mechanisms of CFTR chloride channel dysfunction in cystic fibrosis. *Cell*, **73**, 1251–1254.

Wetsel, R.A., Fleischer, D.T. and Haviland, D.L. (1990). Deficiency of the murine fifth complement component (C5). A 2-base pair gene deletion in a 5′-exon. *J Biol Chem*, **265**, 2435–2440.

Wexler, N.S., Lorimer, J., Porter, J. *et al.* (2004). Venezuelan kindreds reveal that genetic and environmental factors modulate Huntington's disease age of onset. *Proc Natl Acad Sci U S A*, **101**, 3498–3503.

Wheeler, D.A., Srinivasan, M., Egholm, M. *et al.* (2008a). The complete genome of an individual by massively parallel DNA sequencing. *Nature*, **452**, 872–876.

Wheeler, D.L., Barrett, T., Benson, D.A. *et al.* (2008b). Database resources of the National Center for Biotechnology Information. *Nucleic Acids Res*, **36**, D13–21.

Whitcomb, D.C., Gorry, M.C., Preston, R.A. *et al.* (1996). Hereditary pancreatitis is caused by a mutation in the cationic trypsinogen gene. *Nat Genet*, **14**, 141–145.

Whitelaw, E. and Proudfoot, N. (1986). Alpha-thalassaemia caused by a poly(A) site mutation reveals that transcriptional termination is linked to 3′ end processing in the human alpha 2 globin gene. *EMBO J*, **5**, 2915–2922.

Whitlatch, N.L. and Ortel, T.L. (2008). Thrombophilias: when should we test and how does it help? *Semin Respir Crit Care Med*, **29**, 25–39.

Whittaker, J.C., Harbord, R.M., Boxall, N., Mackay, I., Dawson, G. and Sibly, R.M. (2003). Likelihood-based estimation of microsatellite mutation rates. *Genetics*, **164**, 781–787.

Wicker, T., Sabot, F., Hua-Van, A. *et al.* (2007). A unified classification system for eukaryotic transposable elements. *Nat Rev Genet*, **8**, 973–982.

Wicks, S.R., Yeh, R.T., Gish, W.R., Waterston, R.H. and Plasterk, R.H. (2001). Rapid gene mapping in Caenorhabditis elegans using a high density polymorphism map. *Nat Genet*, **28**, 160–164.

Wijmenga, C., Hewitt, J.E., Sandkuijl, L.A. *et al.* (1992). Chromosome 4q DNA rearrangements associated with facioscapulohumeral muscular dystrophy. *Nat Genet*, **2**, 26–30.

Wilder, J. and Hollocher, H. (2001). Mobile elements and the genesis of microsatellites in dipterans. *Mol Biol Evol*, **18**, 384–392.

Wilkie, A.O., Buckle, V.J., Harris, P.C. *et al.* (1990). Clinical features and molecular analysis of the alpha thalassemia/mental retardation syndromes. I. Cases due to deletions involving chromosome band 16p13.3. *Am J Hum Genet*, **46**, 1112–1126.

Wilkie, A.O., Higgs, D.R., Rack, K.A. *et al.* (1991). Stable length polymorphism of up to 260 kb at the tip of the short arm of human chromosome 16. *Cell*, **64**, 595–606.

Willard, H.F. and Waye, J.S. (1987). Chromosome-specific subsets of human alpha satellite DNA: analysis of sequence divergence within and between chromosomal subsets and evidence for an ancestral pentameric repeat. *J Mol Evol*, **25**, 207–214.

Willatt, L., Cox, J., Barber, J. *et al.* (2005). 3q29 microdeletion syndrome: clinical and molecular characterization of a new syndrome. *Am J Hum Genet*, **77**, 154–160.

Willcox, M., Bjorkman, A., Brohult, J., Pehrson, P.O., Rombo, L. and Bengtsson, E. (1983). A case–control study in northern Liberia of Plasmodium falciparum malaria in haemoglobin S and beta-thalassaemia traits. *Ann Trop Med Parasitol*, **77**, 239–246.

Williams, N.M., O'Donovan, M.C. and Owen, M.J. (2005a). Is the dysbindin gene (DTNBP1) a susceptibility gene for schizophrenia? *Schizophr Bull*, **31**, 800–805.

Williams, R.C., Steinberg, A.G., Knowler, W.C. and Pettitt, D.J. (1986). Gm 3;5,13,14 and stated-admixture: independent estimates of admixture in American Indians. *Am J Hum Genet*, **39**, 409–413.

Williams, T.N. (2006). Red blood cell defects and malaria. *Mol Biochem Parasitol*, **149**, 121–127.

Williams, T.N., Maitland, K., Bennett, S. *et al.* (1996). High incidence of malaria in alpha-thalassaemic children. *Nature*, **383**, 522–525.

Williams, T.N., Mwangi, T.W., Roberts, D.J. *et al.* (2005b). An immune basis for malaria protection by the sickle cell trait. *PLoS Med*, **2**, e128.

Williams, T.N., Mwangi, T.W., Wambua, S. *et al.* (2005c). Sickle cell trait and the risk of Plasmodium falciparum malaria and other childhood diseases. *J Infect Dis*, **192**, 178–186.

Williams, T.N., Mwangi, T.W., Wambua, S. *et al.* (2005d). Negative epistasis between the malaria-protective effects of alpha+-thalassemia and the sickle cell trait. *Nat Genet*, **37**, 1253–1257.

Williams, T.N., Wambua, S., Uyoga, S. *et al.* (2005e). Both heterozygous and homozygous alpha+ thalassemias protect against severe and fatal Plasmodium falciparum malaria on the coast of Kenya. *Blood*, **106**, 368–371.

Williams-Blangero, S., Correa-Oliveira, R., Vandeberg, J.L. *et al.* (2004). Genetic influences on plasma cytokine variation in a parasitized population. *Hum Biol*, **76**, 515–525.

Williams-Blangero, S., Subedi, J., Upadhayay, R.P. *et al.* (1999). Genetic analysis of susceptibility to infection with Ascaris lumbricoides. *Am J Trop Med Hyg*, **60**, 921–926.

Williams-Blangero, S., VandeBerg, J.L., Subedi, J. *et al.* (2002). Genes on chromosomes 1 and 13 have significant effects on Ascaris infection. *Proc Natl Acad Sci U S A*, **99**, 5533–5538.

Williamson, D., Langdown, J.V., Myles, T., Mason, C., Henthorn, J.S. and Davies, S.C. (1992). Polycythaemia and microcytosis arising from the combination of a new high oxygen affinity haemoglobin (Hb luton, alpha 89 His-->Leu) and alpha thalassaemia trait. *Br J Haematol*, **82**, 621–622.

Williamson, S.H., Hubisz, M.J., Clark, A.G., Payseur, B.A., Bustamante, C.D. and Nielsen, R. (2007). Localizing recent adaptive evolution in the human genome. *PLoS Genet*, **3**, e90.

Wilson, J.T., Forget, B.G., Wilson, L.B. and Weissman, S.M. (1977). Human globin messenger RNA: importance of cloning for structural analysis. *Science*, **196**, 200–202.

Wilson, R., Ainscough, R., Anderson, K. *et al.* (1994). 2.2 Mb of contiguous nucleotide sequence from chromosome III of C. elegans. *Nature*, **368**, 32–38.

Wilson, W., 3rd, Pardo-Manuel de Villena, F., Lyn-Cook, B.D. *et al.* (2004). Characterization of a common deletion polymorphism of the UGT2B17 gene linked to UGT2B15. *Genomics*, **84**, 707–714.

Winderickx, J., Battisti, L., Hibiya, Y., Motulsky, A.G. and Deeb, S.S. (1993). Haplotype diversity in the human red and green opsin genes: evidence for frequent sequence exchange in exon 3. *Hum Mol Genet*, **2**, 1413–1421.

Winderickx, J., Lindsey, D.T., Sanocki, E., Teller, D.Y., Motulsky, A.G. and Deeb, S.S. (1992). Polymorphism in red photopigment underlies variation in colour matching. *Nature*, **356**, 431–433.

Winkler, C., Modi, W., Smith, M.W. *et al.* (1998). Genetic restriction of AIDS pathogenesis by an SDF-1 chemokine gene variant. ALIVE Study, Hemophilia Growth and Development Study (HGDS), Multicenter AIDS Cohort Study (MACS), Multicenter Hemophilia Cohort Study (MHCS), San Francisco City Cohort (SFCC). *Science*, **279**, 389–393.

Witherspoon, D.J., Marchani, E.E., Watkins, W.S. *et al.* (2006). Human population genetic structure and diversity inferred from polymorphic L1(LINE-1) and Alu insertions. *Hum Hered*, **62**, 30–46.

Wong, A.C., Ning, Y., Flint, J. *et al.* (1997). Molecular characterization of a 130-kb terminal microdeletion at 22q in a child with mild mental retardation. *Am J Hum Genet*, **60**, 113–120.

Wong, K.K., deLeeuw, R.J., Dosanjh, N.S. *et al.* (2007). A comprehensive analysis of common copy-number variations in the human genome. *Am J Hum Genet*, **80**, 91–104.

Wood, B.A. (2005). *Human Evolution: A very Short Introduction.* Oxford University Press, Oxford, UK.

Wood, E.T., Stover, D.A., Slatkin, M., Nachman, M.W. and Hammer, M.F. (2005). The beta -globin recombinational hotspot reduces the effects of strong selection around HbC, a recently arisen mutation providing resistance to malaria. *Am J Hum Genet*, **77**, 637–642.

Wordsworth, S., Buchanan, J., Regan, R. *et al.* (2007). Diagnosing idiopathic learning disability: a cost-effectiveness analysis of microarray technology in the National Health Service of the United Kingdom. *Genomic Med*, **1**, 35–45.

WTCCC (2007). Genome-wide association study of 14,000 cases of seven common diseases and 3,000 shared controls. *Nature*, **447**, 661–678.

Wu, L., Paxton, W.A., Kassam, N. *et al.* (1997). CCR5 levels and expression pattern correlate with infectability by macrophage-tropic HIV-1, in vitro. *J Exp Med*, **185**, 1681–1691.

Xavier, R.J. and Podolsky, D.K. (2007). Unravelling the pathogenesis of inflammatory bowel disease. *Nature*, **448**, 427–434.

Xie, T., Rowen, L., Aguado, B. *et al.* (2003). Analysis of the gene-dense major histocompatibility complex class III region and its comparison to mouse. *Genome Res*, **13**, 2621–2636.

Xing, J., Salem, A.H., Hedges, D.J. *et al.* (2003). Comprehensive analysis of two Alu Yd subfamilies. *J Mol Evol*, **57**, S76–89.

Xing, J., Witherspoon, D.J., Ray, D.A., Batzer, M.A. and Jorde, L.B. (2007). Mobile DNA elements in primate and human evolution. *Am J Phys Anthropol*, **45**, 2–19.

Xing, Y., Yu, T., Wu, Y.N., Roy, M., Kim, J. and Lee, C. (2006). An expectation-maximization algorithm for probabilistic reconstructions of full-length isoforms from splice graphs. *Nucleic Acids Res*, **34**, 3150–3160.

Xu, X., Peng, M. and Fang, Z. (2000). The direction of microsatellite mutations is dependent upon allele length. *Nat Genet*, **24**, 396–399.

Xue, Y., Daly, A., Yngvadottir, B. *et al.* (2006). Spread of an inactive form of caspase-12 in humans is due to recent positive selection. *Am J Hum Genet*, **78**, 659–670.

Xue, Y., Sun, D., Daly, A. *et al.* (2008). Adaptive evolution of UGT2B17 copy-number variation. *Am J Hum Genet*, **83**, 337–346.

Yamada, K. (1992). Population studies of INV(9) chromosomes in 4,300 Japanese: incidence, sex difference and clinical significance. *Jpn J Hum Genet*, **37**, 293–301.

Yamagata, K., Furuta, H., Oda, N. *et al.* (1996). Mutations in the hepatocyte nuclear factor-4alpha gene in maturity-onset diabetes of the young (MODY1). *Nature*, **384**, 458–460.

Yamaguchi, T., Motulsky, A.G. and Deeb, S.S. (1997). Visual pigment gene structure and expression in human retinae. *Hum Mol Genet*, **6**, 981–990.

Yamazaki, K., McGovern, D., Ragoussis, J. *et al.* (2005). Single nucleotide polymorphisms in TNFSF15 confer susceptibility to Crohn's disease. *Hum Mol Genet*, **14**, 3499–3506.

Yan, H., Yuan, W., Velculescu, V.E., Vogelstein, B. and Kinzler, K.W. (2002). Allelic variation in human gene expression. *Science*, **297**, 1143.

Yan, L., Zhang, S., Eiff, B. *et al.* (2000). Thiopurine methyltransferase polymorphic tandem repeat: genotype-phenotype correlation analysis. *Clin Pharmacol Ther*, **68**, 210–219.

Yang, Y., Chung, E.K., Wu, Y.L. *et al.* (2007). Gene copy-number variation and associated polymorphisms of complement component C4 in human systemic lupus erythematosus (SLE): low copy number is a risk factor for and high copy number is a protective factor against SLE susceptibility in European Americans. *Am J Hum Genet*, **80**, 1037–1054.

Yang, Y., Chung, E.K., Zhou, B. *et al.* (2004a). The intricate role of complement component C4 in human systemic lupus erythematosus. *Curr Dir Autoimmun*, **7**, 98–132.

Yang, Y., Lhotta, K., Chung, E.K., Eder, P., Neumair, F. and Yu, C.Y. (2004b). Complete complement components C4A and C4B deficiencies in human kidney diseases and systemic lupus erythematosus. *J Immunol*, **173**, 2803–2814.

Yang, Z., Camp, N.J., Sun, H. *et al.* (2006). A variant of the HTRA1 gene increases susceptibility to age-related macular degeneration. *Science*, **314**, 992–993.

Yang, Z., Mendoza, A.R., Welch, T.R., Zipf, W.B. and Yu, C.Y. (1999). Modular variations of the human major histocompatibility complex class III genes for serine/threonine kinase RP, complement component C4, steroid 21-hydroxylase CYP21, and tenascin TNX (the RCCX module). A mechanism for gene deletions and disease associations. *J Biol Chem*, **274**, 12147–12156.

Yap, M.W., Nisole, S. and Stoye, J.P. (2005). A single amino acid change in the SPRY domain of human Trim5alpha leads to HIV-1 restriction. *Curr Biol*, **15**, 73–78.

Yeager, M., Orr, N., Hayes, R.B. *et al.* (2007). Genome-wide association study of prostate cancer identifies a second risk locus at 8q24. *Nat Genet*, **39**, 645–649.

Yip, Y.L., Scheib, H., Diemand, A.V. *et al.* (2004). The Swiss-Prot variant page and the ModSNP database: a resource for sequence and structure information on human protein variants. *Hum Mutat*, **23**, 464–470.

Yokouchi, Y., Nukaga, Y., Shibasaki, M. *et al.* (2000). Significant evidence for linkage of mite-sensitive childhood asthma to chromosome 5q31-q33 near the inter-leukin 12 B locus by a genome-wide search in Japanese families. *Genomics*, **66**, 152–160.

Yoshikawa, H., Nishimura, T., Nakatsuji, Y. *et al.* (1994). Elevated expression of messenger RNA for periph-eral myelin protein 22 in biopsied peripheral nerves of patients with Charcot–Marie–Tooth disease type 1A. *Ann Neurol*, **35**, 445–450.

Young, I.D. (2005). *Medical Genetics*. Oxford University Press, Oxford, UK.

Yu, C.Y. and Whitacre, C.C. (2004). Sex, MHC and comple-ment C4 in autoimmune diseases. *Trends Immunol*, **25**, 694–699.

Yu, C.Y., Belt, K.T., Giles, C.M., Campbell, R.D. and Porter, R.R. (1986). Structural basis of the polymorphism of human complement components C4A and C4B: gene size, reactivity and antigenicity. *EMBO J*, **5**, 2873–2881.

Yu, X., Yu, Y., Liu, B. *et al.* (2003). Induction of APOBEC3G ubiquitination and degradation by an HIV-1 Vif-Cul5-SCF complex. *Science*, **302**, 1056–1060.

Yvert, G., Brem, R.B., Whittle, J. *et al.* (2003). Trans-acting regulatory variation in Saccharomyces cerevisiae and the role of transcription factors. *Nat Genet*, **35**, 57–64.

Zackai, E.H. and Emanuel, B.S. (1980). Site-specific recipro-cal translocation, t(11;22) (q23;q11), in several unrelated families with 3:1 meiotic disjunction. *Am J Med Genet*, **7**, 507–521.

Zanke, B.W., Greenwood, C.M., Rangrej, J. *et al.* (2007). Genome-wide association scan identifies a colorectal cancer susceptibility locus on chromosome 8q24. *Nat Genet*, **39**, 989–994.

Zaragoza, M.V., Surti, U., Redline, R.W., Millie, E., Chakravarti, A. and Hassold, T.J. (2000). Parental origin and phenotype of triploidy in spontaneous abortions: predominance of diandry and association with the partial hydatidiform mole. *Am J Hum Genet*, **66**, 1807–1820.

Zeder, M.A. and Hesse, B. (2000). The initial domestication of goats (Capra hircus) in the Zagros mountains 10,000 years ago. *Science*, **287**, 2254–2257.

Zeggini, E., Scott, L.J., Saxena, R. *et al.* (2008). Meta-analysis of genome-wide association data and large-scale repli-cation identifies additional susceptibility loci for type 2 diabetes. *Nat Genet*, **40**, 638–645.

Zeggini, E., Weedon, M.N., Lindgren, C.M. *et al.* (2007). Replication of genome-wide association signals in UK samples reveals risk loci for type 2 diabetes. *Science*, **316**, 1336–1341.

Zeidler, M., Stewart, G., Cousens, S.N., Estibeiro, K. and Will, R.G. (1997). Codon 129 genotype and new variant CJD. *Lancet*, **350**, 668.

Zhang, A., Zheng, C., Hou, M. *et al.* (2003). Deletion of the telomerase reverse transcriptase gene and haploinsuffi-ciency of telomere maintenance in Cri du chat syndrome. *Am J Hum Genet*, **72**, 940–948.

Zhang, G., Luo, J., Bruckel, J. *et al.* (2004a). Genetic studies in familial ankylosing spondylitis susceptibility. *Arthritis Rheum*, **50**, 2246–2254.

Zhang, J., Feuk, L., Duggan, G.E., Khaja, R. and Scherer, S.W. (2006). Development of bioinformatics resources for display and analysis of copy number and other structural variants in the human genome. *Cytogenet Genome Res*, **115**, 205–214.

Zhang, L., Lu, H.H., Chung, W.Y., Yang, J. and Li, W.H. (2005a). Patterns of segmental duplication in the human genome. *Mol Biol Evol*, **22**, 135–141.

Zhang, L., Rao, F., Wessel, J. *et al.* (2004b). Functional allelic heterogeneity and pleiotropy of a repeat polymorphism in tyrosine hydroxylase: prediction of catecholamines and response to stress in twins. *Physiol Genomics*, **19**, 277–291.

Zhang, W., Weissfeld, J.L., Romkes, M., Land, S.R., Grandis, J.R. and Siegfried, J.M. (2007). Association of the EGFR intron 1 CA repeat length with lung cancer risk. *Mol Carcinog*, **46**, 372–380.

Zhang, X., Snijders, A., Segraves, R. *et al.* (2005b). High-resolution mapping of genotype-phenotype relation-ships in cri du chat syndrome using array comparative genomic hybridization. *Am J Hum Genet*, **76**, 312–326.

Zhou, X.Y., Tomatsu, S., Fleming, R.E. *et al.* (1998). HFE gene knockout produces mouse model of hereditary hemo-chromatosis. *Proc Natl Acad Sci U S A*, **95**, 2492–2497.

Zhu, X. and Cooper, R.S. (2007). Admixture mapping pro-vides evidence of association of the VNN1 gene with hypertension. *PLoS ONE*, **2**, e1244.

Zhu, Y., Spitz, M.R., Lei, L., Mills, G.B. and Wu, X. (2001). A single nucleotide polymorphism in the matrix metallo-proteinase-1 promoter enhances lung cancer suscepti-bility. *Cancer Res*, **61**, 7825–7829.

Zhu, Y., Strassmann, J.E. and Queller, D.C. (2000). Insertions, substitutions, and the origin of microsatellites. *Genet Res*, **76**, 227–236.

Zijlstra, E.E., el-Hassan, A.M., Ismael, A. and Ghalib, H.W. (1994). Endemic kala-azar in eastern Sudan: a longitu-dinal study on the incidence of clinical and subclinical infection and post-kala-azar dermal leishmaniasis. *Am J Trop Med Hyg*, **51**, 826–836.

Zimmerman, P.A., Buckler-White, A., Alkhatib, G. *et al.* (1997). Inherited resistance to HIV-1 conferred by an inactivating mutation in CC chemokine receptor 5: studies in populations with contrasting clinical phenotypes, defined racial background, and quantified risk. *Mol Med*, **3**, 23–36.

Zimmerman, P.A., Woolley, I., Masinde, G.L. *et al.* (1999). Emergence of FY*A(null) in a Plasmodium vivax-endemic region of Papua New Guinea. *Proc Natl Acad Sci U S A*, **96**, 13973–13977.

Zinn, A.R. and Ross, J.L. (1998). Turner syndrome and haploinsufficiency. *Curr Opin Genet Dev*, **8**, 322–327.

Zivelin, A., Mor-Cohen, R., Kovalsky, V. *et al.* (2006). Prothrombin 20210G>A is an ancestral prothrombotic mutation that occurred in whites approximately 24,000 years ago. *Blood*, **107**, 4666–4668.

Zody, M.C., Garber, M., Sharpe, T. *et al.* (2006). Analysis of the DNA sequence and duplication history of human chromosome 15. *Nature*, **440**, 671–675.

Index